Annual Review of
Neuroscience

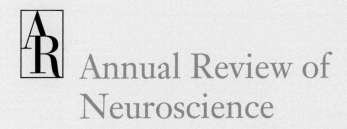

Annual Review of Neuroscience

Volume 37, 2014

www.annualreviews.org • science@annualreviews.org • 650-493-4400

Annual Reviews
4139 El Camino Way • P.O. Box 10139 • Palo Alto, California 94303-0139

Annual Reviews
Palo Alto, California, USA

International Standard Serial Number: 0147-006X
International Standard Book Number: 978-0-8243-2437-7
Library of Congress Control Number: 78643473

TYPESET BY APTARA
PRINTED AND BOUND BY SHERIDAN BOOKS, INC., CHELSEA, MICHIGAN

Contents

Annual Review of
Neuroscience

Volume 37, 2014

Indexes

Errata

An online log of corrections to *Annual Review of Neuroscience* articles may be found at
http://www.annualreviews.org/errata/neuro

Related Articles

Embodied Cognition and Mirror Neurons: A Critical Assessment

Alfonso Caramazza,[1,2] Stefano Anzellotti,[1,2] Lukas Strnad,[1] and Angelika Lingnau[2,3]

[1]Department of Psychology, Harvard University, Cambridge, Massachusetts 02138; email: caramazz@fas.harvard.edu

[2]Center for Mind/Brain Sciences, University of Trento, 38100, Mattarello, Italy

[3]Department of Psychological and Cognitive Sciences, University of Trento, 38068, Rovereto, Trento, Italy

Annu. Rev. Neurosci. 2014. 37:1–15

The *Annual Review of Neuroscience* is online at neuro.annualreviews.org

This article's doi: 10.1146/annurev-neuro-071013-013950

Keywords

concepts, embodied cognition, mirror neurons, simulation, action understanding

Abstract

According to embodied cognition theories, higher cognitive abilities depend on the reenactment of sensory and motor representations. In the first part of this review, we critically analyze the central claims of embodied theories and argue that the existing behavioral and neuroimaging data do not allow investigators to discriminate between embodied cognition and classical cognitive accounts, which assume that conceptual representations are amodal and symbolic. In the second part, we review the main claims and the core electrophysiological findings typically cited in support of the mirror neuron theory of action understanding, one of the most influential examples of embodied cognition theories. In the final part, we analyze the claim that mirror neurons subserve action understanding by mapping visual representations of observed actions on motor representations, trying to clarify in what sense the representations carried by these neurons can be claimed motor.

Contents

INTRODUCTION

Over the past 25 years, numerous theories have been proposed that emphasize the role of perceptual and motor processes for higher cognitive abilities such as language comprehension and action understanding. According to these theories, which we broadly group under the term embodied cognition theories, higher cognitive abilities are achieved in large part or entirely through the reenactment of processes used primarily for sensory input processing or for action execution. This review aims to critically evaluate the central tenets of embodied cognition theories by considering some of the most significant examples of such theories and the evidence supporting them.

Although there are many flavors and varieties of embodied cognition theories, the vast majority of them agree on at least two claims (e.g., Barsalou 2008, Gallese & Lakoff 2005). First, they all converge on the claim that semantic knowledge is carried by sensorimotor representations: The neural systems that are causally involved in forming and retrieving semantic knowledge are the same systems necessary for perceiving different sensory modalities or for producing actions. In line with this claim, studies have proposed that retrieving semantic knowledge of perceptual properties such as the colors of objects critically depends on the neural systems implicated in the perception of these properties (in this example, color perception; Simmons et al. 2007) and that understanding another person's actions requires the contribution from one's own motor system (Rizzolatti & Sinigaglia 2010). Second, most embodied cognition theories emphasize the importance of simulation in conceptual processing (Jeannerod 2001, Zwaan & Taylor 2006). On this account, retrieving semantic knowledge requires neural systems that are involved in perception or action execution and also requires that they perform the same processes utilized during perception or action execution. Semantic processing amounts to a reenactment of stored modality-specific representations in the relevant sensorimotor cortices. For instance, all the semantic knowledge we have about chairs could be exhaustively described as a collection of interacting modality-specific records of what a chair looks like, of the action of sitting, of the somatosensory experiences associated with sitting in a chair, etc. (Barsalou 2008). In this review, we focus on the evidence that investigators have used to support the validity of these two claims.

In both the behavioral and neuroimaging literatures, the arguments offered in support of embodiment are based on numerous interesting findings. However, investigators do not agree on whether the findings actually provide support to the central claims of embodied cognition (Barsalou 2008, Fischer & Zwaan 2008, Glenberg & Kaschak 2002, Kiefer & Pulvermüller 2012) or whether they are orthogonal to such claims, that is, consistent with classical, nonembodied theories of cognition (Caramazza et al. 1990, Chatterjee 2010, Csibra 2008, Jacob & Jeannerod 2005, Mahon & Caramazza 2008). Here we refer to the latter theories as cognitive theories. In the first part of this review, we attempt to clarify the nature of the controversy by presenting the neuroimaging and behavioral results that have been cited in support of embodied theories of semantic knowledge, and we discuss the alternative, cognitive interpretations of these results from the literature. Even though our discussion in this part centers on semantic knowledge, we emphasize that the issues at the core of the controversy are analogous to those in other contexts.

The remainder of the review is then devoted to a critical evaluation of perhaps the most influential embodied theory of cognition: the mirror neuron theory. Mirror neurons were originally discovered in the premotor cortex of the macaque monkey and are characterized by responses produced not only when the animal performs an action but also when it observes a similar action (di Pellegrino et al. 1992, Gallese et al. 1996). For instance, the same neuron would fire at an increased rate both when the monkey grasps an object with its hand and when it passively observes the object being grasped by the hand of the experimenter. This intriguing property inspired a theory that postulates a causal involvement of such motor neurons in action understanding (mirror neuron theory). We review the theory and point to the aspects of the theory that remain debated. In particular, we discuss whether there is sufficient evidence to show that mirror neurons play a causal role in action understanding, and we evaluate the direct-matching hypothesis (one of the central tenets of the mirror neuron theory) in light of the available evidence. Finally, we analyze the central claim that mirror neurons subserve action understanding through the reenactment of motor representations, trying to clarify in what sense the representations carried by mirror neurons can be claimed motor. [For a critical assessment of embodied theories of decision making, see Freedman & Assad (2011)].

EMBODIED COGNITION THEORIES: THE NATURE OF THE CONTROVERSY

Embodiment and Semantic Knowledge: Neuroimaging Studies

A central claim of embodied cognition states that semantic knowledge is represented in sensorimotor systems (Buccino et al. 2005; Goldberg et al. 2006; Pulvermüller et al. 2000, 2005; Simmons et al. 2005, 2007). Numerous studies looked for an overlap between brain areas involved in sensorimotor processes and those involved in the retrieval of semantic knowledge (Hauk et al. 2004, Postle et al. 2008, Simmons et al. 2007; for a recent meta-analysis, see Watson et al. 2013). If semantic knowledge were represented in sensorimotor areas, those areas should be active during sensorimotor processing as well as during retrieval of semantic knowledge. This prediction has been tested for color knowledge and action word comprehension.

The Case of Color Knowledge

Simmons et al. (2007), using fMRI, found evidence for an overlap between areas involved in color perception and those involved in the retrieval of color knowledge. The authors first individuated areas involved in color perception by contrasting activity during a color-discrimination task with activity during a task discriminating between hues of gray. Then, within these areas, they tested whether the activity during retrieval of color knowledge (e.g., TAXI = yellow) was greater than that during a control task requiring subjects to evaluate whether a particular motor property was associated with an object (e.g., HAIR = combed). The authors found a greater signal for the color knowledge task than for the control task in a left fusiform area demonstrated to be more active during color perception than during discrimination of hues of gray. They concluded that this result supported embodied theories of color knowledge. However, their conclusion was too strong. An overlap between brain areas active in two different tasks does not imply an overlap between the neural mechanisms involved in performing those tasks. Several neural populations coexist in a single brain area, and the results cannot rule out that the observed overlap derives from the activity of two different, nonoverlapping networks of neurons (see also Dinstein et al. 2008). Furthermore, cognitive theories of conceptual representation also predict some overlap between activity noted during color perception and that shown during the retrieval of color knowledge. According to cognitive theories, some representations of color do not depend on the specific modality through which the information is accessed, that is, those that are activated when one hears a color word, when one thinks of a color, and when one sees a color. Therefore, these representations would also be active during both color perception and retrieval of color knowledge.

The findings reported by Simmons et al. (2007) may actually be problematic for embodied theories of color knowledge. The double dissociation between color discrimination and color knowledge in cases of brain damage has been well documented (Miceli et al. 2001), ruling out the strong embodied view that would reduce color knowledge entirely to reactivations of mechanisms used primarily for color discrimination. The overlap detected by Simmons et al. (2007) does not occur in areas involved in relatively early stages of color processing (lingual gyrus), as determined by lesion overlap analyses of deficits in color perception (Bouvier & Engel 2006). Instead, and in accord with cognitive theories, the overlap occurs more anteriorly (left fusiform gyrus), in areas that, when damaged, do not affect color perception but impair retrieval of object color knowledge (Miceli et al. 2001).

The Case of Action Words

Another prominent embodied theory in the contemporary literature concerns the understanding of the meaning of action words, which is assumed to depend on the reenactment of motor processes involved in performing those actions (Pulvermüller 2005). The results of many studies have been interpreted to support this theory (Aziz-Zadeh et al. 2006, Boulenger et al. 2009, Pulvermüller et al. 2006, Tettamanti et al. 2005). Here, we discuss a typical example.

In an fMRI study, Hauk et al. (2004) investigated participants' brain activity during passive reading of hand, foot, and mouth action words (e.g., pick, kick, lick) and during the performance of actions with the corresponding body parts. They found that passive reading of action words activates premotor and frontal areas in a somatotopic manner. However, the study did not report a direct analysis of the overlap between the activity in the word and the motor localizer conditions. Although the activity during the two different conditions seems to partially overlap for some effectors, many of the areas of activity during the two conditions are markedly different; therefore, it is difficult to assess whether and to what extent the activations in the two tasks overlap. Nonetheless, the authors interpreted the results as support for the embodied view of action word processing (for similar studies and conclusions, see Aziz-Zadeh et al. 2006, Boulenger et al. 2009, Pulvermüller et al. 2006, Tettamanti et al. 2005).

However, somatotopic activity during action word processing is not by itself evidence supporting an embodied theory of action processing. As we have discussed in the case of color, overlap is predicted by embodied theories, but it is also predicted by nonembodied theories of cognition. Therefore, even in the presence of an overlap, we must ask where the area of overlap is located. Postle and colleagues (2008) investigated the overlap between areas involved in action execution and areas involved in understanding action words more rigorously, and they failed to find reliable somatotopic recruitment of the primary or premotor cortex during the processing of action words (for similarly problematic results, see de Zubicaray et al. 2013, Kemmerer et al. 2008, Kemmerer & Gonzalez-Castillo 2010, Lorey et al. 2013). A recent meta-analysis of fMRI studies on action concepts found no support for the idea that the activation of premotor and motor regions plays a significant role in processing action concepts (Watson et al. 2013; see also Bedny & Caramazza 2011).

In a series of studies, Pulvermüller and colleagues tried to support the embodied view of action word understanding, arguing that when participants read action words their motor cortex is activated rapidly (within 200 ms) and somatotopically (Hauk & Pulvermüller 2004; Pulvermüller et al. 2000, 2005). However, these findings do not address whether such fast and somatotopic activation plays a causal role in semantic processing or is merely the consequence of semantic processing in other, nonmotor areas (Mahon & Caramazza 2008). Studies that used transcranial magnetic stimulation (TMS) over the precentral motor cortex to assess its causal role in semantic processing of action words (Buccino et al. 2004, D'Ausilio et al. 2009, Gerfo et al. 2008, Mottonen & Watkins 2009, Papeo et al. 2009, Willems et al. 2011) have produced inconsistent results (for a review, see Papeo et al. 2013). In contrast, a recent study that used repetitive TMS to interfere with processing in the left posterior middle temporal gyrus (lpMTG), an area known to represent action verb semantics (Peelen et al. 2012), eliminated the action–nonaction verb distinction in the precentral motor cortex (Papeo et al. 2014). This result suggests that activity in the precentral motor cortex during action word comprehension is driven by semantic processing in lpMTG.

In sum, as in the case of color knowledge, the overlap between sensorimotor mechanisms and semantic knowledge does not seem to occur within areas involved in low-level sensorimotor processing, and the activity in precentral motor areas is driven by semantic processing outside the motor system.

Embodiment and Semantic Knowledge: Behavioral Findings

A wide range of behavioral evidence has been produced supporting the claims of embodied cognition theories. All of these studies follow the same general pattern in that they demonstrate various interactions between semantic knowledge and sensorimotor processes. Such interactions are then interpreted as evidence that sensorimotor processes or simulation plays a central role in mediating semantic knowledge.

The relevant evidence comes from various domains. For instance, Hansen et al. (2006) observed that one can sometimes perceive achromatic objects as having a color and that the perceived color is systematically related to the canonical color of the object (e.g., yellow for a banana). These data have been taken to suggest that observers automatically simulate the canonical color of an object as they categorize it. The effects of objects' perceptual properties can become apparent even in tasks that involve a substantial amount of semantic processing. Stanfield & Zwaan (2001) and Zwaan et al. (2002) have shown that in a picture-naming paradigm subjects name an object more quickly if it is preceded by text that implies perceptual properties that match those in the object's depiction. For example, subjects name a picture of an eagle with outstretched wings faster compared with a picture of an eagle with folded wings if the former is preceded by the sentence, "The ranger saw the eagle in the sky." Embodied cognition theory proposes that in order to understand the sentence, subjects simulate the perceptual processes implied by its meaning and are therefore faster at naming a perceptually congruent picture.

Studies have shown similar facilitation effects with other experimental paradigms in the perceptual domain (e.g., Borghi et al. 2004, Bosbach et al. 2005, Meteyard et al. 2008, Solomon & Barsalou 2004) but also in the motor domain (Gentilucci & Gangitano 1998, Glenberg & Kaschak 2002). For example, participants are faster to respond to target words (e.g., "typewriter") following prime words referring to objects that, if manipulated in a typical way, require a similar motor response (e.g., "piano") (Myung et al. 2006; but see Postle et al. 2013). And Rueschemeyer et al. (2010) found that prior planning of motor actions facilitates processing of words denoting objects typically associated with such actions. These results, as explained by embodied cognition theories, are taken to suggest that at least some aspects of semantic knowledge about words and objects are stored in the form of motor representations.

Even though such behavioral evidence is of great interest in its own right, it plays only a very limited role in assessing the two central claims of embodied cognition theories. In particular, the interactions between semantic knowledge and sensorimotor processes do not address whether sensorimotor processes are, in fact, necessary for mediating conceptual representations. Cognitive, nonembodied accounts of semantic knowledge also predict such interactions; however, on these accounts, sensorimotor processes are triggered by retrieving semantic knowledge through association. For example, when one hears the sentence "The ranger saw the eagle in the sky," it is perfectly plausible that one retrieves a visual representation of a flying eagle and is therefore primed to name the picture of a flying eagle faster than a picture of a standing eagle. However, in this case, the activation of sensorimotor representations is a consequence of retrieving semantic knowledge rather than an integral part of it (Chatterjee 2010, Mahon & Caramazza 2008). The mere fact that sensorimotor processes interact with retrieval of semantic knowledge provides no clue about the direction of the causal link between the two. Thus, extant behavioral data do not allow one to discriminate between embodied and cognitive accounts.

THE MIRROR NEURON THEORY OF ACTION UNDERSTANDING

The mirror neuron theory has been immensely influential both as the most complete instantiation of an embodied cognition theory in one particular domain, action understanding, and as the foundation for embodied cognition theories in many other domains, such as language and social cognition. Below, we review the theory's main claims; the core evidence cited as supporting those claims, which stem from monkey physiology and human studies; and some of the problematic issues that some researchers have raised.

Main Claims of the Mirror Neuron Theory

Since its original formulation, several different versions of the mirror neuron theory have been put forth. The basic claim, which has not changed substantially across its various versions, concerns the overlap of neural mechanisms mediating action understanding and action production. It is most clearly expressed in a review by Rizzolatti et al. (2001). These authors maintain that "we understand actions when we map the visual representation of the observed action onto our motor representation of the same action. According to this view, an action is understood when its observation causes the motor system of the observer to resonate. So, when we observe a hand grasping an apple, the same population of neurons that control the execution of grasping movements becomes active in the observer's motor areas. By this approach, the motor knowledge of the observer is used to understand the observed action" (Rizzolatti et al. 2001, p. 661).

The neurons that are active, for instance, both when an individual grasps an apple and when the individual observes someone else grasp an apple are, by definition, mirror neurons. The theory in effect asserts that these neurons constitute a key mechanism shared by action production and action understanding. More specifically, it suggests that populations of mirror neurons are causally involved in mediating both these functions.

The presence of a mechanism that is recruited during both production and understanding does not distinguish the mirror neuron theory from classical cognitive theories according to which central, abstract representations are involved in both comprehension and production of actions and language. However, the mirror neuron theory makes at least three strong claims about the character of the shared mechanism of action production and action understanding, which distinguish it from cognitive, nonembodied theories.

First, the core, novel claim of the mirror neuron theory concerns the motor nature of the representations carried by mirror neurons. The idea that the motor system is involved not only in movement generation but also in understanding actions and intentions is radically different from classical theories for which these processes require the involvement of abstract (or amodal/symbolic) representations.

Second, action understanding mediated by the mirror neuron mechanism is assumed to be direct in the sense that it can be achieved without needing "inferential processing" or other "high-level mental processes" (Rizzolatti & Sinigaglia 2010). The mapping of sensory inputs onto corresponding representations of actions within the motor system is thus postulated to be largely automatic. It presumably does not account for factors such as prior beliefs, specifics of the situation, or the context in which an observed action is carried out because all these likely require the "high-level mental processes" that direct matching between an observed action and a motor representation circumvents. Instead, the matching depends on a "natural response" of the mirror system to the visual input (Rizzolatti & Sinigaglia 2010; but see Cook et al. 2014).

Finally, action understanding involves simulation of the observed actions in the motor system of the observer: Whenever an individual observes an action, his or her understanding is mediated by the same population of premotor neurons that also control his or her own execution of that action. The relevant action is effectively reenacted within the observer's premotor cortex (Rizzolatti & Sinigaglia 2010).

Basic Properties of Macaque Mirror Neurons

Early on after the discovery of mirror neurons, many studies focused on characterizing their basic response properties. At least three important findings emerged from these investigations.

Mirror neurons are activated only by particular kinds of actions. Mirror neurons fire only when the monkey is presented with a natural, transitive action that targets a simultaneously presented object (di Pellegrino et al. 1992), for instance, when the experimenter grasps a piece of food in front of the monkey. Mirror neurons would not fire when the experimenter only moves his hand toward the food but does not grasp it. Furthermore, they would not fire during the presentation of the food alone or when the experimenter performs a grasping movement in absence of an object. Thus, mirror neuron activity during visual observation appears to be triggered by object-directed actions (but see Kraskov et al. 2009).

Mirror neurons have different degrees of congruency. The actions that cause the mirror neuron to fire during both motor production and action observation tend to be congruent

(Gallese et al. 1996). For instance, a mirror neuron that is active when the monkey grasps an object with a precision grip is likely to be activated when the monkey observes the same or a similar action. However, the degree of congruency varies considerably across different mirror neurons. Some mirror neurons exhibit a strict relationship between the performed and observed actions that activate them, such as grasping with a specific type of grip, whereas others fire even when the relationship between the observed and performed action is very loose, such as neurons that respond when an action is performed but are activated by the sight of multiple different actions. Finally, for some mirror neurons there is no clear relationship between observed and performed actions.

Mirror neurons are sensitive to the goal of an action. The observations of congruence between the observed and the executed actions triggering some of the mirror neurons led investigators to propose that mirror neuron activity correlates with action understanding (di Pellegrino et al. 1992, Gallese et al. 1996, Rizzolatti et al. 1996). One of the most influential studies cited in support of such claims is an experiment by Umiltà et al. (2001). In the experiment, two monkeys viewed hand actions performed by an experimenter such as grasping, holding, or placing in two conditions. In one of the conditions, the monkeys observed the actions from start to finish without interruption. In the other condition, the monkeys could only observe the initial stage of the action, but the final stage, during which the hand interacted with the object, was occluded. The researchers found that some neurons in area F5, which showed mirror properties when the monkeys observed the entire hand action sequence, also responded when the final stage of the action was occluded. The authors interpreted this result to mean that, on the basis of the observed part of the action sequence, the monkey understood the action being performed; thus, its understanding was reflected in the activity of the mirror neurons. Because the monkey typically understands the action before it is completed, the firing of these mirror neurons is sustained even if the final part of the action is occluded. Therefore, mirror neuron activity correlates with action understanding.

A study by Fogassi et al. (2005) provided another piece of evidence supporting the correlation between mirror neuron activity and action understanding. The authors recorded neurons in the convexity of the inferior parietal lobule (IPL) of a monkey that responded selectively to reaching actions with extremely similar motor profiles: either reaching for a piece of food and placing it in its mouth or reaching for a piece of food and placing it in a container affixed close to its head. Some of these neurons retained their selectivity for one specific type of action when the monkey was passively viewing the experimenter's actions. Fogassi et al. (2005) take these findings to indicate that the mirror neurons in question selectively encode goals of motor acts and thus facilitate action understanding. They also suggest that intentions are understood by activating one of several possible motor chains (e.g., grasp-to-place versus grasp-to-eat).

The experiments by Umiltà and colleagues (2001) and by Fogassi and colleagues (2005) provide clear cases of mirror neuron activity being sensitive to fairly subtle distinctions between different kinds of observed actions. However, whether these neurons actively contribute to action understanding or whether their activity is only correlated with it is not directly addressed by these experiments. That is, they do not rule out the possibility that mirror neuron activity results from processes that occur in other parts of the brain that mediate action understanding. When a monkey observes an action whose final part is occluded (as in Umiltà et al. 2001), assuming the monkey correctly infers the kind of action being performed, at least two accounts of the mirror neuron activity pattern are equally plausible. On the one hand, the mirror neurons could be actively contributing to the categorization of the observed action. On the other hand, the action could be categorized outside the motor system, and a corresponding nonmotor representation of the action (e.g., crack a nut to get food) could be retrieved (Mahon & Caramazza 2008).

Properties of the Human Mirror System

In the human brain, studies show that the inferior limb of the precentral sulcus/posterior part of the inferior frontal gyrus, the inferior parietal lobe, and the superior temporal sulcus, and recently also the supplementary motor cortex, the primary somatosensory cortex, and visual area MT are recruited during both observation and imitation/execution of actions (Chong et al. 2008, Dinstein et al. 2007, Grèzes et al. 2003, Iacoboni et al. 1999, Kilner et al. 2009, Press et al. 2012; for a recent meta-analysis, see Caspers et al. 2010). Using TMS, many studies have demonstrated that action observation leads to highly effector-specific and even muscle-specific modulations of corticospinal excitability (e.g., Cattaneo et al. 2009, Fadiga et al. 1995, Maeda et al. 2002, Urgesi et al. 2010).

Using multi-voxel pattern analysis (MVPA), Oosterhof et al. (2010) observed above-chance classification of actions across modalities in the left postcentral gyrus and the left anterior parietal cortex. Using a similar approach, Oosterhof et al. (2012) found that the parietal and occipitotemporal cortices contained cross-modal action-specific representations irrespective of the viewpoint of the observed action. By contrast, the ventral premotor cortex contained action-specific representations across modalities for the first- but not the third-person perspective (Caggiano et al. 2011, Maeda et al. 2002). These studies show that high-level representations of actions are not restricted to early sensorimotor areas (but see Cattaneo et al. 2010). Despite various methodological advances, the types of content represented in the various regions of the human mirror system and whether these contents are specifically motor or more abstract remain unclear (Dinstein et al. 2008, Hickok 2009, Oosterhof et al. 2013).

Direct Matching and Simulation

The proposal that conceptual understanding is achieved through sensorimotor simulation is integral to embodied theories of cognition, and in the context of the mirror neuron theory of action understanding, it is intimately linked with the notion of direct matching. On this theory, direct matching is a mechanism through which sensory inputs associated with actions of other individuals are mapped unmediatedly, without involving "higher-level mental processes" such as "inferential processing," onto motor representations in the observer's brain (Rizzolatti & Sinigaglia 2010).

Data from numerous experiments have been interpreted to support the claims of direct matching in action understanding (see especially Fogassi et al. 2005, Gallese et al. 1996, Kohler et al. 2002, Rizzolatti et al. 1996, Umiltà et al. 2001). In one of the most widely cited studies, Kohler and colleagues (2002) report finding mirror neurons in the macaque monkey that become active both when the animal visually observes an action and when it hears a sound that is associated with that action. For example, in one experimental condition, some neurons became active when the monkey cracked a peanut, when visually observing the experimenter crack a peanut, and when hearing the sound of the action alone. The authors interpret these observations as evidence supporting the direct matching hypothesis.

Such an interpretation raises an important question about the nature of the link between incoming sensory representations and the subsequently retrieved motor representations. Mapping the sound of an action onto the motor program corresponding to it requires relatively rich prior knowledge about the action. In the case of visual observation, one could establish a correspondence between the low-level visual inputs and motor representations in the premotor cortex. The information about an action contained in the visual signal allows one to determine which effectors were used and what their position and speed were, among other properties of the action. In contrast, the auditory signal alone provides much less information to establish a correspondence with a motor representation of an action; for example, the sound does not carry information about the effector involved in the action. The triggering of mirror neurons by action sounds represents a learned association that could, in principle, be established between an arbitrary sound and an arbitrary motor representation. The fact that the motor representation of the correct action has been retrieved in the motor system even though the sensory signal alone does not contain sufficient information to determine which motor action was performed implies that the action has already been categorized by the time the motor system is activated. It is not obvious how to reconcile the data about auditory triggering of mirror neurons with the direct matching hypothesis.

ARE MIRROR NEURONS MOTOR?

The centrality of motor representations in action understanding—the claim that mirror neurons are essentially motor—is the defining characteristic of the mirror neuron theory. However, the sense in which mirror neurons can be considered motor, what evidence supports such a claim, and its implications for embodied theories of action understanding are not clear.

Mirror neurons can be considered motor in several ways. First, and most straightforward, is that these neurons fire during active movements, and their responses are selective, responding during certain movements and not others (di Pellegrino et al. 1992, Rizzolatti et al. 1996). However, by definition, mirror neurons are also activated during action observation. Therefore, in this sense, the representations carried by mirror neurons are also visual, and one cannot conclude that actions are understood by reenacting motor representations without also concluding that actions are executed by reenacting visual representations. Thus, the visuomotor character of these neurons does not favor choosing one modality over the other.

A second sense in which the representations carried by mirror neurons are motor is that these neurons were found in areas of the brain that are historically considered motor. Rizzolatti & Sinigaglia (2010) seem to argue that the motor function of mirror neurons depends on their anatomical location. Neurons in area F5, where mirror neurons were found originally (di Pellegrino et al. 1992, Gallese et al. 1996, Rizzolatti et al. 1996), respond during action execution (Kurata & Tanji 1986, Rizzolatti et al. 1981). Thus, mirror neurons could then be considered motor. However, mirror neurons, by definition, also respond to visual stimuli. Area F5 also contains mirror-like neurons that do not fire during action execution and fire only during action observation (di Pellegrino et al. 1992; Gallese et al. 1996, 2002). Accepting the assumption above would lead to the conclusion that even these mirror-like neurons are motor, despite that they are not activated at all during action execution.

In what sense, then, could mirror neuron representations be motor in a way that justifies an embodied theory of action understanding? The motor modality may be predominant over the visual modality in mirror neuron representations in the sense that the informational content of the representations carried by mirror neurons specifies details that are particularly relevant for motor execution (e.g., which muscles are used to perform an action) but not for visual processing (e.g., where they are presented in the visual field). In this case, mirror neurons may be considered predominantly motor in the sense that they carry details specific to the motor modality, but they do not carry other details specific to the visual modality. This determination would allow investigators to interpret mirror neuron activation in terms of reenacting specific motor programs.

A study by Umiltà et al. (2008) is relevant to this issue. These authors investigated the response properties of neurons in areas F5 and F1 of the premotor cortex of monkeys after they were trained to use normal pliers, which require a squeeze action to hold an object, and inverse pliers, which require a squeeze action to release an object. The response pattern of most of the F5 and some of the F1 neurons when monkeys grasped with pliers was extremely similar to the response these neurons exhibited when monkeys grasped objects with reverse pliers. These data thus shed light on how specific the representations mediated by the F5 and a portion of the F1 neurons actually are. The data suggest that neuron activity reflects abstract action properties, such as outcome, rather than just the sequence of motor programs that need to be executed in order to obtain that outcome.

The neurons studied by Umiltà et al. (2008) were not mirror neurons. However, the authors hypothesize that because mirror neurons are found in the same brain regions as those studied in the experiment, one would expect at least some mirror neurons to exhibit the same degree of generalization across different motor actions with the same overall goal. Consistent with this view, Gallese and colleagues (1996) report finding mirror neurons that fire during observation of grasping performed by a monkey either with the hand or with the mouth, a clear indication that mirror neurons represent abstract action goals as opposed to specific motor contents.

Thus far we have adopted a simplified distinction between low-level motor representations and higher-level (abstract, cognitive) representations previously employed in the mirror neuron literature (Rizzolatti & Sinigaglia 2010). Although this distinction can be helpful as a first approximation, it remains unclear on the basis of which criteria the boundary should be drawn. The empirical findings indicate that a richer view is required to appropriately describe the wealth of evidence available in the literature. Action observation and understanding seem to be the outcomes of numerous processing stages at different levels, from early visual areas to the superior temporal sulcus (STS) to the mirror system in the inferior parietal lobe and F5, etc. In a recent study, Mukamel et al. (2010) reported that neurons in the human medial temporal lobe, including the hippocampus and the amygdala, fired both during the execution and the observation of similar actions. These findings indicate that representations active during action execution and action observation are also present outside the regions historically considered motor. The human medial temporal lobe is a highly multimodal brain area known to contain neurons that carry high-level representations of objects that generalize beyond specific views (Kreiman et al. 2000, Quiroga et al. 2005), supporting the hypothesis that these cells store the meaning of a stimulus.

SUMMARY AND CONCLUSIONS

We have provided an overview of the most important empirical results concerning embodied cognition theories and have presented a partial assessment of them, as well. Research motivated by embodied accounts of cognition led to the discovery of many phenomena supporting the close interaction between conceptual processing and sensorimotor representations.

In the context of embodied cognition theories, this body of extraordinarily interesting empirical data has been used by some investigators to argue that conceptual knowledge is mediated primarily by sensorimotor representations and that sensorimotor simulation is an essential part of conceptual processing. We have shown that these claims are unwarranted for two main reasons. First, a substantial part of the evidence cited in support of embodied cognition theories concerns phenomena for which the predictions of embodied and cognitive theories coincide. Therefore, such evidence does not discriminate between embodied and cognitive accounts. In fact, every cognitive theory assumes that perception and action, comprehension and production are bridged through shared, abstract conceptual representations. Cognitive theories would suffer from a strange duality of the mind if there were no possibility for an exchange among perception, action, and conceptual processing. Second, in the field of action understanding, studies on mirror neurons have shown that areas that were thought to carry relatively low-level representations contain neurons that show surprisingly high levels of abstraction (Caggiano et al. 2011, 2012; Ferrari et al. 2005; Gallese et al. 1996; Umiltà et al. 2001, 2008) that, we argue, cannot plausibly be considered motor. At the same time, single-cell recordings in humans individuated neurons located outside the so-called motor system that represent actions with perhaps even greater abstraction (Mukamel et al. 2010). These results suggest that conceptual processing relies on high-level, nonsensorimotor, abstract representations.

DISCLOSURE STATEMENT

The authors are not aware of any affiliations, memberships, funding, or financial holdings that might be perceived as affecting the objectivity of this review.

LITERATURE CITED

Aziz-Zadeh L, Wilson SM, Rizzolatti G, Iacoboni M. 2006. Congruent embodied representations for visually presented actions and linguistic phrases describing actions. *Curr. Biol.* 16:1818–23

Barsalou LW. 2008. Grounded cognition. *Annu. Rev. Psychol.* 59:617–45

Bedny M, Caramazza A. 2011. Perception, action, and word meanings in the human brain: the case from action verbs. *Ann. N. Y. Acad. Sci.* 1224:81–95

Borghi AM, Glenberg AM, Kaschak MP. 2004. Putting words in perspective. *Mem. Cogn.* 32:863–73

Bosbach S, Prinz W, Kerzel D. 2005. Is direction position? Position- and direction-based correspondence effects in tasks with moving stimuli. *Q. J. Exp. Psychol. A* 58:467–506

Boulenger V, Hauk O, Pulvermüller F. 2009. Grasping ideas with the motor system: semantic somatotopy in idiom comprehension. *Cereb. Cortex* 19:1905–14

Bouvier SE, Engel SA. 2006. Behavioral deficits and cortical damage loci in cerebral achromatopsia. *Cereb. Cortex* 16:183–91

Buccino G, Riggio L, Melli G, Binkofski F, Gallese V, Rizzolatti G. 2005. Listening to action-related sentences modulates the activity of the motor system: a combined TMS and behavioral study. *Brain Res. Cogn. Brain Res.* 24:355–63

Buccino G, Vogt S, Ritzl A, Fink GR, Zilles K, et al. 2004. Neural circuits underlying imitation learning of hand actions: an event-related fMRI study. *Neuron* 42:323–34

Caggiano V, Fogassi L, Rizzolatti G, Casile A, Giese MA, Thier P. 2012. Mirror neurons encode the subjective value of an observed action. *Proc. Natl. Acad. Sci. USA* 109:11848–53

Caggiano V, Fogassi L, Rizzolatti G, Pomper JK, Thier P, et al. 2011. View-based encoding of actions in mirror neurons of area F5 in macaque premotor cortex. *Curr. Biol.* 21:144–48

Caramazza A, Hillis AE, Rapp BC, Romani C. 1990. The Multiple Semantics Hypothesis: multiple confusions? *Cogn. Neuropsychol.* 7:161–89

Caspers S, Zilles K, Laird AR, Eickhoff SB. 2010. ALE meta-analysis of action observation and imitation in the human brain. *NeuroImage* 50:1148–67

Cattaneo L, Caruana F, Jezzini A, Rizzolatti G. 2009. Representation of goal and movements without overt motor behavior in the human motor cortex: a transcranial magnetic stimulation study. *J. Neurosci.* 29:11134–38

Cattaneo L, Sandrini M, Schwarzbach J. 2010. State-dependent TMS reveals a hierarchical representation of observed acts in the temporal, parietal, and premotor cortices. *Cereb. Cortex* 20:2252–58

Chatterjee A. 2010. Disembodying cognition. *Lang. Cogn.* 2:79–116

Chong TT, Cunnington R, Williams MA, Kanwisher N, Mattingley JB. 2008. fMRI adaptation reveals mirror neurons in human inferior parietal cortex. *Curr. Biol.* 18:1576–80

Cook R, Bird G, Catmur C, Press C, Heyes C. 2014. Mirror neurons: from origin to function. *Behav. Brain Sci.* In press

Csibra G. 2008. Action mirroring and action understanding: an alternative account. In *Sensorimotor Foundations of Higher Cognition: Attention and Performance XXII*, ed. P Haggard, Y Rossetti, M Kawato, pp. 435–59. New York: Oxford Univ. Press

D'Ausilio A, Pulvermüller F, Salmas P, Bufalari I, Begliomini C, Fadiga L. 2009. The motor somatotopy of speech perception. *Curr. Biol.* 19:381–85 de Zubicaray G, Arciuli J, McMahon K. 2013. Putting an "end" to the motor cortex representations of action words. *J. Cogn. Neurosci.* 25:1957–74

di Pellegrino G, Fadiga L, Fogassi L, Gallese V, Rizzolatti G. 1992. Understanding motor events: a neurophysiological study. *Exp. Brain Res.* 91:176–80

Dinstein I, Hasson U, Rubin N, Heeger DJ. 2007. Brain areas selective for both observed and executed movements. *J. Neurophysiol.* 98:1415–27

Dinstein I, Thomas C, Behrmann M, Heeger DJ. 2008. A mirror up to nature. *Curr. Biol.* 18:R13–18

Fadiga L, Fogassi L, Pavesi G, Rizzolatti G. 1995. Motor facilitation during action observation: a magnetic stimulation study. *J. Neurophysiol.* 73:2608–11

Ferrari PF, Rozzi S, Fogassi L. 2005. Mirror neurons responding to observation of actions made with tools in monkey ventral premotor cortex. *J. Cogn. Neurosci.* 17:212–26

Fischer MH, Zwaan RA. 2008. Embodied language: a review of the role of the motor system in language comprehension. *Q. J. Exp. Psychol.* 61:825–50

Fogassi L, Ferrari PF, Gesierich B, Rozzi S, Chersi F, Rizzolatti G. 2005. Parietal lobe: from action organization to intention understanding. *Science* 308:662–67

Freedman DJ, Assad JA. 2011. A proposed common neural mechanism for categorization and perceptual decisions. *Nat. Neurosci.* 14:143–46

Gallese V, Fadiga L, Fogassi L, Rizzolatti G. 1996. Action recognition in the premotor cortex. *Brain* 119(Pt. 2):593–609

Gallese V, Fadiga L, Fogassi L, Rizzolatti G. 2002. Action representation and the inferior parietal lobule. In *Common Mechanisms in Perception and Action: Attention and Performance*, ed. W Prinz, B Hommel, pp. 334–55. Oxford, UK: Oxford Univ. Press

Gallese V, Lakoff G. 2005. The Brain's concepts: the role of the sensory-motor system in conceptual knowledge. *Cogn. Neuropsychol.* 22:455–79

Gentilucci M, Gangitano M. 1998. Influence of automatic word reading on motor control. *Eur. J. Neurosci.* 10:752–56

Gerfo EL, Oliveri M, Torriero S, Salerno S, Koch G, Caltagirone C. 2008. The influence of rTMS over prefrontal and motor areas in a morphological task: grammatical vs. semantic effects. *Neuropsychologia* 46:764–70

Glenberg AM, Kaschak MP. 2002. Grounding language in action. *Psychon. Bull. Rev.* 9:558–65

Goldberg RF, Perfetti CA, Schneider W. 2006. Perceptual knowledge retrieval activates sensory brain regions. *J. Neurosci.* 26:4917–21

Grèzes J, Armony JL, Rowe J, Passingham RE. 2003. Activations related to "mirror" and "canonical" neurones in the human brain: an fMRI study. *NeuroImage* 18:928–37

Hansen T, Olkkonen M, Walter S, Gegenfurtner KR. 2006. Memory modulates color appearance. *Nat. Neurosci.* 9:1367–68

Hauk O, Johnsrude I, Pulvermüller F. 2004. Somatotopic representation of action words in human motor and premotor cortex. *Neuron* 41:301–7

Hauk O, Pulvermüller F. 2004. Neurophysiological distinction of action words in the fronto-central cortex. *Hum. Brain Mapp.* 21:191–201

Hickok G. 2009. Eight problems for the mirror neuron theory of action understanding in monkeys and humans. *J. Cogn. Neurosci.* 21:1229–43

Iacoboni M, Woods RP, Brass M, Bekkering H, Mazziotta JC, Rizzolatti G. 1999. Cortical mechanisms of human imitation. *Science* 286:2526–28

Jacob P, Jeannerod M. 2005. The motor theory of social cognition: a critique. *Trends Cogn. Sci.* 9:21–25

Jeannerod M. 2001. Neural simulation of action: a unifying mechanism for motor cognition. *NeuroImage* 14:S103–9

Kemmerer D, Castillo JG, Talavage T, Patterson S, Wiley C. 2008. Neuroanatomical distribution of five semantic components of verbs: evidence from fMRI. *Brain Lang.* 107:16–43

Kemmerer D, Gonzalez-Castillo J. 2010. The Two-Level Theory of verb meaning: an approach to integrating the semantics of action with the mirror neuron system. *Brain Lang.* 112:54–76

Kiefer M, Pulvermüller F. 2012. Conceptual representations in mind and brain: theoretical developments, current evidence and future directions. *Cortex* 48:805–25

Kilner JM, Neal A, Weiskopf N, Friston KJ, Frith CD. 2009. Evidence of mirror neurons in human inferior frontal gyrus. *J. Neurosci.* 29:10153–59

Kohler E, Keysers C, Umiltà MA, Fogassi L, Gallese V, Rizzolatti G. 2002. Hearing sounds, understanding actions: action representation in mirror neurons. *Science* 297:846–48

Kraskov A, Dancause N, Quallo MM, Shepherd S, Lemon RN. 2009. Corticospinal neurons in macaque ventral premotor cortex with mirror properties: a potential mechanism for action suppression? *Neuron* 64:922–30

Kreiman G, Koch C, Fried I. 2000. Category-specific visual responses of single neurons in the human medial temporal lobe. *Nat. Neurosci.* 3:946–53

Kurata K, Tanji J. 1986. Premotor cortex neurons in macaques: activity before distal and proximal forelimb movements. *J. Neurosci.* 6:403–11

Lorey B, Naumann T, Pilgramm S, Petermann C, Bischoff M, et al. 2013. How equivalent are the action execution, imagery, and observation of intransitive movements? Revisiting the concept of somatotopy during action simulation. *Brain Cogn.* 81:139–50

Maeda F, Kleiner-Fisman G, Pascual-Leone A. 2002. Motor facilitation while observing hand actions: specificity of the effect and role of observer's orientation. *J. Neurophysiol.* 87:1329–35

Mahon BZ, Caramazza A. 2008. A critical look at the embodied cognition hypothesis and a new proposal for grounding conceptual content. *J. Physiol. Paris* 102:59–70

Meteyard L, Zokaei N, Bahrami B, Vigliocco G. 2008. Visual motion interferes with lexical decision on motion words. *Curr. Biol.* 18:R732–33

Miceli G, Fouch E, Capasso R, Shelton JR, Tomaiuolo F, Caramazza A. 2001. The dissociation of color from form and function knowledge. *Nat. Neurosci.* 4:662–67

Möttönen R, Watkins KE. 2009. Motor representations of articulators contribute to categorical perception of speech sounds. *J. Neurosci.* 29:9819–25

Mukamel R, Ekstrom AD, Kaplan J, Iacoboni M, Fried I. 2010. Single-neuron responses in humans during execution and observation of actions. *Curr. Biol.* 20:750–56

Myung JY, Blumstein SE, Sedivy JC. 2006. Playing on the typewriter, typing on the piano: manipulation knowledge of objects. *Cognition* 98:223–43

Oosterhof NN, Tipper SP, Downing PE. 2012. Viewpoint (in)dependence of action representations: an MVPA study. *J. Cogn. Neurosci.* 24:975–89

Oosterhof NN, Tipper SP, Downing PE. 2013. Crossmodal and action-specific: neuroimaging the human mirror neuron system. *Trends Cogn. Sci.* 17:311–18

Oosterhof NN, Wiggett AJ, Diedrichsen J, Tipper SP, Downing PE. 2010. Surface-based information mapping reveals crossmodal vision-action representations in human parietal and occipitotemporal cortex. *J. Neurophysiol.* 104:1077–89

Papeo L, Lingnau A, Agosta S, Pascual-Leone A, Battelli L, Caramazza A. 2014. The origin of word-related motor activity. *Cereb. Cortex.* In press. doi: 10.1093/cercor/bht423

Papeo L, Pascual-Leone A, Caramazza A. 2013. Disrupting the brain to validate hypotheses on the neurobiology of language. *Front. Hum. Neurosci.* 7:148

Papeo L, Vallesi A, Isaja A, Rumiati RI. 2009. Effects of TMS on different stages of motor and non-motor verb processing in the primary motor cortex. *PLoS ONE* 4:e4508

Peelen M, Romagno D, Caramazza A. 2012. Independent representations of verbs and actions in left temporal cortex. *J. Cogn. Neurosci.* 24:2096–107

Postle N, Ashton R, McFarland K, de Zubicaray GI. 2013. No specific role for the manual motor system in processing the meanings of words related to the hand. *Front. Hum. Neurosci.* 7:11

Postle N, McMahon KL, Ashton R, Meredith M, de Zubicaray GI. 2008. Action word meaning representations in cytoarchitectonically defined primary and premotor cortices. *NeuroImage* 43:634–44

Press C, Weiskopf N, Kilner JM. 2012. Dissociable roles of human inferior frontal gyrus during action execution and observation. *NeuroImage* 60:1671–77

Pulvermüller F. 2005. Brain mechanisms linking language and action. *Nat. Rev. Neurosci.* 6:576–82

Pulvermüller F, Härle M, Hummel F. 2000. Neurophysiological distinction of verb categories. *NeuroReport* 11:2789–93

Pulvermüller F, Huss M, Kherif F, Moscoso del Prado Martin F, Hauk O, Shtyrov Y. 2006. Motor cortex maps articulatory features of speech sounds. *Proc. Natl. Acad. Sci. USA* 103:7865–70

Pulvermüller F, Shtyrov Y, Ilmoniemi R. 2005. Brain signatures of meaning access in action word recognition. *J. Cogn. Neurosci.* 17:884–92

Quiroga RQ, Reddy L, Kreiman G, Koch C, Fried I. 2005. Invariant visual representation by single neurons in the human brain. *Nature* 435:1102–7

Rizzolatti G, Fadiga L, Gallese V, Fogassi L. 1996. Premotor cortex and the recognition of motor actions. *Brain Res. Cogn. Brain Res.* 3:131–41

Rizzolatti G, Fogassi L, Gallese V. 2001. Neurophysiological mechanisms underlying the understanding and imitation of action. *Nat. Rev. Neurosci.* 2:661–70

Rizzolatti G, Scandolara C, Matelli M, Gentilucci M. 1981. Afferent properties of periarcuate neurons in macaque monkeys. I. Somatosensory responses. *Behav. Brain Res.* 2:125–46

Rizzolatti G, Sinigaglia C. 2010. The functional role of the parieto-frontal mirror circuit: interpretations and misinterpretations. *Nat. Rev. Neurosci.* 11:264–74

Rueschemeyer SA, Lindemann O, van Rooij D, van Dam W, Bekkering H. 2010. Effects of intentional motor actions on embodied language processing. *Exp. Psychol.* 57:260–66

Simmons WK, Martin A, Barsalou LW. 2005. Pictures of appetizing foods activate gustatory cortices for taste and reward. *Cereb. Cortex* 15:1602–8

Simmons WK, Ramjee V, Beauchamp MS, McRae K, Martin A, Barsalou LW. 2007. A common neural substrate for perceiving and knowing about color. *Neuropsychologia* 45:2802–10

Solomon KO, Barsalou LW. 2004. Perceptual simulation in property verification. *Mem. Cogn.* 32:244–59

Stanfield RA, Zwaan RA. 2001. The effect of implied orientation derived from verbal context on picture recognition. *Psychol. Sci.* 12:153–56

Tettamanti M, Buccino G, Saccuman MC, Gallese V, Danna M, et al. 2005. Listening to action-related sentences activates fronto-parietal motor circuits. *J. Cogn. Neurosci.* 17:273–81

Umiltà MA, Escola L, Intskirveli I, Grammont F, Rochat M, et al. 2008. When pliers become fingers in the monkey motor system. *Proc. Natl. Acad. Sci. USA* 105:2209–13

Umiltà MA, Kohler E, Gallese V, Fogassi L, Fadiga L, et al. 2001. I know what you are doing: a neurophysiological study. *Neuron* 31:155–65

Urgesi C, Maieron M, Avenanti A, Tidoni E, Fabbro F, Aglioti SM. 2010. Simulating the future of actions in the human corticospinal system. *Cereb. Cortex* 20:2511–21

Watson CE, Cardillo ER, Ianni GR, Chatterjee A. 2013. Action concepts in the brain: an activation likelihood estimation meta-analysis. *J. Cogn. Neurosci.* 25:1191–205

Willems RM, Labruna L, D'Esposito M, Ivry R, Casasanto D. 2011. A functional role for the motor system in language understanding: evidence from theta-burst transcranial magnetic stimulation. *Psychol. Sci.* 22:849–54

Zwaan RA, Stanfield RA, Yaxley RH. 2002. Language comprehenders mentally represent the shapes of objects. *Psychol. Sci.* 13:168–71

Zwaan RA, Taylor LJ. 2006. Seeing, acting, understanding: motor resonance in language comprehension. *J. Exp. Psychol. Gen.* 135:1–11

Translational Control in Synaptic Plasticity and Cognitive Dysfunction

Shelly A. Buffington,* Wei Huang,* and Mauro Costa-Mattioli

Department of Neuroscience, Memory and Brain Research Center, Baylor College of Medicine, Houston, Texas 77030; email: shelly.buffington@bcm.edu, wei.huang@bcm.edu, costamat@bcm.edu

Annu. Rev. Neurosci. 2014. 37:17–38

The *Annual Review of Neuroscience* is online at neuro.annualreviews.org

This article's doi:
10.1146/annurev-neuro-071013-014100

*These authors contributed equally to this work.

Keywords

eIF2α, mTOR, local protein synthesis, memory, autism, neurodegeneration

Abstract

Activity-dependent changes in the strength of synaptic connections are fundamental to the formation and maintenance of memory. The mechanisms underlying persistent changes in synaptic strength in the hippocampus, specifically long-term potentiation and depression, depend on new protein synthesis. Such changes are thought to be orchestrated by engaging the signaling pathways that regulate mRNA translation in neurons. In this review, we discuss the key regulatory pathways that govern translational control in response to synaptic activity and the mRNA populations that are specifically targeted by these pathways. The critical contribution of regulatory control over new protein synthesis to proper cognitive function is underscored by human disorders associated with either silencing or mutation of genes encoding proteins that directly regulate translation. In light of these clinical implications, we also consider the therapeutic potential of targeting dysregulated translational control to treat cognitive disorders of synaptic dysfunction.

Contents

INTRODUCTION

Memory storage is thought to have a physical basis in long-lasting modifications of synaptic function in selective brain circuits. Pioneering studies from the Flexners revealed the requirement for protein synthesis as the first molecular distinction between labile short-term memory (STM), lasting from seconds to several minutes, and more stable long-term memory (LTM), which persists for many hours, days, years, or even a lifetime (reviewed in Dudai 2012, Kandel 2001, McGaugh 2000). Protein synthesis inhibitors selectively block LTM in various species, and these fundamental studies have been supported by more recent genetic manipulations (Costa-Mattioli et al. 2009, Richter & Klann 2009). Hence, the current dogma of the neurobiology of learning posits that the synthesis of specific proteins is what determines whether a synaptic or memory process remains transient or becomes persistently stored in the brain. These newly synthesized proteins are thought either to strengthen a preexisting synaptic connection, by inducing a structural remodeling or a functional change (e.g., insertion of receptors), or to form new synaptic connections. In this article, we focus on recent advances in our understanding of the role of protein synthesis in synaptic plasticity and cognitive dysfunction, emphasizing the different mechanisms by which protein synthesis regulates mnemonic processes, as well as potential pharmacological approaches to the treatment of cognitive disorders where translation is altered. Although translational control is crucial for many biological processes in the central nervous system, we limit our discussion to these specialized areas because we believe they are particularly important for future developments in the field.

TRANSLATIONAL CONTROL IN SYNAPTIC PLASTICITY

STM: short-term memory

LTM: long-term memory

LTP: long-term potentiation

LTD: long-term depression

Synaptic plasticity, the activity-dependent modulation of the strength of synaptic connections, underlies changes in neuronal network dynamics and is therefore thought to be involved in the storage of LTM (Neves et al. 2008). An intriguing aspect of memory is that different types of learning are associated with either strengthening [long-term potentiation (LTP)] or weakening [long-term depression (LTD)] of synaptic efficacy (Malenka & Bear 2004). Repeated activity in a given neural pathway changes the efficacy of its synaptic connections. For instance, with high-frequency activation, an increase in efficacy can last for hours, days, or even weeks (hence LTP). The reverse is also true: Reduced activity lowers synaptic efficacy, resulting in LTD. Both processes are known to require new protein synthesis (Costa-Mattioli et al. 2009, Kandel 2001, Luscher & Huber 2010). Specifically, in ex vivo rodent hippocampal slices, the induction of late-phase LTP

(L-LTP) resulting from 4 trains of 100-Hz high-frequency stimuli (4 × 100 Hz) requires de novo protein synthesis, whereas early-phase LTP (E-LTP) induced by 1 train of 100 Hz stimulus (1 × 100 Hz) is independent of protein synthesis (Kandel 2001). In the same preparation, LTD of CA1 synapses induced by activation of metabotropic glutamate receptors 1/5 (mGluR-LTD), but not N-methyl-D-aspartate receptor–dependent LTD (NMDAR-LTD), also requires new protein synthesis (Huber et al. 2000). We discuss below the translational control mechanisms that underlie both L-LTP and mGluR-LTD.

Mechanisms of Translation

Protein synthesis occurs in three steps: initiation, elongation, and termination. Initiation, the rate-limiting step, is a major target for translational control (Sonenberg & Hinnebusch 2009). Translation initiation begins with the formation of the 43S preinitiation complex, which consists of the small 40S ribosome subunit and the ternary complex; this complex is formed by the initiator methionyl-tRNA (Met-tRNA$_i^{Met}$) and the GTP-bound form of eukaryotic initiation factor eIF2 (**Figure 1**). The 43S preinitiation complex binds to the 5′end of the m^7-G-capped messenger RNA (mRNA), a process that is promoted by eIF3, the poly(A)-binding protein (PABP) and eIF4F. eIF4F consists of the cap-binding protein eIF4E, eIF4A (the DEAD-box RNA helicase), and eIF4G—a scaffold protein that, through binding with eIF3, bridges the 40S ribosome to the 5′end of the mRNA. eIF4G also serves as a scaffold to bind PABP, thus inducing the circularization of the mRNA (Jackson et al. 2010). After attachment, the 43S preinitiation complex scans along the mRNA in a 5′ to 3′ fashion until it encounters the initiation AUG codon in an optimum context [GCC(A/G)CCAUGG, with a purine at the −3 and a G at the +4 positions relative to the A of the AUG codon, which is designated +1] to form the 48S preinitiation complex. During scanning, the eIF4F complex catalyzes the unwinding of secondary structures in the 5′ untranslated region (UTR). The first step in initiation codon recognition is the base pairing between the AUG codon and the anticodon of Met-tRNA$_i^{Met}$ in the peptidyl-tRNA (P) site of the 40S ribosome. Codon-anticodon recognition arrests the scanning ribosome and triggers the activity of the eIF2-specific GTPase activated protein (GAP) eIF5B in the ternary complex. eIF5B hydrolyzes the ternary complex GTP when it is bound to the 40S subunit, a process that reduces eIF2's affinity for Met-tRNA$_i^{Met}$, thus leading eIF2-GDP to dissociate from the ribosome. The translation initiation factors eIF1, eIF1A, eIF3, and eIF2-GDP must dissociate from the complex for the 40S and 60S subunits to join, forming the 80S ribosomal complex, a process facilitated by eIF5B.

After initiation, translation elongation factors are recruited to elongate the polypeptide chain. The translation elongation factors eEF1A and -1B are required for the aminoacyl-tRNA to transfer onto the ribosome. eEF2 is a GTPase that mediates the translocation of the ribosome along the mRNA following peptide bond formation. Upon recognition of a stop codon, the polypeptide chain is released from the mRNA and ribosome through the coordination of termination factors (Jackson et al. 2012).

Translational Control Mechanisms

By activating NMDA or TrkB receptors, synaptic activity leads to changes in general or gene-specific translation (Costa-Mattioli et al. 2009). The mechanisms regulating translation initiation fall into two categories: (*a*) those regulating the recruitment of the ribosome to the 5′end of the mRNA through phosphorylation of translation initiation factors (such as eIF2α and 4E-BPs) and (*b*) those that control translation at the 3′end (PABP and Paips) and impact the mRNA itself. In addition, translation can be regulated at the elongation level.

mGluR-LTD:
a protein synthesis–dependent form of LTD induced by activation of group 1 mGluRs

NMDAR-LTD:
a protein synthesis–independent form of LTD induced by NMDAR activation

Ternary complex:
composed of the initiator methionyl-tRNA (Met-tRNA$_i^{Met}$), eIF2, and GTP

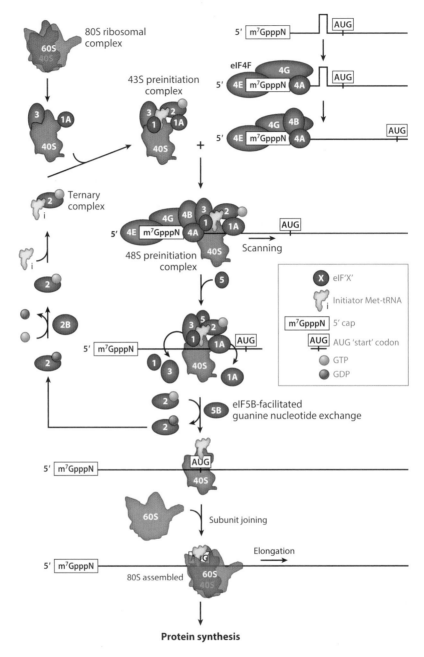

Figure 1

Translation initiation in neurons. Translation intitiation, often the rate-limiting step in protein synthesis, involves multiple fundamental reactions. These include formation of the 43S preinitiation complex, ribosomal scanning along the mRNA, AUG initiation codon recognition, and subunit joining to form the 80S ribosomal complex. Following 80S formation, translation elongation factors are recruited to elongate the forming polypeptide chain. Once the stop codon is reached, the elongation factors are disengaged and release of the newly synthesized protein is orchestrated through the action of translation termination factors.

eIF2α-mediated translational control. Protein synthesis requires the recycling of inactive GDP-bound eIF2 to active GTP-bound eIF2. Phosphorylation of the alpha subunit of eIF2 (eIF2α) at Ser51 blocks the activity of eIF2B, the guanine nucleotide exchange factor (GEF) of eIF2 (Pavitt et al. 1998), thus reducing ternary complex formation and thereby the ability of the cell to synthesize new proteins. Paradoxically, it also results in the translational upregulation of a subset of mRNAs that contain upstream open reading frames (uORFs) in their 5′UTRs. The molecular mechanism underlying this translational upregulation has been explained in great detail for the transcriptional activator GCN4 mRNA in yeast (Hinnebusch 2005) and the transcription factor ATF4 mRNA in mammalian cells (Lu et al. 2004, Vattem & Wek 2004).

The phosphorylation of eIF2α at Ser51 is tightly regulated by kinases and phosphatases (**Figure 2a**). Mammals have four eIF2α kinases: (*a*) heme-regulated kinase HRI (EIF2AK1), which is likely relevant only in erythroid cells; (*b*) the double-strand RNA-dependent kinase PKR (EIF2AK2) activated by viral double-stranded RNA and other stimuli (Garcia et al. 2007); (*c*) the PKR-like endoplasmic reticulum kinase (PERK, EIF2AK3), a transmembrane endoplasmic reticulum (ER) protein kinase enzyme that is activated by ER stress caused by misfolded proteins; and (*d*) the highly conserved eIF2α kinase GCN2 (EIF2AK4), which is activated by amino acid deprivation. Two phosphatase complexes are known to dephosphorylate eIF2α. The first complex, constituted by the catalytic subunit protein phosphatase 1 (PP1) and the regulatory subunit PPP1R15A/GADD34, is induced by phosphorylation of eIF2α (Ron & Harding 2007). The second complex, formed by PP1 and the regulatory protein PPP1R15B/CReP, is constitutively expressed.

Stimuli known to induce a long-lasting change in synaptic strength, such as BDNF or forskolin application, repeated synaptic stimulation, or even behavioral training, all reduce eIF2α phosphorylation (Costa-Mattioli et al. 2009, Takei et al. 2001). In addition, L-LTP, contextual fear, and spatial and gustatory LTM are facilitated in eIF2α knockin heterozygous mice (where the single phosphorylatable Ser51 is replaced by alanine) or mice lacking the eIF2α kinase GCN2 or PKR, in which eIF2α phosphorylation is reduced in the hippocampus (Costa-Mattioli et al. 2005, 2007; Stern et al. 2013; Zhu et al. 2011). By contrast, Sal003, a small molecule inhibitor of eIF2α phosphatases, selectively promotes eIF2α phosphorylation and impairs L-LTP and LTM (Costa-Mattioli et al. 2007). Furthermore, Jiang et al. (2010) showed that L-LTP and LTM were blocked by an independent chemical genetic strategy that selectively activates the phosphorylation of eIF2α in CA1 in vivo. Together, these data demonstrate that eIF2α dephosphorylation is both sufficient and necessary for L-LTP and LTM storage. Consistent with these findings, treatment with integrated stress response inhibitor B (ISRIB), a new compound that blocks the translational effects mediated by eIF2α phosphorylation, or PKRi, a selective inhibitor of PKR, improves spatial and fear-associated LTM (Sidrauski et al. 2013, Zhu et al. 2011). Whether the activity-driven decrease in eIF2α phosphorylation occurs by promoting eIF2α phosphatase activity or by blocking eIF2α kinases remains unknown. Moreover, whether eIF2α phosphorylation regulates other forms of synaptic plasticity has yet to be determined.

How eIF2α phosphorylation controls long-lasting changes in synaptic function has just begun to be elucidated. eIF2α phosphorylation may have a differential role in glutamatergic versus GABAergic neurons. For instance, it is known that the elevation of eIF2α phosphorylation in glutamatergic neurons upregulates the translation of *ATF4* mRNA (Costa-Mattioli et al. 2007, Jiang et al. 2010), which encodes a protein that acts as a repressor of cAMP response element-binding protein (CREB)-mediated L-LTP and LTM (Chen et al. 2003, Costa-Mattioli et al. 2005). By contrast, in GABAergic neurons PKR-mediated phosphorylation of eIF2α locally represses translation of interferon-γ (IFN-γ), resulting in enhanced GABAergic transmission and a consequent depression of neural network excitability (**Figure 2a**) (Zhu et al. 2011). Finally,

Upstream open reading frame (uORF): a short open reading frame found in the 5′UTR that serves as a regulatory unit for expression of its downstream gene

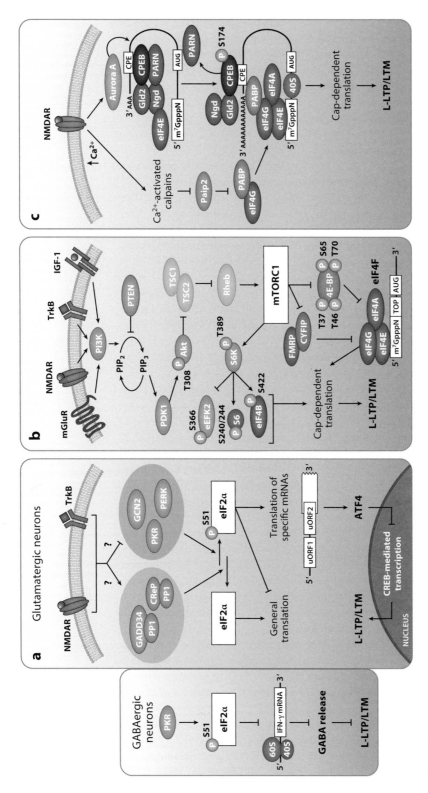

Figure 2

The key signaling pathways that regulate activity-dependent translation initiation in neurons. (*a*) Synaptic activity triggers dephosphorylation of initiation factor eIF2α by either activation of eIF2α-specific phosphatase complexes or inhibition of the eIF2α kinases. Dephosphorylation of eIF2α is both sufficient and necessary to induce L-LTP and LTM formation. In glutamatergic neurons, phosphorylation of eIF2α promotes translation of *ATF4*, a CREB repressor. PKR-mediated phosphorylation of eIF2α in GABAergic neurons negatively regulates translation of the inflammatory cytokine IFN-γ, thus promoting GABA release and maintaining low network rhythmicity in the absence of significant stimulatory inputs. (*b*) Cap-dependent translation is mediated through mTORC1-dependent phosphorylation of its primary downstream effectors, the 4E-BPs and S6Ks. mTORC1 activity is driven by excitatory synaptic inputs, in addition to other cellular signals, that engage the PI3K/Akt signaling pathway. (*c*) Calcium influx following NMDAR activation triggers degradation of Paip2 by calcium-activated calpains, thus releasing PABP to bind eIF4G. The eIF4G-PABP complex then contributes to mRNA circularization by directly bridging the elongated 3′ poly(A) tail to the eIF4F complex at the 5′ cap. Cap-dependent translation is initiated upon subsequent recruitment of the 40S and 60S ribosomal complexes. Abbreviations: GADD34, growth arrest and DNA damage-inducible protein; CReP, constitutive reverter of eIF2α phosphorylation; CREB, cAMP response element-binding protein; PI3K, phosphatidylinositide 3-kinase; PDK1, phosphoinositide-dependent kinase-1.

mice lacking PERK in glutamatergic forebrain neurons display subtle memory deficits (reduced fear extinction) and some behavioral endophenotypes relevant to schizophrenia, as evidenced by behavioral perseveration and decreased prepulse inhibition (Trinh et al. 2012). The phenotype of PERK-deficient mice is remarkably different from that of other mouse models with reduced eIF2α phosphorylation. Thus, whether the schizophrenia-related behaviors in PERK-deficient mice are due to reduced eIF2α phosphorylation or the inability to cope with the ER stress caused by the accumulation of misfolded proteins remains to be determined.

Rapamycin: an immunosuppressant and antiproliferative macrolide

mTORC1-mediated translational control. The mechanistic target of rapamycin complex 1 (mTORC1) regulates translation rates through phosphorylation of its main downstream effectors eIF4E–binding proteins (4E-BPs) and p70 S6 kinases (S6K1/2) (Hay & Sonenberg 2004, Ma & Blenis 2009). The defining component of mTORC1, Raptor, confers substrate specificity to mTORC1 and recruits its downstream targets. Specifically, the 4E-BPs and S6Ks interact with Raptor through a short amino acid sequence called the TOS (mTOR signaling) motif. Rapamycin binds to FKBP-12, and the FKBP-12-rapamycin complex directly binds to and inhibits mTORC1 (Laplante & Sabatini 2012, Wullschleger et al. 2006).

The best-characterized process by which mTORC1 controls translation is by regulating eIF4F complex formation through phosphorylation of 4E-BPs (**Figure 2b**). 4E-BPs (4E-BP1, 4E-BP2, 4E-BP3) are small molecular weight proteins that compete with eIF4G for a common binding site on eIF4E. 4E-BP1, the best characterized 4E-BP, undergoes a hierarchical mTORC1-mediated phosphorylation first at Thr37 and Thr46, which serve as priming sites for the subsequent phosphorylation of Ser65 and Thr70. Phosphorylation of 4E-BP1 at these four sites prevents their binding to eIF4E, allowing it to assemble into the eIF4F complex and thereby stimulating translation rates. 4E-BP2 is the most abundant 4E-BP in the mammalian brain (Bidinosti et al. 2010, Tsukiyama-Kohara et al. 2001). Although compared with 4E-BP2, 4E-BP1 is expressed at lower levels in the brain, it plays an important role in the entrainment and synchrony of the master circadian clock (Cao et al. 2013).

An additional mechanism by which mTORC1 could regulate translation is through phosphorylation of S6Ks (two isoforms exist in vertebrates: S6K1 and S6K2). S6Ks regulate translation initiation, translation elongation, and ribosome biogenesis by phosphorylating eIF4B (a cofactor of eIF4A), eukaryotic elongation factor 2 kinase (eEF2K), and ribosomal protein S6, respectively (reviewed in Ma & Blenis 2009).

mTORC1 activity, as determined by the phosphorylation of its downstream targets S6Ks and 4E-BPs, is stimulated by both L-LTP and LTD-inducing stimuli (Cammalleri et al. 2003, Hou & Klann 2004, Tsokas et al. 2007). Inhibition of mTORC1 by rapamycin blocks L-LTP in rodent hippocampal slices (Cammalleri et al. 2003, Tang et al. 2002), as well as synaptic facilitation in *Aplysia* neuronal cultures (Casadio et al. 1999), highlighting the conserved role of mTORC1 in protein synthesis–dependent long-lasting synaptic potentiation. The contribution of mTORC1 to LTD remains controversial: Although experiments in hippocampal slices originally suggested that rapamycin-mediated blockage of mTORC1 impairs mGluR-LTD (Hou & Klann 2004), recent findings have challenged this view (Bhakar et al. 2012). Removal of upstream negative regulators of mTORC1, however, also blocks mGluR-LTD (Auerbach et al. 2011, Bateup et al. 2011).

Behavioral studies using rapamycin support the idea that mTORC1 is also required for LTM formation in mammals. Specifically, LTM, but not STM, is blocked by rapamycin treatment (Bekinschtein et al. 2007, Blundell et al. 2008). In a recent pharmacogenetic study, a low dose of rapamycin, subthreshold in wild-type mice, was effective in *mTOR* heterozygous mice. These findings demonstrate that direct inhibition of mTORC1 blocks L-LTP and LTM formation and rule out an off-target effect of rapamycin (Stoica et al. 2011).

mTORC1 is also crucial for the reconsolidation of memories associated with electric footshocks or addictive substances (Barak et al. 2013, Blundell et al. 2008, Stoica et al. 2011, Wang et al. 2010b), a process that depends on new protein synthesis (Barak et al. 2013, Milekic & Alberini 2002, Nader et al. 2000). Thus, mTORC1 inhibition or blockade of the translational program directed by mTORC1 holds particular therapeutic potential for conditions plagued by pathological memories, such as post-traumatic stress disorder (PTSD) and drug addiction.

By which mechanism (or mechanisms) does mTORC1 control L-LTP and LTM formation? Studies using mice lacking downstream targets of mTORC1 have begun to answer this question. In mice lacking 4E-BP2, which show enhanced eIF4F complex formation, an E-LTP-inducing protocol elicits L-LTP, but both L-LTP induced by 4×100 Hz and LTM formation are impaired (Banko et al. 2005). These mice exhibit various behavioral abnormalities (Banko et al. 2007), alterations in excitation/inhibition balance, and phenotypes associated with autism spectrum disorders (ASDs) (Gkogkas et al. 2013). Because 4E-BP2 also regulates structural plasticity during development (Ran et al. 2013) and undergoes a different posttranslational modification during adulthood (Bidinosti et al. 2010), an approach to decipher 4E-BP2's function more clearly in the context of L-LTP and LTM may be conditional deletion of 4E-BP2 in the adult brain.

S6Ks are likely the main downstream effectors of mTORC1 activity necessary for long-term facilitation (LTF) in *Aplysia*. Expression of dominant-negative S6K blocks LTF in *Aplysia* sensory neurons, whereas expression of dominant-negative 4E-BP does not (Weatherill et al. 2010). If S6Ks are likewise the major effectors of mTORC1 in the mammalian brain, one would expect that in mice lacking S6Ks, L-LTP and LTM (but not E-LTP and STM) should be blocked. However, mice lacking S6K1 or S6K2 display relatively mild memory deficits, and L-LTP is surprisingly normal (Antion et al. 2008). Although compensation by the remaining S6K could be taking place in the single knockout mice, S6K1/2 double knockout mice have yet to be characterized. Alternatively, studies of conditional S6K1/2 double knockout mice or mutant mice in which all five phosphorylatable serine residues in ribosomal protein S6 are replaced by alanine (Ruvinsky et al. 2005) will determine whether the S6Ks and its major target S6 are major players in mTORC1-mediated plasticity and memory processes in the mammalian brain. That said, recent advances using transcriptome-scale ribosome profiling convincingly show that in nonneuronal cells 4E-BPs are the main effectors by which mTORC1 regulates translation (Hsieh et al. 2012, Thoreen et al. 2012). Specifically, 4E-BPs control translation of mRNAs containing 5′ terminal oligopyrimidine (TOP)- and TOP-like motifs. However, in adult neurons, unlike developing neurons, the major posttranslational modification of 4E-BP2 seems to be deamidation (the conversion of asparagine to aspartic acid), not phosphorylation (Bidinosti et al. 2010). Moreover, in the brain 4E-BPs seem to control translation of *GluA1*, *GluA2*, neuroligins, and *Vip* mRNAs (Cao et al. 2013, Gkogkas et al. 2013, Ran et al. 2013).

These findings raise several interesting conceptual issues. First, is 4E-BP1 or 4E-BP2 the major effector of mTORC1 in the adult brain? Second, do *GluA1*, *GluA2*, neuroligins, and *Vip* mRNAs contain TOP-like motifs? If not, how are they regulated by 4E-BPs? Third, how is 4E-BP2 regulated during L-LTP and LTM formation? Fourth, does mTORC1 control 4E-BP2 deamidation? Finally, given that disruption of eIF4F complex formation inhibits L-LTP and LTM (Hoeffer et al. 2011, 2013), one wonders how mTORC1 regulates 4E-BP2-mediated eIF4F complex formation and translation in the adult brain.

Poly(A)-mediated translational control. The cytoplasmic polyadenylation element-binding protein (CPEB) regulates translation of specific mRNAs containing cytoplasmic polyadenylation elements (CPE) in their 3′UTR (Richter & Klann 2009). Vertebrates have four CPEB paralogs: CPEB1–CPEB4.

CPEB1 localizes at dendrites of mammalian neurons where it regulates translation together with its associated partners: poly(A) polymerase Gld2, deadenylase PARN, and the translation inhibitory factor neuroguidin (Ngd) (Richter & Klann 2009). These proteins comprise a dendritic polyadenylation apparatus (Udagawa et al. 2012) that is subject to regulation by neuronal activity (**Figure 2c**). For instance, NMDAR activation stimulates the phosphorylation of CPEB, which is thought to remove PARN from the RNA complex. Consequently, the poly(A) tail is elongated by Gld2, thus allowing binding of the poly(A)-binding protein (PABP). PABP promotes the recruitment of the eIF4F complex to the 5'end of the mRNA and promotes translation of *CamKIIα* and *NR2A* mRNAs (Huang et al. 2002, Udagawa et al. 2012, Wu et al. 1998). CPEB-1 knockout mice are surprisingly defective in protein synthesis–independent E-LTP and display reduced extinction of hippocampus-dependent spatial and fear memory (Alarcon et al. 2004, Berger-Sweeney et al. 2006). Modulation of CPEB partners also alters LTP: Knockdown of Gld2 impairs LTP at dentate gyrus synapses in vivo; in contrast, this LTP is facilitated in vivo by depletion of the inhibitor Ngd (Udagawa et al. 2012). CPEB2–4, which unlike CPEB1 do not bind to CPE and regulate polyadenylation, bind to another consensus RNA sequence (Huang et al. 2006). These proteins are also expressed in neurons, but their role in mammals remains to be determined.

Significant bodies of work demonstrate that the *Drosophila* and *Aplysia* CPEB homologs, Orb2 and ApCPEB, respectively, also regulate long-term synaptic plasticity and LTM. Orb2 and ApCPEB are thought to act as prion-like proteins that form multimers essential for LTM, but not STM (Kruttner et al. 2012, Majumdar et al. 2012), and the maintenance phase, but not the initiation phase, of LTF (Miniaci et al. 2008; Si et al. 2003a,b, 2010), respectively.

Poly(A)-mediated translational control in neurons can also be regulated by controlling PABP function. The inhibitory PABP-interacting protein 2 (Paip2, of which there are two isoforms Paip2a and Paip2b) controls translation by a dual mechanism: (*a*) decreasing the affinity of PABP for the polyadenylated RNA (Khaleghpour et al. 2001) and (*b*) competing with eIF4G for PABP binding (Karim et al. 2006). Although investigators have not observed functional or mechanistic differences between Paip2A and Paip2B in vitro or in vivo, evidence shows that their tissue distribution in mice differs at both the mRNA and protein levels (Berlanga et al. 2006). In neurons, Paip2a is rapidly degraded by calpains following neuronal stimulation or contextual fear conditioning, and mice lacking Paip2 show a facilitated LTP and LTM after weak behavioral training (Khoutorsky et al. 2013). Paip2a seems to control translation of *CamKIIα* mRNA, an important regulator of LTP and LTM (Silva et al. 1992a,b). Because mRNAs are polyadenylated to different extents in response to synaptic activity (Huang et al. 2002), it will be important to identify the mRNAs regulated by PABP and to determine whether the length of the poly(A) tail makes the mRNA more susceptible or resistant to the PABP–Paip2 ratio.

Translational control during elongation. Translation elongation has also been implicated in synaptic plasticity. One of the best-characterized pathways in translational control at the elongation level is that of eEF2K-eEF2 (Dever & Green 2012). eEF2 is phosphorylated by its kinase eEF2K, a Ca^{2+}-activated protein, in response to neuronal activity. Specifically, activation of α-amino-3-hydroxy-5-methyl-4-isoxazoleproprionic acid receptors (AMPAR), NMDAR, and mGluR1/5 all increase eEF2 phosphorylation (Taha et al. 2013).

Phosphorylation of eEF2 is thought to block general translation by impairing its binding to the ribosome. In cultured hippocampal neurons, eEF2 phosphorylation is maintained at relatively low levels by intrinsic network activity, whereas miniature synaptic excitation by spontaneous glutamate release promotes eEF2 phosphorylation locally at dendrites (Sutton et al. 2007). This observation, together with emerging evidence for compartmentalized regulation of the eEF2K-eEF2 pathway in neurons (Weatherill et al. 2011), points to an intriguing model for synapse-specific modification of translation. In addition, p-eEF2 promotes translation of certain mRNAs by yet unknown mechanisms. In the case of mGluR-LTD, translation of *Map1b* (microtubule associated protein 1B) and *Arc* (activity-regulated cytoskeleton-associated protein) mRNAs, two molecules essential to this process, is selectively upregulated when eEF2 is phosphorylated. RNAi-knockdown of eEF2K in primary neurons (Davidkova & Carroll 2007) and genetic knockout of eEF2K (Park et al. 2008) block the dihydroxyphenylglycine (DHPG)-induced increases in MAP1B and Arc synthesis, respectively, both resulting in an impaired LTD. However, the mechanism by which a particular mRNA can be specifically upregulated by elongating its amino acid chain remains to be determined.

Studies have also documented eEF2's involvement in LTM formation. Phosphorylation of eEF2 in gustatory cortex is elevated within 30 min of novel taste learning during the conditioned taste aversion task (Belelovsky et al. 2005). In eEF2K knockin mice, where eEF2 phosphorylation levels are reduced, associative but not incidental taste learning is impaired (Gildish et al. 2012).

Local translational control at the synapse. Almost a decade after the discovery of LTP (Bliss & Lømo 1973), the identification of synapse-associated polyribosome complexes (Steward & Levy 1982) led to the revolutionary theory that synapses could act as quasi-independent plastic units through local protein synthesis. Colocalization of specific mRNAs and translational machinery at postsynaptic sites suggests that local mRNA translation could play an important role in synaptic processes (Steward 1997, Steward & Schuman 2001, Sutton & Schuman 2006). In a landmark study, Kang & Schuman (1996) observed protein synthesis–dependent LTP in CA1 dendrites that had been surgically severed from their cell bodies. These data provided the first evidence for a functional role of local protein synthesis in synaptic plasticity. Subsequent studies by the Schuman lab and others have not only visualized local translation at dendrites (Sutton & Schuman 2006), but also shown that activity-dependent mRNA translation in dendrites also plays a critical role in other forms of long-lasting synaptic plasticity, such as mGluR-LTD (Huber et al. 2000) and homeostatic plasticity (Sutton et al. 2004, 2006). In the latter, mTORC1 promotes local translation of BDNF in response to AMPAR blockade. BDNF is then released as a retrograde signal to drive the homeostatic upregulation of presynaptic function (Jakawich et al. 2010). This mechanism is conserved between mice and flies (Henry et al. 2012, Penney et al. 2012).

Protein synthesis in dendrites involves trafficking of translationally repressed mRNAs in RNA granules (see Martin & Ephrussi 2009, Swanger & Bassell 2011). The selective derepression of mRNA translation at activated synapses is thought to be fundamental to long-lasting synaptic plasticity (Wang et al. 2010a). Proteins synthesized locally in dendrites could be captured by synapses that have been "tagged" by previous activity, in what is defined as the synaptic tagging hypothesis (Frey & Morris 1997, Martin et al. 1997, Redondo & Morris 2011). Although synaptic tagging occurs at single spines, as recently shown by two-photon imaging in hippocampal slice preparations (Govindarajan et al. 2011), and in behaving animals (Shires et al. 2012), the identities of the locally synthesized proteins and the "tag(s)" remain largely unknown. That said, a limited number of proteins and several dendritically localized candidate mRNAs have been identified. Studies employing in situ hybridization (Eberwine et al. 2002, Lein et al. 2007, Poon et al. 2006) and, more recently, deep RNA sequencing have revealed thousands of dendritically localized mRNAs, including many that encode key regulators of synaptic plasticity (Cajigas et al. 2012).

Dendritic mRNAs translated in response to synaptic stimulation include CaMKIIα, the postsynaptic scaffolding molecule PSD-95, and the GluR1 subunit of AMPA receptors. The importance of local translation of these particular mRNAs is underscored by their collective dys-regulation in fragile X syndrome (FXS) (Muddashetty et al. 2007). In addition, mRNAs encoding components of the translational machinery, such as elongation factor 1 alpha, are also dendritically

Homeostatic plasticity: plasticity initiated at the single cell or network level as a compensatory adaptive response to moderate network excitability

localized and translated in response to stimuli that induce long-term changes in synaptic strength (Huang et al. 2005, Tsokas et al. 2005). Given the number of synaptic mRNAs and local translational machinery, and the spatial limitations of dendritic spines, it will be interesting to determine the number of mRNAs that can be cotranslated at spines and whether this varies between synapses.

Application of novel technologies including high-resolution ribosome profiling (Ingolia et al. 2009) and translating ribosome affinity purification (TRAP) (Doyle et al. 2008) in combination with next-generation sequencing hold great promise for the identification of mRNAs whose translation is specifically upregulated in response to synaptic activation. In fact, simple comparative analysis of data reported in two recent studies employing these technologies reveals that 93 of the 100 mRNAs most sensitive to translational repression by the ATP-competitive inhibitor of mTORC1 Torin 1 (Thoreen et al. 2012) were identified among the neuropil transcriptome (Cajigas et al. 2012). Thus, translational regulation of dendritically localized TOP and TOP-like mRNAs by the mTOR pathway may play an underappreciated role in compartmentalized new protein synthesis and synaptic plasticity.

Translating ribosome affinity purification (TRAP): a biochemical method to isolate polysomal mRNAs from specific cell populations

DYSREGULATION OF TRANSLATIONAL CONTROL IN DISEASE

Dysregulation of the signaling pathways that engage protein synthesis have been linked to the pathophysiology of several neurodegenerative diseases and neurodevelopmental disorders (Bagni et al. 2012, Darnell & Klann 2013, Swanger & Bassell 2011). Indeed, disease progression in Alzheimer's disease (AD) and other neurodegenerative disorders has recently been linked to dysregulation of eIF2α-mediated translational control, whereas several components of the mTORC1 signaling pathway are implicated in the etiology of syndromic ASDs. Finally, loss-of-function mutations in the RNA binding protein and putative translational repressor FMRP cause FXS.

Translational Dysregulation in Neurodegenerative Diseases

Suppression of polysomal mRNA translation in the brains of AD patients was first documented more than 20 years ago (Langstrom et al. 1989). A prevailing hypothesis postulates that the deficits in translational control in AD could be due to aberrant eIF2α phosphorylation. Consistent with this hypothesis, increased phosphorylation of eIF2α has been observed in postmortem analysis of AD patients' brains (Chang et al. 2002a,b; Page et al. 2006) and also in numerous experimental preparations, including cultured neurons treated with synthetic Aβ peptides (Chang et al. 2002a) and brain sections of AD mouse models (Ma et al. 2013, O'Connor et al. 2008, Page et al. 2006). The eIF2α kinases PKR and PERK are activated under similar conditions (Chang et al. 2002a,b; O'Connor et al. 2008; Onuki et al. 2004; Page et al. 2006). As general translation is depressed, translation of ATF4 and β-secretase (BACE1) mRNAs, both of which contain uORFs in the 5′UTR, appear to be elevated by p-eIF2α in AD brains (O'Connor et al. 2008). Because BACE1 promotes amyloidogenesis by initiating cleavage of amyloid precursor protein (APP) to form Aβ, investigators believe that dysregulated translation in AD, via elevated p-eIF2α, establishes a feedforward loop that progressively diminishes the overall structure and function of the affected neuronal population.

Significant efforts have been directed toward restoring physiological translational control in AD in hopes of hindering disease progression. In a recent study, deletion of PERK or GCN2 in APP/PS1 mice (APPswe, PSEN1dE9) not only corrected p-eIF2α levels and the depression in global translation but also rescued the deficits in synaptic plasticity and spatial memory displayed at 10–12 months (Ma et al. 2013). In contrast, however, genetic deletion of GCN2 in the 5XFAD (five familiar AD mutations) mouse model, in which AD progresses faster than in APP/PS1 mice, did not rescue but instead aggravated the memory deficits (Devi & Ohno 2013). Thus, whether blockade of eIF2α kinases could be an effective treatment for AD remains unclear (however, see discussion below).

PKR-eIF2α signaling is also implicated in AD (Morel et al. 2009). PKR activity is upregulated in AD patients' brains and in mouse models of AD (Chang et al. 2002b). In primary neurons, both genetic deletion and pharmacological inhibition of PKR prevented inflammation and apoptosis induced by Aβ peptides (Chang et al. 2002a; Couturier et al. 2010, 2011). Given that genetic or pharmacological inhibition of PKR enhances learning and memory in mice (Zhu et al. 2011), PKR is a promising target for AD treatment. In this regard, a recent study found that exposure of neuronal cultures to β-amyloid oligomers (AβOs) triggered PKR, but not PERK (Laurenco et al. 2013), a process that is attenuated by PKRi. Furthermore, intracerebroventricular infusion of AβOs increased eIF2α phosphorylation and resulted in cognitive impairment only in wild-type mice but not in mice lacking PKR. Hence, in future studies, it would be interesting to determine if direct inhibition of eIF2α, by other means, could also restore the memory deficits in AD mouse models.

mTORC1 Translational Control in Autism Spectrum Disorders

ASDs are a group of neurodevelopmental disorders that share common symptoms, including impaired social interactions, abnormal repetitive behavior, and often intellectual disability (Fombonne 1999, Mefford et al. 2012). Although they are genetically heterogeneous, ASDs are heritable, and some forms are linked to single gene mutations (Zoghbi & Bear 2012). In this regard, human mutations in negative regulators of mTORC1, such as PTEN or tuberous sclerosis complex (TSC) proteins (TSC1/2), are associated with ASD (O'Roak et al. 2012, Sahin 2012). Moreover, mice with mutations in *Pten* or *Tsc1/2* causing hyperactivation of mTORC1, exhibit behaviors analogous to those of human ASD patients (see Costa-Mattioli & Monteggia 2013). Remarkably, the ASD-like phenotypes in mouse models with elevated mTORC1 signaling can be arrested or even reversed by applying rapamycin and/or its synthetic derivatives (termed rapalogs). Thus, considerable recent attention has been directed toward establishing a mechanistic understanding of how the mTORC1 signaling pathway regulates synaptic plasticity and the development of novel, mechanism-based therapies to treat mTORC1-related ASDs.

mTOR consists of two distinct complexes, mTORC1 and mTORC2, each of which critically contributes to synaptic plasticity (Stoica et al. 2011, Huang et al. 2013). Although mTORC2 activity could control some of the ASD-like behaviors in models of mTORC1 hyperactivity, here we focus on what is known about dysregulation of the mTORC1 signaling pathway in relation to disorders of synaptic dysfunction.

mTORC1 integrates a diverse set of extracellular and intracellular inputs to manage cell growth and metabolism, among other essential functions in most mammalian cell types (Laplante & Sabatini 2012). Neurons further exploit the mTORC1 pathway for the integration of activity-dependent signaling. Extracellular neurotrophin or glutamate binding to transmembrane receptors activates the PI3K-Akt-mTORC1 signaling pathway, which is outlined in **Figure 2b**.

Heterozygous loss-of-function mutations in either the *TSC1* or *TSC2* genes cause TSC, a disease that presents with neurological deficits including ASD, epilepsy, and intellectual disability in ∼20–60% of patients (Fombonne 1999, Mefford et al. 2012). Like TSC patients, mice heterozygous for either *Tsc1* or *Tsc2* show deficits in several cognitive tasks including spatial learning in the Morris water maze, context discrimination, and fear-conditioning paradigms (Ehninger et al. 2008, Goorden et al. 2007). *Tsc1*$^{+/-}$ mice are also abnormal in their social behavior (Goorden et al. 2007). In addition, spine density and AMPA/NMDA ratios are increased in *Tsc1*$^{+/-}$ mutants while stimuli that typically induce E-LTP produce L-LTP in *Tsc2*$^{+/-}$ mutants. mGluR-LTD is also impaired in *Tsc1*- and *Tsc2*-deficient neurons (Auerbach et al. 2011, Bateup et al. 2011). At the network level, genetic deletion of Tsc1 results in chronic hyperactivity owing to a loss of inhibitory synaptic transmission (Bateup et al. 2013). These behavioral and synaptic phenotypes are thought to be caused by the hyperactivation of mTORC1 because they are restored by rapamycin.

Inherited mutations in PTEN, which also increase mTORC1 activity, commonly result in multiple neurological phenotypes including ASD, intellectual disability, and macrocephaly in humans (Endersby & Baker 2008). Similarly, conditional knockout of *Pten* in the mouse brain causes impairments in social interaction and learning as well as epilepsy and exaggerated neuronal arborization leading to macrocephaly (Kwon et al. 2006). As in the case of *Tsc2*$^{+/-}$ mice (Bateup et al. 2013), bidirectional plasticity (LTP and LTD) is altered in *Pten*-deficient mice. Analysis of *Nse-Cre Pten* conditional knockout mice revealed that the alterations in synaptic plasticity appear prior to the onset of morphological defects and behavioral abnormalities (Takeuchi et al. 2013). Thus, synaptic dysfunction associated with mTORC1 upregulation could be causally related to both the morphological and the cognitive deficits observed in human patients with loss-of-function mutations in PTEN.

Consistent with a causal role for mTORC1 upregulation in the etiology of ASD-like phenotypes in TSC and/or PTEN-ASD mouse models, rapamycin or rapalog treatment reverses many of the synaptic and behavioral phenotypes in these models (Costa-Mattioli & Monteggia 2013, Ehninger & Silva 2011). The exact mechanisms by which rapamycin ameliorates cognitive dysfunction phenotypes remain unclear as mTORC1 regulates not only translation rates, but also lipid synthesis, autophagy, and mitochondrial function.

The most likely mechanism by which excessive mTORC1 signaling could cause ASD phenotypes is by upregulating cap-dependent translation because hyperactivation of mTORC1 promotes eIF4F complex formation (Gingras et al. 2004). Like mutant mice with hyperactive mTORC1 signaling, mice with elevated eIF4F complex in the brain (4E-BP2 knockout mice or transgenic mice overexpressing eIF4E) also show ASD-like phenotypes, including deficits in social behaviors and stereotypic repetitive behaviors (Gkogkas et al. 2013, Santini et al. 2013). Moreover, disruption of eIF4F complex formation with the protein synthesis inhibitor 4EGI-1 fully rescues the neurophysiological and autistic-like behavioral deficits in both the eIF4E-transgenic mice and the 4E-BP2 knockout mice.

Given that eIF4F complex formation is presumably elevated in TSC and PTEN-ASD mouse models, one would predict that 4EGI-1 treatment could rescue the plasticity and behavioral phenotypes in these models. Conversely, mice overexpressing eIF4E, but not 4E-BP2 knockout mice, should be sensitive to rapamycin-mediated mTORC1 inhibition. These experiments would significantly bolster the notion that the primary mechanism by which mTORC1 regulates synaptic plasticity is through eIF4F-mediated translational control. In addition, whether elevated eIF4F complex formation in the ASD brain leads to an increase in global or specific protein synthesis remains unresolved. In mice overexpressing eIF4E general translation is increased (Santini et al. 2013), whereas in 4E-BP2 knockout mice the translation of specific mRNAs encoding the ASD-associated synaptic adhesion proteins neuroligins 1–4 (*Nlgn1–4*) is upregulated (Gkogkas et al. 2013). As discussed above, whether, and if so how, the mRNA features that dictate mTORC1-mediated regulation of the synthesis of these synaptic proteins differ from those in nonneuronal cells are important questions that have yet to be thoroughly addressed.

Translational Control in Fragile X Syndrome

Dysregulation of translation is also implicated in the pathophysiology of FXS, the most common inherited cause of intellectual disability and ASD (for review see Bassell & Warren 2008, Bhakar et al. 2012, Darnell & Klann 2013, Nelson et al. 2013). FXS is caused by transcriptional silencing of the *Fmr1* gene due to CGG triplet repeat expansion in the 5'UTR of the mRNA (Bassell & Warren 2008, Nelson et al. 2013). *Fmr1* encodes FMRP, an mRNA binding protein that represses translation (Laggerbauer et al. 2001, Li et al. 2001). Accordingly, deletion of *Fmr1* leads to a region-specific increase in general translation (Qin et al. 2005). In addition, protein synthesis inhibitors rescue the LTM phenotype in *Drosophila* FXS mutants (Bolduc et al. 2008). The use of an in vivo UV-crosslinking procedure (CLIP) combined with high throughput sequencing (CLIP-seq) of polysome-associated FMRP showed that FMRP binds to a specific subset of mRNAs. The targets of FMRP include several mRNAs that encode proteins with significant roles in synaptic function, such as NMDAR and mGluR subunits, but also include ASD-linked mRNAs such as *Pten*, *Tsc2*, and other members of the mTOR signaling pathway (Darnell et al. 2011). In this regard, mTORC1 activity seems to be upregulated in subjects (Hoeffer et al. 2012) and mouse models of FXS (Sharma et al. 2010). Some of the phenotypes observed in *Fmr1* knockout mice are restored by rapalog treatment (Busquets-Garcia et al. 2013). In addition, genetic removal of S6K1 also corrects some of the phenotypes in FXS mice (Bhattacharya et al. 2012). How S6K1 regulates translation of FMRP-target mRNAs is unclear. Given that, in one model, FMRP represses translation once associated with polysomes (Khandjian et al. 1996, Stefani et al. 2004), an interesting possibility is that S6K-mediated phosphorylation of eEF2 derepresses translation elongation.

Because eIF4F complex formation seems to be increased in the FXS mouse model (Sharma et al. 2010), FMRP could also repress translation at the initiation level. In this model, FMRP binds to its partner CYFIP1, which acts as a 4E-BP, binding eIF4E through a noncanonical eIF4E motif, thus repressing cap-dependent translation initiation (Napoli et al. 2008).

Several important questions remain regarding the role of translational control in FXS pathophysiology. First, does FMRP inhibit polypeptide elongation or translation initiation? If so, is this a dynamic and reversible process? Second, does disruption of eIF4F complex formation rescue some of the FXS phenotypes? Indeed, would this be specific to eIF4F, or could a more general translation inhibitor, like in FXS *Drosophila* experiments (Bolduc et al. 2008), also correct the FXS phenotypes in mice and/or humans? In this regard, crossing *Tsc2* mice—which are expected to have increased eIF4F but show reduced overall translation rates and several behavioral deficits similar to the *Fmr1* knockout mice—with *Fmr1* knockout mice rescued the mutant phenotypes (Auerbach et al. 2011). Third, if FMRP binds to specific mRNAs, why is general translation increased in *Fmr1* knockout hippocampus? Furthermore, given that modulation of actin polymerization through PAK also corrects the behavioral deficits in the FXS mouse (Dolan et al. 2013, Hayashi et al. 2004), it will be interesting to determine the link between translation and cytoskeletal dynamics in FXS. Recent data demonstrating that, through interaction with Rac1, CYFIP1 also contributes to cytoskeletal remodeling at dendritic spines represent a promising first step in this direction (De Rubeis et al. 2013).

FUTURE PERSPECTIVE

Regulation of protein synthesis is crucial for long-lasting changes in synaptic efficacy and long-term memory storage. Over the past two decades, significant progress has been made in understanding how translation is modulated by synaptic activity. However, the identities of the specific mRNAs translated during L-LTP or LTD remain unknown. One also wonders whether these plasticity-related mRNAs can be exclusively translated at dendrites but not in the soma. If so, what would be the mechanism underlying this process? Furthermore, how many mRNAs can be locally translated in a given spine?

Dysregulated protein synthesis control has been associated with several cognitive disorders, including ASD and Alzheimer's disease. Recent studies targeting different signaling pathways that modulate translation, such as S6K, Tsc, and CPEB, have led to the amelioration of some of the symptoms associated with FXS (Auerbach et al. 2011, Bhattacharya et al. 2012, Udagawa et al. 2013). Although one could therefore envision the dream of personalized medicine for some of these disorders, we should also understand our limitations. For instance, S6K, Tsc, and CPEB each control a different translational program, yet genetic deletion of any one of these in the *Fmr1* knockout rescues some of the FXS symptoms. Thus, in addition to moving toward a much-needed application of these findings in translational medicine, we should also pursue a more thorough understanding of the basic principles of translational control in the nervous system. For instance, how is eIF4F complex formation regulated in the brain? Is neuroguidin, CYFIP, or 4E-BP the primary regulator of cap-dependent translation initiation in neurons? Do CPEB1–4 assume an activity-dependent prion-like conformation in the mammalian brain? If so, how does the aggregated isoform regulate translation rates? Is translation initiation or elongation the principal translational control mechanism in neurons? Are plasticity-related mRNAs degraded or recycled following translation at dendrites? If so, does the mRNA circularize in neurons? Could internal ribosome entry site (IRES)-dependent translation be an alternative mechanism for translational control of some forms of synaptic plasticity (Dyer et al. 2003)? Finally, how does translational control at the pre- and postsynaptic domains, respectively, contribute to the protein synthesis dependence of long-term synaptic plasticity? The next few years are likely to see significant progress toward many of these goals, as well as the development of new treatments for cognitive disorders that target specific translational control mechanisms.

DISCLOSURE STATEMENT

The authors are not aware of any affiliations, memberships, funding, or financial holdings that might be perceived as affecting the objectivity of this review.

ACKNOWLEDGMENTS

The authors thank Drs. K. Krnjević and D. Nelson for insightful comments and discussion and apologize to those whose work was not discussed owing to space limitations. This work was supported by the Whitehall Foundation, the Searle Scholars Program, the U.S. Department of Defense, the National Institute of Neurological Disorders and Stroke, and the National Institute of Mental Health.

NOTE ADDED IN PROOF

After this manuscript was accepted for publication, Viana Di Prisco et al. (2014) reported a new mechanism of translational control of hippocampal mGluR-dependent LTD. The authors provide data highly relevant to the content of this review by identifying neuronal mRNAs that are translationally upregulated by mGluR activation as well as the translational program underlying mGluR-LTD.

LITERATURE CITED

Alarcon JM, Hodgman R, Theis M, Huang Y-S, Kandel ER, Richter JD. 2004. Selective modulation of some forms of Schaffer Collateral-CA1 synaptic plasticity in mice with a disruption of the *CPEB-1* gene. *Learn. Mem.* 11:318–27

Antion MD, Hou L, Wong H, Hoeffer CA, Klann E. 2008. mGluR-dependent long-term depression is associated with increased phosphorylation of S6 and synthesis of elongation factor 1A but remains expressed in S6K-deficient mice. *Mol. Cell. Biol.* 28:2996–3007

Auerbach BD, Osterweil EK, Bear MF. 2011. Mutations causing syndromic autism define an axis of synaptic pathophysiology. *Nature* 480:63–68

Bagni C, Tassone F, Neri G, Hagerman R. 2012. Fragile X syndrome: causes, diagnosis, mechanisms, and therapeutics. *J. Clin. Investig.* 122:4314–22

Banko JL, Merhav M, Stern E, Sonenberg N, Rosenblum K, Klann E. 2007. Behavioral alterations in mice lacking the translation repressor 4E-BP2. *Neurobiol. Learn. Mem.* 87:248–56

Banko JL, Poulin F, Hou L, DeMaria CT, Sonenberg N, Klann E. 2005. The translation repressor 4E-BP2 is critical for eIF4F complex formation, synaptic plasticity, and memory in the hippocampus. *J. Neurosci.* 25:9581–90

Barak S, Liu F, Hamida SB, Yowell QV, Neasta J, et al. 2013. Disruption of alcohol-related memories by mTORC1 inhibition prevents relapse. *Nat. Neurosci.* 16:1111–17

Bassell GJ, Warren ST. 2008. Fragile X syndrome: loss of local mRNA regulation alters synaptic development and function. *Neuron* 60:201–14

Bateup HS, Johnson CA, Denefrio CL, Saulnier JL, Kornacker K, Sabatini BL. 2013. Excitatory/inhibitory synaptic imbalance leads to hippocampal hyperexcitability in mouse models of tuberous sclerosis. *Neuron* 78:510–22

Bateup HS, Takasaki KT, Saulnier JL, Denefrio CL, Sabatini BL. 2011. Loss of Tsc1 in vivo impairs hippocampal mGluR-LTD and increases excitatory synaptic function. *J. Neurosci.* 31:8862–69

Bekinschtein P, Cammarota M, Igaz LM, Bevilaqua LR, Izquierdo I, Medina JH. 2007. Persistence of long-term memory storage requires a late protein synthesis- and BDNF- dependent phase in the hippocampus. *Neuron* 53:261–77

Belelovsky K, Elkobi A, Kaphzan H, Nairn AC, Rosenblum K. 2005. A molecular switch for translational control in taste memory consolidation. *Eur. J. Neurosci.* 22:2560–68

Berger-Sweeney J, Zearfoss NR, Richter JD. 2006. Reduced extinction of hippocampal-dependent memories in CPEB knockout mice. *Learn. Mem.* 13:4–7

Berlanga JJ, Ventoso I, Harding HP, Deng J, Ron D, et al. 2006. Antiviral effect of the mammalian translation initiation factor 2α kinase GCN2 against RNA viruses. *EMBO J.* 25:1730–40

Bhakar AL, Dölen G, Bear MF. 2012. The pathophysiology of fragile X (and what it teaches us about synapses). *Annu. Rev. Neurosci.* 35:417–43

Bhattacharya A, Kaphzan H, Alvarez-Dieppa AC, Murphy JP, Pierre P, Klann E. 2012. Genetic removal of p70 S6 kinase 1 corrects molecular, synaptic, and behavioral phenotypes in fragile X syndrome mice. *Neuron* 76:325–37

Bidinosti M, Ran I, Sanchez-Carbente MR, Martineau Y, Gingras AC, et al. 2010. Postnatal deamidation of 4E-BP2 in brain enhances its association with raptor and alters kinetics of excitatory synaptic transmission. *Mol. Cell* 37:797–808

Bliss TV, Lømo T. 1973. Long-lasting potentiation of synaptic transmission in the dentate area of the anaesthetized rabbit following stimulation of the perforant path. *J. Physiol.* 232:381–89

Blundell J, Kouser M, Powell CM. 2008. Systemic inhibition of mammalian target of rapamycin inhibits fear memory reconsolidation. *Neurobiol. Learn. Mem.* 90:28–35

Bolduc FV, Bell K, Cox H, Broadie KS, Tully T. 2008. Excess protein synthesis in *Drosophila* Fragile X mutants impairs long-term memory. *Nat. Neurosci.* 11:1143–45

Busquets-Garcia A, Gomis-González M, Guegan T, Agustín-Pavón C, Pastor A, et al. 2013. Targeting the endocannabinoid system in the treatment of fragile X syndrome. *Nat. Med.* 19:603–7

Cajigas IJ, Tushev G, Will TJ, Tom Dieck S, Fuerst N, Schuman EM. 2012. The local transcriptome in the synaptic neuropil revealed by deep sequencing and high-resolution imaging. *Neuron* 74:453–66

Cammalleri M, Lütjens R, Berton F, King AR, Simpson C, et al. 2003. Time-restricted role for dendritic activation of the mTOR-p70S6K pathway in the induction of late-phase long-term potentiation in the CA1. *Proc. Natl. Acad. Sci. USA* 100:14368–73

Cao R, Robinson B, Xu H, Gkogkas C, Khoutorsky A, et al. 2013. Translational control of entrainment and synchrony of the suprachiasmatic circadian clock by mTOR/4E-BP1 signaling. *Neuron* 79:712–24

Casadio A, Martin KC, Giustetto M, Zhu H, Chen M, et al. 1999. A transient, neuron-wide form of CREB-mediated long-term facilitation can be stabilized at specific synapses by local protein synthesis. *Cell* 99:221–37

Chang RC, Suen KC, Ma CH, Elyaman W, Ng HK, Hugon J. 2002a. Involvement of double-stranded RNA-dependent protein kinase and phosphorylation of eukaryotic initiation factor-2alpha in neuronal degeneration. *J. Neurochem.* 83:1215–25

Chang RC, Wong AK, Ng HK, Hugon J. 2002b. Phosphorylation of eukaryotic initiation factor-2 α (eIF2 α) is associated with neuronal degeneration in Alzheimer's disease. *Neuroreport* 13:2429–32

Chen A, Muzzio IA, Malleret G, Bartsch D, Verbitsky M, et al. 2003. Inducible enhancement of memory storage and synaptic plasticity in transgenic mice expressing an inhibitor of ATF4 (CREB-2) and C/EBP proteins. *Neuron* 39:655–69

Costa-Mattioli M, Gobert D, Harding H, Herdy B, Azzi M, et al. 2005. Translational control of hippocampal synaptic plasticity and memory by the eIF2 α kinase, GCN2. *Nature* 436:1166–73

Costa-Mattioli M, Gobert D, Stern E, Gamache K, Colina R, et al. 2007. eIF2 α phosphorylation bidirectionally regulates the switch from short- to long-term synaptic plasticity and memory. *Cell* 129:195–206

Costa-Mattioli M, Monteggia LM. 2013. mTOR complexes in neurodevelopmental and neuropsychiatric disorders. *Nat. Neurosci.* 16:1537–43

Costa-Mattioli M, Sossin WS, Klann E, Sonenberg N. 2009. Translational control of long-lasting synaptic plasticity and memory. *Neuron* 61:10–26

Couturier J, Morel M, Pontcharraud R, Gontier V, Fauconneau B, et al. 2010. Interaction of double-stranded RNA-dependent protein kinase (PKR) with the death receptor signaling pathway in amyloid β (Aβ)-treated cells and in APP$_{SL}$PS1 knock-in mice. *J. Biol. Chem.* 285:1272–82

Couturier J, Paccalin M, Morel M, Terro F, Milin S, et al. 2011. Prevention of the β-amyloid peptide-induced inflammatory process by inhibition of double-stranded RNA-dependent protein kinase in primary murine mixed co-cultures. *J. Neuroinflammation* 8:72

Darnell JC, Klann E. 2013. The translation of translational control by FMRP: therapeutic targets for FXS. *Nat. Neurosci.* 16:1530–36

Darnell JC, Van Driesche SJ, Zhang C, Hung KY, Mele A, et al. 2011. FMRP stalls ribosomal translocation on mRNAs linked to synaptic function and autism. *Cell* 146:247–61

Davidkova G, Carroll RC. 2007. Characterization of the role of microtubule-associated protein 1B in metabotropic glutamate receptor-mediated endocytosis of AMPA receptors in hippocampus. *J. Neurosci.* 27:13273–78

De Rubeis S, Pasciuto E, Li KW, Fernández E, Di Marino D, et al. 2013. CYFIP1 coordinates mRNA translation and cytoskeleton remodeling to ensure proper dendritic spine formation. *Neuron* 79:1169–82

Dever TE, Green R. 2012. The elongation, termination, and recycling phases of translation in eukaryotes. *Cold Spring Harb. Perspect. Biol.* 4:a013706

Devi CR, Ohno M. 2013. Deletion of the eIF2 α kinase GCN2 fails to rescue the memory decline associated with Alzheimer's disease. *PLoS ONE* 8:e77335

Dolan BM, Duron SG, Campbell DA, Vollrath B, Shankaranarayana Rao BS, et al. 2013. Rescue of fragile X syndrome phenotypes in *Fmr1* KO mice by the small-molecule PAK inhibitor FRAX486. *Proc. Natl. Acad. Sci. USA* 110:5671–76

Doyle JP, Dougherty JD, Heiman M, Schmidt EF, Stevens TR, et al. 2008. Application of a translational profiling approach for the comparative analysis of CNS cell types. *Cell* 135:749–62

Dudai Y. 2012. The restless engram: consolidations never end. *Annu. Rev. Neurosci.* 35:227–47

Dyer JR, Michel S, Lee W, Castellucci VF, Wayne NL, Sossin WS. 2003. An activity-dependent switch to cap-independent translation triggered by eIF4E dephosphorylation. *Nat. Neurosci.* 6:219–20

Eberwine J, Belt B, Kacharmina JE, Miyashiro K. 2002. Analysis of subcellularly localized mRNAs using in situ hybridization, mRNA amplification, and expression profiling. *Neurochem. Res.* 27:1065–77

Ehninger D, Han S, Shilyansky C, Zhou Y, Li W, et al. 2008. Reversal of learning deficits in a $Tsc2^{+/-}$ mouse model of tuberous sclerosis. *Nat. Med.* 14:843–48

Ehninger D, Silva AJ. 2011. Rapamycin for treating tuberous sclerosis and autism spectrum disorders. *Trends Mol. Med.* 17:78–87

Endersby R, Baker SJ. 2008. PTEN signaling in brain: neuropathology and tumorigenesis. *Oncogene* 27:5416–30

Fombonne E. 1999. The epidemiology of autism: a review. *Psychol. Med.* 29:769–86

Frey U, Morris RG. 1997. Synaptic tagging and long-term potentiation. *Nature* 385:533–36

García MA, Meurs EF, Esteban M. 2007. The dsRNA protein kinase PKR: virus and cell control. *Biochimie* 89:799–811

Gildish I, Manor D, David O, Sharma V, Williams D, et al. 2012. Impaired associative taste learning and abnormal brain activation in kinase-defective eEF2K mice. *Learn. Mem.* 19:116–25

Gingras AC, Raught B, Sonenberg N. 2004. mTOR signaling to translation. *Curr. Top. Microbiol. Immunol.* 279:169–97

Gkogkas CG, Khoutorsky A, Ran I, Rampakakis E, Nevarko T, et al. 2013. Autism-related deficits via dys-regulated eIF4E-dependent translational control. *Nature* 493:371–77

Goorden SM, van Woerden GM, van der Weerd L, Cheadle JP, Elgersma Y. 2007. Cognitive deficits in $Tsc1^{+/-}$ mice in the absence of cerebral lesions and seizures. *Ann. Neurol.* 62:648–55

Govindarajan A, Israely I, Huang S-Y, Tonegawa S. 2011. The dendritic branch is the preferred integrative unit for protein synthesis-dependent LTP. *Neuron* 69:132–46

Hay N, Sonenberg N. 2004. Upstream and downstream of mTOR. *Genes Dev.* 18:1926–45

Hayashi ML, Choi SY, Rao BS, Jung HY, Lee HK, et al. 2004. Altered cortical synaptic morphology and impaired memory consolidation in forebrain-specific dominant-negative PAK transgenic mice. *Neuron* 42:773–87

Henry FE, McCartney AJ, Neely R, Perez AS, Carruthers CJ, et al. 2012. Retrograde changes in presynaptic function driven by dendritic mTORC1. *J. Neurosci.* 32:17128–42

Hinnebusch AG. 2005. Translational regulation of GCN4 and the general amino acid control of yeast. *Annu. Rev. Microbiol.* 59:407–50

Hoeffer CA, Cowansage KK, Arnold EC, Banko JL, Moerke NJ, et al. 2011. Inhibition of the interactions between eukaryotic initiation factors 4E and 4G impairs long-term associative memory consolidation but not reconsolidation. *Proc. Natl. Acad. Sci. USA* 108:3383–88

Hoeffer CA, Sanchez E, Hagerman RJ, Mu Y, Nguyen DV, et al. 2012. Altered mTOR signaling and enhanced CYFIP2 expression levels in subjects with fragile X syndrome. *Genes Brain Behav.* 11:332–41

Hoeffer CA, Santini E, Ma T, Arnold EC, Whelan AM, et al. 2013. Multiple components of eIF4F are required for protein synthesis-dependent hippocampal long-term potentiation. *J. Neurophysiol.* 109:68–76

Hou L, Klann E. 2004. Activation of the phosphoinositide 3-kinase-Akt-mammalian target of rapamycin signaling pathway is required for metabotropic glutamate receptor-dependent long-term depression. *J. Neurosci.* 24:6352–61

Hsieh AC, Liu Y, Edlind MP, Ingolia NT, Janes MR, et al. 2012. The translational landscape of mTOR signalling steers cancer initiation and metastasis. *Nature* 485:55–61

Huang F, Chotiner JK, Steward O. 2005. The mRNA for elongation factor 1α is localized in dendrites and translated in response to treatments that induce long-term depression. *J. Neurosci.* 25:7199–209

Huang W, Zhu PJ, Zhang S, Zhou H, Stoica L, et al. 2013. mTORC2 controls actin polymerization required for consolidation of memory. *Nat. Neurosci.* 16:441–48

Huang YS, Jung MY, Sarkissian M, Richter JD. 2002. *N*-methyl-D-aspartate receptor signaling results in Aurora kinase-catalyzed CPEB phosphorylation and alpha CaMKII mRNA polyadenylation at synapses. *EMBO J.* 21:2139–48

Huang YS, Kan MC, Lin CL, Richter JD. 2006. CPEB3 and CPEB4 in neurons: analysis of RNA-binding specificity and translational control of AMPA receptor GluR2 mRNA. *EMBO J.* 25:4865–76

Huber KM, Kayser MS, Bear MF. 2000. Role for rapid dendritic protein synthesis in hippocampal mGluR-dependent long-term depression. *Science* 288:1254–57

Ingolia NT, Ghaemmaghami S, Newman JRS, Weissman JS. 2009. Genome-wide analysis in vivo of translation with nucleotide resolution using ribosome profiling. *Science* 324:218–23

Jackson RJ, Hellen CU, Pestova TV. 2010. The mechanism of eukaryotic translation initiation and principles of its regulation. *Nat. Rev. Mol. Cell Biol.* 11:113–27

Jackson RJ, Hellen CU, Pestova TV. 2012. Termination and post-termination events in eukaryotic translation. *Adv. Protein Chem. Struct. Biol.* 86:45–93

Jakawich SK, Nasser HB, Strong MJ, McCartney AJ, Perez AS, et al. 2010. Local presynaptic activity gates homeostatic changes in presynaptic function driven by dendritic BDNF synthesis. *Neuron* 68:1143–58

Jiang Z, Belforte JE, Lu Y, Yabe Y, Pickel J, et al. 2010. eIF2α phosphorylation-dependent translation in CA1 pyramidal cells impairs hippocampal memory consolidation without affecting general translation. *J. Neurosci.* 30:2582–94

Kandel ER. 2001. The molecular biology of memory storage: a dialogue between genes and synapses. *Science* 294:1030–38

Kang H, Schuman EM. 1996. A requirement for local protein synthesis in neurotrophin-induced hippocampal synaptic plasticity. *Science* 273:1402–6

Karim MM, Svitkin YV, Kahvejian A, De Crescenzo G, Costa-Mattioli M, Sonenberg N. 2006. A mechanism of translational repression by competition of Paip2 with eIF4G for poly(A) binding protein (PABP) binding. *Proc. Natl. Acad. Sci. USA* 103:9494–99

Khaleghpour K, Kahvejian A, De Crescenzo G, Roy G, Svitkin YV, et al. 2001. Dual interactions of the translational repressor Paip2 with poly(A) binding protein. *Mol. Cell. Biol.* 21:5200–13

Khandjian EW, Corbin F, Woerly S, Rousseau F. 1996. The fragile X mental retardation protein is associated with ribosomes. *Nat. Genet.* 12:91–93

Khoutorsky A, Yanagiya A, Gkogkas CG, Fabian MR, Prager-Khoutorsky M, et al. 2013. Control of synaptic plasticity and memory via suppression of poly(A)-binding protein. *Neuron* 78:298–311

Krüttner S, Stepien B, Noordermeer JN, Mommaas MA, Mechtler K, et al. 2012. *Drosophila* CPEB Orb2A mediates memory independent of its RNA-binding domain. *Neuron* 76:383–95

Kwon CH, Luikart BW, Powell CM, Zhou J, Matheny SA, et al. 2006. Pten regulates neuronal arborization and social interaction in mice. *Neuron* 50:377–88

Laggerbauer B, Ostareck D, Keidel EM, Ostareck-Lederer A, Fischer U. 2001. Evidence that fragile X mental retardation protein is a negative regulator of translation. *Hum. Mol. Genet.* 10:329–38

Langstrom NS, Anderson JP, Lindroos HG, Winblad B, Wallace WC. 1989. Alzheimer's disease-associated reduction of polysomal mRNA translation. *Brain Res. Mol. Brain Res.* 5:259–69

Laplante M, Sabatini DM. 2012. mTOR signaling in growth control and disease. *Cell* 149:274–93

Laurenco MV, Clarke JR, Frozza RL, Bomfim TR, Forny-Germano L, et al. 2013. TNF-α mediates PKR-dependent memory impairment and brain IRS-1 inhibition induced by Alzheimer's β-amyloid oligomers in mice and monkeys. *Cell Metab.* 18:831–43

Lein ES, Hawrylycz MJ, Ao N, Ayres M, Bensinger A, et al. 2007. Genome-wide atlas of gene expression in the adult mouse brain. *Nature* 445:168–76

Li Z, Zhang Y, Ku L, Wilkinson KD, Warren ST, Feng Y. 2001. The fragile X mental retardation protein inhibits translation via interacting with mRNA. *Nucleic Acids Res.* 29:2276–83

Lu PD, Harding HP, Ron D. 2004. Translation reinitiation at alternative open reading frames regulates gene expression in an integrated stress response. *J. Cell Biol.* 167:27–33

Lüscher C, Huber KM. 2010. Group 1 mGluR-dependent synaptic long-term depression: mechanisms and implications for circuitry and disease. *Neuron* 65:445–59

Ma T, Trinh MA, Wexler AJ, Bourbon C, Gatti E, et al. 2013. Suppression of eIF2α kinases alleviates Alzheimer's disease-related plasticity and memory deficits. *Nat. Neurosci.* 16:1299–305

Ma XM, Blenis J. 2009. Molecular mechanisms of mTOR-mediated translational control. *Nat. Rev. Mol. Cell Biol.* 10:307–18

Majumdar A, Cesario WC, White-Grindley E, Jiang H, Ren F, et al. 2012. Critical role of amyloid-like oligomers of *Drosophila* Orb2 in the persistence of memory. *Cell* 148:515–29

Malenka RC, Bear MF. 2004. LTP and LTD: an embarrassment of riches. *Neuron* 44:5–21

Martin KC, Casadio A, Zhu H, Yaping E, Rose JC, et al. 1997. Synapse-specific, long-term facilitation of *Aplysia* sensory to motor synapses: a function for local protein synthesis in memory storage. *Cell* 91:927–38

Martin KC, Ephrussi A. 2009. mRNA localization: gene expression in the spatial dimension. *Cell* 136:719–30

McGaugh JL. 2000. Memory–a century of consolidation. *Science* 287:248–51

Mefford HC, Batshaw ML, Hoffman EP. 2012. Genomics, intellectual disability, and autism. *N. Engl. J. Med.* 366:733–43

Milekic MH, Alberini CM. 2002. Temporally graded requirement for protein synthesis following memory reactivation. *Neuron* 36:521–25

Miniaci MC, Kim JH, Puthanveettil SV, Si K, Zhu H, et al. 2008. Sustained CPEB-dependent local protein synthesis is required to stabilize synaptic growth for persistence of long-term facilitation in *Aplysia*. *Neuron* 59:1024–36

Morel M, Couturier J, Lafay-Chebassier C, Paccalin M, Page G. 2009. PKR, the double stranded RNA-dependent protein kinase as a critical target in Alzheimer's disease. *J. Cell. Mol. Med.* 13:1476–88

Muddashetty RS, Kelić S, Gross C, Xu M, Bassell GJ. 2007. Dysregulated metabotropic glutamate receptor-dependent translation of AMPA receptor and postsynaptic density-95 mRNAs at synapses in a mouse model of fragile X syndrome. *J. Neurosci.* 27:5338–48

Nader K, Schafe GE, Le Doux JE. 2000. Fear memories require protein synthesis in the amygdala for reconsolidation after retrieval. *Nature* 406:722–26

Napoli I, Mercaldo V, Boyl PP, Eleuteri B, Zalfa F, et al. 2008. The fragile X syndrome protein represses activity-dependent translation through CYFIP1, a new 4E-BP. *Cell* 134:1042–54

Nelson DL, Orr HT, Warren ST. 2013. The unstable repeats—three evolving faces of neurological disease. *Neuron* 77:825–43

Neves G, Cooke SF, Bliss TV. 2008. Synaptic plasticity, memory and the hippocampus: a neural network approach to causality. *Nat. Rev. Neurosci.* 9:65–75

O'Connor T, Sadleir KR, Maus E, Velliquette RA, Zhao J, et al. 2008. Phosphorylation of the translation initiation factor eIF2α increases BACE1 levels and promotes amyloidogenesis. *Neuron* 60:988–1009

O'Roak BJ, Vives L, Fu W, Egertson JD, Stanaway IB, et al. 2012. Multiplex targeted sequencing identifies recurrently mutated genes in autism spectrum disorders. *Science* 338:1619–22

Onuki R, Bando Y, Suyama E, Katayama T, Kawasaki H, et al. 2004. An RNA-dependent protein kinase is involved in tunicamycin-induced apoptosis and Alzheimer's disease. *EMBO J.* 23:959–68

Page G, Rioux Bilan A, Ingrand S, Lafay-Chebassier C, Pain S, et al. 2006. Activated double-stranded RNA-dependent protein kinase and neuronal death in models of Alzheimer's disease. *Neuroscience* 139:1343–54

Park S, Park JM, Kim S, Kim JA, Shepherd JD, et al. 2008. Elongation factor 2 and fragile X mental retardation protein control the dynamic translation of Arc/Arg3.1 essential for mGluR-LTD. *Neuron* 59:70–83

Pavitt GD, Ramaiah KV, Kimball SR, Hinnebusch AG. 1998. eIF2 independently binds two distinct eIF2B subcomplexes that catalyze and regulate guanine-nucleotide exchange. *Genes Dev.* 12:514–26

Penney J, Tsurudome K, Liao EH, Elazzouzi F, Livingstone M, et al. 2012. TOR is required for the retrograde regulation of synaptic homeostasis at the *Drosophila* neuromuscular junction. *Neuron* 74:166–78

Poon MM, Choi SH, Jamieson CA, Geschwind DH, Martin KC. 2006. Identification of process-localized mRNAs from cultured rodent hippocampal neurons. *J. Neurosci.* 26:13390–99

Qin M, Kang J, Burlin TV, Jiang C, Smith CB. 2005. Postadolescent changes in regional cerebral protein synthesis: an in vivo study in the FMR1 null mouse. *J. Neurosci.* 25:5087–95

Ran I, Gkogkas CG, Vasuta C, Tartas M, Khoutorsky A, et al. 2013. Selective regulation of GluA subunit synthesis and AMPA receptor-mediated synaptic function and plasticity by the translation repressor 4E-BP2 in hippocampal pyramidal cells. *J. Neurosci.* 33:1872–86

Redondo RL, Morris RGM. 2011. Making memories last: the synaptic tagging and capture hypothesis. *Nat. Rev. Neurosci.* 12:17–30

Richter JD, Klann E. 2009. Making synaptic plasticity and memory last: mechanisms of translational regulation. *Genes Dev.* 23:1–11

Ron D, Harding HP. 2007. eIF2α phosphorylation in cellular stress responses and disease. In *Translational Control in Biology and Medicine*, ed. MB Mathews, N Sonenberg, JWB Hershey, pp. 345–68. Cold Spring Harbor, NY: Cold Spring Harbor Lab. Press

Ruvinsky I, Sharon N, Lerer T, Cohen H, Stolovich-Rain M, et al. 2005. Ribosomal protein S6 phosphorylation is a determinant of cell size and glucose homeostasis. *Genes Dev.* 19:2199–211

Sahin M. 2012. Targeted treatment trials for tuberous sclerosis and autism: no longer a dream. *Curr. Opin. Neurobiol.* 22:895–901

Santini E, Huynh TN, MacAskill AF, Carter AG, Pierre P, et al. 2013. Exaggerated translation causes synaptic and behavioural aberrations associated with autism. *Nature* 493:411–15

Sharma A, Hoeffer CA, Takayasu Y, Miyawaki T, McBride SM, et al. 2010. Dysregulation of mTOR signaling in fragile X syndrome. *J. Neurosci.* 30:694–702

Shires KL, Da Silva BM, Hawthorne JP, Morris RG, Martin SJ. 2012. Synaptic tagging and capture in the living rat. *Nat. Commun.* 3:1246

Si K, Choi Y-B, White-Grindley E, Majumdar A, Kandel ER. 2010. *Aplysia* CPEB can form prion-like multimers in sensory neurons that contribute to long-term facilitation. *Cell* 140:421–35

Si K, Giustetto M, Etkin A, Hsu R, Janisiewicz AM, et al. 2003a. A neuronal isoform of CPEB regulates local protein synthesis and stabilizes synapse-specific long-term facilitation in *Aplysia*. *Cell* 115:893–904

Si K, Lindquist S, Kandel ER. 2003b. A neuronal isoform of the *Aplysia* CPEB has prion-like properties. *Cell* 115:879–91

Sidrauski C, Acosta-Alvear D, Khoutorsky A, Vedantham P, Hearn BR, et al. 2013. Pharmacological brake-release of mRNA translation enhances cognitive memory. *eLife* 2:e00498

Silva AJ, Paylor R, Wehner JM, Tonegawa S. 1992a. Impaired spatial learning in alpha-calcium-calmodulin kinase II mutant mice. *Science* 257:206–11

Silva AJ, Stevens CF, Tonegawa S, Wang Y. 1992b. Deficient hippocampal long-term potentiation in alpha-calcium-calmodulin kinase II mutant mice. *Science* 257:201–6

Sonenberg N, Hinnebusch AG. 2009. Regulation of translation initiation in eukaryotes: mechanisms and biological targets. *Cell* 136:731–45

Stefani G, Fraser CE, Darnell JC, Darnell RB. 2004. Fragile X mental retardation protein is associated with translating polyribosomes in neuronal cells. *J. Neurosci.* 24:7272–76

Stern E, Chinnakkaruppan A, David O, Sonenberg N, Rosenblum K. 2013. Blocking the eIF2α kinase (PKR) enhances positive and negative forms of cortex-dependent taste memory. *J. Neurosci.* 33:2517–25

Steward O. 1997. mRNA localization in neurons: a multipurpose mechanism? *Neuron* 18:9–12

Steward O, Levy WB. 1982. Preferential localization of polyribosomes under the base of dendritic spines in granule cells of the dentate gyrus. *J. Neurosci.* 2:284–91

Steward O, Schuman EM. 2001. Protein synthesis at synaptic sites on dendrites. *Annu. Rev. Neurosci.* 24:299–325

Stoica L, Zhu PJ, Huang W, Zhou H, Kozma SC, Costa-Mattioli M. 2011. Selective pharmacogenetic inhibition of mammalian target of Rapamycin complex I (mTORC1) blocks long-term synaptic plasticity and memory storage. *Proc. Natl. Acad. Sci. USA* 108:3791–96

Sutton MA, Ito HT, Cressy P, Kempf C, Woo JC, Schuman EM. 2006. Miniature neurotransmission stabilizes synaptic function via tonic suppression of local dendritic protein synthesis. *Cell* 125:785–99

Sutton MA, Schuman EM. 2006. Dendritic protein synthesis, synaptic plasticity, and memory. *Cell* 127:49–58

Sutton MA, Taylor AM, Ito HT, Pham A, Schuman EM. 2007. Postsynaptic decoding of neural activity: eEF2 as a biochemical sensor coupling miniature synaptic transmission to local protein synthesis. *Neuron* 55:648–61

Sutton MA, Wall NR, Aakalu GN, Schuman EM. 2004. Regulation of dendritic protein synthesis by miniature synaptic events. *Science* 304:1979–83

Swanger SA, Bassell GJ. 2011. Making and breaking synapses through local mRNA regulation. *Curr. Opin. Genet. Dev.* 21:414–21

Taha E, Gildish I, Gal-Ben-Ari S, Rosenblum K. 2013. The role of eEF2 pathway in learning and synaptic plasticity. *Neurobiol. Learn. Mem.* 105:100–6

Takei N, Kawamura M, Hara K, Yonezawa K, Nawa H. 2001. Brain-derived neurotrophic factor enhances neuronal translation by activating multiple initiation processes: comparison with the effects of insulin. *J. Biol. Chem.* 276:42818–25

Takeuchi K, Gertner MJ, Zhou J, Parada LF, Bennett MV, Zukin RS. 2013. Dysregulation of synaptic plasticity precedes appearance of morphological defects in a Pten conditional knockout mouse model of autism. *Proc. Natl. Acad. Sci. USA* 110:4738–43

Tang SJ, Reis G, Kang H, Gingras AC, Sonenberg N, Schuman EM. 2002. A rapamycin-sensitive signaling pathway contributes to long-term synaptic plasticity in the hippocampus. *Proc. Natl. Acad. Sci. USA* 99:467–72

Thoreen CC, Chantranupong L, Keys HR, Wang T, Gray NS, Sabatini DM. 2012. A unifying model for mTORC1-mediated regulation of mRNA translation. *Nature* 485:109–13

Trinh MA, Kaphzan H, Wek RC, Pierre P, Cavener DR, Klann E. 2012. Brain-specific disruption of the eIF2α kinase PERK decreases ATF4 expression and impairs behavioral flexibility. *Cell Rep.* 1:676–88

Tsokas P, Grace EA, Chan P, Ma T, Sealfon SC, et al. 2005. Local protein synthesis mediates a rapid increase in dendritic elongation factor 1A after induction of late long-term potentiation. *J. Neurosci.* 25:5833–43

Tsokas P, Ma T, Iyengar R, Landau EM, Blitzer RD. 2007. Mitogen-activated protein kinase upregulates the dendritic translation machinery in long-term potentiation by controlling the mammalian target of rapamycin pathway. *J. Neurosci.* 27:5885–94

Tsukiyama-Kohara K, Poulin F, Kohara M, DeMaria CT, Cheng A, et al. 2001. Adipose tissue reduction in mice lacking the translational inhibitor 4E-BP1. *Nat. Med.* 7:1128–32

Udagawa T, Farney NG, Jakovcevski M, Kaphzan H, Alarcon JM, et al. 2013. Genetic and acute CPEB1 depletion ameliorate fragile X pathophysiology. *Nat. Med.* 19:1473–77

Udagawa T, Swanger SA, Takeuchi K, Kim JH, Nalavadi V, et al. 2012. Bidirectional control of mRNA translation and synaptic plasticity by the cytoplasmic polyadenylation complex. *Mol. Cell* 47:253–66

Vattem KM, Wek RC. 2004. Reinitiation involving upstream ORFs regulates ATF4 mRNA translation in mammalian cells. *Proc. Natl. Acad. Sci. USA* 101:11269–74

Viana Di Prisco G, Huang W, Buffington SA, Hsu C-C, Bonnen P, et al. 2014. Translational control of mGluR-dependent long-term depression and object-place learning by eIF2α. *Nat. Neurosci.* In press

Wang DO, Martin KC, Zukin RS. 2010a. Spatially restricting gene expression by local translation at synapses. *Trends Neurosci.* 33:173–82

Wang X, Luo YX, He YY, Li FQ, Shi HS, et al. 2010b. Nucleus accumbens core mammalian target of rapamycin signaling pathway is critical for cue-induced reinstatement of cocaine seeking in rats. *J. Neurosci.* 30:12632–41

Weatherill DB, Dyer J, Sossin WS. 2010. Ribosomal protein S6 kinase is a critical downstream effector of the target of rapamycin complex 1 for long-term facilitation in *Aplysia*. *J. Biol. Chem.* 285:12255–67

Weatherill DB, McCamphill PK, Pethoukov E, Dunn TW, Fan X, Sossin WS. 2011. Compartment-specific, differential regulation of eukaryotic elongation factor 2 and its kinase within *Aplysia* sensory neurons. *J. Neurochem.* 117:841–55

Wu L, Wells D, Tay J, Mendis D, Abbott MA, et al. 1998. CPEB-mediated cytoplasmic polyadenylation and the regulation of experience-dependent translation of α-CaMKII mRNA at synapses. *Neuron* 21:1129–39

Wullschleger S, Loewith R, Hall MN. 2006. TOR signaling in growth and metabolism. *Cell* 124:471–84

Zhu PJ, Huang W, Kalikulov D, Yoo JW, Placzek AN, et al. 2011. Suppression of PKR promotes network excitability and enhanced cognition by interferon-γ-mediated disinhibition. *Cell* 147:1384–96

Zoghbi HY, Bear MF. 2012. Synaptic dysfunction in neurodevelopmental disorders associated with autism and intellectual disabilities. *Cold Spring Harb. Perspect. Biol.* 4(3):a009886

The Perirhinal Cortex

Wendy A. Suzuki[1] and Yuji Naya[2]

[1]Center for Neural Science, New York University, New York, NY 10003;
email: wendy@cns.nyu.edu

[2]Department of Psychology, Peking-Tsinghua Center for Life Sciences and
PKU-IDG/McGovern Institute for Brain Research, Peking University, Beijing 100871,
China; email: yujin@pku.edu.cn

Annu. Rev. Neurosci. 2014. 37:39–53

The *Annual Review of Neuroscience* is online at
neuro.annualreviews.org

This article's doi:
10.1146/annurev-neuro-071013-014207

Keywords

association memory, declarative memory, medial temporal lobe

Abstract

Anatomically, the perirhinal cortex sits at the boundary between the medial temporal lobe and the ventral visual pathway. It has prominent interconnections not only with both these systems, but also with a wide range of unimodal and polymodal association areas. Consistent with these diverse projections, neurophysiological studies reveal a multidimensional set of mnemonic signals that include stimulus familiarity, within- and between-domain associations, associative recall, and delay-based persistence. This wide range of perirhinal memory signals not only includes signals that are largely unique to the perirhinal cortex (i.e., object familiarity), consistent with dual-process theories, but also includes a range of signals (i.e., associative flexibility and recall) that are strongly associated with the hippocampus, consistent with single-process theories. These neurophysiological findings have important implications for bridging the gap between single-process and dual-process models of medial temporal lobe function.

Contents

INTRODUCTION

In primates, the perirhinal cortex lies on the ventral-medial surface of the temporal lobe surrounding the amygdala and anterior hippocampus (**Figure 1**). Brodmann was one of the first to describe the unique cytoarchitectonic features of the perirhinal cortex more than 100 years ago (Brodmann 1909). However, only relatively recently have we started to appreciate the full extent of the contributions of the perirhinal cortex to a range of higher cognitive functions. Findings from a combination of behavioral lesion studies (Murray et al. 1989, Zola-Morgan et al. 1989, Suzuki et al. 1993) and tract-tracing studies (Suzuki & Amaral 1994a,b; Suzuki & Amaral 2003) in nonhuman primates first identified the perirhinal cortex as contributing importantly to visual recognition memory. Since then, extensive work in both humans (Ranganath & Ritchey 2012) and animal model systems (Brown et al. 2010) has focused on specifying the mnemonic functions of the perirhinal cortex with a major emphasis on its contributions to recognition memory.

Recognition memory is thought to comprise two major components: recollection and familiarity. Recollection is defined as memory of the specific contextual details of a particular event or episode, whereas familiarity is defined as awareness of an item having been presented previously without access to additional contextual information about the event. Early physiology studies in monkeys showed that perirhinal cells respond selectively to particular visual stimuli and signal the prior occurrence of that preferred visual stimulus with a decreased response upon stimulus repetition (Brown & Aggleton 2001, Brown et al. 2010, Aggleton et al. 2012). This prominent so-called familiarity signal indicates that the perirhinal cortex is likely involved in a familiarity process as suggested by dual-process models of medial temporal lobe (MTL) function (Davachi 2006, Diana et al. 2007, Eichenbaum et al. 2007, Mayes et al. 2007). These models suggest that familiarity and recollection are distinct forms of memory that are supported by different brain areas; the perirhinal cortex is essential for object familiarity, and the hippocampus is important for recollection that includes contextual and spatial associative memories. Perirhinal-based memories are often conceptualized as important for encoding individual items or objects in memory, whereas the hippocampus is important for associating those items in memory (Murray & Richmond 2001, Davachi 2006, Staresina & Davachi 2008).

Another influential view suggests that the perirhinal cortex contributes to a wide range of functions, including both familiarity and recollection (Squire et al. 2007). This view argues for considerable cooperation between the different MTL structures for both recollection and familiarity (Squire et al. 2007), although the extent to which each area specifically contributes may differ. Moreover, these authors suggest that rather than trying to understand the MTL through the relatively narrow lens of recognition memory, a more fruitful strategy is to focus on specific

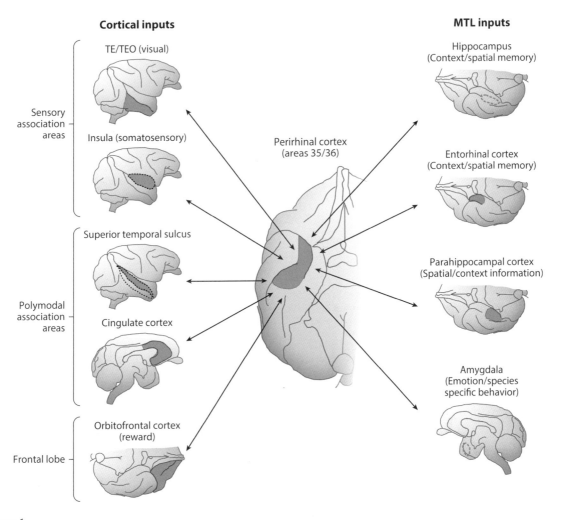

Cortical inputs

Sensory association areas

TE/TEO (visual)

Insula (somatosensory)

Polymodal association areas

Superior temporal sulcus

Cingulate cortex

Frontal lobe

Orbitofrontal cortex (reward)

Perirhinal cortex (areas 35/36)

MTL inputs

Hippocampus (Context/spatial memory)

Entorhinal cortex (Context/spatial memory)

Parahippocampal cortex (Spatial/context information)

Amygdala (Emotion/species specific behavior)

Figure 1

Schematic illustration of the cortical and medial temporal lobe (MTL) connections of the monkey perirhinal cortex (areas 35 and 36).

stimulus attributes of memory and how they are represented in particular MTL areas. We follow this latter approach in our current analysis of the functions of the perirhinal cortex.

In this article, we offer a new view of perirhinal function consistent with its full complement of anatomical projections and neurophysiological response properties. We argue that its prominent and convergent projections from sensory, polymodal, and reward-related cortical areas together with its prominent interconnections with other MTL structures suggest that the perirhinal cortex is designed to associate information in memory across multiple domains and modalities. Consistent with this region's anatomical connections, neurophysiological studies have described not only its prominent visual familiarity responses, but also various associative learning, memory, and recall signals as well as demonstrating striking persistent delay activity across a range of different domains (**Table 1**). We discuss this new view of perirhinal functions in the context of dual-process and single-process models of MTL function. We also address the evidence supporting the role of the perirhinal cortex in visual perception.

Table 1 The memory functions of the perirhinal cortex

Mnemonic category	Neural signal	Behavioral role
Familiarity	Repetitive suppression	Recognition memory[a]
Association		
Within domain (item and item)		
Coding	Un-unitized pair-coding cells (area 36), unitized pair-coding cells (area 35)	Semantic linkage, concepts, higher-order perception[b]
Recall	Pair-recall cells	Flexible associative recall[c]
Between domains		
Item and place, time, or context	Changing cells, object-time cells, object-context signals	Relational memory[d]
Item, configuration, and reward	Object-reward cells, reward condition cells	Expectation of reward[e]
Maintenance over trace interval	Persistent activity for single or multiple items	Conditioning memory, higher-order perception[f]
Information gating	Selective facilitation of signal transfer	Emotional enhancement of memory[g]

[a]Xiang & Brown 1998, Diana et al. 2007, Eichenbaum et al. 2007.
[b]Fujimichi et al. 2010, Bussey & Saksida 2007, Naya et al. 2003a.
[c]Sakai & Miyashita 1991, Naya et al. 2001, Hirabayashi et al. 2013a.
[d]Yanike et al. 2009, Naya & Suzuki 2011, Watson et al. 2012.
[e]Liu & Richmond 2000, Mogami & Tanaka 2006, Ohyama et al. 2012.
[f]Kent & Brown 2012, Naya et al. 2003b, Bussey & Saksida 2007.
[g]Paz et al. 2006, Fernandez & Tendolkar 2006.

THE ANATOMY OF THE PERIRHINAL CORTEX

The perirhinal cortex is composed of two major subdivisions, a smaller and medially situated area 35 and a much larger, laterally situated area 36. Early anatomical studies from the 1970s first identified the perirhinal cortex in monkeys as an area that received inputs from multiple sensory modalities, although the full extent of its connections was not known (Jones & Powell 1970, Van Hoesen & Pandya 1975). Since those early reports, substantial progress has been made in understanding the major cortical and subcortical projections of the perirhinal cortex in monkeys (Suzuki & Amaral 1994a, Lavenex et al. 2004). This region is defined by two major categories of inputs. First, it has major connections with other adjacent regions of the MTL, including, in order of descending strength, the entorhinal cortex, the parahippocampal cortex, the amygdala, and the hippocampus. Second, the perirhinal cortex is defined by its prominent interconnections with a wide range of unimodal and polymodal cortical association areas. The anatomical connections of the perirhinal cortex in rodents are similar though not identical and are not discussed in detail here (Burwell & Amaral 1998, Suzuki 2009a).

Medial Temporal Lobe Connections

One of the strongest MTL connections of the monkey perirhinal cortex is with the entorhinal cortex. The perirhinal projections terminate in a large extent of the entorhinal cortex that includes a wide medial-lateral portion of the anterior entorhinal cortex and more restricted projections to the lateral portions of mid and caudal levels of the entorhinal cortex (Insausti et al. 1987, Suzuki & Amaral 1994b, Mohedano-Moriano et al. 2008). These inputs follow a feedforward projection pattern terminating mainly in superficial layers I–III as well as in layer V (Suzuki &

Amaral 1994b, Mohedano-Moriano et al. 2008). Projections from the entorhinal cortex back to the perirhinal cortex are also prominent and exhibit a medial-lateral topography such that the most lateral portions of the anterior entorhinal cortex project to the most medial portions of the perirhinal cortex (i.e., area 35 and medial portions of area 36), whereas the medial portions of the anterior entorhinal cortex project to the lateral perirhinal cortex (Suzuki & Amaral 1994b, Muñoz & Insausti 2005). These projections follow a classic feedback projection pattern originating mainly from layer V, with weaker involvement of layers VI and III, and terminating in and around perirhinal layer II as well as in layers V and VI (Suzuki & Amaral 1994b, Muñoz & Insausti 2005).

A second major MTL input to the perirhinal cortex comes from the posteriorly adjacent parahippocampal cortex (areas TH and TF). The parahippocampal cortex is an area of polymodal convergence receiving input from ventral stream visual areas TEO and TE and visuospatial input from retrosplenial cortex and posterior parietal cortex, as well as input from auditory association areas (Suzuki & Amaral 1994a). Unlike the rest of the MTL projections to the perirhinal cortex, the projections from the parahippocampal cortex are not strongly reciprocal; the perirhinal cortex receives more prominent projections from the parahippocampal cortex than it sends back (Suzuki & Amaral 1994a, Lavenex et al. 2004).

One major set of MTL projections to the perirhinal cortex that tends to be overlooked in considering the functions of the perirhinal cortex is its prominent interconnections with the amygdala (Stefanacci et al. 1996). These projections exhibit a rostrocaudal gradient whereby the most anterior and ventral regions of the perirhinal cortex receive the strongest amygdala projections and the caudal perirhinal cortex has only weak interconnections with the amygdala. These projections are largely reciprocal and are directed primarily to the lateral and basal nuclei of the amygdala.

The perirhinal cortex also has direct interconnections with the hippocampus and subicular complex. However, the absolute strength of these latter interconnections (estimated using number of labeled cells from retrograde tracer studies) is substantially weaker than its connections with the entorhinal cortex, the parahippocampal cortex, or the amygdala. For nearly its entire rostrocaudal extent, cells in the most distal portions of area CA1 and the most proximal portions of the subiculum project to the perirhinal cortex (Yukie 2000, Insausti & Muñoz 2001). Insausti & Muñoz (2001) have described a weak projection from the presubiculum to the perirhinal cortex. Both these projections are reciprocal (Suzuki & Amaral 1990, Yukie 2000); they provide a direct route by which the perirhinal cortex can influence and be influenced by hippocampal/subicular activity in addition to the prominent indirect connections with these regions provided by the entorhinal cortex.

Cortical Connections

In addition to its major connections with structures within the MTL, the perirhinal cortex is also characterized by its prominent interconnections with a range of both unimodal and polymodal association areas (Suzuki & Amaral 1994a, Lavenex et al. 2002). The strongest cortical input to the perirhinal cortex in monkeys comes from the adjacent unimodal visual areas TE and TEO. Other projections to the perirhinal cortex include input from polymodal regions on the dorsal bank of the superior temporal sulcus, somatosensory input from the insular cortex, and frontal lobe inputs originating from lateral and orbital areas 11, 12, and 13. These orbitofrontal areas have been implicated in reward-related processing (de Araujo et al. 2003, Padoa-Schioppa & Assad 2006, Chaudhry et al. 2009). The perirhinal cortex tends to project back to a much wider extent of the cortex than it receives input from, including some areas that do not project to it at all (Lavenex et al. 2002). This striking asymmetric reciprocity may be related to its prominent role in associative memory recall (see Within-Domain Associations, below).

THE FUNCTIONAL ORGANIZATION OF THE PERIRHINAL CORTEX

Memory

Many previous physiology studies in both monkeys and rodents show that neurons in the perirhinal cortex respond selectively to particular visual objects and demonstrate that this stimulus-selective response is suppressed with repetition (Xiang & Brown 1998, Aggleton et al. 2012) (**Table 1**). This familiarity signal has been described in rodents, monkeys, and humans and reviewed extensively in the literature (Brown et al. 2010, Aggleton et al. 2012). We do not expand on these well-described familiarity signals. Instead, we highlight perirhinal mnemonic signals beyond simple visual object familiarity. In particular, we focus on its role in within-domain (i.e., item-item) and between-domain associations as well as on the growing evidence for its contribution to aspects of conditioning via its interconnections with the amygdala (**Table 1**).

Within-domain associations. A major tenet of dual-process theories is the idea that a unique function of the hippocampus is its ability to encode the relationship between items and events that occur together through flexibly accessible associative links (Eichenbaum & Cohen 2001, Staresina & Davachi 2009). In particular, the hippocampus has the ability to recall the associative link when presented with one item of a learned pair (Polyn et al. 2005, Foster & Wilson 2006, Kuhl et al. 2011). Although extensive evidence supports the important contributions of the hippocampus to associative, relational, and recall functions (Davachi 2006, Eichenbaum et al. 2007), strong neurophysiological evidence also indicates that perirhinal neurons signal both flexible associative memory and recall.

Associative learning and memory signals in the perirhinal cortex have typically been studied using a visual-visual pair-association (VPA) task (Sakai & Miyashita 1991; Erickson & Desimone 1999; Messinger et al. 2001; Naya et al. 2003a,b). In this task, a visual object is presented as a cue stimulus, and after a delay period, a monkey is required to choose a unique, geometrically unrelated visual object that has been paired in memory with the cue stimulus. The perirhinal cortex contains many pair-coding neurons (33% of stimulus-selective neurons) that show significantly ($p < 0.01$) correlated responses to the two visual stimuli in a particular learned pair (Naya et al. 2003a). These striking neurophysiological results are consistent with findings from lesion studies showing that bilateral damage to the perirhinal cortex results in impairment of visual-visual associative memory (Murray et al. 1993, Buckley & Gaffan 1998). Relative to the perirhinal cortex, substantially fewer (4.9%) pair-coding neurons were observed in visual area TE (Naya et al. 2003a). We do not know whether pair-coding perirhinal neurons are generated by converging inputs from individual TE neurons that code the individual stimuli that constitute the paired associate (selective-convergence model) or if they are generated by direct inputs from individual (though sparse) pair-coding neurons in TE (direct-driven model) (Naya et al. 2003a, Hirabayashi et al. 2013b). In either case, the interactions between the perirhinal cortex and area TE are important for item-item association memory (Yoshida et al. 2003). A recent study suggests that pair-coding neurons in area TE are generated by their intrinsic microcircuits (Hirabayashi et al. 2013b). Results from lesion studies in monkeys further suggest that these microcircuits in area TE are influenced by feedback projections from the perirhinal and entorhinal cortices (Higuchi & Miyashita 1996).

Dual-process models allow for an associative function of the perirhinal cortex but suggest that it provides a less flexible unitized kind of associative signal (Davachi 2006, Diana et al. 2007). Although the idea of a unitized representation in the perirhinal cortex is consistent with the description of pair-coding neurons as having a correlated (i.e., similar or unitized) neural response to the two visual stimuli that have been associated in memory (Sakai & Miyashita 1991, Naya et al. 2003a), a more detailed examination of the neurophysiological data suggests a more complex reality. First, despite the fact that perirhinal pair-coding signals show correlated responses to particular learned paired associates, neurons in areas 35 and 36 nonetheless maintain differential response amplitudes to the two different stimuli in the associated pair during the cue presentation period (Fujimichi et al. 2010). This finding suggests that the responses of the perirhinal neurons to the visual stimuli themselves, although significantly correlated, remain distinct and ununitized. However, the story differs for activity during the delay period immediately following the cue presentation. During the delay period of the task, neurons in area 35 (the small subdivision of the perirhinal cortex) represent paired items indiscriminately (i.e., in a unitized fashion), whereas neurons in area 36 (the larger perirhinal subdivision) continue to maintain distinct representations of the associated stimuli. Thus while the entire perirhinal cortex maintains distinct representation of the paired stimuli during the cue presentation period and the largest subdivision (area 36) also maintains this differential response during the delay period, the smaller perirhinal subdivision (area 35) exhibits a clear unitized representation only during the delay period.

Another striking example of associative flexibility in the perirhinal cortex is a unique activity pattern seen during the delay interval of the VPA task. During this period, pair-recall neurons signal the identity of the to-be-remembered paired associate (Sakai & Miyashita 1991; Naya et al. 1996, 2003b), and this neural response represents one of the most striking examples of a long-term associative recall signal ever reported. A subsequent study showed that this recall signal appeared first in area 36 and was propagated backward to area TE, where it was observed at longer latencies (Naya et al. 2001). This physiological finding is consistent with the extensive back projections from area 36 to TE (see The Anatomy of the Perirhinal Cortex, above). Thus, the perirhinal cortex not only signals a long-term representation of the learned paired associates during the cue period of the task (pair-coding neurons), but also provides a highly flexible recollection signal of the sought target during the delay period immediately preceding the choice period of the task (pair-recall neurons).

An important question concerns how the memory retrieval signal develops in the perirhinal cortex. One study showed that area 36 pair-recall signals start as early as the cue period of the VPA task (Naya et al. 2001). However, this study did not directly examine the relationship between these early-responding pair-recall neurons and the pair-coding neurons during the cue period of the task (Naya & Suzuki 2010). One recent study using simultaneous multiunits recording in area 36 used causality analysis to examine the circuits involved in recall during the delay period of the VPA task (Hirabayashi et al. 2013a). This study focused on the relationship between the pair-recall neurons and another category of neurons in the perirhinal cortex known as cue-holding neurons. Cue-holding neurons maintain a representation of a preferred cue stimulus after stimulus offset during the subsequent delay period, which can be considered a form of persistence memory (Naya et al. 2003b). Granger causality analyses suggested a direct coupling from the cue-holding neurons to the pair-recall neurons during the delay period of the VPA task. This result implies that the retrieval of item-item association memory during the delay period may take place through a signal cascade from the cue-holding neurons to the pair-recall neurons. Thus, the perirhinal cortex is deeply involved in the recall/recollection process, and these data are inconsistent with many versions of dual-process models that clearly place recall/recollection functions exclusively in the domain of the hippocampus (Diana et al. 2007, Eichenbaum et al. 2007). It is important to note that no parallel studies of monkey physiology during the VPA task have compared the responses between the perirhinal cortex and hippocampus or examined the interactions between these regions. These kinds of studies will be important to determine, for example, if these regions provide a complementary or a distinct recall/recollection signal during VPA tasks. Other studies described below, however, have directly compared hippocampal and perirhinal neural responses on the same associative learning or memory task.

Between-domain associations. One dual-process model suggests that the perirhinal cortex conveys information for unitized associations and for nonunitized within-domain associations (Mayes et al. 2007). However, this assumption is inconsistent with its convergent projections not only from visual association areas (area TE), but also from areas involved in signaling contextual (hippocampal formation/parahippocampal cortex), spatial (hippocampal formation/parahippocampal cortex), and reward (orbitofrontal cortex) information. For example, perirhinal neural responses have been studied during a conditional motor associative learning task in which animals were required to associate a complex visual scene with a particular rewarded target location (Wirth et al. 2003). This task is also referred to as a location-scene association task and is sensitive to damage to the hippocampus (Gaffan 1992, Murray & Wise 1996) and the perirhinal cortex (Murray et al. 1998). In this task, animals are presented with a complex visual scene with four identical target locations superimposed on it. After a delay interval during which the scene disappears but the four target locations remain visible on the screen, the animal is cued to make an eye movement to a target location, only one of which is rewarded for any given visual scene. With trial and error, animals learn between 2 and 4 novel location-scene associations each session. Yanike et al. (2009) described perirhinal "changing cells" that either came to respond selectively or lost their selective response. This changing neural activity was strongly correlated with the animal's behavioral learning curve for that particular association. Direct and detailed comparison of the timing of the changing cells relative to learning in the hippocampus and the perirhinal cortex suggested that the timing of the changing cells was not different between these two areas (Yanike et al. 2009). Consistent with these findings in monkeys, a parallel study in humans performing a very similar conditional motor association learning task confirmed similar patterns of changing activity in the hippocampus and perirhinal cortex correlated with behavioral performance (Law et al. 2005). These findings show that the perirhinal cortex can signal conjunctions of complex visual stimuli and learned target locations, and they suggest that for new conditional motor associative learning the perirhinal cortex and hippocampus act in parallel to signal the formation of new associations in memory in both monkeys and humans.

Perirhinal and hippocampal responses have also been characterized during the performance of a temporal order memory task. During an encoding phase, animals were shown a sequence of two unique visual items. During a subsequent retrieval phase, these same two items and one distracter stimulus were shown in pseudorandom locations on a video monitor, and the animals were required to touch the first-presented item and then the second (avoiding a distractor image) to obtain a juice reward. Not only did perirhinal neurons provide a prominent visually selective response, but this response was significantly modulated by the temporal order of the stimulus presentation (Naya & Suzuki 2011). Unlike for the conditional motor association task, described above, the hippocampus did not provide a similar item-time association. Instead, a prominent incremental timing signal was seen during the delay period between two item presentations (Pastalkova et al. 2008, MacDonald et al. 2011, Naya & Suzuki 2011). These two examples taken together not only show that perirhinal neurons signal a range of between-domain associations, but also suggest that the perirhinal cortex and hippocampus can respond either very similarly or in distinct ways depending on the particular task demands (i.e., associative learning versus temporal order memory).

Another between-domain association that has been explored in the perirhinal cortex is object-reward associations (Liu et al. 2000, Liu & Richmond 2000, Mogami & Tanaka 2006, Ohyama et al. 2012). Perirhinal neurons can represent the association between particular visual stimuli and particular reward outcomes (Mogami & Tanaka 2006) or cues signaling the imminent arrival of reward (Liu et al. 2000, Liu & Richmond 2000). One recent study reported that perirhinal neurons represent reward information predicted by a configuration of two visual objects (Ohyama et al. 2012). In this task, two visual stimuli are presented sequentially such that the configuration of two cue stimuli predicted upcoming reward. During the second stimulus presentation, substantial numbers of perirhinal neurons showed activity selective to the predicted reward condition regardless of the identity of the second cue stimulus. These findings suggest that the perirhinal cortex signals item-reward associations and can provide a flexible representation for expected reward. Determining which brain region(s) interact with the perirhinal cortex to process information about reward is important. Using a crossed unilateral lesion of the orbitofrontal cortex on one side and the perirhinal and entorhinal cortices on the other, Clark et al. (2013) showed that perirhinal/entorhinal interactions with the orbitofrontal cortex, an area strongly associated with signaling the features of reward (de Araujo et al. 2003, Padoa-Schioppa & Assad 2006, Chaudhry et al. 2009), are essential for monkeys to distinguish between different reward sizes. These findings support the idea that reward information from the orbitofrontal cortex may be combined with associative and contextual information in the perirhinal and entorhinal cortices to estimate expected value.

As highlighted in the anatomy section above, one of the most prominent projections to the perirhinal cortex comes from the adjacent entorhinal and parahippocampal cortices, which both signal aspects of contextual/spatial memory (Hargreaves et al. 2005, Aminoff et al. 2007, Bar et al. 2008). Although dual-process theories typically limit contextual processing to the hippocampus and the parahippocampal cortex, a growing body of reports from fMRI studies have described object-context associations in the perirhinal cortex (Smith et al. 1981, Staresina & Davachi 2008, Staresina et al. 2011), in some cases, similar to contextual associations signaled by the hippocampus, and in other cases, distinct from hippocampal signals. One study scanned subjects as they encoded visual items associated with one of two different contexts (e.g., common versus uncommon items or curved versus straight items) (Watson et al. 2012). Perirhinal activity predicted the accuracy of the retrieval judgments about the behavioral context during the encoding period, and this finding was maintained when memory strength was equated across the items. Thus, even contextual associations, a function typically considered strictly in the domain of the hippocampus in most dual-process models is represented in the perirhinal cortex.

Perirhinal-Amygdala Interactions

A growing body of work has been exploring the functional role of the prominent interconnections between the perirhinal cortex and the amygdala. One study demonstrated simultaneous recording of the perirhinal cortex, the entorhinal cortex, and the basolateral amygdala (BLA) during a trace-conditioning task (Paz et al. 2006). They reported that early in the learning process, BLA activity was associated with increased signal transmission from the perirhinal cortex to the entorhinal cortex, and this activity was increased markedly after reward delivery. Previous studies reported strong inhibition in the projection from the perirhinal cortex to the entorhinal cortex (de Curtis & Pare 2004). In this way, the perirhinal cortex appears to serve as an active gateway of information flow from neocortical areas toward the hippocampus, which can be modulated by the emotional salience of the present situation (Fernandez & Tendolkar 2006, Paz & Pare 2013).

Recent studies show that lesions of rat perirhinal cortex also impaired a range of fear-conditioning paradigms, including trace and delay conditioning (Kent & Brown 2012). Kent & Brown (2012) suggest two ways that the perirhinal cortex may contribute to conditioning. They suggest first that the perirhinal cortex is essential for bridging the temporal gap in the trace-conditioning paradigms because of the prominent persistent activity that has been shown in perirhinal slice preparations, known as endogenous persistent firing (EPF). EPF was first described in entorhinal slice preparations (Egorov et al. 2002, Fransén et al. 2006) and is characterized by persistent neural spiking activity that continues long after the termination of the spike-eliciting current (Navaroli et al. 2012). In fact, this striking cellular phenomenon in the perirhinal cortex is consistent with many other reports of striking perirhinal delay activity. Examples include the recall and cue holding signals seen in the VPA task described above (Naya et al. 2003b), as well as the persistent delay activity seen during a delayed match-to-sample task (Schon et al. 2005), and in situations where implicit learning between temporally adjacent items takes place (Miyashita 1988, Yakovlev et al. 1998, Schapiro et al. 2012). These findings suggest that the perirhinal cortex provides a wide range of transient/persistent memory signals, some of which serve trace conditioning.

Kent & Brown (2012) argue that the perirhinal cortex, in addition to its role in transient memory for the trace interval, contributes to the performance of certain conditioning tasks because of its role in processing particular kinds of conditional stimuli (CS). They argue that the perirhinal cortex is critical when processing CS requiring the ability to treat two or more items or elements as a single entity. Examples of CS requiring an intact perirhinal cortex in rodents include ultrasonic vocalizations, spatial contexts, and discontinuous auditory "pips." Kent & Brown (2012) link this interpretation to theories that emphasize the perceptual functions of the perirhinal cortex (Murray & Wise 2012). However, as we argue below, the data used to support this theory can be interpreted in more than one way.

Perception Although the perirhinal cortex has often been considered to work predominantly for declarative memory (Squire et al. 2007, Wixted & Squire 2010), strong evidence shows that lesions to the perirhinal cortex also impair performances on tasks designed to tap perceptual functions (Buckley et al. 2001, Bussey & Saksida 2007, Graham et al. 2010; but see Shrager et al. 2006). Tasks that show impairment with perirhinal lesions typically cannot be solved on the basis of single feature comparisons. Instead, tasks that show impairment require subjects to choose an odd stimulus out of concurrently presented complex objects or to discriminate between complex objects that contain multiple overlapping elements. These findings suggest that the perirhinal cortex is important for perceptual discriminations when high feature ambiguity is present. These findings also form the basis for a theory that we refer to as the perirhinal perceptual model (PPM), which states that the perirhinal cortex is a perceptual area important for both perception and memory (Bussey et al. 2005, Murray & Wise 2012).

The question of whether the perirhinal cortex should be considered primarily a memory area or an area that participates in both perception and memory has been addressed and debated previously in the literature (Hampton 2005, Baxter 2009, Suzuki 2009b, Suzuki & Baxter 2009, Murray & Wise 2012). The new insight we bring to this debate comes from a detailed consideration of the unique neurophysiological properties of the perirhinal cortex. These unique physiological properties have direct relevance for interpreting the deficit seen on concurrent and oddity discrimination tasks following perirhinal lesions. To solve these tasks successfully, multiple complex stimulus elements must be quickly bound together and then held in memory as they are compared with the other exemplars with multiple overlapping elements. According to the PPM model, the perirhinal cortex solves this problem by encoding conjunctions between sensory features that serve perception (Bussey et al. 2005). The data from the physiological evidence reviewed above suggest that these conjunctions/associations can be encoded by the prominent within-domain (Naya et al. 2003a) and between-domain (Yanike et al. 2009, Naya & Suzuki 2011, Ohyama et al. 2012) associative functions of the perirhinal cortex. It is the broad set of associative memory functions that engage the perirhinal cortex in this particular subset of perceptual tasks that are impaired by perirhinal lesions.

Moreover, once these conjunctive stimuli are represented, they must be maintained in memory during the comparison process. Other physiological evidence described above shows that the perirhinal cortex can represent multiple items through the delay period (Naya et al. 2003b) and can signal the identity of previously presented information over both short and long timescales (Yakovlev et al. 1998, Naya et al. 2003b, Schon et al. 2004). This form of persistent memory may be helpful particularly in situations where large complex amounts of visual information must be held in mind for comparison across visual objects. Thus, we suggest that the impairment seen following perirhinal damage on certain oddity or concurrent discrimination tasks may not be due to perirhinal perceptual functions, but instead to its prominent and diverse mnemonic properties (**Table 1**). How can we definitively distinguish between these two possibilities? We suggest that it will be important to record during oddity or concurrent discrimination as well as during other clear memory-demanding tasks to determine if the perirhinal cortex provides similar mnemonic signals across both categories of tasks (Suzuki 2009b).

SUMMARY

This review shows that the perirhinal cortex is truly a multifaceted memory area that can convey information about stimulus familiarity, within- and between-domain associative learning, memory, and recall and can synergize with the amygdala to modulate information flow to the hippocampus relative to the emotional salience of the situation (**Table 1**). We argue that these mnemonic signals serve declarative memory. Moreover, we argue that the associative learning signals together with persistent delay activity present in this area may underlie the striking deficits seen on oddity and concurrent discrimination tasks, which were previously interpreted as perceptual deficits.

When comparing the diverse mnemonic functions of the perirhinal cortex with those of the hippocampus, neurophysiological studies show that depending on the task demands, investigators see both clear dissociations between the different MTL areas (i.e., temporal order memory task) consistent with dual-process models as well as striking similarities in the signals conveyed by these areas (location-scene association task) consistent with single-process models. This wide range of neurophysiological findings parallel the conflicting pattern of findings cited by the single process and dual process models and suggest that, depending on the situation, both patterns can be seen (Eichenbaum et al. 2007, Wixted & Squire 2010). To address these conflicting models in one comprehensive framework, we suggest a conditionally dynamic model (CD model) of MTL function. This model states that depending on the present task, different MTL structures can express either similar or distinct/dissociable memory signals. Specific task demands as well as the functional interactions between these areas are two key factors that will determine if these structures convey similar or distinct/dissociable signals. An important goal for future studies will be to determine both when and how the interactions between these MTL structures result in either similar or distinct patterns of activity.

DISCLOSURE STATEMENT

The authors are not aware of any affiliations, memberships, funding, or financial holdings that might be perceived as affecting the objectivity of this review.

LITERATURE CITED

Aggleton JP, Brown MW, Albasser MM. 2012. Contrasting brain activity patterns for item recognition memory and associative recognition memory: insights from immediate-early gene functional imaging. *Neuropsychologia* 50(13):3141–55

Aminoff E, Gronau N, Bar M. 2007. The parahippocampal cortex mediates spatial and nonspatial associations. *Cereb. Cortex* 17(7):1493–503

Bar M, Aminoff E, Ishai A. 2008. Famous faces activate contextual associations in the parahippocampal cortex. *Cereb. Cortex* 18(6):1233–38

Baxter MG. 2009. Involvement of medial temporal lobe structures in memory and perception. *Neuron* 61(5):667–77

Brodmann K. 1909. *Vergleichende Lokalisationslehre der Grosshirnrinde.* Leipzig: Johann Ambrosius Barth

Brown MW, Aggleton JP. 2001. Recognition memory: What are the roles of the perirhinal cortex and hippocampus? *Nat. Rev. Neurosci.* 2(1):51–61

Brown MW, Warburton EC, Aggleton JP. 2010. Recognition memory: material, processes, and substrates. *Hippocampus* 20(11):1228–44

Buckley MJ, Booth MCA, Rolls ET, Gaffan D. 2001. Selective perceptual impairments after perirhinal cortex ablation. *J. Neurosci.* 21(24):9824–36

Buckley MJ, Gaffan D. 1998. Perirhinal cortex ablation impairs visual object identification. *J. Neurosci.* 18(6):2268–75

Burwell RD, Amaral DG. 1998. Perirhinal and postrhinal cortices of the rat: interconnectivity and connections with the entorhinal cortex. *J. Comp. Neurol.* 391(3):293–321

Bussey TJ, Saksida LM. 2007. Memory, perception, and the ventral visual-perirhinal-hippocampal stream: thinking outside of the boxes. *Hippocampus* 17(9):898–908

Bussey TJ, Saksida LM, Murray EA. 2005. The perceptual-mnemonic/feature conjunction model of perirhinal cortex function. *Q. J. Exp. Psychol. B* 58(3–4):269–82

Chaudhry AM, Parkinson JA, Hinton EC, Owen AM, Roberts AC. 2009. Preference judgements involve a network of structures within frontal, cingulate and insula cortices. *Eur. J. Neurosci.* 29(5):1047–55

Clark AM, Bouret S, Young AM, Murray EA, Richmond BJ. 2013. Interaction between orbital prefrontal and rhinal cortex is required for normal estimates of expected value. *J. Neurosci.* 33(5):1833–45

Davachi L. 2006. Item, context and relational episodic encoding in humans. *Curr. Opin. Neurobiol.* 16(6):693–700

de Araujo IE, Rolls ET, Kringelbach ML, McGlone F, Phillips N. 2003. Taste-olfactory convergence, and the representation of the pleasantness of flavour, in the human brain. *Eur. J. Neurosci.* 18(7):2059–68

de Curtis M, Paré D. 2004. The rhinal cortices: a wall of inhibition between the neocortex and the hippocampus. *Prog. Neurobiol.* 74(2):101–10

Diana RA, Yonelinas AP, Ranganath C. 2007. Imaging recollection and familiarity in the medial temporal lobe: a three-component model. *Trends Cogn. Sci.* 11(9):379–86

Egorov AV, Hamam BN, Fransén E, Hasselmo ME, Alonso AA. 2002. Graded persistent activity in entorhinal cortex neurons. *Nature* 420:173–78

Eichenbaum H, Cohen NJ. 2001. *From Conditioning to Conscious Recollection.* New York: Oxford Univ. Press

Eichenbaum H, Yonelinas AR, Ranganath C. 2007. The medial temporal lobe and recognition memory. *Annu. Rev. Neurosci.* 30:123–52

Erickson CA, Desimone R. 1999. Responses of macaque perirhinal neurons during and after visual stimulus association learning. *J. Neurosci.* 19(23):10404–16

Fernandez G, Tendolkar I. 2006. The rhinal cortex: 'gatekeeper' of the declarative memory system. *Trends Cogn. Sci.* 10(8):358–62

Foster DJ, Wilson MA. 2006. Reverse replay of behavioural sequences in hippocampal place cells during the awake state. *Nature* 440(7084):680–83

Fransén E, Tahvildari B, Egorov AV, Hasselmo ME, Alonso AA. 2006. Mechanism of graded persistent cellular activity of entorhinal cortex layer V neurons. *Neuron* 49(5):735–46

Fujimichi R, Naya Y, Koyano KW, Takeda M, Takeuchi D, Miyashita Y. 2010. Unitized representation of paired objects in area 35 of the macaque perirhinal cortex. *Eur. J. Neurosci.* 32(4):659–67

Gaffan D. 1992. Amnesia for complex naturalistic scenes and for objects following fornix transection in the rhesus monkey. *Eur. J. Neurosci.* 4:381–88

Graham KS, Barense MD, Lee AC. 2010. Going beyond LTM in the MTL: a synthesis of neuropsychological and neuroimaging findings on the role of the medial temporal lobe in memory and perception. *Neuropsychologia* 48(4):831–53

Hampton RR. 2005. Monkey perirhinal cortex is critical for visual memory, but not for visual perception: reexamination of the behavioural evidence from monkeys. *Q. J. Exp. Psychol. B* 58(3–4):283–99

Hargreaves EL, Rao G, Lee I, Knierim JJ. 2005. Major dissociation between medial and lateral entorhinal input to dorsal hippocampus. *Science* 308:1792–94

Higuchi S-I, Miyashita Y. 1996. Neural code of visual paired associate memory in primate inferotemporal cortex is impaired by perirhinal and entorhinal lesions. *Proc. Natl. Acad. Sci. USA* 93:739–43

Hirabayashi T, Takeuchi D, Tamura K, Miyashita Y. 2013a. Functional microcircuit recruited during retrieval of object association memory in monkey perirhinal cortex. *Neuron* 77(1):192–203

Hirabayashi T, Takeuchi D, Tamura K, Miyashita Y. 2013b. Microcircuits for hierarchical elaboration of object coding across primate temporal areas. *Science* 341:191–95

Insausti R, Amaral DG, Cowan WM. 1987. The entorhinal cortex of the monkey: II. Cortical afferents. *J. Comp. Neurol.* 264:356–95

Insausti R, Muñoz M. 2001. Cortical projections of the non-entorhinal hippocampal formation in the cynomolgus monkey (*Macaca fascicularis*). *Eur. J. Neurosci.* 14(3):435–51

Jones EG, Powell TPS. 1970. An anatomical study of converging sensory pathways within the cerebral cortex of the monkey. *Brain* 93:793–820

Kent BA, Brown TH. 2012. Dual functions of perirhinal cortex in fear conditioning. *Hippocampus* 22(10):2068–79

Kuhl BA, Rissman J, Chun MM, Wagner AD. 2011. Fidelity of neural reactivation reveals competition between memories. *Proc. Natl. Acad. Sci. USA* 108(14):5903–8

Lavenex P, Suzuki WA, Amaral DG. 2002. Perirhinal and parahippocampal cortices of the macaque monkey: projections to the neocortex. *J. Comp. Neurol.* 447(4):394–420

Lavenex P, Suzuki WA, Amaral DG. 2004. Perirhinal and parahippocampal cortices of the macaque monkey: intrinsic projections and interconnections. *J. Comp. Neurol.* 472(3):371–94

Law JR, Flanery MA, Wirth S, Yanike M, Smith AC, et al. 2005. Functional magnetic resonance imaging activity during the gradual acquisition and expression of paired-associate memory. *J. Neurosci.* 25(24):5720–29

Liu Z, Murray EA, Richmond BJ. 2000. Learning motivational significance of visual cues for reward schedules requires rhinal cortex. *Nat. Neurosci.* 3(12):1307–15

Liu Z, Richmond BJ. 2000. Response differences in monkey TE and perirhinal cortex: stimulus association related to reward schedules. *J. Neurophysiol.* 83(3):1677–92

MacDonald CJ, Legage KQ, Eden UT, Eichenbaum H. 2011. Hippocampal "time cells" bridge the gap in memory for discontiguous events. *Neuron* 71(4):737–49

Mayes A, Montaldi D, Migo E. 2007. Associative memory and the medial temporal lobes. *Trends Cogn. Sci.* 11(3):126–35

Messinger A, Squire LR, Zola SM, Albright TD. 2001. Neuronal representations of stimulus associations develop in the temporal lobe during learning. *Proc. Natl. Acad. Sci.* 98(21):12239–44

Miyashita Y. 1988. Neuronal correlate of visual associative long-term memory in the primate temporal cortex. *Nature* 335:817–20

Mogami T, Tanaka K. 2006. Reward association affects neuronal responses to visual stimuli in macaque TE and perirhinal cortices. *J. Neurosci.* 26(25):6761–70

Mohedano-Moriano A, Martinez-Marcos A, Pro-Sistiaga P, Blaizot X, Arroyo-Jimenez MM, et al. 2008. Convergence of unimodal and polymodal sensory input to the entorhinal cortex in the fascicularis monkey. *Neuroscience* 151(1):255–71

Muñoz M, Insausti R. 2005. Cortical efferents of the entorhinal cortex and the adjacent parahippocampal region in the monkey (*Macaca fascicularis*). *Eur. J. Neurosci.* 22(6):1368–88

Murray EA, Bachevalier J, Mishkin M. 1989. Effects of rhinal cortical lesions on visual recognition memory in rhesus monkeys. *Soc. Neurosci. Abstr.* 15:342 (Abstr.)

Murray EA, Baxter MG, Gaffan D. 1998. Monkeys with rhinal cortex damage or neurotoxic hippocampal lesions are impaired on spatial scene learning and object reversals. *Behav. Neurosci.* 112:1291–303

Murray EA, Gaffan D, Mishkin M. 1993. Neural substrates of visual stimulus-stimulus association in rhesus monkeys. *J. Neurosci.* 13:4549–61

Murray EA, Richmond BJ. 2001. Role of perirhinal cortex in object perception, memory, and associations. *Curr. Opin. Neurobiol.* 11(2):188–93

Murray EA, Wise SP. 1996. Role of the hippocampus plus subjacent cortex but not amygdala in visuomotor conditional learning in rhesus monkeys. *Behav. Neurosci.* 110(6):1261–70

Murray EA, Wise SP. 2012. Why is there a special issue on perirhinal cortex in a journal called *Hippocampus*? The perirhinal cortex in historical perspective. *Hippocampus* 22(10):1941–51

Navaroli VL, Zhao Y, Boguszewski P, Brown TH. 2012. Muscarinic receptor activation enables persistent firing in pyramidal neurons from superficial layers of dorsal perirhinal cortex. *Hippocampus* 22(6):1392–404

Naya Y, Sakai K, Miyashita Y. 1996. Activity of primate inferotemporal neurons related to a sought target in pair-association task. *Proc. Natl. Acad. Sci. USA* 93(7):2664–69

Naya Y, Suzuki WA. 2010. Associative memory in the medial temporal lobe. In *Primate Neuroethology*, ed. M Platt, A Ghazanfar, pp. 337–58. Oxford, UK: Oxford Univ. Press

Naya Y, Suzuki WA. 2011. Integrating what and when across the primate medial temporal lobe. *Science* 333:773–76

Naya Y, Yoshida M, Miyashita Y. 2001. Backward spreading of memory-retrieval signal in the primate temporal cortex. *Science* 291:661–64

Naya Y, Yoshida M, Miyashita Y. 2003a. Forward processing of long-term associative memory in monkey inferotemporal cortex. *J. Neurosci.* 23(7):2861–71

Naya Y, Yoshida M, Takeda M, Fujimichi R, Miyashita Y. 2003b. Delay-period activities in two subdivisions of monkey inferotemporal cortex during pair association memory task. *Eur. J. Neurosci.* 18(10):2915–18

Ohyama K, Sugase-Miyamoto Y, Matsumoto N, Shidara M, Sato C. 2012. Stimulus-related activity during conditional associations in monkey perirhinal cortex neurons depends on upcoming reward outcome. *J. Neurosci.* 32(48):17407–19

Padoa-Schioppa C, Assad JA. 2006. Neurons in the orbitofrontal cortex encode economic value. *Nature* 441:223–26

Pastalkova E, Itskov V, Amarasingham A, Buzsáki G. 2008. Internally generated cell assembly sequences in the rat hippocampus. *Science* 321:1322–27

Paz R, Pare D. 2013. Physiological basis for emotional modulation of memory circuits by the amygdala. *Curr. Opin. Neurobiol.* 23(3):381–86

Paz R, Pelletier JG, Bauer EP, Paré D. 2006. Emotional enhancement of memory via amygdala-driven facilitation of rhinal interactions. *Nat. Neurosci.* 9(10):1321–29

Polyn SM, Natu VS, Cohen JD, Norman KA. 2005. Category-specific cortical activity precedes retrieval during memory search. *Science* 310:1963–66

Ranganath C, Ritchey M. 2012. Two cortical systems for memory-guided behaviour. *Nat. Rev. Neurosci.* 13(10):713–26

Sakai K, Miyashita Y. 1991. Neural organization for the long-term memory of paired associates. *Nature* 354:152–55

Schapiro AC, Kustner LV, Turk-Browne NB. 2012. Shaping of object representations in the human medial temporal lobe based on temporal regularities. *Curr. Biol.* 22(17):1622–27

Schon K, Atri A, Hasselmo ME, Tricarico MD, Lopresti ML, Stern CE. 2005. Scopolamine reduces persistent activity related to long-term encoding in the parahippocampal gyrus during delayed matching in humans. *J. Neurosci.* 25(40):9112–23

Schon K, Hasselmo ME, Lopresti ML, Tricarico MD, Stern CE. 2004. Persistence of parahippocampal representation in the absence of stimulus input enhances long-term encoding: a functional magnetic resonance imaging study of subsequent memory after a delayed match-to-sample task. *J. Neurosci.* 24(49):11088–97

Shrager Y, Gold JJ, Hopkins RO, Squire LR. 2006. Intact visual perception in memory-impaired patients with medial temporal lobe lesions. *J. Neurosci.* 26(8):2235–40

Smith EL III, Harwerth RS, Levi DM, Watson JT. 1981. Normal cortical responses in ocularly hypopigmented cats. *Brain Res.* 206:183–86

Squire LR, Wixted JT, Clark RE. 2007. Recognition memory and the medial temporal lobe: a new perspective. *Nat. Rev. Neurosci.* 8:872–83

Staresina BP, Davachi L. 2008. Selective and shared contributions of the hippocampus and perirhinal cortex to episodic item and associative encoding. *J. Cogn. Neurosci.* 20(8):1478–89

Staresina BP, Davachi L. 2009. Mind the gap: binding experiences across space and time in the human hippocampus. *Neuron* 63(2):267–76

Staresina BP, Duncan KD, Davachi L. 2011. Perirhinal and parahippocampal cortices differentially contribute to later recollection of object- and scene-related event details. *J. Neurosci.* 31(24):8739–47

Stefanacci L, Suzuki WA, Amaral DG. 1996. Organization of connections between the amygdaloid complex and the perirhinal and parahippocampal cortices in macaque monkeys. *J. Comp. Neurol.* 375:552–82

Suzuki WA. 2009a. Comparative analysis of the cortical afferents, intrinsic projections and interconnections of the parahippocampal regions in monkeys and rats. In *The Cognitive Neurosciences*, ed. M Gazzaniga, pp. 659–74. Boston, MA: MIT Press

Suzuki WA. 2009b. Perception and the medial temporal lobe: evaluating the current evidence. *Neuron* 61(5):657–66

Suzuki WA, Amaral DG. 1990. Cortical inputs to the CA1 field of the monkey hippocampus originate from the perirhinal and parahippocampal cortex but not from area TE. *Neurosci. Lett.* 115:43–48

Suzuki WA, Amaral DG. 1994a. Perirhinal and parahippocampal cortices of the macaque monkey: cortical afferents. *J. Comp. Neurol.* 350:497–533

Suzuki WA, Amaral DG. 1994b. Topographic organization of the reciprocal connections between monkey entorhinal cortex and the perirhinal and parahippocampal cortices. *J. Neurosci.* 14:1856–77

Suzuki WA, Amaral DG. 2003. The perirhinal and parahippocampal cortices of the macaque monkey: cytoarchitectonic and chemoarchitectonic organization. *J. Comp. Neurol.* 463:67–91

Suzuki WA, Baxter MG. 2009. Memory, perception, and the medial temporal lobe: a synthesis of opinions. *Neuron* 61(5):678–79

Suzuki WA, Zola-Morgan S, Squire LR, Amaral DG. 1993. Lesions of the perirhinal and parahippocampal cortices in the monkey produce long-lasting memory impairment in the visual and tactual modalities. *J. Neurosci.* 13:2430–51

Van Hoesen GW, Pandya DN. 1975. Some connections of the entorhinal (area 28) and perirhinal (area 35) cortices of the rhesus monkey. I. Temporal lobe afferents. *Brain Res.* 95:1–24

Watson HC, Wilding EL, Graham KS. 2012. A role for perirhinal cortex in memory for novel object-context associations. *J. Neurosci.* 32(13):4473–81

Wirth S, Yanike M, Frank LM, Smith AC, Brown EN, Suzuki WA. 2003. Single neurons in the monkey hippocampus and learning of new associations. *Science* 300:1578–81

Wixted JT, Squire LR. 2010. The role of the human hippocampus in familiarity-based and recollection-based recognition memory. *Behav. Brain Res.* 215(2):197–208

Xiang JZ, Brown MW. 1998. Differential neuronal encoding of novelty, familiarity and recency in regions of the anterior temporal lobe. *Neuropharmacology* 37:657–76

Yakovlev V, Fusi S, Berman E, Zohary E. 1998. Inter-trial neuronal activity in inferior temporal cortex: a putative vehicle to generate long-term visual associations. *Nat. Neurosci.* 1(4):310–17

Yanike M, Wirth S, Smith AC, Brown EN, Suzuki WA. 2009. Comparison of associative learning-related signals in the macaque perirhinal cortex and hippocampus. *Cereb. Cortex* 19(5):1064–78

Yoshida M, Naya Y, Miyashita Y. 2003. Anatomical organization of forward fiber projections from area TE to perirhinal neurons representing visual long-term memory in monkeys. *Proc. Natl. Acad. Sci. USA* 100(7):4257–62

Yukie M. 2000. Connections between the medial temporal cortex and the CA1 subfield of the hippocampal formation in the Japanese monkey (*Macaca fuscata*). *J. Comp. Neurol.* 423:282–98

Zola-Morgan S, Squire LR, Amaral DG, Suzuki WA. 1989. Lesions of perirhinal and parahippocampal cortex that spare the amygdala and hippocampal formation produce severe memory impairment. *J. Neurosci.* 9:4355–70

Autophagy and Its Normal and Pathogenic States in the Brain

Ai Yamamoto[1] and Zhenyu Yue[2]

[1]Departments of Neurology, Pathology, and Cell Biology, College of Physicians and Surgeons, Columbia University, New York, NY 10032; email: ay46@cumc.columbia.edu

[2]Departments of Neurology and Neuroscience, Friedman Brain Institute, Icahn School of Medicine at Mount Sinai, New York, NY 10029-6574; email: Zhenyu.yue@mssm.edu

Annu. Rev. Neurosci. 2014. 37:55–78

First published online as a Review in Advance on April 21, 2014

The *Annual Review of Neuroscience* is online at neuro.annualreviews.org

This article's doi: 10.1146/annurev-neuro-071013-014149

Keywords

neurons, glia, macroautophagy, neurodegeneration, neurodevelopment

Abstract

Autophagy is a conserved catabolic process that delivers the cytosol and cytosolic constituents to the lysosome. Its fundamental role is to maintain cellular homeostasis and to protect cells from varying insults, including misfolded proteins and damaged organelles. Beyond these roles, the highly specialized cells of the brain have further adapted autophagic pathways to suit their distinct needs. In this review, we briefly summarize our current understanding of the different forms of autophagy and then offer a closer look at how these pathways impact neuronal and glial functions. The emerging evidence indicates that not only are autophagy pathways essential for neural health, but they have a direct impact on developmental and neurodegenerative processes. Taken together, as we unravel the complex roles autophagy pathways play, we will gain the necessary insight to modify these pathways to protect the human brain and treat neurodegenerative diseases.

Contents

INTRODUCTION

The highly specialized cells of the vertebrate brain depend on catabolic processes not only to eliminate waste, but also to protect against variations in nutrient availability, to promote cellular remodeling, and to defend against invading pathogens. Degradation is achieved primarily by two pathways, the ubiquitin-proteasome system (UPS) and the autophagy–lysosome system. Whereas the proteasome is responsible for most protein turnover (Ciechanover et al. 2000), the lysosome is responsible for recycling not only proteins but also lipids, nucleic acids, and polysaccharides. Identified by de Duve in the 1950s, the lysosome is a membrane-bound organelle that contains a diverse array of hydrolytic enzymes that drive catabolysis under an acidic pH. The delivery of substrates into the lysosomal lumen is achieved by two trafficking pathways: Endocytosis transports extracellular, transmembrane, and membrane-bound constituents; and autophagy transports cytosolic substrates using three distinct pathways, i.e., microautophagy, CMA (chaperone-mediated autophagy), and macroautophagy.

Lysosome-mediated degradation is essential for neural health, and the appreciation of autophagic pathways in the healthy and diseased brain continues to grow. Emerging evidence indicates that the highly specialized neurons and glia not only maintain the basal function of autophagy that is shared across all cells, but also have adapted these processes to suit their specific needs. In this review, we provide a brief overview of our current understanding of the three forms of autophagy and discuss how these pathways impact neurons and glia of the developing and adult brain.

MICROAUTOPHAGY AND CMA

The three major forms of autophagy are distinguished by how substrates are trafficked to the lysosome. Among the three, microautophagy and CMA deliver substrates directly into the lysosomal lumen. Although there is still much to be learned regarding the mechanistic understanding of these pathways, growing evidence indicates that direct substrate delivery to the lysosome, especially via CMA, can have a profound impact on neural function.

Microautophagy

In microautophagy, substrates and the surrounding cytosol are engulfed by the lysosomal membrane. It is the least understood autophagic pathway, although its original description arises from the pioneering ultrastructural studies of starved hepatocytes (reviewed in Mijaljica et al. 2011). There is currently little mechanistic understanding of microautophagy, which gives rise to the suggestion that it occurs only in yeast. Recently, however, a microautophagy-like process was reported to occur in mammalian cells during the formation of multivesicular bodies (MVBs) (Sahu et al. 2011). Substrates are recognized and targeted to late endosomes in an hsc70 (heat shock cognate 70)-dependent manner, a component also required for CMA.

Chaperone-Mediated Autophagy

CMA relies on the directed import of individual proteins into the lysosomal lumen by a concerted effort of chaperone proteins and the single spanning lysosomal membrane protein LAMP2A (lysosome-associated membrane protein 2A) (**Figure 1**) (Kaushik & Cuervo 2012). Proteins targeted for CMA are recognized by a pentapeptide motif similar to KFERQ, which permits its interaction with the cytosolic chaperone, hsc70. Hsc70 traffics the cargo to the lysosome surface, and the subsequent interaction with LAMP2A imports the substrate into the lumen with the help of a lysosome-associated hsc70. CMA has been studied exclusively in mammals, and the lack of conservation in earlier organisms has slowed our mechanistic understanding of this pathway (Bejarano & Cuervo 2010).

First identified as a selective form of lysosomal uptake in the early 1980s (Backer et al. 1983), CMA was shown by Dice and colleagues (1990) to be a selective protein degradative process in response to serum withdrawal. In further studies using paradigms of prolonged starvation, Dice, Knecht, and others found that whereas the degradation of most proteins diminished over time, proteolysis of CMA substrates increased (Cuervo et al. 1995). This finding led to the hypothesis that during periods of starvation, the initial, bulk catabolism is driven by macroautophagy, which is then followed by the more selective degradation of CMA. Consistent with this hypothesis, cross talk between macroautophagy and CMA has been reported, but the underlying mechanism remains unclear (Kaushik et al. 2008). Several disease-related proteins including α-synuclein (Cuervo et al. 2004) have recently been identified as CMA substrates, suggesting that CMA may also play a housekeeping role in neurons (Vogiatzi et al. 2008).

Although deletion of the LAMP2A isoform has not been reported, a primary defect affecting all LAMP2 isoforms leads to Danon disease, a rare X-linked dominant disorder characterized by myopathy and cardiomyopathy (Danon et al. 1981, Nishino et al. 2000). Typically, male patients also present with mild intellectual disability, and although no structural brain abnormalities have been reported (Sugie et al. 2002), brain development still may be affected. Whether a loss of LAMP2 affects the adult brain remains unknown (Eskelinen et al. 2003, Tanaka et al. 2000).

MACROAUTOPHAGY

Macroautophagy first requires the synthesis of a multilamellar vesicle called the autophagosome (AP), which fuses into the endolysosomal system to deliver its cargo (**Figure 1**). In the brain, macroautophagy influences developmental processes as well as degeneration and death; despite

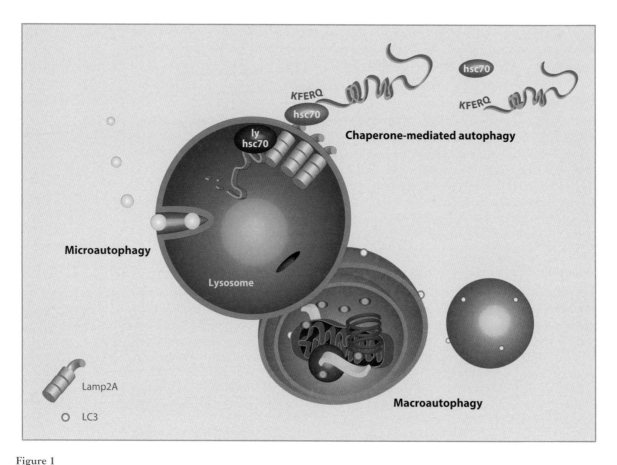

Figure 1

The three forms of autophagy. An illustrated summary of the three forms of autophagy—microautophagy, chaperone-mediated autophagy, and macroautophagy—and their respective interactions with the lysosome.

this importance, however, many fundamental questions about neural macroautophagy still remain unanswered. It may therefore be surprising that neurons were one of the first cell types used in the identification and ultrastructural characterization of this pathway (Dixon 1967, Holtzman & Novikoff 1965). Nonetheless, the key biochemical contributions that followed focused on the liver (reviewed in Yang & Klionsky 2010), driving most of our physiological understanding of macroautophagy today.

In the mid-1990s, Ohsumi and colleagues uncovered that macroautophagy is conserved in yeast, propelling our mechanistic insight into this pathway (Baba et al. 1994, Takeshige et al. 1992). The molecular machinery is currently composed of more than 30 autophagy-related (ATG) genes in yeast, 18 of which have mammalian homologs. Although we briefly describe the molecular underpinnings of macroautophagy here, further details can be found elsewhere (Johnson et al. 2012, Mizushima et al. 2011).

Autophagosome Formation

The primary process in macroautophagy centers on the synthesis of the AP, which is broken down into three steps: induction and nucleation, expansion, and maturation (**Figure 2**). The

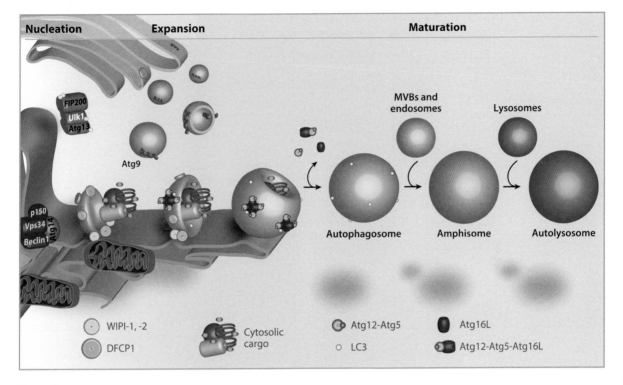

Figure 2

Macroautophagic degradation. An illustration depicting the three steps of autophagosome (AP) formation, nucleation, expansion and maturation. The formation of the AP relies on the concerted effort of the core machinery. The phosphorylation status of the Ulk1 complex attracts different ATG proteins to the formation of the IM. The Beclin1-Atg14L-Vps34 complex enriches the IM and surrounding membrane with PI3P which attracts effector proteins such as WIPI-1 and -2. DFCP1 also cycles from the Golgi to the PI3P-rich IM. As the membrane elongates around the cargo, the tetrameric Atg12-Atg5-Atg16L complex promotes the lipidation of LC3 on PE of the AP growing membrane. Upon completion of the AP, only LC3 remains. During maturation, the AP fuses into the endolysosomal system to form amphisomes and autolysosomes, upon which the contents and inner AP membrane is degraded. Abbreviation: MVB, multivesicular bodies.

ATG proteins contribute toward each of these steps, and overall they can be categorized into six subgroups: the Ulk1 protein kinase complex, the Beclin1-Atg14L-Vps34 lipid kinase complex, the PI3P (phosphatidylinositol-3-monophosphate) binding proteins, Atg9, and the two ubiquitin-like conjugation systems (**Table 1**).

The induction of macroautophagy is regulated through the phosphorylation status of the Ulk1 complex, which drives nucleation of the isolation membrane (IM) (or phagophore). Next, the Beclin1-Atg14L-Vps34 kinase enriches the site with PI3P to recruit the PI3P binding proteins such as WIPI-1 and -2'a DFCP1. At present, the site at which the IM is formed is heavily debated (Axe et al. 2008, Dunn 1990, Ge et al. 2013, Hailey et al. 2010, Hamasaki et al. 2013, Hayashi-Nishino et al. 2009, Ylä-Anttila et al. 2009); it is still conceivable, however, that the membrane origin may differ depending on the type of macroautophagy that is induced or the cell type that is being interrogated.

After the IM forms, the membrane expands and envelopes the cargo. This process requires the final three subgroups: Atg9 and the two ubiquitin-like conjugation systems (Mizushima et al. 2011). How Atg9 promotes elongation is unclear, but as a membrane-spanning protein that shuttles across different vesicles, its significance points toward different membrane sources contributing toward

Table 1 The mammalian ATG gene products and function

Mammalian homologs	Yeast homologs	Functions
Ulk 1, Ulk 2	Atg1	AP initiation
Atg13	Atg13	Member of the Ulk1 complex
FIP200	Atg17	Recruits proteins such as Atg9 and Atg16L to IM
Atg101	None	
Beclin 1	Atg6	AP initiation and elongation
Atg 14/Barkor	Atg14	Member and regulator of the Atg14L-Vps34 kinase complex
WIPI-1, -2	Atg18	AP initiation, elongation, and closure
DFCP1	None	PI3P binding proteins at the IM
Atg2a, 2b	Atg2	
Atg9	Atg9	AP membrane elongation
Atg7	Atg7	E1-like ligase for Atg5 and Atg8 homologs
Atg10	Atg10	E2-like ligase for Atg5
Atg5	Atg5	Acceptor molecule for Ub-like conjugation of Atg12; creates the Atg12-Atg5-Atg16L E3-like ligase for Atg8 conjugation to PE
Atg12	Atg12	Ub-like molecular conjugated to Atg5
Atg16L1, L2	Atg16	Atg12-Atg5-Atg16L E3-like ligase; drives tetramer formation
Atg3	Atg3	E2-like ubiquitin ligase for Atg8 homologs
MAP1LC3A, B,C, GR, GRL1, GRL2	Atg8	Marker (especially LC3B) of the AP membrane; conjugated to PE in a Ub-like manner
Atg4a, b, c	Atg4	Cleave proform of Atg8 homologs to active form; recycles Atg8 from membrane

Abbreviations: AP, autophagosome; IM, isolation membrane; PE, phosphatidylethanolamine; PI3P, phosphatidylinositol-3-monophosphate; Ub, ubiquitin.

AP membrane expansion. The two ubiquitin-like conjugation systems elongate the growing AP membrane by ultimately driving a key reaction: the lipidation of Atg8 and its homologs to the lipid PE (phosphatidylethanolamine) within the growing AP membrane. This result is achieved by two distinct reactions: The first creates an E3-like ligase known as the Atg12-Atg5-Atg16L complex, and the second converts cytosolic Atg8 to its lipidated form (Nishimura et al. 2013). As the membrane closes to form the AP, the Atg12-Atg5-Atg16L leaves the outer membrane, leaving behind an AP membrane labeled with Atg8. The reported presence of Atg16L on vesicles originating from the plasma membrane suggests another membrane source for elongation, but how Atg16L associates with the membrane is uncertain (Ravikumar et al. 2010). Moreover, the lipidation of Atg8 may occur only on membranes from the endoplasmic reticulum (ER)–Golgi intermediate compartment (Ge et al. 2013). It is unclear, however, how these different membrane sources are interrelated if at all.

The Atg8 homolog LC3 (microtubule associated protein 1 light chain 3) is the only protein marker of the AP. Mammalian cells have multiple Atg8 homologs, all of which can potentially label the AP by lipidation, and they are divided into two families. The LC3 family has three isoforms, A, B and C, of which LC3B was the first identified (Kabeya et al. 2000) and LC3C is present only in humans. The GR [GABA(A) receptor associated proteins] family has three members, which include GR, GRL1/GEC1 (GABARAP-like protein 1), and GRL2/GATE-16. Amino acid alignments to yeast Atg8 reveal that it is more similar to GRs than to LC3. Indeed, GRL1 is also present in APs (Chakrama et al. 2010). LC3B, LC3C, and all GR proteins are also highly expressed in the brain. Nonetheless why a large number of mammalian homologs exist is unknown. Weidberg et al. (2010) proposed a functional distinction between the LC3 and GR families in AP formation, but the rationale for this added level of complexity is unclear. As an aside, GR proteins, as their name implies, were initially identified to interact with GABA(A) receptors and to regulate their trafficking in neurons (Leil et al. 2004, Wang et al. 1999). Whether these observations are indicative of a dichotomous function of GR or whether there is a yet-unexplored intersection between neuronal macroautophagy and cell surface protein availability needs to be resolved.

The AP next delivers its cargo to the lysosome, and degradation of the substrates as well as the inner AP membrane ensues. Studies indicate that mammalian APs fuse into the endolysosomal system to become first an amphisome and then an autolysosome [collectively known as autophagic vacuoles (AVs)]. Amphisome formation depends on proteins implicated in the biogenesis of MVBs, a specialized late endosome that sorts endocytic proteins for lysosomal degradation (Rusten et al. 2008). Mutations in ESCRT proteins such as Chmp2b and the AAA+ protein p92/VCP, which are associated with frontotemporal dementias (Lee & Gao 2008, Watts et al. 2004), are characterized by defects in macroautophagic degradation (reviewed in Yamamoto & Simonsen 2011). The dependence on MVBs has promoted the hypothesis that amphisome formation may increase the efficiency of lysosome-mediated degradation. For neurons, amphisome formation may be especially valid because APs must travel long distances before degradation can occur (Hollenbeck 1993, Larsen et al. 2002, Maday et al. 2012).

The fusion of AP membranes raises a vexing question that persists in the field: How are membrane closure and fusion achieved? As have many membrane trafficking events, numerous studies implicate SNARE proteins for both homotypic membrane fusion of the growing AP (Moreau et al. 2011, Nair et al. 2011, Renna et al. 2011) and heterotypic membrane fusion to the lysosome (Hamasaki et al. 2013, Itakura et al. 2012, Pryor et al. 2004, Takáts et al. 2013). If SNAREs are involved, how are they incorporated by the IM or outer AP membrane? Early freeze fracture electron microscopy indicates that the outer AP membrane is smooth and devoid of transmembrane proteins (Fengsrud et al. 2000, Punnonen et al. 1989), an observation that may not support the presence of SNAREs.

Selective Autophagy

Although macroautophagy can lead to bulk degradation of substrates in response to starvation, adaptor proteins known as autophagic receptors and selectivity adaptors can promote selective degradation. Adaptor proteins identified thus far include p62/SQSTM1 (Sequestesome 1), NBR1, Nix, NDP52, Alfy/WDFY3, and OPTN (Optineurin) (reviewed in Mijaljica et al. 2011). These adaptors have been implicated in the selective elimination of ubiquitinated and nonubiquitinated proteins, various organelles such as mitochondria, and invading pathogens.

Selective autophagy in mammals may be related to a pathway in yeast known as Cvt (cytoplasm to vacuole targeting) (Umekawa & Klionsky 2012), which depends on the core AP building machinery, as well as specific adaptor proteins. The Cvt adaptor proteins, Atg11 and Atg19, are required to import vacuolar hydrolases such as aminopeptidase I (ApeI) to the vacuole. ApeI oligomerizes into a large aggregate, which is then recognized sequentially by Atg11 and Atg19 to recruit the AP machinery. Although the building machinery is shared for all forms of macroautophagy, ultrastructural studies reveal that the resulting APs are distinct; for bulk degradation, APs fill with cytoplasm and cytosolic material, whereas for Cvt, APs are filled with only aggregated ApeI (Baba et al. 1997). The tight apposition of the membrane to the cargo is attributed to the direct interaction of Atg11 with Atg8. In mammalian systems, the autophagic receptors directly interact with the Atg8 homologs through an LIR or LRS (LC3 interacting region or recognition sequence) (Birgisdottir et al. 2013). These sequences can differentiate between the LC3 and GR family members, suggesting that the different Atg8 homologs may confer added specificity.

As in Cvt, the priming step in selective autophagy in mammals may be the aggregation of cargo. Whereas the p62-LC3 interaction is the focus of much of its function, p62's alternative name, SQSTM1, aptly describes the ability of p62 to form inclusions by multimerizing through its PB1 domain. Indeed, p62 sequesters ubiquitinated proteins as well as mitochondria (Komatsu et al. 2007a, Narendra et al. 2010, Nezis et al. 2008). This general function of p62 may explain why it is the most pleiotrophic autophagic receptor. As illustrated by aggrephagy and mitophagy, substrate specificity may be acquired through p62's cooperation with other adaptor proteins.

Aggrephagy: The selective degradation of aggregates. The accumulation of ubiquitinated proteins is a hallmark of adult-onset neurodegenerative diseases, and researchers have proposed a role for macroautophagy to intervene in disease progression or prevention (Yamamoto & Simonsen 2011). Aggrephagy relies on the autophagy receptors, p62, NBR1 (Filimonenko et al. 2010, Kirkin et al. 2009), and Alfy (Autophagy linked FYVE protein), the only identified selectivity adaptor (Clausen et al. 2010, Filimonenko et al. 2010). Alfy is a highly conserved member of the BEACH [Beige and Chediak-Higashi syndrome (CHS)] domain proteins, and it mediates the interaction of the p62- and NBR1-positive proteins to Atg12-Atg5 and PI3P (Isakson et al. 2012). Beyond its role in aggrephagy, little else is known about Alfy function in vivo; however, studies in *Drosophila melanogaster* reveal that its homolog Blue Cheese (Bchs) is essential for maintaining a healthy adult brain (Finley et al. 2003), possibly by promoting proper axonal transport of endolysosomal vesicles (Lim & Kraut 2009). More recently, OPTN has been implicated in aggregate turnover as well (Korac et al. 2013).

Although studies in the brain indicate that preformed intraneuronal inclusions can be degraded, whether this process is driven by aggrephagy remains uncertain (Yamamoto & Simonsen 2011). Alfy is most highly expressed in the brain (Isakson et al. 2012), and its overexpression can promote the elimination of aggregates in primary cortical neurons (Filimonenko et al. 2010). Moreover, a recent study using cortico-striatal slice cultures revealed that protein aggregates can be cleared in a macroautophagy-dependent manner and that aggregating proteins are preferentially degraded over bulk cytoplasm (Proenca et al. 2013). Together these studies strongly suggest that aggrephagy can occur in the brain, but long-term examination in vivo is still required.

Mitophagy: The selective degradation of mitochondria. Similar to aggrephagy, mitophagy relies on the sequestration then targeted degradation of mitochondria (de Vries & Przedborski 2013, Narendra et al. 2012). One means by which this may be achieved is through the interplay between a mitochondrial kinase PINK1 (P10 inducible kinase 1) and an E3 ubiquitin ligase Parkin (Narendra et al. 2012). When mitochondria become damaged, PINK1 stabilizes in the outer mitochondrial membrane, which recruits then phosphorylates proteins, possibly including cytosolic Parkin. Subsequently, Parkin ubiquitinates different outer mitochondrial membrane proteins such as Miro, which in turn attracts the autophagic receptors p62 and Nix (Ding et al. 2010, Novak et al. 2010, Schweers et al. 2007).

As with aggrephagy, the role of mitophagy in the brain is unclear. Mutations in PINK1 and Parkin cause familial Parkinson's disease (fPD), and disease-causing mutations can lead to defects in mitophagy (Geisler et al. 2010). Nonetheless, mitochondrial abnormalities or dopamine (DA)-related deficits have not been reported in mice deficient for p62, PINK1, or Parkin (Kitada et al. 2007, Komatsu et al. 2007a, Perez & Palmiter 2005, Wooten et al. 2008). In contrast, mice deficient for members of the AP machinery accumulate abnormal mitochondria in the brain (Liang et al. 2010). These findings suggest that although mitochondria are eliminated by macroautophagy, bulk macroautophagy is sufficient for mitochondrial turnover in the developing brain. PINK1-Parkin-dependent mitophagy, however, may demonstrate more relevance in aging DA neurons. For example, Dawson and colleagues have found that after a ten-month period, depletion of Parkin in adult substantia nigra leads to neuronal loss (Shin et al. 2011). Nonetheless, it remains to be seen whether the toxicity observed is due to a defect in mitophagy or other Parkin-related functions.

NEURONAL AUTOPHAGY

All forms of autophagy were discovered under conditions of starvation, and the ensuing biochemical studies revealed that at the cellular level, starvation is translated into signaling through the large, serine/threonine kinase mTOR (mammalian target of rapamycin). mTOR is found in two different complexes (reviewed in Efeyan et al. 2012) and in the form of mTORC1 (mTOR complex I) acts as a negative regulator of macroautophagy. How starvation and mTORC1 inhibition can promote CMA and microautophagy is less certain.

Although the autophagic response to mTORC1 inhibition is robust in most organs, the response in the brain has been controversial. Acute starvation in transgenic mice expressing GFP-LC3 (green fluorescent protein-tagged LC3) triggers a rapid increase in GFP-LC3-positive APs in the liver, muscle, and heart but rarely in the brain, despite strong expression in many neuronal subtypes and glia (Mizushima et al. 2004). Biochemical measures indicating LC3 lipidation are also difficult to detect in neural tissue. In contrast, other studies indicate that intermittent fasting and mTOR inhibition can induce macroautophagy in neurons (Alirezaei et al. 2010, Kaushik et al. 2011, Proenca et al. 2013). Moreover, TSC1 (tubular sclerosis complex I) mutations that lead to hyperactive mTOR appear to chronically inactivate autophagy (McMahon et al. 2012). The same mutations, however, do not mimic the neurodegenerative phenotype of mice that completely lack the AP machinery (Hara et al. 2006, Liang et al. 2010), indicating that mTOR-independent autophagic events continue to occur.

This varied response to mTORC1 may reflect the physiology of the vertebrate brain versus peripheral tissues such as the liver. The primary energy source for vertebrates is glucose, and the brain is its highest consumer. To maintain brain activity during periods of starvation, gluconeogenesis drives the breakdown of glycogen, adipose tissue, and proteins. Under conditions of nutrient deprivation, hepatocytes can account for as much as 75% of the ongoing protein degradation (Mortimore et al. 1989, Seglen & Bohley 1992), and thus the liver can lose significant mass (>40%) yet the brain loses much less (Goodman et al. 1984, Wagenmakers et al. 1984). Thus unlike the liver, from which we have derived most of our knowledge of autophagy, the brain's drive to degrade proteins in response to mTORC1 inhibition may be blunted. Moreover, depending on how neurons interact with the periphery, how they respond to mTORC1 will vary.

Although mTORC1-dependent autophagy in the brain is controversial, it is clear that autophagy can be induced in the developing and adult brain. In neurons, mechanical stress such as axotomy or nerve crush, excitotoxic stress, and drug-induced toxicity can evoke AP accumulation across neuronal subtypes (Yue et al. 2009). Activation may be achieved by mTORC1-independent mechanisms relying on AMP kinase, increasing PI3P levels, and changes in intracellular calcium (reviewed in Johnson et al. 2012). Further mechanistic understanding of how autophagy is induced in the brain will provide a framework for therapeutic interventions but will also help provide much needed tools that will dissect how autophagic events influence neural function and health.

Genetic studies have firmly established that basal macroautophagy is essential for the development of a healthy brain (Hara et al. 2006, Komatsu et al. 2006, Liang et al. 2010); however, it can also play a deleterious role, promoting neuronal damage and loss. For example, inhibition of macroautophagy during excitotoxic stress can be protective and potentially promote recovery

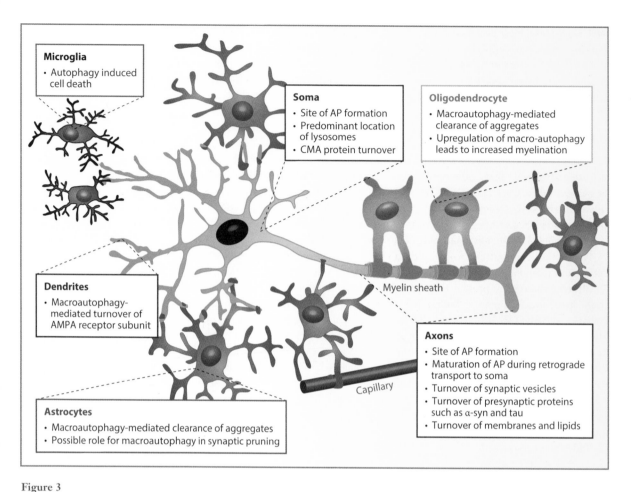

Figure 3

Neural Autophagy: A summary. A schematic representation summarizing the different autophagic events in different regions of the neuron and the three different glial subtypes discussed in this review. Abbreviations: AMPA, α-amino-3-hydroxy-5-methyl-4-isoxazolepropionic acid; AP, autophagosome; CMA, chaperone-mediated autophagy.

(Dong et al. 2012, Koike et al. 2008, Yue et al. 2002). This dichotomy suggests that the impact of autophagy can be influenced not only by neuronal subtype, but also by the neuron's age and where within the neuron that macroautophagy is activated. Moreover, the yet unknown contribution of glial autophagy may also be a key factor in how autophagy affects neural health (**Figure 3**).

Neuronal Subtype

Neurons are extremely diverse in form and function and can be readily distinguished by morphology, biochemistry, and activity; it thus may be unsurprising that different neuronal subtypes depend on basal macroautophagy in various ways. This is readily revealed when macroautophagy is conditionally disrupted using *Nestin-cre*: The pattern of protein accumulation, dysfunction, and death varies across neuronal subtypes (Hara et al. 2006, Komatsu et al. 2006). For example, Purkinje cells are among the most vulnerable neurons and degenerate within 12 postnatal weeks. In contrast, midbrain DA neurons are largely spared until 9 months of age (Friedman et al. 2012).

Although Purkinje cells require macroautophagy, hyperactivation of macroautophagy can also be neurotoxic (Yue et al. 2002), indicating that neuronal dependence on this pathway is more complex than the initial studies implied.

Not only is the dependence on macroautophagy subtype specific, but its significance can also vary. Mechanical stressors such as axotomy lead to a rapid induction and accumulation of AVs in neuronal axons and soma. Studies in nigrostriatal DA neurons versus RGCs (retinal ganglia cells) of the optic nerve reveal that the induction of macroautophagy in DA neurons is deleterious (Cheng et al. 2011), but in RGCs it is beneficial (Rodríguez-Muela et al. 2012). This finding would suggest that in the former, AV accumulation promotes axonal fragmentation and swelling, whereas for the latter, macroautophagy clears and recycles the damaged material to promote an environment that permits regrowth. Why these differences occur is unclear. Nonetheless, the subtype-specific responsiveness and reliance on autophagy indicate that augmenting these pathways systemically may be problematic. Understanding the basis of these differences will play an important role in our understanding of neuronal autophagy overall.

Neuronal Compartments

Soma. The size and specialization of neurons permit macroautophagy to occur independently within different regions of the same cell. Moreover, adaptations of the macroautophagic pathways to support specific neural activities have been observed, particularly for synaptic transmission. APs are synthesized in the soma (Bains et al. 2009, Lee et al. 2010b, Young et al. 2009), but the presence of ER in axons and terminals throughout the neuron (Broadwell & Cataldo 1984) suggests that AP formation can occur broadly. This is particularly the case for axons, but for dendrites it remains unknown. Although AP formation may occur in different regions of the cell, the primary location of lysosomes in adult neurons is the soma (Roberts & Gorenstein 1987); thus trafficking events will strongly dictate the efficiency of degradation.

Axons. The ability of APs to form distally in axons has been indicated since the earliest studies with axon injury models (Matthews & Raisman 1972). Further studies revealed that upon sequestration, the APs mature and form amphisomes while being transported retrograde back to the soma in a dynein/dynactin-dependent manner (Hollenbeck 1993, Maday et al. 2012). Many of these studies were performed on neurons of the peripheral nervous system; however, recent examinations of CNS-derived neurons reveal similar findings in culture and in the brain (Wong & Holzbaur 2014, Yang et al. 2013).

The axonal autophagic cargoes remain to be characterized, but limited evidence suggests that they are presynaptic proteins and membranous structures such as synaptic vesicles (Yang et al. 2013). Tau protein, which stabilizes axonal microtubules, may also be degraded by autophagy (Wang et al. 2009, 2012). A deficiency in macroautophagy causes Purkinje cell axons to accumulate aberrant membrane organelles, suggesting that axonal macroautophagy may help to balance membrane networks within the axon to prevent axon dystrophy (Komatsu et al. 2007b). Similarly, macroautophagy may play a key role in axon outgrowth (Ban et al. 2013, Coupé et al. 2012).

Although autophagy in axons is required for neuronal function, a profound accumulation of AVs can also be observed in axons of dying or degenerating neurons. Pathological or pharmacological conditions may disrupt maturation directly, but similar observations can be made if retrograde transport is affected (Yang et al. 2013). A growing number of studies also implicate macroautophagy in axonopathies such as pcd (Purkinje cell death) (Yang et al. 2013); however, studies of peripheral axons reveal that macroautophagy is induced not only during degeneration but also during regeneration (Mohseni 2011), suggesting that the stimulus for macroautophagy may promote distinct functions.

Synapse development and synaptic activity. A local role for macroautophagy in axons raises questions about its role in the synapse. Studies in *D. melanogaster* show that macroautophagy may promote neuromuscular junction growth by reducing the levels of Highwire, an E3 ubiquitin ligase (Shen & Ganetzky 2009). In rodents, disruption of *Atg7* in midbrain DA neurons causes an altered number of synaptic vesicles and aberrantly evoked DA transmission (prior to the onset of degeneration), consistent with a presynaptic function of macroautophagy in axon terminals (Hernandez et al. 2012). Early evidence also indicates that macroautophagy acts postsynaptically; cultured neurons treated with *N*-methyl-D-aspartate (NMDA) demonstrate a profound induction of AP formation (Borsello et al. 2003). A more recent report demonstrates that depolarization and NMDA-based long-term depression stimulate macroautophagic activity in spines and dendrites, accompanied by the enhanced degradation of GluR1, an AMPA (α-amino-3-hydroxy-5-methyl-4-isoxazolepropionic acid) receptor subunit (Shehata et al. 2012). In summary, although there is still much for investigators to learn, neurons have adapted the flexibility of macroautophagy to suit their complex needs. Further study examining the role of macroautophagy in neurons will shed more light on how its role may be further extended to plastic changes in the adult brain.

GLIAL AUTOPHAGY

Although we know the fate of neurons upon early deletion of core *ATG* genes, the fate of glial cells remains largely unexamined. Like neurons, glial cells represent a diverse and specialized collection of cell types and thus may also demonstrate temporal- and subtype-dependent reliance on autophagy. Unlike neurons, however, glial cells replicate and respond more robustly to nutrient changes and thus may also demonstrate characteristics similar to peripheral cells. The limited studies available indicate that glia such as astrocytes and oligodendrocytes can mount a macroautophagic response to accumulated proteins in an mTORC1-dependent and -independent manner (Janen et al. 2010, Schwarz et al. 2012, Tang et al. 2008, Zschocke et al. 2011), but the efficaciousness of this response is unclear. Autophagy-induced cell death in microglial cells has also been reported (Arroyo et al. 2013).

Glial subtypes may also have adapted macroautophagy to promote specialized functions. Oligodendroctyes increase myelination in response to an upregulation of macroautophagy in both a dysmyelination mutant and control animals (Smith et al. 2013). The peripheral Schwann cells show a similar responsiveness to macroautophagy (Fortun et al. 2003, Rangaraju et al. 2010). Moreover, Schwann cells associated with regenerating axons may upregulate macroautophagy (Mohseni 2011). These data suggest that glia may reclaim membranes and proteins by macroautophagy to promote myelinating events. Astrocytes may rely on macroautophagy to prune axons (Song et al. 2008). Thus, in light of the tight interrelationship shared between neurons and glia, the disruption of macroautophagy in glia may have a profound non-cell-autonomous influence on neuronal health and development. How other forms of autophagy impact glial health and function also need to be examined.

AUTOPHAGY AND NEURODEVELOPEMENT

Macroautophagy is highly active during development from early preimplantation onward (Mizushima & Levine 2010); generally, however, beyond the postnatal degenerative phenotype that has been reported (Hara et al. 2006, Komatsu et al. 2006, Liang et al. 2010), the loss of ATG proteins in early development does not lead to gross structural changes in the brain (Kuma et al. 2004). A moderate exception thus far has been hypothalamic neurons; macroautophagy is essential for axonal projections to mature (Coupe et al. 2012). Hypothalamic neurons project their axons postnatally, which may account for the differential reliance on axonal macroautophagy. Intrinsically different macroautophagic activity is also reflected in a recent study, which shows that macroautophagy maintains postnatal neural stem cells and promotes neurogenesis but that it is dispensable for these purposes during the embryonic period (Wang et al. 2013).

In contrast, disruption of proteins involved in modulating macroautophagy reveals a more striking impact on neurodevelopment. Ambra1 is a vertebrate protein that acts as a positive regulator of Beclin1 (Fimia et al. 2007). During development, it is highly expressed in the neural tube, and its deficiency leads to neural tube closure defects, possibly due to unchecked hyperproliferation of cells at the onset of neurulation. Moreover, several autosomal recessive disorders, Vici syndrome (Cullup et al. 2013), and two different hereditary spastic paraparesis (Oz-Levi et al. 2012, Vantaggiato et al. 2013) demonstrate motor and cognitive delays, intellectual disability, and callosal agenesis or hypogenesis, which suggests that degradation by macroautophagy is essential for neuronal connectivity and axonal outgrowth. The proteins involved, EPG5, ZFYVE26, and TECPR2, have been implicated in AP maturation (Behrends et al. 2010, Vantaggiato et al. 2013, Zhao et al. 2013). In mice deficient for *Epg5*, the callosal phenotype was replicated, but other symptoms associated with Vici syndrome were not observed (Zhao et al. 2013). A deletion of *ZFYVE26* or *TECPR2* has not yet been reported. Why proteins involved in AP maturation, rather than the core ATG proteins, are more essential to CNS development is unclear. Unlike ATG proteins, these proteins may help to balance autophagic versus endocytic degradation; thus the neurodevelopmental defects may reflect a more generalized disruption of vesicle trafficking.

AUTOPHAGY, AGING, AND NEURODEGENERATION

Studies of postmortem brains of AD (Alzheimer's disease), PD (Parkinson's disease), and HD (Huntington's disease) gave some of the earliest evidence that macroautophagy may contribute to neurodegeneration. Recent studies have shown that macroautophagy plays an important role in ALS (amyotrophic lateral sclerosis). Moreover, the dysregulation of CMA has been implicated in the pathogenesis of adult-onset disorders, particularly PD (Koga & Cuervo 2011). Efforts to decipher how autophagy impacts neurodegeneration has highlighted the complex role that autophagy plays in the brain.

Alzheimer's Disease

Although the classic pathological hallmarks of AD include intraneuronal neurofibrillary tangles and the extracellular senile plaques, another prominent feature is the presence of "autophagic neurites," dystrophic neurites characterized by swollen axonal or dendritic varicosities filled with AVs (Nixon et al. 2005). Although the pathogenic significance of these structures remains elusive, several reports suggest that upregulated protein synthesis or transcription of lysosomal components and cellular factors promoting autophagy is upregulated in AD brains (Ginsberg et al. 2010, Lipinski et al. 2010, Nixon & Cataldo 2006). Although this upregulation may be an attempt by the neurons to clear amyloid precursor protein (APP) products such as Abeta and CTF (C-terminal fragment) (Steele et al. 2013b, Vingtdeux et al. 2011), the accumulation of AVs indicates that AV maturation fails (Nixon & Cataldo 2006). The lack of degradation and completion of the autophagic process promotes a state of autophagic stress (Chu 2006), which contributes towards pathogenesis. Other studies in postmortem AD brains, however, show that mTORC1 is hyperactivated (Li et al. 2005) and Beclin 1 protein levels are decreased (Pickford et al. 2008), indicating that autophagic activation may be compromised in AD brains. As mentioned above, these apparently contradictory findings may be influenced largely by from where and when the samples were collected. Nonetheless, the aberrant accumulation of AV structures in postmortem samples is indicative that at the end stage of disease, degradation by macroautophagy is failing. When and how macroautophagy begins to exert toxicity are still uncertain.

In early-onset FAD (familial AD), defects in macroautophagy may be an underlying cause. The most common cause of FAD is mutations in PS1 (Presenilin 1), a multispanning intramembrane protease best known as a member of γ-secretase, a multimeric enzyme complex responsible for the cleavage of several substrates including APP and Notch (De Strooper et al. 2012). A growing number of studies indicate that PS1 also has functions outside γ-secretase (De Strooper et al. 2012), such as lysosomal degradation and AP maturation (Lee et al. 2010b, Neely et al. 2011). PS1 mutations have been implicated in disruption of endolysosomal transport and function (Lee et al. 2010a), which are essential for AP maturation and degradation. The exact mechanism, however, remains to be clarified (Zhang et al. 2012).

Parkinson's Disease

Genetic and experimental evidence strongly implicates a role for both macroautophagy and CMA in PD. PD is characterized by the degeneration of DA neurons in the nigrostriatal pathway and the presence of Lewy bodies, an eosinophilic cytoplasmic inclusion comprised largely of α-syn (α-synuclein) fibrils, and Lewy neurites, similar proteinaceous structures composed of abnormal α-syn fibrils and granular material. Beyond the recessive PINK1 and Parkin mutations that diminish mitophagy (see Mitophagy, above), aberrant levels or activity of autosomal dominant mutations in α-syn and LRRK2 may compromise basal macroautophagy and CMA, likely underlying the pathogenesis of PD.

α-Syn is linked to familial as well as sporadic PD and may either modulate or be a target for macroautophagy. Although the physiologic function of α-syn is unclear, the available evidence suggests that it is involved in vesicle trafficking (Cooper et al. 2006, Nemani et al. 2010). Along these lines, excessive α-syn may impair macroautophagy by inhibiting the small GTPase Rab1a and altering the localization of Atg9 (Winslow et al. 2010). α-syn fibrils also reduce AP turnover (Tanik et al. 2013). In contrast, aggregates containing disease-causing mutations in α-syn can be cleared by an upregulation of macroautophagy (Sarkar et al. 2007, Steele et al. 2013a), and clearance can be dependent on aggrephagy (Filimonenko et al. 2010).

Genetic mutations in LRRK2 are linked to the most common familial forms of PD, and several studies find that LRRK2 participates in the regulation of macroautophagy (reviewed in Gómez-Suaga et al. 2012). Although the molecular mechanism is not well understood, the pharmacological inhibition of LRRK2 kinase activity promotes macroautophagy. Nonetheless, despite the autophagy-associated defects in LRRK2 animal models, these deficiencies may be several steps downstream from LRRK2. Whether members of the AP machinery are a direct target remains to be seen.

Both wildtype α-syn and LRRK2 proteins are CMA substrates, and their PD-linked mutations impair CMA-mediated degradation (Cuervo et al. 2004, Orenstein et al. 2013). Moreover, the interaction between α-syn and oxidized DA also interferes with CMA activity, broadening the impact of dysfunctional CMA to idiopathic PD (Martinez-Vicente et al. 2008). The mechanisms of CMA inhibition differ between the two proteins: monomeric α-syn blocks internalization, whereas LRRK2 evokes a more complex self-perpetuating inhibitory effect (Orenstein et al. 2013). In light of the cross talk between CMA and macroautophagy, it would be interesting to determine if a corresponding upregulation of macroautophagy occurs in neurons with compromised CMA. Future in vivo studies are anticipated.

Huntington's Disease

Currently, one of the primary therapeutic questions in HD is how to preferentially eliminate the disease-causing protein (Sah & Aronin 2011), and one approach investigators are exploring is the activation of macroautophagy. Unlike other adult-onset neurodegenerative disorders, all known cases of this hereditary disorder are caused by a trinucleotide repeat expansion in the gene *HD*. Elimination of the resulting mutant protein product in experimental models not only halts the progression of pathogenesis, but also promotes amelioration of the disease (Kordasiewicz et al. 2012, Régulier et al. 2003, Yamamoto et al. 2000).

Whether the activation of macroautophagy in the HD brain will be beneficial, however, still remains uncertain. The inhibition of mTORC1 in vivo showed early promise (Ravikumar et al. 2004), but the complexity of mTORC1 inhibition and neural autophagy suggests alternative therapeutic approaches may be necessary (Fox et al. 2010, King et al. 2008, Tsvetkov et al. 2010). Despite the numerous studies examining macroautophagy and HD, a fundamental question remains: Can the full-length mutant htt protein be eliminated by macroautophagy? Htt is a 350-kDa protein; the pathogenic mutation is present at the very N-terminus. Early studies indicated that transgenic expression of a short N-terminus fragment was sufficient to recapitulate key aspects of HD (Mangiarini et al. 1996), and these fragments have been studied primarily in the context of autophagy. Although subsequent biochemical studies suggest that this fragment is pathogenically relevant (Landles et al. 2010), questions remain whether the fragment alone truly represents HD (Bowles et al. 2012).

Although a role for autophagic stress has not been demonstrated in HD, some studies have reported a potential dysfunction in basal macroautophagy (Heng et al. 2010, Martinez-Vicente et al. 2010). In a curious twist, complete deletion of the polyglutamine stretch, even nonpathogenic lengths, leads to enhanced neuronal macroautophagy (Zheng et al. 2010). These data imply that Htt, and specifically its polyglutamine stretch, can modulate this pathway. The regulation may likely be due to Htt's role in retrograde transport, which can be impeded in a polyglutamine-length-dependent manner (Colin et al. 2008, Liot et al. 2013, Wong & Holzbaur 2014). Phosphorylation of Htt can overcome the polyglutamine-dependent inhibition, thus providing an added avenue by which macroautophagy deficiencies may be overcome.

Amyotrophic Lateral Sclerosis

An emerging hypothesis for ALS pathogenesis is defective protein degradation, and of particular interest is the functional convergence on macroautophagy (Robberecht & Philips 2013, Thomas et al. 2013). ALS is a degenerative disease that affects motor neurons in the motor cortex, the brain stem, and the spinal cord. Although research has not identified a mutation in core *ATG* genes, two selective macroautophagy proteins, p62/SQSTM1 (Hirano et al. 2013, Shimizu et al. 2013) and OPTN (Maruyama et al. 2010), have recently been linked to sporadic and familial ALS, suggesting defects in aggrephagy. Both proteins can be found in ALS-associated inclusions (Mizuno et al. 2006, Osawa et al. 2011), but why motor neurons may rely on aggrephagy still needs to be established.

Upregulation of macroautophagy has been both promising (Castillo et al. 2013, Feng et al. 2008) and disappointing for ALS (Gill et al. 2009, Pizzasegola et al. 2009, Zhang et al. 2011). In all of these studies, ALS was modeled using transgenic overexpression of mutant SOD1 (superoxide dismutase-1). These models show a profound accumulation of AVs in motor neurons (Li et al. 2008, Tian et al. 2011), and thus the strength of the induction may have profoundly influenced the outcome. Whether SOD1-dependent ALS and the rare OPTN- and p62-dependent forms of ALS converge remains to be determined. But if there is convergence, upregulation of basal macroautophagy may not suffice if substrates cannot be properly targeted for degradation.

CONCLUSION

Understanding the role of autophagy in neurodegeneration reignited our interest in neural autophagy, and over the past decade, the complexity of these essential pathways was revealed. Although many outstanding questions still remain, it is incontrovertible that autophagy plays a vital role in neurons and glia, far beyond their role of eliminating nonessential proteins. As scientists continue to delve deeper into the regulation and import of autophagy, we will gain further insight into how these specialized cells use autophagic pathways, ultimately helping to determine how modulation of these pathways influences neural development and contributes toward neural disease.

DISCLOSURE STATEMENT

The authors are not aware of any affiliations, memberships, funding, or financial holdings that might be perceived as affecting the objectivity of this review.

ACKNOWLEDGMENTS

Funding is provided by the NIH [RO1 NS077111 (A.Y.), RO1 NS063973 (A.Y.), RO1 NS060123 (Z.Y.)], The Parkinson Disease Foundation (A.Y.), and Cure Huntington's Disease Initiative Foundation (Z.Y.).

LITERATURE CITED

Alirezaei M, Kemball CC, Flynn CT, Wood MR, Whitton JL, Kiosses WB. 2010. Short-term fasting induces profound neuronal autophagy. *Autophagy* 6:702–10

Arroyo DS, Soria JA, Gaviglio EA, Garcia-Keller C, Cancela LM, et al. 2013. Toll-like receptor 2 ligands promote microglial cell death by inducing autophagy. *FASEB J.* 27:299–312

Axe EL, Walker SA, Manifava M, Chandra P, Roderick HL, et al. 2008. Autophagosome formation from membrane compartments enriched in phosphatidylinositol 3-phosphate and dynamically connected to the endoplasmic reticulum. *J. Cell Biol.* 182:685–701

Baba M, Osumi M, Scott SV, Klionsky DJ, Ohsumi Y. 1997. Two distinct pathways for targeting proteins from the cytoplasm to the vacuole/lysosome. *J. Cell Biol.* 139:1687–95

Baba M, Takeshige K, Baba N, Ohsumi Y. 1994. Ultrastructural analysis of the autophagic process in yeast: detection of autophagosomes and their characterization. *J. Cell Biol.* 124:903–13

Backer JM, Bourret L, Dice JF. 1983. Regulation of catabolism of microinjected ribonuclease A requires the amino-terminal 20 amino acids. *Proc. Natl. Acad. Sci. USA* 80:2166–70

Bains M, Florez-McClure ML, Heidenreich KA. 2009. Insulin-like growth factor-I prevents the accumulation of autophagic vesicles and cell death in Purkinje neurons by increasing the rate of autophagosome-to-lysosome fusion and degradation. *J. Biol. Chem.* 284:20398–407

Ban BK, Jun MH, Ryu HH, Jang DJ, Ahmad ST, Lee JA. 2013. Autophagy negatively regulates early axon growth in cortical neurons. *Mol. Cell. Biol.* 33:3907–19

Behrends C, Sowa ME, Gygi SP, Harper JW. 2010. Network organization of the human autophagy system. *Nature* 466:68–76

Bejarano E, Cuervo AM. 2010. Chaperone-mediated autophagy. *Proc. Am. Thorac. Soc.* 7:29–39

Birgisdottir AB, Lamark T, Johansen T. 2013. The Lir motif—crucial for selective autophagy. *J. Cell Sci.* 126:3237–47

Borsello T, Croquelois K, Hornung JP, Clarke PG. 2003. *N*-methyl-D-aspartate-triggered neuronal death in organotypic hippocampal cultures is endocytic, autophagic and mediated by the c-Jun N-terminal kinase pathway. *Eur. J. Neurosci.* 18:473–85

Bowles KR, Brooks SP, Dunnett SB, Jones L. 2012. Gene expression and behaviour in mouse models of HD. *Brain Res. Bull.* 88:276–84

Broadwell RD, Cataldo AM. 1984. The neuronal endoplasmic reticulum: its cytochemistry and contribution to the endomembrane system. II. Axons and terminals. *J. Comp. Neurol.* 230:231–48

Castillo K, Nassif M, Valenzuela V, Rojas F, Matus S, et al. 2013. Trehalose delays the progression of amyotrophic lateral sclerosis by enhancing autophagy in motoneurons. *Autophagy* 9:1308–20

Chakrama FZ, Seguin-Py S, Le Grand JN, Fraichard A, Delage-Mourroux R, et al. 2010. GABARAPL1 (GEC1) associates with autophagic vesicles. *Autophagy* 6:495–505

Cheng HC, Kim SR, Oo TF, Kareva T, Yarygina O, et al. 2011. Akt suppresses retrograde degeneration of dopaminergic axons by inhibition of macroautophagy. *J. Neurosci.* 31:2125–35

Chu CT. 2006. Autophagic stress in neuronal injury and disease. *J. Neuropathol. Exp. Neurol.* 65:423–32

Ciechanover A, Orian A, Schwartz AL. 2000. Ubiquitin-mediated proteolysis: biological regulation via destruction. *Bioessays* 22:442–51

Clausen TH, Lamark T, Isakson P, Finley K, Larsen KB, et al. 2010. p62/SQSTM1 and ALFY interact to facilitate the formation of p62 bodies/ALIS and their degradation by autophagy. *Autophagy* 6:330–44

Colin E, Zala D, Liot G, Rangone H, Borrell-Pagès M, et al. 2008. Huntingtin phosphorylation acts as a molecular switch for anterograde/retrograde transport in neurons. *EMBO J.* 27:2124–34

Cooper AA, Gitler AD, Cashikar A, Haynes CM, Hill KJ, et al. 2006. Alpha-synuclein blocks ER-Golgi traffic and Rab1 rescues neuron loss in Parkinson's models. *Science* 313:324–28

Coupé B, Ishii Y, Dietrich MO, Komatsu M, Horvath TL, Bouret SG. 2012. Loss of autophagy in proopiomelanocortin neurons perturbs axon growth and causes metabolic dysregulation. *Cell Metab.* 15:247–55

Cuervo AM, Knecht E, Terlecky SR, Dice JF. 1995. Activation of a selective pathway of lysosomal proteolysis in rat liver by prolonged starvation. *Am. J. Physiol.* 269:C1200–8

Cuervo AM, Stefanis L, Fredenburg R, Lansbury PT, Sulzer D. 2004. Impaired degradation of mutant alpha-synuclein by chaperone-mediated autophagy. *Science* 305:1292–95

Cullup T, Kho AL, Dionisi-Vici C, Brandmeier B, Smith F, et al. 2013. Recessive mutations in Epg5 cause Vici syndrome, a multisystem disorder with defective autophagy. *Nat. Genet.* 45:83–87

Danon MJ, Oh SJ, DiMauro S, Manaligod JR, Eastwood A, et al. 1981. Lysosomal glycogen storage disease with normal acid maltase. *Neurology* 31:51–57

De Strooper B, Iwatsubo T, Wolfe MS. 2012. Presenilins and γ-secretase: structure, function, and role in Alzheimer disease. *Cold Spring Harb. Perspect. Med.* 2:A006304

De Vries RL, Przedborski S. 2013. Mitophagy and Parkinson's disease: be eaten to stay healthy. *Mol. Cell Neurosci.* 55:37–43

Dice JF, Terlecky SR, Chiang HL, Olson TS, Isenman LD, et al. 1990. A selective pathway for degradation of cytosolic proteins by lysosomes. *Semin. Cell Biol.* 1:449–55

Ding WX, Ni HM, Li M, Liao Y, Chen X, et al. 2010. Nix is critical to two distinct phases of mitophagy, reactive oxygen species-mediated autophagy induction and Parkin-ubiquitin-p62-mediated mitochondrial priming. *J. Biol. Chem.* 285:27879–90

Dixon JS. 1967. "Phagocytic" lysosomes in chromatolytic neurones. *Nature* 215:657–58

Dong XX, Wang YR, Qin S, Liang ZQ, Liu BH, et al. 2012. p53 mediates autophagy activation and mitochondria dysfunction in kainic acid-induced excitotoxicity in primary striatal neurons. *Neuroscience* 207:52–64

Dunn WA Jr. 1990. Studies on the mechanisms of autophagy: formation of the autophagic vacuole. *J. Cell Biol.* 110:1923–33

Efeyan A, Zoncu R, Sabatini DM. 2012. Amino acids and mTORC1: from lysosomes to disease. *Trends Mol. Med.* 18:524–33

Eskelinen EL, Tanaka Y, Saftig P. 2003. At the acidic edge: emerging functions for lysosomal membrane proteins. *Trends Cell Biol.* 13:137–45

Feng HL, Leng Y, Ma CH, Zhang J, Ren M, Chuang DM. 2008. Combined lithium and valproate treatment delays disease onset, reduces neurological deficits and prolongs survival in an amyotrophic lateral sclerosis mouse model. *Neuroscience* 155:567–72

Fengsrud M, Erichsen ES, Berg TO, Raiborg C, Seglen PO. 2000. Ultrastructural characterization of the delimiting membranes of isolated autophagosomes and amphisomes by freeze-fracture electron microscopy. *Eur. J. Cell Biol.* 79:871–82

Filimonenko M, Isakson P, Finley KD, Anderson M, Jeong H, et al. 2010. The selective macroautophagic degradation of aggregated proteins requires the Pi3p-binding protein Alfy. *Mol. Cell* 38:265–79

Fimia GM, Stoykova A, Romagnoli A, Giunta L, Di Bartolomeo S, et al. 2007. Ambra1 regulates autophagy and development of the nervous system. *Nature* 447:1121–25

Finley KD, Edeen PT, Cumming RC, Mardahl-Dumesnil MD, Taylor BJ, et al. 2003. *blue cheese* mutations define a novel, conserved gene involved in progressive neural degeneration. *J. Neurosci.* 23:1254–64

Fortun J, Dunn WA Jr, Joy S, Li J, Notterpek L. 2003. Emerging role for autophagy in the removal of aggresomes in Schwann cells. *J. Neurosci.* 23:10672–80

Fox JH, Connor T, Chopra V, Dorsey K, Kama JA, et al. 2010. The mTOR kinase inhibitor Everolimus decreases S6 kinase phosphorylation but fails to reduce mutant huntingtin levels in brain and is not neuroprotective in the R6/2 mouse model of Huntington's disease. *Mol. Neurodegener.* 5:26

Friedman LG, Lachenmayer ML, Wang J, He L, Poulose SM, et al. 2012. Disrupted autophagy leads to dopaminergic axon and dendrite degeneration and promotes presynaptic accumulation of α-synuclein and LRRK2 in the brain. *J. Neurosci.* 32:7585–93

Ge L, Melville D, Zhang M, Schekman R. 2013. The ER–Golgi intermediate compartment is a key membrane source for the LC3 lipidation step of autophagosome biogenesis. *eLife* 2:e00947

Geisler S, Holmstrom KM, Treis A, Skujat D, Weber SS, et al. 2010. The PINK1/Parkin-mediated mitophagy is compromised by PD-associated mutations. *Autophagy* 6:871–78

Gill A, Kidd J, Vieira F, Thompson K, Perrin S. 2009. No benefit from chronic lithium dosing in a sibling-matched, gender balanced, investigator-blinded trial using a standard mouse model of familial ALS. *PLoS ONE* 4:E6489

Ginsberg SD, Alldred MJ, Counts SE, Cataldo AM, Neve RL, et al. 2010. Microarray analysis of hippocampal CA1 neurons implicates early endosomal dysfunction during Alzheimer's disease progression. *Biol. Psychiatry* 68:885–93

Gómez-Suaga P, Fdez E, Blanca Ramírez M, Hilfiker S. 2012. A link between autophagy and the pathophysiology of LRRK2 in Parkinson's disease. *Park. Dis.* 2012:324521

Goodman MN, Lowell B, Belur E, Ruderman NB. 1984. Sites of protein conservation and loss during starvation: influence of adiposity. *Am. J. Physiol.* 246:E383–90

Hailey DW, Rambold AS, Satpute-Krishnan P, Mitra K, Sougrat R, et al. 2010. Mitochondria supply membranes for autophagosome biogenesis during starvation. *Cell* 141:656–67

Hamasaki M, Furuta N, Matsuda A, Nezu A, Yamamoto A, et al. 2013. Autophagosomes form at ER-mitochondria contact sites. *Nature* 495:389–93

Hara T, Nakamura K, Matsui M, Yamamoto A, Nakahara Y, et al. 2006. Suppression of basal autophagy in neural cells causes neurodegenerative disease in mice. *Nature* 441:885–89

Hayashi-Nishino M, Fujita N, Noda T, Yamaguchi A, Yoshimori T, Yamamoto A. 2009. A subdomain of the endoplasmic reticulum forms a cradle for autophagosome formation. *Nat. Cell Biol.* 11:1433–37

Heng MY, Duong DK, Albin RL, Tallaksen-Greene SJ, Hunter JM, et al. 2010. Early autophagic response in a novel knock-in model of Huntington disease. *Hum. Mol. Genet.* 19:3702–20

Hernandez D, Torres CA, Setlik W, Cebrián C, Mosharov EV, et al. 2012. Regulation of presynaptic neurotransmission by macroautophagy. *Neuron* 74:277–84

Hirano M, Nakamura Y, Saigoh K, Sakamoto H, Ueno S, et al. 2013. Mutations in the gene encoding p62 in Japanese patients with amyotrophic lateral sclerosis. *Neurology* 80:458–63

Hollenbeck PJ. 1993. Products of endocytosis and autophagy are retrieved from axons by regulated retrograde organelle transport. *J. Cell Biol.* 121:305–15

Holtzman E, Novikoff AB. 1965. Lysomes in the rat sciatic nerve following crush. *J. Cell Biol.* 27:651–69

Isakson P, Holland P, Simonsen A. 2012. The role of ALFY in selective autophagy. *Cell Death Differ.* 20:12–20

Itakura E, Kishi-Itakura C, Mizushima N. 2012. The hairpin-type tail-anchored SNARE syntaxin 17 targets to autophagosomes for fusion with endosomes/lysosomes. *Cell* 151:1256–69

Janen SB, Chaachouay H, Richter-Landsberg C. 2010. Autophagy is activated by proteasomal inhibition and involved in aggresome clearance in cultured astrocytes. *Glia* 58:1766–74

Johnson CW, Melia TJ, Yamamoto A. 2012. Modulating macroautophagy: a neuronal perspective. *Future Med. Chem.* 4:1715–31

Kabeya Y, Mizushima N, Ueno T, Yamamoto A, Kirisako T, et al. 2000. Lc3, a mammalian homologue of yeast Apg8p, is localized in autophagosome membranes after processing. *EMBO J.* 19:5720–28

Kaushik S, Cuervo AM. 2012. Chaperone-mediated autophagy: a unique way to enter the lysosome world. *Trends Cell Biol.* 22:407–17

Kaushik S, Massey AC, Mizushima N, Cuervo AM. 2008. Constitutive activation of chaperone-mediated autophagy in cells with impaired macroautophagy. *Mol. Biol. Cell* 19:2179–92

Kaushik S, Rodriguez-Navarro JA, Arias E, Kiffin R, Sahu S, et al. 2011. Autophagy in hypothalamic AgRP neurons regulates food intake and energy balance. *Cell Metab.* 14:173–83

King MA, Hands S, Hafiz F, Mizushima N, Tolkovsky AM, Wyttenbach A. 2008. Rapamycin inhibits polyglutamine aggregation independently of autophagy by reducing protein synthesis. *Mol. Pharmacol.* 73:1052–63

Kirkin V, Lamark T, Sou YS, Bjorkoy G, Nunn JL, et al. 2009. A role for NBR1 in autophagosomal degradation of ubiquitinated substrates. *Mol. Cell* 33:505–16

Kitada T, Pisani A, Porter DR, Yamaguchi H, Tscherter A, et al. 2007. Impaired dopamine release and synaptic plasticity in the striatum of PINK1-deficient mice. *Proc. Natl. Acad. Sci. USA* 104:11441–46

Koga H, Cuervo AM. 2011. Chaperone-mediated autophagy dysfunction in the pathogenesis of neurodegeneration. *Neurobiol. Dis.* 43:29–37

Koike M, Shibata M, Tadakoshi M, Gotoh K, Komatsu M, et al. 2008. Inhibition of autophagy prevents hippocampal pyramidal neuron death after hypoxic-ischemic injury. *Am. J. Pathol.* 172:454–69

Komatsu M, Waguri S, Chiba T, Murata S, Iwata J, et al. 2006. Loss of autophagy in the central nervous system causes neurodegeneration in mice. *Nature* 441:880–84

Komatsu M, Waguri S, Koike M, Sou YS, Ueno T, et al. 2007a. Homeostatic levels of p62 control cytoplasmic inclusion body formation in autophagy-deficient mice. *Cell* 131:1149–63

Komatsu M, Wang QJ, Holstein GR, Friedrich VL Jr, Iwata J, et al. 2007b. Essential role for autophagy protein Atg7 in the maintenance of axonal homeostasis and the prevention of axonal degeneration. *Proc. Natl. Acad. Sci. USA* 104:14489–94

Korac J, Schaeffer V, Kovacevic I, Clement AM, Jungblut B, et al. 2013. Ubiquitin-independent function of optineurin in autophagic clearance of protein aggregates. *J. Cell Sci.* 126:580–92

Kordasiewicz HB, Stanek LM, Wancewicz EV, Mazur C, McAlonis MM, et al. 2012. Sustained therapeutic reversal of Huntington's disease by transient repression of huntingtin synthesis. *Neuron* 74:1031–44

Kuma A, Hatano M, Matsui M, Yamamoto A, Nakaya H, et al. 2004. The role of autophagy during the early neonatal starvation period. *Nature* 432:1032–36

Landles C, Sathasivam K, Weiss A, Woodman B, Moffitt H, et al. 2010. Proteolysis of mutant huntingtin produces an exon 1 fragment that accumulates as an aggregated protein in neuronal nuclei in Huntington disease. *J. Biol. Chem.* 285:8808–23

Larsen KE, Fon EA, Hastings TG, Edwards RH, Sulzer D. 2002. Methamphetamine-induced degeneration of dopaminergic neurons involves autophagy and upregulation of dopamine synthesis. *J. Neurosci.* 22:8951–60

Lee JA, Gao FB. 2008. ESCRT, autophagy, and frontotemporal dementia. *BMB Rep.* 41:827–32

Lee JH, Yu WH, Kumar A, Lee S, Mohan PS, et al. 2010a. Lysosomal proteolysis and autophagy require presenilin 1 and are disrupted by Alzheimer-related PS1 mutations. *Cell* 141:1146–58

Lee JY, Koga H, Kawaguchi Y, Tang W, Wong E, et al. 2010b. HDAC6 controls autophagosome maturation essential for ubiquitin-selective quality-control autophagy. *EMBO J.* 29:969–80

Leil TA, Chen ZW, Chang CS, Olsen RW. 2004. GABAA receptor-associated protein traffics GABAA receptors to the plasma membrane in neurons. *J. Neurosci.* 24:11429–38

Li L, Zhang X, Le W. 2008. Altered macroautophagy in the spinal cord of SOD1 mutant mice. *Autophagy* 4:290–93

Li X, Alafuzoff I, Soininen H, Winblad B, Pei JJ. 2005. Levels of mTOR and its downstream targets 4E-BP1, eEF2, and eEF2 kinase in relationships with tau in Alzheimer's disease brain. *FEBS J.* 272:4211–20

Liang CC, Wang C, Peng X, Gan B, Guan JL. 2010. Neural-specific deletion of FIP200 leads to cerebellar degeneration caused by increased neuronal death and axon degeneration. *J. Biol. Chem.* 285:3499–509

Lim A, Kraut R. 2009. The *Drosophila* BEACH family protein, Blue Cheese, links lysosomal axon transport with motor neuron degeneration. *J. Neurosci.* 29:951–63

Liot G, Zala D, Pla P, Mottet G, Piel M, Saudou F. 2013. Mutant Huntingtin alters retrograde transport of TrkB receptors in striatal dendrites. *J. Neurosci.* 33:6298–309

Lipinski MM, Zheng B, Lu T, Yan Z, Py BF, et al. 2010. Genome-wide analysis reveals mechanisms modulating autophagy in normal brain aging and in Alzheimer's disease. *Proc. Natl. Acad. Sci. USA* 107:14164–69

Maday S, Wallace KE, Holzbaur EL. 2012. Autophagosomes initiate distally and mature during transport toward the cell soma in primary neurons. *J. Cell Biol.* 196:407–17

Mangiarini L, Sathasivam K, Seller M, Cozens B, Harper A, et al. 1996. Exon 1 of the HD gene with an expanded CAG repeat is sufficient to cause a progressive neurological phenotype in transgenic mice. *Cell* 87:493–506

Martinez-Vicente M, Talloczy Z, Kaushik S, Massey AC, Mazzulli J, et al. 2008. Dopamine-modified α-synuclein blocks chaperone-mediated autophagy. *J. Clin. Investig.* 118:777–88

Martinez-Vicente M, Talloczy Z, Wong E, Tang G, Koga H, et al. 2010. Cargo recognition failure is responsible for inefficient autophagy in Huntington's disease. *Nat. Neurosci.* 13:567–76

Maruyama H, Morino H, Ito H, Izumi Y, Kato H, et al. 2010. Mutations of optineurin in amyotrophic lateral sclerosis. *Nature* 465:223–26

Matthews MR, Raisman G. 1972. A light and electron microscopic study of the cellular response to axonal injury in the superior cervical ganglion of the rat. *Proc. R. Soc. Lond. B* 181:43–79

McMahon J, Huang X, Yang J, Komatsu M, Yue Z, et al. 2012. Impaired autophagy in neurons after disinhibition of mammalian target of rapamycin and its contribution to epileptogenesis. *J. Neurosci.* 32:15704–14

Mijaljica D, Prescott M, Devenish RJ. 2011. Microautophagy in mammalian cells: revisiting a 40-year-old conundrum. *Autophagy* 7:673–82

Mizuno Y, Amari M, Takatama M, Aizawa H, Mihara B, Okamoto K. 2006. Immunoreactivities of p62, an ubiqutin-binding protein, in the spinal anterior horn cells of patients with amyotrophic lateral sclerosis. *J. Neurol. Sci.* 249:13–18

Mizushima N, Levine B. 2010. Autophagy in mammalian development and differentiation. *Nat. Cell Biol.* 12:823–30

Mizushima N, Yamamoto A, Matsui M, Yoshimori T, Ohsumi Y. 2004. In vivo analysis of autophagy in response to nutrient starvation using transgenic mice expressing a fluorescent autophagosome marker. *Mol. Biol. Cell* 15:1101–11

Mizushima N, Yoshimori T, Ohsumi Y. 2011. The role of Atg proteins in autophagosome formation. *Annu. Rev. Cell Dev. Biol.* 27:107–32

Mohseni S. 2011. Autophagy in insulin-induced hypoglycaemic neuropathy. *Pathology* 43:254–60

Moreau K, Ravikumar B, Renna M, Puri C, Rubinsztein DC. 2011. Autophagosome precursor maturation requires homotypic fusion. *Cell* 146:303–17

Mortimore GE, Pösö AR, Lardeux BR. 1989. Mechanism and regulation of protein degradation in liver. *Diabetes Metab. Rev.* 5:49–70

Nair U, Jotwani A, Geng J, Gammoh N, Richerson D, et al. 2011. SNARE proteins are required for macroautophagy. *Cell* 146:290–302

Narendra D, Kane LA, Hauser DN, Fearnley IM, Youle RJ. 2010. p62/SQSTM1 is required for Parkin-induced mitochondrial clustering but not mitophagy; VDAC1 is dispensable for both. *Autophagy* 6:1090–106

Narendra D, Walker JE, Youle R. 2012. Mitochondrial quality control mediated by PINK1 and Parkin: links to parkinsonism. *Cold Spring Harb. Perspect. Biol.* 4:a011338

Neely KM, Green KN, Laferla FM. 2011. Presenilin is necessary for efficient proteolysis through the autophagy-lysosome system in a γ-secretase-independent manner. *J. Neurosci.* 31:2781–91

Nemani VM, Lu W, Berge V, Nakamura K, Onoa B, et al. 2010. Increased expression of α-synuclein reduces neurotransmitter release by inhibiting synaptic vesicle reclustering after endocytosis. *Neuron* 65:66–79

Nezis IP, Simonsen A, Sagona AP, Finley K, Gaumer S, et al. 2008. Ref(2)P, the *Drosophila melanogaster* homologue of mammalian p62, is required for the formation of protein aggregates in adult brain. *J. Cell Biol.* 180:1065–71

Nishimura T, Kaizuka T, Cadwell K, Sahani MH, Saitoh T, et al. 2013. Fip200 regulates targeting of Atg16l1 to the isolation membrane. *EMBO Rep.* 14:284–91

Nishino I, Fu J, Tanji K, Yamada T, Shimojo S, et al. 2000. Primary LAMP-2 deficiency causes X-linked vacuolar cardiomyopathy and myopathy (Danon disease). *Nature* 406:906–10

Nixon RA, Cataldo AM. 2006. Lysosomal system pathways: genes to neurodegeneration in Alzheimer's disease. *J. Alzheimers Dis.* 9:277–89

Nixon RA, Wegiel J, Kumar A, Yu WH, Peterhoff C, et al. 2005. Extensive involvement of autophagy in Alzheimer disease: an immuno-electron microscopy study. *J. Neuropathol. Exp. Neurol.* 64:113–22

Novak I, Kirkin V, McEwan DG, Zhang J, Wild P, et al. 2010. Nix is a selective autophagy receptor for mitochondrial clearance. *EMBO Rep.* 11:45–51

Orenstein SJ, Kuo SH, Tasset I, Arias E, Koga H, et al. 2013. Interplay of LRRK2 with chaperone-mediated autophagy. *Nat. Neurosci.* 16:394–406

Osawa T, Mizuno Y, Fujita Y, Takatama M, Nakazato Y, Okamoto K. 2011. Optineurin in neurodegenerative diseases. *Neuropathology* 31:569–74

Oz-Levi D, Ben-Zeev B, Ruzzo EK, Hitomi Y, Gelman A, et al. 2012. Mutation in TECPR2 reveals a role for autophagy in hereditary spastic paraparesis. *Am. J. Hum. Genet.* 91:1065–72

Perez FA, Palmiter RD. 2005. Parkin-deficient mice are not a robust model of parkinsonism. *Proc. Natl. Acad. Sci. USA* 102:2174–79

Pickford F, Masliah E, Britschgi M, Lucin K, Narasimhan R, et al. 2008. The autophagy-related protein beclin 1 shows reduced expression in early Alzheimer disease and regulates amyloid β accumulation in mice. *J. Clin. Investig.* 118:2190–99

Pizzasegola C, Caron I, Daleno C, Ronchi A, Minoia C, et al. 2009. Treatment with lithium carbonate does not improve disease progression in two different strains of SOD1 mutant mice. *Amyotroph. Lateral Scler.* 10:221–28

Proenca CC, Stoehr N, Bernhard M, Seger S, Genoud C, et al. 2013. Atg4b-dependent autophagic flux alleviates Huntington's disease progression. *PLoS ONE* 8:E68357

Pryor PR, Mullock BM, Bright NA, Lindsay MR, Gray SR, et al. 2004. Combinatorial SNARE complexes with VAMP7 or VAMP8 define different late endocytic fusion events. *EMBO Rep.* 5:590–95

Punnonen EL, Pihakaski K, Mattila K, Lounatmaa K, Hirsimäki P. 1989. Intramembrane particles and filipin labelling on the membranes of autophagic vacuoles and lysosomes in mouse liver. *Cell Tissue Res.* 258:269–76

Rangaraju S, Verrier JD, Madorsky I, Nicks J, Dunn WA Jr, Notterpek L. 2010. Rapamycin activates autophagy and improves myelination in explant cultures from neuropathic mice. *J. Neurosci.* 30:11388–97

Ravikumar B, Moreau K, Jahreiss L, Puri C, Rubinsztein DC. 2010. Plasma membrane contributes to the formation of pre-autophagosomal structures. *Nat. Cell Biol.* 12:747–57

Ravikumar B, Vacher C, Berger Z, Davies JE, Luo S, et al. 2004. Inhibition of mTOR induces autophagy and reduces toxicity of polyglutamine expansions in fly and mouse models of Huntington disease. *Nat. Genet.* 36:585–95

Régulier E, Trottier Y, Perrin V, Aebischer P, Déglon N. 2003. Early and reversible neuropathology induced by tetracycline-regulated lentiviral overexpression of mutant huntingtin in rat striatum. *Hum. Mol. Genet.* 12:2827–36

Renna M, Schaffner C, Winslow AR, Menzies FM, Peden AA, et al. 2011. Autophagic substrate clearance requires activity of the syntaxin-5 SNARE complex. *J. Cell Sci.* 124:469–82

Robberecht W, Philips T. 2013. The changing scene of amyotrophic lateral sclerosis. *Nat. Rev. Neurosci.* 14:248–64

Roberts VJ, Gorenstein C. 1987. Examination of the transient distribution of lysosomes in neurons of developing rat brains. *Dev. Neurosci.* 9:255–64

Rodríguez-Muela N, Germain F, Mariño G, Fitze PS, Boya P. 2012. Autophagy promotes survival of retinal ganglion cells after optic nerve axotomy in mice. *Cell Death Differ.* 19:162–69

Rusten TE, Filimonenko M, Rodahl LM, Stenmark H, Simonsen A. 2008. ESCRTing autophagic clearance of aggregating proteins. *Autophagy* 4:233–36

Sah DW, Aronin N. 2011. Oligonucleotide therapeutic approaches for Huntington disease. *J. Clin. Investig.* 121:500–7

Sahu R, Kaushik S, Clement CC, Cannizzo ES, Scharf B, et al. 2011. Microautophagy of cytosolic proteins by late endosomes. *Dev. Cell* 20:131–39

Sarkar S, Davies JE, Huang Z, Tunnacliffe A, Rubinsztein DC. 2007. Trehalose, a novel mTOR-independent autophagy enhancer, accelerates the clearance of mutant huntingtin and alpha-synuclein. *J. Biol. Chem.* 282:5641–52

Schwarz L, Goldbaum O, Bergmann M, Probst-Cousin S, Richter-Landsberg C. 2012. Involvement of macroautophagy in multiple system atrophy and protein aggregate formation in oligodendrocytes. *J. Mol. Neurosci.* 47:256–66

Schweers RL, Zhang J, Randall MS, Loyd MR, Li W, et al. 2007. Nix is required for programmed mitochondrial clearance during reticulocyte maturation. *Proc. Natl. Acad. Sci. USA* 104:19500–5

Seglen PO, Bohley P. 1992. Autophagy and other vacuolar protein degradation mechanisms. *Experientia* 48:158–72

Shehata M, Matsumura H, Okubo-Suzuki R, Ohkawa N, Inokuchi K. 2012. Neuronal stimulation induces autophagy in hippocampal neurons that is involved in AMPA receptor degradation after chemical long-term depression. *J. Neurosci.* 32:10413–22

Shen W, Ganetzky B. 2009. Autophagy promotes synapse development in *Drosophila*. *J. Cell Biol.* 187:71–79

Shimizu H, Toyoshima Y, Shiga A, Yokoseki A, Arakawa K, et al. 2013. Sporadic ALS with compound heterozygous mutations in the SQSTM1 gene. *Acta Neuropathol.* 126:453–59

Shin J-H, Ko HS, Kang H, Lee Y, Lee Y-I, et al. 2011. PARIS (ZNF746) repression of PGC-1a contributes to neurodegeneration in Parkinson's disease. *Cell* 144:689–702

Smith CM, Mayer JA, Duncan ID. 2013. Autophagy promotes oligodendrocyte survival and function following dysmyelination in a long-lived myelin mutant. *J. Neurosci.* 33:8088–100

Song JW, Misgeld T, Kang H, Knecht S, Lu J, et al. 2008. Lysosomal activity associated with developmental axon pruning. *J. Neurosci.* 28:8993–9001

Steele JW, Ju S, Lachenmayer ML, Liken J, Stock A, et al. 2013a. Latrepirdine stimulates autophagy and reduces accumulation of α-synuclein in cells and in mouse brain. *Mol. Psychiatry* 18:882–88

Steele JW, Lachenmayer ML, Ju S, Stock A, Liken J, et al. 2013b. Latrepirdine improves cognition and arrests progression of neuropathology in an Alzheimer's mouse model. *Mol. Psychiatry* 18:889–97

Sugie K, Yamamoto A, Murayama K, Oh SJ, Takahashi M, et al. 2002. Clinicopathological features of genetically confirmed Danon disease. *Neurology* 58:1773–78

Takáts S, Nagy P, Varga Á, Pircs K, Kárpáti M, et al. 2013. Autophagosomal Syntaxin17-dependent lysosomal degradation maintains neuronal function in *Drosophila*. *J. Cell Biol.* 201:531–39

Takeshige K, Baba M, Tsuboi S, Noda T, Ohsumi Y. 1992. Autophagy in yeast demonstrated with proteinase-deficient mutants and conditions for its induction. *J. Cell Biol.* 119:301–11

Tanaka Y, Guhde G, Suter A, Eskelinen EL, Hartmann D, et al. 2000. Accumulation of autophagic vacuoles and cardiomyopathy in LAMP-2-deficient mice. *Nature* 406:902–6

Tang G, Yue Z, Talloczy Z, Hagemann T, Cho W, et al. 2008. Autophagy induced by Alexander disease-mutant GFAP accumulation is regulated by p38/MAPK and mTOR signaling pathways. *Hum. Mol. Genet.* 17:1540–55

Tanik SA, Schultheiss CE, Volpicelli-Daley LA, Brunden KR, Lee VM. 2013. Lewy body-like α-synuclein aggregates resist degradation and impair macroautophagy. *J. Biol. Chem.* 288:15194–210

Thomas M, Alegre-Abarrategui J, Wade-Martins R. 2013. RNA dysfunction and aggrephagy at the centre of an amyotrophic lateral sclerosis/frontotemporal dementia disease continuum. *Brain* 136:1345–60

Tian F, Morimoto N, Liu W, Ohta Y, Deguchi K, et al. 2011. In vivo optical imaging of motor neuron autophagy in a mouse model of amyotrophic lateral sclerosis. *Autophagy* 7:985–92

Tsvetkov AS, Miller J, Arrasate M, Wong JS, Pleiss MA, Finkbeiner S. 2010. A small-molecule scaffold induces autophagy in primary neurons and protects against toxicity in a Huntington disease model. *Proc. Natl. Acad. Sci. USA* 107:16982–87

Umekawa M, Klionsky DJ. 2012. The cytoplasm-to-vacuole targeting pathway: a historical perspective. *Int. J. Cell Biol.* 2012:142634

Vantaggiato C, Crimella C, Airoldi G, Polishchuk R, Bonato S, et al. 2013. Defective autophagy in spastizin mutated patients with hereditary spastic paraparesis type 15. *Brain* 136:3119–39

Vingtdeux V, Chandakkar P, Zhao H, d'Abramo C, Davies P, Marambaud P. 2011. Novel synthetic small-molecule activators of AMPK as enhancers of autophagy and amyloid-β peptide degradation. *FASEB J.* 25:219–31

Vogiatzi T, Xilouri M, Vekrellis K, Stefanis L. 2008. Wild type α-synuclein is degraded by chaperone-mediated autophagy and macroautophagy in neuronal cells. *J. Biol. Chem.* 283:23542–56

Wagenmakers AJ, Schepens JT, Veerkamp JH. 1984. Increase of the activity state and loss of total activity of the branched-chain 2-oxo acid dehydrogenase in rat diaphragm during incubation. *Biochem. J.* 224:491–96

Wang C, Liang C-C, Bian ZC, Zhu Y, Guan J-L. 2013. FIP200 is required for maintenance and differentiation of postnatal neural stem cells. *Nat. Neurosci.* 16:532–42

Wang H, Bedford FK, Brandon NJ, Moss SJ, Olsen RW. 1999. GABA(A)-receptor-associated protein links GABA(A) receptors and the cytoskeleton. *Nature* 397:69–72

Wang JT, Medress ZA, Barres BA. 2012. Axon degeneration: molecular mechanisms of a self-destruction pathway. *J. Cell Biol.* 196:7–18

Wang Y, Martinez-Vicente M, Krüger U, Kaushik S, Wong E, et al. 2009. Tau fragmentation, aggregation and clearance: the dual role of lysosomal processing. *Hum. Mol. Genet.* 18:4153–70

Watts GD, Wymer J, Kovach MJ, Mehta SG, Mumm S, et al. 2004. Inclusion body myopathy associated with Paget disease of bone and frontotemporal dementia is caused by mutant valosin-containing protein. *Nat. Genet.* 36:377–81

Weidberg H, Shvets E, Shpilka T, Shimron F, Shinder V, Elazar Z. 2010. LC3 and GATE-16/GABARAP subfamilies are both essential yet act differently in autophagosome biogenesis. *EMBO J.* 29:1792–802

Winslow AR, Chen CW, Corrochano S, Acevedo-Arozena A, Gordon DE, et al. 2010. α-Synuclein impairs macroautophagy: implications for Parkinson's disease. *J. Cell Biol.* 190:1023–37

Wong YC, Holzbaur EL. 2014. The regulation of autophagosome dynamics by Huntingtin and HAP1 is disrupted by expression of mutant Huntingtin, leading to defective cargo degradation. *J. Neurosci.* 34(4):1293–305

Wooten MW, Geetha T, Babu JR, Seibenhener ML, Peng J, et al. 2008. Essential role of sequestosome 1/p62 in regulating accumulation of Lys63-ubiquitinated proteins. *J. Biol. Chem.* 283:6783–89

Yamamoto A, Lucas JJ, Hen R. 2000. Reversal of neuropathology and motor dysfunction in a conditional model of Huntington's disease. *Cell* 101:57–66

Yamamoto A, Simonsen A. 2011. The elimination of accumulated and aggregated proteins: a role for aggrephagy in neurodegeneration. *Neurobiol. Dis.* 43:17–28

Yang Y, Coleman M, Zhang L, Zheng X, Yue Z. 2013. Autophagy in axonal and dendritic degeneration. *Trends Neurosci.* 36:418–28

Yang Z, Klionsky DJ. 2010. Eaten alive: a history of macroautophagy. *Nat. Cell Biol.* 12:814–22

Ylä-Anttila P, Vihinen H, Jokitalo E, Eskelinen EL. 2009. 3D tomography reveals connections between the phagophore and endoplasmic reticulum. *Autophagy* 5:1180–85

Young JE, Martinez RA, La Spada AR. 2009. Nutrient deprivation induces neuronal autophagy and implicates reduced insulin signaling in neuroprotective autophagy activation. *J. Biol. Chem.* 284:2363–73

Yue Z, Friedman L, Komatsu M, Tanaka K. 2009. The cellular pathways of neuronal autophagy and their implication in neurodegenerative diseases. *Biochim. Biophys. Acta* 1793:1496–507

Yue Z, Horton A, Bravin M, DeJager PL, Selimi F, Heintz N. 2002. A novel protein complex linking the delta 2 glutamate receptor and autophagy: implications for neurodegeneration in lurcher mice. *Neuron* 35:921–33

Zhang X, Garbett K, Veeraraghavalu K, Wilburn B, Gilmore R, et al. 2012. A role for presenilins in autophagy revisited: normal acidification of lysosomes in cells lacking PSEN1 and PSEN2. *J. Neurosci.* 32:8633–48

Zhang X, Li L, Chen S, Yang D, Wang Y, et al. 2011. Rapamycin treatment augments motor neuron degeneration in SOD1(G93a) mouse model of amyotrophic lateral sclerosis. *Autophagy* 7:412–25

Zhao YG, Zhao H, Sun H, Zhang H. 2013. Role of Epg5 in selective neurodegeneration and Vici syndrome. *Autophagy* 9:1258–62

Zheng S, Clabough EB, Sarkar S, Futter M, Rubinsztein DC, Zeitlin SO. 2010. Deletion of the huntingtin polyglutamine stretch enhances neuronal autophagy and longevity in mice. *PLoS Genet.* 6:E1000838

Zschocke J, Zimmermann N, Berning B, Ganal V, Holsboer F, Rein T. 2011. Antidepressant drugs diversely affect autophagy pathways in astrocytes and neurons—dissociation from cholesterol homeostasis. *Neuropsychopharmacology* 36:1754–68

Apolipoprotein E in Alzheimer's Disease: An Update

Jin-Tai Yu,[1,2] Lan Tan,[1,2] and John Hardy[3]

[1] Department of Neurology, Qingdao Municipal Hospital, School of Medicine, Qingdao University, Qingdao 266071, China; email: yu-jintai@163.com, dr.tanlan@163.com

[2] College of Medicine and Pharmaceutics, Ocean University of China, Qingdao 266003, China

[3] Reta Lila Weston Laboratories and Department of Molecular Neuroscience, University College London Institute of Neurology, London WC1N 3BG, United Kingdom; email: j.hardy@ucl.ac.uk

Annu. Rev. Neurosci. 2014. 37:79–100

First published online as a Review in Advance on April 21, 2014

The *Annual Review of Neuroscience* is online at neuro.annualreviews.org

This article's doi: 10.1146/annurev-neuro-071013-014300

Keywords

Alzheimer's disease, apolipoprotein E, polymorphism, amyloid-β, tau, pathogenesis, therapy

Abstract

The vast majority of Alzheimer's disease (AD) cases are late onset (LOAD), which is genetically complex with heritability estimates up to 80%. Apolipoprotein E (*APOE*) has been irrefutably recognized as the major genetic risk factor, with semidominant inheritance, for LOAD. Although the mechanisms that underlie the pathogenic nature of APOE in AD are still not completely understood, emerging data suggest that APOE contributes to AD pathogenesis through both amyloid-β (Aβ)-dependent and Aβ-independent pathways. Given the central role for APOE in the modulation of AD pathogenesis, many therapeutic strategies have emerged, including converting APOE conformation, regulating APOE expression, mimicking APOE peptides, blocking the APOE/Aβ interaction, modulating APOE lipidation state, and gene therapy. Accumulating evidence also suggests the utility of *APOE* genotyping in AD diagnosis, risk assessment, prevention, and treatment response.

Contents

INTRODUCTION

Alzheimer's disease (AD) is the leading cause of dementia in the elderly and the fourth leading cause of death in developed countries, characterized pathologically by the loss of synapses and neurons as well as the formation of extracellular amyloid plaques and intracellular neurofibrillary tangles (Association 2013). Both early-onset AD and late-onset AD (LOAD) have a genetic component. Twin studies predicted the heritability of LOAD to be as high as 80% (Gatz et al. 2006). Thus far, apolipoprotein E (*APOE*) has been established unequivocally as the most important susceptibility gene for LOAD (Genin et al. 2011), which influences susceptibility for <50% of common LOAD. Although recent large genome-wide association studies (GWAS) have identified many other new loci for LOAD, the risk-increasing effects of these genes to AD are much smaller than those of *APOE* (Bettens et al. 2013). As the largest-established genetic risk factor for LOAD, *APOE* has been the primary target of numerous studies investigating this disease's underlying molecular neuropathology, pharmacological therapy, clinical progression, diagnosis, prevention, and treatment response. In this review, we aim to assess the possible mechanisms by which this protein exerts its modulatory effect on AD and provide new perspectives for AD therapy by targeting APOE. We also present the recent advances and challenges on the utility of *APOE* genotyping in AD diagnosis, risk assessment, prevention, and therapeutic response.

STRUCTURE AND NEUROBIOLOGY OF APOE

Human APOE protein is a lipoprotein of 299 amino acids with two structural domains: The N-terminal domain contains the receptor-binding region, and the C-terminal domain contains the lipid-binding region (**Figure 1**) (Chen et al. 2011b). Human *APOE* gene is located on chromosome

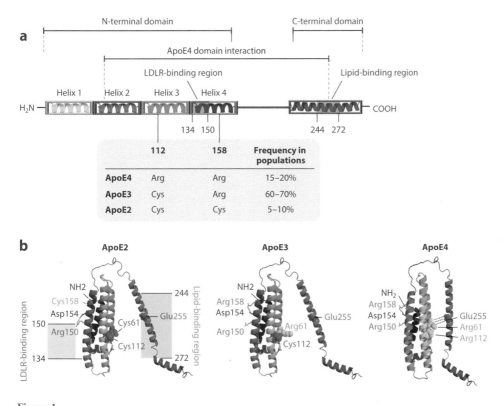

Figure 1

(*a*) Linear diagram of the domain structure of human apolipoprotein E (APOE), a lipoprotein of 299 amino acids with three structural domains (Chen et al. 2011b). The N-terminal domain contains the two polymorphic positions, 112 and 158, that distinguish the three common APOE isoforms. (*b*) Model of the structure of lipid-free APOE. The N-terminal domain is shown as the structure obtained from X-ray crystallography of the isolated domain, whereas the C-terminal domain and linker are modeled as helices on the basis of structure prediction and circular dichroism spectroscopy (Hatters et al. 2006). In APOE3 and APOE4, Arg158 forms a salt bridge with Asp154. In APOE2, an alternative salt bridge forms between Arg150 and Asp154, which lies in the basic LDLR-binding region (Dong et al. 1996). In APOE4, Arg112 orients the side chain of Arg61 into the aqueous environment where it can interact with Glu255, resulting in an interaction between the N- and C-terminal domains (Zhong & Weisgraber 2009). Domain interaction occurs to a significantly lesser extent in APOE2 and APOE3 because both contain Cys112, resulting in a different conformation of Arg61. Abbreviation: LDLR, low-density lipoprotein receptor.

19q13 and encodes three major APOE isoforms: APOE2, APOE3, and APOE4. The differences among the three APOE isoforms are limited to amino acid residues 112 and 158, which have a major effect on the structure and function of APOE at the molecular and cellular levels and are also associated with neuropathological conditions (**Figure 1**) (Zhong & Weisgraber 2009). The recently published solution NMR structure of APOE3 (Chen et al. 2011b) differs greatly from the suggested domain-domain interactions in earlier models (**Figure 1**) (Zhong & Weisgraber 2009). The Arg/Cys change at position 112 may affect the movement of Arg114, which, in turn, perturbs the ionization of His140: This affects the charge distribution of helix 4, finally resulting in the structural difference between APOE3 and APOE4 (Frieden & Garai 2012).

APOE is expressed in many tissues, with the highest expression in the liver followed by the brain. In the brain, astrocytes have been recognized as the primary source of APOE, although microglia and neurons also synthesize this protein (Xu et al. 2006). In the central nervous system (CNS), APOE takes up lipids generated after neuronal degeneration and redistributes them to cells requiring lipids for proliferation, membrane repair, or remyelination of new axons (Hauser et al. 2011). In addition, APOE modulates glutamate receptor function and synaptic plasticity by regulating APOE receptor recycling in neurons (Chen et al. 2010). Recent studies suggest that APOE controls cerebrovascular integrity via cyclophilin A (Bell et al. 2012).

GENETICS OF *APOE* IN ALZHEIMER'S DISEASE

In 1993, genetic analysis identified *APOE4* as the major risk factor for AD (Corder et al. 1993) and immediately thereafter identified *APOE2* as being protective against disease (Chartier-Harlin et al. 1994). In general terms, one allele of *APOE4* shifts the risk curve for the disease to be 5 years earlier, two copies of *APOE4* shift it 10 years earlier, and one copy of the *APOE2* allele shifts it 5 years later; these general rules also hold true in APP and presenilin mutation carriers (Noguchia et al. 1993, Pastor et al. 2003). Although the majority of the risk at the locus of the *APOE* gene is encoded by amino acid changes, there has been ongoing controversy about whether other risks may also be at the locus, which is encoded by genetic variability in expression. This controversy has not been fully resolved, but expression analysis suggests that *APOE4/APOE3* heterozygotes with a high APOE4/APOE3 expression ratio are at higher risk than those with a low ratio (Lambert et al. 1997). For 15 years, *APOE* was the only established locus for LOAD. However, with the advent of GWAS and whole-exome/genome sequencing, many other loci have now been identified. Loci identified by GWAS have much smaller effect sizes than that of *APOE*, typically with odds ratios between 1.1 and 1.3 in contrast to a ratio of $\sim$4 for *APOE4* (Lambert et al. 2013). In addition, loci identified by sequencing can have similar odds ratios, but because they are much rarer than *APOE4* is in the population, they have a much smaller population attributable risk (Guerreiro et al. 2013, Jonsson et al. 2013). Of particular note, however, and relevant to the below discussions of APOE's role in disease, all of the loci identified to date map to pathways involved in immunity and inflammation, cholesterol metabolism, or endosomal vesicle recycling (Jones et al. 2010). This information is consistent given that these pathways relate to the role of APOE in AD.

AMYLOID-β-DEPENDENT ROLES FOR APOE IN ALZHEIMER'S DISEASE

Effects of APOE Receptors on APP Trafficking and Amyloid-β Production

APOE receptors are a group of transmembrane proteins known as the LDLR (low-density lipoprotein receptor) family that mediate endocytosis of ligands and then recycle back to the cell surface (Holtzman et al. 2012). Numerous adaptors that bind to APOE receptors interact with the NPxY motif of APP (Hauser et al. 2011). Several studies have confirmed that APOE receptors play an important role in modulating APP trafficking and amyloid-β (Aβ) production (**Figure 2a**) (Bu 2009). The effects of APOE isoforms on APP processing and Aβ production have been investigated in cell-culture systems. Although some studies suggest that lipid-poor and lipid-free APOE4 enhance Aβ production by increasing LRP1- and APOER2-mediated APP endocytosis (Ye et al. 2005), others found no clear evidence for isoform-specific effects on APP processing (Irizarry et al. 2004). Recently, differential coexpression correlation network analysis of the *APOE4* and LOAD transcriptomic changes identified a set of candidate core *APOE4* effectors, including several genes encoding known or novel modulators of APP processing and endocytic trafficking, such as *APBA2*, *ITM2B*, *FYN*, *RNF219*, and *SV2A* (Rhinn et al. 2013). Further investigation of the possible roles of APOE isoforms in Aβ production in vivo is warranted.

Effects of APOE on Amyloid-β Aggregation

Several studies have reported human APOE isoform-dependent differences in Aβ deposition (APOE4 ≫ APOE3 > APOE2) (**Figure 2b**) (Bales et al. 2009, Castellano et al. 2011, Youmans et al. 2012). Neuropathological examination in LOAD suggests that APOE ε4 allele dosage is associated with increased Aβ, Aβ oligomers, and plaque accumulation in the brain (Christensen et al. 2010, Hashimoto et al. 2012, Koffie et al. 2012). Using AD transgenic animal models, several laboratories replicated the APOE isoform-dependent effects (APOE4>APOE3>APOE2) on both brain Aβ oligomers formation (Hashimoto et al. 2012, Youmans et al. 2012) and Aβ plaque deposition (Bales et al. 2009). Recently, hAPP transgenic mice expressing human *APOE3* or *APOE4* demonstrated a gene dose-dependent effect and an isoform-dependent effect of *APOE* on Aβ accumulation (APOE4>APOE3>APOE2) (Bien-Ly et al. 2012). Further elucidating the underlying in vivo mechanisms of APOE isoform-mediated differences in Aβ accumulation is critical for relating these findings to AD pathogenesis. Studies using positron emission tomography (PET) imaging with [11C] Pittsburgh compound B (PIB) have also shown that the *APOE* ε4 allele increases brain amyloid burden in a dose-dependent manner both in cognitively normal elderly patients (Reiman et al. 2009) and in prodromal AD patients (Villemagne et al. 2011), although PIB studies in AD patients have produced conflicting results (Ossenkoppele et al. 2013, Villemagne et al. 2011). The increase in fibrillar Aβ deposition occurs concomitantly with decreased cerebrospinal fluid (CSF) Aβ42 levels, supporting the notion that APOE4 promotes Aβ deposition in brain (Morris et al. 2010). Recently, a florbetapir PET GWAS also identified APOE as the top modulator of cerebral amyloid deposition (Ramanan et al. 2013).

Effects of APOE on Amyloid-β Clearance

LOAD is characterized by an overall impairment in Aβ clearance (Mawuenyega et al. 2010). APOE is the best-characterized Aβ chaperone (Bu 2009). Although the exact molecular and cellular processes by which the APOE isoform modulates Aβ clearance is unclear, several APOE/Aβ-based mechanisms have been proposed to function in an isoform-dependent manner (APOE2>APOE3≫APOE4) (**Figure 2b**). The primary effects of such mechanisms include the following (Tai et al. 2013): (*a*) Aβ clearance via microglia (Lee et al. 2012), astrocytes (Basak et al. 2012), and neurons (Li et al. 2012); (*b*) Aβ clearance via the blood-brain barrier (BBB) (Bachmeier et al. 2013); (*c*) Aβ clearance via enzymatic degradation (Jiang et al. 2008); (*d*) Aβ oligomerization (Hashimoto et al. 2012); and (*e*) Aβ clearance by drainage via the interstitial fluid (Castellano et al. 2011) or perivasculature (Hawkes et al. 2012). APOE may facilitate the uptake of APOE/Aβ complexes via APOE receptors, such as LDLR, LRP1, and sortilin (Bu 2009, Carlo et al. 2013). Furthermore, APOE isoform-specific APOE/Aβ levels may affect the dynamic compartmentalization of Aβ in the CNS (Hong et al. 2011). This suggestion is supported by findings that levels of soluble APOE/Aβ are lower and oligomeric Aβ levels are higher with APOE4 in AD transgenic mouse brain and in human synaptosomes and CSF (Tai et al. 2013). Moreover, the observed APOE isoform-specific effect on Aβ clearance is also attributed to its lipidation status, and optimal lipidation of APOE2 versus APOE3 and APOE4 makes an important contribution to Aβ clearance in AD synapses (Arold et al. 2012). APOE2 and APOE3 are more highly lipidated than the APOE4 isoform (Jiang et al. 2008). Interestingly, APOE isoforms also regulate

a Effects of APOE receptors on APP trafficking and Aβ production

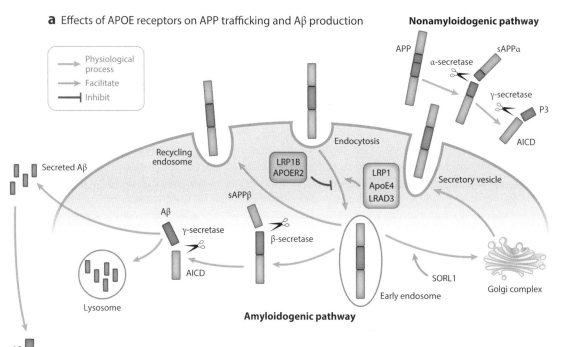

b Effects of APOE on Aβ aggregation and clearance

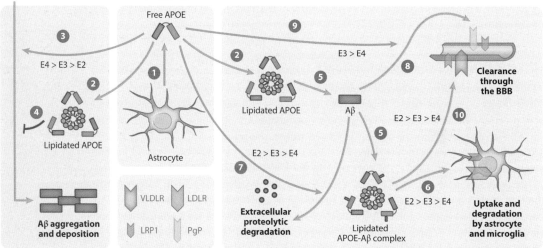

the metabolism of soluble Aβ by competing with the LRP1-dependent cellular uptake pathway in astrocytes (Verghese et al. 2013).

AMYLOID-β-INDEPENDENT ROLES FOR APOE IN ALZHEIMER'S DISEASE

The Effect of APOE on Tau Pathology

APOE has also been reported to affect tau neuropathological changes seen in AD brains in vitro and in animal models (**Figure 3**). APOE3 may inhibit the abnormal hyperphosphorylation and the destabilization of the neuronal cytoskeleton in AD (Strittmatter et al. 1994), whereas the C-terminal-truncated form of APOE4 is neurotoxic and stimulates tau phosphorylation, leading to pre-NFTs (neurofibrillary tangles) (Harris et al. 2003). In agreement with this suggestion, 3xTG-ApoE4 mice have more total human tau accumulation in both somatodendritic and intra-axonal compartments (Bennett et al. 2013). Unlike in vitro and animal models, human studies of the relationship between APOE and tau show contradictory results (Kim et al. 2009). Interestingly, the *APOE* genotype appears to interact with AD CSF biomarkers, total tau, and phosphorylated tau (Leoni 2011). APOE4 carriers are consistently found with higher tau, phosphorylated tau, and tau/Aβ ratios at every disease stage (Mattsson et al. 2009, Vemuri et al. 2010).

Numerous studies have reported a link between APOE and glycogen synthase kinase 3 (GSK3), one of the main kinases that phosphorylates tau. Studies have reported that APOE regulates GSK3 activity and that APOE4 has the greatest effect (Hoe et al. 2006). The mechanism linking APOE, GSK3, and tau phosphorylation is based on the cell surface LDLR family to which APOE binds (Hoe et al. 2006). Most of these studies were performed in cell culture. Therefore, further in vivo studies are needed to prove an APOE isoform-dependent effect on tau hyperphosphorylation and its underlying mechanism.

←

Figure 2

Aβ-dependent roles for APOE in AD. (*a*) Effects of APOE receptors on APP trafficking and Aβ production. In the nonamyloidogenic pathway, APP is cleaved by α-secretase and then by γ-secretase to produce a minimally toxic peptide called p3. In the amyloidogenic pathway, APP is internalized and delivered to endosomes. APP is cleaved first by β-secretase and then by γ-secretase to generate the toxic Aβ peptide. Although most Aβ peptides are secreted to the extracellular space, some can aggregate in the lysosomes and contribute to intraneuronal Aβ accumulation. LRP1 fast endocytosis enhances APP endocytosis and processing to Aβ, whereas LRP1B decreases APP processing to Aβ and APOER2 has differential effects depending on ligand binding. LRAD3 as a novel APOE receptor can interact with APP and affect APP processing by decreasing sAPPα and increasing Aβ production. SORL1 may shuttle APP to the Golgi compartments and reduce its processing via β-secretase in the early endosome, thus decreasing processing to Aβ. Additionally, APOE4 may promote APP amyloidogenic processing in a manner that depends on LRP1 function. (*b*) Effects of APOE on Aβ aggregation and clearance. In the brain, APOE is primarily produced by astrocytes (❶) and is subsequently lipidated by ABCA1 to form lipoprotein particles (❷). APOE could accelerate Aβ aggregation and deposition in an isoform-dependent manner (❸). In contrast, increasing APOE lipidation status decreases brain Aβ levels in APP transgenic animals (❹). In the extracellular space, lipidated APOE is able to bind to soluble Aβ and to transport Aβ within the central nervous system (❺). APOE may also facilitate the cellular uptake and degradation of Aβ by astrocytes or microglias through the endocytosis of the lipidated APOE-Aβ complex, which is mediated by different members of the LDL family, such as LDLR and LRP1 (❻). APOE facilitates extracellular proteolytic degradation of Aβ via IDE and other related enzymes in an isoform-dependent manner (❼). At the BBB, soluble Aβ is predominantly transported from the interstitial fluid into the bloodstream via LRP1 and P-glycoprotein (❽). Free APOE (i.e., not bound to Aβ) in the brain can facilitate Aβ BBB transit, both in vitro and in vivo in an isoform-dependent manner (❾). APOER2/Aβ and APOE3/Aβ complexes are cleared at the BBB via both VLDLR and LRP1 at a substantially faster rate than is the APOE4/Aβ complex (❿). Abbreviations: Aβ, amyloid-β; AD, Alzheimer's disease; AICD, APP intracellular domain; APOE, apolipoprotein E; APOER2, APOE receptor 2; BBB, blood-brain barrier; LDL, low-density lipoprotein; LRAD3, LDL receptor class A domain containing 3; SORL1, sortilin-related receptor 1 (also known as SORLA or LR11); VLDLR, very-low-density lipoprotein receptor.

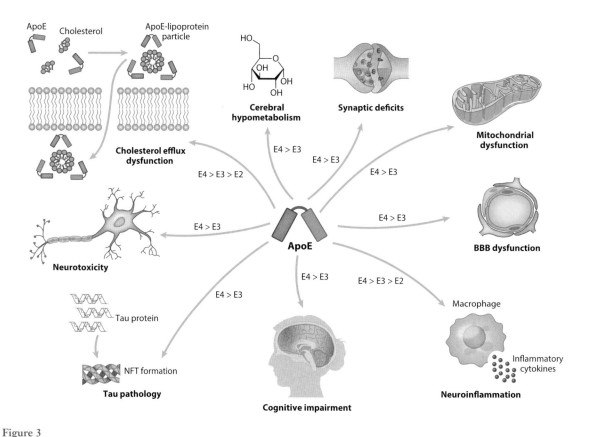

Figure 3

Summary of Aβ-independent roles for ApoE in the pathogenesis of Alzheimer's disease. The isoform-dependent effects of ApoE are indicated. Abbreviations: ApoE, apolipoprotein E; BBB, blood-brain barrier; NFT, neurofibrillary tangle.

The Effect of APOE on Synaptic Plasticity, Dendritic Spine Integrity, and Cognitive Impairment

As the major lipid transport protein in the CNS, APOE plays an important role in promoting synapse formation and synaptic plasticity (Hauser et al. 2011, Klein et al. 2010). APOE4 is a cofactor that enhances the toxicity of oligomeric Aβ by directing it to synapses, thus providing a link between the *APOE* ε4 genotype and synapse loss (Koffie et al. 2012). Several in vitro studies have illustrated the requirement of APOE-lipoprotein complexes in neurite outgrowth and synaptogenesis (Mauch et al. 2001). APOE also differentially regulates dendritic spine density and morphology in an isoform- and age-dependent manner (Dumanis et al. 2009, Klein et al. 2010). APOE4 TR mice have significantly fewer and shorter dendritic spines in cortical neurons compared with APOE3 and APOE2 TR mice, and these differences increase gradually with age (Dumanis et al. 2009, Klein et al. 2010).

APOE also affects long-term potentiation (i.e., the current in vitro memory model) in an isoform-specific manner (via induction of long-term potentiation: APOE4>APOE3>APOE2) (Korwek et al. 2009). APOE isoforms may differentially affect synaptic plasticity and dendritic spine integrity via the modulation of spine elimination (Basak & Kim 2010); as a result of isoform-dependent levels of APOE (Basak & Kim 2010, Dumanis et al. 2009); and by acting as signaling molecules through interactions with APOE receptors (Chen et al. 2010), NMDA receptors

(Korwek et al. 2009), and a novel intracellular PKCε pathway (Sen et al. 2012). Further studies are warranted to determine the exact molecular and cellular mechanisms of these phenomena.

The Role of APOE in Neuroinflammation

Increasing evidence suggests that neuroinflammation contributes to AD neuropathogenesis (Broussard et al. 2012). APOE modulates inflammatory and immune responses in an isoform-dependent manner (APOE4>APOE3>APOE2) (Keene et al. 2011, Zhang et al. 2011). Although APOE3 is critical for suppressing inflammation, APOE4 is associated with an overactive proinflammatory immune phenotype (Keene et al. 2011, Zhang et al. 2011). Activation of primary microglia from *APOE* transgenic mice with lipopolysaccharide led to genotype-dependent differences in cytokine secretion (APOE4>APOE3>APOE2) (Vitek et al. 2009). Several in vivo studies also provide support regarding the role of dramatically increased proinflammatory cytokines in *APOE4/4* mice (Zhu et al. 2012). Moreover, a human study demonstrated that the *APOE* genotype is related to the levels of the plasma inflammatory markers (APOE4>APOE3>APOE2) (Ringman et al. 2012). Additionally, anti-inflammatory interventions in AD also show greater benefit in *APOE* ε4 carriers as opposed to noncarriers (Szekely et al. 2008). Through interactions with APOE receptors, various APOE isoforms differently modulate the components of the innate immune system via multiple signaling pathways, including NF-κB signaling (Bell et al. 2012), p38 MAP kinase signaling (Baitsch et al. 2011), JNK signaling (Pocivavsek et al. 2009), and endoplasmic reticulum stress signaling (Cash et al. 2012).

The Role of APOE in Neurotoxicity

APOE4 (>APOE3>APOE2) may also have direct neurotoxic effects and induce neuropathology through various cellular pathways (Mahley & Huang 2012). APOE4 is more susceptible to proteolytic cleavage than is APOE3, thus resulting in the generation of neurotoxic fragments (Brecht et al. 2004, Harris et al. 2003). The N-terminal fragment may play a role in neurotoxicity, possibly involving GTPase, CREB activation, calcium dysregulation, and/or activation of the AKT pathway. By contrast, the C-terminal fragment is more likely associated with AD-like neuropathological changes (Mahley & Huang 2012). A recent study shows that C-terminal-truncated APOE4 inefficiently clears Aβ and acts in concert with Aβ to elicit neuronal and behavioral deficits in transgenic mice (Bien-Ly et al. 2011). This C-terminal fragment also stimulates tau phosphorylation and the formation of NFT-like inclusions in vitro and in vivo (Brecht et al. 2004, Harris et al. 2003). Furthermore, some studies also strongly suggest that APOE4 causes age- and tau-dependent impairment of hilar GABAergic interneurons (Andrews-Zwilling et al. 2010).

The Role of APOE in Lipid Metabolism

Growing evidence suggests that dysregulated lipid metabolism may be involved in AD pathogenesis (Di Paolo & Kim 2011). APOE is also a potent modulator of plasma lipoprotein and cholesterol levels (Hauser et al. 2011). APOE is the major component of lipoprotein particles in the brain that mediate transport of cholesterol and other lipids between neurons and glial cells, indicating that the effects of APOE4 in AD pathogenesis are mediated via mechanisms related to brain lipid metabolism (Hauser et al. 2011). In cultured neurons, APOE4 is less efficient than APOE2 and APOE3 in transporting brain cholesterol (Rapp et al. 2006). Moreover, APOE-induced intracellular Aβ degradation is mediated by the cholesterol efflux function of APOE, which lowers cellular cholesterol levels and subsequently facilitates the intracellular trafficking of Aβ to lysosomes for degradation (Lee et al. 2012). Because the cholesterol efflux activity is APOE isoform dependent (APOE2>APOE3>APOE4) (Hara et al. 2003), the higher risk of AD observed in APOE4 carriers may be due to the poorer efficiency of cholesterol efflux by APOE4.

The Effect of APOE on Metabolic and Structural Alterations in the Brain

Brain energy metabolism in human subjects is most often studied using 18-fluoro-2-deoxyglucose (18F-FDG) PET. Several studies have shown that cognitively normal elderly (Protas et al. 2013), prodromal AD patients (Mosconi et al. 2008), and AD patients (Ossenkoppele et al. 2013) who are *APOE* ε4 carriers have reduced posterior cingulate glucose metabolism in a dose-dependent manner. Moreover, functional MRI studies have reported that brain activity in the default mode network is increased in young, middle-aged, and elderly carriers of *APOE* ε4 (Filippini et al. 2009, Sheline et al. 2010). Whether APOE isoform-specific lipid dysregulation or cellular signaling contributes specifically to cerebral hypometabolism in AD patients or whether the hypometabolism is secondary to brain injury, altered Aβ metabolism, or another AD-related dysfunction remains under investigation. Moreover, many MRI studies have found that *APOE*-mediated cerebral structural alterations may contribute to a predisposition to AD. High-resolution structural MRI of a pool of older subjects showed a selective, dose-dependent association of *APOE* ε4 with hippocampal CA1 apical neuropil atrophy and episodic memory across the cognitive spectrum (Kerchner et al. 2014). Recently, a study of infants showed that *APOE* ε4 carriers had decreased cortical gray matter volume in the precuneus, posterior-middle cingulate, lateral temporal, and medial occipitotemporal regions, all areas preferentially affected by AD (Dean et al. 2014). These findings indicate that *APOE* ε4 carriers have brain structural and developmental alterations that provide a foothold for the neuropathological changes associated with the subsequent course of AD.

The Effect of APOE on Mitochondrial Dysfunction

Mitochondrial dysfunction appears to be a precipitating or exacerbating factor in both familial AD and LOAD (Colca & Feinstein 2012). Mitochondrial dysfunction in AD patients varies with *APOE* genotype, and the effects are greater in *APOE4* carriers than noncarriers (Gibson et al. 2000). Moreover, in the postmortem cortex of younger individuals, *APOE* ε4 carriers have greater oxidative mitochondrial dysfunction than do noncarriers (Conejero-Goldberg et al. 2011, Valla et al. 2010). Transgenic *APOE4* mice also exhibit structural (Shenk et al. 2009) and functional (Strum et al. 2007) mitochondrial abnormalities.

Several studies have implicated mitochondria as a potential target for APOE or its proteolytic fragments. The APOE4(1-272) fragment can bind to subunits of mitochondrial respiratory complexes and perturb their activities (Nakamura et al. 2009). In both N2A cells and cortical neurons, full-length APOE4 reduced the expression of subunits of mitochondrial respiratory complexes, resulting in a reduction in mitochondrial respiratory function (Chen et al. 2011a). A recent study demonstrates clear *APOE* genotype differences in the proteomic response of mitochondria (James et al. 2012). It would be of interest to determine whether upregulating the expression of mitochondrial energy metabolism genes can protect against APOE4-induced mitochondrial dysfunction and mitigate AD pathology.

The Effect of APOE on Blood-Brain Barrier Permeability

Recent data from brain imaging studies in AD patients and animal models suggest that BBB dysfunction may precede cognitive decline and onset of neurodegenerative changes, thereby contributing to the onset and progression of dementia (Zlokovic 2011). Increasing evidence suggests that APOE plays an important role in maintaining BBB integrity in an isoform-dependent manner and that lack of APOE leads to BBB leakage (Bell et al. 2012, Nishitsuji et al. 2011). Moreover, APOE regulates the integrity of tight junctions in an isoform-dependent manner in in vitro BBB models (Nishitsuji et al. 2011). Similarly, BBB permeability is increased in *APOE4* knockin mice compared with *APOE3* knockin mice (Bell et al. 2012, Nishitsuji et al. 2011). By activating a proinflammatory CypA–nuclear factor-κB–matrix metalloproteinase-9 pathway in pericytes using a lipoprotein receptor, researchers also found that animals carrying *APOE4*, but not *APOE2* or *APOE3*, have a disrupted BBB (Bell et al. 2012). These vascular defects in *APOE*-deficient and *APOE4*-expressing mice precede neuronal dysfunction and eventually cause neurodegenerative changes.

APOE-TARGETED THERAPEUTICS

Converting APOE4 into an APOE3-Like Conformation

The involvement of APOE4 in AD mandates the development of APOE4-specific therapies (**Figure 4**). Evidence suggests that APOE4 domain interaction mediates its detrimental effects in AD pathogenesis (Brodbeck et al. 2011, Chen et al. 2012, Zhong & Weisgraber 2009).

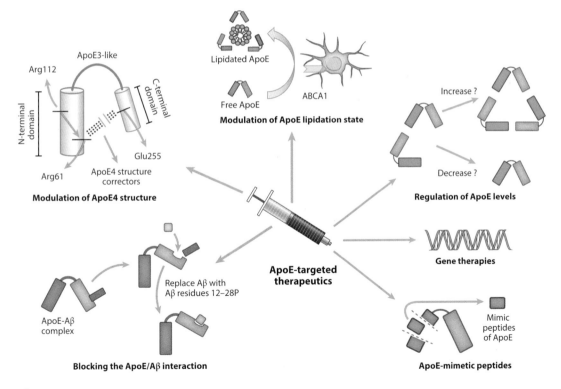

Figure 4

Summary of ApoE-based therapeutic approaches for Alzheimer's disease. Desired therapeutic strategies include modulation of the ApoE4 structure, regulation of ApoE levels, mimicking of the protective effects of endogenous ApoE3, modulation of ApoE lipidation states, blockade of the ApoE/Aβ interaction, and ApoE-targeted gene therapies. Abbreviations: Aβ, amyloid-β; ApoE, apolipoprotein E.

However, mutant APOE4 (APOE4-R61T) lacking domain interaction behaves like APOE3 and does not cause these detrimental effects (Chen et al. 2011a). Hence, disrupting APOE4 domain interaction with small molecules (Brodbeck et al. 2011, Chen et al. 2012) may represent a new therapeutic strategy. Notably, using high-throughput screening, researchers have found several small molecules, such as GIND-25, PH-002, and CB9032258, that inhibit domain interaction and thus alter the detrimental effects of APOE4 in neuronal cell culture (Brodbeck et al. 2011, Chen et al. 2012). Recent studies have also noted out that at least three peptides (peptides 15–30, 116–123, and 271–279) differ between APOE3 and APOE4 (Frieden & Garai 2012). Thus, stabilizing these regions may alter the properties of APOE4 to become more similar to those of APOE3. Site-directed mutagenesis within these peptides or small-molecule screening should also enable identification of residues that change the behavior of APOE4 to resemble that of APOE3.

Regulation of APOE Expression

In addition to isoform status, the amount of APOE also plays a role in AD pathogenesis. Despite some conflicting studies (Bachmeier et al. 2013, Korwek et al. 2009), several groups have reported that APOE concentration in the brains of APOE knockin mice varies in an APOE isoform-dependent manner (APOE2>APOE3>APOE4) (Bales et al. 2009). Recent human studies have revealed that plasma APOE levels are reduced in AD and correlate with brain amyloid load through PIB-PET (Gupta et al. 2011). Moreover, CSF APOE protein levels are associated with lower CSF $A\beta42$ levels, independent of the APOE genotype (Cruchaga et al. 2012). Therefore, targeting proteins in the brain that modulate APOE levels represents an attractive pathway for decreasing amyloid deposition in AD therapy.

The expression of APOE is transcriptionally regulated by the ligand-activated nuclear receptors PPARγ and LXRs, which form obligate heterodimers with RXRs (Cramer et al. 2012). Several studies have shown that activation of LXRs by the agonists GW3965 and T0901317 (T0) increases murine APOE expression levels, promotes $A\beta$ clearance, decreases brain $A\beta$ load, and reserves memory deficits in transgenic mice (Donkin et al. 2010, Jiang et al. 2008, Namjoshi et al. 2013). Recently, pharmacological activation of RXRs with bexarotene reversed $A\beta$-induced deficits in neural function and diminished behavioral abnormalities in APPPS1-21 mice in an APOE-dependent way (Cramer et al. 2012). These studies are controversial, although other groups have reported no change in plaque numbers in treated mice even while confirming changes in APP metabolism (Fitz et al. 2013, Landreth et al. 2013, Price et al. 2013, Tesseur et al. 2013, Veeraraghavalu et al. 2013).

Interestingly, some recent studies directly suggest that decreasing, rather than increasing, human APOE levels regardless of isoform or age at treatment may be a better therapeutic approach for AD (Bien-Ly et al. 2012, Kim et al. 2011, Kuszczyk et al. 2013) Additionally, decreasing APOE4, or increasing APOE2, may generate beneficial effects such as increasing $A\beta$ clearance through the BBB, decreasing BBB susceptibility to injury, and attenuating $A\beta$ neurotoxicity (Bell et al. 2012, Bu 2009).

APOE-Mimetic Peptides

In addition to blocking the potentially deleterious effects of the APOE4 isoform, another therapeutic strategy is to mimic the protective effects of endogenous APOE3. However, this approach remains controversial. APOE mimetic peptides are short synthetic peptides that share structural as well as biological features of native APOE (Osei-Hwedieh et al. 2011). In different AD models, some APOE-mimetic peptides, such as COG112, COG133, and COG1410, inhibit neurodegeneration, improve cognitive functions, decrease $A\beta$ levels, and reduce neuroinflammation and hyperphosphorylation of tau (Ghosal et al. 2013, Sarantseva et al. 2009, Vitek et al. 2012). Additionally, a double or triple tandem repeat of amino acids 141–149 located in the APOE receptor-binding region has increased cell surface APP and decreased Aβ levels in in vitro cell culture (Minami et al. 2010). Thus, the use of APOE-mimetic peptides appears to be a possibility in the development of AD treatments.

Gene Therapy Directed Toward APOE

The first study of gene delivery of human *APOE* in mice provided interesting and promising results (Dodart et al. 2005). Expression of human APOE4 led to increased Aβ burden, indicating that APOE4 is deleterious. In addition, compared with APOE3 and APOE4 in PDAPP mice, lentivirus-mediated APOE2 expression significantly decreased both Aβ42 concentration and Aβ burden (Dodart et al. 2005). This proof-of-concept study indicated that gene delivery of *APOE2* may have a disease-modifying effect on amyloid pathology. Moreover, as discussed above, some recent studies suggest that decreasing, rather than increasing, human APOE levels, regardless of isoform, may be a better therapeutic approach (Bien-Ly et al. 2012, Kim et al. 2011). Hence, gene-inhibition strategies, such as antisense oligonucleotides and RNA interference, may present attractive alternative approaches for decreasing *APOE* expression in AD.

Blocking the APOE/Amyloid-β Interaction

Pharmacological blockade or neutralization of the APOE/Aβ interaction may also provide an alternative therapeutic strategy for AD. Aβ12-28P, a synthetic peptide mimicking Aβ12–28, reduces intraneuronal Aβ accumulation, inhibits loss of synaptic proteins, and ameliorates memory deficits in APP/PS1 transgenic mice (Kuszczyk et al. 2013, Sadowski et al. 2006). Recently, a study tried to map the primary components of the APOE/Aβ interaction to smaller protein segments more suited to guiding the development of small-molecule inhibitors (Liu et al. 2011). Three peptides (APOE249-256[251-Leu], Aβ17–21[20-Ile], and Aβ17–21), Congo red dye, and X-34 are efficient blockers of APOE/Aβ binding. Moreover, Aβ20–29 may also potentially block the APOE/Aβ interaction and play an effective role in reducing Aβ fibrillization (Hao et al. 2010). Additionally, an APOE-specific antibody, HJ6.3, inhibits amyloid accumulation in APPswe/PS1ΔE9 mice, suggesting anti-APOE immunotherapy may represent a novel AD therapeutic strategy (Kim et al. 2012). These approaches should be assessed carefully because they could disrupt APOE-lipid interactions and the associated beneficial functions of APOE.

Modulation of APOE Lipidation State

The lipidation status of APOE is an important functional parameter (Hauser et al. 2011). Membrane-bound ABCA1 is the primary factor in the brain that regulates APOE lipidation (Wahrle et al. 2004). Poorly lipidated APOE promotes amyloidogenesis in AD mouse models (Wahrle et al. 2005), whereas selective ABCA1 overexpression is sufficient to block amyloid plaque formation (Wahrle et al. 2008). ABC1 transcription is controlled primarily by LXRs, RXRs, or PPARγ (Hauser et al. 2011). In addition, LXRs, RXRs, or PPARγ agonists, such as T0901317, GW3965, bexarotene, and pioglitazone, could reduce amyloid levels and improve cognitive performance in AD mice (Cramer et al. 2012, Donkin et al. 2010, Jiang et al. 2008, Mandrekar-Colucci et al. 2012, Namjoshi et al. 2013). However, besides modulating APOE lipidation, other mechanisms, such as anti-inflammatory pathways and direct enhancement of APOE expression triggered by these agonists, may also contribute (Cramer et al. 2012, Cui et al. 2012, Terwel et al. 2011).

APOE GENOTYPING FOR ALZHEIMER'S DISEASE RISK ASSESSMENT, DIAGNOSIS, AND PREVENTION

APOE is the most robust risk marker available for LOAD: Individuals carrying the ε4 allele are at increased risk of AD compared with those carrying the more common ε3 allele. By contrast, the ε2 allele decreases risk (Bettens et al. 2013, Liu et al. 2013). In *APOE* ε4 carriers, lifetime risk of AD at 85 years is estimated to be as high as 68% and 35% for *APOE* ε4ε4 and *APOE* ε3ε4 female carriers, respectively (Genin et al. 2011). Moreover, *APOE* ε4 also affects age-related memory trajectories: In presymptomatic individuals, accelerating declines occur in a gene-dose manner (Caselli et al. 2009), which is in agreement with a recent GWAS that also identified *APOE4* as the only significant gene associated with age-related cognitive decline in humans (De Jager et al. 2012).

Although an association between the *APOE* ε4 allele and increased AD risk has been established, the utility of *APOE* genotyping in clinical diagnosis of AD is still unclear (McKhann et al. 2011). Current practice guidelines advocate *APOE* genotyping in cases of dementia and mild cognitive impairment and also in asymptomatic participants within the context of clinical/epidemiological research, but *APOE* genotyping has not been recommended for prognostication in cognitively intact persons outside the research area (Schipper 2011).

Recently, increasing evidence indicates that *APOE* interacts with potentially modifiable risk and protective factors for AD, including exercise (Lautenschlager et al. 2008), type 2 diabetes mellitus and prediabetic states (Dore et al. 2009, Ronnemaa et al. 2008), nutrition (Samieri et al. 2011), hyperlipidemia (Hall et al. 2006), and traumatic brain injury (Wilson & Montgomery 2007). Moreover, the beneficial effects of some preventive strategies may be more prominent in patients positive for *APOE* ε4 (Kariv-Inbal et al. 2012, Samieri et al. 2011). Additionally, several epidemiological and prospective studies suggest that long-term use of NSAIDs was associated with a significantly lower risk of AD in *APOE* ε4 subjects (in 't Veld et al. 2001, Szekely et al. 2008). Delineation of the influences of the *APOE* genotype on modifiable AD risk factors and prevention may spur consideration of *APOE* testing for presymptomatic individuals seeking to define their personal risk (Schipper 2011). The Risk Evaluation and Education for Alzheimer's Disease (REVEAL) study is a series of multisite randomized clinical trials that examine the impact of *APOE* genetic susceptibility testing on asymptomatic individuals. REVEAL I showed that an AD genetic risk assessment with *APOE* genotyping can be given to relatives of people with AD without causing severe adverse psychological or behavioral effects (Green et al. 2009).

APOE GENOTYPES IN ALZHEIMER'S DISEASE THERAPEUTICS

Roughly 2% of the population bears the *APOE* ε4/ε4 genotype, and the major risk conferred by this genotype indicates approximately 30% onset by age 75 and >50% by age 85 (Genin et al. 2011). As such, it would be appropriate to prioritize targeting of these individuals, as well as presenilins or APP mutation carriers, in clinical trials aimed at developing novel preventive therapeutics. In more than 100 clinical trials for dementia, *APOE* has been used as the only gene of reference for the pharmacogenomics of AD (Cacabelos et al. 2012). Although it remains a controversial area, several studies have suggested that the presence of the *APOE* ε4 allele differentially affects the quality and extent of drug responsiveness in AD patients treated with both cholinesterase inhibitors (Patterson et al. 2011, Petersen et al. 2005) and other noncholinergic strategies (Risner et al. 2006). The *APOE* ε4 allele could be useful as an explanatory variable or covariate if warranted by a drug's action (Kennedy et al. 2014).

Because the *APOE* ε4 allele is associated with an increased amyloid burden, it may have a greater influence on Aβ-directed therapies. Interestingly, in a clinical trial in mild to moderate

AD of a monoclonal antiamyloid antibody (bapineuzumab), analyses revealed positive effects on cognitive and functional endpoints only in *APOE* ε4 noncarriers, and vasogenic edema and sulcal effusions (ARIA-E) were more frequent in *APOE* ε4 carriers (Salloway et al. 2009, Sperling et al. 2012). The potential benefit found in secondary analyses in a phase II study was encouraging enough for Janssen and Pfizer to progress to two phase III trial tests, with both manufacturers conducting separate trials in *APOE* ε4 carriers and noncarriers.

CONCLUDING REMARKS

APOE is the major genetic risk factor for AD. The single amino acid differences among the APOE isoforms modulate APOE structure to profoundly affect its functions. Although the mechanisms that underlie the pathogenic nature of APOE in AD are not completely understood, emerging data suggest that APOE may contribute to AD pathogenesis through both Aβ-dependent and Aβ-independent pathways in an isoform-dependent manner (**Figures 2** and **3**). Further studies will be important to clarify whether the *APOE* ε4 allele influences AD pathogenesis by a gain of toxic function, a loss of protective function, or a combination of both. Given the central role for APOE in the modulation of AD processes, many therapeutic strategies targeting APOE have emerged (**Figure 4**). However, the effects of these therapeutic interventions on the human apoE isoforms are still unknown. Moreover, although current practice guidelines advocate *APOE* genotyping in cases of dementia and mild cognitive impairment and also in asymptomatic participants within the context of clinical/epidemiological research, further delineation of the influences of the *APOE* genotype on modifiable AD risk factors and prevention may prompt consideration of presymptomatic *APOE* testing for individuals actively seeking to define their personal risk. Additionally, emerging evidence suggests that drugs designed for AD treatment have the potential to interact with the *APOE* genotype, which may be an important factor to consider in future trial designs for AD.

DISCLOSURE STATEMENT

The authors are not aware of any affiliations, memberships, funding, or financial holdings that might be perceived as affecting the objectivity of this review.

ACKNOWLEDGMENTS

We thank Dr. Teng Jiang and Dr. Chuandong Jia for preparing the figures. This work was supported by grants to J.-T.Y. and L.T. from the National Natural Science Foundation of China (81000544, 81171209) and by an anonymous foundation (J.H.).

LITERATURE CITED

Alzheimer's Assoc. 2013. Alzheimer's disease facts and figures. *Alzheimer's Dement.* 9:208–45

Andrews-Zwilling Y, Bien-Ly N, Xu Q, Li G, Bernardo A, et al. 2010. Apolipoprotein E4 causes age- and tau-dependent impairment of GABAergic interneurons, leading to learning and memory deficits in mice. *J. Neurosci.* 30:13707–17

Arold S, Sullivan P, Bilousova T, Teng E, Miller CA, et al. 2012. Apolipoprotein E level and cholesterol are associated with reduced synaptic amyloid β in Alzheimer's disease and apoE TR mouse cortex. *Acta Neuropathol.* 123:39–52

Bachmeier C, Paris D, Beaulieu-Abdelahad D, Mouzon B, Mullan M, Crawford F. 2013. A multifaceted role for apoE in the clearance of β-amyloid across the blood-brain barrier. *Neurodegener. Dis.* 11:13–21

Baitsch D, Bock HH, Engel T, Telgmann R, Muller-Tidow C, et al. 2011. Apolipoprotein E induces anti-inflammatory phenotype in macrophages. *Arterioscler. Thromb. Vasc. Biol.* 31:1160–68

Bales KR, Liu F, Wu S, Lin S, Koger D, et al. 2009. Human APOE isoform-dependent effects on brain β-amyloid levels in PDAPP transgenic mice. *J. Neurosci.* 29:6771–79

Basak JM, Kim J. 2010. Differential effects of ApoE isoforms on dendritic spines in vivo: linking an Alzheimer's disease risk factor with synaptic alterations. *J. Neurosci.* 30:4526–27

Basak JM, Verghese PB, Yoon H, Kim J, Holtzman DM. 2012. Low-density lipoprotein receptor represents an apolipoprotein E-independent pathway of Aβ uptake and degradation by astrocytes. *J. Biol. Chem.* 287:13959–71

Bell RD, Winkler EA, Singh I, Sagare AP, Deane R, et al. 2012. Apolipoprotein E controls cerebrovascular integrity via cyclophilin A. *Nature* 485:512–16

Bennett RE, Esparza TJ, Lewis HA, Kim E, Mac Donald CL, et al. 2013. Human apolipoprotein E4 worsens acute axonal pathology but not amyloid-β immunoreactivity after traumatic brain injury in 3xTG-AD mice. *J. Neuropathol. Exp. Neurol.* 72:396–403

Bettens K, Sleegers K, Van Broeckhoven C. 2013. Genetic insights in Alzheimer's disease. *Lancet Neurol.* 12:92–104

Bien-Ly N, Andrews-Zwilling Y, Xu Q, Bernardo A, Wang C, Huang Y. 2011. C-terminal-truncated apolipoprotein (apo) E4 inefficiently clears amyloid-beta (Aβ) and acts in concert with Aβ to elicit neuronal and behavioral deficits in mice. *Proc. Natl. Acad. Sci. USA* 108:4236–41

Bien-Ly N, Gillespie AK, Walker D, Yoon SY, Huang Y. 2012. Reducing human apolipoprotein E levels attenuates age-dependent Aβ accumulation in mutant human amyloid precursor protein transgenic mice. *J. Neurosci.* 32:4803–11

Brecht WJ, Harris FM, Chang S, Tesseur I, Yu GQ, et al. 2004. Neuron-specific apolipoprotein e4 proteolysis is associated with increased tau phosphorylation in brains of transgenic mice. *J. Neurosci.* 24:2527–34

Brodbeck J, McGuire J, Liu Z, Meyer-Franke A, Balestra ME, et al. 2011. Structure-dependent impairment of intracellular apolipoprotein E4 trafficking and its detrimental effects are rescued by small-molecule structure correctors. *J. Biol. Chem.* 286:17217–26

Broussard GJ, Mytar J, Li RC, Klapstein GJ. 2012. The role of inflammatory processes in Alzheimer's disease. *Inflammopharmacology* 20:109–26

Bu G. 2009. Apolipoprotein E and its receptors in Alzheimer's disease: pathways, pathogenesis and therapy. *Nat. Rev. Neurosci.* 10:333–44

Cacabelos R, Martinez R, Fernandez-Novoa L, Carril JC, Lombardi V, et al. 2012. Genomics of dementia: APOE- and CYP2D6-related pharmacogenetics. *Int. J. Alzheimer's Dis.* 2012:518901

Carlo AS, Gustafsen C, Mastrobuoni G, Nielsen MS, Burgert T, et al. 2013. The pro-neurotrophin receptor sortilin is a major neuronal apolipoprotein E receptor for catabolism of amyloid-β peptide in the brain. *J. Neurosci.* 33:358–70

Caselli RJ, Dueck AC, Osborne D, Sabbagh MN, Connor DJ, et al. 2009. Longitudinal modeling of age-related memory decline and the APOE ε4 effect. *N. Engl. J. Med.* 361:255–63

Cash JG, Kuhel DG, Basford JE, Jaeschke A, Chatterjee TK, et al. 2012. Apolipoprotein E4 impairs macrophage efferocytosis and potentiates apoptosis by accelerating endoplasmic reticulum stress. *J. Biol. Chem.* 287:27876–84

Castellano JM, Kim J, Stewart FR, Jiang H, DeMattos RB, et al. 2011. Human apoE isoforms differentially regulate brain amyloid-β peptide clearance. *Sci. Transl. Med.* 3:89ra57

Chartier-Harlin MC, Parfitt M, Legrain S, Perez-Tur J, Brousseau T, et al. 1994. Apolipoprotein E, ε4 allele as a major risk factor for sporadic early and late-onset forms of Alzheimer's disease: analysis of the 19q13.2 chromosomal region. *Hum. Mol. Genet.* 3:569–74

Chen HK, Ji ZS, Dodson SE, Miranda RD, Rosenblum CI, et al. 2011a. Apolipoprotein E4 domain interaction mediates detrimental effects on mitochondria and is a potential therapeutic target for Alzheimer disease. *J. Biol. Chem.* 286:5215–21

Chen HK, Liu Z, Meyer-Franke A, Brodbeck J, Miranda RD, et al. 2012. Small molecule structure correctors abolish detrimental effects of apolipoprotein E4 in cultured neurons. *J. Biol. Chem.* 287:5253–66

Chen J, Li Q, Wang J. 2011b. Topology of human apolipoprotein E3 uniquely regulates its diverse biological functions. *Proc. Natl. Acad. Sci. USA* 108:14813–18

Chen Y, Durakoglugil MS, Xian X, Herz J. 2010. ApoE4 reduces glutamate receptor function and synaptic plasticity by selectively impairing ApoE receptor recycling. *Proc. Natl. Acad. Sci. USA* 107:12011–16

Christensen DZ, Schneider-Axmann T, Lucassen PJ, Bayer TA, Wirths O. 2010. Accumulation of intraneuronal Aβ correlates with ApoE4 genotype. *Acta Neuropathol.* 119:555–66

Colca JR, Feinstein DL. 2012. Altering mitochondrial dysfunction as an approach to treating Alzheimer's disease. *Adv. Pharmacol.* 64:155–76

Conejero-Goldberg C, Hyde TM, Chen S, Dreses-Werringloer U, Herman MM, et al. 2011. Molecular signatures in post-mortem brain tissue of younger individuals at high risk for Alzheimer's disease as based on APOE genotype. *Mol. Psychiatry* 16:836–47

Corder EH, Saunders AM, Strittmatter WJ, Schmechel DE, Gaskell PC, et al. 1993. Gene dose of apolipoprotein E type 4 allele and the risk of Alzheimer's disease in late onset families. *Science* 261:921–23

Cramer PE, Cirrito JR, Wesson DW, Lee CY, Karlo JC, et al. 2012. ApoE-directed therapeutics rapidly clear β-amyloid and reverse deficits in AD mouse models. *Science* 335:1503–6

Cruchaga C, Kauwe JS, Nowotny P, Bales K, Pickering EH, et al. 2012. Cerebrospinal fluid APOE levels: an endophenotype for genetic studies for Alzheimer's disease. *Hum. Mol. Genet.* 21:4558–71

Cui W, Sun Y, Wang Z, Xu C, Peng Y, Li R. 2012. Liver X receptor activation attenuates inflammatory response and protects cholinergic neurons in APP/PS1 transgenic mice. *Neuroscience* 210:200–10

De Jager PL, Shulman JM, Chibnik LB, Keenan BT, Raj T, et al. 2012. A genome-wide scan for common variants affecting the rate of age-related cognitive decline. *Neurobiol. Aging* 33:1017e1–15

Dean DC 3rd, Jerskey BA, Chen K, Protas H, Thiyyagura P, et al. 2014. Brain differences in infants at differential genetic risk for late-onset Alzheimer disease: a cross-sectional imaging study. *JAMA Neurol.* 71:11–22

Di Paolo G, Kim TW. 2011. Linking lipids to Alzheimer's disease: cholesterol and beyond. *Nat. Rev. Neurosci.* 12:284–96

Dodart JC, Marr RA, Koistinaho M, Gregersen BM, Malkani S, et al. 2005. Gene delivery of human apolipoprotein E alters brain Aβ burden in a mouse model of Alzheimer's disease. *Proc. Natl. Acad. Sci. USA* 102:1211–16

Dong LM, Parkin S, Trakhanov SD, Rupp B, Simmons T, et al. 1996. Novel mechanism for defective receptor binding of apolipoprotein E2 in type III hyperlipoproteinemia. *Nat. Struct. Biol.* 3:718–22

Donkin JJ, Stukas S, Hirsch-Reinshagen V, Namjoshi D, Wilkinson A, et al. 2010. ATP-binding cassette transporter A1 mediates the beneficial effects of the liver X receptor agonist GW3965 on object recognition memory and amyloid burden in amyloid precursor protein/presenilin 1 mice. *J. Biol. Chem.* 285:34144–54

Dore GA, Elias MF, Robbins MA, Elias PK, Nagy Z. 2009. Presence of the APOE ε4 allele modifies the relationship between type 2 diabetes and cognitive performance: the Maine-Syracuse Study. *Diabetologia* 52:2551–60

Dumanis SB, Tesoriero JA, Babus LW, Nguyen MT, Trotter JH, et al. 2009. ApoE4 decreases spine density and dendritic complexity in cortical neurons in vivo. *J. Neurosci.* 29:15317–22

Filippini N, MacIntosh BJ, Hough MG, Goodwin GM, Frisoni GB, et al. 2009. Distinct patterns of brain activity in young carriers of the APOE-ε4 allele. *Proc. Natl. Acad. Sci. USA* 106:7209–14

Fitz NF, Cronican AA, Lefterov I, Koldamova R. 2013. Comment on "ApoE-directed therapeutics rapidly clear β-amyloid and reverse deficits in AD mouse models." *Science* 340:924

Frieden C, Garai K. 2012. Structural differences between apoE3 and apoE4 may be useful in developing therapeutic agents for Alzheimer's disease. *Proc. Natl. Acad. Sci. USA* 109:8913–18

Gatz M, Reynolds CA, Fratiglioni L, Johansson B, Mortimer JA, et al. 2006. Role of genes and environments for explaining Alzheimer disease. *Arch. Gen. Psychiatry* 63:168–74

Genin E, Hannequin D, Wallon D, Sleegers K, Hiltunen M, et al. 2011. APOE and Alzheimer disease: a major gene with semi-dominant inheritance. *Mol. Psychiatry* 16:903–7

Ghosal K, Stathopoulos A, Thomas D, Phenis D, Vitek MP, Pimplikar SW. 2013. The apolipoprotein-E-mimetic COG112 protects amyloid precursor protein intracellular domain-overexpressing animals from Alzheimer's disease-like pathological features. *Neurodegener. Dis.* 12:51–58

Gibson GE, Haroutunian V, Zhang H, Park LC, Shi Q, et al. 2000. Mitochondrial damage in Alzheimer's disease varies with apolipoprotein E genotype. *Ann. Neurol.* 48:297–303

Green RC, Roberts JS, Cupples LA, Relkin NR, Whitehouse PJ, et al. 2009. Disclosure of APOE genotype for risk of Alzheimer's disease. *N. Engl. J. Med.* 361:245–54

Guerreiro R, Wojtas A, Bras J, Carrasquillo M, Rogaeva E, et al. 2013. TREM2 variants in Alzheimer's disease. *N. Engl. J. Med.* 368:117–27

Gupta VB, Laws SM, Villemagne VL, Ames D, Bush AI, et al. 2011. Plasma apolipoprotein E and Alzheimer disease risk: the AIBL study of aging. *Neurology* 76:1091–98

Hall K, Murrell J, Ogunniyi A, Deeg M, Baiyewu O, et al. 2006. Cholesterol, APOE genotype, and Alzheimer disease: an epidemiologic study of Nigerian Yoruba. *Neurology* 66:223–27

Hao J, Zhang W, Zhang P, Liu R, Liu L, et al. 2010. Aβ20-29 peptide blocking apoE/Aβ interaction reduces full-length Aβ42/40 fibril formation and cytotoxicity in vitro. *Neuropeptides* 44:305–13

Hara M, Matsushima T, Satoh H, Iso-o N, Noto H, et al. 2003. Isoform-dependent cholesterol efflux from macrophages by apolipoprotein E is modulated by cell surface proteoglycans. *Arterioscler. Thromb. Vasc. Biol.* 23:269–74

Harris FM, Brecht WJ, Xu Q, Tesseur I, Kekonius L, et al. 2003. Carboxyl-terminal-truncated apolipoprotein E4 causes Alzheimer's disease-like neurodegeneration and behavioral deficits in transgenic mice. *Proc. Natl. Acad. Sci. USA* 100:10966–71

Hashimoto T, Serrano-Pozo A, Hori Y, Adams KW, Takeda S, et al. 2012. Apolipoprotein E, especially apolipoprotein E4, increases the oligomerization of amyloid β peptide. *J. Neurosci.* 32:15181–92

Hatters D, Peters-Libeu C, Weisgraber KH. 2006. Apolipoprotein E structure: insights into function. *Trends Biochem. Sci.* 31:445–54

Hauser PS, Narayanaswami V, Ryan RO. 2011. Apolipoprotein E: from lipid transport to neurobiology. *Prog. Lipid Res.* 50:62–74

Hawkes CA, Sullivan PM, Hands S, Weller RO, Nicoll JA, Carare RO. 2012. Disruption of arterial perivascular drainage of amyloid-β from the brains of mice expressing the human APOE ε4 allele. *PLoS ONE* 7:e41636

Hoe HS, Freeman J, Rebeck GW. 2006. Apolipoprotein E decreases tau kinases and phospho-tau levels in primary neurons. *Mol. Neurodegener.* 1:18

Holtzman DM, Herz J, Bu G. 2012. Apolipoprotein E and apolipoprotein E receptors: normal biology and roles in Alzheimer disease. *Cold Spring Harb. Perspect. Med.* 2:a006312

Hong S, Quintero-Monzon O, Ostaszewski BL, Podlisny DR, Cavanaugh WT, et al. 2011. Dynamic analysis of amyloid β protein in behaving mice reveals opposing changes in ISF versus parenchymal Aβ during age-related plaque formation. *J. Neurosci.* 31:15861–69

in 't Veld BA, Ruitenberg A, Hofman A, Launer LJ, van Duijn CM, et al. 2001. Nonsteroidal antiinflammatory drugs and the risk of Alzheimer's disease. *N. Engl. J. Med.* 345:1515–21

Irizarry MC, Deng A, Lleo A, Berezovska O, Von Arnim CA, et al. 2004. Apolipoprotein E modulates γ-secretase cleavage of the amyloid precursor protein. *J. Neurochem.* 90:1132–43

James R, Searcy JL, Le Bihan T, Martin SF, Gliddon CM, et al. 2012. Proteomic analysis of mitochondria in APOE transgenic mice and in response to an ischemic challenge. *J. Cereb. Blood Flow Metab.* 32:164–76

Jiang Q, Lee CY, Mandrekar S, Wilkinson B, Cramer P, et al. 2008. ApoE promotes the proteolytic degradation of Aβ. *Neuron* 58:681–93

Jones L, Holmans PA, Hamshere ML, Harold D, Moskvina V, et al. 2010. Genetic evidence implicates the immune system and cholesterol metabolism in the aetiology of Alzheimer's disease. *PLoS ONE* 5:e13950

Jonsson T, Stefansson H, Steinberg S, Jonsdottir I, Jonsson PV, et al. 2013. Variant of TREM2 associated with the risk of Alzheimer's disease. *N. Engl. J. Med.* 368:107–16

Kariv-Inbal Z, Yacobson S, Berkecz R, Peter M, Janaky T, et al. 2012. The isoform-specific pathological effects of apoE4 in vivo are prevented by a fish oil (DHA) diet and are modified by cholesterol. *J. Alzheimer's Dis.* 28:667–83

Keene CD, Cudaback E, Li X, Montine KS, Montine TJ. 2011. Apolipoprotein E isoforms and regulation of the innate immune response in brain of patients with Alzheimer's disease. *Curr. Opin. Neurobiol.* 21:920–28

Kennedy RE, Cutter GR, Schneider LS. 2014. Effect of *APOE* genotype status on targeted clinical trials outcomes and efficiency in dementia and mild cognitive impairment resulting from Alzheimer's disease. *Alzheimer's Dement.* doi: 10.1016/j.jalz.2013.03.003. In press

Kerchner GA, Berdnik D, Shen JC, Bernstein JD, Fenesy MC, et al. 2014. APOE ε4 worsens hippocampal CA1 apical neuropil atrophy and episodic memory. *Neurology* 82:691–97

Kim J, Basak JM, Holtzman DM. 2009. The role of apolipoprotein E in Alzheimer's disease. *Neuron* 63:287–303

Kim J, Eltorai AE, Jiang H, Liao F, Verghese PB, et al. 2012. Anti-apoE immunotherapy inhibits amyloid accumulation in a transgenic mouse model of Aβ amyloidosis. *J. Exp. Med.* 209:2149–56

Kim J, Jiang H, Park S, Eltorai AE, Stewart FR, et al. 2011. Haploinsufficiency of human APOE reduces amyloid deposition in a mouse model of amyloid-β amyloidosis. *J. Neurosci.* 31:18007–12

Klein RC, Mace BE, Moore SD, Sullivan PM. 2010. Progressive loss of synaptic integrity in human apolipoprotein E4 targeted replacement mice and attenuation by apolipoprotein E2. *Neuroscience* 171:1265–72

Koffie RM, Hashimoto T, Tai HC, Kay KR, Serrano-Pozo A, et al. 2012. Apolipoprotein E4 effects in Alzheimer's disease are mediated by synaptotoxic oligomeric amyloid-β. *Brain* 135:2155–68

Korwek KM, Trotter JH, Ladu MJ, Sullivan PM, Weeber EJ. 2009. ApoE isoform-dependent changes in hippocampal synaptic function. *Mol. Neurodegener.* 4:21

Kuszczyk MA, Sanchez S, Pankiewicz J, Kim J, Duszczyk M, et al. 2013. Blocking the interaction between apolipoprotein E and Aβ reduces intraneuronal accumulation of Aβ and inhibits synaptic degeneration. *Am. J. Pathol.* 182:1750–68

Lambert JC, Ibrahim-Verbaas CA, Harold D, Naj AC, Sims R, et al. 2013. Meta-analysis of 74,046 individuals identifies 11 new susceptibility loci for Alzheimer's disease. *Nat. Genet.* 45:1452–58

Lambert JC, Perez-Tur J, Dupire MJ, Galasko D, Mann D, et al. 1997. Distortion of allelic expression of apolipoprotein E in Alzheimer's disease. *Hum. Mol. Genet.* 6:2151–54

Landreth GE, Cramer PE, Lakner MM, Cirrito JR, Wesson DW, et al. 2013. Response to comments on "ApoE-directed therapeutics rapidly clear β-amyloid and reverse deficits in AD mouse models." *Science* 340:924

Lautenschlager NT, Cox KL, Flicker L, Foster JK, van Bockxmeer FM, et al. 2008. Effect of physical activity on cognitive function in older adults at risk for Alzheimer disease: a randomized trial. *J. Am. Med. Assoc.* 300:1027–37

Lee CY, Tse W, Smith JD, Landreth GE. 2012. Apolipoprotein E promotes β-amyloid trafficking and degradation by modulating microglial cholesterol levels. *J. Biol. Chem.* 287:2032–44

Leoni V. 2011. The effect of apolipoprotein E (ApoE) genotype on biomarkers of amyloidogenesis, tau pathology and neurodegeneration in Alzheimer's disease. *Clin. Chem. Lab. Med.* 49:375–83

Li J, Kanekiyo T, Shinohara M, Zhang Y, LaDu MJ, et al. 2012. Differential regulation of amyloid-β endocytic trafficking and lysosomal degradation by apolipoprotein E isoforms. *J. Biol. Chem.* 287:44593–601

Liu CC, Kanekiyo T, Xu H, Bu G. 2013. Apolipoprotein E and Alzheimer disease: risk, mechanisms and therapy. *Nat. Rev. Neurol.* 9:106–18

Liu Q, Wu WH, Fang CL, Li RW, Liu P, et al. 2011. Mapping ApoE/Aβ binding regions to guide inhibitor discovery. *Mol. Biosyst.* 7:1693–700

Mahley RW, Huang Y. 2012. Apolipoprotein E sets the stage: response to injury triggers neuropathology. *Neuron* 76:871–85

Mandrekar-Colucci S, Karlo JC, Landreth GE. 2012. Mechanisms underlying the rapid peroxisome proliferator-activated receptor-γ-mediated amyloid clearance and reversal of cognitive deficits in a murine model of Alzheimer's disease. *J. Neurosci.* 32:10117–28

Mattsson N, Zetterberg H, Hansson O, Andreasen N, Parnetti L, et al. 2009. CSF biomarkers and incipient Alzheimer disease in patients with mild cognitive impairment. *J. Am. Med. Assoc.* 302:385–93

Mauch DH, Nagler K, Schumacher S, Goritz C, Muller EC, et al. 2001. CNS synaptogenesis promoted by glia-derived cholesterol. *Science* 294:1354–57

Mawuenyega KG, Sigurdson W, Ovod V, Munsell L, Kasten T, et al. 2010. Decreased clearance of CNS β-amyloid in Alzheimer's disease. *Science* 330:1774

McKhann GM, Knopman DS, Chertkow H, Hyman BT, Jack CR Jr, et al. 2011. The diagnosis of dementia due to Alzheimer's disease: recommendations from the National Institute on Aging-Alzheimer's Association workgroups on diagnostic guidelines for Alzheimer's disease. *Alzheimer's Dement.* 7:263–69

Minami SS, Cordova A, Cirrito JR, Tesoriero JA, Babus LW, et al. 2010. ApoE mimetic peptide decreases Aβ production in vitro and in vivo. *Mol. Neurodegener.* 5:16

Morris JC, Roe CM, Xiong C, Fagan AM, Goate AM, et al. 2010. APOE predicts amyloid-β but not tau Alzheimer pathology in cognitively normal aging. *Ann. Neurol.* 67:122–31

Mosconi L, De Santi S, Brys M, Tsui WH, Pirraglia E, et al. 2008. Hypometabolism and altered cerebrospinal fluid markers in normal apolipoprotein E E4 carriers with subjective memory complaints. *Biol. Psychiatry* 63:609–18

Nakamura T, Watanabe A, Fujino T, Hosono T, Michikawa M. 2009. Apolipoprotein E4 (1–272) fragment is associated with mitochondrial proteins and affects mitochondrial function in neuronal cells. *Mol. Neurodegener.* 4:35

Namjoshi DR, Martin G, Donkin J, Wilkinson A, Stukas S, et al. 2013. The liver X receptor agonist GW3965 improves recovery from mild repetitive traumatic brain injury in mice partly through apolipoprotein E. *PLoS ONE* 8:e53529

Nishitsuji K, Hosono T, Nakamura T, Bu G, Michikawa M. 2011. Apolipoprotein E regulates the integrity of tight junctions in an isoform-dependent manner in an in vitro blood-brain barrier model. *J. Biol. Chem.* 286:17536–42

Noguchi S, Murakami K, Yamada N, Payami H, Kaye J, et al. 1993. Apolipoprotein E genotype and Alzheimer's disease. *Lancet* 342:737–38

Osei-Hwedieh DO, Amar M, Sviridov D, Remaley AT. 2011. Apolipoprotein mimetic peptides: mechanisms of action as anti-atherogenic agents. *Pharmacol. Ther.* 130:83–91

Ossenkoppele R, van der Flier WM, Zwan MD, Adriaanse SF, Boellaard R, et al. 2013. Differential effect of APOE genotype on amyloid load and glucose metabolism in AD dementia. *Neurology* 80:359–65

Pastor P, Roe CM, Villegas A, Bedoya G, Chakraverty S, et al. 2003. Apolipoprotein E ε4 modifies Alzheimer's disease onset in an E280A PS1 kindred. *Ann. Neurol.* 54:163–69

Patterson CE, Todd SA, Passmore AP. 2011. Effect of apolipoprotein E and butyrylcholinesterase genotypes on cognitive response to cholinesterase inhibitor treatment at different stages of Alzheimer's disease. *Pharmacogenomics J.* 11:444–50

Petersen RC, Thomas RG, Grundman M, Bennett D, Doody R, et al. 2005. Vitamin E and donepezil for the treatment of mild cognitive impairment. *N. Engl. J. Med.* 352:2379–88

Pocivavsek A, Burns MP, Rebeck GW. 2009. Low-density lipoprotein receptors regulate microglial inflammation through c-Jun N-terminal kinase. *Glia* 57:444–53

Price AR, Xu G, Siemienski ZB, Smithson LA, Borchelt DR, et al. 2013. Comment on "ApoE-directed therapeutics rapidly clear β-amyloid and reverse deficits in AD mouse models." *Science* 340:924

Protas HD, Chen K, Langbaum JB, Fleisher AS, Alexander GE, et al. 2013. Posterior cingulate glucose metabolism, hippocampal glucose metabolism, and hippocampal volume in cognitively normal, late-middle-aged persons at 3 levels of genetic risk for Alzheimer disease. *JAMA Neurol.* 70:320–25

Ramanan VK, Risacher SL, Nho K, Kim S, Swaminathan S, et al. 2013. APOE and BCHE as modulators of cerebral amyloid deposition: a florbetapir PET genome-wide association study. *Mol. Psychiatry* 19:351–57

Rapp A, Gmeiner B, Huttinger M. 2006. Implication of apoE isoforms in cholesterol metabolism by primary rat hippocampal neurons and astrocytes. *Biochimie* 88:473–83

Reiman EM, Chen K, Liu X, Bandy D, Yu M, et al. 2009. Fibrillar amyloid-β burden in cognitively normal people at 3 levels of genetic risk for Alzheimer's disease. *Proc. Natl. Acad. Sci. USA* 106:6820–25

Rhinn H, Fujita R, Qiang L, Cheng R, Lee JH, Abeliovich A. 2013. Integrative genomics identifies APOE ε4 effectors in Alzheimer's disease. *Nature* 500:45–50

Ringman JM, Elashoff D, Geschwind DH, Welsh BT, Gylys KH, et al. 2012. Plasma signaling proteins in persons at genetic risk for Alzheimer disease: influence of APOE genotype-plasma biomarkers and AD genetic risk. *Arch. Neurol.* 69:757–64

Risner ME, Saunders AM, Altman JF, Ormandy GC, Craft S, et al. 2006. Efficacy of rosiglitazone in a genetically defined population with mild-to-moderate Alzheimer's disease. *Pharmacogenomics J.* 6:246–54

Ronnemaa E, Zethelius B, Sundelof J, Sundstrom J, Degerman-Gunnarsson M, et al. 2008. Impaired insulin secretion increases the risk of Alzheimer disease. *Neurology* 71:1065–71

Sadowski MJ, Pankiewicz J, Scholtzova H, Mehta PD, Prelli F, et al. 2006. Blocking the apolipoprotein E/amyloid-β interaction as a potential therapeutic approach for Alzheimer's disease. *Proc. Natl. Acad. Sci. USA* 103:18787–92

Salloway S, Sperling R, Gilman S, Fox NC, Blennow K, et al. 2009. A phase 2 multiple ascending dose trial of bapineuzumab in mild to moderate Alzheimer disease. *Neurology* 73:2061–70

Samieri C, Feart C, Proust-Lima C, Peuchant E, Dartigues JF, et al. 2011. Omega-3 fatty acids and cognitive decline: modulation by ApoEε4 allele and depression. *Neurobiol. Aging* 32:2317e13–22

Sarantseva S, Timoshenko S, Bolshakova O, Karaseva E, Rodin D, et al. 2009. Apolipoprotein E-mimetics inhibit neurodegeneration and restore cognitive functions in a transgenic *Drosophila* model of Alzheimer's disease. *PLoS ONE* 4:e8191

Schipper HM. 2011. Presymptomatic apolipoprotein E genotyping for Alzheimer's disease risk assessment and prevention. *Alzheimer's Dement.* 7:e118–23

Sen A, Alkon DL, Nelson TJ. 2012. Apolipoprotein E3 (ApoE3) but not ApoE4 protects against synaptic loss through increased expression of protein kinase C epsilon. *J. Biol. Chem.* 287:15947–58

Sheline YI, Morris JC, Snyder AZ, Price JL, Yan Z, et al. 2010. *APOE4* allele disrupts resting state fMRI connectivity in the absence of amyloid plaques or decreased CSF Aβ42. *J. Neurosci.* 30:17035–40

Shenk JC, Liu J, Fischbach K, Xu K, Puchowicz M, et al. 2009. The effect of acetyl-L-carnitine and R-α-lipoic acid treatment in ApoE4 mouse as a model of human Alzheimer's disease. *J. Neurol. Sci.* 283:199–206

Sperling R, Salloway S, Brooks DJ, Tampieri D, Barakos J, et al. 2012. Amyloid-related imaging abnormalities in patients with Alzheimer's disease treated with bapineuzumab: a retrospective analysis. *Lancet Neurol.* 11:241–49

Strittmatter WJ, Weisgraber KH, Goedert M, Saunders AM, Huang D, et al. 1994. Hypothesis: microtubule instability and paired helical filament formation in the Alzheimer disease brain are related to apolipoprotein E genotype. *Exp. Neurol.* 125:163–71; discuss. 72–74

Strum JC, Shehee R, Virley D, Richardson J, Mattie M, et al. 2007. Rosiglitazone induces mitochondrial biogenesis in mouse brain. *J. Alzheimer's Dis.* 11:45–51

Szekely CA, Breitner JC, Fitzpatrick AL, Rea TD, Psaty BM, et al. 2008. NSAID use and dementia risk in the Cardiovascular Health Study: role of APOE and NSAID type. *Neurology* 70:17–24

Tai LM, Bilousova T, Jungbauer L, Roeske SK, Youmans KL, et al. 2013. Levels of soluble apolipoprotein E/amyloid-beta (Aβ) complex are reduced and oligomeric Aβ increased with APOE4 and Alzheimer disease in a transgenic mouse model and human samples. *J. Biol. Chem.* 288:5914–26

Terwel D, Steffensen KR, Verghese PB, Kummer MP, Gustafsson JA, et al. 2011. Critical role of astroglial apolipoprotein E and liver X receptor-α expression for microglial Aβ phagocytosis. *J. Neurosci.* 31:7049–59

Tesseur I, Lo AC, Roberfroid A, Dietvorst S, Van Broeck B, et al. 2013. Comment on "ApoE-directed therapeutics rapidly clear β-amyloid and reverse deficits in AD mouse models." *Science* 340:924

Valla J, Yaari R, Wolf AB, Kusne Y, Beach TG, et al. 2010. Reduced posterior cingulate mitochondrial activity in expired young adult carriers of the APOE ε4 allele, the major late-onset Alzheimer's susceptibility gene. *J. Alzheimer's Dis.* 22:307–13

Veeraraghavalu K, Zhang C, Miller S, Hefendehl JK, Rajapaksha TW, et al. 2013. Comment on "ApoE-directed therapeutics rapidly clear β-amyloid and reverse deficits in AD mouse models." *Science* 340:924

Vemuri P, Wiste HJ, Weigand SD, Knopman DS, Shaw LM, et al. 2010. Effect of apolipoprotein E on biomarkers of amyloid load and neuronal pathology in Alzheimer disease. *Ann. Neurol.* 67:308–16

Verghese PB, Castellano JM, Garai K, Wang Y, Jiang H, et al. 2013. ApoE influences amyloid-beta (Aβ) clearance despite minimal apoE/Aβ association in physiological conditions. *Proc. Natl. Acad. Sci. USA* 110:E1807–16

Villemagne VL, Pike KE, Chetelat G, Ellis KA, Mulligan RS, et al. 2011. Longitudinal assessment of Aβ and cognition in aging and Alzheimer disease. *Ann. Neurol.* 69:181–92

Vitek MP, Brown CM, Colton CA. 2009. APOE genotype-specific differences in the innate immune response. *Neurobiol. Aging* 30:1350–60

Vitek MP, Christensen DJ, Wilcock D, Davis J, Van Nostrand WE, et al. 2012. APOE-mimetic peptides reduce behavioral deficits, plaques and tangles in Alzheimer's disease transgenics. *Neurodegener. Dis.* 10:122–26

Wahrle SE, Jiang H, Parsadanian M, Hartman RE, Bales KR, et al. 2005. Deletion of *Abca1* increases Aβ deposition in the PDAPP transgenic mouse model of Alzheimer disease. *J. Biol. Chem.* 280:43236–42

Wahrle SE, Jiang H, Parsadanian M, Kim J, Li A, et al. 2008. Overexpression of ABCA1 reduces amyloid deposition in the PDAPP mouse model of Alzheimer disease. *J. Clin. Investig.* 118:671–82

Wahrle SE, Jiang H, Parsadanian M, Legleiter J, Han X, et al. 2004. ABCA1 is required for normal central nervous system ApoE levels and for lipidation of astrocyte-secreted apoE. *J. Biol. Chem.* 279:40987–93

Wilson M, Montgomery H. 2007. Impact of genetic factors on outcome from brain injury. *Br. J. Anaesth.* 99:43–48

Xu Q, Bernardo A, Walker D, Kanegawa T, Mahley RW, Huang Y. 2006. Profile and regulation of apolipoprotein E (ApoE) expression in the CNS in mice with targeting of green fluorescent protein gene to the ApoE locus. *J. Neurosci.* 26:4985–94

Ye S, Huang Y, Mullendorff K, Dong L, Giedt G, et al. 2005. Apolipoprotein (apo) E4 enhances amyloid β peptide production in cultured neuronal cells: apoE structure as a potential therapeutic target. *Proc. Natl. Acad. Sci. USA* 102:18700–5

Youmans KL, Tai LM, Nwabuisi-Heath E, Jungbauer L, Kanekiyo T, et al. 2012. APOE4-specific changes in Aβ accumulation in a new transgenic mouse model of Alzheimer disease. *J. Biol. Chem.* 287:41774–86

Zhang H, Wu LM, Wu J. 2011. Cross-talk between apolipoprotein E and cytokines. *Mediators Inflamm.* 2011:949072

Zhong N, Weisgraber KH. 2009. Understanding the association of apolipoprotein E4 with Alzheimer disease: clues from its structure. *J. Biol. Chem.* 284:6027–31

Zhu Y, Nwabuisi-Heath E, Dumanis SB, Tai LM, Yu C, et al. 2012. APOE genotype alters glial activation and loss of synaptic markers in mice. *Glia* 60:559–69

Zlokovic BV. 2011. Neurovascular pathways to neurodegeneration in Alzheimer's disease and other disorders. *Nat. Rev. Neurosci.* 12:723–38

Function and Dysfunction of Hypocretin/Orexin: An Energetics Point of View

Xiao-Bing Gao and Tamas Horvath

Yale Program in Integrative Cell Signaling and Neurobiology of Metabolism, Section of Comparative Medicine, Yale University School of Medicine, New Haven, Connecticut 06520; email: xiao-bing.gao@yale.edu, tamas.horvath@yale.edu

Annu. Rev. Neurosci. 2014. 37:101–16

First published online as a Review in Advance on April 24, 2014

The *Annual Review of Neuroscience* is online at neuro.annualreviews.org

This article's doi: 10.1146/annurev-neuro-071013-013855

Keywords

hypocretin/orexin, hypocretin, orexin, energy balance, sleep/wake regulation, motivational behavior

Abstract

The basic elements of animal behavior that are critical to survival include energy, arousal, and motivation: Energy intake and expenditure are fundamental to all organisms for the performance of any type of function; according to the Yerkes-Dodson law, an optimal level of arousal is required for animals to perform normal functions; and motivation is critical to goal-oriented behaviors in higher animals. The brain is the primary organ that controls these elements and, through evolution, has developed specialized structures to accomplish this task. The orexin/hypocretin system in the perifornical/lateral hypothalamus, which was discovered 15 years ago, is one such specialized area. This review summarizes a fast-growing body of evidence discerning how the orexin/hypocretin system integrates internal and external cues to regulate energy intake that can then be used to generate sufficient arousal for animals to perform innate and goal-oriented behaviors.

Contents

INTRODUCTION

Locomotion is a fundamental characteristic of the kingdom Animalia and is essential for individuals and species within the kingdom to survive. For example, movement is required for most animals to perform such tasks as foraging for food, escaping predators, and reproducing, among others. Several elements are essential to execute locomotive functions, first among these elements being energy. Energy is required for biochemical reactions and biophysical activities to sustain the normal functionality of every cell, but it is also needed for animals to overcome environmental factors (e.g., gravity, air or water resistance) to execute locomotor activity. Second among the essential elements is the maintenance of a sufficient level of arousal for animals to perform daily activities. The shift between rest (sleep) and active (wake) states occurs in both lower animals such as *C. elegans* and higher animals such as mammals (Mackiewicz et al. 2008). According to the Yerkes-Dodson law, both hypo- and hyperarousal will compromise normal animal behaviors (Yerkes & Dodson 1908). The third important component is motivation (desire to achieve a certain behavior) to initiate and sustain locomotion in animals. This is particularly true in goal-oriented behaviors.

In the mammalian brain, many neuronal systems are responsible for one or all of the elements required for locomotor activity in animals. A small number of nerve cells in the perifornical/lateral hypothalamus (LH) that synthesize the neuropeptide hypocretin/orexin participate in neural mechanisms underlying all three elements critical to animal behavior. Hypocretin/orexin (Hcrt) was discovered independently by two groups of investigators 15 years ago (de Lecea et al. 1998, Sakurai et al. 1998). Hcrt comprises two peptides, Hcrt-1 (orexin A) and Hcrt-2 (orexin B) derived from a common 130–amino acid precursor peptide (preprohypocretin) by proteolytic cleavage (de Lecea et al. 1998, Sakurai et al. 1998). In mammals, Hcrt is synthesized only by neurons in the LH, and investigators have estimated that rodents have a few thousand Hcrt cells and humans have about 50,000–90,000 (Peyron et al. 1998, Thannickal et al. 2000, Sutcliffe & de Lecea 2002). Even though this population of neurons is small, studies have detected the nerve fibers containing Hcrt all over the central nervous system and even in peripheral organs (Peyron et al. 1998, Trivedi et al. 1998, van den Pol 1999, Marcus et al. 2001, Adeghate et al. 2010). At the cellular level, Hcrt generally acts as an excitatory transmitter to enhance synaptic transmission, neuronal activity, and intracellular calcium levels in central neurons (van den Pol et al. 1998, Hagan et al. 1999, Davis et al. 2003, Korotkova et al. 2003).

BIDIRECTIONAL ROLES OF Hcrt IN THE REGULATION OF ENERGY BALANCE

The intake and use of energy are fundamental characteristics of all organisms and are essential to their performance of basic functions. The hypothalamus of the brain is a classic structure that participates in the regulation of food intake in animals. Hetherington & Ranson (1940) incidentally found that lesions of the LH resulted in decreased food intake. Their work was extended by Anand & Brobeck (1951a,b), who demonstrated that circumscribed lesions in the LH at tuberal levels of the hypothalamus led to a remarkable reduction in food intake and to death by starvation. These pioneering observations led investigators to consider the LH as a feeding center (Stellar 1954, Sawchenko 1998). A large body of evidence subsequently supported this concept. Delgado & Anand (1953) showed that electrical stimulation of the LH induced food intake. Ungerstedt (1971) demonstrated that selective chemical lesions of dopamine neurons projecting rostrally through the LH led to hypophagia. Stricker et al. (1978) found that chemical lesions with glutamate agonists depressed feeding and body weight regulation. And still other studies showed that neurons in the LH fired spontaneously during naturally occurring feeding behavior and hypoglycemia (Katafuchi et al. 1985, Himmi et al. 1988). However, the exact identity of the cells responsible for the role of the LH in positive energy balance was not clear until the end of the twentieth century.

Hcrt Promotes Energy Intake in Animals

Hcrt's role in food intake originates back to when it was first discovered (Sakurai et al. 1998). Investigators found that the application of Hcrt and antagonists of Hcrt receptors modulates food intake. The intracerebroventricular administration of Hcrt (either Hcrt-1 or -2) leads to a short-term increase in food consumption in rats (Sakurai et al. 1998). Local administration of Hcrt-1 to brain areas such as the paraventricular nucleus, the dorsomedial nucleus, or the LH/perifonical area triggers food intake in rats (Dube et al. 1999, Thorpe et al. 2003). Applying the selective Hcrt-1 receptor antagonist, SB-334867, suppresses food intake (Haynes et al. 2000, Rodgers et al. 2002). Additionally, the activity level of the Hcrt neurons is closely related to the amount of food intake. Fasting enhances expression of Hcrt mRNA and peptides in rats (Sakurai et al. 1998, Mondal et al. 1999, Yamamoto et al. 2000). Acute hypoglycemia induces upregulation of Hcrt mRNA and c-Fos expression in Hcrt neurons in rats when food is not available (Cai et al. 1999, Griffond et al. 1999, Moriguchi et al. 1999), whereas leptin treatment decreases the concentration of Hcrt in the LH and fasting-induced upregulation of Hcrt mRNA (López et al. 2000, Beck et al. 2001). Hcrt may also be involved in intense hyperphagia produced by $GABA_A$ receptor stimulation in the nucleus accumbens shell (Baldo et al. 2004). Finally, Hcrt helps control peripheral organs essential to feeding behavior. Some evidence indicates that Hcrt neurons innervate the dorsal motor nucleus of the vagus (DMV) (Peyron et al. 1998), which is a key region that controls gastric acid secretion and gut motility. The activation of Hcrt neurons by hypoglycemia triggers c-Fos expression in the DMV (Cai et al. 2001), and the central application of Hcrt increases both gastric acid secretion and gastric motor function in rats (Takahashi et al. 1999, Krowicki et al. 2002). The current consensus is that Hcrt may promote positive energy balance by promoting arousal or mediating the rewarding aspects of feeding (Yamanaka et al. 2003, Harris et al. 2005). However, the mechanisms underlying these two processes are still largely unclear.

Hcrt Regulates Energy Expenditure

Hcrt's ability to increase energy expenditure in animals was reported shortly after the discovery of the peptide. Lubkin & Stricker-Krongrad (1998) were the first to show that microinjection of

Hcrt-1 into the third ventricle stimulated the metabolic rate in mice. They also showed that the effect of Hcrt-1 was more potent in the dark cycle than in the light cycle. This observation was later confirmed in mice and rats (Asakawa et al. 2002, Wang et al. 2003, Semjonous et al. 2009). By microinjecting Hcrt-1 into various brain structures, Wang and colleagues (2003) showed that a direct infusion of Hcrt-1 into the arcuate nucleus (ARC) led to an enhancement in whole-body O_2 consumption (VO_2) in urethane-anesthetized rats, whereas Kiwaki et al. (2004) showed that Hcrt-1 microinjection into the paraventricular nucleus of hypothalamus (PVH) and LH increased thermogenesis in conscious rats. The discrepancy in action sites of Hcrt in rats may reveal new aspects of Hcrt-mediated effects on energy expenditure because Hcrt promotes physical activity when microinjected into many brain areas, including the PVH and LH (España et al. 2001; Kotz et al. 2002, 2006, 2008; Kiwaki et al. 2004). Research clearly shows that Hcrt enhances energy expenditure by promoting cardiovascular functions and thermogenesis, which is mediated by the sympathetic nervous system (see reviews by Samson et al. 2005, Székely 2006, Teske et al. 2010). In addition, new evidence has shown that Hcrt is required to mediate the mobilization of brown adipose tissue (BAT) (Sellayah et al. 2011). Compared with their wild-type littermates, Hcrt knockout mice expressed higher levels of preadipocyte markers, fewer mitochondria, and lowered levels of BAT thermogenic proteins (such as PPAR-$\gamma1/\gamma2$, PGC-1α/β, and UCP-1) in brown fat cells (Sellayah et al. 2011). These defects led to a compromised thermogenesis induced by high-fat diet and cold exposure (Sellayah et al. 2011, Sellayah & Sikder 2012). These results are supported by a recent study showing the direct innervation of the raphe pallidus by Hcrt-containing nerve fibers of the LH to promote BAT thermogenesis in rats (Tupone et al. 2011). The most recent study supporting this argument is by Kotz and colleagues (2012), who found that Hcrt is critical in obesity resistance in animals.

Earlier studies have shown that Hcrt may be an important player in the regulation of glucose metabolism (Cai et al. 1999, 2001; Jöhren et al. 2006; Paranjape et al. 2007; Tsuneki et al. 2010). Most recent developments further reveal how the Hcrt system may be involved in the regulation of glucose metabolism and the development of diabetes. First, both Hcrt and its receptors have been found in peripheral tissues, including the endocrine pancreas, and studies in vitro and in vivo have demonstrated that Hcrt modulates insulin and glucose levels (reviewed by Heinonen et al. 2008, Chandra & Liddle 2009). Second, locally applying Hcrt into brain structures directly affects glucose metabolism in the peripheral organs. During the light phase, disinhibition of Hcrt neurons by microinjecting bicuculline, the GABA$_A$ receptor antagonist, into the perifornical area increased basal endogenous glucose production (EGP) in rats, which was prevented by pretreating the HcrtR1 antagonist and hepatic sympathetic denervation (Yi et al. 2009). In addition, plasma insulin clamped at severalfold of the basal level did not counteract the effects of disinhibition of Hcrt neurons on glucose production, suggesting hepatic insulin resistance (Yi et al. 2009). The local infusion of Hcrt into the ventromedial hypothalamus (VMH) enhanced glucose uptake in skeletal muscles by activating beta-2-adrenergic receptors (Shiuchi et al. 2009). Third, a role for Hcrt and the HcrtR1 pathway in the regulation of glucose metabolism and the development of diabetes mellitus (DM) is strengthened by a report that HcrtR1-immunopositive nerves were found in the pancreas of normal and DM rats (Adeghate et al. 2010). The number of HcrtR1-positive cells increased significantly in the islets after DM was induced by streptozotocin (STZ) in this species (Adeghate et al. 2010). In line with these results, Hcrt knockout mice showed more efficient glucose use in glucose tolerance tests and reduced hyperglycemia after the STZ treatment (Adeghate et al. 2010).

Hcrt System Serves as a Sensor of Energy Status in Animals

Unlike many other neuropeptides, which promote energy intake and decrease energy expenditure [such as neuropeptide Y (NPY) and agouti-related protein (AgRP) or vice versa (such as α-MSH)

(Semjonous et al. 2009)], the actions of the Hcrt system are bidirectional. The mechanisms through which Hcrt modulates energy homeostasis are not clear and are still emerging. To promote energy intake, the Hcrt system must be able to sense the animal's energy status. The Hcrt system should be activated to promote food/energy intake when the energy state is low, whereas it should be inhibited when the energy status is high, as it would be after a meal. However, as a promoter of energy expenditure, the Hcrt system is expected to do the opposite. Although it seems paradoxical, the evidence supports the existence of both situations. Yamanaka et al. (2003) originally showed that the activity in isolated Hcrt neurons could be inhibited by molecules encoding cues for energy status, such as glucose, insulin, and leptin. However, this report suffered from limitations originating from the approaches used in their experiments. The acutely isolated Hcrt neurons led to a compromised internal cellular content and external environment. The concentration of glucose used in the study was not at physiological levels either.

Later, Burdakov and colleagues (2006) showed in brain slices that Hcrt neurons could be inhibited by an elevated glucose level resembling the physiological levels in the cerebrospinal fluid (CSF) after a meal through the opening of tandem-pore K^+ channels. However, a later report argued that in mice with a deficiency in tandem-pore K^+ channels, Hcrt neurons could still be inhibited by elevated glucose levels in the artificial cerebrospinal fluid (Guyon et al. 2009). Surprisingly, two of the most recent reports demonstrated that with an intact intracellular content, the Hcrt neurons did not respond to an increase in the ambient glucose levels when cell-attached extracellular or perforated whole-cell recordings were performed in Hcrt neurons (Parsons & Hirasawa 2010, Liu et al. 2011). In fact, a lowered level of ambient glucose inhibited Hcrt neurons, owing to lowered intracellular ATP levels and the opening of K-ATP channels in the Hcrt neurons (Liu et al. 2011). The difference between the responses to extracellular glucose may lie within the experimental conditions used in the above experiments. When conventional whole-cell recording was used, in which the original intracellular content was compromised by the pipette solution, Hcrt neurons were inhibited by high glucose levels and activated by low glucose amounts (Yamanaka et al. 2003; Burdakov et al. 2006). When the intracellular content was kept intact with extracellular or perforated whole-cell recording, Hcrt neurons were inhibited by low levels of glucose (Parsons & Hirasawa 2010, Liu et al. 2011). These results suggest that the Hcrt neurons' response to the changes in ambient glucose levels may involve heterogeneous mechanisms. Identifying key molecule(s) missing during the conventional whole-cell recording may provide critical insight into the understanding of mechanisms underlying the sensing of ambient energy status by Hcrt neurons.

Hcrt AS A POTENT PLAYER IN AROUSAL/WAKE MAINTENANCE

The role of the hypothalamus in sleep regulation was originally proposed by von Economo on the basis of his observations in human patients that lesions in the posterior hypothalamus and midbrain junction lead to sleepiness, whereas anterior hypothalamic inflammation leads to insomnia and chorea (von Economo 1930). Later, the sleep-promoting effect of lesions and inhibition of the posterior LH was confirmed in monkeys, rats, and cats (Ranson 1939, Nauta 1946, Swett & Hobson 1968, Lin et al. 1989). How the LH participates in sleep regulation, however, remained elusive until Hcrt was discovered in this brain area.

Hcrt is a potent arousal/wake promoter in the brain (reviewed by de Lecea & Sutcliffe 2005; Sakurai 2005; Saper 2006, 2013). First, a deficiency in Hcrt itself and its receptor (HcrtR2 or OXR2) leads to narcolepsy in dogs, mice, and human patients (Chemelli et al. 1999, Lin et al. 1999, Nishino et al. 2000, Thannickal et al. 2000, Ripley et al. 2001). Second, the concentration of Hcrt fluctuates in animals depending on the behavioral state. The Hcrt-1 level in CSF is high during the active period and low during the inactive period in rodents and squirrel monkeys (Fujiki et al. 2001, Yoshida et al. 2001, Zeitzer et al. 2003). The activity in these neurons changes along with the rats' behavioral state. The c-Fos expression increases in Hcrt neurons during the dark phase and sleep deprivation (Estabrooke et al. 2001). Also, Hcrt neurons are generally active in the dark phase (awake) and silent in the light phase (sleep) (Lee et al. 2005, Mileykovskiy et al. 2005). Direct and selective stimulation of Hcrt neurons with the optogenetic approach increases the probability of the subject's transition to wakefulness from either slow-wave sleep or rapid-eye-movement sleep (Adamantidis et al. 2007). Last, by projecting to major arousal areas such as the locus coeruleus and the basal forebrain, Hcrt neurons promote arousal and antagonize sleep and muscle atonia by integrating various sensory and homeostatic inputs (Bourgin et al. 2000, España et al. 2001, van den Pol et al. 2002, Lee et al. 2005, Yoshida et al. 2006). Thus, in the flip-flop switch model of sleep regulation, Hcrt neurons consolidate wakefulness by setting the threshold for state transitions (Saper 2006, 2013; Selbach & Haas 2006). Under the physiological condition, the Hcrt system is closely regulated by a network of negative feedbacks to maintain an optimal output (Burt et al. 2011). It is not clear, however, whether the closely regulated output of the Hcrt system and/or other arousal system underlies the optimal arousal levels required for animals to perform normal functions as described by the Yerkes-Dodson law, an interesting phenomenon that has not been well understood.

The latest development in understanding the role of Hcrt in the regulation of the sleep and wake cycles provides a more complicated picture than previously described. One of the hypothesized functions of sleep is to preserve energy, which is crucial to animals in the natural environment where food is not always accessible. Therefore, the study of the integration of the sleep/wake cycle and energy balance regulation is essential to the overall understanding of the homeostatic regulatory processes in animals. The original reports by Yamanaka et al. (2003) on the regulation of Hcrt neuron activity by molecules encoding ambient energy supplies such as glucose provided a critical clue that the Hcrt system may be a node where the neuronal systems controlling energy balance and sleep/wake status converge. Liu & Gao (2007) later showed that adenosine, a substance resulting from cellular activity and metabolism, inhibited Hcrt neurons. On the one hand, adenosine is usually considered a potent sleep-promoting substance (Porkka-Heiskanen & Kalinchuk 2011). On the other hand, a main source of extracellular adenosine comes from the energy metabolism of cells (Porkka-Heiskanen & Kalinchuk 2011). This particular piece of evidence not only shows the sleep-promoting effects of adenosine on Hcrt neurons, but also suggests that energy use can be translated into a mechanism used by the Hcrt system as a cue to limit energy expenditure and modulate the behavioral state. Shulman et al. (2003) showed that the animals' energy status is highly correlated with their behavioral state. The brain utilizes more energy in the activated state than it does in the anesthetized state, and a low-energy state leads to unconsciousness in animals (Shulman et al. 1999, 2009). Insufficient energy may attenuate wakefulness and promote sleep in animals (Benington & Heller 1995, Scharf et al. 2008). The evidence supporting this hypothesis is emerging, particularly from the studies on Hcrt neurons. Parsons & Hirasawa (2010) have shown that lactate, but not glucose, is required to maintain the firing of Hcrt neurons in vitro, and Gao and colleagues (Liu et al. 2011) have further shown that the intracellular levels of ATP are a key factor to maintain the membrane potential and firing of action potential in Hcrt neurons. More importantly, the intracellular levels of ATP are lower in Hcrt neurons in sleeping mice than in sleep-deprived animals (Liu et al. 2011). This finding is consistent with the results indicating that the cerebral consumption of glucose and oxygen and glucose uptake fluctuate across the sleep/wake cycle in humans and animals (Boyle et al. 1994, Vyazovskiy et al. 2008). The correlation between activity and the intracellular levels of ATP in Hcrt neurons may serve as a mechanism underlying the energy hypothesis of sleep; i.e., the decrease in ATP levels may limit arousal/wakefulness and promote sleep in animals (Benington & Heller 1995, Scharf et al. 2008). In summary, research on the relationship between energy and behavioral states is revealing new avenues for understanding the sleep/wake cycle and behavioral states in the context of energy metabolism in animals. The role of the Hcrt system is no doubt critical to deciphering the mystery behind these processes.

Hcrt IN REWARDING/MOTIVATIONAL BEHAVIORS

The perifornical/LH area is a brain structure known to be responsible for reward-seeking behaviors. In rats, electrical stimulation of the LH induces marked reinforcement activity: Investigators observed a robust self-administration of electric current when the stimulating electrodes were implanted within the LH area (Olds & Milner 1954, Olds 1958). Many drugs of abuse, such as morphine and amphetamine, modulate the self-reinforcement induced by stimulation of the LH area (Adams et al. 1972, Goodall & Carey 1975). Drugs of abuse, in addition to electrical stimulation, induce self-administration in rats when applied to the LH directly. For instance, administering D-Ala2-Met-enkephalin, a long-lasting analogue of encephalin, through a cannula implanted in the LH area triggers self-administration in rats, which can be blocked by naloxone (Olds & Williams 1980). Similar results have been reported in self-administration of morphine and morphine-induced place preference in mice (Cazala et al. 1987).

Hcrt neurons are clearly required in the development of drug addiction in animal models and human patients (see reviews by Baimel & Borgland 2012, España 2012, Mahler et al. 2012). First, Hcrt neurons are activated when rodents are exposed to drugs of abuse, and the expression of c-Fos in Hcrt neurons is enhanced by opiates, cocaine, amphetamine, and nicotine in various animal models of drug-seeking behavior (Georgescu et al. 2003, Harris et al. 2005, Pasumarthi et al. 2006, McPherson et al. 2007, Plaza-Zabala et al. 2011). Second, direct applications of Hcrt to reward centers in the brain promote drug-induced plasticity in these brain areas and drug-seeking behavior in animals. Activation of Hcrt neurons or administration of Hcrt directly into the ventral tegmental area (VTA) reinstates extinguished morphine-seeking behavior and increases the break point in a progressive ratio task for cocaine (Boutrel et al. 2005, Harris et al. 2005, Hamlin et al. 2008, España et al. 2011). Also, a central application of Hcrt [intracerebroventricular (i.c.v.)] reinstates extinguished nicotine-seeking behavior in mice (Plaza-Zabala et al. 2011). The findings that Hcrt induces synaptic plasticity in the VTA and that Hcrt receptor antagonists abolish cocaine or amphetamine-induced locomotor sensitization, potentiation of glutamatergic currents, and expression levels of genes associated with synaptic plasticity in the VTA in rats may provide mechanisms underlying the role that Hcrt plays in reward centers of the brain (Borgland et al. 2006, Winrow et al. 2010). Third, the disruption of Hcrt receptor–mediated pathways attenuates or blocks drug-seeking behavior in animals. The development of morphine dependence is attenuated by systemic application of the selective Hcrt-1 receptor antagonist, SB334867 (Georgescu et al. 2003). SB334867 blocks the acquisition of cocaine sensitization, attenuates cocaine and amphetamine conditioned place preference (CPP), and reduces break points for cocaine under a progressive ratio (PR) schedule of reinforcement (Harris et al. 2005; Borgland et al. 2006, 2009; España et al. 2010). SB334867 significantly decreases nicotine reinforcement in rats as well (Hollander et al. 2008, LeSage et al. 2010). Drug-seeking behaviors are attenuated or abolished in animals with a deficiency in Hcrt peptide or receptors (Georgescu et al. 2003, Hollander et al. 2012). In human narcoleptic patients (who have a deficiency in Hcrt peptide or neurons), the tendency toward drug abuse is significantly low (Guilleminault et al. 1974).

Despite the growing body of evidence on the Hcrt system's participation in reward-seeking behavior, the Hcrt system's role in the development of motivational behaviors is not yet clear.

Harris et al. (2005) proposed that the Hcrt system may be responsible for cue-induced seeking conduct (i.e., CPP) for cocaine and morphine. The Hcrt system may contribute to psychostimulatory effects of drugs (such as cocaine and amphetamine) (Borgland et al. 2006). Borgland et al. (2009) recently suggested that the Hcrt system may be involved not in the rewarding aspect of drugs of abuse but rather in the motivational aspect of drug-seeking behavior, as indicated in cocaine self-administration experiments. Ho & Berridge (2013) showed that activating Hcrt signaling in the ventral pallidum (VP) generated hedonic (liking) responses to sweetness in rats. The Hcrt system may also be responsible for sleep disorders in drug addicts (Rao et al. 2013). Additionally, the Hcrt system may mediate the effects of metabolic states on rewarding and motivational behaviors in animals. Carroll et al. (1979) showed that food restriction increases sensitivity to drug reward. According to the current framework, stimulation of D1 receptors, upregulation of protein kinases, and activation of epigenetic mechanisms in the NAc are responsible for the effects of food restriction on drug reward (reviewed by Carr 2011). The latest progress in the field suggests that not only food restriction but also metabolic status may shape drug reward in animals. Consistent with the effects of food restriction on drug use, overnutrition or obesity decreases the subject's sensitivity to drug reward (reviewed by Kenny 2011). Emerging evidence proposes that D2 dopamine receptor–mediated mechanisms in striatal neurons underlie this process (Kenny 2011). Because food and calorie restriction and diet-induced obesity represent two opposite sides of the same coin, a common mechanism is likely responsible for the effects of metabolic status on drug reward. Is the Hcrt system one of the converging points where metabolic status interacts with motivational behaviors in animals? Studies have demonstrated that cues encoding hunger and food restriction increase the expression of Hcrt mRNA in animals (Sakurai et al. 1998; Cai et al. 1999, 2001; Kurose et al. 2002), whereas obesity leads to the downregulation of Hcrt expression in rats and mice (Cai et al. 2000, Beck et al. 2001, Stricker-Krongrad et al. 2002). We have also shown previously that an acute food restriction leads to the reorganization of the neural circuitry onto Hcrt neurons (Horvath & Gao 2005). These data clearly implicate the adaptation of Hcrt neurons induced by changes in metabolic status. Therefore, an important question arises about whether the effects of the metabolic status on motivational behaviors in animals require the Hcrt system, an issue that has not yet been thoroughly explored.

DYSFUNCTION OF Hcrt IN ALTERED ANIMAL BEHAVIORS AND HUMAN DISEASES

Deficiency in the Hcrt system in narcolepsy is well addressed in human patients and animal models (Chemelli et al. 1999, Lin et al. 1999, Nishino et al. 2000, Thannickal et al. 2000, Ripley et al. 2001). Although in animal models, the loss of Hcrt peptide, its receptors (particularly OXR2), and Hcrt-containing neurons produce a narcolepsy-like phenotype (Lin et al. 1999, Hara et al. 2001, Willie et al. 2003), in most human narcolepsy cases, the mutation does not occur in either the Hcrt ligand or Hcrt receptor genes. The low (or undetectable) level of Hcrt in the CSF in patients with narcolepsy–cataplexy is likely due to the loss of Hcrt-containing neurons (Thannickal et al. 2000). The mechanisms underlying the selective loss of Hcrt-containing neurons in narcoleptic patients are not yet clear, but the latest evidence suggests an autoimmune process may be involved in this pathological condition (Han 2012, Mahlios et al. 2013). The replacement of Hcrt in narcoleptic animal models is currently being tested as a potential treatment for narcolepsy (Blanco-Centurion et al. 2013, Kantor et al. 2013).

In an animal model of narcolepsy, obesity is one of the phenotypes observed in the study (Hara et al. 2001). In fact, historic records have shown an increased body weight in narcoleptic patients (Schuld et al. 2000, Akinnusi et al. 2012). The exact link between obesity and narcolepsy in human patients is uncertain, but in mice with genetically ablated Hcrt neurons, hypophagia was observed in conjunction with obesity. This piece of evidence implies that the balance may tilt toward energy expenditure when the Hcrt system functions normally in intact animals.

Although research has not yet established a direct association, dysfunction in the Hcrt system has been seen in many other diseases and pathological conditions. For instance, the dual Hcrt receptor antagonists SB649868 and suvorexant have been effective treatments for insomnia in clinical trials (Winrow & Renger 2013). This evidence may shed new light on the development of treatments for insomnia. In patients with Prader–Willi syndrome (PWS), the loss of muscle tone in infancy and sleep disturbances with excessive daytime sleepiness (EDS) at later developmental stages imply a similarity between this disease and narcolepsy caused by a deficiency in the Hcrt system (Camfferman et al. 2008). Some evidence indicates that impaired levels of Hcrt-1 in the CSF correlate with the severity of EDS in PWS patients (Nevsimalova et al. 2005), whereas the total number of Hcrt neurons is not significantly different in PWS patients as compared with age-matched controls (Fronczek et al. 2005). Thus, the reduced CSF levels of Hcrt in PWS patients may very likely be due to impaired functions in neurons expressing this neuropeptide.

PERSPECTIVE

The Hcrt system is only one part of an ever-evolving network controlling the most fundamental functions in animals. The intake and use of energy are the basis of all biological activities. The Hcrt system may be a critical hub of the brain positioned to control energy intake and expenditure and, at the same time, may contribute significantly to the control of behavioral (rest versus active) states and motivational behaviors in animals (**Figure 1**). Several lines of research are still needed. First, investigators must discern just how the Hcrt system monitors the ambient energy state and nutritional stores. To this end, new evidence must materialize and inconsistencies in published reports must still be addressed. Determining the molecular and cellular mechanisms that underlie the sensing of cues encoding energy status and nutritional stores is also essential to improving our

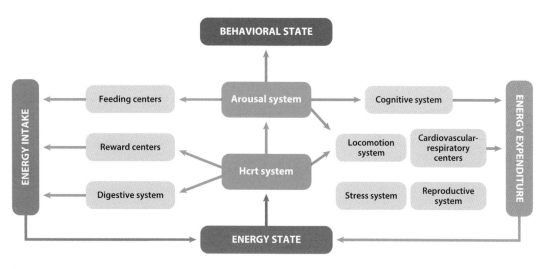

Figure 1

A theoretical scheme of how the Hcrt system regulates animal behaviors depending on animals' energy states. As proposed in this diagram, the Hcrt system promotes both energy intake and expenditure. The energy state based on the available energy store in animals may fine-tune the activity in the Hcrt system and help determine animals' behavioral states. In this simplified diagram, the interaction of the Hcrt system with other homeostatic centers and feedbacks from environmental cues are not included.

understanding. Second, we need to understand the connectomics of Hcrt neurons with an emphasis on the brain structures and peripheral organs responsible for energy intake and expenditure. Interactions between the Hcrt neurons and the NPY, melanin-concentrating hormone (MCH) and other neuronal populations have been reported previously (Horvath et al. 1999, Guan et al. 2002, van den Pol et al. 2004, Rao et al. 2008); however, the contribution of these interactions in the context of bidirectional effects of Hcrt on energy homeostasis has not been shown. Third, and most important, a comprehensive understanding of how an animal's energy state shapes the Hcrt system, and in turn modulates other homeostatic, motivational, and cognitive functions, is critical.

DISCLOSURE STATEMENT

The authors are not aware of any affiliations, memberships, funding, or financial holdings that might be perceived as affecting the objectivity of this review.

LITERATURE CITED

Adamantidis AR, Zhang F, Aravanis AM, Deisseroth K, de Lecea L. 2007. Neural substrates of awakening probed with optogenetic control of hypocretin neurons. *Nature* 450:420–24

Adams WJ, Lorens SA, Mitchell CL. 1972. Morphine enhances lateral hypothalamic self-stimulation in the rat. *Proc. Soc. Exp. Biol. Med.* 140:770–71

Adeghate E, Fernandez-Cabezudo M, Hameed R, El-Hasasna H, El Wasila M, et al. 2010. Orexin-1 receptor co-localizes with pancreatic hormones in islet cells and modulates the outcome of streptozotocin-induced diabetes mellitus. *PLoS ONE* 5(1):e8587

Akinnusi ME, Saliba R, Porhomayon J, El-Solh AA. 2012. Sleep disorders in morbid obesity. *Eur. J. Intern. Med.* 23(3):219–26

Anand BK, Brobeck JR. 1951a. Hypothalamic control of food intake in rats and cats. *Yale J. Biol. Med.* 24:123–40

Anand BK, Brobeck JR. 1951b. Localization of a "feeding center" in the hypothalamus of the rat. *Proc. Soc. Exp. Biol. Med.* 77:323–24

Asakawa A, Inui A, Goto K, Yuzuriha H, Takimoto Y, et al. 2002. Effects of agouti-related protein, orexin and melanin-concentrating hormone on oxygen consumption in mice. *Int. J. Mol. Med.* 10:523–25

Baimel C, Borgland SL. 2012. Hypocretin modulation of drug-induced synaptic plasticity. *Prog. Brain Res.* 198:123–31

Baldo BA, Gual-Bonilla L, Sijapati K, Daniel RA, Landry CF, Kelley AE. 2004. Activation of a subpopulation of orexin/hypocretin-containing hypothalamic neurons by GABA$_A$ receptor-mediated inhibition of the nucleus accumbens shell, but not by exposure to a novel environment. *Eur. J. Neurosci.* 19:376–86

Beck B, Richy S, Dimitrov T, Stricker-Krongrad A. 2001. Opposite regulation of hypothalamic orexin and neuropeptide Y receptors and peptide expressions in obese Zucker rats. *Biochem. Biophys. Res. Commun.* 286:518–23

Benington JH, Heller HC. 1995. Restoration of brain energy metabolism as the function of sleep. *Prog. Neurobiol.* 45:347–60

Blanco-Centurion C, Liu M, Konadhode R, Pelluru D, Shiromani PJ. 2013. Effects of orexin gene transfer in the dorsolateral pons in orexin knockout mice. *Sleep* 36(1):31–40

Borgland SL, Chang S-J, Bowers MS, Thompson J, Vittoz N, et al. 2009. Orexin A/hypocretin-1 selectively promotes motivation for positive reinforcers. *J. Neurosci.* 29:11215–25

Borgland SL, Taha SA, Sarti F, Fields HL, Bonci A. 2006. Orexin A in the VTA is critical for the induction of synaptic plasticity and behavioral sensitization to cocaine. *Neuron* 49:589–601

Bourgin P, Huitrón-Résendiz S, Spier AD, Fabre V, Morte B, et al. 2000. Hypocretin-1 modulates rapid eye movement sleep through activation of locus coeruleus neurons. *J. Neurosci.* 20:7760–65

Boutrel B, Kenny PJ, Specio SE, Martin-Fardon R, Markou A, et al. 2005. Role for hypocretin in mediating stress-induced reinstatement of cocaine-seeking behavior. *Proc. Natl. Acad. Sci. USA* 102:19168–73

Boyle PJ, Scott JC, Krentz AJ, Nagy RJ, Comstock E, Hoffman C. 1994. Diminished brain glucose metabolism is a significant determinant for falling rates of systemic glucose utilization during sleep in normal humans. *J. Clin. Invest.* 93:529–35

Burdakov D, Jensen LT, Alexopoulos H, Williams RH, Fearon IM, et al. 2006. Tandem-pore K$^+$ channels mediate inhibition of orexin neurons by glucose. *Neuron* 50(5):711–22

Burt J, Alberto CO, Parsons MP, Hirasawa M. 2011. Local network regulation of orexin neurons in the lateral hypothalamus. *Am. J. Physiol. Regul. Integr. Comp. Physiol.* 301(3):R572–80

Cai XJ, Evans ML, Lister CA, Leslie RA, Arch JRS, et al. 2001. Hypoglycemia activates orexin neurons and selectively increases hypothalamic orexin-B levels. Responses inhibited by feeding and possibly mediated by the nucleus of the solitary tract. *Diabetes* 50:105–12

Cai XJ, Lister CA, Buckingham RE, Pickavance L, Wilding J, et al. 2000. Down-regulation of orexin gene expression by severe obesity in the rats: studies in Zucker fatty and zucker diabetic fatty rats and effects of rosiglitazone. *Brain Res. Mol. Brain Res.* 77:131–37

Cai XJ, Widdowson PS, Harrold J, Wilson S, Buckingham RE, et al. 1999. Hypothalamic orexin expression: modulation by blood glucose and feeding. *Diabetes* 48:2132–37

Camfferman D, McEvoy RD, O'Donoghue F, Lushington K. 2008. Prader Willi Syndrome and excessive daytime sleepiness. *Sleep Med. Rev.* 12:65–75

Carr KD. 2011. Food scarcity, neuroadaptations, and the pathogenic potential of dieting in an unnatural ecology: binge eating and drug abuse. *Physiol. Behav.* 104:162–67

Carroll ME, France CP, Meisch RA. 1979. Food deprivation increases oral and intravenous drug intake in rats. *Science* 205:319–21

Cazala P, Darracq C, Saint-Marc M. 1987. Self-administration of morphine into the lateral hypothalamus in the mouse. *Brain Res.* 416:283–88

Chandra R, Liddle RA. 2009. Neural and hormonal regulation of pancreatic secretion. *Curr. Opin. Gastroenterol.* 25:441–46

Chemelli RM, Willie JT, Sinton CM, Elmquist JK, Scammell T, et al. 1999. Narcolepsy in orexin knockout mice: molecular genetics of sleep regulation. *Cell* 98:437–51

Davis SF, Williams KW, Xu W, Glatzer NR, Smith BN. 2003. Selective enhancement of synaptic inhibition by hypocretin (orexin) in rat vagal motor neurons: implications for autonomic regulation. *J. Neurosci.* 23:3844–54

de Lecea L, Kilduff TS, Peyron C, Gao X, Foye PE, et al. 1998. The hypocretins: hypothalamus-specific peptides with neuroexcitatory activity. *Proc. Natl. Acad. Sci. USA* 95:322–27

de Lecea L, Sutcliffe JG. 2005. The hypocretins and sleep. *FEBS J.* 272:5675–88

Delgado J, Anand BK. 1953. Increase of food intake induced by electrical stimulation of the lateral hypothalamus. *Am. J. Physiol.* 172:162–68

Dube MG, Kalra SP, Kalra PS. 1999. Food intake elicited by central administration of orexins/hypocretins: identification of hypothalamic sites of action. *Brain Res.* 842:473–77

España RA. 2012. Hypocretin/orexin involvement in reward and reinforcement. *Vitam. Horm.* 89:185–208

España RA, Baldo BA, Kelley AE, Berridge CW. 2001. Wake-promoting and sleep-suppressing actions of hypocretin (orexin): basal forebrain sites of action. *Neuroscience* 106:699–715

España RA, Melchior JR, Roberts DC, Jones SR. 2011. Hypocretin 1/orexin A in the ventral tegmental area enhances dopamine responses to cocaine and promotes cocaine self-administration. *Psychopharmacology* 214:415–26

España RA, Oleson EB, Locke JL, Brookshire BR, Roberts DC, Jones SR. 2010. The hypocretin-orexin system regulates cocaine self-administration via actions on the mesolimbic dopamine system. *Eur. J. Neurosci.* 31:336–48

Estabrooke IV, McCarthy MT, Ko E, Chou TC, Chemelli RM, et al. 2001. Fos expression in orexin neurons varies with behavioral state. *J. Neurosci.* 21:1656–62

Fronczek R, Lammers GJ, Balesar R, Unmehopa UA, Swaab DF. 2005. The number of hypothalamic hypocretin (orexin) neurons is not affected in Prader-Willi syndrome. *J. Clin. Endocrinol. Metab.* 90(9):5466–70

Fujiki N, Yoshida Y, Ripley B, Honda K, Mignot E, Nishino S. 2001. Changes in CSF hypocretin-1 (orexin A) levels in rats across 24 hours and in response to food deprivation. *Neuroreport* 17:993–97

Georgescu D, Zachariou V, Barrot M, Mieda M, Willie JT, et al. 2003. Involvement of the lateral hypothalamic peptide orexin in morphine dependence and withdrawal. *J. Neurosci.* 23:3106–11

Goodall EB, Carey RJ. 1975. Effects of D- versus L-amphetamine, food deprivation, and current intensity on self-stimulation of the lateral hypothalamus, substantia nigra, and medial frontal cortex of the rat. *J. Comp. Physiol. Psychol.* 89:1029–45

Griffond B, Risold PY, Jacquemard C, Colard C, Fellman D. 1999. Insulin-induced hypoglycemia increases prehypocretin (orexin) mRNA in the rat lateral hypothalamic area. *Neurosci. Lett.* 262:77–80

Guan JL, Uehara K, Lu S, Wang QP, Funahashi H, et al. 2002. Reciprocal synaptic relationships between orexin- and melanin-concentrating hormone-containing neurons in the rat lateral hypothalamus: a novel circuit implicated in feeding regulation. *Int. J. Obes. Relat. Metab. Disord.* 26(12):1523–32

Guilleminault C, Carskadon M, Dement WC. 1974. On the treatment of rapid eye movement narcolepsy. *Arch. Neurol.* 30:90–93

Guyon A, Tardy MP, Rovère C, Nahon JL, Barhanin J, Lesage F. 2009. Glucose inhibition persists in hypothalamic neurons lacking tandem-pore K^+ channels. *J. Neurosci.* 29(8):2528–33

Hagan JJ, Leslie RA, Patel S, Evans ML, Wattam TA, et al. 1999. Orexin A activates locus coeruleus cell firing and increases arousal in the rat. *Proc. Natl. Acad. Sci. USA* 96:10911–16

Hamlin AS, Clemens KJ, McNally GP. 2008. Renewal of extinguished cocaine-seeking. *Neuroscience* 151:659–70

Han F. 2012. Sleepiness that cannot be overcome: narcolepsy and cataplexy. *Respirology* 17(8):1157–65

Hara J, Beuckmann CT, Nambu T, Willie JT, Chemelli RM, et al. 2001. Genetic ablation of orexin neurons in mice results in narcolepsy, hypophagia, and obesity. *Neuron* 30(2):345–54

Harris GC, Wimmer M, Aston-Jones G. 2005. A role for lateral hypothalamic orexin neurons in reward seeking. *Nature* 437(7058):556–59

Haynes AC, Jackson B, Chapman H, Tadayyon M, Johns A, et al. 2000. A selective orexin-1 receptor antagonist reduces food consumption in male and female rats. *Regul. Pept.* 96:45–51

Heinonen MV, Purhonen AK, Mäkelä KA, Herzig KH. 2008. Functions of orexins in peripheral tissues. *Acta Physiol.* 192:471–85

Hetherington AW, Ranson SW. 1940. Hypothalamic lesions and adiposity in the rat. *Anat. Rec.* 78:149–72

Himmi T, Boyer A, Orsini JC. 1988. Changes in lateral hypothalamic neuronal activity accompanying hyper- and hypoglycemias. *Physiol. Behav.* 44:347–54

Ho CY, Berridge KC. 2013. An orexin hotspot in ventral pallidum amplifies hedonic 'liking' for sweetness. *Neuropsychopharmacology* 38(9):1655–64

Hollander JA, Lu Q, Cameron MD, Kamenecka TM, Kenny PJ. 2008. Insular hypocretin transmission regulates nicotine reward. *Proc. Natl. Acad. Sci. USA* 105:19480–85

Hollander JA, Pham D, Fowler CD, Kenny PJ. 2012. Hypocretin-1 receptors regulate the reinforcing and reward-enhancing effects of cocaine: pharmacological and behavioral genetics evidence. *Front. Behav. Neurosci.* 6:47

Horvath TL, Diano S, van den Pol AN. 1999. Synaptic interaction between hypocretin (orexin) and neuropeptide Y cells in the rodent and primate hypothalamus: a novel circuit implicated in metabolic and endocrine regulations. *J Neurosci.* 19(3):1072–87

Horvath TL, Gao XB. 2005. Input organization and plasticity of hypocretin neurons: possible clues to obesity's association with insomnia. *Cell Metab.* 1:279–86

Jöhren O, Gremmels JA, Qadri F, Dendorfer A, Dominiak P. 2006. Adrenal expression of orexin receptor subtypes is differentially regulated in experimental streptozotocin induced type-1 diabetes. *Peptides* 27:2764–69

Kantor S, Mochizuki T, Lops SN, Ko B, Clain E, et al. 2013. Orexin gene therapy restores the timing and maintenance of wakefulness in narcoleptic mice. *Sleep* 36(8):1129–38

Katafuchi T, Oomura Y, Yoshimatsu H. 1985. Single unit activity in the rat lateral hypothalamus during 2-deoxy-D-glucose induced and natural feeding behavior. *Brain Res.* 359:1–9

Kenny PJ. 2011. Common cellular and molecular mechanisms in obesity and drug addiction. *Nat. Rev. Neurosci.* 12(11):638–51

Kiwaki K, Kotz CM, Wang C, Lanningham-Foster L, Levine JA. 2004. Orexin A (hypocretin 1) injected into hypothalamic paraventricular nucleus and spontaneous physical activity in rats. *Am. J. Physiol. Endocrinol. Metab.* 286:E551–59

Korotkova TM, Sergeeva OA, Eriksson KS, Haas HL, Brown RE. 2003. Excitation of ventral tegmental area dopaminergic and nondopaminergic neurons by orexins/hypocretins. *J. Neurosci.* 23:7–11

Kotz C, Nixon J, Butterick T, Perez-Leighton C, Teske J, Billington C. 2012. Brain orexin promotes obesity resistance. *Ann. N. Y. Acad. Sci.* 1264(1):72–86

Kotz CM, Teske JA, Billington CJ. 2008. Neuroregulation of nonexercise activity thermogenesis and obesity resistance. *Am. J. Physiol. Regul. Integr. Comp. Physiol.* 294:R699–710

Kotz CM, Teske JA, Levine JA, Wang C. 2002. Feeding and activity induced by orexin A in the lateral hypothalamus in rats. *Regul. Pept.* 104:27–32

Kotz CM, Wang C, Teske JA, Thorpe AJ, Novak CM, et al. 2006. Orexin A mediation of time spent moving in rats: neural mechanisms. *Neuroscience* 142:29–36

Krowicki ZK, Burmeister MA, Berthoud HR, Scullion RT, Fuchs K, Hornby PJ. 2002. Orexins in rat dorsal motor nucleus of the vagus potently stimulate gastric motor function. *Am. J. Physiol. Gastrointest. Liver Physiol.* 283:G465–72

Kurose T, Ueta Y, Yamamoto Y, Serino R, Ozaki Y, et al. 2002. Effects of restricted feeding on the activity of hypothalamic Orexin (OX)-A containing neurons and OX2 receptor mRNA level in the paraventricular nucleus of rats. *Regul. Pept.* 104:145–51

Lee MG, Hassani OK, Jones BE. 2005. Discharge of identified orexin/hypocretin neurons across the sleep-waking cycle. *J. Neurosci.* 25:6716–20

LeSage MG, Perry JL, Kotz CM, Shelley D, Corrigall WA. 2010. Nicotine self-administration in the rat: effects of hypocretin antagonists and changes in hypocretin mRNA. *Psychopharmacology* 209:203–12

Lin JS, Sakai K, Vanni-Mercier G, Jouvet M. 1989. A critical role of the posterior hypothalamus in the mechanisms of wakefulness determined by microinjection of muscimol in freely moving cats. *Brain Res.* 479:225–40

Lin L, Faraco J, Li R, Kadotani H, Rogers W, et al. 1999. The sleep disorder canine narcolepsy is caused by a mutation in the *hypocretin (orexin) receptor 2* gene. *Cell* 98(3):365–76

Liu Z-W, Gan G, Suyama S, Gao X-B. 2011. Intracellular energy status regulates activity in hypocretin/orexin neurones: a link between energy and behavioural states. *J. Physiol.* 589(Pt. 17):4157–66

Liu Z-W, Gao X-B. 2007. Adenosine inhibits activity of hypocretin/orexin neurons by the A1 receptor in the lateral hypothalamus: a possible sleep-promoting effect. *J. Neurophysiol.* 97(1):837–48

López M, Seoane L, García MC, Lago F, Casanueva FF, et al. 2000. Leptin regulation of prepro-orexin and orexin receptor mRNA levels in the hypothalamus. *Biochem. Biophys. Res. Commun.* 269(1):41–45

Lubkin M, Stricker-Krongrad A. 1998. Independent feeding and metabolic actions of orexins in mice. *Biochem. Biophys. Res. Commun.* 253(2): 241–45

Mackiewicz M, Naidoo N, Zimmerman JE, Pack AI. 2008. Molecular mechanisms of sleep and wakefulness. *Ann. N. Y. Acad. Sci.* 1129:335–49

Mahler SV, Smith RJ, Moorman DE, Sartor GC, Aston-Jones G. 2012. Multiple roles for orexin/hypocretin in addiction. *Prog. Brain Res.* 198:79–121

Mahlios J, De la Herrán-Arita AK, Mignot E. 2013. The autoimmune basis of narcolepsy. *Curr. Opin. Neurobiol.* 23:767–73

Marcus JN, Aschkenasi CJ, Lee CE, Chemelli RM, Saper CB, et al. 2001. Differential expression of orexin receptors 1 and 2 in the rat brain. *J. Comp. Neurol.* 435(1):6–25

McPherson CS, Featherby T, Krstew E, Lawrence AJ. 2007. Quantification of phosphorylated cAMP-response element-binding protein expression throughout the brain of amphetamine-sensitized rats: activation of hypothalamic orexin A-containing neurons. *J. Pharmacol. Exp. Ther.* 323:805–12

Mileykovskiy BY, Kiyashchenko LI, Siegel JM. 2005. Behavioral correlates of activity in identified hypocretin/orexin neurons. *Neuron* 46:787–98

Mondal MS, Nakazato M, Date Y, Murakami N, Yanagisawa M, Matsukura S. 1999. Widespread distribution of orexin in rat brain and its regulation upon fasting. *Biochem. Biophys. Res. Commun.* 256(3):495–99

Moriguchi T, Sakurai T, Nambu T, Yanagisawa M, Goto K. 1999. Neurons containing orexin in the lateral hypothalamic area of the adult rat brain are activated by insulin-induced acute hypoglycemia. *Neurosci. Lett.* 264:101–4

Nauta WJH. 1946. Hypothalamic regulation of sleep in rats: an experimental study. *J. Neurophysiol.* 9:285–316

Nevsimalova S, Vankova J, Stepanova I, Seemanova E, Mignot E, Nishino S. 2005. Hypocretin deficiency in Prader-Willi syndrome. *Eur. J. Neurol.* 12(1):70–72

Nishino S, Ripley B, Overeem S, Lammers GJ, Mignot E. 2000. Hypocretin (orexin) deficiency in human narcolepsy. *Lancet* 355:39–40

Olds J. 1958. Self-stimulation of the brain; its use to study local effects of hunger, sex, and drugs. *Science* 127:315–24

Olds J, Milner P. 1954. Positive reinforcement produced by electrical stimulation of septal area and other regions of rat brain. *J. Comp. Physiol. Psychol.* 47:419–27

Olds ME, Williams KN. 1980. Self-administration of D-Ala^2-Met-enkephalinamide at hypothalamic self-stimulation sites. *Brain Res.* 194:155–70

Paranjape S, Vavaiya K, Kale A, Briski K. 2007. Role of dorsal vagal motor nucleus orexin-receptor-1 in glycemic responses to acute versus repeated insulin administration. *Neuropeptides* 41:111–16

Parsons MP, Hirasawa M. 2010. ATP-sensitive potassium channel-mediated lactate effect on orexin neurons: implications for brain energetics during arousal. *J. Neurosci.* 30(24):8061–70

Pasumarthi RK, Reznikov LR, Fadel J. 2006. Activation of orexin neurons by acute nicotine. *Eur. J. Pharmacol.* 535:172–76

Peyron C, Tighe DK, van den Pol AN, de Lecea L, Heller HC, et al. 1998. Neurons containing hypocretin (orexin) project to multiple neuronal systems. *J. Neurosci.* 18:9996–10015

Plaza-Zabala A, Flores Á, Maldonado R, Berrendero F. 2011. Hypocretin/orexin signaling in the hypothalamic paraventricular nucleus is essential for the expression of nicotine withdrawal. *Biol. Psychiatry* 71:214–23

Porkka-Heiskanen T, Kalinchuk AV. 2011. Adenosine, energy metabolism and sleep homeostasis. *Sleep Med. Rev.* 15(2):123–35

Ranson SW. 1939. Somnolence caused by hypothalamic lesions in the monkey. *Arch. Neurol. Psychiatry* 41:1–23

Rao Y, Lu M, Ge F, Marsh DJ, Qian S, et al. 2008. Regulation of synaptic efficacy in hypocretin/orexin-containing neurons by melanin concentrating hormone in the lateral hypothalamus. *J. Neurosci.* 28(37):9101–10

Rao Y, Mineur YS, Gan G, Wang AH, Liu ZW, et al. 2013. Repeated in vivo exposure of cocaine induces long-lasting synaptic plasticity in hypocretin/orexin-producing neurons in the lateral hypothalamus in mice. *J. Physiol.* 591:1951–66

Ripley B, Overeem S, Fujiki N, Nevsimalova S, Uchino M, et al. 2001. CSF hypocretin/orexin levels in narcolepsy and other neurological conditions. *Neurology* 57:2253–58

Rodgers RJ, Ishii Y, Halford JC, Blundell JE. 2002. Orexins and appetite regulation. *Neuropeptides* 36:303–25

Sakurai T. 2005. Roles of orexin/hypocretin in regulation of sleep/wakefulness and energy homeostasis. *Sleep Med. Rev.* 9:231–41

Sakurai T, Amemiya A, Ishii M, Matsuzaki I, Chemelli RM, et al. 1998. Orexins and orexin receptors: a family of hypothalamic neuropeptides and G protein-coupled receptors that regulate feeding behavior. *Cell* 92:573–85

Samson WK, Taylor MM, Ferguson AV. 2005. Non-sleep effects of hypocretin/orexin. *Sleep Med. Rev.* 9:243–52

Saper CB. 2006. Staying awake for dinner: hypothalamic integration of sleep, feeding, and circadian rhythms. *Prog. Brain Res.* 153:243–52

Saper CB. 2013. The neurobiology of sleep. *Continuum* 19(1 Sleep Disord.):19–31

Sawchenko PE. 1998. Toward a new neurobiology of energy balance, appetite, and obesity: the anatomists weigh in. *J. Comp. Neurol.* 402:435–41

Scharf MT, Naidoo N, Zimmerman JE, Pack AI. 2008. The energy hypothesis of sleep revisited. *Prog. Neurobiol.* 86:264–80

Schuld A, Hebebrand J, Geller F, Pollmächer T. 2000. Increased body-mass index in patients with narcolepsy. *Lancet* 355:1274–75

Selbach O, Haas HL. 2006. Hypocretins: the timing of sleep and waking. *Chronobiol. Int.* 23(1–2):63–70

Sellayah D, Bharaj P, Sikder D. 2011. Orexin is required for brown adipose tissue development, differentiation, and function. *Cell Metab.* 14:478–90

Sellayah D, Sikder D. 2012. Orexin receptor-1 mediates brown fat developmental differentiation. *Adipocyte* 1:58–63

Semjonous NM, Smith KL, Parkinson JR, Gunner DJ, Liu YL, et al. 2009. Coordinated changes in energy intake and expenditure following hypothalamic administration of neuropeptides involved in energy balance. *Int. J. Obes.* 33:775–85

Shiuchi T, Haque MS, Okamoto S, Inoue T, Kageyama H, et al. 2009. Hypothalamic orexin stimulates feeding-associated glucose utilization in skeletal muscle via sympathetic nervous system. *Cell Metab.* 10:466–80

Shulman RG, Hyder F, Rothman DL. 2003. Cerebral metabolism and consciousness. *C. R. Biol.* 326:253–73

Shulman RG, Hyder F, Rothman DL. 2009. Baseline brain energy supports the state of consciousness. *Proc. Natl. Acad. Sci. USA* 106:11096–101

Shulman RG, Rothman DL, Hyder F. 1999. Stimulated changes in localized cerebral energy consumption under anesthesia. *Proc. Natl. Acad. Sci. USA* 96:3245–50

Stellar E. 1954. The physiology of motivation. *Psychol. Rev.* 61:5–22

Stricker EM, Swerdloff AF, Zigmond MJ. 1978. Intrahypothalamic injections of kainic acid produce feeding and drinking deficits in rats. *Brain Res.* 158:470–73

Stricker-Krongrad A, Richy S, Beck B. 2002. Orexins/hypocretins in the ob/ob mouse: hypothalamic gene expression, peptide content and metabolic effects. *Regul. Pept.* 104:11–20

Sutcliffe JG, de Lecea L. 2002. The hypocretins: setting the arousal threshold. *Nat. Rev. Neurosci.* 3:339–49

Swett C, Hobson J. 1968. The effects of posterior hypothalamic lesions on behavioral and electrographic manifestations of sleep and waking in cats. *Arch. Ital. Biol.* 106:283–93

Székely M. 2006. Orexins, energy balance, temperature, sleep-wake cycle. *Am. J. Physiol. Regul. Integr. Comp. Physiol.* 291:R530–32

Takahashi N, Okumura T, Yamada H, Kohgo Y. 1999. Stimulation of gastric acid secretion by centrally administered orexin-A in conscious rats. *Biochem. Biophys. Res. Commun.* 254:623–27

Teske JA, Billington CJ, Kotz CM. 2010. Hypocretin/orexin and energy expenditure. *Acta Physiol.* 198:303–12

Thannickal TC, Moore RY, Nienhuis R, Ramanathan L, Gulyani S, et al. 2000. Reduced number of hypocretin neurons in human narcolepsy. *Neuron* 27:469–74

Thorpe AJ, Mullett MA, Wang C, Kotz CM. 2003. Peptides that regulate food intake: regional, metabolic, and circadian specificity of lateral hypothalamic orexin A feeding stimulation. *Am. J. Physiol. Regul. Integr. Comp. Physiol.* 284:R1409–17

Trivedi P, Yu H, MacNeil DJ, Van der Ploeg LH, Guan XM. 1998. Distribution of orexin receptor mRNA in the rat brain. *FEBS Lett.* 438(1–2):71–75

Tsuneki H, Wada T, Sasaoka T. 2010. Role of orexin in the regulation of glucose homeostasis. *Acta Physiol.* 198:335–48

Tupone D, Madden CJ, Cano G, Morrison SF. 2011. An orexinergic projection from perifornical hypothalamus to raphe pallidus increases rat brown adipose tissue thermogenesis. *J. Neurosci.* 31:15944–55

Ungerstedt U. 1971. Adipsia and aphagia after 6-hydroxydopamine induced degeneration of the nigro-striatal dopamine system. *Acta Physiol. Scand.* 367:95–122

van den Pol AN. 1999. Hypothalamic hypocretin (orexin): robust innervation of the spinal cord. *J. Neurosci.* 19:3171–82

van den Pol AN, Acuna-Goycolea C, Clark KR, Ghosh PK. 2004. Physiological properties of hypothalamic MCH neurons identified with selective expression of reporter gene after recombinant virus infection. *Neuron* 42(4):635–52

van den Pol AN, Gao X-B, Obrietan K, Kilduff T, Belousov AB. 1998. Presynaptic and postsynaptic actions and modulation of neuroendocrine neurons by a new hypothalamic peptide, hypocretin/orexin. *J. Neurosci.* 18:7962–71

van den Pol AN, Ghosh PK, Liu RJ, Li Y, Aghajanian GK, Gao XB. 2002. Hypocretin (orexin) enhances neuron activity and cell synchrony in developing mouse GFP-expressing locus coeruleus. *J. Physiol.* 541:169–85

von Economo C. 1930. Sleep as a problem of localization. *J. Nerv. Ment. Dis.* 71:249–59

Vyazovskiy VV, Cirelli C, Tononi G, Tobler I. 2008. Cortical metabolic rates as measured by 2-deoxyglucose-uptake are increased after waking and decreased after sleep in mice. *Brain Res. Bull.* 75:591–97

Wang J, Osaka T, Inoue S. 2003. Orexin-A-sensitive site for energy expenditure localized in the arcuate nucleus of the hypothalamus. *Brain Res.* 971:128–34

Willie JT, Chemelli RM, Sinton CM, Tokita S, Williams SC, et al. 2003. Distinct narcolepsy syndromes in *orexin receptor-2* and *orexin* null mice: molecular genetic dissection of non-REM and REM sleep regulatory processes. *Neuron* 38(5):715–30

Winrow CJ, Renger JJ. 2013. Discovery and development of orexin receptor antagonists as therapeutics for insomnia. *Br. J. Pharmacol.* 171:283–93

Winrow CJ, Tanis KQ, Reiss DR, Rigby AM, Uslaner JM, et al. 2010. Orexin receptor antagonism prevents transcriptional and behavioral plasticity resulting from stimulant exposure. *Neuropharmacology* 58:185–94

Yamamoto Y, Ueta Y, Serino R, Nomura M, Shibuya I, Yamashita H. 2000. Effects of food restriction on the hypothalamic prepro-orexin gene expression in genetically obese mice. *Brain Res. Bull.* 51:515–21

Yamanaka A, Beuckmann CT, Willie JT, Hara J, Tsujino N, et al. 2003. Hypothalamic orexin neurons regulate arousal according to energy balance in mice. *Neuron* 38(5):701–13

Yerkes RM, Dodson JD. 1908. The relation of strength of stimulus to rapidity of habit-formation. *J. Comp. Neurol. Psychol.* 18:459–82

Yi CX, Serlie MJ, Ackermans MT, Foppen E, Buijs RM, et al. 2009. A major role for perifornical orexin neurons in the control of glucose metabolism in rats. *Diabetes* 58:1998–2005

Yoshida K, McCormack S, España RA, Crocker A, Scammell TE. 2006. Afferents to the orexin neurons of the rat brain. *J. Comp. Neurol.* 494:845–61

Yoshida Y, Fujiki N, Nakajima T, Ripley B, Matsumura H, et al. 2001. Fluctuation of extracellular hypocretin-1 (orexin A) levels in the rat in relation to the light-dark cycle and sleep-wake activities. *Eur. J. Neurosci.* 14:1075–81

Zeitzer JM, Buckmaster CL, Parker KJ, Hauck CM, Lyons DM, Mignot E. 2003. Circadian and homeostatic regulation of hypocretin in a primate model: implications for the consolidation of wakefulness. *J. Neurosci.* 23:3555–60

Reassessing Models of Basal Ganglia Function and Dysfunction

Alexandra B. Nelson[1,2] and Anatol C. Kreitzer[1,3]

[1]The Gladstone Institutes, San Francisco, California 94158;
email: akreitzer@gladstone.ucsf.edu

[2]Department of Neurology, University of California, San Francisco, California 94122

[3]Departments of Anatomy and Physiology, University of California, San Francisco,
California 94143; email: alexandra.nelson@ucsf.edu

Annu. Rev. Neurosci. 2014. 37:117–35

The *Annual Review of Neuroscience* is online at
neuro.annualreviews.org

This article's doi:
10.1146/annurev-neuro-071013-013916

Keywords

striatum, Parkinson's disease, dopamine

Abstract

The basal ganglia are a series of interconnected subcortical nuclei. The function and dysfunction of these nuclei have been studied intensively in motor control, but more recently our knowledge of these functions has broadened to include prominent roles in cognition and affective control. This review summarizes historical models of basal ganglia function, as well as findings supporting or conflicting with these models, while emphasizing recent work in animals and humans directly testing the hypotheses generated by these models.

Contents

INTRODUCTION

The basal ganglia are a set of deep forebrain nuclei consisting of the striatum (caudate and putamen in primates), the globus pallidus (internal and external segments), the subthalamic nucleus, and the substantia nigra (pars reticulata and pars compacta). Along with other brain regions, including the cerebral cortex, the thalamus, and several brain stem nuclei, the basal ganglia form a network with both open- and closed-loop circuitry (see **Figure 1**). The basal ganglia are evolutionarily conserved: Their basic anatomy and connectivity are preserved across most vertebrates, from the lamprey to the human (Reiner et al. 1998, Stephenson-Jones et al. 2012). Although the structure and basic units of neural computation are likely to be similar across species, the breadth of functions may differ.

Deciphering the basic neural computations that the basal ganglia perform, as well as the overall function this system serves, has been a long-standing area of intense research. Much of the ongoing interest in these nuclei derives from their role in human disease and the striking symptoms humans with lesions of these structures experience. Although inferring the normal function of a brain region or circuit from the disease state is challenging and in some cases perilous, examination of important human basal ganglia diseases gives a sense of the potential functions of the basal ganglia. Findings in humans with these diseases, combined with neuroanatomical and neurochemical studies in animals, led to the development of the rate model. Evolving over several years, this model was formulated by a number of investigators, including Albin, Penney, Young, and DeLong (Albin et al. 1989; DeLong 1983, 1990; Penney & Young 1983). The rate model describes the basic anatomical and neurochemical connections between basal ganglia nuclei and postulates that some human movement disorders are caused by imbalanced activity in basal ganglia nuclei. At the time, the authors highlighted limitations and caveats of the rate model, but the simplicity of the model made it extraordinarily attractive for clinicians and basic scientists alike: It generated testable hypotheses regarding the physiology and connectivity of basal ganglia nuclei in health and

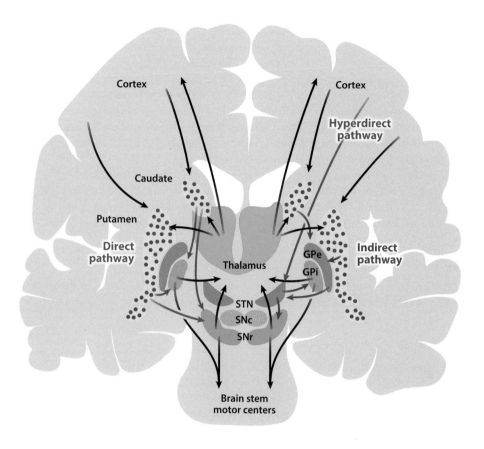

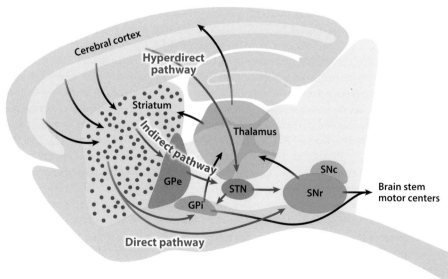

Figure 1

Simplified basal
ganglia circuit
diagram. Basal ganglia
nuclei and their major
connections in
primates (*above*),
shown in coronal view,
and rodents (*below*),
shown in sagittal view.
Many additional
connections between
nuclei are omitted for
simplicity; see text for
details. In both panels,
the direct pathway is
shown in blue, the
indirect pathway in
red, and the
hyperdirect pathway in
green. Black arrows
represent connections
shared by multiple
pathways. Blue and red
dots in the primate
caudate/putamen and
rodent striatum
represent direct
pathway–forming and
indirect
pathway–forming
medium spiny
neurons, respectively.
Abbreviations: GPe,
globus pallidus, pars
externa; GPi, globus
pallidus, pars interna;
SNc, substantia nigra,
pars compacta; SNr,
substantia nigra, pars
reticulata; STN,
subthalamic nucleus.

disease. As evidence has accumulated, however, it has become clear that the basal ganglia encode information using a more sophisticated algorithm than firing rate alone. Nevertheless, we argue that although current investigators continue to identify deficiencies in the rate model, it has been a remarkable foundation and catalyst to develop a greater understanding of basal ganglia function in healthy and diseased brains.

In this review, we begin with an overview of the functional neuroanatomy of the basal ganglia. We describe the key components and predictions of the rate model and discuss findings that support, question, or refute the model, primarily in the context of motor control. Finally, we examine some of the most recent experiments designed to test the rate model and discuss how these experiments have revised our sense of how the basal ganglia circuit drives normal motor behavior as well as disease manifestations.

This review does not cover in detail the related role of the basal ganglia in reinforcement learning and habit formation, but the reader is directed to two excellent recent reviews of this topic (Graybiel 2008, Redgrave et al. 2010). A complementary recent review examines the evidence for and against the rate model in animal models of Parkinson's disease (Ellens & Leventhal 2013).

THE BASAL GANGLIA CIRCUIT

The primate and rodent basal ganglia circuit is illustrated in simplified form in **Figure 1**. The composition of the individual nuclei and their more detailed connectivity are described below.

Input Nuclei

The striatum and subthalamic nucleus (STN) are the primary input nuclei of the basal ganglia. The striatum receives input from nearly every cortical area, which can be grouped into regions subserving sensorimotor, cognitive, and affective functions (Redgrave et al. 2010). As originally schematized (Alexander et al. 1986), and more recently reviewed (Redgrave et al. 2010), cortical areas of related behavioral function send excitatory inputs to subregions of the striatum, forming distinct channels. Although some evidence indicates that these channels have distinct functions, additional evidence shows that they overlap at the level of the striatum and may serve an integrative function between cognitive, motor, and limbic signals deriving from cortical areas (Mailly et al. 2013). The striatum also receives extensive excitatory input from the thalamus (reviewed in Smith et al. 2004). The thalamostriatal projection is also topographically organized into parallel but overlapping motor, cognitive, and limbic circuits (Berendse & Groenewegen 1990, Elena Erro et al. 2002, Ragsdale & Graybiel 1991, Smith & Parent 1986). Much of this input derives from a small region within the intralaminar nuclei of the thalamus, the centromedian and parafascicular (CM/Pf) complex (Jones & Leavitt 1974, Sadikot et al. 1992), but input also derives from the mediodorsal (Parent 1976, Royce 1978, Sato et al. 1979, Wall et al. 2013) and ventrolateral nuclei of the thalamus (Elena Erro et al. 2002; McFarland & Haber 2000, 2001). The striatum also receives a dense projection from midbrain dopaminergic neurons in the substantia nigra, pars compacta (SNc) and the ventral tegmental area (VTA).

The striatum is composed primarily of γ-aminobutyric acid (GABA)-ergic spiny projection neurons, also called medium spiny neurons for their characteristic medium size and numerous dendritic spines. These neurons receive excitatory inputs from the cortex, the thalamus, and in more ventral regions, the amygdala and the hippocampus, and are the sole striatal projection neurons. Based on their projection patterns, medium spiny neurons can be divided into two groups, which is a critical component of the rate model. Approximately half of the medium spiny neurons project directly to basal ganglia output nuclei (SNr or GPi), forming the direct pathway, whereas the other half project to the globus pallidus pars externa (GPe) and polysynaptically to the output nuclei, forming the indirect pathway (see **Figure 1**). The striatum also contains cholinergic and GABAergic interneurons, whose role in overall circuit function is as yet unknown.

The other basal ganglia input nucleus, the STN, receives relatively restricted cortical input from primary motor, supplementary motor, and premotor cortices (Groenewegen & Berendse 1990, Kitai & Deniau 1981, Monakow et al. 1978, Nambu et al. 1996). The STN also receives input from the GPe through the indirect pathway (Carpenter et al. 1981, Groenewegen & Berendse 1990). Composed primarily of glutamatergic neurons, the STN sends an excitatory projection to the basal ganglia output nuclei (Carpenter et al. 1981, Parent & Hazrati 1995, Smith et al. 1990). The cortex-STN-SNr/GPi projection, also known as the hyperdirect pathway, has been of growing interest (Nambu et al. 2000, 2002). Recent articles have highlighted the involvement of this pathway in the development of action canceling or stop signals for motor and cognitive programs (Sano et al. 2013, Schmidt et al. 2013). In addition to the canonical feedforward connection between STN and the output nuclei, anatomical studies have identified STN projections back to the GPe (Carpenter et al. 1981, Sato et al. 2000b) and the striatum (Kita & Kitai 1987, Smith et al. 1990, Wall et al. 2013). The function of these connections is unclear, but some have postulated that this loop forms an intrinsic pacemaker (Plenz & Kital 1999).

Intrinsic and Output Nuclei

The globus pallidus is composed of two regions, the intrinsic basal ganglia nucleus GPe and the basal ganglia output nucleus GPi (an analogous structure in rodents is embedded in the internal capsule and is termed the entopeduncular nucleus). Both GPe and GPi make numerous connections within the basal ganglia (reviewed in Jaeger & Kita 2011). GPe neurons receive GABAergic inputs from striatal projection neurons of the indirect pathway, as well as excitatory inputs from the STN. GABAergic GPe neurons (Jessell et al. 1978) send their axons to almost every basal ganglia nucleus, including the STN (forming the canonical indirect pathway), the GPi and SNr, and also the striatum itself (Bevan et al. 1998, Kita & Kita 2001, Mallet et al. 2012, Sato et al. 2000a). The GPi and SNr, the basal ganglia output nuclei, are also composed of GABAergic neurons. GPi receives inhibitory inputs from striatum (the direct pathway) and excitatory inputs from the STN (the canonical indirect pathway), as well as the thalamus (Deschênes et al. 1996), the pedunculopontine tegmental nucleus (PPTg) (Edley & Graybiel 1983), and the GPe (Sato et al. 2000a). The GPi projects to the thalamus, particularly the ventral nuclei and the CM/Pf (Parent & Parent 2004). Basal ganglia output is also directed at several brain stem nuclei, including the superior colliculus and the PPTg. Superior colliculus output pathways are important for regulation of eye movements and orienting behaviors (Hikosaka 2007). PPTg output pathways are believed to participate in both motor and attentional control mechanisms (Benarroch 2013). These basal ganglia–brain stem connections form a direct output pathway for regulation of motor behavior, as distinguished from the basal ganglia–thalamocortical pathway, which requires looping back to the cortex to regulate behavior.

Dopaminergic Neurons

The SNc and VTA are located in the midbrain and contain GABAergic and dopaminergic neurons. Studies recently showed that some midbrain dopamine neurons also release glutamate and GABA, although the behavioral role of corelease is unclear (Stuber et al. 2010, Tritsch et al. 2012). Dopaminergic midbrain neurons project widely to many brain regions, including prominent projections to the frontal cortex and striatum, where through G protein–coupled receptors, dopamine release modulates excitability and synaptic function in target structures. SNc dopaminergic neurons project primarily to the more dorsal/lateral (sensorimotor and cognitive) portions of the striatum, whereas VTA dopaminergic neurons project primarily to the more medial and ventral (cognitive/affective/limbic) portions of the striatum (Smith & Kieval 2000).

BASAL GANGLIA DISEASES

Human basal ganglia diseases are a major motivation for ongoing research, as well as a source of much of the data for and against the rate model, so we briefly describe some of the major disease entities and how they relate to basal ganglia circuit function.

Huntington's Disease

Huntington's disease (HD) is a human neurodegenerative disease caused by autosomal dominant inheritance of an expanded trinucleotide repeat in the gene encoding huntingtin. Anatomically, patients with HD initially develop synaptic changes (Ferrante et al. 1991) and later develop cell loss within the striatum (reviewed in Reiner et al. 2011), with preferential early loss of indirect pathway striatal neurons (Albin et al. 1990, Deng et al. 2004, Reiner et al. 1988). As the disease progresses, neurodegeneration is observed in both the striatum and the cortex, as demonstrated by reduced striatal and cortical volume in living patients (Jernigan et al. 1991, Rosas et al. 2001) and at autopsy (Halliday et al. 1998). Presymptomatic carriers of the HD gene show similar findings (Aylward et al. 2004) and behavioral abnormalities are present during early symptomatic phases (Tabrizi et al. 2013). Since the striatum is implicated in early HD, we presume behavioral deficits are the result of basal ganglia dysfunction. At a microcircuit level, evidence indicates that the psychiatric disease manifestations correlate with more prominent degeneration within the striosome compartment of the striatum (Tippett et al. 2007). HD-related cognitive deficits include impairments in processing speed, task-set switching, sequencing, and concentration (reviewed in Paulsen 2011), suggesting a role for the basal ganglia in many cognitive functions traditionally attributed to the frontal cortex.

Parkinson's Disease

Parkinson's disease (PD) is a neurodegenerative disease characterized by progressive cell loss in multiple brain regions, particularly brain stem nuclei and, most prominently as it pertains to motor symptoms, the SNc. In late stages of the disease, there can be more widespread neuropathology, including involvement of the cerebral cortex (Irwin et al. 2012). PD, like HD, produces psychiatric, cognitive, and motor symptoms. It is difficult to tease apart the relationship of these symptoms to basal ganglia dysfunction, however, because there is neurodegeneration in other areas. Despite this difficulty, considerable evidence shows that basal ganglia dysfunction due to loss of striatal dopamine does contribute to nonmotor symptoms, including mood and cognitive symptoms (Cools 2006).

Dystonia

Characterized by abnormal involuntary twisting movements and postures caused by coactivation of normally antagonistic muscle groups, dystonia has been historically characterized as a hyperkinetic movement disorder. As a symptom, it is a component of several different disease entities, including both Huntington's disease and Parkinson's disease. However, in primary dystonias, little to no brain pathology is observed, suggesting circuit dysfunction, rather than neurodegeneration, may be the cause. These observations imply that dystonia may not reflect a single pathophysiologic process, but is rather a manifestation of convergent circuit mechanisms.

THE RATE MODEL

Developed in the 1980s and early 1990s, the rate model incorporated a growing body of literature describing the connectivity, neurochemistry, and physiology of the basal ganglia (Albin et al. 1989, DeLong 1990). In large part, it sprang from observations about deviations in basal ganglia structure and function in disease, particularly Huntington's disease, Parkinson's disease, and hemiballism, a movement disorder seen with acute lesions of the subthalamic nucleus. It postulated that the basal ganglia process cortical input through parallel pathways from the striatum through to basal ganglia output nuclei and feed it back to the cortex via a thalamic relay. Increases or decreases in firing rate of different basal ganglia nuclei were postulated to regulate basal ganglia output and behavior. Some of the key components and conclusions of the rate model are listed below:

1. The basal ganglia form an interconnected network.
2. The basal ganglia are involved in both motor and cognitive function.
3. The cortex may generate motor or cognitive commands, but the execution and/or maintenance of these commands relies on the integrity of the basal ganglia as a positive feedback system to sustain activity.
4. The network can be subdivided into two major pathways, the direct and indirect pathways.
5. Dopamine differentially acts on the two major pathways at the level of the striatum.
6. Among several possible actions, a subset are selected by the striatum; competing actions are suppressed by lateral inhibition at several levels of the circuit.
7. Network output is integrated at the level of the basal ganglia output nuclei, the GPi and SNr, which inhibit or disinhibit thalamocortical and/or brain stem areas to suppress or promote specific actions.

These key concepts, in particular the idea that activity in the direct and indirect pathways has opposing effects on behavioral output, led to numerous physiological, pathological, and behavioral predictions. Some of the key predictions, as well as references to experiments supporting or refuting these predictions, are listed in **Table 1**.

The simplest formulation of the rate model treats the direct and indirect pathways as groups of neurons with uniform responses. However, given the somatotopic, functional, and synaptic organization of the basal ganglia, varying responses are more likely. Physiological recordings from normal nonhuman primates during motor and cognitive tasks show a wide variety of responses during a single task, even within the same anatomic region. This idea, often referred to as the action selection model, was articulated and advanced in a series of articles by Jonathan Mink (Mink 1996, 2001; Mink & Thach 1993) and Okihide Hikosaka (Hikosaka 1991, 1998). This model shares some of the key concepts of the rate model but suggests that different neural ensembles within a basal ganglia pathway may activate or inhibit individual motor programs.

SUPPORT FOR THE RATE MODEL

Since the rate model was first developed, evidence from both humans and animal models has tended to support some, but not all, of its major predictions, summarized below.

Humans

The surgical treatment of movement disorders, including deep brain stimulation (DBS), has allowed for physiological recordings of the human basal ganglia. One of the major caveats of such

Table 1 Key predictions of the rate model. For each key prediction, important findings supporting or refuting the rate model are listed, as well as limitations of these findings and critical future experiments

Rate model predictions	Supportive findings	Conflicting findings	Critical experiments/limitations
Dopamine has opposing effects on direct and indirect pathway neurons.	Dopamine increases the excitability of direct, more than indirect, pathway neurons (Planert et al. 2013). Dopamine has bidirectional effects on synaptic function in direct and indirect pathways (Andre et al. 2010).	The firing rate of essentially all striatal neurons increases after MPTP treatment (Liang et al. 2008, Rothblat & Schneider 1993).	Most experiments have been performed in vitro with bath application of dopamine, which does not reflect phasic release of dopamine. It will be critical to test effects of phasic dopamine release on the activity of identified direct and indirect pathway neurons in vivo and on intrinsic excitability versus synaptic inputs in vitro.
Ablation or inactivation of direct pathway neurons leads to decreased movement.	Ablation of direct pathway neurons results in bradykinesia (Drago et al. 1998) or ipsilateral turning (Hikida et al. 2010).		Ablation experiments are limited in that they are often chronic manipulations (thus adaptation may occur) and do not necessarily reflect normal function of neuronal ensembles. Optogenetic or pharmacogenetic inhibition of direct or indirect pathway neurons would demonstrate bidirectional control of movement.
Ablation or inactivation of indirect pathway neurons leads to increased or excessive movement.	Ablation of indirect pathway neurons results in hyperactivity (Durieux et al. 2009, Sano et al. 2003) or contralateral turning (Hikida et al. 2010).		Ablation experiments are limited in that they are often chronic manipulations (thus adaptation may occur) and do not necessarily reflect normal function of neuronal ensembles. Optogenetic or pharmacogenetic inhibition of direct or indirect pathway neurons would demonstrate bidirectional control of movement.
Activation of direct pathway neurons leads to increased or involuntary movement.	Optogenetic activation of direct pathway neurons increases locomotion (Kravitz et al. 2010); direct pathway neurons show increased activity during initiation and ongoing performance of trained motor sequences (Jin et al. 2014).	Both pathways are active in movement initiation (Cui et al. 2013, Jin et al. 2014).	It would be critical to perform additional recordings of identified direct pathway neurons during normal movement and pathological states of involuntary movement, with focus on both firing rate and pattern.
Activation of indirect pathway neurons leads to decreased movement or bradykinesia.	Optogenetic activation of indirect pathway neurons decreases locomotion (Kravitz et al. 2010).	Both pathways are active in movement initiation (Cui et al. 2013, Jin et al. 2014).	It would be critical to perform additional recordings of identified indirect pathway neurons during normal movement and pathological states of reduced movement, with focus on both firing rate and pattern.

(Continued)

Decreases in basal ganglia output (at the GPi or SNr level) correlate with increased movement.	Decreased GPi firing rate is observed in HD (Starr et al. 2008, Cubo et al. 2000), dystonia (Starr et al. 2005, Vitek et al. 1999), and levodopa- or graft-induced dyskinesias (Lozano et al. 2000, Papa et al. 1999, Richardson et al. 2011). Decreased SNr firing predicts movement initiation in mice (Freeze et al. 2013) or primates (Hikosaka & Wurtz 1983).	Normal GPi firing rates have been observed in HD (Tang et al. 2005). Reductions in oscillatory activity correlate with reductions in bradykinesia (Rosin et al. 2011).	Correlation does establish causal relationships; rate changes may not be causal. It is important to identify and monitor the activity of functionally distinct ensembles within basal ganglia nuclei, rather than random sampling of neurons within each structure. It will be important to initiate (or correct) firing rate changes versus abnormal patterned activity in normal animals (or both) to determine which feature is causative.
Increases in basal ganglia output (at the GPi or SNr level) should correlate with decreased movement.	Increased SNr/GPi firing rates have been observed in PD patients (Filion & Tremblay 1991, Hutchison et al. 1998), and increased SNr firing rates occur with stopping/canceling movement (Schmidt et al. 2013). Increased SNr firing predicts decreased movement in mice (Freeze et al. 2013).	No change in firing rates of SNr neurons is observed in MPTP-treated monkeys (Wichmann et al. 1999).	Correlation does establish causal relationships; rate changes may not be causal. It is important to identify and monitor the activity of functionally distinct ensembles within basal ganglia nuclei. It will be important to initiate (or correct) firing rate changes versus abnormal patterned activity in normal animals (or both) to determine which feature is causative.

Abbreviations: GPi, globus pallidus pars interna; HD, Huntington's disease; MPTP, 1-methyl-4-phenyl-1,2,3,6-tetrahydropyridine; PD, Parkinson's disease; SNr, substantia nigra pars reticulata.

research is the lack of control subjects undergoing DBS: Comparisons are often made to non-human primates or to patients with another disease. However, numerous key observations from human recordings support the rate model.

In HD, neuropathological studies have shown preferential degeneration of striatal indirect pathway neurons, as measured by immunoreactivity for enkephalin and other markers (Deng et al. 2004, Reiner et al. 1988). These data led to the prediction that in HD the GPe will be disinhibited, leading to decreased activity in the STN and output nuclei. A small number of human HD patients have undergone DBS implantation or lesion of the GPi (Cubo et al. 2000, Kang et al. 2011, Moro et al. 2004). One study showed no significant change in the firing rate or pattern of GPi neurons in HD (Tang et al. 2005), whereas two other studies found significant increases in GPe firing and significant decreases in GPi firing in HD as compared with PD, consistent with the rate model (Cubo et al. 2000, Starr et al. 2008).

Dystonia was not discussed at length in the original articulation of the rate model; however, if dystonia is considered with chorea as a hyperkinetic movement disorder, the rate model would predict that an imbalance in direct and indirect pathway activity (with net loss of indirect pathway activity) would lead to disinhibition of the GPe and increased inhibition of the STN and GPi. Decreased STN and GPi firing rates would disinhibit thalamocortical and brain stem motor circuits, triggering aberrant coactivation of muscle groups. Several investigators have observed lowered firing rates in the GPi of dystonia patients as measured during intraoperative recordings, as compared with firing rates in control nonhuman primates or patients with PD (Starr et al. 2005, Vitek et al. 1999). In addition, some have observed lower STN firing rates in dystonia as compared with PD (Schrock et al. 2009), although these rates were somewhat higher than those observed in essential tremor (Steigerwald et al. 2008) and in normal nonhuman primates (Bergman et al. 1994).

The rate model postulates that dopamine has opposing effects on the activity of direct pathway–versus indirect pathway–forming striatal neurons (Albin et al. 1989). A corollary is that in Parkinson's disease, the loss of nigrostriatal dopamine would lead to an increase in indirect pathway activity (resulting in increased STN and GPi firing rates) and decreases in direct pathway activity (resulting in increased GPi firing rate). Indeed, the firing rate of STN and GPi neurons is increased in PD patients (Benazzouz et al. 2002, Hutchison et al. 1998). Intraoperative administration of a dopamine agonist reduces the firing rate of GPi neurons (Levy et al. 2001a), and the development of levodopa-induced dyskinesias is associated with an even more profound decrease in GPi firing (Lozano et al. 2000). A corollary is that decreasing STN or GPi firing rates should be therapeutic. In fact, local inactivation of STN neurons with muscimol or lidocaine produces short-latency antiparkinsonian effects in humans undergoing DBS surgery (Levy et al. 2001b). In the same study, at later time points of the muscimol infusion, patients developed dyskinesias, suggesting that soon after infusion, a normal firing rate in the STN was achieved, relieving parkinsonian symptoms; however, later firing rates were further depressed, producing hyperkinesias. In the case of dyskinesias associated with grafting of fetal midbrain neurons, lower firing rates, as well as bursting, were observed in the GPi (Richardson et al. 2011), which also supports the rate model.

Animal Models

In the 1980s, a small group of people inadvertently self-administered an impure designer drug, which resulted in the acute development of parkinsonism. The responsible contaminant was later identified as 1-methyl-4-phenyl-1,2,3,6-tetrahydropyridine (MPTP), which triggers the death of dopaminergic neurons of the SNc (Langston et al. 1983). This discovery led to the development of an MPTP-lesioned primate model of Parkinson's disease, which facilitated a direct examination of basal ganglia nuclei in the context of parkinsonism. The rate model predicts that MPTP treatment would decrease striatal direct pathway activity and increase indirect pathway activity, causing inhibition of GPe, disinhibition of STN, and increased basal ganglia output activity. GPe recordings show decreases in firing rate following MPTP treatment (Filion & Tremblay 1991), which reverse following dopamine agonist treatment (Filion et al. 1991). STN firing rates increase following dopamine depletion with MPTP (Bergman et al. 1994), and improvements in parkinsonism result from inhibition or lesion of the STN (Bergman et al. 1990, Guridi et al. 1994, Wichmann et al. 1994). The output nuclei of the basal ganglia, GPi and SNr, showed overall higher rates of discharge (Filion & Tremblay 1991) as well as increased firing in response to movements. These firing abnormalities at the level of the output nuclei also reversed with dopamine agonist treatment (Filion et al. 1991), resulting in profound decreases in firing rate in the context of levodopa-induced dyskinesias (Papa et al. 1999).

In rodents, investigators have reported differential activity of the striatal direct and indirect pathways in parkinsonism using a backpropagating action potential method for identifying striatal neurons of each pathway. They observed a decrease in direct pathway and an increase in indirect pathway firing rates (Mallet et al. 2006), which supports the rate model. The advent of genetic techniques for selectively targeting direct and indirect pathway neurons has permitted more rigorous testing of rate model hypotheses. Direct and indirect pathway striatal neurons have several distinct, almost nonoverlapping markers (Gerfen et al. 1990), which can form the basis for cell-type-specific manipulations. Direct pathway neurons express Substance P and the D1 dopamine receptor, whereas indirect pathway neurons express the D2 dopamine receptor, A2a adenosine receptor, and enkephalin (Schiffmann & Vanderhaeghen 1993). Ablation of indirect pathway neurons by selective expression of an immunotoxin under the D2 dopamine receptor promoter resulted in hyperlocomotor phenotypes (Sano et al. 2003), much as would be expected in early Huntington's disease. Ablation of direct pathway neurons by selective expression of the diphtheria toxin resulted in slowing of movement (bradykinesia) (Drago et al. 1998), whereas ablation of indirect pathway neurons led to an increase in locomotor activity (Durieux et al. 2009, 2012). Inducible and cell-type-selective block of neurotransmission with tetanus toxin in direct or indirect pathway striatal neurons produced opposing effects on rotational behavior as well as psychostimulant-induced locomotion (Hikida et al. 2010). Likewise, deletion of the key signaling protein DARPP-32 from either direct or indirect pathway neurons also resulted in divergent motor behaviors, including loss of hyperkinetic behaviors in the former and increased locomotion in the latter (Bateup et al. 2010).

Neuronal ablation or blocking neurotransmission, even with rapid techniques, may lead to plasticity in basal ganglia microcircuits. To determine whether acutely altering the balance of activity in the two pathways could achieve similar results, several groups have employed optogenetics, which allows millisecond timescale manipulations. Selective activation of striatal direct pathway neurons increases locomotor activity, whereas activation of indirect pathway neurons suppresses movement (Kravitz et al. 2010). Likewise, stimulation of the direct pathway or inactivation of the indirect pathway (via high-frequency optogenetic stimulation of STN inputs) in parkinsonian animals ameliorates bradykinesia (Gradinaru et al. 2009, Kravitz et al. 2010). However, like electrical stimulation, optogenetic manipulations impose intense, probably supraphysiologic levels of activity on the basal ganglia circuit. They also show that such imbalances in activity are sufficient to produce behavior, but they do not prove that they are necessary.

INCONSISTENCIES IN AND LIMITATIONS OF THE RATE MODEL

Numerous experimental observations, highlighted both by the original papers and by subsequent studies, indicate the rate model does not entirely explain the relationship between basal ganglia neural activity and behavior.

Humans

Many clinicians and investigators have been puzzled by the observation that lesioning or stimulation of the GPi is therapeutic in both hypokinetic (Parkinson's disease) and hyperkinetic movement disorders (dystonia, chorea). This paradox was included in initial articulations of the model, and the authors implied that there might be some shared, rather than directly opposing, circuit mechanisms involved in diseases such as Parkinson's disease and dystonia. In some ways, this is not surprising, given the overlap of symptoms between classically hypokinetic and hyperkinetic disorders, such as the presence of dystonia and parkinsonism in both PD and HD. An interesting related observation is the fact that temporary inactivation of basal ganglia output in normal experimental animals does not cause marked behavioral dysfunction (Desmurget & Turner 2008). The mild parkinsonism observed in these primates mirrors what is observed in some dystonic patients with GPi DBS (Berman et al. 2009). Thus, loss of basal ganglia output (as in pallidotomy or temporary inactivation) does not cause severe behavioral disturbances in normal animals, but gain of abnormal basal ganglia output (as in the disease state) can.

Although firing rate changes have been observed in disease states, these are often small in magnitude, and there are more salient changes in patterning (bursting, oscillatory activity) and synchrony, both within and across nodes in the basal ganglia network. Abnormal activity patterns have been studied extensively in Parkinson's disease (Eusebio et al. 2009, Litvak et al. 2011) and dystonia (Chen et al. 2006, Starr et al. 2005, Weinberger et al. 2012) and show correlations between such phenomena and disease manifestations. However, such observations cannot demonstrate causality (Eusebio & Brown 2009). The fact that similar oscillations are seen in Parkinson's disease and dystonia suggests that oscillations in and of themselves may not cause the motor phenotype. Human recordings have identified additional disease-specific physiologic signatures (de Hemptinne et al. 2013, Shimamoto et al. 2013), and these may help us develop hypotheses about the causal role of neural activity in movement disorders. Less is known about how such bursting, synchrony, or oscillations contribute to normal basal ganglia function in humans, but this is an area of ongoing research, which can be most effectively carried out using animal models (Leventhal et al. 2012).

Animal Models

Recordings from the striatum have been less supportive of the rate model, which predicts that loss of dopamine will produce reciprocal decreases and increases in the firing of direct and indirect pathway neurons. In chronically parkinsonian animals, the overall firing rate of striatal neurons increases markedly (Liang et al. 2008, Rothblat & Schneider 1993). This observation may be due to oversampling of high firing rate indirect pathway neurons; however, using rate responses to levodopa administration as a means to identify neurons, Liang et al. (2008) observed that both direct and indirect pathway neurons have markedly elevated firing rates.

As in human studies, nonhuman primate studies have often shown minimal changes in overall firing rate but suggest that patterned activity may be a driver of abnormal behavior. SNr, for example, shows essentially no firing rate change in MPTP-treated monkeys but an increase in bursting (Wichmann et al. 1999). STN shows increases in both rate and bursting (Bergman et al. 1994). GPe and GPi also show increased levels of synchrony and oscillations after MPTP treatment (Raz et al. 2000, Wichmann et al. 1994). Although definitive evidence that these oscillations are necessary and sufficient to produce parkinsonian motor behavior is still lacking, reductions in abnormal patterned activity in MPTP-treated monkeys correlate with behavioral improvements with several therapies, including levodopa (Heimer et al. 2006, Tachibana et al.

2011), pharmacologic inactivations of basal ganglia nuclei (Tachibana et al. 2011), and DBS (Hahn et al. 2008, McCairn & Turner 2009, Vitek et al. 2012).

Although optogenetic studies in rodents have shown that direct and indirect pathway activation is sufficient to cause opposing behaviors (Kravitz et al. 2010, Tai et al. 2012), they have not shown that such imbalances are necessary to produce behavior. Subsequent studies using observational (rather than interventional) methods have told a more complicated story. A minority of unidentified striatal neurons modulate firing during specific movements or tasks (Hikosaka et al. 1989, Hollerman et al. 1998, Kawagoe et al. 1998, Kimchi & Laubach 2009). Combining cell-type-specific genetic methods with in vivo imaging, a recent paper demonstrated that both direct and indirect pathway striatal neurons are activated simultaneously during a motor task (Cui et al. 2013). A similar study using single-unit recordings and juxtacellular labeling of striatal neurons also found cooperative activity of the direct and indirect pathways during voluntary movements (Isomura et al. 2013). Using optogenetics and single-unit recordings, our laboratory has found coactivation of both direct and indirect pathway neurons during spontaneous locomotion (A.V. Kravitz, M.M. Finucane, G. Cui, D.M. Friend, K.H. LeBlanc, D.M. Lovinger, R.M. Costa, K.S. Pollard, and A.C. Kreitzer, submitted). Although these results seem, at first glance, to contradict the rate model, upon further examination, they support an action selection form of the model. If activation of the direct and indirect pathways is important for selecting certain actions, and suppressing others, it would stand to reason that for every direct pathway ensemble activated to choose a particular action, several indirect pathway ensembles would be activated to suppress competing actions. In fact, these conclusions are further supported by the observation in rodents that SNr shows both excitatory and inhibitory responses during locomotion (Fan et al. 2012; Gulley et al. 2002, 1999; Jin & Costa 2010). Recent work in our laboratory (Freeze et al. 2013) has also identified excitatory and inhibitory SNr responses to optogenetic activation of the direct pathway, but the rate model–predicted inhibitory responses had the strongest correlation with gating of movement. Excited neurons may suppress competing actions, supporting the action selection model.

CONCLUSIONS

The Rate Model and subsequent elaborations have helped propel a large body of hypothesis-driven research on basal ganglia function. Many of its basic assumptions have been validated by both physiologic and behavioral studies, although these studies have suggested that the system is more complicated than initially assumed. The rate model suggested that firing rates correlate with behavior. Subsequent studies in humans with basal ganglia disorders, as well as in animal models, suggest that different groups of neurons within a particular nucleus may be activated or suppressed in order to release certain motor programs but not others, yielding a mixture of responses when physiological responses are measured. In addition, basal ganglia nuclei may use both the rate and timing codes contained in spike trains. Although the direct causal relationship of such patterned activity to behavior has not yet been explored in detail, we hope it will be a focus of second-generation optogenetic, physiologic, and behavioral studies.

DISCLOSURE STATEMENT

The authors are not aware of any affiliations, memberships, funding, or financial holdings that might be perceived as affecting the objectivity of this review.

LITERATURE CITED

Albin RL, Young AB, Penney JB. 1989. The functional anatomy of basal ganglia disorders. *Trends Neurosci.* 12:366–75

Albin RL, Young AB, Penney JB, Handelin B, Balfour R, et al. 1990. Abnormalities of striatal projection neurons and *N*-methyl-D-aspartate receptors in presymptomatic Huntington's disease. *N. Engl. J. Med.* 322:1293–98

Alexander GE, DeLong MR, Strick PL. 1986. Parallel organization of functionally segregated circuits linking basal ganglia and cortex. *Annu. Rev. Neurosci.* 9:357–81

Andre VM, Cepeda C, Cummings DM, Jocoy EL, Fisher YE, et al. 2010. Dopamine modulation of excitatory currents in the striatum is dictated by the expression of D1 or D2 receptors and modified by endocannabinoids. *Eur. J. Neurosci.* 31:14–28

Aylward EH, Sparks BF, Field KM, Yallapragada V, Shpritz BD, et al. 2004. Onset and rate of striatal atrophy in preclinical Huntington disease. *Neurology* 63:66–72

Bateup HS, Santini E, Shen W, Birnbaum S, Valjent E, et al. 2010. Distinct subclasses of medium spiny neurons differentially regulate striatal motor behaviors. *Proc. Natl. Acad. Sci. USA* 107:14845–50

Benarroch EE. 2013. Pedunculopontine nucleus: functional organization and clinical implications. *Neurology* 80:1148–55

Benazzouz A, Breit S, Koudsie A, Pollak P, Krack P, Benabid AL. 2002. Intraoperative microrecordings of the subthalamic nucleus in Parkinson's disease. *Mov. Disord.* 17(Suppl. 3):S145–49

Berendse HW, Groenewegen HJ. 1990. Organization of the thalamostriatal projections in the rat, with special emphasis on the ventral striatum. *J. Comp. Neurol.* 299:187–228

Bergman H, Wichmann T, DeLong MR. 1990. Reversal of experimental parkinsonism by lesions of the subthalamic nucleus. *Science* 249:1436–38

Bergman H, Wichmann T, Karmon B, DeLong MR. 1994. The primate subthalamic nucleus. II. Neuronal activity in the MPTP model of parkinsonism. *J. Neurophysiol.* 72:507–20

Berman BD, Starr PA, Marks WJ Jr, Ostrem JL. 2009. Induction of bradykinesia with pallidal deep brain stimulation in patients with cranial-cervical dystonia. *Stereotact. Funct. Neurosurg.* 87:37–44

Bevan MD, Booth PA, Eaton SA, Bolam JP. 1998. Selective innervation of neostriatal interneurons by a subclass of neuron in the globus pallidus of the rat. *J. Neurosci.* 18:9438–52

Carpenter MB, Carleton SC, Keller JT, Conte P. 1981. Connections of the subthalamic nucleus in the monkey. *Brain Res.* 224:1–29

Chen CC, Kühn AA, Hoffmann KT, Kupsch A, Schneider GH, et al. 2006. Oscillatory pallidal local field potential activity correlates with involuntary EMG in dystonia. *Neurology* 66:418–20

Cools R. 2006. Dopaminergic modulation of cognitive function-implications for L-DOPA treatment in Parkinson's disease. *Neurosci. Biobehav. Rev.* 30:1–23

Cubo E, Shannon KM, Penn RD, Kroin JS. 2000. Internal globus pallidotomy in dystonia secondary to Huntington's disease. *Mov. Disord.* 15:1248–51

Cui G, Jun SB, Jin X, Pham MD, Vogel SS, et al. 2013. Concurrent activation of striatal direct and indirect pathways during action initiation. *Nature* 494:238–42

de Hemptinne C, Ryapolova-Webb ES, Air EL, Garcia PA, Miller KJ, et al. 2013. Exaggerated phase-amplitude coupling in the primary motor cortex in Parkinson disease. *Proc. Natl. Acad. Sci. USA* 110:4780–85

DeLong MR. 1983. The neurophysiologic basis of abnormal movements in basal ganglia disorders. *Neurobehav. Toxicol. Teratol.* 5:611–16

DeLong MR. 1990. Primate models of movement disorders of basal ganglia origin. *Trends Neurosci.* 13:281–85

Deng YP, Albin RL, Penney JB, Young AB, Anderson KD, Reiner A. 2004. Differential loss of striatal projection systems in Huntington's disease: a quantitative immunohistochemical study. *J. Chem. Neuroanat.* 27:143–64

Deschênes M, Bourassa J, Doan VD, Parent A. 1996. A single-cell study of the axonal projections arising from the posterior intralaminar thalamic nuclei in the rat. *Eur. J. Neurosci.* 8:329–43

Desmurget M, Turner RS. 2008. Testing basal ganglia motor functions through reversible inactivations in the posterior internal globus pallidus. *J. Neurophysiol.* 99:1057–76

Drago J, Padungchaichot P, Wong JY, Lawrence AJ, McManus JF, et al. 1998. Targeted expression of a toxin gene to D1 dopamine receptor neurons by *Cre*-mediated site-specific recombination. *J. Neurosci.* 18:9845–57

Durieux PF, Bearzatto B, Guiducci S, Buch T, Waisman A, et al. 2009. D_2R striatopallidal neurons inhibit both locomotor and drug reward processes. *Nat. Neurosci.* 12:393–95

Durieux PF, Schiffmann SN, de Kerchove d'Exaerde A. 2012. Differential regulation of motor control and response to dopaminergic drugs by D1R and D2R neurons in distinct dorsal striatum subregions. *EMBO J.* 31:640–53

Edley SM, Graybiel AM. 1983. The afferent and efferent connections of the feline nucleus tegmenti pedunculopontinus, pars compacta. *J. Comp. Neurol.* 217:187–215

Elena Erro M, Lanciego JL, Gimenez-Amaya JM. 2002. Re-examination of the thalamostriatal projections in the rat with retrograde tracers. *Neurosci. Res.* 42:45–55

Ellens DJ, Leventhal DK. 2013. Review: electrophysiology of basal ganglia and cortex in models of Parkinson disease. *J. Parkinsons Dis.* 3:241–54

Eusebio A, Brown P. 2009. Synchronisation in the beta frequency-band—the bad boy of parkinsonism or an innocent bystander? *Exp. Neurol.* 217:1–3

Eusebio A, Pogosyan A, Wang S, Averbeck B, Gaynor LD, et al. 2009. Resonance in subthalamo-cortical circuits in Parkinson's disease. *Brain* 132:2139–50

Fan D, Rossi MA, Yin HH. 2012. Mechanisms of action selection and timing in substantia nigra neurons. *J. Neurosci.* 32:5534–48

Ferrante RJ, Kowall NW, Richardson EP Jr. 1991. Proliferative and degenerative changes in striatal spiny neurons in Huntington's disease: a combined study using the section-Golgi method and calbindin D28k immunocytochemistry. *J. Neurosci.* 11:3877–87

Filion M, Tremblay L. 1991. Abnormal spontaneous activity of globus pallidus neurons in monkeys with MPTP-induced parkinsonism. *Brain Res.* 547:142–51

Filion M, Tremblay L, Bédard PJ. 1991. Effects of dopamine agonists on the spontaneous activity of globus pallidus neurons in monkeys with MPTP-induced parkinsonism. *Brain Res.* 547:152–61

Freeze BS, Kravitz AV, Hammack N, Berke JD, Kreitzer AC. 2013. Control of basal ganglia output by direct and indirect pathway projection neurons. *J. Neurosci.* 33:18531–39

Gerfen CR, Engber TM, Mahan LC, Susel Z, Chase TN, et al. 1990. D1 and D2 dopamine receptor-regulated gene expression of striatonigral and striatopallidal neurons. *Science* 250:1429–32

Gradinaru V, Mogri M, Thompson KR, Henderson JM, Deisseroth K. 2009. Optical deconstruction of parkinsonian neural circuitry. *Science* 324:354–59

Graybiel AM. 2008. Habits, rituals, and the evaluative brain. *Annu. Rev. Neurosci.* 31:359–87

Groenewegen HJ, Berendse HW. 1990. Connections of the subthalamic nucleus with ventral striatopallidal parts of the basal ganglia in the rat. *J. Comp. Neurol.* 294:607–22

Gulley JM, Kosobud AE, Rebec GV. 2002. Behavior-related modulation of substantia nigra pars reticulata neurons in rats performing a conditioned reinforcement task. *Neuroscience* 111:337–49

Gulley JM, Kuwajima M, Mayhill E, Rebec GV. 1999. Behavior-related changes in the activity of substantia nigra pars reticulata neurons in freely moving rats. *Brain Res.* 845:68–76

Guridi J, Herrero MT, Luquin R, Guillen J, Obeso JA. 1994. Subthalamotomy improves MPTP-induced parkinsonism in monkeys. *Stereotact. Funct. Neurosurg.* 62:98–102

Hahn PJ, Russo GS, Hashimoto T, Miocinovic S, Xu W, et al. 2008. Pallidal burst activity during therapeutic deep brain stimulation. *Exp. Neurol.* 211:243–51

Halliday GM, McRitchie DA, Macdonald V, Double KL, Trent RJ, McCusker E. 1998. Regional specificity of brain atrophy in Huntington's disease. *Exp. Neurol.* 154:663–72

Heimer G, Rivlin-Etzion M, Bar-Gad I, Goldberg JA, Haber SN, Bergman H. 2006. Dopamine replacement therapy does not restore the full spectrum of normal pallidal activity in the 1-methyl-4-phenyl-1,2,3,6-tetra-hydropyridine primate model of Parkinsonism. *J. Neurosci.* 26:8101–14

Hikida T, Kimura K, Wada N, Funabiki K, Nakanishi S. 2010. Distinct roles of synaptic transmission in direct and indirect striatal pathways to reward and aversive behavior. *Neuron* 66:896–907

Hikosaka O. 1991. Basal ganglia—possible role in motor coordination and learning. *Curr. Opin. Neurobiol.* 1:638–43

Hikosaka O. 1998. Neural systems for control of voluntary action—a hypothesis. *Adv. Biophys.* 35:81–102

Hikosaka O. 2007. GABAergic output of the basal ganglia. *Prog. Brain Res.* 160:209–26

Hikosaka O, Sakamoto M, Usui S. 1989. Functional properties of monkey caudate neurons. III. Activities related to expectation of target and reward. *J. Neurophysiol.* 61:814–32

Hikosaka O, Wurtz RH. 1983. Visual and oculomotor functions of monkey substantia nigra pars reticulata. I. Relation of visual and auditory responses to saccades. *J. Neurophysiol.* 49:1230–53

Hollerman JR, Tremblay L, Schultz W. 1998. Influence of reward expectation on behavior-related neuronal activity in primate striatum. *J. Neurophysiol.* 80:947–63

Hutchison WD, Allan RJ, Opitz H, Levy R, Dostrovsky JO, et al. 1998. Neurophysiological identification of the subthalamic nucleus in surgery for Parkinson's disease. *Ann. Neurol.* 44:622–28

Irwin DJ, White MT, Toledo JB, Xie SX, Robinson JL, et al. 2012. Neuropathologic substrates of Parkinson disease dementia. *Ann. Neurol.* 72:587–98

Isomura Y, Takekawa T, Harukuni R, Handa T, Aizawa H, et al. 2013. Reward-modulated motor information in identified striatum neurons. *J. Neurosci.* 33:10209–20

Jaeger D, Kita H. 2011. Functional connectivity and integrative properties of globus pallidus neurons. *Neuroscience* 198:44–53

Jernigan TL, Salmon DP, Butters N, Hesselink JR. 1991. Cerebral structure on MRI, Part II: Specific changes in Alzheimer's and Huntington's diseases. *Biol. Psychiatry* 29:68–81

Jessell TM, Emson PC, Paxinos G, Cuello AC. 1978. Topographic projections of substance P and GABA pathways in the striato- and pallido-nigral system: a biochemical and immunohistochemical study. *Brain Res.* 152:487–98

Jin X, Costa RM. 2010. Start/stop signals emerge in nigrostriatal circuits during sequence learning. *Nature* 466:457–62

Jin X, Tecuapetla F, Costa RM. 2014. Basal ganglia subcircuits distinctively encode the parsing and concatenation of action sequences. *Nat. Neurosci.* 17:423–30

Jones EG, Leavitt RY. 1974. Retrograde axonal transport and the demonstration of non-specific projections to the cerebral cortex and striatum from thalamic intralaminar nuclei in the rat, cat and monkey. *J. Comp. Neurol.* 154:349–77

Kang GA, Heath S, Rothlind J, Starr PA. 2011. Long-term follow-up of pallidal deep brain stimulation in two cases of Huntington's disease. *J. Neurol. Neurosurg. Psychiatry* 82:272–77

Kawagoe R, Takikawa Y, Hikosaka O. 1998. Expectation of reward modulates cognitive signals in the basal ganglia. *Nat. Neurosci.* 1:411–16

Kimchi EY, Laubach M. 2009. Dynamic encoding of action selection by the medial striatum. *J. Neurosci.* 29:3148–59

Kita H, Kita T. 2001. Number, origins, and chemical types of rat pallidostriatal projection neurons. *J. Comp. Neurol.* 437:438–48

Kita H, Kitai ST. 1987. Efferent projections of the subthalamic nucleus in the rat: light and electron microscopic analysis with the PHA-L method. *J. Comp. Neurol.* 260:435–52

Kitai ST, Deniau JM. 1981. Cortical inputs to the subthalamus: intracellular analysis. *Brain Res.* 214:411–15

Kravitz AV, Freeze BS, Parker PRL, Kay K, Thwin MT, et al. 2010. Regulation of parkinsonian motor behaviours by optogenetic control of basal ganglia circuitry. *Nature* 466:622–26

Langston JW, Ballard P, Tetrud JW, Irwin I. 1983. Chronic Parkinsonism in humans due to a product of meperidine-analog synthesis. *Science* 219:979–80

Leventhal DK, Gage GJ, Schmidt R, Pettibone JR, Case AC, Berke JD. 2012. Basal ganglia beta oscillations accompany cue utilization. *Neuron* 73:523–36

Levy R, Dostrovsky JO, Lang AE, Sime E, Hutchison WD, Lozano AM. 2001a. Effects of apomorphine on subthalamic nucleus and globus pallidus internus neurons in patients with Parkinson's disease. *J. Neurophysiol.* 86:249–60

Levy R, Lang AE, Dostrovsky JO, Pahapill P, Romas J, et al. 2001b. Lidocaine and muscimol microinjections in subthalamic nucleus reverse Parkinsonian symptoms. *Brain* 124:2105–18

Liang L, DeLong MR, Papa SM. 2008. Inversion of dopamine responses in striatal medium spiny neurons and involuntary movements. *J. Neurosci.* 28:7537–47

Litvak V, Jha A, Eusebio A, Oostenveld R, Foltynie T, et al. 2011. Resting oscillatory cortico-subthalamic connectivity in patients with Parkinson's disease. *Brain* 134:359–74

Lozano AM, Lang AE, Levy R, Hutchison W, Dostrovsky J. 2000. Neuronal recordings in Parkinson's disease patients with dyskinesias induced by apomorphine. *Ann. Neurol.* 47:S141–46

Mailly P, Aliane V, Groenewegen HJ, Haber SN, Deniau JM. 2013. The rat prefrontostriatal system analyzed in 3D: evidence for multiple interacting functional units. *J. Neurosci.* 33:5718–27

Mallet N, Ballion B, Le Moine C, Gonon F. 2006. Cortical inputs and GABA interneurons imbalance projection neurons in the striatum of parkinsonian rats. *J. Neurosci.* 26:3875–84

Mallet N, Micklem BR, Henny P, Brown MT, Williams C, et al. 2012. Dichotomous organization of the external globus pallidus. *Neuron* 74:1075–86

McCairn KW, Turner RS. 2009. Deep brain stimulation of the globus pallidus internus in the parkinsonian primate: local entrainment and suppression of low-frequency oscillations. *J. Neurophysiol.* 101:1941–60

McFarland NR, Haber SN. 2000. Convergent inputs from thalamic motor nuclei and frontal cortical areas to the dorsal striatum in the primate. *J. Neurosci.* 20:3798–813

McFarland NR, Haber SN. 2001. Organization of thalamostriatal terminals from the ventral motor nuclei in the macaque. *J. Comp. Neurol.* 429:321–36

Mink JW. 1996. The basal ganglia: focused selection and inhibition of competing motor programs. *Prog. Neurobiol.* 50:381–425

Mink JW. 2001. Neurobiology of basal ganglia circuits in Tourette syndrome: faulty inhibition of unwanted motor patterns? *Adv. Neurol.* 85:113–22

Mink JW, Thach WT. 1993. Basal ganglia intrinsic circuits and their role in behavior. *Curr. Opin. Neurobiol.* 3:950–57

Monakow KH, Akert K, Künzle H. 1978. Projections of the precentral motor cortex and other cortical areas of the frontal lobe to the subthalamic nucleus in the monkey. *Exp. Brain Res.* 33:395–403

Moro E, Lang AE, Strafella AP, Poon YY, Arango PM, et al. 2004. Bilateral globus pallidus stimulation for Huntington's disease. *Ann. Neurol.* 56:290–94

Nambu A, Takada M, Inase M, Tokuno H. 1996. Dual somatotopical representations in the primate subthalamic nucleus: evidence for ordered but reversed body-map transformations from the primary motor cortex and the supplementary motor area. *J. Neurosci.* 16:2671–83

Nambu A, Tokuno H, Hamada I, Kita H, Imanishi M, et al. 2000. Excitatory cortical inputs to pallidal neurons via the subthalamic nucleus in the monkey. *J. Neurophysiol.* 84:289–300

Nambu A, Tokuno H, Takada M. 2002. Functional significance of the cortico-subthalamo-pallidal 'hyperdirect' pathway. *Neurosci. Res.* 43:111–17

Papa SM, Desimone R, Fiorani M, Oldfield EH. 1999. Internal globus pallidus discharge is nearly suppressed during levodopa-induced dyskinesias. *Ann. Neurol.* 46:732–38

Parent A. 1976. Striatal afferent connections in the turtle (*Chrysemys picta*) as revealed by retrograde axonal transport of horseradish peroxidase. *Brain Res.* 108:25–36

Parent A, Hazrati LN. 1995. Functional anatomy of the basal ganglia. II. The place of subthalamic nucleus and external pallidum in basal ganglia circuitry. *Brain Res. Brain Res. Rev.* 20:128–54

Parent M, Parent A. 2004. The pallidofugal motor fiber system in primates. *Park. Relat. Disord.* 10:203–11

Paulsen JS. 2011. Cognitive impairment in Huntington disease: diagnosis and treatment. *Curr. Neurol. Neurosci. Rep.* 11:474–83

Penney JB Jr, Young AB. 1983. Speculations on the functional anatomy of basal ganglia disorders. *Annu. Rev. Neurosci.* 6:73–94

Planert H, Berger TK, Silberberg G. 2013. Membrane properties of striatal direct and indirect pathway neurons in mouse and rat slices and their modulation by dopamine. *PloS ONE* 8:e57054

Plenz D, Kital ST. 1999. A basal ganglia pacemaker formed by the subthalamic nucleus and external globus pallidus. *Nature* 400:677–82

Ragsdale CW Jr, Graybiel AM. 1991. Compartmental organization of the thalamostriatal connection in the cat. *J. Comp. Neurol.* 311:134–67

Raz A, Vaadia E, Bergman H. 2000. Firing patterns and correlations of spontaneous discharge of pallidal neurons in the normal and the tremulous 1-methyl-4-phenyl-1,2,3,6-tetrahydropyridine vervet model of parkinsonism. *J. Neurosci.* 20:8559–71

Redgrave P, Rodriguez M, Smith Y, Rodriguez-Oroz MC, Lehericy S, et al. 2010. Goal-directed and habitual control in the basal ganglia: implications for Parkinson's disease. *Nat. Rev. Neurosci.* 11:760–72

Reiner A, Albin RL, Anderson KD, D'Amato CJ, Penney JB, Young AB. 1988. Differential loss of striatal projection neurons in Huntington disease. *Proc. Natl. Acad. Sci. USA* 85:5733–37

Reiner A, Dragatsis I, Dietrich P. 2011. Genetics and neuropathology of Huntington's disease. *Int. Rev. Neurobiol.* 98:325–72

Reiner A, Medina L, Veenman CL. 1998. Structural and functional evolution of the basal ganglia in vertebrates. *Brain Res. Brain Res. Rev.* 28:235–85

Richardson RM, Freed CR, Shimamoto SA, Starr PA. 2011. Pallidal neuronal discharge in Parkinson's disease following intraputamenal fetal mesencephalic allograft. *J. Neurol. Neurosurg. Psychiatry* 82:266–71

Rosas HD, Goodman J, Chen YI, Jenkins BG, Kennedy DN, et al. 2001. Striatal volume loss in HD as measured by MRI and the influence of CAG repeat. *Neurology* 57:1025–28

Rosin B, Slovik M, Mitelman R, Rivlin-Etzion M, Haber SN, et al. 2011. Closed-loop deep brain stimulation is superior in ameliorating parkinsonism. *Neuron* 72:370–84

Rothblat DS, Schneider JS. 1993. Response of caudate neurons to stimulation of intrinsic and peripheral afferents in normal, symptomatic, and recovered MPTP-treated cats. *J. Neurosci.* 13:4372–78

Royce GJ. 1978. Cells of origin of subcortical afferents to the caudate nucleus: a horseradish peroxidase study in the cat. *Brain Res.* 153:465–75

Sadikot AF, Parent A, Smith Y, Bolam JP. 1992. Efferent connections of the centromedian and parafascicular thalamic nuclei in the squirrel monkey: a light and electron microscopic study of the thalamostriatal projection in relation to striatal heterogeneity. *J. Comp. Neurol.* 320:228–42

Sano H, Chiken S, Hikida T, Kobayashi K, Nambu A. 2013. Signals through the striatopallidal indirect pathway stop movements by phasic excitation in the substantia nigra. *J. Neurosci.* 33:7583–94

Sano H, Yasoshima Y, Matsushita N, Kaneko T, Kohno K, et al. 2003. Conditional ablation of striatal neuronal types containing dopamine D2 receptor disturbs coordination of basal ganglia function. *J. Neurosci.* 23:9078–88

Sato F, Lavallée P, Lévesque M, Parent A. 2000a. Single-axon tracing study of neurons of the external segment of the globus pallidus in primate. *J. Comp. Neurol.* 417:17–31

Sato F, Parent M, Levesque M, Parent A. 2000b. Axonal branching pattern of neurons of the subthalamic nucleus in primates. *J. Comp. Neurol.* 424:142–52

Sato M, Itoh K, Mizuno N. 1979. Distribution of thalamo-caudate neurons in the cat as demonstrated by horseradish peroxidase. *Exp. Brain Res. Exp.* 34:143–53

Schiffmann SN, Vanderhaeghen JJ. 1993. Adenosine A2 receptors regulate the gene expression of striatopallidal and striatonigral neurons. *J. Neurosci.* 13:1080–87

Schmidt R, Leventhal DK, Mallet N, Chen F, Berke JD. 2013. Canceling actions involves a race between basal ganglia pathways. *Nat. Neurosci.* 16:1118–24

Schrock LE, Ostrem JL, Turner RS, Shimamoto SA, Starr PA. 2009. The subthalamic nucleus in primary dystonia: single-unit discharge characteristics. *J. Neurophysiol.* 102:3740–52

Shimamoto SA, Ryapolova-Webb ES, Ostrem JL, Galifianakis NB, Miller KJ, Starr PA. 2013. Subthalamic nucleus neurons are synchronized to primary motor cortex local field potentials in Parkinson's disease. *J. Neurosci.* 33:7220–33

Smith Y, Hazrati LN, Parent A. 1990. Efferent projections of the subthalamic nucleus in the squirrel monkey as studied by the PHA-L anterograde tracing method. *J. Comp. Neurol.* 294:306–23

Smith Y, Kieval JZ. 2000. Anatomy of the dopamine system in the basal ganglia. *Trends Neurosci.* 23:S28–33

Smith Y, Parent A. 1986. Differential connections of caudate nucleus and putamen in the squirrel monkey (*Saimiri sciureus*). *Neuroscience* 18:347–71

Smith Y, Raju DV, Pare JF, Sidibe M. 2004. The thalamostriatal system: a highly specific network of the basal ganglia circuitry. *Trends Neurosci.* 27:520–27

Starr PA, Kang GA, Heath S, Shimamoto S, Turner RS. 2008. Pallidal neuronal discharge in Huntington's disease: support for selective loss of striatal cells originating the indirect pathway. *Exp. Neurol.* 211:227–33

Starr PA, Rau GM, Davis V, Marks WJ Jr, Ostrem JL, et al. 2005. Spontaneous pallidal neuronal activity in human dystonia: comparison with Parkinson's disease and normal macaque. *J. Neurophysiol.* 93:3165–76

Steigerwald F, Pötter M, Herzog J, Pinsker M, Kopper F, et al. 2008. Neuronal activity of the human subthalamic nucleus in the parkinsonian and nonparkinsonian state. *J. Neurophysiol.* 100:2515–24

Stephenson-Jones M, Ericsson J, Robertson B, Grillner S. 2012. Evolution of the basal ganglia: dual-output pathways conserved throughout vertebrate phylogeny. *J. Comp. Neurol.* 520:2957–73

Stuber GD, Hnasko TS, Britt JP, Edwards RH, Bonci A. 2010. Dopaminergic terminals in the nucleus accumbens but not the dorsal striatum corelease glutamate. *J. Neurosci.* 30:8229–33

Tabrizi SJ, Scahill RI, Owen G, Durr A, Leavitt BR, et al. 2013. Predictors of phenotypic progression and disease onset in premanifest and early-stage Huntington's disease in the TRACK-HD study: analysis of 36-month observational data. *Lancet Neurol.* 12:637–49

Tachibana Y, Iwamuro H, Kita H, Takada M, Nambu A. 2011. Subthalamo-pallidal interactions underlying parkinsonian neuronal oscillations in the primate basal ganglia. *Eur. J. Neurosci.* 34:1470–84

Tai LH, Lee AM, Benavidez N, Bonci A, Wilbrecht L. 2012. Transient stimulation of distinct subpopulations of striatal neurons mimics changes in action value. *Nat. Neurosci.* 15:1281–89

Tang JK, Moro E, Lozano AM, Lang AE, Hutchison WD, et al. 2005. Firing rates of pallidal neurons are similar in Huntington's and Parkinson's disease patients. *Exp. Brain Res.* 166:230–36

Tippett LJ, Waldvogel HJ, Thomas SJ, Hogg VM, van Roon-Mom W, et al. 2007. Striosomes and mood dysfunction in Huntington's disease. *Brain* 130:206–21

Tritsch NX, Ding JB, Sabatini BL. 2012. Dopaminergic neurons inhibit striatal output through non-canonical release of GABA. *Nature* 490:262–66

Vitek JL, Chockkan V, Zhang JY, Kaneoke Y, Evatt M, et al. 1999. Neuronal activity in the basal ganglia in patients with generalized dystonia and hemiballismus. *Ann. Neurol.* 46:22–35

Vitek JL, Zhang J, Hashimoto T, Russo GS, Baker KB. 2012. External pallidal stimulation improves parkinsonian motor signs and modulates neuronal activity throughout the basal ganglia thalamic network. *Exp. Neurol.* 233:581–86

Wall NR, De La Parra M, Callaway EM, Kreitzer AC. 2013. Differential innervation of direct- and indirect-pathway striatal projection neurons. *Neuron* 79:347–60

Weinberger M, Hutchison WD, Alavi M, Hodaie M, Lozano AM, et al. 2012. Oscillatory activity in the globus pallidus internus: comparison between Parkinson's disease and dystonia. *Clin. Neurophysiol.* 123:358–68

Wichmann T, Bergman H, DeLong MR. 1994. The primate subthalamic nucleus. III. Changes in motor behavior and neuronal activity in the internal pallidum induced by subthalamic inactivation in the MPTP model of parkinsonism. *J. Neurophysiol.* 72:521–30

Wichmann T, Bergman H, Starr PA, Subramanian T, Watts RL, DeLong MR. 1999. Comparison of MPTP-induced changes in spontaneous neuronal discharge in the internal pallidal segment and in the substantia nigra pars reticulata in primates. *Exp. Brain Res.* 125:397–409

A Mitocentric View of Parkinson's Disease

Nele A. Haelterman,[1] Wan Hee Yoon,[2,3]
Hector Sandoval,[2] Manish Jaiswal,[2,3]
Joshua M. Shulman,[1,2,4] and Hugo J. Bellen[1,2,3,5]

[1]Program in Developmental Biology, [2]Department of Molecular and Human Genetics,
[3]Howard Hughes Medical Institute, [4]Department of Neurology, [5]Department of Neuroscience,
Jan and Dan Duncan Neurological Research Institute, Baylor College of Medicine, Houston,
Texas 77030; email: nhaelter@bcm.edu, whyoon@bcm.edu, hectors@bcm.edu,
jaiswal@bcm.edu, Joshua.Shulman@bcm.edu, hbellen@bcm.edu

Annu. Rev. Neurosci. 2014. 37:137–59

First published online as a Review in Advance on
May 5, 2014

The *Annual Review of Neuroscience* is online at
neuro.annualreviews.org

This article's doi:
10.1146/annurev-neuro-071013-014317

Keywords

PD genes, reactive oxygen species, mitochondria, electron transport chain,
mitochondrial unfolded protein response, mitochondrial dynamics

Abstract

Parkinson's disease (PD) is a common neurodegenerative disease, yet the
underlying causative molecular mechanisms are ill defined. Numerous observations based on drug studies and mutations in genes that cause PD
point to a complex set of rather subtle mitochondrial defects that may be
causative. Indeed, intensive investigation of these genes in model organisms
has revealed roles in the electron transport chain, mitochondrial protein homeostasis, mitophagy, and the fusion and fission of mitochondria. Here, we
attempt to synthesize results from experimental studies in diverse systems
to define the precise function of these PD genes, as well as their interplay
with other genes that affect mitochondrial function. We propose that subtle
mitochondrial defects in combination with other insults trigger the onset
and progression of disease, in both familial and idiopathic PD.

Contents

INTRODUCTION

Parkinson's disease (PD) is a progressive and incurable neurodegenerative disorder affecting 1% of the older adult population. The defining motor characteristics of PD, known collectively as parkinsonism, include tremor, increased muscle tone, slow movements, and impaired gait and balance (Fahn 2003, Lees et al. 2009). In autopsy studies, PD is characterized by the loss of dopamine (DA) neurons in the midbrain substantia nigra (SN) in association with α-synuclein protein aggregates, termed Lewy bodies (Goedert et al. 2013, Jellinger 2009). The SN is the source of DA for other basal ganglia nuclei, which have an essential role in initiating and facilitating movement. In addition, a range of nonmotor manifestations are observed (Chaudhuri & Schapira 2009), including neuropsychiatric symptoms (e.g., depression, cognitive impairment), autonomic nervous system dysfunction, sleep disturbance, pain, and constipation, which likely relate to more widespread synuclein pathology and neurodegeneration throughout the nervous system.

Although PD pathogenesis has been studied extensively, the primary cause(s) remain(s) elusive. Alterations in numerous cellular processes have been implicated, including oxidative stress (Fariello 1988), excitotoxicity (Olney et al. 1990), the ubiquitin-proteasome system (Lowe et al. 1988), the endolysosomal compartment, and mitochondrial dysfunction (Schapira et al. 1989). All these may be directly or indirectly related to mechanisms of neurodegeneration in PD. However, mitochondrial toxins potently induce SN degeneration, and genes that have been linked to familial PD and related disorders are increasingly being shown to affect mitochondrial function (**Table 1**). Based on recent studies, we focus here on a mitocentric view of PD, suggesting that mitochondrial dysfunction is in fact a primary cause of PD.

Mitochondria are double-walled, dynamic, filamentous organelles that constitute the cell's major source of adenosine triphosphate (ATP) production, the chemical energy of the cell. Within mitochondria, ATP is produced by the citric acid cycle (Krebs cycle) in the matrix via the action of four respiratory complexes (CI, CII, CIII, CIV) and ATP synthase in the mitochondrial inner membrane (MIM) (**Figure 1**). Mitochondrial mechanisms have been implicated in diverse human disorders (Vafai & Mootha 2012). A severe impairment of mitochondrial function typically has dramatic consequences. For instance, mutations in genes encoding subunits of the respiratory complexes lead to Leigh syndrome, a fatal multisystem disorder of childhood onset. In contrast, more subtle mitochondrial impairment is associated with comparatively insidious onset and heterogeneous clinical manifestations, including cancer, diabetes, cardiomyopathy, anemia,

PD: Parkinson's disease

Substantia nigra (SN): pigmented region of the midbrain (due to melanin) containing dopaminergic neurons; participates with other basal ganglia nuclei in movement control

Ubiquitin-proteasome system: the system that tags misfolded or damaged proteins with ubiquitin and degrades them in the proteasome

Table 1 Mitochondrial phenotypes associated with selected genes linked to PD and related Parkinsonian disorders[a]

Gene	Function	Mitochondrial phenotypes			References
		Fly	**Rodent**	**Human**	
α-*Synuclein* (*SNCA*) Dominant	Synaptic vesicle formation/ membrane fusion	**Mito function** *Mutant O/E*[b]: (↑) level of CV subunits	**Mito function** *Mutant O/E*[b]: (↓) CIV activity	**Mito function** *WT* and *mutant O/E*[c]: (↓) CI activity; (↓) Δψ$_m$ **Mito dynamics** *WT* and *mutant O/E*[c]: (↑) fragmentation	Devi et al. 2008, Kamp et al. 2010, Martin et al. 2006, Parihar et al. 2009, Polymeropoulos et al. 1997, Uéda et al. 1993, Xun et al. 2007
Parkin (*PARK2*) Recessive	E3 ubiquitin ligase	**Mito morphology/ dynamics** *Lof*[d]: abnormal; (↓) fission	**Mito function** *Lof*[d]: (↓) level of CI and CIV subunits; (↓) ETC activity **Mito morphology/ dynamics** *Conditional lof*[d]: (↓) biogenesis	**Mito function** *Lof*[e]: (↓) CI activity **Mito dynamics** *Lof*[e]: (↑) fragmentation	Greene et al. 2003, Kitada et al. 1998, Lutz et al. 2009, Müftüoglu et al. 2004, Palacino et al. 2004, Pesah et al. 2004, Shin et al. 2011
PINK1 Recessive	Kinase	**Mito function** *Lof*[d]: (↓) CI activity; (↓) ATP level; (↓) mtDNA level; (↓) Δψ$_m$ **Mito morphology** *Lof*[d]: abnormal	**Mito function** *Lof*[d]: (↓) CI activity; (↓) respiration; (↓) mitochondrial preprotein import; (↓) ATP *Lof*[e]: (↓) Δψ$_m$; (↓) ATP **Mito morphology/ dynamics** *Lof*[e]: (↑) fragmentation *Mutant O/E*[c]: (↑) fragmentation; (↑) fission	**Mito function** *Lof*[e]: (↓) CI activity; (↓) Δψ$_m$; (↓) respiration; (↓) mtDNA level/synthesis; (↓) ATP *Mutant O/E*[c]: (↓) Δψ$_m$; (↓) CI activity **Mito morphology/ dynamics** *Lof*[e]: (↑) fragmentation *Lof*[e]: (↑) biogenesis	Amo et al. 2011, Bonifati et al. 2005, Clark et al. 2006, Cui et al. 2010, Dagda et al. 2009, Exner et al. 2007, Gautier et al. 2008, Gegg et al. 2009, Gispert et al. 2009, Heeman et al. 2011, Hoepken et al. 2007, Morais et al. 2009, Park et al. 2006, Piccoli et al. 2008, Seibler et al. 2011, Unoki & Nakamura 2001, Vos et al. 2012, Yuan et al. 2010
DJ-1 (*Park7*) Recessive	Cysteine protease/redox-regulated chaperone	**Mito function** *Lof*[d]: (↓) mtDNA level; (↓) respiration; (↓) ATP level **Mito morphology** *Lof*[d]: abnormal	**Mito function** *Lof*[e]: (↓) CI activity; (↓) CI assembly; (↓) ATP; (↓) respiration; (↓) Δψ$_m$; (↓) UCPs expression **Mito morphology/ dynamics** *Lof*[e]: (↑) fragmentation; (↓) fusion *Mutant O/E*[c]: (↑) fragmentation	**Mito function** *Lof*[e]: (↓) Δψ$_m$ **Mito morphology/ dynamics** *Lof*[e]: (↑) fragmentation *Mutant O/E*[c]: (↑) fragmentation *WT O/E*[c]: (↑) elongated mitochondria	Bonifati et al. 2003, Guzman et al. 2010, Hao et al. 2010, Heo et al. 2010, Irrcher et al. 2010, Krebiehl et al. 2010, Kwon et al. 2011, Nagakubo et al. 1997, Thomas et al. 2011, Wang et al. 2012a

(*Continued*)

Table 1 *(Continued)*

Gene	Function	Mitochondrial phenotypes			References
		Fly	**Rodent**	**Human**	**References**
LRRK2 Dominant	Kinase/GTPase activity	**Mito morphology** *Mutant O/E[b]:* abnormal	**Mito functions** *Mutant O/E[c]:* (↓) Δψ$_m$ **Mito morphology/ dynamics** *WT* and *mutant O/E[c]:* (↑) fission	**Mito function** *Lof[e]:* (↓) Δψ$_m$; (↓) ATP; (↓) CI, CII, CIV; (↑) UCPs expression *Mutant O/E[c]:* (↓) Δψ$_m$; (↑) UCPs expression **Mito morphology/ dynamics** *Mutant O/E[c]:* (↑) fragmentation	Cherra et al. 2013, Mortiboys et al. 2010, Niu et al. 2012, Paisán-Ruíz et al. 2004, Papkovskaia et al. 2012, Wang et al. 2012b
ATP13A2 Recessive	Lysosomal ATPase	NA	**Mito morphology** *Lof[e]:* (↑) fragmentation	**Mito function** *Lof[e]:* (↓) ATP; (↑) mtDNA level; (↑) oxygen consumption rate **Mito morphology** *Lof[e]:* (↑) fragmentation	Grünewald et al. 2012, Ramirez et al. 2006, Ramonet et al. 2012, Schultheis et al. 2004
FBXO7 Recessive	E3 ubiquitin protein ligase subunit	NA	NA	**Mito function** *Lof[e]:* (↓) Parkin recruitment to mitochondria	Burchell et al. 2013, Ilyin et al. 2000, Shojaee et al. 2008
Vps35 Dominant	Subunit of the retromer complex	NA	NA	**Mito function** *Lof[e]:* (↓) delivery of MAPL from mitochondria to peroxisomes	Braschi et al. 2010, Edgar & Polak 2000, Vilariño-Güell et al. 2011

[a]Abbreviations: CI, complex I; CIV, complex IV; CV, complex V; ETC, electron transport chain; *Gof, gain-of-function; Lof, loss-of-function;* MAPL, mitochondrial-anchored protein ligase; mtDNA, mitochondrial DNA; NA, not available; O/E, overexpression; UCPs, mitochondrial uncoupling proteins; WT, wild type; Δψ$_m$, mitochondrial membrane potential.
[b]Overexpression in vivo animal model.
[c]Overexpression in vitro cell culture.
[d]Loss of function in vivo animal model.
[e]Loss of function in vitro cell culture.

Complex I (CI): large protein complex, located in the inner mitochondrial membrane, that oxidizes nicotinamide adenine dinucleotide (NADH) to transfer electrons from NADH to ubiquinone

and neurodegeneration (Schapira 2012, Schon & Przedborski 2011). In fact, numerous studies have documented a mitochondrial complex I (CI) deficiency in the SN of PD patients (Keeney et al. 2006, Mizuno et al. 1989, Parker et al. 1989, Schapira et al. 1989). Moreover, a recent report similarly implicated early dysregulation in the mitochondrial electron transport chain (ETC), on the basis of transcriptional profiling of the SN from PD patients (Zheng et al. 2010).

Below, we first address the environmental and genetic evidence that links the etiology of PD to mitochondria. Next, we address how the many genes implicated in PD affect distinct mitochondrial functions such as the ETC, the mitochondrial unfolded protein response (UPR^{mt}), and mitochondrial dynamics. In the final section, we describe models that explain the relative vulnerability of certain cell types, such as SN DA neurons, despite a global mitochondrial dysfunction,

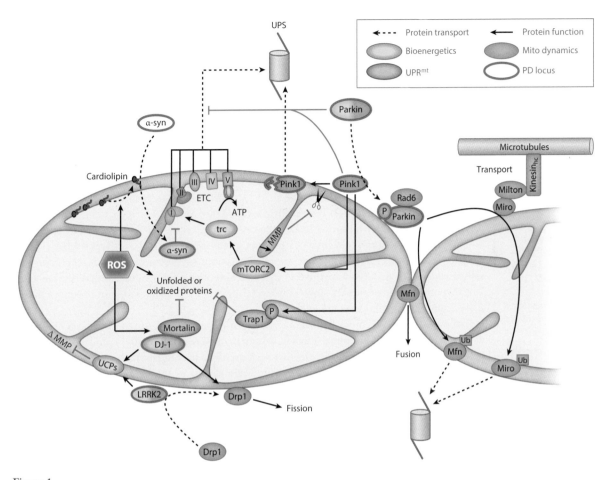

Figure 1

Proteins implicated in Parkinson's disease (PD) maintain a healthy pool of mitochondria. The effects of PD-linked proteins (*thick green line*) on mitochondrial function can be divided into three different groups. Bioenergetics (*yellow*): Pink1 and Parkin regulate the turnover of several subunits of ETC complexes by the ubiquitin proteasome system (UPS). In addition, Pink1 supports complex I activity through a kinase cascade involving *tricornered* (*trc*) and mTORC2. Mitochondrially localized α-synuclein (α-syn), however, inhibits complex I. Oxidation of the mitochondrial phospholipid cardiolipin translocates this lipid from the inner to the outer mitochondrial membrane where it serves as a receptor for α-synuclein. Hence, in the presence of elevated levels of reactive oxygen species (ROS), mitochondrial α-synuclein import is increased. LRRK2 and DJ-1 regulate the activity of uncoupling proteins (UCPs), which preserve the mitochondrial membrane potential (MMP). Both PD proteins therefore indirectly maintain ATP production. Mitochondrial unfolded protein response (UPR^mt, *blue*): When activated, Pink1 phosphorylates the chaperone Trap1, a protein that protects mitochondrial proteins from ROS. In addition, Trap1 works with Hsp60 to refold imported proteins, sustaining mitochondrial protein homeostasis. In the presence of ROS, DJ-1 is translocated into mitochondria, where it interacts with the chaperone Mortalin to salvage oxidized proteins. Mitochondrial dynamics (*orange*): LRRK2 directs mitochondrial fission as it interacts with and recruits the mitochondrial fission protein Drp1. DJ-1 increases mitochondrial fission by regulating Drp1 levels, although the exact mechanism through which the chaperone executes this function is not clear. The E3 ubiquitin ligase Parkin inhibits mitochondrial fusion through ubiquitin-mediated degradation of the fusion protein Mitofusin (Mfn). Mitochondrial transport is regulated by the Pink1/Parkin pathway. Here, Parkin-mediated degradation of the adaptor protein Miro is thought to detach a dysfunctional mitochondrion from kinesin motor proteins, halting its transport. Finally, when a dysfunctional mitochondrion cannot be repaired, it is cleared through mitophagy. Under basal conditions, the protein kinase Pink1 is imported into the mitochondrial intermembrane space, where it is cleaved by 2 proteases and subsequently degraded. However, upon mitigation of the MMP, Pink1 is no longer cleaved. The protein then phosphorylates and activates the E3 ubiquitin ligase Parkin. Parkin, along with E2 ligases such as Rad6, initiates mitophagy by ubiquitinating target proteins.

and we further attempt to generalize lessons from rare familial forms of PD to the more common sporadic disease.

MITOCHONDRIAL DYSFUNCTION UNDERLIES PARKINSON'S DISEASE

The Environmental Link

Electron transport chain (ETC): 5 protein complexes that produce ATP via an electrochemical proton gradient across the inner mitochondrial membrane by coupling electron transfer between electron donors and acceptors

Mitochondrial unfolded protein response (UPR^{mt}): mitochondrial stress response to an accumulation of unfolded or misfolded proteins in the matrix of mitochondria

Autosomal recessive juvenile parkinsonism (AR-JP): familial form of early-onset parkinsonism, usually characterized at autopsy by substantia nigra degeneration in the absence of Lewy bodies

A wave of interest in a potential role for mitochondria in PD came in 1976, when investigators found that drug users who took opioid analogs (MPPP) laced with 1-methyl-4-phenyl-1,2,3,6-tetrahydropyridine (MPTP) developed PD-like motor symptoms resulting from degeneration of DA neurons in the SN (Langston et al. 1983, Martinez & Greenamyre 2012). Subsequent investigations showed that MPTP is readily converted into MPP$^+$, which crosses the blood brain barrier and is selectively imported in DA neurons (Nicklas et al. 1985). MPP$^+$ binds to CI and inhibits electron transport, reducing ATP production and increasing reactive oxygen species (ROS) levels (Nicklas et al. 1985). Together, these observations suggested that SN degeneration and subsequent motor manifestations in PD may similarly result from mitochondrial dysfunction. This hypothesis gained further support when CI defects were documented in postmortem SN tissue from PD patients (Schapira et al. 1989).

Following on these initial findings, a variety of pesticides and toxins have been associated with SN degeneration in animal models (Martinez & Greenamyre 2012), and evidence from human epidemiologic studies suggests that such exposure may in fact increase PD risk (Tanner et al. 2011). Several of the implicated toxins were shown to inhibit mitochondrial function, and such compounds have facilitated the development of useful animal model systems for PD studies (Martinez & Greenamyre 2012). For instance, the widely used herbicide paraquat, which is structurally similar to MPTP, is a potent redox cycler that accepts electrons from CI. This process converts the molecule to a free radical, which interacts with O_2 to generate superoxide anions and other ROS. Similarly, the naturally occurring pesticide rotenone systemically inhibits CI and leads to the degeneration of DA neurons in the SN. Finally, the fungicide maneb contains the mitotoxin manganese ethylene-bisdithiocarbamate, which preferentially inhibits complex III of the ETC and causes proteasomal inhibition, oxidative stress, and cytoplasmic α-synuclein aggregation in DA neuron cell lines (Zhou et al. 2004).

In summary, compelling data from animal models coupled with evidence from epidemiologic studies have led to the hypothesis that environmental exposure to mitochondrial toxins may be an important contributor to PD in the population (Martinez & Greenamyre 2012). Large meta-analyses suggest an approximately twofold increased risk of PD due to pesticide exposure (Priyadarshi et al. 2000). Thus additional environmental and/or genetic insults, in combination with aging, would likely be required to trigger PD in most affected individuals.

The Genetic Link

Rapid advances in genetics over the past two decades have transformed our thinking about PD from a primarily environmentally influenced disorder to one with a substantial genetic contribution (Shulman et al. 2011, Trinh & Farrer 2013). In particular, many of the genes responsible for familial forms of PD and related disorders have strong links to mitochondrial function. In 1998, mutations in the *parkin* gene were discovered to cause autosomal recessive, juvenile parkinsonism (AR-JP) (Kitada et al. 1998), and subsequent work has proven *parkin* to be the most important cause of PD in children and young adults (Lücking et al. 2000, Periquet et al. 2003). As detailed below, studies of *parkin* as well as *pink1* and *DJ-1*, which are less common causes of AR-JP (Bonifati et al. 2003, 2005; Rogaeva et al. 2004; Valente et al. 2004), have strongly implicated these genes in mitochondrial function. At presentation, AR-JP is often clinically indistinguishable from typical, late-onset PD cases. As the disease progresses unusual features emerge, including prominent dystonia, diurnal fluctuations, and an overall benign course without development of characteristic PD nonmotor symptoms, suggesting a more anatomically restricted pathology. In fact, most autopsies of PD patients who carry mutations in *parkin* show SN degeneration in the absence of Lewy body pathology (Poulopoulos et al. 2012), and some have argued that AR-JP and PD should be classified as distinct clinicopathologic entities (Doherty et al. 2013). Nevertheless, AR-JP and other Mendelian disorders with prominent parkinsonism highlight numerous genes with established links to mitochondrial function that appear to be similarly required for DA neurons to survive (**Table 1**). Therefore, studies of these loci will likely provide important mechanistic and perhaps therapeutic insights relevant to PD.

Although only a minority of idiopathic PD cases report a family history (10–30%), these figures likely underestimate the genetic contributions because of the strong influence of age on disease manifestation. Indeed, late-onset PD is now recognized to be influenced by a large number of genetic susceptibility variants, including both common and rare alleles with a range of potency (Shulman et al. 2011, Trinh & Farrer 2013). Families with autosomal dominant PD have led to the discovery of mutations or gene multiplication at the α-*synuclein* (*SNCA*) locus (Polymeropoulos et al. 1997, Singleton et al. 2003), as well as mutations in *LRRK2* (Paisán-Ruíz et al. 2004, Zimprich et al. 2004) and *VPS35* (Vilariño-Güell et al. 2011, Zimprich et al. 2011). Whereas *SNCA* mutations are associated with earlier-onset and more aggressive familial disease, clinical presentations in families with *LRRK2* and *VPS35* mutations closely overlap with idiopathic PD. In addition to rare, dominant mutations with high penetrance, investigators have identified additional common susceptibility variants at both the *LRRK2* and *SNCA* loci (Healy et al. 2008, Maraganore et al. 2006).

Table 1 highlights selected genetic causes of PD and related disorders with established connections to mitochondrial biology. As expanded upon below, many such genes have been directly or indirectly associated with mitochondrial defects including abnormal mitochondria morphology, decreased ETC-activity, and altered mitochondrial membrane potential. **Table 1** also highlights a number of genes, including *ATP13A2* and *FBXO7*, that cause recessive disorders characterized by prominent parkinsonism along with additional neurologic features not seen in PD. We have limited our discussion to Mendelian causes of parkinsonism in which the causal genes are clearly established. However, large genome-wide association studies have recently identified numerous common polymorphisms with strong statistical evidence of association with PD susceptibility (Int. Parkinson's Dis. Genomics Consort. 2011). Although these genetic loci individually have modest effects on disease risk, they likely have a major effect on PD at the population level because they are common. In the coming years, it will be of great interest to confirm the genes responsible for associations at these loci and to explore their potential links to mitochondrial and/or other cellular pathways.

Idiopathic PD: most common form of PD, characterized by late-onset, non-Mendelian inheritance and pathologically defined by nigral degeneration and Lewy bodies

Genome-wide association study: unbiased analysis of polymorphisms throughout the human genome to identify variants associated with traits at the population level

PROTEINS ENCODED BY PD LOCI AFFECT DIVERSE MITOCHONDRIAL FUNCTIONS

As described above, mutations in a plethora of genes cause PD (**Table 1**). Over the past decade, many of these genes were found to directly or indirectly affect mitochondria. Studies in different model organisms revealed their roles in mitochondrial biogenesis, physiology, stucture, and dynamics, as well as in quality control.

Deficits in the Electron Transport Chain

Mitophagy: form of autophagy that degrades mitochondria

Ndi1p: *S. cerevisiae* protein that catalyzes the oxidation of NADH and the reduction of ubiquinone and can hence bypass CI

E3 ubiquitin ligase: enzyme that transfers ubiquitin from E2 ubiquitin-conjugating enzymes to target proteins

Mitochondrial membrane potential (MMP): electrochemical gradient across the inner mitochondrial membrane that is essential to generate ATP

Chaperone: protein that interacts with unfolded or partially folded proteins to facilitate correct folding or unfolding

Pink1 and Parkin regulate the activity of different ETC complexes. Mutations in *parkin* and *pink1* are two important genetic causes of AR-JP, and the encoded proteins appear to have pivotal roles in regulating mitochondrial function. Much of our understanding originates from pioneering studies on *parkin* and *pink1* homologs in the fruit fly, *Drosophila melanogaster*, where mutations lead to prominent mitochondrial phenotypes in muscles and sperm (Clark et al. 2006, Greene et al. 2003, Park et al. 2006, Pesah et al. 2004). Subsequent genetic analyses have established convincingly that the proteins function in a linear pathway to regulate mitochondrial morphology and to induce clearance of dysfunctional mitochondria, a process termed mitophagy (**Figure 1** and see paragraph below).

In addition, recent studies showed that both proteins directly regulate mitochondrial function. Under basal conditions, low levels of the Pink1 kinase span the mitochondrial outer and inner membranes (MOM and MIM), where it acts on respiratory complexes. Wu et al. (2013) recently suggested that Pink1 regulates CI activity in flies by activating a kinase pathway that involves tricornered and mTORC2 (**Figure 1**). Loss of Pink1 also leads to a reduction in CI activity in flies, mice, and humans (Hoepken et al. 2007, Morais et al. 2009). Moreover, phenotypes observed in *Drosophila pink1* mutants are rescued by Ndi1p, a yeast protein that bypasses CI by transferring electrons from NADH to ubiquinone (Vilain et al. 2012). Although *pink1* mutants display a mild defect in CI activity, Vincow et al. (2013) recently documented that Pink1 regulates the protein turnover of multiple subunits of all 5 ETC complexes (**Figure 1**).

The E3 ubiquitin ligase Parkin normally localizes to the cytosol and is recruited only to dysfunctional mitochondria. Similar to *pink1* mutants, *parkin* mutant flies display reduced ATP levels (Vos et al. 2012). However, unlike *pink1* mutants, *parkin* mutants do not display CI defects and hence cannot be rescued by Ndi1p expression (Vilain et al. 2012). This finding suggests that Parkin may differentially regulate the activity of various ETC complexes, compared with Pink1. Indeed, Parkin was recently shown to ubiquitinate and thereby regulate the protein levels of many of the ETC subunits in a human cell line (Sarraf et al. 2013, Vincow et al. 2013). These studies suggest that, in addition to displaying CI deficiency, *parkin* mutant flies may display dysfunction of other subunits of the ETC, similar to results from studies of leukocytes from patients with *parkin* mutations (Müftüoglu et al. 2004).

A screen for mutants in flies that enhance the *pink1*-associated phenotypes led to the isolation of a modifier, *heixuedian* (*heix*), which encodes a protein that synthesizes vitamin K_2 (Vos et al. 2012). Vitamin K_2 functions as an electron carrier, downstream of complex II (CII), and can therefore compensate for defects in upstream respiratory complexes. Loss of *heix* enhances the *pink1* defects associated with loss of the mitochondrial membrane potential (MMP), ATP production, and motility. Overexpression of *heix* or food supplementation of vitamin K_2 rescues the energy defects and the mitochondrial morphological defects of both *pink1* and *parkin* mutants. Together, these findings suggest that Pink1 and Parkin are required to support ATP production, although both proteins affect the activity of different complexes within the ETC and their functions may turn out to be more complicated than the simple linear pathway suggested here.

DJ-1 supports ATP production. Mutations in DJ-1, which encodes a protein related to a family of molecular chaperones, are a rare cause of AR-JP (**Table 1**) (Bonifati et al. 2003, Klein & Lohmann-Hedrich 2007). DJ-1 is expressed in most cell types and localizes to the cytosol, mitochondrial matrix, and intermembrane space (Moore et al. 2006, Zhang et al. 2005). We previously reviewed DJ-1's role in protection against oxidative stress and in PD pathogenesis (Jaiswal et al. 2012). Unlike vertebrates, who have only one copy of *DJ-1*, the *Drosophila* genome encodes two paralogs (*DJ-1α* and *DJ-1β*). When both are deleted, flies display an age-dependent decline in mitochondrial DNA and respiration, leading to reduced ATP levels (Hao et al. 2010). Fly or human *DJ-1* overexpression rescues the phenotypes of *pink1* mutants, suggesting that the two proteins possess at least partially overlapping functions (Hao et al. 2010). In vertebrates, *DJ-1* null mouse embryonic fibroblasts and DA primary neurons display reduced mitochondrial respiration and MMP (Heo et al. 2012, Krebiehl et al. 2010). These cells also display reduced CI assembly and an increased sensitivity to oxidative stress owing to downregulation of mitochondrial uncoupling proteins (UCPs) (Guzman et al. 2010, Heo et al. 2012). In sum, data from various systems suggest that DJ-1 plays a role in maintaining respiratory complex stability and/or MMP, similar to the roles of other genes implicated in PD and related disorders.

Uncoupling proteins (UCPs): members of the family of mitochondrial anion carrier proteins that dissipate the proton gradient across the inner mitochondrial membrane

Cardiolipin: mitochondria-specific phospholipid that regulates many processes such as mitochondrial dynamics, in addition to various signaling pathways

α-Synuclein inhibits complex I. As introduced above, aggregation of α-synuclein comprises the defining neuropathology of PD (Lewy bodies). Furthermore, common polymorphisms and rare mutations in *α-synuclein* have been genetically linked to PD risk (**Table 1**). Although most studies have focused on the cytoplasmic role of α-synuclein in PD pathogenesis, recent evidence also links the protein to mitochondria (reviewed in Nakamura 2013). α-Synuclein contains a cryptic mitochondrial targeting signal and the protein accumulates in the SN and striatal mitochondria of PD patients, where it inhibits CI activity (Butler et al. 2012, Devi et al. 2008). Moreover, several groups have reported that overexpression of wild-type or mutant α-synuclein increases ROS levels in various mammalian cell lines (Junn & Mouradian 2002, Parihar et al. 2009). Elevated ROS levels may enhance mitochondrial translocation of α-synuclein, which is dependent on the phospholipid cardiolipin. Under basal conditions, cardiolipin only localizes to the MIM, but the lipid becomes more abundant on the MOM under conditions of high oxidative stress (Cole et al. 2008). One possible interpretation is that aggregated or misfolded α-synuclein can inhibit CI activity, increase ROS, and thereby potentiate its own mitochondrial translocation in a vicious cycle that further disrupts ETC function (**Figure 1**). In this context, it is intriguing that several mitochondrial toxins have also induced α-synuclein aggregation in animal models (Martinez & Greenamyre 2012).

LRRK2 affects ETC activity. Mutations in *LRRK2* are a common cause of familial PD (Healy et al. 2008). The LRRK2 protein is localized primarily to the cytoplasm and membranes, including the mitochondrial membrane. It functions as a kinase, a GTPase, and a scaffolding protein (Papkovskaia et al. 2012, West et al. 2005). Several lines of evidence support a role for LRRK2 in mitochondria. The MMP and total intracellular ATP levels were reduced both in cells derived from skin biopsies of patients with LRRK2 mutations as well as in human cell lines expressing mutant *LRRK2* (Mortiboys et al. 2010). Furthermore, LRRK2's kinase activity regulates UCPs to maintain MMP and ATP production (**Figure 1**) (Papkovskaia et al. 2012). In *Drosophila*, overexpression of human, PD-associated mutant *LRRK2* reduces life span and increases sensitivity to rotenone, suggesting a potentially conserved mitochondrial role (Ng et al. 2009). Coexpression of *parkin* substantially rescues these *LRRK2*-induced phenotypes. Moreover, activating AMP kinase, a key cellular regulator of energy metabolism, significantly improves mitochondrial function in both *LRRK2* and *parkin* mutant flies (Ng et al. 2012). Together, these results suggest that Parkin and LRRK2 may operate in a common pathway. Wild-type *LRRK2* overexpression enhanced the viability of DA neurons in *Caenorhabditis elegans* exposed to rotenone (Saha et al. 2009), suggesting that LRRK2 protects against mitochondrial damage. In conclusion, numerous studies implicate mitochondrial impairment in LRRK2-associated PD caused by *LRRK2* mutations. However, additional work is needed to better define the relevant mechanisms and their relationships with other known LRRK2 functions (Kett & Dauer 2012).

Defects in Protein Homeostasis: The Mitochondrial Unfolded Protein Response

Translocase of the inner membrane (TIM) complex: protein complex that facilitates the translocation of nuclear-encoded mitochondrial proteins from the intermembrane space into the matrix

When mitochondria receive a toxic insult that impairs protein folding, they induce a mitochondrial-specific unfolded protein response (UPR^{mt}) (Haynes & Ron 2010). This response involves a signal that is sent from mitochondria to the nucleus, leading to expression of mitochondrially targeted chaperones and proteases. Although the UPR^{mt} signaling pathway is not well understood, recent studies link mitochondrial chaperones to PD pathogenesis.

The chaperones Mortalin and DJ-1 in the response to oxidative stress. An initial clue linking impaired UPR^{mt} to PD came from the finding that levels of Mortalin (HSPA9), a conserved mitochondrial chaperone of the Hsp70 protein family, are reduced in the SN of PD patients (De Mena et al. 2009, Jin et al. 2006). Mortalin is a multifunctional protein that localizes predominantly to mitochondria and interacts with the translocase of the inner membrane (TIM) complex (D'Silva et al. 2004, Kang et al. 1990). Mortalin may function with another interaction partner, Hsp60, to help refold matrix proteins following TIM-mediated import (Wadhwa et al. 2005). In addition to playing a role in protein import, Mortalin is implicated in the response to oxidative stress. Reducing Mortalin expression in mammalian cells leads to a collapse of the MMP and a spike in ROS production (Burbulla et al. 2010). Furthermore, in DA primary neurons, Jin et al. identified a physical interaction between Mortalin and DJ-1, which is enhanced upon rotenone treatment (**Figure 1**) (Jin et al. 2007). Although this finding could support a direct role for Mortalin in the oxidative stress response, the enhanced interaction may also be a consequence of increased mitochondrial DJ-1 levels induced by ROS.

Identification of rare variants in PD cohorts produced genetic evidence that potentially links *Mortalin* and PD (Burbulla et al. 2010, De Mena et al. 2009). Although these results require confirmation, it is notable that mitochondrial phenotypes, induced by knockdown of *Mortalin* expression in mammalian cells, were rescued by wild-type *Mortalin* but not by forms of *Mortalin* harboring PD-associated variants (Burbulla et al. 2010). All mitochondrial impairments could also be rescued by overexpressing *Parkin* (Yang et al. 2011). In addition, Pink1 interacts with Mortalin in vitro (Rakovic et al. 2011), suggesting that the Pink1/Parkin pathway may impinge on regulating this chaperone. Further evidence supporting involvement of the UPR^{mt} in PD etiology came from the finding that α-synuclein physically interacts with Mortalin and that *Mortalin* expression is reduced upon overexpression of wild-type and mutant α-synuclein (Jin et al. 2007). In conclusion, although the precise effects of *Mortalin* loss on neuronal function and maintenance remain unclear, changes in the expression or function of this protein will likely promote sensitivity to subsequent mitochondrial insults.

The chaperone Trap-1 functions with Pink1. Another mitochondrial chaperone that has been linked to PD is TNF-receptor associated protein 1 (Trap1), a member of the Hsp90 family that localizes predominantly to mitochondria (Felts et al. 2000). This chaperone was the first in vivo target identified for Pink1, and evidence proposed that Pink1's function in protecting the cell from ROS-induced cell death depends on Trap1 phosphorylation (**Figure 1**) (Pridgeon et al. 2007). Similar to a loss of *pink1* or *parkin*, *Drosophila trap1* null mutants display dysfunctional mitochondria as well as an age-dependent decline in motor performance (Costa et al. 2013). Overexpression of Trap1 rescues *pink1*, but not *parkin*-mutant phenotypes, suggesting that Trap1 and Parkin may function in parallel to protect the cell from dysfunctional mitochondria (Zhang et al. 2013b).

In addition to rescuing *pink1* mutant phenotypes in *Drosophila*, *Trap1* expression in rat primary cortical neurons and a human DA cell culture protected DA neurons from α-synuclein-induced toxicity (Butler et al. 2012). Although investigators initially interpreted these data as linking α-synuclein overexpression and mitochondrial dysfunction, further studies showed that Trap1 can also induce a more general ER-mediated UPR. Thus, the observed protective effect of Trap1 may, in fact, be mediated by an increased ability of the cell to cope with unfolded cytoplasmic proteins (Takemoto et al. 2011).

Mitochondrial Hsp60 may play a role in the Pink1/Parkin pathway. The third mitochondrial chaperone that may be associated with PD is mitochondrial Hsp60. This key player of the UPR^{mt} interacts in vitro with both Pink1 and Parkin (Davison et al. 2009, Rakovic et al. 2011). In addition, mitochondrial Hsp60 is downregulated by about 40% upon loss of Pink1 in a DA neuronal cell culture (Kim et al. 2012). Although this observation requires confirmation, Parkin and/or Pink1 may interact with Hsp60 to shut down the UPR^{mt} when a mitochondrion is deemed too damaged to save. Alternatively, if the UPR^{mt} is extended, Hsp60 might interact with both proteins to induce mitophagy.

Taken together, because PD has been associated with mitochondrial dysfunction and a concomitant increase in ROS, one might expect to find the UPR^{mt} to be activated in PD model systems. However, several components of the UPR^{mt} have instead been found to be either reduced or dysfunctional in such models, and consistent observations have been made in PD patients. Furthermore, experiments in mammalian cell cultures showed that the three chaperones (mitochondrial Hsp60, Mortalin, and Trap1) control the gating of the mitochondrial permeability transition pore (Ghosh et al. 2010, Qu et al. 2012). Hence, their loss may sensitize a cell for apoptotic death, owing to a reduced threshold for pore opening and release of cytochrome c. *Pink1*-mutant mice similarly show a lower threshold for opening of the mitochondrial transition pore and are indeed more susceptible to toxic insults (Morais et al. 2009). Whether this observation is due to a change in the activity or levels of any of the aforementioned chaperones remains to be determined.

Defects in Mitochondrial Dynamics

As mitochondria age or are exposed to environmental toxins, they accumulate numerous mitochondrial DNA mutations and protein insults that impair their function and may lead to cell death if repair does not occur. It is therefore important for a cell to maintain a healthy pool of mitochondria. Several protective measures exist to restore or eliminate dysfunctional mitochondria. These processes include mitochondrial fusion and fission, mitochondrial trafficking, and the clearance of dysfunctional mitochondria via mitophagy. These three processes play a critical role in mitochondrial quality control, especially in cells that consume much energy and do not divide, such as neuronal cells (Itoh et al. 2012, Schon & Przedborski 2011). Several genes linked to PD and related parkinsonian disorders have been implicated in the regulation of mitochondrial dynamics, suggesting that failure of these mechanisms may promote disease pathogenesis.

Tipping the balance toward demise: most PD loci alter fusion/fission. In response to mitochondrial dysfunction, fusion mixes key constituents, potentially complementing deficiencies from damaged proteins, DNA, and/or membrane lipids. Among the required cellular machinery, mitofusin (Mfn) and optic atrophy 1 (Opa 1) mediate fusion of the outer and inner mitochondrial membranes, respectively. Conversely, mitochondrial fission requires dynamin related protein 1 (Drp1). Loss of Mfn or Opa1 leads to small, fragmented mitochondria, whereas the loss of Drp1 causes a network of large interconnected mitochondria (Chan 2012, Youle & van der Bliek 2012). Hence, a constant balance of fission and fusion is required to maintain a healthy pool of mitochondria. Mutations in Mfn2 lead to Charcot-Marie-Tooth type 2A (Züchner et al. 2004), an inherited peripheral neuropathy, and loss of Opa1 leads to autosomal dominant optic atrophy (Eiberg et al.

1994). Loss of Drp1 causes a rare infantile mitochondrial encephalopathy, consisting of early neurologic failure and death (Waterham et al. 2007). Although these genes are not known to be genetically linked with PD themselves, they do show convincing interactions with established PD genes in model organisms.

In *Drosophila*, mutations in *pink1* or *parkin* lead to aberrant mitochondria that are swollen and lose their typical cristae structure (Clark et al. 2006, Greene et al. 2003, Park et al. 2006). Reducing mitochondrial fusion or increasing fission in either *pink1* or *parkin* mutant flies restores the morphological defects (Deng et al. 2008, Poole et al. 2008). However, although adding one copy of *drp1* partially restores energy levels in a *pink1* mutant, it fails to restore CI-activity (Liu et al. 2011, Vilain et al. 2012). Finally, whereas *pink1* or *parkin* mutant flies are viable, removing only a single copy of *drp1* in a *pink1* or *parkin* mutant is fatal (Poole et al. 2008). These genetic interactions strongly suggest that the functions of Parkin and Pink1 and mediators of mitochondrial fusion/fission may be linked.

Several groups have reported that round, fragmented mitochondria appear upon overexpression of either wild-type or mutant forms of α-synuclein, both in vitro as well as in vivo (Kamp et al. 2010, Nakamura et al. 2011). The ability of α-synuclein to induce membrane fragmentation is specific for mitochondrial membranes and appears to be independent of the fission protein Drp1 (Nakamura et al. 2011). Indeed, an age-dependent decrease in both Mfn1 and Mfn2 was observed in neurons of mice overexpressing mutant α-synuclein (Xie & Chung 2012), suggesting a possible pathological role for α-synuclein in reducing mitochondrial fusion.

Studies recently demonstrated that LRRK2 plays a role in regulating mitochondrial dynamics in *Drosophila* DA neurons. Overexpression of mutant, but not wild-type, human LRRK2 resulted in enlarged mitochondria resembling those seen in *pink1* or *parkin* mutants (Ng et al. 2012). These defects were significantly rescued by *parkin* coexpression, suggesting that these proteins may function in a common pathway (Ng et al. 2009). Skin biopsies from patients with *LRRK2* mutations displayed increased mitochondrial length and interconnectivity, decreased ATP levels, and MMP (Mortiboys et al. 2010). In mouse cortical primary neurons, LRRK2 also interacts with and phosphorylates the fission protein Drp1 (Niu et al. 2012). Overexpressing wild-type and mutant *LRRK2* in this system, as well as in primary DA neurons, results in recruitment of Drp1 to the mitochondrial membrane leading to mitochondrial fragmentation and clearance (Niu et al. 2012, Wang et al. 2012b). Similar to *LRRK2*, mutations in *DJ-1* result in increased Drp1 levels and increased mitochondrial fragmentation, although the underlying mechanisms remain elusive (Wang et al. 2012a). These studies suggest that LRRK2 and DJ-1 play important roles in regulating mitochondrial dynamics and quality control, although additional work is needed to understand the mechanisms (**Figure 1**).

In sum, altered mitochondrial fusion/fission is a feature observed in many experimental systems relevant to PD. However, because enhanced mitochondrial fusion and/or fission can be a response to injury, many of the observations described here could be secondary in nature.

Mitochondrial trafficking is controlled by Pink1 and Parkin. Mitochondrial trafficking delivers a mobile source of ATP to subcellular locations that have high energy demands, such as the neuronal synapse (Saxton & Hollenbeck 2012, Sheng & Cai 2012). Newly formed mitochondria travel along microtubules using kinesins and adaptor proteins that link mitochondria to the motor proteins (Saxton & Hollenbeck 2012, Sheng & Cai 2012). Specifically, in flies, Pink1 phosphorylates Miro, a rho-like GTPase that resides on the MOM, where it functions as an adaptor to microtubule motor proteins (**Figure 1**) (Liu et al. 2012, Wang et al. 2011). Phosphorylation of Miro triggers its degradation via a Parkin-dependent ubiquitination pathway and causes mitochondria to detach from cognate kinesin motors. Because Pink1 is activated upon mitochondrial dysfunction (see next section), this action may serve as a mechanism to block dysfunctional mitochondria from moving along the axon.

Miro and the adaptor protein Milton, which links mitochondria to the heavy chain of kinesin, form a complex with the fusion protein Mfn, which is also targeted for degradation by Parkin in both flies and mammals (Glater et al. 2006, Glauser et al. 2011, Górska-Andrzejak et al. 2003, Ziviani et al. 2010). In addition, genetically altering the mitochondrial fission/fusion balance in a wild-type background impedes mitochondrial trafficking, leading to synapses and dendrites that are devoid of mitochondria (Liu et al. 2012, Sheng & Cai 2012). Together, these findings suggest a tight link between regulators of mitochondrial dynamics and intracellular trafficking. Pink1 and Parkin affect both processes, possibly to ensure the presence of sufficient functional mitochondria at subcellular locations with energy-intensive demands. Because LRRK2 and DJ-1 each regulate Drp1 levels, they may indirectly regulate mitochondrial trafficking as well (Niu et al. 2012, Wang et al. 2012a).

Impaired clearance of dysfunctional mitochondria in PD. When mitochondria can no longer be repaired, they are cleared through a form of autophagy known as mitophagy (Narendra et al. 2010). For degradation, mitochondria initially undergo fragmentation, allowing autophagosomes to engulf them. The protein machinery that mediates fission and fusion is essential for this process, and genetic manipulations that inhibit fission or enhance fusion interfere with mitophagy (Gomes et al. 2011).

Under normal conditions, the ubiquitously expressed Pink1 spans both the MOM and the MIM (Jin et al. 2010, Weihofen et al. 2009). Within the MIM, it is cleaved by two proteins [mitochondrial processing peptidase (Greene et al. 2012) and presenilin-associated rhomboid-like protease (Jin et al. 2010)], after which it is thought to be degraded (Matsuda et al. 2013). Under basal conditions, Pink1 protein levels are therefore kept low. Once a mitochondrion becomes dysfunctional and loses its MMP, the import of Pink1 into the MIM is blocked, and it is no longer cleaved. Pink1 therefore quickly accumulates on the MOM, where it initiates mitophagy (Jin & Youle 2012, Youle & Narendra 2011). It does so by recruiting and activating the E3 ubiquitin ligase Parkin, a process that requires Pink1's kinase activity (Kim et al. 2008). However, whether Pink1 acts directly on Parkin or whether it indirectly recruits Parkin to the MOM is still under debate (Iguchi et al. 2013, Kim et al. 2008, Narendra et al. 2010). In our opinion, the data are most consistent with a model where Pink1 phosphorylates several proteins, one of which would subsequently recruit Parkin. One potential candidate would be the Pink1-target MFN2, which is required to recruit Parkin to damaged mitochondria (Chen & Dorn 2013).

Upon activation, Parkin functions with E2 ubiquitin-conjugating enzymes, such as the recently identified Rad6, to ubiquitinate its targets (Haddad et al. 2013). Over the past few years, investigators have identified many substrates for parkin-dependent ubiquitination, including numerous mitochondrial proteins (**Figure 1**) (Sarraf et al. 2013). Overall, activation of the Pink/Parkin pathway halts the trafficking of a dysfunctional mitochondrion, initiates its fragmentation, and leads to its subsequent degradation. In addition, the Pink1/Parkin pathway appears to support mitochondrial biogenesis. Shin et al. recently found that Parkin-activation leads to the degradation of Paris, which represses the transcription of a key regulator of mitochondrial biogenesis, PGC1-α (Shin et al. 2011).

Because most of the mitophagy-related research has focused on Pink1 and Parkin, the question arises whether impaired mitophagy is a general theme in the pathogenesis of PD. A recent study in mice found that overexpressing wild-type or mutant α-synuclein induces mitophagy (Sampaio-Marques et al. 2012). Similarly, overexpressing mutant LRRK2 in mouse cortical neurons results in autophagic degradation of mitochondria (Cherra et al. 2013). In conclusion, multiple aspects of mitochondrial dynamics appear to be affected to various extents in mutants for PD-related loci (**Table 1**).

MITOCHONDRIA AND PD: AN INTEGRATED MODEL

As summarized in this review, numerous genes associated with familial forms of PD and related disorders with prominent parkinsonism can be linked to mitochondrial mechanisms. In addition, mitochondrial toxins cause SN degeneration in both humans and various other model systems. These data support models in which mitochondria play a central role in DA neuronal loss in PD. It is remarkable that even in cases of AR-JP caused by mutations in *parkin* or *pink1*, where mitochondrial dysfunction seems to be the primary insult, the overall effect on mitochondrial function appears subtle. This conclusion is based in part on studies of model organisms, where loss of homologous genes causes a mild reduction of CI activity and does not significantly impact viability (Morais et al. 2009, Vilain et al. 2012). A similar partial CI deficiency has been documented in postmortem SN tissue from PD patients (Bindoff et al. 1991; Parker et al. 1989, 2008; Schapira et al. 1989). In humans, mutations that more severely affect ETC function are associated with correspondingly more aggressive disease. For instance, Leigh syndrome, a mitochondrial encephalopathy, has an infantile onset, affects multiple organ systems, and is often fatal (Vafai & Mootha 2012). Consistent with these observations, mutations in Leigh syndrome–associated gene homologs in mice or flies are lethal [e.g., Sco2 (Porcelli et al. 2010, Yang et al. 2010), LRPPRC (Ruzzenente et al. 2012), C8ORF38 (Zhang et al. 2013a), SURF1 (Agostino et al. 2003, Zordan et al. 2006), among others].

Although the overall extent of mitochondrial dysfunction may be modest, one important lesson from studies of Mendelian forms of PD and related parkinsonian disorders is that the responsible genetic lesions appear to have distributed effects on several core features of mitochondrial biology, including (*a*) the ETC, (*b*) the UPR^{mt}, and (*c*) mitochondrial dynamics. As shown in **Figure 2**, we propose that simultaneous disruptions in these core systems interact and overwhelm the capacity of the mitochondria to compensate for additional insults, promoting a vicious cycle that may ultimately result in cell death. However, given the near-universal requirement of mitochondria in all eukaryotic cells, this model does not explain the selective vulnerability of certain neuronal subtypes in PD, such as DA neurons in the SN. Therefore, apart from affecting these core mitochondrial processes (a, b, and c), an additional insult ("second hit") is likely required, tipping the balance toward severe dysfunction in distinct cellular contexts. In the case of DA neurons, elevated endogenous ROS levels may fulfill the second-hit requirement (see sidebar, Why Are DA Neurons More Susceptible To Stress Than Are Other Neurons?). For example, in an individual with juvenile

---

Figure 2

The multiple hit model of Parkinson's disease (PD). *Top panel*: In familial parkinsonism, for example due to loss of function of *Parkin*, several key aspects of mitochondrial homeostasis are mildly affected, including energy production, protein folding (UPR^{mt}), or dynamics (fission and fusion). These defects interact and amplify one another, but a second hit, such as elevated endogenous reactive oxygen species (ROS) or cellular activity, is likely required to trigger neuronal dysfunction and loss in certain vulnerable cell types, such as in DA neurons. *Bottom panel*: In idiopathic PD, multiple hits, including both common and rare genomic variations at diverse susceptibility loci, may in combination cause similar, subtle, and distributed defects in mitochondrial function. α-Synuclein pathology, ROS, potential environmental factors, and the widespread cellular effects of aging further degrade mitochondrial activity. In a potential feedback mechanism, mitochondrial dysfunction may also promote α-synuclein aggregation, which in turn may further amplify mitochondrial defects, ultimately leading to neuronal dysfunction and cell death.

parkinsonism due to *parkin* mutations (hit 1), endogenous ROS levels (hit 2) may be augmented to toxic levels owing to baseline deficiencies in the ETC. This action may cause oxidized, misfolded, and dysfunctional proteins to accumulate, which then overwhelms the already-strained UPR^{mt}. As mitochondrial function further deteriorates, clearance mechanisms fail to respond properly, owing to impaired mitochondrial dynamics. Ultimately, injured mitochondria accumulate within neurons, causing progressive functional derangements and eventually cell death.

Compared with the genetic causes of AR-JP, our understanding of the potential mitochondrial impact of dominant causes of PD, such as that due to *SNCA* and *LRRK2* mutations, is less well

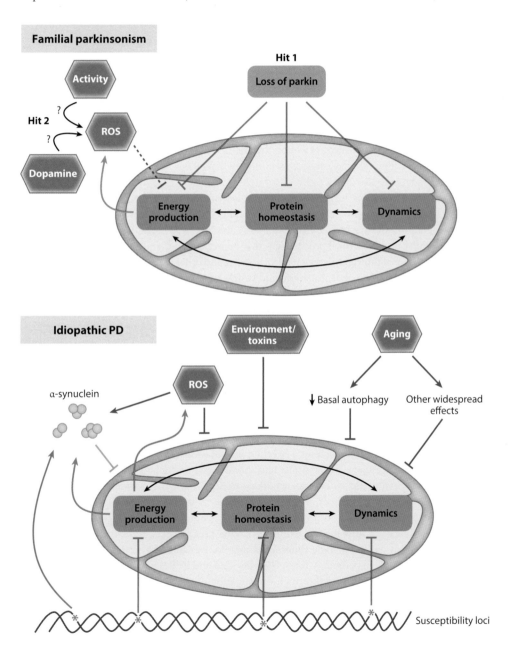

WHY ARE DA NEURONS MORE SUSCEPTIBLE TO STRESS THAN ARE OTHER NEURONS?

What makes DA neurons more vulnerable to oxidative stress than other neurons has been a conundrum for PD. One view is that oxidation of cytosolic DA and its metabolites leads to the generation of cytotoxic free radicals. Oxidized DA forms covalent bonds with several proteins, including mitochondrial Trap1, a CI subunit, DJ-1, and even α-synuclein, rendering the latter more prone to aggregation (Van Laar et al. 2009). Potentially inconsistent with this model, however, levodopa (a form of DA) is among the safest and most effective treatments for PD, and large clinical trials do not support enhanced disease progression (Fahn et al. 2004). Alternatively, a cell's activity pattern may render it more susceptible to subsequent toxic insults. Indeed, DA neurons contain pacemaker properties (Guzman et al. 2010). This continuous activity requires high amounts of ATP and therefore imparts a demand on the cell to function properly. Pacemaking activity is found in DA neurons of the SN but not in those of the adjacent ventral tegmental area, which also use DA but are relatively spared in PD. Thus, both ROS production via DA oxidation and pacemaker firing properties may converge to explain the susceptibility of SN DA neurons in PD.

developed. However, accumulating evidence (detailed above) indicates that α-synuclein aggregates and/or LRRK2 dysfunction may cause similar, distributed effects on mitochondria, in which case the two-hit model may also apply.

A key remaining question is whether in the absence of known Mendelian mutations, as is true in the vast majority of idiopathic PD cases, primary mitochondrial failure is also a primary cause of neurodegeneration. Alternatively, mitochondrial dysfunction may secondarily follow from other upstream, cellular insults. A third possibility is that neuronal injury and death result from a more complex, bidirectional interplay between mitochondria and other cellular processes, including oxidative metabolism, lysosomal degradation, and vesicle trafficking. Nevertheless, our mitocentric view favors a central, causal role in all PD cases, and we further suggest that the multi-hit model provides an excellent, generalizable framework (**Figure 2**). Idiopathic PD is currently best understood as a complex genetic disorder that is likely due to the interaction between numerous genetic susceptibility factors and that is modified by the environment and aging. These varied risk factors may therefore constitute multiple hits, and we speculate that as the genetic and nongenetic factors are better defined, many will be found to impinge on the core mitochondrial impairments (ETC, UPR^{mt}, and dynamics) that we have come to understand from the investigation of familial PD.

Thus, we speculate that, collectively, many inherited or de novo PD risk variants are analogous to the single genetic lesion owing to loss of *parkin* or *pink1* in AR-JP, resulting in broad effects on mitochondrial function. Such PD risk variants are predicted to cumulatively degrade mitochondrial reserve mechanisms, establishing a baseline increased susceptibility profile. Although it is necessary, this context alone is likely insufficient. Additional hits, including environmental factors such as mitochondrial toxins and aging, are likely needed to set the disease process in motion. α-Synuclein is likely a key player, given the evidence of a potential feedback loop between mitochondrial toxicity and enhanced aggregation (see above and **Figure 2**). Furthermore, because α-synuclein pathology is now recognized to have widespread effects in many cell types beyond the SN, key clinical and pathologic differences may be explained between idiopathic PD and AR-JP, where α-synuclein pathology is not typically seen and neurodegeneration does not appear to extend outside the SN. A more thorough investigation of mitochondrial mechanisms may not only help us identify additional risk factors, but ultimately help explain how such factors interact to trigger the onset and progression of disease, providing important clues to new and successful therapeutic strategies for the future.

DISCLOSURE STATEMENT

The authors are not aware of any affiliations, memberships, funding, or financial holdings that might be perceived as affecting the objectivity of this review.

LITERATURE CITED

Agostino A, Invernizzi F, Tiveron C, Fagiolari G, Prelle A, et al. 2003. Constitutive knockout of *Surf1* is associated with high embryonic lethality, mitochondrial disease and cytochrome *c* oxidase deficiency in mice. *Hum. Mol. Genet.* 12:399–413

Amo T, Sato S, Saiki S, Wolf AM, Toyomizu M, et al. 2011. Mitochondrial membrane potential decrease caused by loss of PINK1 is not due to proton leak, but to respiratory chain defects. *Neurobiol. Dis.* 41:111–18

Bindoff LA, Birch-Machin MA, Cartlidge NE, Parker WD Jr, Turnbull DM. 1991. Respiratory chain abnormalities in skeletal muscle from patients with Parkinson's disease. *J. Neurol. Sci.* 104:203–8

Bonifati V, Rizzu P, van Baren MJ, Schaap O, Breedveld GJ, et al. 2003. Mutations in the *DJ-1* gene associated with autosomal recessive early-onset parkinsonism. *Science* 299:256–59

Bonifati V, Rohé CF, Breedveld GJ, Fabrizio E, De Mari M, et al. 2005. Early-onset parkinsonism associated with *PINK1* mutations: frequency, genotypes, and phenotypes. *Neurology* 65:87–95

Braschi E, Goyon V, Zunino R, Mohanty A, Xu L, McBride HM. 2010. Vps35 mediates vesicle transport between the mitochondria and peroxisomes. *Curr. Biol.* 20:1310–15

Burbulla L, Schelling C, Kato H, Rapaport D, Woitalla D, et al. 2010. Dissecting the role of the mitochondrial chaperone mortalin in Parkinson's disease: functional impact of disease-related variants on mitochondrial homeostasis. *Hum. Mol. Genet.* 19:4437–52

Burchell VS, Nelson DE, Sanchez-Martinez A, Delgado-Camprubi M, Ivatt RM, et al. 2013. The Parkinson's disease-linked proteins Fbxo7 and Parkin interact to mediate mitophagy. *Nat. Neurosci.* 16:1257–65

Butler EK, Voigt A, Lutz AK, Toegel JP, Gerhardt E, et al. 2012. The mitochondrial chaperone protein TRAP1 mitigates α-synuclein toxicity. *PLoS Genet.* 8:e1002488

Chan DC. 2012. Fusion and fission: interlinked processes critical for mitochondrial health. *Annu. Rev. Genet.* 46:265–87

Chaudhuri KR, Schapira AH. 2009. Non-motor symptoms of Parkinson's disease: dopaminergic pathophysiology and treatment. *Lancet Neurol.* 8:464–74

Chen Y, Dorn GW 2nd. 2013. PINK1-phosphorylated mitofusin 2 is a Parkin receptor for culling damaged mitochondria. *Science* 340:471–75

Cherra SJ III, Steer E, Gusdon AM, Kiselyov K, Chu CT. 2013. Mutant LRRK2 elicits calcium imbalance and depletion of dendritic mitochondria in neurons. *Am. J. Pathol.* 182:474–84

Clark IE, Dodson MW, Jiang C, Cao JH, Huh JR, et al. 2006. *Drosophila pink1* is required for mitochondrial function and interacts genetically with *parkin. Nature* 441:1162–66

Cole NB, Dieuliis D, Leo P, Mitchell DC, Nussbaum RL. 2008. Mitochondrial translocation of α-synuclein is promoted by intracellular acidification. *Exp. Cell Res.* 314:2076–89

Costa AC, Loh SH, Martins LM. 2013. *Drosophila Trap1* protects against mitochondrial dysfunction in a PINK1/parkin model of Parkinson's disease. *Cell Death Dis.* 4:e467

Cui M, Tang X, Christian WV, Yoon Y, Tieu K. 2010. Perturbations in mitochondrial dynamics induced by human mutant PINK1 can be rescued by the mitochondrial division inhibitor mdivi-1. *J. Biol. Chem.* 285:11740–52

D'Silva P, Liu Q, Walter W, Craig E. 2004. Regulated interactions of mtHsp70 with Tim44 at the translocon in the mitochondrial inner membrane. *Nat. Struct. Mol. Biol.* 11:1084–91

Dagda RK, Cherra SJ 3rd, Kulich SM, Tandon A, Park D, Chu CT. 2009. Loss of PINK1 function promotes mitophagy through effects on oxidative stress and mitochondrial fission. *J. Biol. Chem.* 284:13843–55

Davison E, Pennington K, Hung C-C, Peng J, Rafiq R, et al. 2009. Proteomic analysis of increased Parkin expression and its interactants provides evidence for a role in modulation of mitochondrial function. *Proteomics* 9:4284–97

De Mena L, Coto E, Sánchez-Ferrero E, Ribacoba R, Guisasola LM, et al. 2009. Mutational screening of the mortalin gene (HSPA9) in Parkinson's disease. *J. Neural Transm.* 116:1289–93

Deng H, Dodson MW, Huang H, Guo M. 2008. The Parkinson's disease genes *pink1* and *parkin* promote mitochondrial fission and/or inhibit fusion in *Drosophila*. *Proc. Natl. Acad. Sci. USA* 105:14503–8

Devi L, Raghavendran V, Prabhu BM, Avadhani NG, Anandatheerthavarada HK. 2008. Mitochondrial import and accumulation of alpha-synuclein impair complex I in human dopaminergic neuronal cultures and Parkinson disease brain. *J. Biol. Chem.* 283:9089–100

Doherty KM, Silveira-Moriyama L, Parkkinen L, Healy DG, Farrell M, et al. 2013. Parkin disease: a clinico-pathologic entity? *JAMA Neurol.* 70:571–79

Edgar AJ, Polak JM. 2000. Human homologues of yeast vacuolar protein sorting 29 and 35. *Biochem. Biophys. Res. Commun.* 277:622–30

Eiberg H, Kjer B, Kjer P, Rosenberg T. 1994. Dominant optic atrophy (OPA1) mapped to chromosome 3q region. I. Linkage analysis. *Hum. Mol. Genet.* 3:977–80

Exner N, Treske B, Paquet D, Holmström K, Schiesling C, et al. 2007. Loss-of-function of human PINK1 results in mitochondrial pathology and can be rescued by parkin. *J. Neurosci.* 27:12413–18

Fahn S. 2003. Description of Parkinson's disease as a clinical syndrome. *Ann. N. Y. Acad. Sci.* 991:1–14

Fahn S, Oakes D, Shoulson I, Kieburtz K, Rudolph A, et al. 2004. Levodopa and the progression of Parkinson's disease. *N. Engl. J. Med.* 351:2498–508

Fariello RG. 1988. Experimental support for the implication of oxidative stress in the genesis of parkinsonian syndromes. *Funct. Neurol.* 3:407–12

Felts SJ, Owen BA, Nguyen P, Trepel J, Donner DB, Toft DO. 2000. The hsp90-related protein TRAP1 is a mitochondrial protein with distinct functional properties. *J. Biol. Chem.* 275:3305–12

Gautier CA, Kitada T, Shen J. 2008. Loss of PINK1 causes mitochondrial functional defects and increased sensitivity to oxidative stress. *Proc. Natl. Acad. Sci. USA* 105:11364–69

Gegg ME, Cooper JM, Schapira AH, Taanman JW. 2009. Silencing of PINK1 expression affects mitochondrial DNA and oxidative phosphorylation in dopaminergic cells. *PLoS ONE* 4:e4756

Ghosh JC, Siegelin MD, Dohi T, Altieri DC. 2010. Heat shock protein 60 regulation of the mitochondrial permeability transition pore in tumor cells. *Cancer Res.* 70:8988–93

Gispert S, Ricciardi F, Kurz A, Azizov M, Hoepken HH, et al. 2009. Parkinson phenotype in aged PINK1-deficient mice is accompanied by progressive mitochondrial dysfunction in absence of neurodegeneration. *PLoS ONE* 4:e5777

Glater EE, Megeath LJ, Stowers RS, Schwarz TL. 2006. Axonal transport of mitochondria requires milton to recruit kinesin heavy chain and is light chain independent. *J. Cell Biol.* 173:545–57

Glauser L, Sonnay S, Stafa K, Moore DJ. 2011. Parkin promotes the ubiquitination and degradation of the mitochondrial fusion factor mitofusin 1. *J. Neurochem.* 118:636–45

Goedert M, Spillantini MG, Del Tredici K, Braak H. 2013. 100 years of Lewy pathology. *Nat. Rev. Neurol.* 9:13–24

Gomes LC, Di Benedetto G, Scorrano L. 2011. During autophagy mitochondria elongate, are spared from degradation and sustain cell viability. *Nat. Cell Biol.* 13:589–98

Górska-Andrzejak J, Stowers RS, Borycz J, Kostyleva R, Schwarz TL, Meinertzhagen IA. 2003. Mitochondria are redistributed in *Drosophila* photoreceptors lacking milton, a kinesin-associated protein. *J. Comp. Neurol.* 463:372–88

Greene AW, Grenier K, Aguileta MA, Muise S, Farazifard R, et al. 2012. Mitochondrial processing peptidase regulates PINK1 processing, import and Parkin recruitment. *EMBO Rep.* 13:378–85

Greene JC, Whitworth AJ, Kuo I, Andrews LA, Feany MB, Pallanck LJ. 2003. Mitochondrial pathology and apoptotic muscle degeneration in *Drosophila parkin* mutants. *Proc. Natl. Acad. Sci. USA* 100:4078–83

Grünewald A, Arns B, Seibler P, Rakovic A, Münchau A, et al. 2012. ATP13A2 mutations impair mitochondrial function in fibroblasts from patients with Kufor-Rakeb syndrome. *Neurobiol. Aging* 33:1843.e1–7

Guzman JN, Sanchez-Padilla J, Wokosin D, Kondapalli J, Ilijic E, et al. 2010. Oxidant stress evoked by pacemaking in dopaminergic neurons is attenuated by DJ-1. *Nature* 468:696–700

Haddad DM, Vilain S, Vos M, Esposito G, Matta S, et al. 2013. Mutations in the intellectual disability gene *Ube2a* cause neuronal dysfunction and impair Parkin-dependent mitophagy. *Mol. Cell* 50:831–43

Hao L-Y, Giasson BI, Bonini NM. 2010. DJ-1 is critical for mitochondrial function and rescues PINK1 loss of function. *Proc. Natl. Acad. Sci. USA* 107:9747–52

Haynes C, Ron D. 2010. The mitochondrial UPR – protecting organelle protein homeostasis. *J. Cell Sci.* 123:3849–55

Healy DG, Falchi M, O'Sullivan SS, Bonifati V, Durr A, et al. 2008. Phenotype, genotype, and worldwide genetic penetrance of *LRRK2*-associated Parkinson's disease: a case-control study. *Lancet Neurol.* 7:583–90

Heeman B, Van den Haute C, Aelvoet SA, Valsecchi F, Rodenburg RJ, et al. 2011. Depletion of PINK1 affects mitochondrial metabolism, calcium homeostasis and energy maintenance. *J. Cell Sci.* 124:1115–25

Heo JY, Park JH, Kim SJ, Seo KS, Han JS, et al. 2012. DJ-1 null dopaminergic neuronal cells exhibit defects in mitochondrial function and structure: involvement of mitochondrial complex I assembly. *PLoS ONE* 7:e32629

Hoepken HH, Gispert S, Morales B, Wingerter O, Del Turco D, et al. 2007. Mitochondrial dysfunction, peroxidation damage and changes in glutathione metabolism in PARK6. *Neurobiol. Dis.* 25:401–11

Iguchi M, Kujuro Y, Okatsu K, Koyano F, Kosako H, et al. 2013. Parkin-catalyzed ubiquitin-ester transfer is triggered by PINK1-dependent phosphorylation. *J. Biol. Chem.* 288:22019–32

Ilyin GP, Rialland M, Pigeon C, Guguen-Guillouzo C. 2000. cDNA cloning and expression analysis of new members of the mammalian F-box protein family. *Genomics* 67:40–47

Int. Parkinson's Dis. Genomics Consort. (IPDGC), Wellcome Trust Case Control Consort. 2 (WTCCC2). 2011. A two-stage meta-analysis identifies several new loci for Parkinson's disease. *PLoS Genet.* 7:e1002142

Irrcher I, Aleyasin H, Seifert EL, Hewitt SJ, Chhabra S, et al. 2010. Loss of the Parkinson's disease-linked gene DJ-1 perturbs mitochondrial dynamics. *Hum. Mol. Genet.* 19:3734–46

Itoh K, Nakamura K, Iijima M, Sesaki H. 2012. Mitochondrial dynamics in neurodegeneration. *Trends Cell Biol.* 23:64–71

Jaiswal M, Sandoval H, Zhang K, Bayat V, Bellen HJ. 2012. Probing mechanisms that underlie human neurodegenerative diseases in *Drosophila. Annu. Rev. Genet.* 46:371–96

Jellinger KA. 2009. A critical evaluation of current staging of α-synuclein pathology in Lewy body disorders. *Biochim. Biophys. Acta* 1792:730–40

Jin J, Hulette C, Wang Y, Zhang T, Pan C, et al. 2006. Proteomic identification of a stress protein, mortalin/mthsp70/GRP75: relevance to Parkinson disease. *Mol. Cell. Proteomics* 5:1193–204

Jin J, Li G, Davis J, Zhu D, Wang Y, et al. 2007. Identification of novel proteins associated with both α-synuclein and DJ-1. *Mol. Cell. Proteomics* 6:845–59

Jin SM, Lazarou M, Wang C, Kane LA, Narendra DP, Youle RJ. 2010. Mitochondrial membrane potential regulates PINK1 import and proteolytic destabilization by PARL. *J. Cell Biol.* 191:933–42

Jin SM, Youle RJ. 2012. PINK1- and Parkin-mediated mitophagy at a glance. *J. Cell Sci.* 125:795–99

Junn E, Mouradian MM. 2002. Human α-synuclein over-expression increases intracellular reactive oxygen species levels and susceptibility to dopamine. *Neurosci. Lett.* 320:146–50

Kamp F, Exner N, Lutz AK, Wender N, Hegermann J, et al. 2010. Inhibition of mitochondrial fusion by α-synuclein is rescued by PINK1, Parkin and DJ-1. *EMBO J.* 29:3571–89

Kang P, Ostermann J, Shilling J, Neupert W, Craig E, Pfanner N. 1990. Requirement for hsp70 in the mitochondrial matrix for translocation and folding of precursor proteins. *Nature* 348:137–43

Keeney PM, Xie J, Capaldi RA, Bennett JP Jr. 2006. Parkinson's disease brain mitochondrial complex I has oxidatively damaged subunits and is functionally impaired and misassembled. *J. Neurosci.* 26:5256–64

Kett LR, Dauer WT. 2012. Leucine-rich repeat kinase 2 for beginners: six key questions. *Cold Spring Harb. Perspect. Med.* 2:a009407

Kim K-H, Song K, Yoon S-H, Shehzad O, Kim Y-S, Son J. 2012. Rescue of PINK1 protein null-specific mitochondrial complex IV deficits by ginsenoside Re activation of nitric oxide signaling. *J. Biol. Chem.* 287:44109–20

Kim Y, Park J, Kim S, Song S, Kwon SK, et al. 2008. PINK1 controls mitochondrial localization of Parkin through direct phosphorylation. *Biochem. Biophys. Res. Commun.* 377:975–80

Kitada T, Asakawa S, Hattori N, Matsumine H, Yamamura Y, et al. 1998. Mutations in the *parkin* gene cause autosomal recessive juvenile parkinsonism. *Nature* 392:605–8

Klein C, Lohmann-Hedrich K. 2007. Impact of recent genetic findings in Parkinson's disease. *Curr. Opin. Neurol.* 20:453–64

Krebiehl G, Ruckerbauer S, Burbulla LF, Kieper N, Maurer B, et al. 2010. Reduced basal autophagy and impaired mitochondrial dynamics due to loss of Parkinson's disease-associated protein DJ-1. *PLoS ONE* 5:e9367

Kwon HJ, Heo JY, Shim JH, Park JH, Seo KS, et al. 2011. DJ-1 mediates paraquat-induced dopaminergic neuronal cell death. *Toxicol. Lett.* 202:85–92

Langston JW, Ballard P, Tetrud JW, Irwin I. 1983. Chronic Parkinsonism in humans due to a product of meperidine-analog synthesis. *Science* 219:979–80

Lees AJ, Hardy J, Revesz T. 2009. Parkinson's disease. *Lancet* 373:2055–66

Liu S, Sawada T, Lee S, Yu W, Silverio G, et al. 2012. Parkinson's disease-associated kinase PINK1 regulates Miro protein level and axonal transport of mitochondria. *PLoS Genet.* 8:e1002537

Liu W, Acin-Peréz R, Geghman KD, Manfredi G, Lu B, Li C. 2011. Pink1 regulates the oxidative phosphorylation machinery via mitochondrial fission. *Proc. Natl. Acad. Sci. USA* 108:12920–24

Lowe J, Blanchard A, Morrell K, Lennox G, Reynolds L, et al. 1988. Ubiquitin is a common factor in intermediate filament inclusion bodies of diverse type in man, including those of Parkinson's disease, Pick's disease, and Alzheimer's disease, as well as Rosenthal fibres in cerebellar astrocytomas, cytoplasmic bodies in muscle, and Mallory bodies in alcoholic liver disease. *J. Pathol.* 155:9–15

Lücking CB, Dürr A, Bonifati V, Vaughan J, De Michele G, et al. 2000. Association between early-onset Parkinson's disease and mutations in the *parkin* gene. *N. Engl. J. Med.* 342:1560–67

Lutz AK, Exner N, Fett ME, Schlehe JS, Kloos K, et al. 2009. Loss of parkin or PINK1 function increases Drp1-dependent mitochondrial fragmentation. *J. Biol. Chem.* 284:22938–51

Maraganore DM, de Andrade M, Elbaz A, Farrer MJ, Ioannidis JP, et al. 2006. Collaborative analysis of α-synuclein gene promoter variability and Parkinson disease. *JAMA* 296:661–70

Martin LJ, Pan Y, Price AC, Sterling W, Copeland NG, et al. 2006. Parkinson's disease alpha-synuclein transgenic mice develop neuronal mitochondrial degeneration and cell death. *J. Neurosci.* 26:41–50

Martinez TN, Greenamyre JT. 2012. Toxin models of mitochondrial dysfunction in Parkinson's disease. *Antioxid. Redox Signal.* 16:920–34

Matsuda S, Kitagishi Y, Kobayashi M. 2013. Function and characteristics of PINK1 in mitochondria. *Oxid. Med. Cell. Longev.* 2013:601587

Mizuno Y, Ohta S, Tanaka M, Takamiya S, Suzuki K, et al. 1989. Deficiencies in complex I subunits of the respiratory chain in Parkinson's disease. *Biochem. Biophys. Res. Commun.* 163:1450–55

Moore DJ, Dawson VL, Dawson TM. 2006. Lessons from *Drosophila* models of DJ-1 deficiency. *Sci. Aging Knowl. Environ.* 2006:pe2

Morais VA, Verstreken P, Roethig A, Smet J, Snellinx A, et al. 2009. Parkinson's disease mutations in PINK1 result in decreased Complex I activity and deficient synaptic function. *EMBO Mol. Med.* 1:99–111

Mortiboys H, Johansen KK, Aasly JO, Bandmann O. 2010. Mitochondrial impairment in patients with Parkinson disease with the G2019S mutation in LRRK2. *Neurology* 75:2017–20

Müftüoglu M, Elibol B, Dalmizrak O, Ercan A, Kulaksiz G, et al. 2004. Mitochondrial complex I and IV activities in leukocytes from patients with parkin mutations. *Mov. Disord.* 19:544–48

Nagakubo D, Taira T, Kitaura H, Ikeda M, Tamai K, et al. 1997. DJ-1, a novel oncogene which transforms mouse NIH3T3 cells in cooperation with *ras*. *Biochem. Biophys. Res. Commun.* 231:509–13

Nakamura K. 2013. α-Synuclein and mitochondria: partners in crime? *Neurotherapeutics* 10:391–99

Nakamura K, Nemani VM, Azarbal F, Skibinski G, Levy JM, et al. 2011. Direct membrane association drives mitochondrial fission by the Parkinson disease-associated protein α-synuclein. *J. Biol. Chem.* 286:20710–26

Narendra DP, Jin SM, Tanaka A, Suen DF, Gautier CA, et al. 2010. PINK1 is selectively stabilized on impaired mitochondria to activate Parkin. *PLoS Biol.* 8:e1000298

Ng CH, Guan MS, Koh C, Ouyang X, Yu F, et al. 2012. AMP kinase activation mitigates dopaminergic dysfunction and mitochondrial abnormalities in *Drosophila* models of Parkinson's disease. *J. Neurosci.* 32:14311–17

Ng CH, Mok SZS, Koh C, Ouyang X, Fivaz ML, et al. 2009. Parkin protects against LRRK2 G2019S mutant-induced dopaminergic neurodegeneration in *Drosophila*. *J. Neurosci.* 29:11257–62

Nicklas WJ, Vyas I, Heikkila RE. 1985. Inhibition of NADH-linked oxidation in brain mitochondria by 1-methyl-4-phenyl-pyridine, a metabolite of the neurotoxin, 1-methyl-4-phenyl-1,2,5,6-tetrahydropyridine. *Life Sci.* 36:2503–8

Niu J, Yu M, Wang C, Xu Z. 2012. Leucine-rich repeat kinase 2 disturbs mitochondrial dynamics via Dynamin-like protein. *J. Neurochem.* 122:650–58

Olney JW, Zorumski CF, Stewart GR, Price MT, Wang GJ, Labruyere J. 1990. Excitotoxicity of L-dopa and 6-OH-dopa: implications for Parkinson's and Huntington's diseases. *Exp. Neurol.* 108:269–72

Paisán-Ruíz C, Jain S, Evans EW, Gilks WP, Simón J, et al. 2004. Cloning of the gene containing mutations that cause PARK8-linked Parkinson's disease. *Neuron* 44:595–600

Palacino JJ, Sagi D, Goldberg MS, Krauss S, Motz C, et al. 2004. Mitochondrial dysfunction and oxidative damage in *parkin*-deficient mice. *J. Biol. Chem.* 279:18614–22

Papkovskaia TD, Chau KY, Inesta-Vaquera F, Papkovsky DB, Healy DG, et al. 2012. G2019S leucine-rich repeat kinase 2 causes uncoupling protein-mediated mitochondrial depolarization. *Hum. Mol. Genet.* 21:4201–13

Parihar MS, Parihar A, Fujita M, Hashimoto M, Ghafourifar P. 2009. Alpha-synuclein overexpression and aggregation exacerbates impairment of mitochondrial functions by augmenting oxidative stress in human neuroblastoma cells. *Int. J. Biochem. Cell Biol.* 41:2015–24

Park J, Lee SB, Lee S, Kim Y, Song S, et al. 2006. Mitochondrial dysfunction in *Drosophila PINK1* mutants is complemented by *parkin*. *Nature* 441:1157–61

Parker WD Jr, Boyson SJ, Parks JK. 1989. Abnormalities of the electron transport chain in idiopathic Parkinson's disease. *Ann. Neurol.* 26:719–23

Parker WD Jr, Parks JK, Swerdlow RH. 2008. Complex I deficiency in Parkinson's disease frontal cortex. *Brain Res.* 1189:215–18

Periquet M, Latouche M, Lohmann E, Rawal N, De Michele G, et al. 2003. *Parkin* mutations are frequent in patients with isolated early-onset parkinsonism. *Brain* 126:1271–78

Pesah Y, Pham T, Burgess H, Middlebrooks B, Verstreken P, et al. 2004. *Drosophila parkin* mutants have decreased mass and cell size and increased sensitivity to oxygen radical stress. *Development* 131:2183–94

Piccoli C, Sardanelli A, Scrima R, Ripoli M, Quarato G, et al. 2008. Mitochondrial respiratory dysfunction in familiar parkinsonism associated with PINK1 mutation. *Neurochem. Res.* 33:2565–74

Polymeropoulos MH, Lavedan C, Leroy E, Ide SE, Dehejia A, et al. 1997. Mutation in the α-synuclein gene identified in families with Parkinson's disease. *Science* 276:2045–47

Poole AC, Thomas RE, Andrews LA, McBride HM, Whitworth AJ, Pallanck LJ. 2008. The PINK1/Parkin pathway regulates mitochondrial morphology. *Proc. Natl. Acad. Sci. USA* 105:1638–43

Porcelli D, Oliva M, Duchi S, Latorre D, Cavaliere V, et al. 2010. Genetic, functional and evolutionary characterization of scox, the *Drosophila melanogaster* ortholog of the human *SCO1* gene. *Mitochondrion* 10:433–48

Poulopoulos M, Levy OA, Alcalay RN. 2012. The neuropathology of genetic Parkinson's disease. *Mov. Disord.* 27:831–42

Pridgeon JW, Olzmann JA, Chin L-S, Li L. 2007. PINK1 protects against oxidative stress by phosphorylating mitochondrial chaperone TRAP1. *PLoS Biol.* 5:e172

Priyadarshi A, Khuder SA, Schaub EA, Shrivastava S. 2000. A meta-analysis of Parkinson's disease and exposure to pesticides. *Neurotoxicology* 21:435–40

Qu M, Zhou Z, Chen C, Li M, Pei L, et al. 2012. Inhibition of mitochondrial permeability transition pore opening is involved in the protective effects of mortalin overexpression against beta-amyloid-induced apoptosis in SH-SY5Y cells. *Neurosci. Res.* 72:94–102

Rakovic A, Grünewald A, Voges L, Hofmann S, Orolicki S, et al. 2011. PINK1-interacting proteins: proteomic analysis of overexpressed PINK1. *Park. Dis.* 2011:153979

Ramirez A, Heimbach A, Gründemann J, Stiller B, Hampshire D, et al. 2006. Hereditary parkinsonism with dementia is caused by mutations in ATP13A2, encoding a lysosomal type 5 P-type ATPase. *Nat. Genet.* 38:1184–91

Ramonet D, Podhajska A, Stafa K, Sonnay S, Trancikova A, et al. 2012. PARK9-associated ATP13A2 localizes to intracellular acidic vesicles and regulates cation homeostasis and neuronal integrity. *Hum. Mol. Genet.* 21:1725–43

Rogaeva E, Johnson J, Lang AE, Gulick C, Gwinn-Hardy K, et al. 2004. Analysis of the PINK1 gene in a large cohort of cases with Parkinson disease. *Arch. Neurol.* 61:1898–904

Ruzzenente B, Metodiev MD, Wredenberg A, Bratic A, Park CB, et al. 2012. LRPPRC is necessary for polyadenylation and coordination of translation of mitochondrial mRNAs. *EMBO J.* 31:443–56

Saha S, Guillily MD, Ferree A, Lanceta J, Chan D, et al. 2009. LRRK2 modulates vulnerability to mitochondrial dysfunction in *Caenorhabditis elegans*. *J. Neurosci.* 29:9210–18

Sampaio-Marques B, Felgueiras C, Silva A, Rodrigues M, Tenreiro S, et al. 2012. SNCA (α-synuclein)-induced toxicity in yeast cells is dependent on sirtuin 2 (Sir2)-mediated mitophagy. *Autophagy* 8:1494–509

Sarraf SA, Raman M, Guarani-Pereira V, Sowa ME, Huttlin EL, et al. 2013. Landscape of the PARKIN-dependent ubiquitylome in response to mitochondrial depolarization. *Nature* 496:372–76

Saxton WM, Hollenbeck PJ. 2012. The axonal transport of mitochondria. *J. Cell Sci.* 125:2095–104

Schapira AHV. 2012. Mitochondrial diseases. *Lancet* 379:1825–34

Schapira AH, Cooper JM, Dexter D, Jenner P, Clark JB, Marsden CD. 1989. Mitochondrial complex I deficiency in Parkinson's disease. *Lancet* 1:1269

Schon EA, Przedborski S. 2011. Mitochondria: the next (neurode)generation. *Neuron* 70:1033–53

Schultheis PJ, Hagen TT, O'Toole KK, Tachibana A, Burke CR, et al. 2004. Characterization of the P5 subfamily of P-type transport ATPases in mice. *Biochem. Biophys. Res. Commun.* 323:731–38

Seibler P, Graziotto J, Jeong H, Simunovic F, Klein C, Krainc D. 2011. Mitochondrial Parkin recruitment is impaired in neurons derived from mutant PINK1 induced pluripotent stem cells. *J. Neurosci.* 31:5970–76

Sheng Z-H, Cai Q. 2012. Mitochondrial transport in neurons: impact on synaptic homeostasis and neurodegeneration. *Nat. Rev. Neurosci.* 13:77–93

Shin JH, Ko HS, Kang H, Lee Y, Lee YI, et al. 2011. PARIS (ZNF746) repression of PGC-1α contributes to neurodegeneration in Parkinson's disease. *Cell* 144:689–702

Shojaee S, Sina F, Banihosseini SS, Kazemi MH, Kalhor R, et al. 2008. Genome-wide linkage analysis of a Parkinsonian-pyramidal syndrome pedigree by 500 K SNP arrays. *Am. J. Hum. Genet.* 82:1375–84

Shulman JM, De Jager PL, Feany MB. 2011. Parkinson's disease: genetics and pathogenesis. *Annu. Rev. Pathol. Mech. Dis.* 6:193–222

Singleton AB, Farrer M, Johnson J, Singleton A, Hague S, et al. 2003. α-Synuclein locus triplication causes Parkinson's disease. *Science* 302:841

Takemoto K, Miyata S, Takamura H, Katayama T, Tohyama M. 2011. Mitochondrial TRAP1 regulates the unfolded protein response in the endoplasmic reticulum. *Neurochem. Int.* 58:880–87

Tanner CM, Kamel F, Ross GW, Hoppin JA, Goldman SM, et al. 2011. Rotenone, paraquat, and Parkinson's disease. *Environ. Health Perspect.* 119:866–72

Thomas KJ, McCoy MK, Blackinton J, Beilina A, van der Brug M, et al. 2011. DJ-1 acts in parallel to the PINK1/parkin pathway to control mitochondrial function and autophagy. *Hum. Mol. Genet.* 20:40–50

Trinh J, Farrer M. 2013. Advances in the genetics of Parkinson disease. *Nat. Rev. Neurol.* 9:445–54

Uéda K, Fukushima H, Masliah E, Xia Y, Iwai A, et al. 1993. Molecular cloning of cDNA encoding an unrecognized component of amyloid in Alzheimer disease. *Proc. Natl. Acad. Sci. USA* 90:11282–86

Unoki M, Nakamura Y. 2001. Growth-suppressive effects of BPOZ and EGR2, two genes involved in the PTEN signaling pathway. *Oncogene* 20:4457–65

Vafai SB, Mootha VK. 2012. Mitochondrial disorders as windows into an ancient organelle. *Nature* 491:374–83

Valente EM, Salvi S, Ialongo T, Marongiu R, Elia AE, et al. 2004. PINK1 mutations are associated with sporadic early-onset parkinsonism. *Ann. Neurol.* 56:336–41

Van Laar V, Mishizen A, Cascio M, Hastings TG. 2009. Proteomic identification of dopamine-conjugated proteins from isolated rat brain mitochondria and SH-SY5Y cells. *Neurobiol. Dis.* 34:487–500

Vilain S, Esposito G, Haddad D, Schaap O, Dobreva MP, et al. 2012. The yeast complex I equivalent NADH dehydrogenase rescues *pink1* mutants. *PLoS Genet.* 8:e1002456

Vilariño-Güell C, Wider C, Ross OA, Dachsel JC, Kachergus JM, et al. 2011. VPS35 mutations in Parkinson disease. *Am. J. Hum. Genet.* 89:162–67

Vincow ES, Merrihew G, Thomas RE, Shulman NJ, Beyer RP, et al. 2013. The PINK1-Parkin pathway promotes both mitophagy and selective respiratory chain turnover in vivo. *Proc. Natl. Acad. Sci. USA* 110:6400–5

Vos M, Esposito G, Edirisinghe JN, Vilain S, Haddad DM, et al. 2012. Vitamin K2 is a mitochondrial electron carrier that rescues *pink1* deficiency. *Science* 336:1306–10

Wadhwa R, Takano S, Kaur K, Aida S, Yaguchi T, et al. 2005. Identification and characterization of molecular interactions between mortalin/mtHsp70 and HSP60. *Biochem. J.* 391:185–90

Wang X, Petrie TG, Liu Y, Liu J, Fujioka H, Zhu X. 2012a. Parkinson's disease-associated DJ-1 mutations impair mitochondrial dynamics and cause mitochondrial dysfunction. *J. Neurochem.* 121:830–39

Wang X, Winter D, Ashrafi G, Schlehe J, Wong YL, et al. 2011. PINK1 and Parkin target Miro for phosphorylation and degradation to arrest mitochondrial motility. *Cell* 147:893–906

Wang X, Yan MH, Fujioka H, Liu J, Wilson-Delfosse A, et al. 2012b. LRRK2 regulates mitochondrial dynamics and function through direct interaction with DLP1. *Hum. Mol. Genet.* 21:1931–44

Waterham HR, Koster J, van Roermund CW, Mooyer PA, Wanders RJ, Leonard JV. 2007. A lethal defect of mitochondrial and peroxisomal fission. *N. Engl. J. Med.* 356:1736–41

Weihofen A, Thomas KJ, Ostaszewski BL, Cookson MR, Selkoe DJ. 2009. Pink1 forms a multiprotein complex with Miro and Milton, linking Pink1 function to mitochondrial trafficking. *Biochemistry* 48:2045–52

West AB, Moore DJ, Biskup S, Bugayenko A, Smith WW, et al. 2005. Parkinson's disease-associated mutations in leucine-rich repeat kinase 2 augment kinase activity. *Proc. Natl. Acad. Sci. USA* 102:16842–47

Wu Z, Sawada T, Shiba K, Liu S, Kanao T, et al. 2013. Tricornered/NDR kinase signaling mediates PINK1-directed mitochondrial quality control and tissue maintenance. *Genes Dev.* 27:157–62

Xie W, Chung KK. 2012. Alpha-synuclein impairs normal dynamics of mitochondria in cell and animal models of Parkinson's disease. *J. Neurochem.* 122:404–14

Xun Z, Sowell RA, Kaufman TC, Clemmer DE. 2007. Lifetime proteomic profiling of an A30P alpha-synuclein *Drosophila* model of Parkinson's disease. *J. Proteome Res.* 6:3729–38

Yang H, Brosel S, Acin-Perez R, Slavkovich V, Nishino I, et al. 2010. Analysis of mouse models of cytochrome *c* oxidase deficiency owing to mutations in *Sco2*. *Hum. Mol. Genet.* 19:170–80

Yang H, Zhou X, Liu X, Yang L, Chen Q, et al. 2011. Mitochondrial dysfunction induced by knockdown of mortalin is rescued by Parkin. *Biochem. Biophys. Res. Commun.* 410:114–20

Youle RJ, Narendra DP. 2011. Mechanisms of mitophagy. *Nat. Rev. Mol. Cell Biol.* 12:9–14

Youle RJ, van der Bliek AM. 2012. Mitochondrial fission, fusion, and stress. *Science* 337:1062–65

Yuan XL, Guo JF, Shi ZH, Xiao ZQ, Yan XX, et al. 2010. R492X mutation in PTEN-induced putative kinase 1 induced cellular mitochondrial dysfunction and oxidative stress. *Brain Res.* 1351:229–37

Zhang K, Li Z, Jaiswal M, Bayat V, Xiong B, et al. 2013a. The C8ORF38 homologue Sicily is a cytosolic chaperone for a mitochondrial complex I subunit. *J. Cell Biol.* 200:807–20

Zhang L, Karsten P, Hamm S, Pogson JH, Müller-Rischart AK, et al. 2013b. TRAP1 rescues PINK1 loss-of-function phenotypes. *Hum. Mol. Genet.* 22:2829–41

Zhang L, Shimoji M, Thomas B, Moore DJ, Yu SW, et al. 2005. Mitochondrial localization of the Parkinson's disease related protein DJ-1: implications for pathogenesis. *Hum. Mol. Genet.* 14:2063–73

Zheng B, Liao Z, Locascio JJ, Lesniak KA, Roderick SS, et al. 2010. PGC-1α, a potential therapeutic target for early intervention in Parkinson's disease. *Sci. Transl. Med.* 2:52ra73

Zhou Y, Shie FS, Piccardo P, Montine TJ, Zhang J. 2004. Proteasomal inhibition induced by manganese ethylene-bis-dithiocarbamate: relevance to Parkinson's disease. *Neuroscience* 128:281–91

Zimprich A, Benet-Pagès A, Struhal W, Graf E, Eck SH, et al. 2011. A mutation in VPS35, encoding a subunit of the retromer complex, causes late-onset Parkinson disease. *Am. J. Hum. Genet.* 89:168–75

Zimprich A, Biskup S, Leitner P, Lichtner P, Farrer M, et al. 2004. Mutations in *LRRK2* cause autosomal-dominant parkinsonism with pleomorphic pathology. *Neuron* 44:601–7

Ziviani E, Tao RN, Whitworth AJ. 2010. *Drosophila* Parkin requires PINK1 for mitochondrial translocation and ubiquitinates mitofusin. *Proc. Natl. Acad. Sci. USA* 107:5018–23

Zordan MA, Cisotto P, Benna C, Agostino A, Rizzo G, et al. 2006. Post-transcriptional silencing and functional characterization of the *Drosophila melanogaster* homolog of human *Surf1*. *Genetics* 172:229–41

Züchner S, Mersiyanova IV, Muglia M, Bissar-Tadmouri N, Rochelle J, et al. 2004. Mutations in the mitochondrial GTPase mitofusin 2 cause Charcot-Marie-Tooth neuropathy type 2A. *Nat. Genet.* 36:449–51

Coupling Mechanism and Significance of the BOLD Signal: A Status Report

Elizabeth M.C. Hillman

Departments of Biomedical Engineering and Radiology and the Kavli Institute for Brain Science, Columbia University, New York, NY 10027; email: elizabeth.hillman@columbia.edu

Annu. Rev. Neurosci. 2014. 37:161–81

The *Annual Review of Neuroscience* is online at neuro.annualreviews.org

This article's doi:
10.1146/annurev-neuro-071013-014111

Keywords

neurovascular coupling, fMRI, astrocytes, pericytes, vascular endothelium

Abstract

Functional magnetic resonance imaging (fMRI) provides a unique view of the working human mind. The blood-oxygen-level-dependent (BOLD) signal, detected in fMRI, reflects changes in deoxyhemoglobin driven by localized changes in brain blood flow and blood oxygenation, which are coupled to underlying neuronal activity by a process termed neurovascular coupling. Over the past 10 years, a range of cellular mechanisms, including astrocytes, pericytes, and interneurons, have been proposed to play a role in functional neurovascular coupling. However, the field remains conflicted over the relative importance of each process, while key spatiotemporal features of BOLD response remain unexplained. Here, we review current candidate neurovascular coupling mechanisms and propose that previously overlooked involvement of the vascular endothelium may provide a more complete picture of how blood flow is controlled in the brain. We also explore the possibility and consequences of conditions in which neurovascular coupling may be altered, including during postnatal development, pathological states, and aging, noting relevance to both stimulus-evoked and resting-state fMRI studies.

Contents

INTRODUCTION

Noninvasive assessment of human brain function is a major challenge. Techniques such as electroencephalography (EEG) and magnetoencephalography (MEG) provide direct measurement of neuronal activity but are faced with sensitivity, localization, and resolution issues and are rarely used for functional brain imaging research. All other current methods of human functional brain imaging rely on proxy measures of neuronal activity that are related to local blood flow, oxygenation, or metabolism (Raichle 1998). The most common of these functional brain imaging techniques is functional magnetic resonance imaging (fMRI).

WHAT IS THE fMRI BOLD SIGNAL?

fMRI relies upon the measurement of $T2^*$ relaxation, which is sensitive primarily to local concentrations of paramagnetic deoxyhemoglobin (HbR) (Ogawa et al. 1990). The so-called fMRI blood-oxygen-level-dependent (BOLD) signal increases with decreasing HbR and can be analyzed to produce localized maps of functional activity in the human brain. However, despite widespread use for almost 20 years, the fMRI BOLD signal is still poorly understood (Attwell et al. 2010, Girouard & Iadecola 2006, Logothetis 2010, Sirotin & Das 2009).

Interpretation of the fMRI BOLD signal is intrinsically linked to understanding the underlying physiological and metabolic processes in the brain that modulate blood flow. A prevailing misconception is that BOLD provides a direct measurement of neuronal oxygen consumption. However, this is generally not the case; classic positive BOLD signals, seen in response to functional stimuli, represent a decrease in HbR and thus an overoxygenation of the responding region (Attwell & Iadecola 2002). These positive BOLD responses correspond to a local, actively actuated, increase in blood flow and volume, which brings blood in sufficient excess to increase local oxygenation levels (Raichle 1998). This response typically begins within ~500 ms and peaks 3–5 s after stimulus onset (**Figure 1**), even for short stimuli lasting less than 1 s (Hirano et al. 2011, Martindale et al. 2003, Yesilyurt et al. 2008), with more complex dynamics for prolonged stimuli (**Figures 2** and

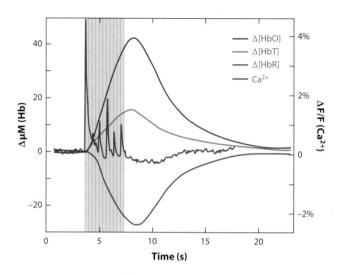

Figure 1

A typical stimulus-evoked response in the rat somatosensory cortex. Stimulus was 4 s of ~1 mA, 3 Hz forepaw stimulation. Data were acquired using multispectral optical intrinsic signal imaging of the exposed cortex, averaged over the responding region. Dark gray trace shows calcium response to the same stimulation, measured using bulk cortical injection of calcium-sensitive fluorophore Oregon green 488 BAPTA-1 AM (Bouchard et al. 2009). Figure reproduced from Hillman (2007). Notably, there is a distinct increase in total hemoglobin (HbT) corresponding to vessel dilation and an increase in the number of red blood cells per unit volume of cortex, consistent with an increase in blood flow. Oxyhemoglobin (HbO) increases while deoxyhemoglobin (HbR) decreases, indicating a net overoxygenation of the region. The fMRI BOLD signal is sensitive to changes in HbR, where stimulus-evoked 'positive BOLD' corresponds to the decrease in HbR shown here. The response begins within ~500 ms of stimulus onset and peaks at 3–5 s before slowly returning to baseline. Note that the first large calcium response (corresponding to neuronal activity) precedes marked hemodynamic changes.

4) (Martindale et al. 2005). The active process linking local neuronal activity to an orchestrated increase in local blood flow is termed neurovascular coupling. The spatiotemporal properties of the BOLD response to stimulus are therefore strongly dependent upon neurovascular coupling and the dynamic properties of the vasculature that alters blood flow through its physical dilation and constriction.

For classical interpretation of fMRI BOLD signals (Boynton et al. 1996), it is assumed that neurovascular coupling is so robust that any increase in neuronal activity generates a proportional increase in local blood flow, irrespective of brain region, brain development, and pathological state (Logothetis 2010). We return to the possible implications of these conditions in later sections. However, we begin by reviewing what is currently known about the cause, manifestations, and mechanisms of neurovascular coupling that drive stimulus-evoked functional hyperemia leading to positive BOLD in the normal brain.

WHY DOES BLOOD FLOW INCREASE?

The BOLD response is clearly an essential component of normal brain function (Girouard & Iadecola 2006, Mogi & Horiuchi 2011, Schroeter et al. 2007), yet its relationship to neuronal metabolism is less clear. Questioning whether neurovascular coupling simply matches supply to demand, several studies have maintained animals in hyperoxic or hyperglycemic states, with the surprising result that all continued to exhibit functional hyperemia in response to stimulation despite plentiful availability of oxygen and/or glucose (Lindauer et al. 2010, Wolf et al. 1997). Hypoglycemia in humans produced similar results (Powers et al. 1996). These studies demonstrate that blood flow increases are not triggered simply by local sensing of depleted nutrients.

The fact that blood flow increases are high enough to generate local hyperoxygenation, far exceeding oxygen consumption, also suggests an indirect relation between oxygen supply and demand. While one recent study suggested that high oxygen gradients are required to supply all active cells (Devor et al. 2011), others have inferred that the role of the hemodynamic response is to provide higher levels of glucose rather than oxygen (Fox & Raichle 1986, Fox et al. 1988, Heeger & Ress 2002, Paulson et al. 2009). Roles for functional hyperemia in waste removal and heat regulation have also been proposed (Yablonskiy et al. 2000).

The relative delay in the peak of increased blood flow (**Figure 1**) further confirms that neurons do not rely upon functional hyperemia to meet their initial needs for increased oxygen and glucose, since neuronal firing may have ended prior to measurable changes in blood flow. One explanation proposed for this discrepancy is that neurons, or associated support cells, may maintain sufficient stores of nutrients such as glycogen to support initial neuronal responses (Brown & Ransom 2007, Heeger & Ress 2002), which may be necessary if the speed of the vascular response is physically limited. A later blood flow peak may serve to replenish supplies or, in the case of prolonged stimulation, could be required to sustain neuronal responses once initial stores have been depleted. One model that fits with this view is the astrocyte-neuron lactate shuttle (Pellerin et al. 2007, Pellerin & Magistretti 1994), which posits that astrocytes undergo anaerobic glycolysis in response to elevated glutamate levels, preserving local oxygen supplies for neuronal use, while supplying lactate as an energy substrate to neurons (Schurr 2005). Astrocytes, located between neurons and blood vessels, are well positioned to serve this role, although the lactate shuttle model itself remains controversial (Hertz 2004, Kasischke et al. 2004).

So while the underlying purpose of functional hyperemia remains unresolved, the observations above point toward functional hyperemia being driven by an actively triggered process, initiated soon after stimulus onset, but which is not, at least initially, causally dependent on local metabolic needs (Attwell & Iadecola 2002).

SPATIOTEMPORAL PROPERTIES OF THE HEMODYNAMIC RESPONSE

The sometimes counterintuitive properties of the BOLD response described above provide an important framework for determining the mechanisms initiating and sustaining functional hyperemia. Adding to this picture, recent in vivo optical imaging and two-photon microscopy studies (see sidebar) have sought to further define the precise vascular and spatiotemporal dynamics of the cortical hemodynamic response to stimulation (Blinder et al. 2013, Chen et al. 2011, Drew et al. 2011, Hillman et al. 2007, Sirotin et al. 2009, Stefanovic et al. 2007, Tian et al. 2010). The properties of functional hyperemia that have emerged are illustrated in **Figure 2**.

Vascular Features of Functional Hyperemia

Capillary hyperemia. Capillaries are increasingly thought to physically expand during functional hyperemia (Stefanovic et al. 2007, Villringer et al. 1994). Whether active or passive, this increase in capillary diameter could explain observed increases in parenchymal [HbT] (Chen et al. 2011, Culver et al. 2005, Hillman et al. 2007, Sirotin et al. 2009), as well as the deeper location of cerebral blood volume (CBV) changes compared with BOLD in MION-MRI (Zhao et al. 2006). Localized parenchymal HbT increases occur very early, prior to dilation of pial arteries, and may be the underlying cause of observations initially interpreted as the initial dip (see sidebar, The Initial Dip) (Chen et al. 2011, Sirotin et al. 2009).

Arteriolar/arterial dilation. Penetrating arterioles dilate during functional hyperemia (Tian et al. 2010), although their relative timing in relation to capillary dilations has yet to be shown. Specific branches of pial arteries also dilate rapidly, extending up to >1 mm away from the center of the responding region (Li et al. 2003, Ngai & Winn 2002) with recent evidence supporting high-speed retrograde propagation of vasodilation (Chen et al. 2011, Iadecola et al. 1997, Chen et al. 2014).

Venous changes. Increases in arterial, arteriolar, and possibly capillary diameters decrease resistance and increase blood flow. Veins are less likely to exhibit diameter increases, but instead increase their speed of blood flow (Bouchard et al. 2009, Drew et al. 2011, Hillman et al. 2007). Venous oxygenation levels increase after a short delay, as increased blood flow increases the net oxygenation level of blood leaving the capillary beds. The onset delay of venous oxygenation increases is usually attributed to capillary transit time but may also have contributions from oxygen consumption dynamics (Hillman et al. 2007).

Short- and long-duration stimuli. Short stimuli generate responses with a similar temporal shape irrespective of stimulus duration, and with nonlinear scaling of amplitude (up to stimuli 3–4 s long) (Hirano et al. 2011, Martindale et al. 2003, Yesilyurt et al. 2008). Longer-duration stimuli tend to show a peak-then-plateau pattern (**Figures 2c** and **4a**) (Dunn et al. 2005). More distant pial arteries tend to return toward baseline after the initial peak, whereas the plateau phase is more localized to the central parenchyma (Berwick et al. 2008) and can be more variable in amplitude than the initial peak (Drew et al. 2011, Martindale et al. 2005). Often in fMRI, the plateau is higher than the initial peak (Mandeville et al. 1999).

Cessation of stimulus. As blood flow decreases, there is some evidence for residual high HbT in the capillary beds, which combined with frequently observed post-stimulus arterial vasoconstriction (Devor et al. 2008) could play a role in the poststimulus undershoot observed in fMRI (Buxton 2012, Hillman et al. 2007).

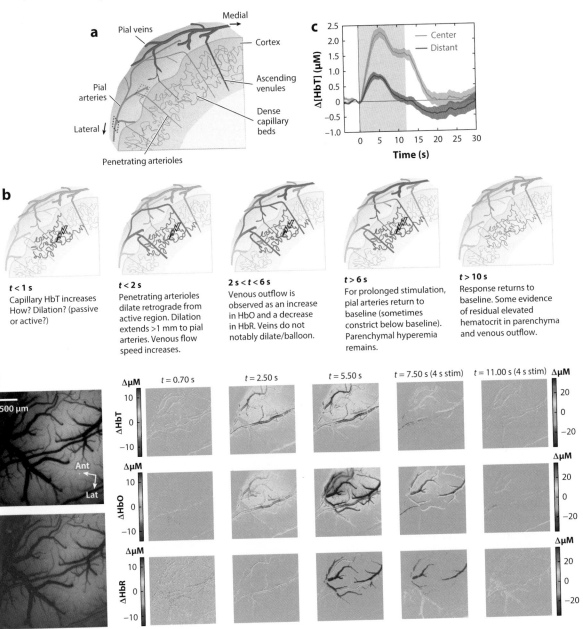

b

t < 1 s
Capillary HbT increases How? Dilation? (passive or active?)

t < 2 s
Penetrating arterioles dilate retrograde from active region. Dilation extends >1 mm to pial arteries. Venous flow speed increases.

2 s < t < 6 s
Venous outflow is observed as an increase in HbO and a decrease in HbR. Veins do not notably dilate/balloon.

t > 6 s
For prolonged stimulation, pial arteries return to baseline (sometimes constrict below baseline). Parenchymal hyperemia remains.

t > 10 s
Response returns to baseline. Some evidence of residual elevated hematocrit in parenchyma and venous outflow.

THE INITIAL DIP

Early optical imaging studies reported a transient increase in cortical HbR shortly after stimulus onset (corresponding to an "initial dip" in the fMRI BOLD signal) (Malonek & Grinvald 1996). This "dip" was found to localize to a more discrete region of the cortex than changes at later time points and was interpreted as a direct marker of neuronal oxygen consumption. Optical intrinsic signal imaging studies using 610–630-nm light (assumed to be wholly sensitive to HbR) routinely observed this dip (Chen et al. 2005, Das & Gilbert 1997). As fMRI became more widely used, however, relatively few researchers reported an initial dip in the BOLD signal (Hu & Yacoub 2012, Raichle 1998). Optical imaging studies employing additional wavelengths similarly found less evidence for an initial dip (Chen et al. 2011, Lindauer et al. 2001), instead observing localized, early increases in [HbT] rather than increases in [HbR] (Chen et al. 2011). A recent study in awake behaving primates demonstrated that early HbT increases (assessed at isosbestic wavelength 530 nm) could in fact account for initial absorption increases at 610–630 nm, suggesting contamination and misinterpretation of earlier optical initial dip intrinsic signal data (Sirotin et al. 2009). While the mechanistic basis of rapid, localized [HbT] increases is still being explored, it is likely a result of active neurovascular coupling. Therefore, although an initial dip may still occur in the brain, for example if functional hyperemia is delayed, optical intrinsic signal imaging maps produced using only 610–630 nm are unlikely to represent early oxygen consumption. Studies that have directly measured brief, local decreases in PO_2 prior to increases in blood flow (Ances et al. 2001, Li et al. 2010, Parpaleix et al. 2013) are consistent with this conclusion because local oxygen depletion need not have measurable effects on the oxygenation level of red blood cells within adjacent capillaries. Few recent fMRI and optical imaging studies have noted marked initial dips in their data (Chen et al. 2011, Devor et al. 2008, Harris et al. 2010, Hirano et al. 2011, Martin et al. 2013).

BOLD Signal Representations of Vascular Changes

In relation to fMRI BOLD, it is important to consider the manifestation of the dynamics described above in terms of changes in [HbR]. Although the dilation of arteries and arterioles can be significant (5–25%) (Tian et al. 2010), their small volume and high oxygenation means that arteries themselves contribute relatively little to the fMRI BOLD signal (Hillman et al. 2007).

←

Figure 2

Vascular evolution of normal stimulus-evoked functional hyperemia. (*a*) Schematic cut through of the mammalian cortex. Major cortical blood vessels are located on the pial surface with penetrating arterioles diving perpendicularly into the cortex and branching into dense capillary beds within the cortical layers. Blood drains from these capillaries via perpendicularly oriented ascending venules, which join a network of large draining veins on the pial surface. (*b*) Schematic sequence of the vascular dynamic response to functional stimulation (see text for citations). (*c*) Dynamics of the Δ[HbT] response to a 12-s duration, 3 Hz, ~1 mA electrical hindpaw stimulation recorded using optical imaging of the rat somatosensory cortex. 'Center' and 'Distant' vessels sampled are indicated in panel *a*. Time courses reveal that the response is more sustained within the central region, while more distant arteries return to baseline earlier, after a peak at 3–5 seconds. Δ[HbT] is independent of oxygenation dynamics and here represents only hyperemia due to arterial/arteriolar dilation and increased concentrations of red blood cells in the capillary beds. (*d*) A sequence of optical intrinsic signal imaging (OISI) data acquired on the rat somatosensory cortex in response to 4-s, 3 Hz hindpaw stimulation. *Left*: images showing the field of view under green (530 nm) illumination. *Below*: a composite based on oxygenation-dependent baseline reflectance highlighting arteries (*red*), veins (*blue*), and parenchyma (*green*). Time sequence: [HbT] changes show an initial increase in parenchymal signal by 0.7 s after stimulus onset (color scale for 0.7 s maps shown at *left*; all other time points use color scale at *right*). Increased contrast of pial arteries corresponds to dilation (confirmed by full width half maximum calculation). Increased contrast of the intervening parenchymal space corresponds to an increase in the number of red blood cells per unit volume within the capillary beds (as well as diving and ascending arterioles and venules). [HbR] (deoxyhemoglobin) changes show a distinctly different pattern and can be seen to localize to the shape of the draining veins, with decreases delayed relative to initial changes in [HbT]. Data reproduced from Chen et al. (2011, 2014) and Bouchard et al. (2009).

Veins, however, exhibit large increases in [HbR] that contribute significantly to the fMRI BOLD signal. However, these increases can be expected to be both delayed and superficially weighted with respect to the active capillary beds and may even be shifted medially as a result of the drainage patterns of the venous system (Turner 2002). These features are visible in the optical imaging sequence shown in **Figure 2d**.

Linearity versus Nonlinearity of the BOLD Response

Initial approaches to modeling and interpreting the fMRI BOLD signal assumed linear relations in which a simple hemodynamic response function (HRF) could be convolved with some driving function representing neuronal activity (Boynton et al. 1996, Buxton et al. 1998, Heeger & Ress 2002). The linearity of the BOLD response was subsequently questioned by numerous reports (Devor et al. 2003; Friston et al. 2000; Hewson-Stoate et al. 2005; Hirano et al. 2011; Martin et al. 2006, 2013; Martindale et al. 2005; Sheth et al. 2004; Yesilyurt et al. 2008), which identified nonlinear scaling in the amplitude and duration of the hemodynamic response to stimuli of different frequencies, amplitudes, and durations. A 5-ms visual stimulus, for example, yields a BOLD response almost identical to that from a 250-ms stimulus, both peaking at 3–5 s after stimulus onset. A linear relation would predict a response up to 50 times larger in the latter case (Yesilyurt et al. 2008). The peak-plateau properties of the response for longer stimulus durations add further complexity to HRF-based modeling and quantification (Kennerley et al. 2012, Martin et al. 2013, Martindale et al. 2005). The vascular properties of the hemodynamic response (**Figure 2**) indicate a more complex spatiotemporal neurovascular relationship than originally assumed (Buxton 2012, Hwan Kim et al. 2013).

In a seminal study that acquired neuronal and fMRI BOLD data simultaneously in primates, linear modeling required site-specific HRFs and found fit discrepancies for long-duration stimuli (Logothetis et al. 2001). This study highlighted the importance of considering which measure of neuronal activity should or could be expected to correlate with the BOLD signal and found that local field potentials (LFPs) better predicted BOLD than postsynaptic multiunit spiking activity (MUA) (Cardoso et al. 2012, Logothetis 2010, Sirotin & Das 2009). Such questions will remain unanswered until a comprehensive picture of the cellular and vascular mechanisms of neurovascular coupling is identified.

COUPLING MECHANISMS: HOW DOES BLOOD FLOW CHANGE?

Early investigators proposed that local metabolic factors modulate local blood flow in the brain (Friedland & Iadecola 1991, Roy & Sherrington 1890). The possibility of direct action of neuronally derived substances such as glutamate and nitric oxide (NO) on the vasculature persisted (Attwell & Iadecola 2002) until the past 10 years, when a number of seminal studies introduced additional possible cellular mediators of neurovascular coupling, including astrocytes (Takano et al. 2006), interneurons (Cauli et al. 2004), and pericytes (Peppiatt et al. 2006). Each of these cellular candidates is described below (pathways illustrated in **Figure 3**).

Astrocytes

Astrocyte end-feet ensheath diving arterioles, capillaries, and ascending venules throughout the cortex, and almost all astrocytes, have a process in contact with a blood vessel (McCaslin et al. 2011). Astrocyte involvement in neurovascular coupling has become widely accepted following a study that performed in vivo uncaging of calcium in astrocytic end-feet resulting in dilation of adjacent penetrating arterioles. The response to uncaging was blocked by SC-560 (COX1), methyl arachidonyl fluorophosphonate (MAFP) (PLA$_2$), and indomethacin (COX1 and -2) but was unaffected by NS-398 (COX2), L-NAME (NO), MS-PPOH and miconazole (EETS), and caffeine (adenosine). Researchers proposed that a buildup of glutamate (implying local neuronal activity) is sensed by metabotropic glutamate receptors (mGluR5s) on astrocytes, leading to increased intracellular calcium and generation of arachidonic acid derivatives such as prostaglandins capable of causing vasodilation through direct action on perivascular smooth muscle.

However, recent studies have begun to question the role of astrocytes in neurovascular coupling. Although increases in astrocytic intracellular calcium have been observed to correlate to stimulation in many cases (Schummers et al. 2008, Wang et al. 2006, Winship et al. 2007), a recent comprehensive in vivo two-photon study suggested that measurable increases in astrocytic calcium occur after the onset of arteriolar dilation (Nizar et al. 2013). The same study also showed that mice lacking astrocytic inositol trisphosphate (IP$_3$) type-2 receptors (necessary for generating intracellular calcium increases, which in turn would trigger the release of vasoactive arachidonic acid derivatives) still exhibited normal stimulus-evoked functional hyperemia (confirmed by Takata et al. 2013). A further recent study found that astrocytes do not express mGluR5s in the adult brain, the presumed pathway for glutamate-mediated astrocyte activation (Sun et al. 2013). Finally, astrocytes do not directly contact pial arteries above the cortical surface, making astrocytes unlikely mediators of pial artery dilations (Chen et al. 2011, McCaslin et al. 2011). While an additional recent study has countered some of these concerns with new observations of higher speed changes in astrocytic calcium (Lind et al. 2013), the role of astrocytes as the primary mediators of functional hyperemia in the cortex has become less certain.

Pericytes

Pericytes are small cells that wrap around vessels, particularly capillaries, and are involved in vasculogenesis and the blood–brain barrier (Armulik et al. 2005, Winkler et al. 2011). In 2006, Peppiatt et al. showed that pericytes in acute in vitro retina and cerebellar slices evoked capillary vasoconstriction in response to electrical stimulation, GABA, and ATP in the retina, and noradrenaline in the cerebellum, and that this preconstriction could be reversed by glutamate. Another in vitro whole-retina study observed capillary dilation in response to bradykinin and cholinergic agonists, as well as histamine (Schönfelder et al. 1998). Cultured pericytes relax in response to prostacyclin PGI$_2$, NO (endothelial or neuronal), vasoactive intestinal peptide (VIP), and low pH, whereas constriction (wrinkling) has been observed in response to endothelial-derived ET-1, thromboxane A$_2$, and angiotensin II and the catecholamines serotonin, histamine, and noradrenaline (Hamilton et al. 2010). The presence of contractile proteins in pericytes suggests that active (rather than passive) capillary dilation could play a role in functional hyperemia.

However, discriminating active from passive effects in vivo can be very challenging, and the interplay between capillary, arteriolar, and arterial diameters, and their relative influence on vascular resistance and flow is not yet well understood. One recent in vivo study examining pericyte function failed to observe capillary dilations in response to bicuculline despite upstream dilation (Fernández-Klett et al. 2010). Further work is needed to define both the behavior of capillaries during stimulus-evoked functional hyperemia and the potential role of pericytes in mediating capillary dilation.

Neuronal Networks and Interneurons

Afferents from the basal forebrain are known to modulate regional blood flow via acetylcholine (ACh) release (Arnerić et al. 1988, Chédotal et al. 1994, Hotta et al. 2011, Kocharyan et al. 2007, Takata et al. 2013, Vaucher & Hamel 1995). Researchers have also demonstrated the release of vasoactive substances by cortical interneurons including VIP and NO (dilation) and neuropeptide Y (NPY) and somatostatin (SOM) (constriction) (Cauli et al. 2004). It has been proposed that such local and distributed neuronal networks play a role in neurovascular coupling, potentially providing an alternate route to local coupling within the cortex (Piché et al. 2010, Sato & Sato 1995). More conservative models suggest that interneurons may fine-tune local hemodynamics, with astrocytes or pericytes perhaps acting as intermediaries (Attwell & Iadecola 2002, Cauli & Hamel 2010). However, a recent study returned to the possible importance of deep brain regions, showing that norepinephrine release from locus coeruleus afferents could generate large-scale cerebral vasoconstrictions, demonstrating that this effect may play a role in controlling and constraining functional hyperemia (Bekar et al. 2012, Kozberg et al. 2013).

Propagated Vasodilation

Overall, because no single coupling mechanism has been demonstrated incontrovertibly, the prevailing view is that some combination of the mechanisms above must work together to generate the blood flow response (Attwell et al. 2010). However, one component of neurovascular coupling that has been overlooked in recent years is the vasculature itself, and the potential importance of propagated vasodilation in the generation of functional hyperemia.

Retrograde propagation of vasodilation was first explored in the brain in 1997 (Iadecola et al. 1997) but was only recently demonstrated to occur in the cortex during functional hyperemia (Chen et al. 2011, 2014; Tian et al. 2010). Early studies explored propagation of vasodilation in vessels within the hamster cheek pouch but observed speeds that were too slow to account for the properties of the brain's hemodynamic response (Duling & Berne 1970). However, recent work has identified a more rapidly propagated vasodilation mechanism, mediated via endothelial hyperpolarization (Bagher & Segal 2011, Figueroa & Duling 2009). This hyperpolarization can propagate electrically within the vascular endothelium itself, traveling distances exceeding 1 mm with limited attenuation (Wölfle et al. 2011) and causing self-dilation via myoendothelial coupling to encircling smooth muscle cells (SMCs) via myoendothelial gap junctions or some endothelium-derived hyperpolarizing factor (EDHF). This EDHF-type dilation has been shown to be independent of NO and COX pathways, persisting in the presence of indomethacin and L-NAME. A second form of endothelially propagated vasodilation has also been characterized, which is linked to a slowly moving endothelial calcium wave (Tallini et al. 2007) that can propagate up to 500 microns and cause dilation of SMCs via endothelial release of NO and prostanoids such as prostacyclin (whose production in endothelial cells depends primarily on COX1 but can also involve COX2) (de Wit & Griffith 2010).

Both forms of endothelium-dependent vasodilation can be initiated by IP_3-mediated increases in intracellular endothelial calcium and can be generated experimentally by iontophoresis of ACh into the vascular endothelium. Short applications (<0.2–1 s) result in vasodilation closely resembling the BOLD HRF (Wölfle et al. 2011). Other identified initiators of EDHF-type propagated vasodilation include adenosine, ATP, UTP, K^+ ions, and bradykinin (Marrelli et al. 2003, Rosenblum 1986, Winter & Dora 2007, You et al. 1997). Studies in isolated rat middle cerebral arteries have demonstrated that EDHF-type propagated vasodilation requires an endothelial intracellular calcium increase exceeding 340 nM, whereas slower NO/prostanoid-dependent vasodilation requires a lower threshold of 220 nM (Marrelli 2001). These endothelial and SMC pathways are incorporated into **Figure 3**.

An Integrated Model?

Despite having been extensively observed and characterized in peripheral vessels, endothelially mediated vasodilation in the brain has not been considered in recent models of functional neurovascular coupling. Incorporating propagated vasodilation presents a potentially more complete picture of how functional hyperemia may be generated (**Figure 4**). EDHF-type propagated vasodilation provides an elegant mechanism to explain the rapid dilation of distant pial arteries. The possibility that endothelial signaling is initiated at the capillary level, close to active neurons, and travels retrograde along an integrative vascular route would explain the selective recruitment of specific arterial branches and the generation of an optimally localized increase in blood flow (Erinjeri & Woolsey 2002, Sarelius & Pohl 2010). Most neurovascular coupling research in recent years has focused on extravascular interactions with vascular SMCs (Cauli & Hamel 2010, Takano et al. 2006) and mechanisms by which vasoactive prostanoids, EETs, and NO could be generated by perivascular cells such as neurons and astrocytes. Recognizing that these same substances can be released by the vascular endothelium itself removes the requirement for extravascular signaling at the level of diving arterioles. Initiation of functional hyperemia could thus occur at capillary endothelial cells, perhaps via pericytes, astrocytes, or interneurons as intermediaries, or potentially via direct ACh or other neurogenic signaling to endothelial receptors (Arnerić et al. 1988).

We further hypothesize that fast and slow components of propagated vasodilation (Tallini et al. 2007) could provide a physical basis for the nonlinearities of the BOLD response. As shown in **Figure 4**, rapid and transient EDHF-dependent dilation of distant pial arteries combined with a slower, more spatially restricted NO/prostanoid-dependent dilation sustaining capillary hyperemia could elegantly explain the spatiotemporal patterns illustrated in **Figure 2**. Differing intracellular calcium thresholds (Marrelli 2001) may trigger fast EDHF-type dilation in response to large, initial neuronal responses, whereas slower NO and prostanoid-dependent dilation may be sustained by lower levels of ongoing activity. The possibly differing latencies and regenerative properties of these fast and slow responses could explain the influence of stimulus frequency, amplitude, and anesthesia on the resultant hemodynamic response. Such two-phase models have previously been proposed, based empirically on the spatiotemporal dynamics of the hemodynamic response (Cauli & Hamel 2010, Martindale et al. 2005, Tomita 2007); however, until now these models have not had a defined cellular and mechanistic basis. The involvement of at least two distinct endothelial mechanisms (EDHF, NO and prostanoid-dependent) may explain pharmacology results that have, almost without exception, failed to completely eliminate functional hyperemia through blockade of a single pathway (Lecrux et al. 2011, Takano et al. 2006). These same studies may need to be reinterpreted if the sensitivity of the vascular endothelium to the same pharmacological manipulations was not considered. Although further work is needed to develop this integrative model, one recent study has already confirmed a role for endothelial propagation of vasodilation in the brain during functional hyperemia (Chen et al. 2014). Endothelial involvement in neurovascular coupling would be highly consistent with the established importance of endothelium-dependent vasodilation in the rest of the body (Andresen et al. 2006, Marrelli 2001).

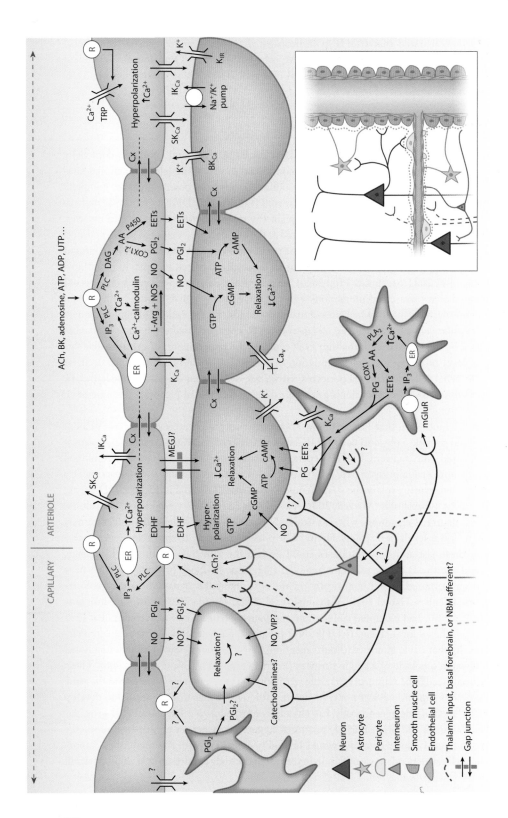

POSITIVE VERSUS NEGATIVE BOLD: STIMULUS-EVOKED VERSUS RESTING STATE

If oxygen consumption were to increase in the absence of functional hyperemia (e.g., if neurovascular coupling were impaired or neuronal activity failed to reach some threshold level), the resulting increase in [HbR] would produce a negative BOLD signal (**Figure 5*b***). Such a region would be classically interpreted as exhibiting a decrease in neuronal activity or even local inhibition (Devor et al. 2007). Vasoconstriction could equally result in an increase in [HbR] and thus negative BOLD (**Figure 5*c***), although here, the resultant changes in [HbO] and [HbR] will depend on the properties of the vascular compartment that changed volume (see **Figure 5** caption). It is important to consider that the evidence presented above suggests that vasodilation (leading to positive BOLD) is a triggered response to a specific stimulation event. This could mean that a reduction in neuronal activity would not necessarily be expected to evoke a coupled vasoconstriction by analogous mechanisms. While our above discussions of mechanisms focused on mediators of vasodilation more than constriction, a primary candidate to explain observed vasoconstrictions might be the effect of a global influence such as the norepinephrine network, uniformly driving vasoconstriction in the absence of counteracting vasodilation (Bekar et al. 2012, Kozberg et al. 2013). A reactive and possibly threshold-dependent model of functional blood flow regulation raises questions regarding whether resting-state fMRI, in which only baseline fluctuations in the BOLD signal are measured, can be assumed to represent the properties of underlying neuronal activity in the same way as stimulus-evoked responses.

A more detailed understanding of neurovascular coupling mechanisms could directly address the question of whether fMRI BOLD responses are truly neurogenic, and explain many other effects that continue to confound fMRI interpretation such as "negative BOLD" (see sidebar, Positive versus Negative BOLD, and **Figure 5**). Understanding the mechanistic basis of spatiotemporal non-linearities in the BOLD response could permit the design of fMRI stimulation paradigms that might better probe the timing and spatial extent of neuronal activity.

Figure 3

Candidate neurovascular coupling pathways [modified from Félétou & Vanhoutte (2004) and Attwell et al. (2010)]. Astrocytes can sense glutamate via metabotropic glutamate receptors (mGluR) and increase their intracellular calcium (Ca^{2+}), which can generate arachidonic acid (AA) from phospholipase A_2 (PLA_2) which is converted by COX1 (or 3) to prostaglandins (PG) and by P450 epoxygenase to epoxyeicosatrienoic acid (EETs). Both PGs and EETs can relax smooth muscle cells (SMCs) through conversion of adenosine triphosphate (ATP) to cyclic adenosine monophosphate (cAMP). Endothelial cells can increase their intracellular calcium through transient receptor potential (TRP) cation channels, and in response to receptor (R) binding, through IP_3-mediated release of calcium from intracellular stores [endoplasmic reticulum (ER)]. Endothelial receptor targets include acetylcholine (ACh), bradykinin (BK), adenosine diphosphate (ADP), ATP, uridine triphosphate (UTP), and adenosine. Receptor binding can activate phospholipase C (PLC) (or PLA_2), which via diacyl-glycerol (DAG) can also produce EETs and AA derivatives including prostacyclin (PGI_2), both of which can drive SMC relaxation via cAMP, while increased intracellular calcium can drive the production of endothelial nitric oxide (NO), which can affect SMC relaxation through conversion of guanosine triphosphate (GTP) to cyclic guanosine monophosphate (cGMP). Intracellular calcium increases also lead to endothelial hyperpolarization through opening of calcium-dependent potassium channels (K_{Ca}). Endothelial hyperpolarization could be coupled to adjacent SMCs through myoendothelial gap junctions (MEGJs) or some other endothelium-derived hyperpolarizing factor (EDHF) such as K^+ efflux through endothelial SKCa and IKCa channels by activating KIR and/or the Na^+/K^+ ATPase. SMC hyperpolarization causes relaxation through inactivation of voltage-dependent calcium channels (Ca_v). Endothelial hyperpolarization can spread rapidly to adjacent endothelial cells, likely via gap junctions. Pericytes possess many SMC-like properties and could relax in response to NO and PGI_2 from astrocytes, neurons, or endothelial cells or in response to neuropeptides such as vasointenstinal peptides (VIPs). Pericytes or astrocytes could also be involved in signaling *to* endothelial cells. Question marks represent many other potential signaling pathways yet to be identified. Additional abbreviation: NMB, nucleus basalis of Meynert.

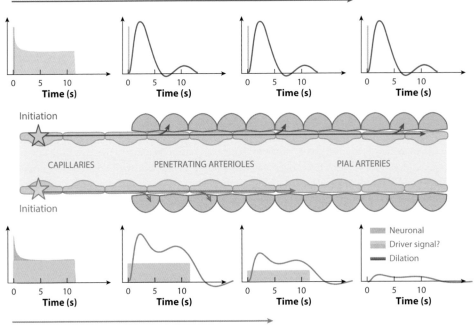

Fast, long-range endothelial hyperpolarization

Initiation

CAPILLARIES   PENETRATING ARTERIOLES   PIAL ARTERIES

Initiation

Neuronal
Driver signal?
Dilation

Slower, short-range, calcium wave mediating NO/prostanoid-dependent vasodilation

Figure 4

Proposed model of nonlinear neurovascular coupling incorporating fast and slow propagated vasodilation. Neuronal activity at the capillary level could either directly or indirectly cause an increase in endothelial intracellular calcium. An initial large-amplitude increase in endothelial calcium could initiate endothelial hyperpolarization, which would be rapidly propagated with minimal attenuation to drive relaxation of perivascular SMCs all the way up to the pial arteries (Wölfle et al. 2011). The same initial increase in endothelial calcium could also drive a slower propagating wave of increased calcium within the endothelium (Tallini et al. 2007), bringing NO and prostanoid-dependent vasodilation over a shorter distance. Combined, these two effects would generate spatiotemporal nonlinearities consistent with the properties of functional hyperemia. A lower threshold in endothelial calcium (Marrelli 2001) for slow propagation might explain continued parenchymal hyperemia, but only transient (initial) long-range dilation as shown in **Figure 2**.

IS NEUROVASCULAR COUPLING ALWAYS NORMAL?

A range of recent reports have explored conditions under which the stimulus-evoked fMRI BOLD response is altered or abnormal, including during postnatal development (Kozberg et al. 2013), in diseases such as Alzheimer's and stroke (Girouard & Iadecola 2006, Hamilton et al. 2010), diabetes (Mogi & Horiuchi 2011), aging, and drug use (D'Esposito et al. 2003, Schroeter et al. 2007). While most of these studies inferred that altered BOLD responses were a reflection of altered underlying neuronal activity, it is becoming increasingly recognized that in some situations, neurovascular coupling itself could be altered.

The emerging technique of resting state functional connectivity mapping (RS-FCM) has provided additional interesting findings in relation to the diseased brain. RS-FCM records a subject's brain in the absence of specific stimuli. Resulting BOLD data show seemingly random fluctuations in signal throughout the brain; however, analysis of this data can reveal distinct spatial patterns corresponding to regions where signals are temporally correlated (Fox et al. 2005). These regions are inferred to be functionally connected. Similar functional connectivity networks have been found

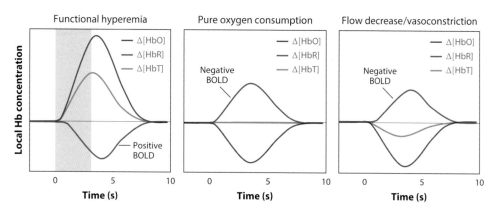

Figure 5

Conditions for positive and negative BOLD. *Left*: Normal 'positive BOLD' in which functional hyperemia increases [HbT] and [HbO]. A decrease in [HbR] occurs because of the wash-in of oxygenated blood. *Middle*: Possible response when oxygen consumption occurs in the absence of functional hyperemia. This sign of increased metabolic activity would be measured as 'negative BOLD'. *Right*: Arteriolar vasoconstriction would decrease [HbT] and [HbO], assuming high arterial oxygen saturation. The consequent decrease in flow would cause deoxygenation, even if oxygen consumption does not change. If there were an associated decrease in the volume of the capillary beds (where oxygen saturation is <98%), the HbR decrease caused by this volume decrease would compete with increasing HbR owing to increasing relative oxygen extraction.

across subjects; however, abnormal networks have also been identified in a wide range of neurological and psychological disorders including Alzheimer's, schizophrenia, depression, autism, attention deficit hyperactivity disorder, glioma, and multiple sclerosis (Greicius et al. 2004, Rocca et al. 2010, Zhang & Raichle 2010). It is routinely assumed that RS-FCM is detecting alterations in the neuronal connectivity or processing of the brain in these conditions. However, since fMRI measures are purely hemodynamic in origin, it is important to consider that alterations in apparent functional connectivity networks could, in some cases, be the result of altered neurovascular coupling. Although this possibility retains the potential utility of RS-FCM for clinical diagnosis, the possibility that neurovascular impairment could be a symptom or cause of a pathological state exhibiting abnormal functional connectivity could provide new insights into the condition and could help to elucidate new therapies (Nicolakakis & Hamel 2011).

We should not overlook that intact neurovascular coupling is essential for normal brain function. Neurovascular deficits could feasibly manifest as attentional or psychological in the short term and neurodegenerative in the long term (Roy & Sherrington 1890, Schroeter et al. 2007). Improving our understanding of neurovascular coupling in the resting, developing, and diseased brain is a critical next step. Development of fMRI-based neurovascular assessment tools and neurovascular therapies could be exciting new clinical frontiers.

SUMMARY POINTS

1. Stimulus-evoked positive BOLD is an actively driven process governed by neurovascular coupling mechanisms and the properties of the brain's vasculature.

2. Although a range of candidate mechanisms for neurovascular coupling have been proposed, a consensus has yet to be reached in the field.

3. Incorporation of propagated vasodilation and endothelial signaling into existing models may provide new explanations for the sensitivities and properties of the fMRI BOLD response.

4. Resting-state functional connectivity mapping has brought fMRI to clinical populations where neurovascular coupling may be impaired. Abnormal coupling may be the cause of alterations in functional connectivity networks in some pathologies.

5. The continuing utility and significance of both stimulus-evoked and resting-state fMRI BOLD are inextricably tied to improving our understanding of the cellular and vascular bases of neurovascular coupling in health and disease.

DISCLOSURE STATEMENT

The author is not aware of any affiliations, memberships, funding, or financial holdings that might be perceived as affecting the objectivity of this review.

ACKNOWLEDGMENTS

The author thanks Mariel Kozberg, Mohammed Shaik, Brenda Chen, Aleksandr Rayshubskiy, Costantino Iadecola, David Attwell, Karl Kasischke, Ingrid Sarelius, and Caryl Hill for their comments and contributions. The author acknowledges support from NINDS 1R01NS063226, 1R01NS076628, and R21NS053684, the Human Frontier Science Program (HFSP), NSF (CAREER 0954796 and student fellowships), and NCATS UL1 RR024156.

LITERATURE CITED

Akerboom J, Chen T-W, Wardill TJ, Tian L, Marvin JS, et al. 2012 . Optimization of a GCaMP calcium indicator for neural activity imaging. *J. Neurosci.* 32:13819–40

Ances BM, Buerk DG, Greenberg JH, Detre JA. 2001. Temporal dynamics of the partial pressure of brain tissue oxygen during functional forepaw stimulation in rats. *Neurosci. Lett.* 306:106–10

Andresen J, Shafi NI, Bryan RM Jr. 2006. Endothelial influences on cerebrovascular tone. *J. Appl. Physiol.* 100:318–27

Armulik A, Abramsson A, Betsholtz C. 2005. Endothelial/pericyte interactions. *Circ. Res.* 97:512–23

Arnerić SP, Honig MA, Milner TA, Greco S, Iadecola C, Reis DJ. 1988. Neuronal and endothelial sites of acetylcholine synthesis and release associated with microvessels in rat cerebral cortex: ultrastructural and neurochemical studies. *Brain Res.* 454:11–30

Attwell D, Buchan AM, Charpak S, Lauritzen M, MacVicar BA, Newman EA. 2010. Glial and neuronal control of brain blood flow. *Nature* 468:232–43

Attwell D, Iadecola C. 2002. The neural basis of functional brain imaging signals. *Trends Neurosci.* 25:621–25

Bagher P, Segal SS. 2011. Regulation of blood flow in the microcirculation: role of conducted vasodilation. *Acta Physiol.* 202:271–84

Bekar LK, Wei HS, Nedergaard M. 2012. The locus coeruleus-norepinephrine network optimizes coupling of cerebral blood volume with oxygen demand. *J. Cereb. Blood Flow Metab.* 32:2135–45

Berwick J, Johnston D, Jones M, Martindale J, Martin C, et al. 2008. Fine detail of neurovascular coupling revealed by spatiotemporal analysis of the hemodynamic response to single whisker stimulation in rat barrel cortex. *J. Neurophysiol.* 99:787–98

Blinder P, Tsai PS, Kaufhold JP, Knutsen PM, Suhl H, Kleinfeld D. 2013. The cortical angiome: an interconnected vascular network with noncolumnar patterns of blood flow. *Nat. Neurosci.* 16:889–97

Boorman L, Kennerley AJ, Johnston D, Jones M, Zheng Y, et al. 2010. Negative blood oxygen level dependence in the rat: a model for investigating the role of suppression in neurovascular coupling. *J. Neurosci.* 30:4285–94

Bouchard MB, Chen BR, Burgess SA, Hillman EMC. 2009. Ultra-fast multispectral optical imaging of cortical oxygenation, blood flow, and intracellular calcium dynamics. *Opt. Express* 17:15670–78

Boynton GM, Engel SA, Glover GH, Heeger DJ. 1996. Linear systems analysis of functional magnetic resonance imaging in human V1. *J. Neurosci.* 16:4207–21

Brown AM, Ransom BR. 2007. Astrocyte glycogen and brain energy metabolism. *Glia* 55:1263–71

Buxton RB. 2012. Dynamic models of BOLD contrast. *NeuroImage* 62:953–61

Buxton RB, Wong EC, Frank LR. 1998. Dynamics of blood flow and oxygenation changes during brain activation: the balloon model. *Magn. Reson. Med.* 39:855–64

Cardoso MMB, Sirotin YB, Lima B, Glushenkova E, Das A. 2012. The neuroimaging signal is a linear sum of neurally distinct stimulus- and task-related components. *Nat. Neurosci.* 15:1298–306

Cauli B, Hamel E. 2010. Revisiting the role of neurons in neurovascular coupling. *Front. Neuroenerg.* 2:9

Cauli B, Tong X-K, Rancillac A, Serluca N, Lambolez B, et al. 2004. Cortical GABA interneurons in neurovascular coupling: relays for subcortical vasoactive pathways. *J. Neurosci.* 24:8940–49

Chédotal A, Umbriaco D, Descarries L, Hartman BK, Hamel E. 1994. Light and electron microscopic immunocytochemical analysis of the neurovascular relationships of choline acetyltransferase and vasoactive intestinal polypeptide nerve terminals in the rat cerebral cortex. *J. Comp. Neurol.* 343:57–71

Chen BR, Bouchard MB, McCaslin AFH, Burgess SA, Hillman EMC. 2011. High-speed vascular dynamics of the hemodynamic response. *NeuroImage* 54:1021–30

Chen BR, Kozberg MG, Bouchard MB, Shaik MA, Hillman EMC. 2014. A critical role for the vascular endothelium in functional neurovascular coupling in the brain. *J. Am. Heart Assoc.* In press

Chen LM, Friedman RM, Roe AW. 2005. Optical imaging of SI topography in anesthetized and awake squirrel monkeys. *J. Neurosci.* 25:7648–59

Culver JP, Siegel AM, Franceschini MA, Mandeville JB, Boas DA. 2005. Evidence that cerebral blood volume can provide brain activation maps with better spatial resolution than deoxygenated hemoglobin. *NeuroImage* 27:947–59

D'Esposito M, Deouell LY, Gazzaley A. 2003. Alterations in the BOLD fMRI signal with ageing and disease: a challenge for neuroimaging. *Nat. Rev. Neurosci.* 4:863–72

Das A, Gilbert CD. 1997. Distortions of visuotopic map match orientation singularities in primary visual cortex. *Nature* 387:594–98

de Wit C, Griffith T. 2010. Connexins and gap junctions in the EDHF phenomenon and conducted vasomotor responses. *Pflügers Arch. - Eur. J. Physiol.* 459:897–914

Devor A, Dunn AK, Andermann ML, Ulbert I, Boas DA, Dale AM. 2003. Coupling of total hemoglobin concentration, oxygenation, and neural activity in rat somatosensory cortex. *Neuron* 39:353–59

Devor A, Hillman EMC, Tian P, Waeber C, Teng IC, et al. 2008. Stimulus-induced changes in blood flow and 2-deoxyglucose uptake dissociate in ipsilateral somatosensory cortex. *J. Neurosci.* 28:14347–57

Devor A, Sakadžić S, Saisan PA, Yaseen MA, Roussakis E, et al. 2011. "Overshoot" of O_2 is required to maintain baseline tissue oxygenation at locations distal to blood vessels. *J. Neurosci.* 31:13676–81

Devor A, Tian P, Nishimura N, Teng IC, Hillman EMC, et al. 2007. Suppressed neuronal activity and concurrent arteriolar vasoconstriction may explain negative blood oxygenation level-dependent signal. *J. Neurosci.* 27:4452–59

Drew PJ, Shih AY, Kleinfeld D. 2011. Fluctuating and sensory-induced vasodynamics in rodent cortex extend arteriole capacity. *Proc. Natl. Acad. Sci. USA* 108:8473–78

Duling BR, Berne RM. 1970. Propagated vasodilation in the microcirculation of the hamster cheek pouch. *Circ. Res.* 26:163–70

Dunn AK, Devor A, Bolay H, Andermann M, Moskowitz M, et al. 2003. Simultaneous imaging of total cerebral hemoglobin concentration, oxygenation, and blood flow during functional activation. *Opt. Lett.* 28:28–30

Dunn AK, Devor A, Dale AM, Boas DA. 2005. Spatial extent of oxygen metabolism and hemodynamic changes during functional activation of the rat somatosensory cortex. *NeuroImage* 27:279–90

Eggebrecht AT, White BR, Ferradal SL, Chen C, Zhan Y, et al. 2012. A quantitative spatial comparison of high-density diffuse optical tomography and fMRI cortical mapping. *NeuroImage* 61:1120–28

Erinjeri JP, Woolsey TA. 2002. Spatial integration of vascular changes with neural activity in mouse cortex. *J. Cereb. Blood Flow Metab.* 22:353–60

Félétou M, Vanhoutte PM. 2004. EDHF: new therapeutic targets? *Pharmacol. Res.* 49:565–80

Fernández-Klett F, Offenhauser N, Dirnagl U, Priller J, Lindauer U. 2010. Pericytes in capillaries are contractile in vivo, but arterioles mediate functional hyperemia in the mouse brain. *Proc. Natl. Acad. Sci. USA* 107:22290–95

Figueroa XF, Duling BR. 2009. Gap junctions in the control of vascular function. *Antioxid. Redox Signal.* 11:251–66

Fox MD, Snyder AZ, Vincent JL, Corbetta M, Van Essen DC, Raichle ME. 2005. The human brain is intrinsically organized into dynamic, anticorrelated functional networks. *Proc. Natl. Acad. Sci. USA* 102(27):9673–78

Fox PT, Raichle ME. 1986. Focal physiological uncoupling of cerebral blood flow and oxidative metabolism during somatosensory stimulation in human subjects. *Proc. Natl. Acad. Sci. USA* 83:1140–44

Fox PT, Raichle ME, Mintun MA, Dence C. 1988. Nonoxidative glucose consumption during focal physiologic neural activity. *Science* 241:462–64

Friedland RP, Iadecola C. 1991. Roy and Sherrington (1890): a centennial reexamination of "On the Regulation of the Blood-Supply of the Brain." *Neurology* 41:10–14

Friston KJ, Mechelli A, Turner R, Price CJ. 2000. Nonlinear responses in fMRI: the balloon model, Volterra kernels, and other hemodynamics. *NeuroImage* 12:466–77

Girouard H, Iadecola C. 2006. Neurovascular coupling in the normal brain and in hypertension, stroke, and Alzheimer disease. *J. Appl. Physiol.* 100:328–35

Greicius MD, Srivastava G, Reiss AL, Menon V. 2004. Default-mode network activity distinguishes Alzheimer's disease from healthy aging: evidence from functional MRI. *Proc. Natl. Acad. Sci. USA* 101:4637–42

Hamilton NB, Attwell D, Hall CN. 2010. Pericyte-mediated regulation of capillary diameter: a component of neurovascular coupling in health and disease. *Front. Neuroenerg.* 2:5

Harris S, Jones M, Zheng Y, Berwick J. 2010. Does neural input or processing play a greater role in the magnitude of neuroimaging signals? *Front. Neuroenerg.* 2:15

Heeger DJ, Ress D. 2002. What does fMRI tell us about neuronal activity? *Nat. Rev. Neurosci.* 3:142–51

Hertz L. 2004. The astrocyte-neuron lactate shuttle: a challenge of a challenge. *J. Cereb. Blood Flow Metab.* 24:1241–48

Hewson-Stoate N, Jones M, Martindale J, Berwick J, Mayhew J. 2005. Further nonlinearities in neurovascular coupling in rodent barrel cortex. *NeuroImage* 24:565–74

Hillman EMC. 2007. Optical brain imaging in-vivo: techniques and applications from animal to man. *J. Biomed. Opt.* 12:051402

Hillman EMC, Devor A, Bouchard M, Dunn AK, Krauss GW, et al. 2007. Depth-resolved optical imaging and microscopy of vascular compartment dynamics during somatosensory stimulation. *NeuroImage* 35:89–104

Hirano Y, Stefanovic B, Silva AC. 2011. Spatiotemporal evolution of the functional magnetic resonance imaging response to ultrashort stimuli. *J. Neurosci.* 31:1440–47

Hotta H, Uchida S, Kagitani F, Maruyama N. 2011. Control of cerebral cortical blood flow by stimulation of basal forebrain cholinergic areas in mice. *J. Physiol. Sci.* 61:201–9

Hu X, Yacoub E. 2012. The story of the initial dip in fMRI. *NeuroImage* 62:1103–8

Hwan Kim J, Khan R, Thompson JK, Ress D. 2013. Model of the transient neurovascular response based on prompt arterial dilation. *J. Cereb. Blood Flow Metab.* 33:1429–39

Iadecola C, Yang G, Ebner TJ, Chen G. 1997. Local and propagated vascular responses evoked by focal synaptic activity in cerebellar cortex. *J. Neurophysiol.* 78:651–59

Kasischke KA, Vishwasrao HD, Fisher PJ, Zipfel WR, Webb WW. 2004. Neural activity triggers neuronal oxidative metabolism followed by astrocytic glycolysis. *Science* 305:99–103

Kennerley AJ, Harris S, Bruyns-Haylett M, Boorman L, Zheng Y, et al. 2012. Early and late stimulus-evoked cortical hemodynamic responses provide insight into the neurogenic nature of neurovascular coupling. *J. Cereb. Blood Flow Metab.* 32:468–80

Kocharyan A, Fernandes P, Tong X-K, Vaucher E, Hamel E. 2007. Specific subtypes of cortical GABA interneurons contribute to the neurovascular coupling response to basal forebrain stimulation. *J. Cereb. Blood Flow Metab.* 28:221–31

Kozberg MG, Chen BR, DeLeo SE, Bouchard MB, Hillman EMC. 2013. Resolving the transition from negative to positive blood oxygen level-dependent responses in the developing brain. *Proc. Natl. Acad. Sci. USA* 110:4380–85

Lecrux C, Toussay X, Kocharyan A, Fernandes P, Neupane S, et al. 2011. Pyramidal neurons are "neurogenic hubs" in the neurovascular coupling response to whisker stimulation. *J. Neurosci.* 31:9836–47

Li B, Freeman RD. 2010. Neurometabolic coupling in the lateral geniculate nucleus changes with extended age. *J. Neurophysiol.* 104(1):414–25

Li P, Luo Q, Luo W, Chen S, Cheng H, Zeng S. 2003. Spatiotemporal characteristics of cerebral blood volume changes in rat somatosensory cortex evoked by sciatic nerve stimulation and obtained by optical imaging. *J. Biomed. Opt.* 8:629–35

Lind BL, Brazhe AR, Jessen SB, Tan FCC, Lauritzen MJ. 2013. Rapid stimulus-evoked astrocyte Ca2+ elevations and hemodynamic responses in mouse somatosensory cortex in vivo. *Proc. Natl. Acad. Sci. USA* 110(48):E4678–87

Lindauer U, Leithner C, Kaasch H, Rohrer B, Foddis M, et al. 2010. Neurovascular coupling in rat brain operates independent of hemoglobin deoxygenation. *J. Cereb. Blood Flow Metab.* 30:757–68

Lindauer U, Royl G, Leithner C, Kühl M, Gold L, et al. 2001. No evidence for early decrease in blood oxygenation in rat whisker cortex in response to functional activation. *NeuroImage* 13:988–1001

Logothetis NK. 2010. Neurovascular uncoupling: much ado about nothing. *Front. Neuroenerg.* 2:2

Logothetis NK, Pauls J, Augath M, Trinath T, Oeltermann A. 2001. Neurophysiological investigation of the basis of the fMRI signal. *Nature* 412:150–57

Malonek D, Grinvald A. 1996. Interactions between electrical activity and cortical microcirculation revealed by imaging spectroscopy: implications for functional brain mapping. *Science* 272:551–54

Mandeville JB, Marota JJA, Ayata C, Zaharchuk G, Moskowitz MA, et al. 1999. Evidence of a cerebrovascular postarteriole Windkessel with delayed compliance. *J. Cereb. Blood Flow Metab.* 19:679–89

Marrelli SP. 2001. Mechanisms of endothelial P2Y$_1$- and P2Y$_2$-mediated vasodilatation involve differential [Ca2+]$_i$ responses. *Am. J. Physiol. - Heart Circ. Physiol.* 281:H1759–66

Marrelli SP, Eckmann MS, Hunte MS. 2003. Role of endothelial intermediate conductance KCa channels in cerebral EDHF-mediated dilations. *Am. J. Physiol. - Heart Circ. Physiol.* 285:H1590–99

Martin C, Martindale J, Berwick J, Mayhew J. 2006. Investigating neural-hemodynamic coupling and the hemodynamic response function in the awake rat. *NeuroImage* 32:33–48

Martin C, Zheng Y, Sibson NR, Mayhew JEW, Berwick J. 2013. Complex spatiotemporal haemodynamic response following sensory stimulation in the awake rat. *NeuroImage* 66:1–8

Martindale J, Berwick J, Martin C, Kong Y, Zheng Y, Mayhew J. 2005. Long duration stimuli and nonlinearities in the neural-haemodynamic coupling. *J. Cereb. Blood Flow Metab.* 25:651–61

Martindale J, Mayhew J, Berwick J, Jones M, Martin C, et al. 2003. The hemodynamic impulse response to a single neural event. *J. Cereb. Blood Flow Metab.* 23:546–55

Mayhew J, Hu D, Zheng Y, Askew S, Hou Y, et al. 1998. An evaluation of linear model analysis techniques for processing images of microcirculation activity. *NeuroImage* 7:49–71

McCaslin AFH, Chen BR, Radosevich AJ, Cauli B, Hillman EMC. 2011. In-vivo 3D morphology of astrocyte-vasculature interactions in the somatosensory cortex: implications for neurovascular coupling. *J. Cereb. Blood Flow Metab.* 31:795–806

Mogi M, Horiuchi M. 2011. Neurovascular coupling in cognitive impairment associated with diabetes mellitus. *Circ. J.* 75:1042–48

Ngai AC, Winn HR. 2002. Pial arteriole dilation during somatosensory stimulation is not mediated by an increase in CSF metabolites. *Am. J. Physiol. Heart Circ. Physiol.* 282:H902–7

Nicolakakis N, Hamel E. 2011. Neurovascular function in Alzheimer's disease patients and experimental models. *J. Cereb. Blood Flow Metab.* 31:1354–70

Nielsen AN, Lauritzen M. 2001. Coupling and uncoupling of activity-dependent increases of neuronal activity and blood flow in rat somatosensory cortex. *J. Physiol.* 533:773–85

Nizar K, Uhlirova H, Tian P, Saisan PA, Cheng Q, et al. 2013. In vivo stimulus-induced vasodilation occurs without IP3 receptor activation and may precede astrocytic calcium increase. *J. Neurosci.* 33:8411–22

Ogawa S, Lee TM, Kay AR, Tank DW. 1990. Brain magnetic resonance imaging with contrast dependent on blood oxygenation. *Proc. Natl. Acad. Sci. USA* 87:9868–72

Parpaleix A, Houssen YG, Charpak S. 2013. Imaging local neuronal activity by monitoring PO_2 transients in capillaries. *Nat. Med.* 19:241–46

Paulson OB, Hasselbalch SG, Rostrup E, Knudsen GM, Pelligrino D. 2009. Cerebral blood flow response to functional activation. *J. Cereb. Blood Flow Metab.* 30:2–14

Pellerin L, Bouzier-Sore AK, Aubert A, Serres S, Merle M, et al. 2007. Activity-dependent regulation of energy metabolism by astrocytes: an update. *Glia* 55:1251–56

Pellerin L, Magistretti PJ. 1994. Glutamate uptake into astrocytes stimulates aerobic glycolysis: a mechanism coupling neuronal activity to glucose utilization. *Proc. Natl. Acad. Sci. USA* 91:10625–29

Peppiatt CM, Howarth C, Mobbs P, Attwell D. 2006. Bidirectional control of CNS capillary diameter by pericytes. *Nature* 443:700–4

Piché M, Uchida S, Hara S, Aikawa Y, Hotta H. 2010. Modulation of somatosensory-evoked cortical blood flow changes by GABAergic inhibition of the nucleus basalis of Meynert in urethane-anaesthetized rats. *J. Physiol.* 588:2163–71

Powers WJ, Hirsch IB, Cryer PE. 1996. Effect of stepped hypoglycemia on regional cerebral blood flow response to physiological brain activation. *Am. J. Physiol. - Heart Circ. Physiol.* 270:H554–59

Raichle ME. 1998. Behind the scenes of functional brain imaging: a historical and physiological perspective. *Proc. Natl. Acad. Sci. USA* 95:765–72

Rayshubskiy A, Wojtasiewicz TJ, Mikell CB, Bouchard MB, Timerman D, et al. 2013. Direct, intraoperative observation of ~0.1 Hz hemodynamic oscillations in awake human cortex: implications for fMRI. *NeuroImage* 87:323–31

Rocca MA, Valsasina P, Absinta M, Riccitelli G, Rodegher ME, et al. 2010. Default-mode network dysfunction and cognitive impairment in progressive MS. *Neurology* 74:1252–59

Rosenblum WI. 1986. Endothelial dependent relaxation demonstrated in vivo in cerebral arterioles. *Stroke* 17:494–97

Roy CS, Sherrington CS. 1890. On the regulation of the blood-supply of the brain. *J. Physiol.* 11:85–108

Sakadzić S, Roussakis E, Yaseen MA, Mandeville ET, Srinivasan VJ, et al. 2010. Two-photon high-resolution measurement of partial pressure of oxygen in cerebral vasculature and tissue. *Nat. Meth.* 7:755–59

Sarelius I, Pohl U. 2010. Control of muscle blood flow during exercise: local factors and integrative mechanisms. *Acta Physiol.* 199:349–65

Sato A, Sato Y. 1995. Cholinergic neural regulation of regional cerebral blood flow. *Alzheimer Dis. Assoc. Disord.* 9:28–38

Schönfelder U, Hofer A, Paul M, Funk RHW. 1998. In situ observation of living pericytes in rat retinal capillaries. *Microvasc. Res.* 56:22–29

Schroeter ML, Cutini S, Wahl MM, Scheid R, Yves von Cramon D. 2007. Neurovascular coupling is impaired in cerebral microangiopathy—an event-related Stroop study. *NeuroImage* 34:26–34

Schummers J, Yu H, Sur M. 2008. Tuned responses of astrocytes and their influence on hemodynamic signals in the visual cortex. *Science* 320:1638–43

Schurr A. 2006. Lactate: the ultimate cerebral oxidative energy substrate? *J. Cereb. Blood Flow Metab.* 26:142–52

Sheth SA, Nemoto M, Guiou M, Walker M, Pouratian N, Toga AW. 2004. Linear and nonlinear relationships between neuronal activity, oxygen metabolism, and hemodynamic responses. *Neuron* 42:347–55

Shibuki K, Hishida R, Murakami H, Kudoh M, Kawaguchi T, et al. 2003. Dynamic imaging of somatosensory cortical activity in the rat visualized by flavoprotein autofluorescence. *J. Physiol.* 549:919–27

Sirotin YB, Das A. 2009. Anticipatory haemodynamic signals in sensory cortex not predicted by local neuronal activity. *Nature* 457:475–79

Sirotin YB, Hillman EMC, Bordier C, Das A. 2009. Spatiotemporal precision and hemodynamic mechanism of optical point spreads in alert primates. *Proc. Natl. Acad. Sci. USA* 106:18390–95

Stefanovic B, Hutchinson E, Yakovleva V, Schram V, Russell JT, et al. 2007. Functional reactivity of cerebral capillaries. *J. Cereb. Blood Flow Metab.* 28:961–72

Sun W, McConnell E, Pare J-F, Xu Q, Chen M, et al. 2013. Glutamate-dependent neuroglial calcium signaling differs between young and adult brain. *Science* 339:197–200

Takano T, Tian G-F, Peng W, Lou N, Libionka W, et al. 2006. Astrocyte-mediated control of cerebral blood flow. *Nat. Neurosci.* 9:260–67

Takata N, Nagai T, Ozawa K, Oe Y, Mikoshiba K, Hirase H. 2013. Cerebral blood flow modulation by basal forebrain or whisker stimulation can occur independently of large cytosolic Ca^{2+} signaling in astrocytes. *PLoS One* 8:e66525

Tallini YN, Brekke JF, Shui B, Doran R, Hwang S-M, et al. 2007. Propagated endothelial Ca^{2+} waves and arteriolar dilation in vivo: measurements in $Cx40^{BAC}$–GCaMP2 transgenic mice. *Circ. Res.* 101:1300–9

Tian P, Teng IC, May LD, Kurz R, Lu K, et al. 2010. Cortical depth-specific microvascular dilation underlies laminar differences in blood oxygenation level-dependent functional MRI signal. *Proc. Natl. Acad. Sci. USA* 107:15246–51

Tomita M. 2007. Blood flow control in the brain: possible biphasic mechanism of functional hyperemia. *Asian Biomed.* 1(1):17–32

Turner R. 2002. How much cortex can a vein drain? Downstream dilution of activation-related cerebral blood oxygenation changes. *NeuroImage* 16:1062–67

Vanzetta I, Grinvald A. 1999. Increased cortical oxidative metabolism due to sensory stimulation: implications for functional brain imaging. *Science* 286:1555–58

Vanzetta I, Hildesheim R, Grinvald A. 2005. Compartment-resolved imaging of activity-dependent dynamics of cortical blood volume and oximetry. *J. Neurosci.* 25:2233–44

Vaucher E, Hamel E. 1995. Cholinergic basal forebrain neurons project to cortical microvessels in the rat: electron microscopic study with anterogradely transported *Phaseolus vulgaris* leucoagglutinin and choline acetyltransferase immunocytochemistry. *J. Neurosci.* 15:7427–41

Villringer A, Them A, Lindauer U, Einhäupl K, Dirnagl U. 1994. Capillary perfusion of the rat brain cortex. An in vivo confocal microscopy study. *Circ. Res.* 75:55–62

Wang X, Lou N, Xu Q, Tian G, Peng WG, et al. 2006. Astrocytic Ca^{2+} signaling evoked by sensory stimulation in vivo. *Nat. Neurosci.* 9:816–23

Winkler EA, Bell RD, Zlokovic BV. 2011. Central nervous system pericytes in health and disease. *Nat. Neurosci.* 14:1398–405

Winship IR, Plaa N, Murphy TH. 2007. Rapid astrocyte calcium signals correlate with neuronal activity and onset of the hemodynamic response in vivo. *J. Neurosci.* 27:6268–72

Winter P, Dora KA. 2007. Spreading dilatation to luminal perfusion of ATP and UTP in rat isolated small mesenteric arteries. *J. Physiol.* 582:335–47

Wolf T, Lindauer U, Villringer A, Dirnagl U. 1997. Excessive oxygen or glucose supply does not alter the blood flow response to somatosensory stimulation or spreading depression in rats. *Brain Res.* 761:290–99

Wölfle SE, Chaston DJ, Goto K, Sandow SL, Edwards FR, Hill CE. 2011. Non-linear relationship between hyperpolarisation and relaxation enables long distance propagation of vasodilatation. *J. Physiol.* 589:2607–23

Yablonskiy DA, Ackerman JJH, Raichle ME. 2000. Coupling between changes in human brain temperature and oxidative metabolism during prolonged visual stimulation. *Proc. Natl. Acad. Sci. USA* 97:7603–8

Yeşilyurt B, Uğurbil K, Uludağ K. 2008. Dynamics and nonlinearities of the BOLD response at very short stimulus durations. *Magn. Reson. Imaging* 26:853–62

You J, Johnson TD, Childres WF, Bryan RM Jr. 1997. Endothelial-mediated dilations of rat middle cerebral arteries by ATP and ADP. *Am. J. Physiol. - Heart Circ. Physiol.* 273:H1472–77

Zhang D, Raichle ME. 2010. Disease and the brain's dark energy. *Nat. Rev. Neurol.* 6:15–28

Zhao F, Wang P, Hendrich K, Ugurbil K, Kim S-G. 2006. Cortical layer-dependent BOLD and CBV responses measured by spin-echo and gradient-echo fMRI: insights into hemodynamic regulation. *NeuroImage* 30:1149–60

Cortical Control of Whisker Movement

Carl C.H. Petersen

Laboratory of Sensory Processing, Brain Mind Institute, Faculty of Life Sciences,
École Polytechnique Fédérale de Lausanne (EPFL), CH-1015 Lausanne, Switzerland;
email: carl.petersen@epfl.ch

Annu. Rev. Neurosci. 2014. 37:183–203

First published online as a Review in Advance on
May 9, 2014

The *Annual Review of Neuroscience* is online at
neuro.annualreviews.org

This article's doi:
10.1146/annurev-neuro-062012-170344

Keywords

motor cortex, somatosensory cortex, whisker motor control, neural circuits

Abstract

Facial muscles drive whisker movements, which are important for active tactile sensory perception in mice and rats. These whisker muscles are innervated by cholinergic motor neurons located in the lateral facial nucleus. The whisker motor neurons receive synaptic inputs from premotor neurons, which are located within the brain stem, the midbrain, and the neocortex. Complex, distributed neural circuits therefore regulate whisker movement during behavior. This review focuses specifically on cortical whisker motor control. The whisker primary motor cortex (M1) strongly innervates brain stem reticular nuclei containing whisker premotor neurons, which might form a central pattern generator for rhythmic whisker protraction. In a parallel analogous pathway, the whisker primary somatosensory cortex (S1) strongly projects to the brain stem spinal trigeminal interpolaris nucleus, which contains whisker premotor neurons innervating muscles for whisker retraction. These anatomical pathways may play important functional roles, since stimulation of M1 drives exploratory rhythmic whisking, whereas stimulation of S1 drives whisker retraction.

Contents

INTRODUCTION

Mice and rats use their whiskers extensively to gather tactile sensory information about their immediate environment. The whiskers are typically moved backward and forward at high frequencies during exploratory behavior. As a moving whisker encounters an object, it bends and exerts forces within the whisker follicle. These mechanical forces are thought to open stretch-activated ion channels driving action potential firing in whisker sensory neurons. Self-generated movements of the whiskers, driven by the animal's own muscles, are therefore responsible for an important component of sensory input. Conversely, sensory input from whisker-object contacts alters whisker movements (Mitchinson et al. 2007, Grant et al. 2009, Crochet et al. 2011), presumably to optimize the quality of incoming sensory information. Interactions between sensory and motor components of the whisker system are therefore prominent.

The primary whisker sensory neurons have cell bodies in the trigeminal ganglion, and they innervate the principal trigeminal nucleus and the spinal trigeminal nuclei. These brain stem nuclei in turn send whisker sensory information to various downstream pathways, including the trigemino-thalamo-cortical (Brecht 2007, Petersen 2007, Feldmeyer et al. 2013); trigemino-tectal/-collicular (Steindler 1985, Cohen et al. 2008); trigemino-facial (Nguyen & Kleinfeld 2005); and trigemino-pontine/-olivary/-cerebellar pathways (Swenson et al. 1984, Molinari et al. 1996, Yatim et al. 1996). Sensory signals from the whiskers are thus processed in a large number of interconnected brain areas, which together underlie whisker sensorimotor coordination and sensory perception. Because the tactile whisker sensory signals are often actively gathered, they relate closely to self-generated whisker and head movements. To derive a self-consistent percept of the tactile environment, rats and mice must therefore integrate sensory information and motor commands. The neocortex is a prominent region for sensorimotor integration. Recent reviews cover cortical processing of sensory signals from whiskers (Brecht 2007, Petersen 2007, Diamond et al. 2008, Bosman et al. 2011, Feldmeyer et al. 2013). This review focuses on the cortical control of whisker movement.

The cortex has long been known to play an important role in motor control, beginning with the pioneering experiments of Fritsch & Hitzig (1870). These first investigations of movements evoked by electrical stimulation of the cortex in dogs were later extended to monkeys by Ferrier (1874) and Sherrington (1906). The important experiments of Penfield & Boldrey (1937) defined motor maps of awake human patients undergoing neurosurgical interventions, finding an organization broadly similar to that observed in the previous animal experiments. Electrical stimulation of a wide variety of brain regions evoked diverse sensations and movements (Penfield & Boldrey 1937). A region anterior to the central sulcus evoked robust movements, defining the primary motor cortex (M1). Stimulation of medial M1 evoked primarily lower limb movements, whereas stimulation of more lateral M1 evoked upper limb movements and, further laterally, head and face movements. The maps of evoked movements suggested a motor homunculus, whereby movements of nearby body parts were controlled by nearby regions of the cortex, mirroring the somatotopic organization of the primary somatosensory cortex (S1). However, even in the initial report (Penfield & Boldrey 1937) it was clear that movements could also be evoked from stimulation of other cortical areas, such as by stimulating the somatosensory cortex lying posterior to the central sulcus. More recent experiments in monkeys confirm that motor-related neurons are widely distributed in the monkey brain, with transneuronal retrogradely labeled neurons from muscles being found in both M1 and S1 (Rathelot & Strick 2006). This study agrees with data from monkeys showing that neurons in both M1 and S1 project to the spinal cord (Coulter & Jones 1977). Together these data raise questions about the definition of M1. Further questions about the overall organization of motor maps in the monkey cortex have been raised by the work of Graziano and colleagues (Graziano et al. 2002, Cooke et al. 2003, Graziano & Aflalo 2007). Rather than the motor cortex containing a map of muscles as implied by the motor homunculus, Graziano suggests that the motor cortex is divided into areas controlling different types of behaviors. Through applying intracortical microstimulation for long durations compatible with the timescale of behavior, Graziano and colleagues found that complex sequences of movements were evoked by stimulating localized regions of the frontal and parietal cortices. Stimulation of specific cortical regions evoked reproducible behaviors, including climbing, feeding, defensive movements, and reaching to defined locations in space. Whether motor maps in monkeys are organized as a homunculus, or rather in terms of behavioral repertoire, remains controversial. As one thinks about the general role of the cortex in controlling movement, it may also be useful to consider an evolutionary perspective. The neocortex, as its name implies, was added at a late time in the evolution of the brain. In terms of motor control, the cortex therefore may serve as a high-order sensorimotor circuit interacting with a series of lower-level sensorimotor circuits in basal ganglia, the midbrain, and the brain stem, all of which are evolutionarily older. Such considerations suggest that cortical activity is likely to affect motor output through interactions with many different subcortical pathways.

Investigation into the detailed mechanisms underlying cortical motor control has begun in rats and mice. Early experiments used intracortical microstimulation to evoke movements, which were only qualitatively evaluated. Similar to the results in humans and monkeys, in rodents, movements could be evoked by stimulating a large fraction of the neocortex, including M1 and S1 (Hall & Lindholm 1974, Donoghue & Wise 1982, Neafsey et al. 1986). Motor maps based on surface stimulation (Woolsey 1958) and intracortical microstimulation (Hall & Lindholm 1974, Donoghue & Wise 1982, Gioanni & Lamarche 1985, Neafsey et al. 1986, Miyashita et al. 1994, Brecht et al. 2004a) reveal forelimb and hindlimb representations bordering (and overlapping) with their S1 representations. Head, whisker, and eye movements map to more anterior and medial locations. Although these motor maps are reproducible, the experimental methods suffer from numerous important limitations. Intracortical microstimulation affects neurons near the electrode but also stimulates axons of passage, limiting the spatial resolution of these motor-mapping experiments and reducing the possibility for causal interpretations of the data. New stimulation techniques are therefore desirable. In addition, most motor-mapping experiments have been carried out in lightly anesthetized animals, a condition designed to minimize movement and sensation. The overall brain state may influence how neural circuits in the neocortex connect to motor output (Tandon et al. 2008), and thus motor maps in awake animals may differ substantially from those obtained under anesthesia. Quantitative analysis of movements evoked by precise stimulation in awake animals is therefore necessary. Powerful molecular and genetic tools have now been developed for investigating the mouse brain, raising interest in defining cortical circuits for motor control in the mouse. Recent advances in optogenetics (Nagel et al. 2003, Boyden et al. 2005, Zhang et al. 2007, Chow et al. 2010, Zhang et al. 2011) have helped improve the specificity of stimulation, which is useful for defining more precise neocortical motor maps (Matyas et al. 2010, Harrison et al. 2012). Furthermore, new transsynaptic viral methods can now label and genetically manipulate specifically connected synaptic circuits (Wickersham et al. 2007, Osakada et al. 2011). These new methods have recently been applied to the mouse whisker motor system, providing new insights reviewed here in the context of existing knowledge.

MUSCLES DRIVING WHISKER MOVEMENT

The whiskers are moved back and forth through large angles ($\sim$90°) and at high frequencies ($\sim$5–20 Hz) during active exploratory whisking (**Figure 1a**). Although it is important to note that the whiskers can move in complex multidimensional ways (Knutsen et al. 2008), a simple one-dimensional variable, the whisker angle, accounts for much of the overall movement of whiskers. Most of the time, the whiskers all move bilaterally in synchrony, but whiskers can also move independently (Sachdev et al. 2002), which occurs extensively during whisker-object contact. The whisker can be thought of as pivoting around its insertion point in the mystacial pad. Intrinsic muscles within the mystacial pad (Dörfl 1982, Hill et al. 2008) attach superficially on one whisker and form a sling around the base (deep in the pad) of the immediately anterior whisker follicle. Each whisker is served by its own intrinsic muscle. Contraction of an intrinsic muscle causes the base of the whisker follicle (located deep in the mystical pad) to move toward the posterior. This contraction rotates the external part of the whisker forward, the pad insertion point acting as the pivot. Intrinsic muscles therefore drive whisker protraction (**Figure 1b**). Extrinsic muscles nasolabialis and maxillolabialis act on the superficial part of the whisker pad attaching to external anchor points posteriorly. Contraction of these extrinsic muscles drives whisker retraction by translating the whole whisker pad and posteriorly rotating individual follicles (**Figure 1b**). Electromyograph (EMG) recordings have defined the timing of contraction of different whisker muscles during whisking and have found that the intrinsic muscles are active during whisker protraction and that the extrinsic muscles nasolabialis and maxillolabialis are often active during whisker retraction (Hill et al. 2008, Moore et al. 2013). In addition, several other muscles acting on the whiskers and mystacial pad have been described (Hill et al. 2008, Haidarliu et al. 2012). The whisker muscles are unusual in at least two ways: First, the intrinsic whisker muscles are distinguished from skeletal muscles by predominance of a fast-contracting, fast-fatigable muscle type (Jin et al. 2004); and, second, the whisker muscles do not appear to have spindles and thus there is no direct proprioceptive feedback (Rice et al. 1997).

WHISKER MOTOR NEURONS IN THE FACIAL NUCLEUS

Both the intrinsic and extrinsic whisker muscles are innervated by the facial nerve, with the cell bodies of the cholinergic motor neurons lying in the lateral facial nucleus (**Figure 2a**). Retrograde labeling from injections into whisker muscles has shown that the motor neurons innervating the intrinsic muscles are located more ventrally within the lateral facial nucleus, whereas the motor neurons that innervate the extrinsic muscle are located more dorsally within the lateral facial nucleus (Takatoh et al. 2013). The lateral facial nucleus therefore appears to have a well-ordered

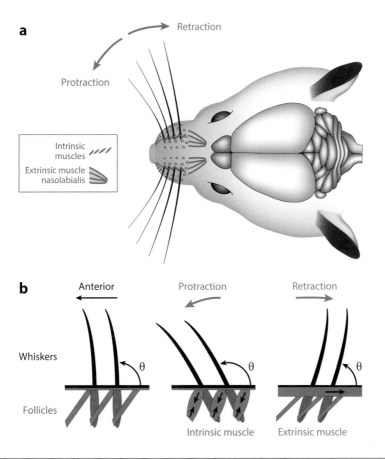

a

Retraction

Protraction

Intrinsic muscles ////

Extrinsic muscle nasolabialis

b

Anterior

Protraction

Retraction

Whiskers

θ

θ

θ

Follicles

Intrinsic muscle

Extrinsic muscle

Figure 1

Muscles controlling whisker movement. (*a*) The whiskers on the snout of mice and rats are arranged in a stereotypical highly ordered manner in the mystacial pad. Although whiskers have several degrees of freedom, the most important whisker movement is one-dimensional: forward (protraction) and backward (retraction). (*b*) The whisker follicle inserts into the pad, and movements of the whisker are generated by two types of muscle (*left*). Intrinsic muscle (*green*) forms a sling around the deep base of one whisker follicle and attaches to the upper part of the immediately posterior whisker follicle. Contraction of intrinsic muscle pulls the base of the whisker posteriorly, generating rotation of the whisker such that it protracts (*middle*). Extrinsic muscles (*red*) nasolabialis and maxillolabialis attach posteriorly to bone and act superficially on the whisker pad. Contraction of these extrinsic muscles causes whisker retraction by pulling the whisker pad backward and also causing backward rotation of whiskers (*right*).

map, with ventral protraction motor neurons innervating intrinsic whisker muscles and dorsal retraction motor neurons innervating extrinsic whisker muscles.

In an elegant series of experiments, Herfst & Brecht (2008) made whole-cell membrane potential recordings from whisker motor neurons in the lateral facial nucleus of the anesthetized rat. Injection of depolarizing current triggered action potential firing in the single neuron being recorded, which evoked reliable whisker movements with short latency (**Figure 2b**). Most whisker motor neurons evoked protraction of only a single whisker, presumably through contraction of the intrinsic muscle attached to an individual whisker. A smaller fraction of neurons recorded in the facial nucleus evoked whisker retraction, typically involving multiple whiskers and likely resulting from contraction of extrinsic muscle acting on the whole whisker pad. Single action potentials in different motor neurons of the facial nucleus evoked whisker movements with very

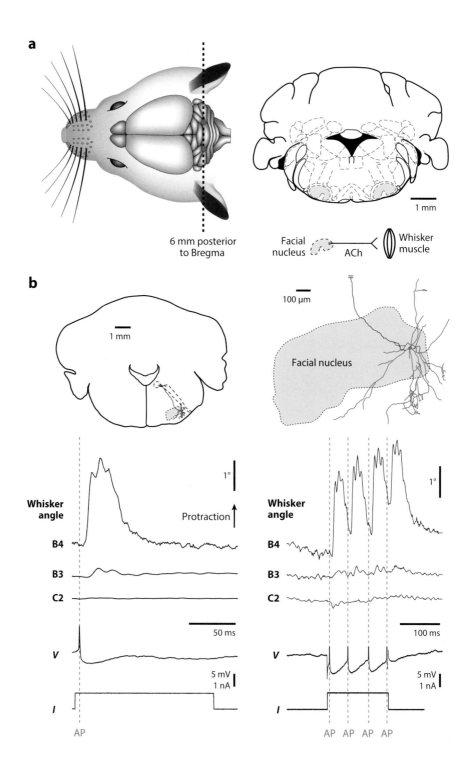

a

6 mm posterior
to Bregma

1 mm

Facial
nucleus — ACh — Whisker
muscle

b

1 mm

100 μm

Facial nucleus

Whisker
angle

1°

Protraction ↑

B4

B3

C2

50 ms

V

5 mV
1 nA

I

AP

Whisker
angle

1°

B4

B3

C2

100 ms

V

5 mV
1 nA

I

AP AP AP AP

different amplitudes (ranging from −0.6° to 5.6°), but trial-to-trial variability for an individual motor neuron was low. Latencies from action potential to onset of whisker movement were short (ranging from 4.0 to 11.1 ms). The measurements of Herfst & Brecht (2008) therefore define a fast and reliable pathway for controlling whisker movements by motor neurons located in the lateral facial nucleus.

The firing patterns of whisker motor neurons during behavior are unknown, and future experiments are needed to measure their activity directly during exploratory whisking. It will be of great interest to examine the relative timing of action potential firing in different whisker motor neurons during the whisking cycle. Furthermore, the recruitment patterns of different motor neurons under different behavioral circumstances will be important to study because the whiskers are moved in different ways depending on task requirements. A first step toward such data was made by Cramer & Keller (2006), who recorded from whisker motor neurons during fictive whisking driven by stimulation of whisker M1. Their data suggest that phasic action potential firing of motor neurons occurs shortly before each whisker protraction and that larger-amplitude movements are associated with increases in firing rate of individual motor neurons, as well as with recruitment of additional motor neurons.

WHISKER PREMOTOR NEURONS

The whisker motor neurons located in the lateral facial nucleus therefore directly drive whisker movement. These motor neurons receive synaptic inputs distributed across their somatodendritic compartments, which determine when action potentials are fired. To understand the mechanisms controlling whisker movement, we therefore need to learn about the whisker premotor neurons, which innervate the whisker motor neurons. Hattox et al. (2002) injected the retrograde anatomical tracer cholera toxin subunit B into the lateral facial nucleus and found retrogradely labeled cell bodies in a large number of brain areas including the brain stem reticular formation, the nucleus ambiguus, the pedunculopontine tegmental nucleus, the Kölliker-Fuse nucleus, the parabrachial nuclei, the superior colliculus, the red nucleus, the periaqueductal gray, the mesencephalon, the pons, and several nuclei involved in oculomotor behaviors. However, the specificity of classical retrograde labeling methods is limited. Retrograde transsynaptic viral methods based on modified rabies virus have recently been developed, which promise to reveal specific monosynaptic neural circuits (Wickersham et al. 2007). Rabies virus is well known to spread across neurons in the nervous system, apparently crossing synapses in an exclusively retrograde manner (Ugolini 1995, Kelly & Strick 2000). However, intact rabies virus will replicate and spread sequentially across

←——————————————————————————————

Figure 2

Whisker motor neurons are located in the lateral facial nucleus. (*a*) Schematic drawing to indicate the location of the facial nucleus, ~6 mm posterior to Bregma in the mouse (*left*). A schematic drawing of a coronal section of the mouse brain (Paxinos & Franklin 2001) showing the ventral location of the facial nucleus (*right*). The lateral facial nucleus contains the cholinergic motor neurons controlling whisker movement. (*b*) Whole-cell membrane potential recording of an anatomically identified neuron in the lateral facial nucleus in an anesthetized rat (*above*; axon in *blue*, dendrites and soma in *red*) (Herfst & Brecht 2008). Injection of depolarizing current through the recording electrode evoked a single action potential (AP) (*lower left*) or a train of four action potentials (*lower right*). Each action potential in the motor neuron drove a brief forward protraction of the B4 whisker but had little impact on other nearby whiskers B3 and C2. Panel *a* (*right*) is modified from Paxinos & Franklin (2001) and reprinted with permission from Academic Press. Panel *b* is modified from Herfst & Brecht (2008) and reprinted with permission from the American Physiological Society. Other abbreviation: ACh, acetylcholine.

many synapses, which complicates the interpretation of data. A critical step to map monosynaptically connected neurons is thus to restrict the spread of rabies virus so that it can cross only one synapse. One gene in the rabies virus genome encodes for the glycoprotein G, which is essential for infection. Rabies virus lacking G (ΔG-rabies) can then be transcomplemented by expression of rabies G in specific cell types, from which the virus can then spread retrogradely (Wickersham et al. 2007). Because the spreading ΔG-rabies virus does not encode G in its genome, it cannot make its glycoprotein in the upstream infected neurons, and therefore it cannot spread beyond the first-order presynaptic neurons (Wickersham et al. 2007). Replacing the gene encoding G by GFP (ΔG-GFP rabies) allows investigators to visualize infected neurons using fluorescence imaging. Such monosynaptic rabies-based circuit-mapping methods have now been applied to study the organization of premotor neurons in the mouse spinal cord (Stepien et al. 2010) and in whisker premotor neurons (Takatoh et al. 2013).

Monosynaptic rabies virus tracing from intrinsic and extrinsic whisker muscles has provided a comprehensive map of whisker premotor neurons (**Figure 3a**). Takatoh et al. (2013) injected ΔG-GFP rabies into whisker muscles of transgenic mice expressing rabies G in the cholinergic motor neurons. Motor neurons infected with ΔG-GFP rabies could therefore complement the G-deficient rabies with rabies G expressed transgenically from the mouse genome. The rabies thus spread one synapse retrogradely to label premotor neurons with high GFP levels. The locations of whisker premotor neurons found with monosynaptic rabies (Takatoh et al. 2013) generally agreed with previous retrograde labeling (Hattox et al. 2002). Whisker premotor neurons were prominently labeled in the dorsal medullary reticular nucleus of the brain stem, the intermediate reticular nucleus of the brain stem (IRt, **Figure 3a**), the gigantocellular reticular nucleus of the brain stem (GiRt, **Figure 3a**), the Kölliker-Fuse nucleus, the pre-Bötzinger and Bötzinger complexes, the rostral part of the lateral paragigantocellular nucleus, the rostral part of the spinal trigeminal interpolaris nucleus (SP5i), the spinal trigeminal oralis nucleus, the superior colliculus, and the mesencephalic reticular nucleus. The rabies-based tracing showed some differences in premotor circuits for intrinsic and extrinsic muscles. Motor neurons controlling intrinsic muscles

---

Figure 3

Whisker premotor neurons. (*a*) Transsynaptic modified rabies virus can be used to label premotor neurons (Takatoh et al. 2013) (*above*). The glycoprotein G was replaced by GFP in the rabies genome, making ΔG-GFP rabies virus. This virus was injected into whisker muscles to infect motor neurons. The motor neurons of the transgenic mouse specifically express rabies G, generated by Cre recombinase expressed in cholinergic neurons (Chat-Cre) acting on loxP-stop-loxP elements to drive expression of rabies G from the RΦGT transgene. The transgenic rabies G transcomplements the ΔG-GFP rabies virus in whisker motor neurons, making a new infectious virus that can retrogradely specifically infect the presynaptic whisker premotor neurons. Only monosynaptically connected neurons are labeled because the viral genome remains G-deficient and therefore cannot propagate. Infected neurons express GFP and can therefore be visualized through fluorescence microscopy. Premotor neurons expressing GFP are found in many brain locations including the brain stem, the midbrain, and the neocortex. When the rabies virus is injected into intrinsic muscle, many premotor neurons are labeled in brain stem reticular (Rt) nuclei. These Rt nuclei are located posterior to the facial nucleus. The schematic drawings indicate a plane 7 mm posterior to Bregma (*lower left*), which contains brain stem Rt nuclei, shown in the coronal section (*lower middle*) (Paxinos & Franklin 2001). GFP-labeled whisker premotor neurons are evident in intermediate reticular nucleus (IRt) and gigantocellularis (GiRt) (*lower right*) (Takatoh et al. 2013). (*b*) A schematic drawing (*left*) of a horizontal section of the brain stem stained with cytochrome oxidase (*middle*). The section includes the facial nucleus (FN), the Rt nuclei, and the spinal trigeminal interpolaris nucleus (SP5i). The lesion sites, labeled Rt and SP5i, show the locations that had previously been electrically stimulated in the awake head-restrained mouse. Stimulation of the Rt nuclei drove whisker protraction (*green*), whereas stimulation of SP5i drove whisker retraction (*red*) (*right*). Panel *a* (*upper, lower right*) is reprinted from Takatoh et al. (2013) with permission from Cell Press. Panel *a* (*lower middle*) is modified from Paxinos & Franklin (2001) and reprinted with permission from Academic Press. Panel *b* is modified from Matyas et al. (2010) and reprinted with permission from the American Association for the Advancement of Science. Other abbreviations: GFP, green fluorescent protein; NA, nucleus ambiguus; PCRt, parvicellular reticular nucleus.

were more strongly innervated by IRt. Motor neurons controlling extrinsic muscle, however, were more strongly innervated by the rostral part of SP5i. In agreement with this spatial difference in the location of premotor neurons for intrinsic and extrinsic muscles, microstimulation of the brain stem reticular nuclei (Rt) evokes whisker protraction presumably by contracting intrinsic whisker muscles, whereas microstimulation of SP5i evokes whisker retraction, presumably by contracting extrinsic muscles (**Figure 3b**) (Matyas et al. 2010).

Further investigations into the specific neural circuits controlling intrinsic and extrinsic muscles will be of great interest. A key goal is to record the activity of defined whisker premotor neurons during different whisker behaviors. Future experiments could also utilize rabies virus expressing

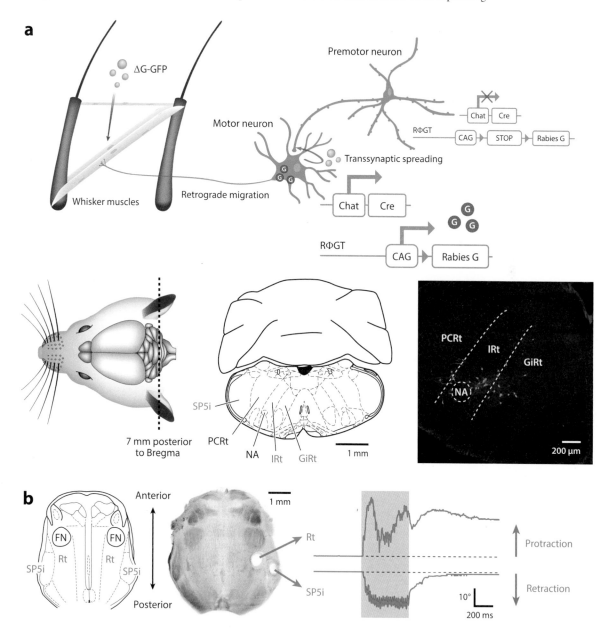

channelrhodopsin-2 (ChR2) (Boyden et al. 2005) to stimulate specific premotor circuits and measure the evoked movements. Of equal importance would be optogenetic inactivation experiments to investigate the role of the different premotor circuits for specific aspects of whisker behavior.

Recently, Moore et al. (2013) made lesions at various locations in the whisker-related brain stem, finding that a ventral region of the intermediate reticular nucleus of the brain stem (vIRt) lying medial to the nucleus ambiguus was essential for whisking. Furthermore, action potential firing of neurons near vIRt was phase-locked to whisker protraction, and whisking could be induced by pharmacological stimulation of neurons near vIRt. Whisking premotor neurons in vIRt might therefore form a central pattern generator driving rhythmic whisker protraction (Moore et al. 2013).

In addition to the complex distribution of premotor neurons in the brain stem and midbrain, monosynaptic rabies virus injected in whisker muscles also labels a very sparse population of layer-5 pyramidal neurons in the neocortex, with a few premotor neurons apparently residing in both M1 and S1 (see Takatoh et al. 2013, supplemental figure 1). Some neocortical neurons therefore appear to be whisker premotor neurons, directly innervating whisker motor neurons.

INNERVATION OF BRAIN STEM BY THE SENSORIMOTOR CORTEX

Injection of anterograde tracers into M1 and S1 reveals the direct long-range axonal projections from glutamatergic pyramidal neurons in these cortical areas (White & DeAmicis 1977, Wise & Jones 1977, Porter & White 1983, Welker et al. 1988, Miyashita et al. 1994, Grinevich et al. 2005, Aronoff et al. 2010, Matyas et al. 2010, Mao et al. 2011). Both M1 and S1 project to a wide variety of brain regions that could directly or indirectly cause whisker movement, including the striatum, the thalamus, the superior colliculus, the pons, the red nucleus, and various brain stem nuclei. Here, we focus on the extensive cortical innervation of the brain stem, which is among the more direct pathways in which the cortex can drive whisker movement. However, it is important to note that the cortex can affect whisker movement using many alternative routes, notably including pathways via the superior colliculus (Hemelt & Keller 2008) and cerebellar circuits (Legg et al. 1989).

Injection of lentivirus-expressing GFP into whisker M1 served as a viral-based anterograde tracer, labeling axonal output to different brain stem nuclei (Grinevich et al. 2005). Grinevich et al. (2005) found some direct innervation of the lateral facial nucleus from M1 (**Figure 4a,b**), in agreement with the monosynaptic rabies experiments indicating a few premotor neurons in the sensorimotor cortex (Takatoh et al. 2013). However, the most prominent axonal projection to the brain stem from M1 is the strong innervation of the Rt nuclei, including the dorsal medullary reticular nucleus (MDd), the IRt, and the GiRt (**Figure 4a,c**) (Grinevich et al. 2005, Matyas et al. 2010). Monosynaptic rabies tracing from whisker motor neurons retrogradely labeled these brain stem Rt nuclei, so these cortical projections from M1 could directly innervate whisker premotor neurons. In particular, M1 axons innervate vIRt, the region proposed to be the whisking central pattern generator (Moore et al. 2013). Among other possible pathways, M1 might thus drive whisker movement through direct innervation of motor neurons in the facial nucleus and through premotor neurons located in the brain stem reticular nuclei.

Analysis of the axonal projections from whisker S1 shows a pattern of subcortical connectivity very similar to that found from whisker M1 (Matyas et al. 2010). S1 and M1 project to neighboring regions of the striatum, the thalamus, the superior colliculus, the pons, the red nucleus, and the brain stem (Matyas et al. 2010). In the brain stem, S1 strongly innervates SP5i (Matyas et al. 2010). According to monosynaptic rabies tracing, SP5i is supposed to contain whisker premotor neurons preferentially innervating extrinsic whisker muscles (Takatoh et al. 2013). The axonal projection of S1 neurons to the brain stem SP5i could therefore drive whisker retraction via extrinsic whisker muscles.

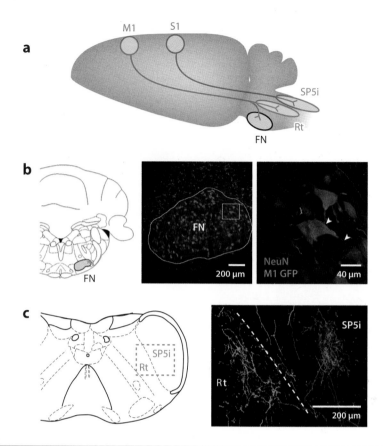

Figure 4

Innervation of the brain stem by the sensorimotor cortex. (*a*) A schematic drawing showing axonal projections from M1 (*green*) innervating the facial nucleus (FN) and brain stem reticular (Rt) nuclei. Axonal projections from S1 (*red*) innervate spinal trigeminal interpolaris (SP5i). (*b*) Lentivirus-expressing GFP was injected into M1, and some labeled axons (*green*) were found in the FN with neurons stained for NeuN (*red*) (Grinevich et al. 2005, an M1 axon in close proximity to a FN cell is highlighted by *arrowheads*). (*c*) Anterograde tracing of axons from M1 (*green*) and S1 (*red*) reveals dense axonal labeling in the brain stem. M1 strongly innervates Rt, whereas S1 strongly innervates SP5i (Matyas et al. 2010). Panels *a* and *c* are modified from Matyas et al. (2010) and reprinted with permission from the American Association for the Advancement of Science. Panel *b* is modified from Grinevich et al. (2005) and reprinted with permission from the Society for Neuroscience.

WHISKER MOVEMENTS EVOKED BY STIMULATION OF THE SENSORIMOTOR CORTEX

The similarity of axonal projections from M1 and S1 to the brain stem suggests that they could equally drive whisker movement through their apparently parallel, analogous projections from M1 to Rt and from S1 to SP5i. Consistent with this hypothesis, stimulation of either M1 or S1 was found to evoke whisker movements (Matyas et al. 2010). Before stimulating, the sensorimotor neocortex was functionally mapped through voltage-sensitive dye imaging (Grinvald & Hildesheim 2004, Ferezou et al. 2007, Matyas et al. 2010). The C2 whisker representation in S1 was defined as the location of the earliest sensory response evoked by whisker deflection (**Figure 5a**). Neurons in the C2 barrel column project directly to whisker M1 (Ferezou et al. 2007, Matyas et al. 2010, Mao et al. 2011). Whisker M1 therefore receives sensory input, and its location can thus also be

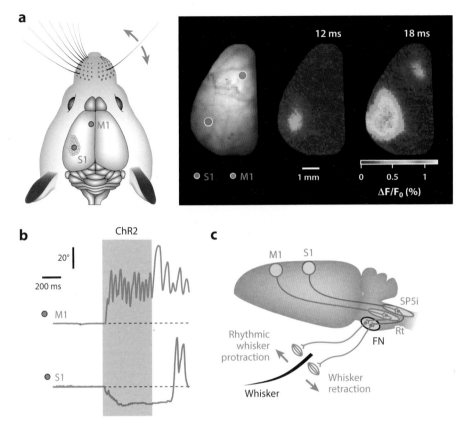

Figure 5

Whisker movements evoked by stimulating the sensorimotor cortex. (*a*) Functional localization of the sensory responses evoked by brief 1-ms deflection of the C2 whisker imaged using voltage-sensitive dye. At 12 ms after whisker deflection (*center right*) a localized depolarization reveals the location of the C2 whisker representation in S1. A few milliseconds later at 18 ms (*far right*), the depolarization has spread within S1 and a second localized hot spot of depolarization appears in the frontal cortex, identifying the location of whisker M1. (*b*) Optogenetic stimulation of M1 with channelrhodopsin-2 (ChR2) drives rhythmic whisker protraction (*green*, with S1 inactivated). Optogenetic stimulation of S1 drives whisker retraction (*red*, with M1 inactivated). (*c*) Schematic summary of the signaling pathway through which M1 and S1 might evoke whisker movements. M1 (*green*) projects to brain stem reticular (Rt) nuclei, exciting Rt premotor neurons, which evoke activity in protraction motor neurons of the facial nucleus (FN) driving contraction of intrinsic whisker muscles. Neural circuits within Rt nuclei may form a central pattern generator underlying rhythmic protraction. S1 (*red*) projects to the spinal trigeminal interpolaris nucleus (SP5i), exciting SP5i premotor neurons, which in turn would evoke activity in FN retraction motor neurons to drive contraction of extrinsic whisker muscles. All panels are modified from Matyas et al. (2010) and reproduced with permission from the American Association for the Advancement of Science.

functionally localized through voltage-sensitive dye imaging (Ferezou et al. 2007, Matyas et al. 2010). Approximately 6 ms after the initial depolarization in S1, a secondary localized hot spot of activity was found in the frontal cortex, defining the location of whisker M1 (**Figure 5a**). Intracortical microstimulation was then targeted to the functionally identified regions of M1 and S1. To prevent complications induced by cortico-cortical signaling, S1 was inactivated when M1 was stimulated, and vice versa: M1 was inactivated when stimulating S1 (Matyas et al. 2010). These intracortical microstimulation experiments revealed that M1 drives rhythmic whisker protraction, whereas S1 drives whisker retraction. Latencies for evoking movement were shorter for S1 (14.8 ± 2.8 ms) than for M1 (21.1 ± 5.8 ms) (Matyas et al. 2010). Optogenetic stimulation showed the same whisker motor map, S1 driving short-latency whisker retraction and M1 driving rhythmic whisker protraction (Matyas et al. 2010, Mateo et al. 2011) (**Figure 5*b***). These results are in good agreement with the anatomical connectivity of neural circuits described above. Activity in M1 neurons projecting to reticular nuclei in the brain stem may excite whisker premotor neurons (perhaps in vIRt), which in turn innervate whisker motor neurons in the facial nucleus that preferentially drive whisker protraction (**Figure 5*c***). On the other hand, activity in S1 neurons projecting to the brain stem may drive firing of premotor neurons in SP5i, which preferentially innervate retraction motor neurons (**Figure 5*c***). Future studies must directly test this hypothesis by applying the increasingly precise molecular methods that have been developed, such as rabies virus and optogenetic interventions. Different types of neocortical neurons will likely make different impacts on whisker movement. Action potential firing in some individual cortical neurons appears to evoke a measurable whisker movement, albeit with long latencies (Brecht et al. 2004b). It will therefore be important to study neocortical cell-type specificity in the context of connected neural circuits for motor control.

There is some uncertainty about the overall structure and function of whisker M1 (Brecht 2011). In the discussion above, whisker M1 is defined through its sensory map, as a localized hot spot of whisker-deflection evoked activity, which colocalizes with the axonal projection from S1 to M1. Deflection of different whiskers evokes somatotopically organized hot spots of activity in both S1 and M1 (Ferezou et al. 2007). Thus a well-defined sensory whisker map in M1 has been defined functionally using voltage-sensitive dye imaging (Ferezou et al. 2007) and anatomically through tracing of axonal projections from S1 to M1 (Mao et al. 2011). Intracortical microstimulation experiments suggest well-ordered whisker motor maps in M1 (Brecht et al. 2004a), and in future studies it will therefore be interesting to investigate if the sensory map in M1 aligns and colocalizes precisely with the motor map in M1. This proposed mapping is currently under debate; some studies suggest that the sensory map in M1 does not match the location of the motor map (Smith & Alloway 2013). Intracortical microstimulation studies also suggest that M1 might contain distinct whisker-related subregions, including a whisker retraction area that is supposed to be spatially separated from a rhythmic whisking area (Haiss & Schwarz 2005). However, Matyas et al. (2010) found that the whisker motor maps in M1 change dramatically upon inactivation of S1, the retraction area in M1 becoming a protraction area after S1 inactivation. M1 motor maps are therefore not trivial to interpret, owing to strong cortico-cortical connectivity. Currently, the simplest possibility is that there is only one whisker M1 region, which directly drives exploratory rhythmic whisker protraction and which can also indirectly drive whisker retraction via cortico-cortical connectivity to S1 (Matyas et al. 2010). Areas surrounding whisker M1 also appear to drive whisker protraction, suggesting an overall broad tuning of M1 (Matyas et al. 2010).

CORRELATION OF WHISKER MOVEMENT AND CORTICAL ACTIVITY IN M1 AND S1

So far we have established some possible neural circuits that allow activity in M1 and S1 to drive distinct whisker movements when these cortical regions are directly stimulated. However, to investigate possible physiological roles of these cortical brain regions with respect to whisker motor control, studies must record the activity of neurons in M1 and S1 and correlate their activity with whisker movements during behavior. The neocortex contains a large diversity of cell types, and one might anticipate that the most direct impact of cortical neurons upon whisker movement is via neurons projecting to the whisker motor regions of the brain stem. Unfortunately, measurements have not yet been made of the activity of these specific brain stem–projecting neocortical neurons during whisker-related behavior. Recordings have, however, been made from unlabeled neurons and other types of neurons in M1 and S1 during whisker-related behavior.

Extracellular recordings of action potential firing of neurons in whisker M1 during quantified whisker movements have been reported in only a small number of studies (Carvell et al. 1996, Hill et al. 2011, Friedman et al. 2012), and there are no published measurements of membrane potential from M1 of awake mice during whisking behavior. In agreement with Carvell et al. (1996), Friedman et al. (2012) found that M1 neurons increase firing rate during whisking compared with nonwhisking periods (**Figure 6a**). In addition, Friedman et al. (2012) reported that the increased firing in M1 preceded the onset of whisker movement. Increased action potential firing in M1 neurons could therefore contribute to the initiation of whisking, perhaps through the previously discussed projections to the reticular brain stem, for example onto vIRt neurons driving rhythmic whisker protraction. However, in probably the most detailed study of M1 activity during whisking to date, Hill et al. (2011) show that firing rates of M1 neurons both increase and decrease so that the average rate across the population is little changed during whisking. In future studies, it will therefore be important to distinguish different cell types in M1, which could have different activity patterns during whisking. Hill et al. (2011) do, however, find that M1 activity is modulated in important ways during whisking. They report that M1 neurons change firing rate with respect to the amplitude of whisking and the midpoint of whisking, and some cells also showed rapid modulations in firing rate at specific phases during the whisking cycle (**Figure 6b**). The modulation of M1 activity during whisking was not changed when the sensory whisker nerve (the infraorbital nerve, ION) was cut (Hill et al. 2011). The activity of M1 neurons may therefore relate primarily to motor commands rather than to sensory information.

Researchers have also begun to study the activity of neurons in S1 during whisker-related behavior. Juxtasomal recordings from anatomically identified excitatory neurons in S1 reveal that the overall spike rate is not different when comparing epochs of whisking and no whisking (de Kock & Sakmann 2009). However, slender-tufted pyramidal neurons located in L5A of the S1

Figure 6

Correlation of activity in sensorimotor cortex with whisker movement. (*a*) Example extracellular recording of spiking activity in M1 (Friedman et al. 2012), with a spike raster of five whisking onset epochs (*above*) and the spike time histogram from many epochs (*below*). Time zero is aligned across trials to be the onset time of whisking. Action potential firing increases shortly before onset of whisking and remains elevated during the first second of whisking. (*b*) In a different example extracellular recording of spiking activity in M1 (Hill et al. 2011), the firing rate is modulated by the amplitude of whisking (*left*), by the midpoint of the whisker position during whisking (*middle*), and by the phase of the whisker position within the whisking cycle (*right*). (*c*) Simultaneous recording of local field potential (LFP) and whole-cell recording of membrane potential (V_m) in layer 2/3 of the C2 barrel column of S1 in an awake behaving mouse (Poulet & Petersen 2008). During quiet wakefulness, when the whisker angle (*green*) is not changing, the LFP and V_m show slow, large-amplitude, synchronous fluctuations. When the mouse is actively whisking, the LFP and V_m reduce variance, reduce slow fluctuations, and on average depolarize. Whisking therefore induces an important change in cortical state, which does not depend on sensory input from the whisker. (*d*) When the membrane potential across different whisking cycles is aligned to the peak of protraction and averaged, then there is an obvious fast phase-locked V_m fluctuation (Poulet & Petersen 2008). Averaging at random times (Shuffled, S; Normal unshuffled, N) shows the noise level. The same analysis carried out in mice with cut sensory infraorbital nerves (IONs) reveals that the fast phase-locked V_m modulation during whisking depends on sensory input because it is absent after cutting IONs. Panel *a* is modified from Friedman et al. (2012) and reproduced with permission from the American Physiological Society. Panel *b* is modified from Hill et al. (2011) and reproduced with permission from Cell Press. Panels *c* and *d* are modified from Poulet & Petersen (2008) and reproduced with permission from the Nature Publishing Group.

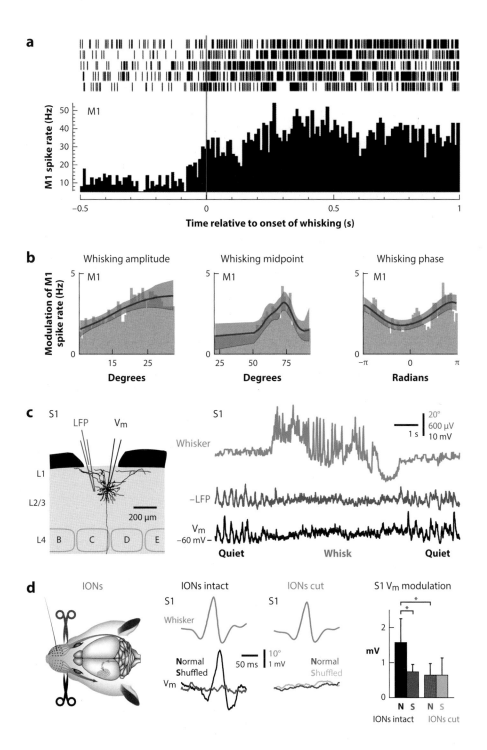

a

M1

M1 spike rate (Hz)

Time relative to onset of whisking (s)

b

Modulation of M1 spike rate (Hz)

Whisking amplitude

M1

Degrees

Whisking midpoint

M1

Degrees

Whisking phase

M1

Radians

c

S1

LFP Vm

L1

L2/3

L4 B C D E

200 µm

S1

Whisker

20°
600 µV
10 mV
1 s

−LFP

Vm
−60 mV −

Quiet Whisk Quiet

d

IONs

IONs intact

S1
Whisker

IONs cut

S1
Whisker

S1 Vm modulation

Normal
Shuffled

Vm

10°
1 mV
50 ms

Normal
Shuffled

mV

N S   N S
IONs intact IONs cut

barrel cortex increased their firing rate during whisking (de Kock & Sakmann 2009). Whole-cell membrane potential recordings have been carried out from identified neurons in layer 2/3 during whisking (Crochet & Petersen 2006; Poulet & Petersen 2008; Gentet et al. 2010, 2012). In agreement with extracellular recordings, the whole-cell recordings of excitatory neurons do not, on average, change spike rate during whisking. However, the spike rates of inhibitory γ-aminobutyric acid (GABA)ergic neurons change in a cell-type-specific manner. Parvalbumin-expressing fast-spiking GABAergic neurons reduce firing rate during whisking, non-fast-spiking presumed $5HT_{3A}$-receptor expressing neurons increase firing rate during whisking, and somatostatin-expressing neurons decrease firing rate during whisking (Gentet et al. 2010, 2012; Petersen & Crochet 2013). Thus GABAergic inhibition is significantly reorganized in S1 during whisking. In addition, an important change in brain state that accompanies whisking is clearly observed in S1. During quiet wakefulness (when the whiskers are not moving), the local field potential and membrane potential of neurons in S1 often show slow, large-amplitude fluctuations (Petersen et al. 2003, Crochet & Petersen 2006, Poulet & Petersen 2008, Okun et al. 2010, Zagha et al. 2013). During active whisking, the membrane potential of excitatory neurons depolarizes, the membrane potential variance decreases, and the slow membrane potential fluctuations are suppressed (Crochet & Petersen 2006, Poulet & Petersen, 2008) (**Figure 6c**). These changes in cortical state do not depend on sensory input because whisking induces a similar change in S1 cortical state when the sensory nerve (ION) is cut (Poulet & Petersen 2008). S1 therefore appears to encode motor-related signals, similar to results in mouse V1 (Niell & Stryker 2010, Keller et al. 2012). Increases in thalamic firing (Poulet et al. 2012), motor cortex firing (Zagha et al. 2013) or neuromodulation (Lee & Dan 2012) likely contribute to driving the brain state changes in S1 during whisking.

Analyzed at higher temporal resolution, extracellular recordings from S1 reveal rapid cycle-by-cycle modulation during whisking, with individual units firing at specific phases of the whisking cycle (Fee et al. 1997, Curtis & Kleinfeld 2009). These firing rate modulations are likely driven by membrane potential fluctuations phase-locked to the whisking cycle (Crochet & Petersen 2006, Poulet & Petersen 2008, Crochet et al. 2011) (**Figure 6d**). Different neurons depolarize and fire at different phases of the whisking cycle, thus encoding whisker position on the millisecond timescale. This fast phase-locked activity in S1 does depend on sensory signals from the periphery (Fee et al. 1997, Poulet & Petersen 2008) (**Figure 6d**). This finding contrasts with the phase-locked activity in M1, which is independent of sensory reafference signals. Whereas the fast phase-locked signals in S1 are largely related to sensory signals, the signals in M1 are more likely related to motor commands, which agrees with the overall notions of cortical organization. However, S1 activity may nonetheless impact whisker movement through direct and indirect pathways. S1 activity is rapidly signaled to M1 (Ferezou et al. 2007) and so the whisking-induced changes in S1 activity could also be relayed to M1. S1 activity might also impact whisker movement more directly via brain stem projections and other subcortical projections, as discussed earlier.

IMPACT OF S1 AND M1 INACTIVATION ON WHISKER MOVEMENT

Whereas cortical stimulation reveals pathways for motor control and electrophysiological measurements reveal correlations of neural activity and movement, inactivation experiments investigate the necessity of the neural activity. Precise inactivation experiments with cell-type specificity and spatiotemporal control will be enormously informative in future experiments. Experiments until now have largely been limited to lesions or pharmacological inactivations of S1 or M1, but these experiments have nonetheless revealed some important general insights into whisker motor control. Most importantly, rodents can whisk in a relatively normal way after lesion of M1 (Welker 1964, Semba & Komisaruk 1984, Gao et al. 2003). Thus although M1 activity can evoke whisking, it appears that other pathways normally drive spontaneous exploratory whisking, including pathways signaling via serotonin (Hattox et al. 2003). Pharmacological inactivation of S1 reduces whisker retraction evoked by high-frequency whisker stimulation (Matyas et al. 2010), but it otherwise appears to make little impact on spontaneous whisker movement. Various brain areas and signaling pathways will likely play diverse roles during different behaviors. In the future, it will therefore be important to investigate the roles of S1 and M1 during specific learned behaviors. Huber et al. (2012) trained mice to locate objects with whiskers and found that pharmacological inactivation of M1 changes whisker movement and behavioral performance in this task.

CONCLUSIONS AND FUTURE PERSPECTIVES

Current evidence suggests that the cortex can drive whisker movements using at least two distinct pathways. Whisker M1 projects to brain stem Rt, which contains whisker premotor neurons and includes the proposed whisking central pattern generator (vIRt), preferentially innervating protraction motor neurons of intrinsic whisker muscles. This anatomical pathway could therefore account for the rhythmic protraction movements evoked by stimulating M1. In an analogous parallel pathway, whisker S1 projects to the brain stem SP5i, which contains whisker premotor neurons preferentially innervating retraction motor neurons of extrinsic whisker muscles. This pathway could therefore account for the whisker retraction evoked by stimulating S1.

These two different cortical regions drive qualitatively different whisker movements. M1 drives rhythmic whisker protraction, which resembles exploratory whisking. M1 activity therefore appears well-suited to increase the amount of sensory information arriving from the whiskers during active sensing. S1, however, drives whisker retraction, which might serve as a negative feedback signal, preventing overstimulation of the whisker system. M1 and S1 therefore appear to play fundamentally different roles in whisker motor control. A region close to, or overlapping with, whisker M1 has been suggested to be a frontal orienting field (Erlich et al. 2011), homologous to the frontal eye field of primates. Whisker M1 may therefore be involved in multiple brain functions, including whisker motor control, spatial attention, and preparation of orienting responses. How actions and action plans are mapped onto the cortex therefore remains poorly understood in mice, rats, and monkeys (Graziano & Aflalo 2007).

Although it is clear from stimulation experiments that the cortex can drive whisker movements, these experiments do not necessarily indicate the physiological role of M1 and S1 in regulating normal whisker movement during behavior. Recordings of M1 and S1 activity in behaving animals have not yet been targeted to the cell types likely to be most directly related to whisker movement, so we currently know rather little about the causal influences of normal patterns of cortical activity upon whisker movement. Future experiments in behaving animals must therefore record the activity of defined types of neurons across the whisker motor control pathways, including the cortical neurons projecting to the brain stem and other motor-related brain regions. To test whether the activity in these neural circuits accounts for the movements, future experimental work must specifically interfere with the proposed synaptic pathways, perhaps through combining retrograde transsynaptic virus (Wickersham et al. 2007) and optogenetic inhibition (Zhang et al. 2007, Chow et al. 2010).

In conclusion, although the synaptic pathways for the cortex to control whisker movement are beginning to be mapped, we still understand little about the normal physiological role of the cortex in whisker motor control. Furthermore, in this review, we have focused only on the simplest pathways from the cortex to whisker motor neurons via brain stem, but it is important to remember that there are many more complex signaling pathways, for example through the basal ganglia, the superior colliculus, and the cerebellum, all of which likely contribute during behavior.

DISCLOSURE STATEMENT

The author is not aware of any affiliations, memberships, funding, or financial holdings that might be perceived as affecting the objectivity of this review.

ACKNOWLEDGMENTS

This work was funded by grants from the Swiss National Science Foundation, the Human Frontier Science Program and the European Research Council. I thank Varun Sreenivasan and Alexandros Kyriakatos for valuable discussions.

LITERATURE CITED

Aronoff R, Matyas F, Mateo C, Ciron C, Schneider B, Petersen CCH. 2010. Long-range connectivity of mouse primary somatosensory barrel cortex. *Eur. J. Neurosci.* 31:2221–33

Bosman LW, Houweling AR, Owens CB, Tanke N, Shevchouk OT, et al. 2011. Anatomical pathways involved in generating and sensing rhythmic whisker movements. *Front. Integr. Neurosci.* 5:53

Boyden ES, Zhang F, Bamberg E, Nagel G, Deisseroth K. 2005. Millisecond-timescale, genetically targeted optical control of neural activity. *Nat. Neurosci.* 8:1263–68

Brecht M. 2007. Barrel cortex and whisker-mediated behaviors. *Curr. Opin. Neurobiol.* 17:408–16

Brecht M. 2011. Movement, confusion, and orienting in frontal cortices. *Neuron* 72:193–96

Brecht M, Krauss A, Muhammad S, Sinai-Esfahani L, Bellanca S, Margrie TW. 2004a. Organization of rat vibrissa motor cortex and adjacent areas according to cytoarchitectonics, microstimulation, and intracellular stimulation of identified cells. *J. Comp. Neurol.* 479:360–73

Brecht M, Schneider M, Sakmann B, Margrie TW. 2004b. Whisker movements evoked by stimulation of single pyramidal cells in rat motor cortex. *Nature* 427:704–10

Carvell GE, Miller SA, Simons DJ. 1996. The relationship of vibrissal motor cortex unit activity to whisking in the awake rat. *Somatosens. Mot. Res.* 13:115–27

Chow BY, Han X, Dobry AS, Qian X, Chuong AS, et al. 2010. High-performance genetically targetable optical neural silencing by light-driven proton pumps. *Nature* 463:98–102

Cohen JD, Hirata A, Castro-Alamancos MA. 2008. Vibrissa sensation in superior colliculus: wide-field sensitivity and state-dependent cortical feedback. *J. Neurosci.* 28:11205–20

Cooke DF, Taylor CSR, Moore T, Graziano MSA. 2003. Complex movements evoked by microstimulation of the ventral intraparietal area. *Proc. Natl. Acad. Sci. USA* 100:6163–68

Coulter JD, Jones EG. 1977. Differential distribution of corticospinal projections from individual cytoarchitectonic fields in the monkey. *Brain Res.* 129:335–40

Cramer NP, Keller A. 2006. Cortical control of a whisking central pattern generator. *J. Neurophysiol.* 96:209–17

Crochet S, Petersen CCH. 2006. Correlating whisker behavior with membrane potential in barrel cortex of awake mice. *Nat. Neurosci.* 9:608–10

Crochet S, Poulet JFA, Kremer Y, Petersen CCH. 2011. Synaptic mechanisms underlying sparse coding of active touch. *Neuron* 69:1160–75

Curtis JC, Kleinfeld D. 2009. Phase-to-rate transformations encode touch in cortical neurons of a scanning sensorimotor system. *Nat. Neurosci.* 12:492–501

de Kock CP, Sakmann B. 2009. Spiking in primary somatosensory cortex during natural whisking in awake head-restrained rats is cell-type specific. *Proc. Natl. Acad. Sci. USA* 106:16446–50

Diamond ME, von Heimendahl M, Knutsen PM, Kleinfeld D, Ahissar E. 2008. 'Where' and 'what' in the whisker sensorimotor system. *Nat. Rev. Neurosci.* 9:601–12

Donoghue JP, Wise SP. 1982. The motor cortex of the rat: cytoarchitecture and microstimulation mapping. *J. Comp. Neurol.* 212:76–88

Dörfl J. 1982. The musculature of the mystacial vibrissae of the white mouse. *J. Anat.* 135:147–54

Erlich JC, Bialek M, Brody CD. 2011. A cortical substrate for memory-guided orienting in the rat. *Neuron* 72:330–43

Fee MS, Mitra PP, Kleinfeld D. 1997. Central versus peripheral determinants of patterned spike activity in rat vibrissa cortex during whisking. *J. Neurophysiol.* 78:1144–49

Feldmeyer D, Brecht M, Helmchen F, Petersen CCH, Poulet JFA, et al. 2013. Barrel cortex function. *Prog. Neurobiol.* 103:3–27

Ferezou I, Haiss F, Gentet LJ, Aronoff R, Weber B, Petersen CCH. 2007. Spatiotemporal dynamics of cortical sensorimotor integration in behaving mice. *Neuron* 56:907–23

Ferrier D. 1874. Experiments on the brain of monkeys—No. 1. *Proc. R. Soc. Lond.* 23:409–30

Friedman WA, Zeigler HP, Keller A. 2012. Vibrissae motor cortex unit activity during whisking. *J. Neurophysiol.* 107:551–63

Fritsch G, Hitzig E. 1870. Über die elektrische Erregbarkeit des Grosshirns. *Arch. Anat. Physiol. Wiss. Med.* 37:300–32

Gao P, Hattox AM, Jones LM, Keller A, Zeigler HP. 2003. Whisker motor cortex ablation and whisker movement patterns. *Somatosens. Mot. Res.* 20:191–98

Gentet LJ, Avermann M, Matyas F, Staiger JF, Petersen CCH. 2010. Membrane potential dynamics of GABAergic neurons in the barrel cortex of behaving mice. *Neuron* 65:422–35

Gentet LJ, Kremer Y, Taniguchi H, Huang ZJ, Staiger JF, Petersen CCH. 2012. Unique functional properties of somatostatin-expressing GABAergic neurons in mouse barrel cortex. *Nat. Neurosci.* 15:607–12

Gioanni Y, Lamarche M. 1985. A reappraisal of rat motor cortex organization by intracortical microstimulation. *Brain Res.* 344:49–61

Grant RA, Mitchinson B, Fox CW, Prescott TJ. 2009. Active touch sensing in the rat: anticipatory and regulatory control of whisker movements during surface exploration. *J. Neurophysiol.* 101:862–74

Graziano MS, Aflalo TN. 2007. Mapping behavioral repertoire onto the cortex. *Neuron* 56:239–51

Graziano MS, Taylor CS, Moore T, Cooke DF. 2002. The cortical control of movement revisited. *Neuron* 36:349–62

Grinevich V, Brecht M, Osten P. 2005. Monosynaptic pathway from rat vibrissa motor cortex to facial motor neurons revealed by lentivirus-based axonal tracing. *J. Neurosci.* 25:8250–58

Grinvald A, Hildesheim R. 2004. VSDI: a new era in functional imaging of cortical dynamics. *Nat. Rev. Neurosci.* 5:874–85

Haidarliu S, Golomb D, Kleinfeld D, Ahissar E. 2012. Dorsorostral snout muscles in the rat subserve coordinated movement for whisking and sniffing. *Anat. Rec.* 295:1181–91

Haiss F, Schwarz C. 2005. Spatial segregation of different modes of movement control in the whisker representation of rat primary motor cortex. *J. Neurosci.* 25:1579–87

Hall RD, Lindholm EP. 1974. Organization of motor and somatosensory neocortex in the albino rat. *Brain Res.* 66:23–38

Harrison TC, Ayling OG, Murphy TH. 2012. Distinct cortical circuit mechanisms for complex forelimb movement and motor map topography. *Neuron* 74:397–409

Hattox A, Li Y, Keller A. 2003. Serotonin regulates rhythmic whisking. *Neuron* 39:343–52

Hattox AM, Priest CA, Keller A. 2002. Functional circuitry involved in the regulation of whisker movements. *J. Comp. Neurol.* 442:266–76

Hemelt ME, Keller A. 2008. Superior colliculus control of vibrissa movements. *J. Neurophysiol.* 100:1245–54

Herfst LJ, Brecht M. 2008. Whisker movements evoked by stimulation of single motor neurons in the facial nucleus of the rat. *J. Neurophysiol.* 99:2821–32

Hill DN, Bermejo R, Zeigler HP, Kleinfeld D. 2008. Biomechanics of the vibrissa motor plant in rat: rhythmic whisking consists of triphasic neuromuscular activity. *J. Neurosci.* 28:3438–55

Hill DN, Curtis JC, Moore JD, Kleinfeld D. 2011. Primary motor cortex reports efferent control of vibrissa motion on multiple timescales. *Neuron* 72:344–56

Huber D, Gutnisky DA, Peron S, O'Connor DH, Wiegert JS, et al. 2012. Multiple dynamic representations in the motor cortex during sensorimotor learning. *Nature* 484:473–78

Jin TE, Witzemann V, Brecht M. 2004. Fiber types of the intrinsic whisker muscle and whisking behavior. *J. Neurosci.* 24:3386–93

Keller GB, Bonhoeffer T, Hübener M. 2012. Sensorimotor mismatch signals in primary visual cortex of the behaving mouse. *Neuron* 74:809–15

Kelly RM, Strick PL. 2000. Rabies as a transneuronal tracer of circuits in the central nervous system. *J. Neurosci. Methods* 103:63–71

Knutsen PM, Biess A, Ahissar E. 2008. Vibrissal kinematics in 3D: tight coupling of azimuth, elevation, and torsion across different whisking modes. *Neuron* 59:35–42

Lee S-H, Dan Y. 2012. Neuromodulation of brain states. *Neuron* 76:209–22

Legg CR, Mercier B, Glickstein M. 1989. Corticopontine projection in the rat: the distribution of labelled cortical cells after large injections of horseradish peroxidase in the pontine nuclei. *J. Comp. Neurol.* 286:427–41

Mao T, Kusefoglu D, Hooks BM, Huber D, Petreanu L, Svoboda K. 2011. Long-range neuronal circuits underlying the interaction between sensory and motor cortex. *Neuron* 72:111–23

Mateo C, Avermann M, Gentet LJ, Zhang F, Deisseroth K, Petersen CCH. 2011. In vivo optogenetic stimulation of neocortical excitatory neurons drives brain-state-dependent inhibition. *Curr. Biol.* 21:1593–602

Matyas F, Sreenivasan V, Marbach F, Wacongne C, Barsy B, et al. 2010. Motor control by sensory cortex. *Science* 330:1240–43

Mitchinson B, Martin CJ, Grant RA, Prescott TJ. 2007. Feedback control in active sensing: Rat exploratory whisking is modulated by environmental contact. *Proc. Biol. Sci.* 274:1035–41

Miyashita E, Keller A, Asanuma H. 1994. Input–output organization of the rat vibrissal motor cortex. *Exp. Brain Res.* 99:223–32

Molinari HH, Schultze KE, Strominger NL. 1996. Gracile, cuneate, and spinal trigeminal projections to inferior olive in rat and monkey. *J. Comp. Neurol.* 375:467–80

Moore JD, Deschênes M, Furuta T, Huber D, Smear MC, et al. 2013. Hierarchy of orofacial rhythms revealed through whisking and breathing. *Nature* 497:205–10

Nagel G, Szellas T, Huhn W, Kateriya S, Adeishvili N, et al. 2003. Channelrhodopsin-2, a directly light-gated cation-selective membrane channel. *Proc. Natl. Acad. Sci. USA* 100:13940–45

Neafsey EJ, Bold EL, Haas G, Hurley-Gius KM, Quirk G, et al. 1986. The organization of the rat motor cortex: a microstimulation mapping study. *Brain Res.* 396:77–96

Niell CM, Stryker MP. 2010. Modulation of visual responses by behavioral state in mouse visual cortex. *Neuron* 65:472–79

Nguyen QT, Kleinfeld D. 2005. Positive feedback in a brainstem tactile sensorimotor loop. *Neuron* 45:447–57

Okun M, Naim A, Lampl I. 2010. The subthreshold relation between cortical local field potential and neuronal firing unveiled by intracellular recordings in awake rats. *J. Neurosci.* 30:4440–48

Osakada F, Mori T, Cetin AH, Marshel JH, Virgen B, Callaway EM. 2011. New rabies virus variants for monitoring and manipulating activity and gene expression in defined neural circuits. *Neuron* 71:617–31

Paxinos G, Franklin KBJ. 2001. *The Mouse Brain in Stereotaxic Coordinates.* San Diego: Academic. 2nd ed.

Penfield W, Boldrey E. 1937. Somatic motor and sensory representation in the cerebral cortex of man as studied by electrical stimulation. *Brain* 60:389–443

Petersen CCH. 2007. The functional organization of the barrel cortex. *Neuron* 56:339–55

Petersen CCH, Crochet S. 2013. Synaptic computation and sensory processing in neocortical layer 2/3. *Neuron* 78:28–48

Petersen CCH, Hahn TTG, Mehta M, Grinvald A, Sakmann B. 2003. Interaction of sensory responses with spontaneous depolarization in layer 2/3 barrel cortex. *Proc. Natl. Acad. Sci. USA* 100:13638–43

Porter LL, White EL. 1983. Afferent and efferent pathways of the vibrissal region of primary motor cortex in the mouse. *J. Comp. Neurol.* 214:279–89

Poulet JFA, Fernandez LM, Crochet S, Petersen CCH. 2012. Thalamic control of cortical states. *Nat. Neurosci.* 15:370–72

Poulet JFA, Petersen CCH. 2008. Internal brain state regulates membrane potential synchrony in barrel cortex of behaving mice. *Nature* 454:881–85

Rathelot JA, Strick PL. 2006. Muscle representation in the macaque motor cortex: an anatomical perspective. *Proc. Natl. Acad. Sci. USA* 103:8257–62

Rice FL, Fundin BT, Arvidsson J, Aldskogius H, Johansson O. 1997. Comprehensive immunofluorescence and lectin binding analysis of vibrissal follicle sinus complex innervation in the mystacial pad of the rat. *J. Comp. Neurol.* 385:149–84

Sachdev RN, Sato T, Ebner FF. 2002. Divergent movement of adjacent whiskers. *J. Neurophysiol.* 87:1440–48

Semba K, Komisaruk BR. 1984. Neural substrates of two different rhythmical vibrissal movements in the rat. *Neuroscience* 12:761–74

Sherrington CS. 1906. *The Integrative Action of the Nervous System*. New Haven, CT: Yale Univ. Press

Smith JB, Alloway KD. 2013. Rat whisker motor cortex is subdivided into sensory-input and motor-output areas. *Front. Neural Circuits* 7:4

Steindler DA. 1985. Trigeminocerebellar, trigeminotectal, and trigeminothalamic projections: a double retrograde axonal tracing study in the mouse. *J. Comp. Neurol.* 237:155–75

Stepien AE, Tripodi M, Arber S. 2010. Monosynaptic rabies virus reveals premotor network organization and synaptic specificity of cholinergic partition cells. *Neuron* 68:456–72

Swenson RS, Kosinski RJ, Castro AJ. 1984. Topography of spinal, dorsal column nuclear, and spinal trigeminal projections to the pontine gray in rats. *J. Comp. Neurol.* 222:301–11

Takatoh J, Nelson A, Zhou X, Bolton MM, Ehlers MD, et al. 2013. New modules are added to vibrissal premotor circuitry with the emergence of exploratory whisking. *Neuron* 77:346–60

Tandon S, Kambi N, Jain N. 2008. Overlapping representations of the neck and whiskers in the rat motor cortex revealed by mapping at different anaesthetic depths. *Eur. J. Neurosci.* 27:228–37

Ugolini G. 1995. Specificity of rabies virus as a transneuronal tracer of motor networks: transfer from hypoglossal motoneurons to connected second-order and higher order central nervous system cell groups. *J. Comp. Neurol.* 356:457–80

Welker E, Hoogland PV, Van der Loos H. 1988. Organization of feedback and feedforward projections of the barrel cortex: a PHA-L study in the mouse. *Exp. Brain Res.* 73:411–35

Welker WI. 1964. Analysis of sniffing of the albino rat. *Behaviour* 22:223–44

White EL, DeAmicis RA. 1977. Afferent and efferent projections of the region of mouse SmI cortex which contains the posteromedial barrel subfield. *J. Comp. Neurol.* 175:455–82

Wickersham IR, Lyon DC, Barnard RJ, Mori T, Finke S, et al. 2007. Monosynaptic restriction of transsynaptic tracing from single, genetically targeted neurons. *Neuron* 53:639–47

Wise SP, Jones EG. 1977. Cells of origin and terminal distribution of descending projections of the rat somatic sensory cortex. *J. Comp. Neurol.* 175:129–57

Woolsey CN. 1958. Organization of somatic sensory and motor areas of the cerebral cortex. In *Biological and Biochemical Bases of Behavior*, ed. HF Harlow, CN Woolsy, pp. 63–81. Madison: Univ. Wis. Press

Yatim N, Billig I, Compoint C, Buisseret P, Buisseret-Delmas C. 1996. Trigeminocerebellar and trigemino-olivary projections in rats. *Neurosci. Res.* 25:267–83

Zagha E, Casale AE, Sachdev RNS, McGinley MJ, McCormick DA. 2013. Motor cortex feedback influences sensory processing by modulating network state. *Neuron* 79:567–78

Zhang F, Vierock J, Yizhar O, Fenno LE, Tsunoda S, et al. 2011. The microbial opsin family of optogenetic tools. *Cell* 147:1446–57

Zhang F, Wang LP, Brauner M, Liewald JF, Kay K, et al. 2007. Multimodal fast optical interrogation of neural circuitry. *Nature* 446:633–39

Neural Coding of Uncertainty and Probability

Wei Ji Ma[1] and Mehrdad Jazayeri[2]

[1]Center for Neural Science and Department of Psychology, New York University, New York, New York 10003; email: weijima@nyu.edu

[2]McGovern Institute for Brain Research and Department of Brain and Cognitive Sciences, Massachusetts Institute of Technology, Cambridge, Massachusetts 02139; email: mjaz@mit.edu

Annu. Rev. Neurosci. 2014. 37:205–20

The *Annual Review of Neuroscience* is online at neuro.annualreviews.org

This article's doi: 10.1146/annurev-neuro-071013-014017

Keywords

Bayesian inference, decision making, perception, population encoding

Abstract

Organisms must act in the face of sensory, motor, and reward uncertainty stemming from a pandemonium of stochasticity and missing information. In many tasks, organisms can make better decisions if they have at their disposal a representation of the uncertainty associated with task-relevant variables. We formalize this problem using Bayesian decision theory and review recent behavioral and neural evidence that the brain may use knowledge of uncertainty, confidence, and probability.

Contents

PROBABILITY, BELIEFS, AND UNCERTAINTY

The information that organisms have available to make decisions is remarkably limited and impoverished. For example, visual signals are degraded in the dark, other individuals' internal states are not directly accessible, and the amount of food available in food sources may vary depending on many unknown factors. Moreover, our nervous system is not perfect. For example, three-dimensional visual objects are projected onto two-dimensional sensors, causing ambiguity (Kersten et al. 2004), and responses to the same sensory event are inherently variable (Faisal et al. 2008).

Knowing the nature of internal and external stochastic processes will generally help organisms to make better decisions. In sensory processing, to infer the state of the world from variable and ambiguous sensory inputs, the brain would benefit from knowing the probabilistic relationships between stimuli and the sensory responses they evoke. On the motor side, knowing the statistics of one's motor variability can enhance the effectiveness of movements. Knowledge of the variability of rewards and costs could also be highly informative. Higher faculties such as anticipation, planning, decision making, and thinking can all benefit from knowledge about the probabilistic contingencies in the environment and stochasticity within the nervous system.

"Belief" is a term used to describe an agent's knowledge of probabilistic information about variables that describe the state of the world, the state of the body, or a mental state. Mathematically, belief can be viewed as a "subjective probability": The stronger one's belief in a particular proposition, the higher the corresponding subjective probability. For example, when trying to cross a road, you could maintain a belief distribution over the speed of an approaching car. Some beliefs may be hardwired from birth, whereas others could be acquired flexibly through experience or vary from trial to trial as the sensory input varies. For an optimal observer, beliefs are based on the actual probabilistic relationships between variables, but in general they need not be. Uncertainty is typically specified by some measure of the width of the belief distribution. In the road-crossing example, sensory uncertainty could be defined as the standard deviation of the belief distribution over the car speed (e.g., 30 ± 2 km/h): In fog, the uncertainty may be higher (e.g., 30 ± 10 km/h).

The primary alternative to using belief distributions and uncertainty is to use point estimates of variables, e.g., "The car's speed is 30 km/h." However, in most realistic conditions, uncertainty information is relevant for decision making: You might cross the road if a car's speed is 30 ± 2 km/h but not if it is 30 ± 10 km/h. The same holds in higher cognition: When you hear an outrageous statement but are uncertain whether the speaker is joking, you may seek clarification instead of getting angry. In this paper, we review recent studies that have shown that, under suitable conditions, humans and animals behave as if they do make use of belief distributions and uncertainty. We also provide a critical view of what it means to use belief distributions and discuss outstanding questions related to the neural basis of such probabilistic computation.

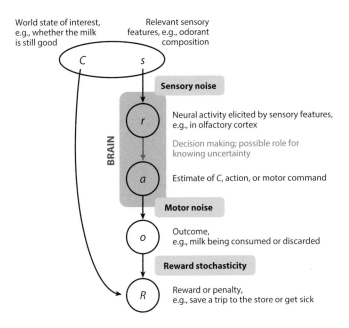

World state of interest, e.g., whether the milk is still good

Relevant sensory features, e.g., odorant composition

C   *s*

BRAIN

Sensory noise

r — Neural activity elicited by sensory features, e.g., in olfactory cortex

Decision making; possible role for knowing uncertainty

a — Estimate of *C*, action, or motor command

Motor noise

o — Outcome, e.g., milk being consumed or discarded

Reward stochasticity

R — Reward or penalty, e.g., save a trip to the store or get sick

Figure 1

A probabilistic formalization of different classes of decision-making tasks. To simplify notation, we ignore temporal dynamics. Each node contains a random variable, which is described by a probability distribution that is conditioned on the variables from which arrows point. For example, the statistics of the outcome *o* depend on the action taken, *a* (we ignore potential other dependencies for simplicity). The brain's goal is to realize an appropriate mapping from neural activity *r* to motor commands *a*. This can involve taking into account uncertainty/probabilistic information.

THEORY OF PROBABILISTIC COMPUTATION

Decision-making studies have advanced along three relatively independent tracks: (*a*) Perceptual decisions are concerned with judgments about a stimulus; (*b*) sensorimotor decisions incorporate sensorimotor contingencies to specify when, where, and how to make movements; and (*c*) reward-based decisions focus on behavioral responses that are based on an assessment of the utility of choice options. **Figure 1** shows a schematic that encompasses these different classes of decision making. The so-called forward or generative model characterizes the true probabilistic relationships between different variables. Suppose there is a particular aspect of the world (the world includes one's own body), denoted *C*, that is of interest, for example, whether just-expired milk is still good. Relevant to *C* are certain sensory features, denoted *s*, for example, the odorants of the milk. Each value of *C* and each value of *s* have a certain frequency of occurring together in the world, which is represented by a joint probability distribution $p(C,s)$. Upon interrogating the environment, the sensory feature *s* evokes a neural population activity, denoted *r*, for example, the olfactory response to the odorants. Because of the inherent variability in neural responses, the relationship between *r* and *s* is stochastic and is best described by a probability distribution $p(r|s)$. Decision making or acting is the process of mapping the neural activity *r* to a decision or action (motor command), denoted *a*, for example, to drink or not to drink the milk. In perceptual tasks, *a* is simply the observer's estimate of the world state *C*. The action *a* leads to an outcome *o* in the external world, for example, the milk being ingested. This mapping may be subject to motor noise and can be formulated by $p(o|a)$. Finally, the state of the world and the outcome may lead to reward or punishment, denoted *R*, for example, getting sick from drinking bad milk. In general, reward may be stochastic for a given world state C and outcome o; for example, there might be 50% chance of getting sick after drinking bad milk.

We can use **Figure 1** to define more rigorously different types of belief and uncertainty. Let us first consider an optimal (ideal) observer, for whom beliefs follow directly from the true probability distributions. Sensory or perceptual beliefs are over C and s and are captured by the distribution $p(C,s|r)$; in our examples, these would be distributions over car speed or milk quality. Outcome or motor beliefs are expressed by $p(o|a)$, which represents the agent's knowledge of the nature of the noise that will influence a given motor command. Finally, the belief distribution over reward is $p(R|o,C)$: For example, one option might give you a 50% chance of receiving $20, while the other option yields $8 for sure. In perceptual and sensorimotor tasks, the contingency between outcome and reward is typically deterministic, which reduces $p(R|o,C)$ to a deterministic reward or cost function; for example, when o is an estimate of C and both are continuous, the cost function can be defined as the squared estimation error $(o - C)^2$. These three types of belief distributions and their corresponding measure of uncertainty are distinct and may have different neural underpinnings.

The fundamental question we ask is whether and how the brain incorporates belief distributions or uncertainty in the mapping from response r to action a. At one extreme, the brain would not use belief distributions at all, but instead use point estimates of stimuli, even when all the dependencies shown in **Figure 1** are stochastic. At the other extreme, the brain acts as an optimal observer that knows everything about the generative model shown in **Figure 1** and uses this knowledge in the best possible way. In between lie many possibilities, including cases in which the brain utilizes either belief distributions over a subset of variables or belief distributions that deviate from the true distributions.

Bayesian decision theory specifies how an optimal observer would utilize belief distributions. The agent first computes the probability of receiving reward R when sensory activity is r and the planned action is a. This probability is

$$p(R|r, a) = \iint p(R|o, C)p(o|a)p(C|r)\,do\,dC. \qquad 1.$$

The three factors in the integrand correspond to the three belief distributions: reward, outcome, and sensory beliefs, respectively. The optimal observer uses these belief distributions to average over the world state C and outcome o that are not known (such integration is also called marginalization). Optimality is defined as executing actions a that maximize utility under the distribution $p(R|r,a)$. Utility could simply be the expected value of R, or it could be a complicated nonlinear function of the distribution $p(R|r,a)$, for example in order to account for risk aversion (Kahneman & Tversky 1979, Glimcher et al. 2008).

The distribution $p(C|r)$ is known as the posterior distribution over C and represents knowledge about the world state C after combining sensory information with prior expectations. The posterior can be further broken down (using Bayes' rule) as the normalized product of a likelihood $p(r|C)$ and a prior $p(C)$. The likelihood $p(r|C)$ is computed by integrating the product of $p(r|s)$ (the likelihood over s) and $p(s|C)$ over s (another marginalization); this operation transforms beliefs over s into beliefs over C. Let us consider a simple example of a sensory decision in which an observer receives a unit reward ($R = 1$) for correctly estimating the state of the world C. In this case, because the outcome o and action a are determined by the observer's estimate of C, denoted $\hat{C}$, Equation 1 can be simplified to $p(R = 1|r, \hat{C}) = p(C = \hat{C}|r)$. Then, maximizing reward reduces to maximizing the posterior, i.e., the strength of the observer's belief that $\hat{C}$ is the true state of the world.

Bayesian decision theory provides a rigorous definition for confidence, namely the belief associated with the proposition that the observer has chosen or intends to choose. For example, in the sensory decision task described above, confidence would be the posterior probability of the observer's estimate of the state of the world, $p(C = \hat{C} \,|\, r)$ (or a monotonic function of it). More generally, confidence can be defined as the observer's belief that the chosen action maximizes utility (de Martino et al. 2013). A basic way to measure confidence in humans is using a discrete rating scale (Peirce & Jastrow 1884), but we review several other methods below (see Neural Signatures of Uncertainty). Confidence generally correlates with task performance, and the strength of this correlation—measuring the accuracy of one's knowledge of the quality of one's decisions—is itself typically correlated with, although sometimes dissociable from, performance (Fleming & Dolan 2012).

Using belief distributions in decision making, also called probabilistic computation, is neither necessary nor sufficient for optimal performance (Ma 2012). This is because the Bayesian decision strategy ultimately amounts to a specific deterministic mapping from neural activity r to an action a: $a = F(r)$. For example, in simple two-alternative detection or discrimination tasks, this mapping reduces to a comparison of a point estimate of the stimulus with a fixed criterion, without any need to represent uncertainty (Green & Swets 1966). Indeed, observers in signal detection theory models do not commonly utilize an internal measure of uncertainty. However, in most tasks of realistic complexity, optimal performance does require keeping track of entire belief distributions, and this fact is typically used to connect optimal performance to probabilistic computation. Conversely, using belief distributions or uncertainty does not guarantee optimality. The behavior of an agent with incorrect beliefs $q(R\,|\,o,C)$, $q(o\,|\,a)$, $q(r\,|\,C)$, or $q(C)$ or of an agent who does not compute Equation 1 correctly may be suboptimal even though the agent performs probabilistic computation. Here, our focus is not on characterizing the extent to which observers behave optimally, but rather on the ways in which belief distributions or uncertainty information may be used to guide behavior.

BEHAVIORAL EVIDENCE FOR PROBABILISTIC COMPUTATION

A typical behavioral experiment specifies the task contingencies (distributions over C, s, and R) and analyzes behavioral outcomes (o) to infer the mapping from r to a. To know whether the brain uses uncertainty information, tasks have been designed in which maximization of accuracy or reward requires the observer to take trial-to-trial variations in uncertainty into account. For example, we can vary the likelihood over C by changing the stimulus s or the reliability of the information r provides about s, the prior over C by changing the statistics of C in the world, or reward beliefs by changing the probabilities of reward.

We first consider a study in which trial-to-trial feedback was provided and sensory reliability was not varied. When trial-to-trial feedback is provided, it may be possible for the brain to gradually learn to implement an optimal stimulus-to-response strategy without performing true probabilistic computation, i.e., without using belief distributions on every trial. Yang & Shadlen (2007) trained monkeys to choose between two options using evidence provided by four visual shapes. Each shape was associated with a log likelihood ratio (log LR), and the cumulative log LR specified the odds with which the option would be rewarded. They found that monkeys combined the evidence from individual shapes and made decisions based on the cumulative log LR. Although it is possible that the animals used uncertainty information to perform the task, it is also possible that the animals used the extensive training period to establish a stimulus-response mapping that assigned a suitable weight to each shape for or against the two options.

One line of evidence in support of probabilistic computation and against stimulus-response mapping strategy comes from studies in which the reliability of a stimulus was varied on a trial-by-trial basis. These studies have shown that subjects adjust their behavior based on the reliabilities of stimuli even when the point estimates of the stimuli remain unchanged. For example, in a set of cue combination experiments, Angelaki and colleagues showed that monkeys optimally combine an optic flow (visual) cue of varying reliability with a self-motion (vestibular) cue to make judgments about heading direction (Gu et al. 2008, Morgan et al. 2008, Fetsch et al. 2009), suggesting that the animal's decisions are based on a trial-to-trial estimate of stimulus reliability. Similarly, subjects take stimulus reliability into account in gaze direction perception (Mareschal et al. 2013), coincidence detection (Miyazaki et al. 2005), time interval reproduction (Jazayeri & Shadlen 2010, Acerbi et al. 2012), speeded reaching movements (Tassinari et al. 2006, Landy et al. 2012), and dynamic sensorimotor tasks (Faisal & Wolpert 2009, Turnham et al. 2011, O'Reilly et al. 2013), as well as when the number of reliable stimuli (Van den Berg et al. 2012) or the reward contingencies (Feng et al. 2009, Kiani & Shadlen 2009) are varied.

Another line of evidence in support of the probabilistic computation comes from studies in which subjects were shown to take sensory uncertainty into account even when trial-to-trial feedback was randomized or completely withheld. One study had human subjects perform a discrimination task with asymmetric rewards but without feedback until after many trials (Whiteley & Sahani 2008). Subjects approximately maximized reward, suggesting that they used trial-to-trial sensory uncertainty information. Analogous conclusions were reached in a comparable auditory task (Maiworm et al. 2011), a spatial reasoning task in natural scenes (D'Antona et al. 2013), an auditory-visual causal inference task (Kording et al. 2007), and speech perception studies (Ma et al. 2009, Bejjanki et al. 2011). A Bayesian integration model of sound localization by the barn owl (Fischer & Pena 2011) was based on data obtained using random rewards (Hausmann et al. 2009). These results provide stronger evidence for the hypothesis that the brain computes with sensory uncertainty.

Some studies combined both solutions: They varied the reliability of the sensory information from trial to trial and withheld trial-to-trial feedback. This approach has been used to study cue combination (reviewed in Trommershauser et al. 2011) as well as the integration of sensory inputs with a prior distribution in domains as diverse as speed perception (Stocker & Simoncelli 2006), orientation perception (Girshick et al. 2011), duration estimation (Ahrens & Sahani 2011, Cicchini et al. 2012), and reaching movements (Kording & Wolpert 2004, Battaglia & Schrater 2007). An additional reason for withholding feedback in some of these studies was to make subjects use prior distributions derived from natural statistics rather than from experimental statistics. The strategy of withholding feedback and varying reliability has been extended to multiple-item categorical tasks, namely visual search (Ma et al. 2011), change detection (Keshvari et al. 2012), oddity detection (Hillis et al. 2002; as modeled by Hospedales & Vijayakumar 2009), and simultaneity judgment (Magnotti et al. 2013). In an orientation categorization task, it was shown that observers used stimulus reliability to adjust category boundaries from trial to trial in a near-optimal manner (Qamar et al. 2013). Finally, in a novel design, subjects selected one of two images on which they wanted to perform an orientation identification task (Barthelme & Mamassian 2009). Subjects chose the most informative image most of the time, and their performance was higher than when they were not allowed to choose. Together, these studies provide converging evidence that in many behavioral settings the brain has trial-to-trial access to sensory uncertainty information.

THE FLEXIBILITY OF PROBABILISTIC COMPUTATION

The concern about feedback highlights a more general question regarding probabilistic computation: To what extent do subjects utilize various types of belief distributions flexibly across tasks, modalities, and motor effectors? The degree of flexibility has implications for how the brain

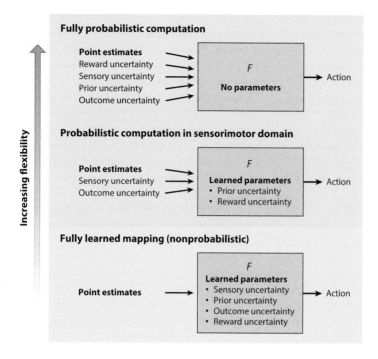

Figure 2

Three possible levels of flexibility of probabilistic computation. The boxes marked F indicate the mapping from neural activity r to action a, as implemented in the network architecture. Feeding into these is the information extracted from r that is used by the brain on a trial-to-trial basis. Information that is not available or used has to be incorporated through the parameters of the mapping function F (shown inside the box) that cannot be modified flexibly. By "point estimates" we refer to point estimates of stimulus features of interest (but not of stimulus features that correlate with sensory uncertainty), of outcomes, and of rewards.

may implement probabilistic computation in neural circuits. At the most flexible level (**Figure 2**, *top row*), the sensory likelihoods (say, over s and over C), prior, outcome belief distribution, and reward belief distribution (or cost function) are all independently accessible, presumably each through the activity of a population of neurons, and are used on each trial through Equation 1. In this case, the mapping function F from r to a is realized through a transformation that does not depend on beliefs and uncertainties. This strategy requires a complicated neural machinery but has the advantage of being able to instantly accommodate any change in sensory and motor variability and to integrate any known priors and cost functions. At the other extreme (**Figure 2**, *bottom row*), on each trial the subject has access only to point estimates of the sensory stimuli. As such, optimal behavior is possible only if the width of the likelihood function, prior, outcome beliefs, and reward beliefs are learned as parameters of the mapping function F. Therefore, any change in any of these beliefs requires a modification of the mapping function. When feedback is provided, the subject can learn F through trial and error (Law & Gold 2008), but this may take many trials. Withholding feedback makes it easier to assess the flexibility of probabilistic computation, although, in simple tasks, belief distributions can also be learned in an unsupervised fashion (Raphan & Simoncelli 2011). In principle, an observer's behavior may show intermediate levels of flexibility where some—but not all—belief distributions are readily accessible. For example, the brain may be able to flexibly accommodate changes in sensory or motor uncertainty, but require continuous learning to accommodate priors and reward beliefs (**Figure 2**, *middle row*).

One way to measure the flexibility of probabilistic computations is to test whether previously learned belief distributions would transfer to other conditions (Seydell et al. 2008, Maloney & Mamassian 2009). The most convincing case for probabilistic computation could be made if subjects utilize belief distributions flexibly across tasks, environmental statistics, sensory modalities, motor effectors, and cost functions. Varying sensory uncertainty from trial to trial, as described above, addresses one aspect of this flexibility (**Figure 2**, *middle row*), but the question of flexibility with respect to other belief distributions (**Figure 2**, *top row*) remains unanswered. Two studies that had subjects take into account prior information that varied from trial to trial found probabilistic but suboptimal behavior (Hudson et al. 2007, Acerbi et al. 2014). One human fMRI study found that changes in the likelihood and prior can be attributed to different brain areas (Vilares et al. 2012), but these signals could also be correlates of other factors such as attention or task difficulty.

Trommershäuser and colleagues conducted a series of studies (reviewed in Trommershauser et al. 2008) in which human subjects made speeded hand movements to a green disc on a screen while trying to avoid a red disc. The cost function was varied by changing the point value of a disc or the degree of overlap. In these experiments, subjects obtained near-maximal reward given their level of motor noise. The investigators found no evidence of learning, suggesting that subjects instantly incorporated previously acquired implicit knowledge of their motor uncertainty. Seydell et al. (2008) tested transfer in a similar task by comparing performance under a particular cost function between observers trained on the same cost function and observers trained on a different cost function; no difference was found, suggesting that outcome uncertainty was encoded and used. Fleming et al. (2013) asked whether humans could optimally combine sensory uncertainty information with knowledge of motor uncertainty and a cost function that changed from trial to trial. Subjects overweighted sensory uncertainty relative to their motor uncertainty, suggesting that sensory beliefs were not properly taken into account.

NEURAL REALIZATIONS OF PROBABILISTIC COMPUTATION

If organisms use information about sensory uncertainty in decision making on a trial-by-trial basis, then the question arises as to how sensory uncertainty is represented in neural populations. Theoretical schemes proposed to answer this question have been reviewed elsewhere (Jazayeri 2008, Ma et al. 2008, Vilares & Kording 2011, Pouget et al. 2013), and we only briefly summarize them here. In probabilistic population codes (Foldiak 1993, Sanger 1996, Pouget et al. 2003, Jazayeri & Movshon 2006, Ma et al. 2006), the brain has knowledge of the generative process $p(r|s)$ and thus automatically possesses a likelihood function over s when given a pattern of activity r. Uncertainty is implicitly represented in the population activity; for example, more uncertainty may correspond to a lower total spike count. Neural operations performed on input populations correspond to manipulations of the corresponding likelihood functions. To give this framework predictive power, one needs to assume a specific form of neural variability, $p(r|s)$. One family of distributions that has been proposed is "Poisson-like" (Ma et al. 2006). Poisson-like variability is well-established at the level of single neurons (Dean 1981, Tolhurst et al. 1983, Britten et al. 1993, Softky & Koch 1993), but recent developments have also found evidence in support of this functional description at the population level (Graf et al. 2011, Berens et al. 2012). Under the Poisson-like assumption, the logarithm of the likelihood function is linear in population activity (Jazayeri & Movshon 2006, Ma et al. 2006). As a corollary, the log LR in a two-alternative discrimination task can be straightforwardly written as a linear combination of neural activity (Jazayeri & Movshon 2006, Ma et al. 2006) or, in the reduced case of two neurons, as a function of the difference between those two neurons (Beck et al. 2008). Assuming Poisson-like probabilistic population codes, Fetsch et al. (2012) obtained likelihoods from neural activity in cortical area MSTd, where visual and vestibular cues for self-motion are integrated, and accurately predicted the monkey's cue integration behavior, including a moderate deviation from optimality.

Another class of models is based on the assumption that the activity of a neuron at a given point in time is a sample from the belief distribution that is to be represented (Hoyer & Hyvarinen 2003, Paulin 2005, Fiser et al. 2010, Shi et al. 2010, Griffiths et al. 2012). In these so-called sampling codes, the entire probability distribution and the corresponding uncertainty are represented across time or across a neural population. These schemes hold that the probability of a variable of interest is directly mapped onto firing rate. In support of this proposal, the spontaneous activity in the ferret visual cortex may reflect the statistics of the environment (Berkes et al. 2011). Moreover, there is some behavioral evidence that observers sample from the posterior distribution (Moreno-Bote et al. 2011, Gershman et al. 2012). Sampling codes of this kind have not been thoroughly formalized, but certain versions of its formulation may be implausible. For example, if single-neuron firing rate were a sample of a sensory belief distribution, then firing rate variability should increase with uncertainty, which is inconsistent with the variance-reducing effect that decreasing contrast has on visual cortical neurons (Tolhurst et al. 1983).

Explicit probability codes compose a third class of neural codes for uncertainty (Barlow & Levick 1969, Anderson 1994, Anastasio et al. 2000, Barber et al. 2003, Lee & Mumford 2003, Rao 2004, Deneve 2008). In this class, the activity of a neuron tuned to a stimulus feature is monotonically related to the probability density of that feature (typically through a linear or logarithmic transformation). Higher uncertainty is then represented by a wider activation pattern across the population.

More physiological evidence is needed to distinguish between these schemes. However, any scheme must address how basic Bayesian computations can be implemented using biologically plausible neural operations. This is still a work in progress. An important computation is to combine a likelihood function with a prior. Several recent studies have proposed that priors over sensory variables are encoded through the organization and distribution of tuning curves in the sensory representation. For example, more neurons may be dedicated to stimuli that have higher probability (Fischer & Pena 2011, Ganguli & Simoncelli 2011, Girshick et al. 2011). If the density of neurons encodes the prior and the sensory tuning function is proportional to the likelihood, a simple population vector decoder can estimate the most probable stimulus (Fischer & Pena 2011). More sophisticated representations of the prior are also possible, if the decoder properly weights the sensory responses (Ganguli & Simoncelli 2011). Finally, if the logarithm of the prior is encoded by the tuning functions (Simoncelli 2009), then an optimal linear decoder could integrate the prior and likelihood information (Jazayeri & Movshon 2006). In a different view, integration of the prior and likelihood information is mediated by interactions between spontaneous activity and stimulus-evoked sensory responses (Berkes et al. 2011). Priors may also be encoded by neurons downstream of sensory representations (Basso & Wurtz 1997, Platt & Glimcher 1999, Janssen & Shadlen 2005, Churchland et al. 2008). In this scenario, the prior could exert its effect on behavior either through feedback mechanisms akin to attentional modulation (Ghose & Maunsell 2002) or through linear operations that exploit Poisson-like sensory variability (Ma et al. 2006).

Computation can also consist of combining multiple likelihoods over the same variable, as in cue combination, or of transforming a likelihood over one variable (say, s in **Figure 1**) into a likelihood over another variable (say, C in **Figure 1**), as in categorization tasks. Most work on the implementation of these computations has been done within a Poisson-like probabilistic population code framework; herein, cue combination and evidence accumulation are implemented through linear operations on neural activity (Jazayeri & Movshon 2006, Ma et al. 2006, Beck et al. 2008), but Kalman filtering (as used in motor control and visual tracking) and many forms of categorization require quadratic operations and divisive normalization (Beck et al. 2011, Ma et al. 2011, Qamar et al. 2013). Virtually no work has been done on how mid-level and high-level visual computations, such as inferring Gestalt or obtaining viewpoint invariance, are performed probabilistically by a neural network. Finally, any probabilistic representation is bound to change at some point downstream, because the organism must make an estimate or execute an action. Depending on task demands, training regimen, and species, this process may take place in different brain areas, including the premotor and association areas in cortex (Jazayeri & Movshon 2006), in subcortical areas (Beck et al. 2008), or at the level of the muscles (Simoncelli 2009).

NEURAL SIGNATURES OF UNCERTAINTY

Recent studies have begun to examine the neural representation of uncertainty in animal models. Kepecs et al. (2008) trained rats to categorize a mixture of two odors and controlled sensory uncertainty by varying the proportion of each odor and the category boundary. Recording from neurons in the orbitofrontal cortex established a neural correlate of the animal's confidence (decision certainty), which was consistent with the predictions of a computational model. The authors also used a separate task to ensure that rats were able to use the uncertainty information: When given the option of initiating a new trial instead of waiting for a reward, rats resorted to this option more often when uncertainty was high. However, because orbitofrontal activity was not recorded during the task in which animals had to use uncertainty information, whether the recorded orbitofrontal signals contributed to the animal's measure of confidence remains unclear. In another study, Kiani & Shadlen (2009) trained monkeys to perform a motion discrimination task in which the monkeys could opt out of the decision on some trials by choosing a small but certain reward. The animals did so when sensory evidence was weak, and firing rates of LIP neurons on those trials were in between the activity levels of when the monkey chose either category. These findings confirm earlier work that LIP responses in this task vary monotonically with the log-LR (Gold & Shadlen 2007). Whether animals use this signal to make their trial-by-trial confidence judgments remains undetermined. In another study, monkeys used saccades to make a bet on whether the decision they made on a prior visual search task was correct (Middlebrooks & Sommer 2012). The animal's confidence, which was inferred from the magnitude of the bet, was reflected in a simultaneous recording of neural activity in the supplementary eye field. Most recently, Komura et al. (2013) trained monkeys in a direction discrimination task in which the animal either could discriminate the direction of a cloud of moving dots and receive a large reward for correct judgments or could opt out of the decision task altogether to receive a smaller reward. They found that neural activity in the pulvinar nucleus of the thalamus decreased with sensory uncertainty. Moreover, reversible inactivation of the region of interest in the dorsal pulvinar increased the proportion of opt-out responses, suggesting that signals in this region of the pulvinar contribute to the animal's assessment of uncertainty.

In reward-based decision making, recent neural studies have begun to examine the neural correlates of expected reward (mean of R given a) and risk (variance of R given a). A thorough discussion of this literature can be found elsewhere (Rangel et al. 2008, Rushworth & Behrens 2008, Lee et al. 2012); we highlight only two studies. Preuschoff et al. (2006) asked subjects to bet which of two randomly drawn integers m and n between 1 and 10 was greater. After placing the bet, subjects were informed of the value of m. This manipulation dissociated expected reward from risk: Expected reward was monotonically related to m, whereas risk was highest at intermediate values of m. Using fMRI in humans, this study and similar ones found correlates of risk in the striatum, insula, and lateral orbitofrontal cortex. Behrens et al. (2007) used a task in which humans chose between two options and gained information about the probability of each leading to reward by observing either a stable or a volatile history of rewards. The authors found activation in the anterior cingulate cortex correlated with the subject's estimate of volatility.

Animal studies have also investigated the topic of expected reward and reward uncertainty (Schultz 2000, Hikosaka et al. 2008, Rangel et al. 2008). Early studies found a crucial role for dopamine in reward probability and risk (Fiorillo et al. 2003). Others have examined the ways in which the probability or utility of reward influences neural activity and choice behavior (Platt & Glimcher 1999, Sugrue et al. 2004, Lau & Glimcher 2008, So & Stuphorn 2010, Chen et al. 2013) as well as how trial-to-trial variations of expected reward control behavior in stochastic environments (Barraclough et al. 2004, Dorris & Glimcher 2004).

Taken together, these studies provide evidence for widespread representation of different kinds of uncertainty and open the door to several novel lines of inquiry. For example, what are the differences and similarities between the neural codes for the three different types of beliefs? What are the algorithms and neural mechanisms by which neurons measure or represent uncertainty? How do the computational principles for estimating confidence generalize across brain areas and behavioral contexts? Do the observed neural correlates of confidence causally contribute to an animal's ability to assess sensory uncertainty and decision confidence?

FUTURE DIRECTIONS

The study of probabilistic computation by the brain is still in its early stages, and many open questions remain. At the behavioral level, experiments must establish the extent to which belief distributions are encoded and utilized in a flexible manner, as schematized in **Figure 2**. One strategy is to assess how organisms act in the face of multiple types of uncertainty, as was done in a recent study by Fleming et al. (2013). Such tasks are also ecologically relevant: As the example of spoiled milk illustrates, natural behavior can often improve by combining knowledge about multiple types of belief. It is often fruitful to study sensory, outcome, or reward uncertainty in isolation, but we believe progress can be made by combining established paradigms from these individual domains. In any experimental work, particular care should be taken to address the impact of feedback on claims of probabilistic computation.

Recent work has explored the possibility that neural computation is probabilistic but suboptimal (Beck et al. 2012, Whiteley & Sahani 2012, Orhan et al. 2014, Acerbi et al. 2014). Suboptimality can arise from at least two sources: (a) wrong or incompletely learned beliefs and (b) neural networks implementing approximate computations. It will be important to develop behavioral paradigms to tease apart and characterize the factors that contribute to suboptimal performance. This line of work may shed light on disorders of higher brain function that are associated with faulty probabilistic computations.

A direction for physiological investigation is to move beyond correlation and establish a causal link between behavior and neural activity associated with belief and uncertainty. Also important are understanding the learning mechanisms at multiple levels of analysis—from synapses to circuits—that enable the brain to encode beliefs as well as assessing if different types of belief are used and represented differently. Mechanisms of integration are also poorly understood: For example, we do not know how neurons combine various types of belief. Finally, it will be valuable to design experiments to test the theoretical schemes for the neural representation of probability.

On the theoretical side, a challenge is to propose a flexible and general framework for the neural implementation of Equation 1. Theoretical schemes have focused predominantly on relatively simple sensory problems, for example, the probabilistic representation of a single, one-dimensional stimulus feature. Future studies should expand the reach of probabilistic models to more complex and naturalistic stimuli. They should address how a particular representational scheme can be used to perform complex computations, for example, categorization in a high-dimensional space or realization of the Gestalt "principles" of perception. An intriguing idea is that in dynamic natural scenes, such as when trying to predict whether a stack of blocks will topple, people build beliefs over possible futures by mentally simulating the dynamics, in this example, the laws of physics (Battaglia et al. 2013). The application of probabilistic models to the problem of object recognition (Kersten et al. 2004), which traditional models seek to explain without reference to beliefs (DiCarlo et al. 2012), needs to be explored. A higher-order form of inference that has barely been studied at the neural level is structure learning (Tenenbaum & Griffiths 2001, Kemp & Tenenbaum 2008, Braun et al. 2010, Pouget et al. 2013), the process of learning generalizable rules for categorizing stimuli or performing actions. Study of structure learning, generalization, and model selection could help to bridge the gap between simple psychophysical tasks and more cognitive domains. Finally, capacity limitations in the encoding stage have largely been ignored in probabilistic models of decision making but deserve attention (Palmer et al. 1990, Keshvari et al. 2013, Mazyar et al. 2013).

The idea that the brain computes with belief distributions has already had a profound impact on neuroscience, psychology, and cognitive science. In the coming years, we foresee greater convergence among these fields and progress in applying the concepts of belief and uncertainty to understand the computational and neural underpinnings of more complex and more natural behaviors.

DISCLOSURE STATEMENT

The authors are not aware of any affiliations, memberships, funding, or financial holdings that might be perceived as affecting the objectivity of this review.

LITERATURE CITED

Acerbi L, Vijayakumar S, Wolpert DM. 2014. On the origins of suboptimality in human probabilistic inference. *PLOS Comp. Biol.* In press

Acerbi L, Wolpert DM, Vijayakumar S. 2012. Internal representations of temporal statistics and feedback calibrate motor-sensory interval timing. *PLoS Comput. Biol.* 8(11):e1002771

Ahrens M, Sahani M. 2011. Observers exploit stochastic models of sensory change to help judge the passage of time. *Curr. Biol.* 21:1–7

Anastasio TJ, Patton PE, Belkacem-Boussaid K. 2000. Using Bayes' rule to model multisensory enhancement in the superior colliculus. *Neural Comput.* 12(5):1165–87

Anderson C. 1994. Neurobiological computational systems. In *Computational Intelligence Imitating Life*, pp. 213–22. New York: IEEE Press

Barber MJ, Clark JW, Anderson CH. 2003. Neural representation of probabilistic information. *Neural Comput.* 15(8):1843–64

Barlow HB, Levick WR. 1969. Three factors limiting the reliable detection of light by retinal ganglion cells of the cat. *J. Physiol.* 200:1–24

Barraclough DJ, Conroy ML, et al. 2004. Prefrontal cortex and decision making in a mixed-strategy game. *Nat. Neurosci.* 7:404–10

Barthelme S, Mamassian P. 2009. Evaluation of objective uncertainty in the visual system. *PLoS Comput. Biol.* 5(9):e1000504

Basso MA, Wurtz RH. 1997. Modulation of neuronal activity by target uncertainty. *Nature* 389:66–69

Battaglia PW, Hamrick JB, Tenenbaum JB. 2013. Simulation as an engine of physical scene understanding. *Proc. Natl. Acad. Sci. USA* 110(45):18327–32

Battaglia PW, Schrater PR. 2007. Humans trade off viewing time and movement duration to improve visuo-motor accuracy in a fast reaching task. *J. Neurosci.* 27(26):6984–94

Beck J, Ma WJ, Kiani R, Hanks T, Churchland AK, et al. 2008. Bayesian decision making with probabilistic population codes. *Neuron* 60(6):1142–52

Beck JM, Latham PE, Pouget A. 2011. Marginalization in neural circuits with divisive normalization. *J. Neurosci.* 31(43):15310–19

Beck JM, Ma WJ, Pitkow X, Latham PE, Pouget A. 2012. Not noisy, just wrong: the role of suboptimal inference in behavioral variability. *Neuron* 74(1):30–39

Behrens TEJ, Woolrich MW, Walton ME, Rushworth MFS. 2007. Learning the value of information in an uncertain world. *Nat. Neurosci.* 10(9):1214–21

Bejjanki VR, Clayards M, Knill DC, Aslin RN. 2011. Cue integration in categorical tasks: insights from audio-visual speech perception. *PLoS ONE* 6(5):e19812

Berens P, Ecker AS, Cotton RJ, Ma WJ, Bethge M, Tolias AS. 2012. A fast and simple population code for orientation in primate V1. *J. Neurosci.* 32(31):10618–26

Berkes P, Orban G, Lengyel M, Fiser J. 2011. Spontaneous cortical activity reveals hallmarks of an optimal internal model of the environment. *Science* 331(6013):83–87

Braun DA, Mehring C, Wolpert DM. 2010. Structure learning in action. *Behav. Brain Res.* 206(2):157–65

Britten KH, Shadlen MN, Newsome WT, Movshon JA. 1993. Responses of neurons in macaque MT to stochastic motion signals. *Vis. Neurosci.* 10(6):1157–69

Chen X, Mihalas S, Niebur E, Stuphorn V. 2013. Mechanisms underlying the influence of saliency on value-based decisions. *J. Vis.* 13(12):18

Churchland AK, Kiani R, Shadlen MN. 2008. Decision-making with multiple alternatives. *Nat. Neurosci.* 11(6):693–702

Cicchini GM, Arrighi R, Cecchetti L, Giusti M, Burr DC. 2012. Optimal encoding of interval timing in expert percussionists. *J. Neurosci.* 32(3):1056–60

D'Antona AD, Perry JS, Geisler WS. 2013. Humans make efficient use of natural image statistics when performing spatial interpolation. *J. Vis.* 13(14):11

de Martino B, Fleming SM, Garrett N, Dolan RJ. 2013. Confidence in value-based choice. *Nat. Neurosci.* 16:105–10

Dean AF. 1981. The variability of discharge of simple cells in the cat striate cortex. *Exp. Brain Res.* 44:437–40

Deneve S. 2008. Bayesian spiking neurons I: inference. *Neural Comput.* 20(1):91–117

DiCarlo JJ, Zoccolan D, Rust NC. 2012. How does the brain solve visual object recognition? *Neuron* 73(3):415–34

Dorris MC, Glimcher PW. 2004. Activity in posterior parietal cortex is correlated with the relative subjective desirability of action. *Neuron* 44(2):365–78

Faisal A, Selen LPJ, Wolpert DM. 2008. Noise in the nervous system. *Nat. Rev. Neurosci.* 9(4):292–303

Faisal AA, Wolpert DM. 2009. Near optimal combination of sensory and motor uncertainty in time during a naturalistic perception-action task. *J. Neurophysiol.* 101(4):1901–12

Feng S, Holmes P, Rorie A, Newsome WT. 2009. Can monkeys choose optimally when faced with noisy stimuli and unequal rewards? *PLoS Comput. Biol.* 5(2):e1000284

Fetsch CR, Pouget A, DeAngelis GC, Angelaki DE. 2012. Neural correlates of reliability-based cue weighting during multisensory integration. *Nat. Neurosci.* 15:146–54

Fetsch CR, Turner AH, DeAngelis GC, Angelaki DE. 2009. Dynamic reweighting of visual and vestibular cues during self-motion perception. *J. Neurosci.* 29(49):15601–12

Fiorillo CD, Tobler PN, Schultz W. 2003. Discrete coding of reward probability and uncertainty by dopamine neurons. *Science* 299(5614):1898–902

Fischer BJ, Pena JL. 2011. Owl's behavior and neural representation predicted by Bayesian inference. *Nat. Neurosci.* 14(8):1061–66

Fiser J, Berkes P, Orbán G, Lengyel M. 2010. Statistically optimal perception and learning: from behavior to neural representations. *Trends Cogn. Sci.* 14(3):119–30

Fleming SM, Dolan RJ. 2012. The neural basis of metacognitive ability. *Phil. Trans. R. Soc. B* 367(1594):1338–49

Fleming SM, Maloney LT, Daw ND. 2013. The irrationality of categorical perception. *J. Neurosci.* 33(49):19060–70

Foldiak P. 1993. The 'ideal homunculus': statistical inference from neural population responses. In *Computation and Neural Systems*, ed. F Eeckman, J Bower, pp. 55–60. Norwell, MA: Kluwer Acad.

Ganguli D, Simoncelli EP. 2011. Implicit encoding of prior probabilities in optimal neural populations. In *Advances in Neural Information Processing Systems*, ed. J Shawe-Taylor, RS Zemel, P Bartlett, F Pereira, KW Weinberger, pp. 658–66. Cambridge, MA: MIT Press

Gershman SJ, Vul E, Tenenbaum JB. 2012. Multistability and perceptual inference. *Neural Comput.* 24:1–24

Ghose GM, Maunsell JH. 2002. Attentional modulation in visual cortex depends on task timing. *Nature* 419(6907):616–20

Girshick AR, Landy MS, Simoncelli EP. 2011. Cardinal rules: visual orientation perception reflects knowledge of environmental statistics. *Nat. Neurosci.* 14:926–32

Glimcher PW, Fehr E, Camerer C, Poldrack RA, eds. 2008. *Neuroeconomics: Decision Making and the Brain.* New York: Academic

Gold JI, Shadlen MN. 2007. The neural basis of decision making. *Annu. Rev. Neurosci.* 30:535–74

Graf AB, Kohn A, Jazayeri M, Movshon JA. 2011. Decoding the activity of neuronal populations in macaque primary visual cortex. *Nat. Neurosci.* 14(2):239–45

Green DM, Swets JA. 1966. *Signal Detection Theory and Psychophysics.* Los Altos, CA: Wiley

Griffiths TL, Vul E, Sanborn AN. 2012. Bridging levels of analysis for probabilistic models of cognition. *Curr. Dir. Psychol. Sci.* 21(4):263–68

Gu Y, Angelaki DE, DeAngelis GC. 2008. Neural correlates of multisensory cue integration in macaque MSTd. *Nat. Neurosci.* 11(10):1201–10

Hausmann L, von Campenhausen M, Endler F, Singheiser M, Wagner H. 2009. Improvements of sound localization abilities by the facial ruff of the barn owl (*Tyto alba*) as demonstrated by virtual ruff removal. *PLoS ONE* 4:e7721

Hikosaka O, Bromberg-Martin E, Hong S, Matsumoto M. 2008. New insights on the subcortical representation of reward. *Curr. Opin. Neurobiol.* 18(2):203–8

Hillis JM, Ernst MO, Banks MS, Landy MS. 2002. Combining sensory information: mandatory fusion within, but not between, senses. *Science* 298(5598):1627–30

Hospedales T, Vijayakumar S. 2009. Multisensory oddity detection as Bayesian inference. *PLoS ONE* 4(1):e4205

Hoyer PO, Hyvärinen A. 2003. *Interpreting Neural Response Variability as Monte Carlo Sampling of the Posterior.* Cambridge, MA: MIT Press

Hudson TE, Maloney LT, Landy MS. 2007. Movement planning with probabilistic target information. *J. Neurophysiol.* 98(5):3034–46

Janssen P, Shadlen MN. 2005. A representation of the hazard rate of elapsed time in macaque area LIP. *Nat. Neurosci.* 8(2):234–41

Jazayeri M. 2008. Probabilistic sensory recoding. *Curr. Opin. Neurobiol.* 18(4):431–37

Jazayeri M, Movshon JA. 2006. Optimal representation of sensory information by neural populations. *Nat. Neurosci.* 9(5):690–96

Jazayeri M, Shadlen MN. 2010. Temporal context calibrates interval timing. *Nat. Neurosci.* 13(8):1020–26

Kahneman D, Tversky A. 1979. Prospect theory: an analysis of decision under risk. *Econometrica* 47:263–91

Kemp C, Tenenbaum JB. 2008. The discovery of structural form. *Proc. Natl. Acad. Sci. USA* 105(31):10687–92

Kepecs A, Uchida N, Zariwala HA, Mainen ZF. 2008. Neural correlates, computation and behavioural impact of decision confidence. *Nature* 455:227–33

Kersten D, Mamassian P, Yuille A. 2004. Object perception as Bayesian inference. *Annu. Rev. Psychol.* 55:271–304

Keshvari S, Van den Berg R, Ma WJ. 2012. Probabilistic computation in human perception under variability in encoding precision. *PLoS ONE* 7(6):e40216

Kiani R, Shadlen MN. 2009. Representation of confidence associated with a decision by neurons in the parietal cortex. *Science* 324:759–64

Komura Y, Nikkuni A, Hirashima N, Uetake T, Miyamoto A. 2013. Responses of pulvinar neurons reflect a subject's confidence in visual categorization. *Nat. Neurosci.* 16(6):749–55

Körding KP, Beierholm U, Ma WJ, Quartz S, Tenenbaum JB, Shams L. 2007. Causal inference in multisensory perception. *PLoS ONE* 2(9):e943

Kording KP, Wolpert DM. 2004. Bayesian integration in sensorimotor learning. *Nature* 427(6971):244–47

Landy MS, Trommershauser J, Daw ND. 2012. Dynamic estimation of task-relevant variance in movement under risk. *J. Neurosci.* 32(37):12702–11

Lau B, Glimcher PW. 2008. Value representations in the primate striatum during matching behavior. *Neuron* 58(3):451–63

Law CT, Gold JI. 2008. Neural correlates of perceptual learning in a sensory-motor, but not a sensory, cortical area. *Nat. Neurosci.* 11(4):505–13

Lee D, Seo H, Jung MW. 2012. Neural basis of reinforcement learning and decision making. *Annu. Rev. Neurosci.* 35:287–308

Lee TS, Mumford D. 2003. Hierarchical Bayesian inference in the visual cortex. *J. Opt. Soc. Am.* 20(7):1434–48

Ma WJ. 2012. Organizing probabilistic models of perception. *Trends Cogn. Sci.* 16(10):511–18

Ma WJ, Beck JM, Latham PE, Pouget A. 2006. Bayesian inference with probabilistic population codes. *Nat. Neurosci.* 9(11):1432–38

Ma WJ, Beck JM, Pouget A. 2008. Spiking networks for Bayesian inference and choice. *Curr. Opin. Neurobiol.* 18:217–22

Ma WJ, Navalpakkam V, Beck JM, van den Berg R, Pouget A. 2011. Behavior and neural basis of near-optimal visual search. *Nat. Neurosci.* 14(6):783–90

Ma WJ, Zhou X, Ross LA, Foxe JJ, Parra LC. 2009. Lip-reading aids word recognition most in moderate noise: a Bayesian explanation using high-dimensional feature space. *PLoS ONE* 4(3):e4638

Magnotti JF, Ma WJ, Beauchamp MS. 2013. Causal inference of asynchronous audiovisual speech. *Front. Psychol.* 4:798

Maiworm M, König R, Röder B. 2011. Integrative processing of perception and reward in an auditory localization paradigm. *Exp. Psychol.* 58(3):217–26

Maloney LT, Mamassian P. 2009. Bayesian decision theory as a model of human visual perception: testing Bayesian transfer. *Vis. Neurosci.* 26(1):147–55

Mareschal I, Calder AJ, Clifford CWG. 2013. Humans have an expectation that gaze is directed toward them. *Curr. Bio.* 23(8):717–21

Mazyar H, van den Berg R, Seilheimer RL, Ma WJ. 2013. Independence is elusive: set size effects on encoding precision in visual search. *J. Vis.* 13(5). doi: 10.1167/13.5.8

Middlebrooks PG, Sommer MA. 2012. Neuronal correlates of metacognition in primate frontal cortex. *Neuron* 75(3):517–30

Miyazaki M, Nozaki D, Nakajima Y. 2005. Testing Bayesian models of human coincidence timing. *J. Neurophysiol.* 94(1):395–99

Moreno-Bote R, Knill DC, Pouget A. 2011. Bayesian sampling in visual perception. *Proc. Natl. Acad. Sci. USA* 108(30):12491–96

Morgan ML, DeAngelis GC, Angelaki DE. 2008. Multisensory integration in macaque visual cortex depends on cue reliability. *Neuron* 59:662–73

Najemnik J, Geisler WS. 2005. Optimal eye movement strategies in visual search. *Nature* 434(7031):387–91

O'Reilly JX, Jbabdi S, Rushworth MFS, Behrens TEJ. 2013. Brain systems for probabilistic and dynamic prediction: computational specificity and integration. *PLoS Biol.* 11(9):e1001662

Orhan AE, Sims CR, Jacobs RA, Knill DC. 2014. The adaptive nature of visual working memory. *Curr. Dir. Psychol. Sci.* In press

Palmer J. 1990. Attentional limits on the perception and memory of visual information. *J. Exp. Psychol. Hum. Percept. Perform.* 16:332–50

Paulin MG. 2005. Evolution of the cerebellum as a neuronal machine for Bayesian state estimation. *J. Neural Eng.* 2(3):S219–34

Peirce CS, Jastrow J. 1884. On small differences of sensation. *Mem. Natl. Acad. Sci.* 3:73–83

Platt ML, Glimcher PW. 1999. Neural correlates of decision variables in parietal cortex. *Nature* 400(6741):233–38

Pouget A, Beck JM, Ma WJ, Latham PE. 2013. Probabilistic brains: knowns and unknowns. *Nat. Neurosci.* 16:1170–78

Pouget A, Dayan P, Zemel RS. 2003. Inference and computation with population codes. *Annu. Rev. Neurosci.* 26:381–410

Preuschoff K, Bossaerts P, Quartz SR. 2006. Neural differentiation of expected reward and risk in human subcortical structures. *Neuron* 51(3):381–90

Qamar AT, Cotton RJ, George RG, Beck JM, Prezhdo E, et al. 2013. Trial-to-trial, uncertainty-based adjustment of decision boundaries in visual categorization. *Proc. Natl. Acad. Sci. USA* 110(50):20332–37

Rangel A, Camerer C, Montague PR. 2008. A framework for studying the neurobiology of value-based decision making. *Nat. Rev. Neurosci.* 9(7):545–56

Rao RP. 2004. Bayesian computation in recurrent neural circuits. *Neural Comput.* 16(1):1–38

Raphan M, Simoncelli EP. 2011. Least squares estimation without priors or supervision. *Neural Comput.* 23(2):374–420

Rushworth MFS, Behrens TEJ. 2008. Choice, uncertainty and value in prefrontal and cingulate cortex. *Nat. Neurosci.* 11(4):389–97

Sanger T. 1996. Probability density estimation for the interpretation of neural population codes. *J. Neurophysiol.* 76(4):2790–93

Schultz W. 2000. Multiple reward signals in the brain. *Nat. Rev. Neurosci.* 1(3):199–207

Seydell A, McCann BC, Trommershäuser J, Knill DC. 2008. Learning stochastic reward distributions in a speeded pointing task. *J. Neurosci.* 28(17):4356–67

Shi L, Griffiths TL, Feldman NH, Sanborn AN. 2010. Exemplar models as a mechanism for performing Bayesian inference. *Psychon. Bull. Rev.* 17(4):443–64

Simoncelli EP. 2009. Optimal estimation in sensory systems. In *The Cognitive Neurosciences*, Vol. 4, ed. M Gazzanig, pp. 525–35. Cambridge, MA: MIT Press

So NY, Stuphorn V. 2010. Supplementary eye field encodes option and action value for saccades with variable reward. *J. Neurophysiol.* 104(5):2634–53

Softky WR, Koch C. 1993. The highly irregular firing of cortical cells is inconsistent with temporal integration of random EPSPs. *J. Neurosci.* 13(1):334–50

Stocker AA, Simoncelli EP. 2006. Noise characteristics and prior expectations in human visual speed perception. *Nat. Neurosci.* 9(4):578–85

Sugrue LP, Corrado GS, Newsome WT. 2004. Matching behavior and the representation of value in the parietal cortex. *Science* 304(5678):1782–87

Tassinari H, Hudson TE, Landy MS. 2006. Combining priors and noisy visual cues in a rapid pointing task. *J. Neurosci.* 26:10154–63

Tenenbaum JB, Griffiths TL. 2001. Generalization, similarity, and Bayesian inference. *Behav. Brain Sci.* 24(4):629–40; discuss. 652–791

Tolhurst D, Movshon J, Dean AF. 1983. The statistical reliability of signals in single neurons in cat and monkey visual cortex. *Vis. Res.* 23:775–85

Trommershauser J, Kording K, Landy MS, eds. 2011. *Sensory Cue Integration.* New York: Oxford Univ. Press

Trommershäuser J, Maloney L, Landy MS. 2008. Decision making, movement planning and statistical decision theory. *Trends Cogn. Sci.* 12(8):291–97

Turnham EJ, Braun DA, Wolpert DM. 2011. Inferring visuomotor priors for sensorimotor learning. *PLoS Comput. Biol.* 7(3):e1001112

van den Berg R, Vogel M, Josić K, Ma WJ. 2012. Optimal inference of sameness. *Proc. Natl. Acad. Sci. USA* 109(8):3178–83

Vilares I, Howard JD, Fernandes HL, Gottfried JA, Kording KP. 2012. Differential representations of prior and likelihood uncertainty in the human brain. *Curr. Biol.* 22(18):1641–48

Vilares I, Kording KP. 2011. Bayesian models: the structure of the world, uncertainty, behavior, and the brain. *Ann. NY Acad. Sci.* 1224:22–39

Whiteley L, Sahani M. 2008. Implicit knowledge of visual uncertainty guides decisions with asymmetric outcomes. *J. Vis.* 8(3):1–15

Whiteley L, Sahani M. 2012. Attention in a Bayesian framework. *Front. Hum. Neurosci.* 6:100

Yang T, Shadlen MN. 2007. Probabilistic reasoning by neurons. *Nature* 447:1075–80

Neural Tube Defects

Nicholas D.E. Greene and Andrew J. Copp

Newlife Birth Defects Research Center, Institute of Child Health, University College London, WC1N 1EH, United Kingdom; email: n.greene@ucl.ac.uk

Annu. Rev. Neurosci. 2014. 37:221–42

The *Annual Review of Neuroscience* is online at neuro.annualreviews.org

This article's doi: 10.1146/annurev-neuro-062012-170354

Keywords

anencephaly, spina bifida, folic acid, genetics

Abstract

Neural tube defects (NTDs), including spina bifida and anencephaly, are severe birth defects of the central nervous system that originate during embryonic development when the neural tube fails to close completely. Human NTDs are multifactorial, with contributions from both genetic and environmental factors. The genetic basis is not yet well understood, but several nongenetic risk factors have been identified as have possibilities for prevention by maternal folic acid supplementation. Mechanisms underlying neural tube closure and NTDs may be informed by experimental models, which have revealed numerous genes whose abnormal function causes NTDs and have provided details of critical cellular and morphological events whose regulation is essential for closure. Such models also provide an opportunity to investigate potential risk factors and to develop novel preventive therapies.

Contents

INTRODUCTION

Neural tube defects (NTDs) are severe birth defects of the central nervous system that originate during embryogenesis and result from failure of the morphogenetic process of neural tube closure (see sidebar). In higher vertebrates, the neural tube is generated by the processes that shape, bend, and fuse the neural plate, and fusion in the dorsal midline progressively seals the neural tube as it forms. If closure is not completed, the neuroepithelium remains exposed to the environment and consequently subject to degeneration and neuronal deficit. The type and severity of these open NTDs vary with the level of the body axis affected. Thus, failure of closure in the prospective brain and spinal cord results in anencephaly and open spina bifida (myelomeningocele), respectively.

NTDs: Neural tube defects

NEURAL TUBE CLOSURE

Neural tube closure is discontinuous: Closure initiates de novo at specific sites along the body axis, and progression of fusion in the dorsal midline extends the closed region to close the neuropores in the brain (anterior and hindbrain neuropores) and spine (posterior neuropore).

Although the unifying feature of open NTDs is incomplete neural tube closure, evidence points to many different possible causes, both genetic and environmental. In humans, it appears that most NTDs are multifactorial, resulting from an additive contribution of several risk factors, which are each individually insufficient to disrupt neural tube closure (the multifactorial threshold model) (Harris & Juriloff 2007). The challenge of identifying the primary cause of NTDs in individual patients is highlighted by the numerous candidate genes and environmental factors indicated by epidemiologic studies and experimental models. Moreover, the potential for gene-gene and gene-environment interactions introduces further potential complexity.

UNDERSTANDING THE EMBRYONIC BASIS OF NTDs: NEURAL TUBE CLOSURE

Determining the specific causes of NTDs is best achieved in the context of an understanding of the mechanisms underlying neural tube closure (reviewed by Copp & Greene 2013, Greene & Copp 2009). Given the inaccessibility of the neurulation-stage human embryo, our knowledge of the key principles of neural tube closure comes mainly from analysis of experimental models, particularly other mammals, amphibians, and birds, in which primary neural tube closure is achieved through folding and fusion of the neuroepithelium.

Primary Neurulation: Subtypes of NTDs Relate to Stages of Closure

In the prospective brain and most of the spinal cord, neural tube formation essentially involves the bending of the neuroepithelium at the midline to generate neural folds that elevate, meet, and fuse in the dorsal midline (primary neurulation). Rather than simultaneously rolling up along the extent of the rostrocaudal axis, neural tube closure is discontinuous with distinct sites of initiation located at characteristic axial levels. Moreover, the morphological and molecular requirements for closure vary along the body axis, such that an individual NTD usually affects only a portion of the neural tube. NTDs can thus be attributed to failure of particular initiation events or disruption of the progression of closure between these sites (**Figure 1**).

In mice, closure is first achieved on embryonic day 8.5 at the level of the hindbrain/cervical boundary (closure 1) (**Figure 1***a*), and failure of this event leads to craniorachischisis (Copp et al. 2003). Closure initiates at a second site on embryonic day 9, closure 2, in the caudal forebrain or forebrain/midbrain boundary. Once initial contact and fusion have been established between the tips of the neural folds, closure spreads bidirectionally from the sites of closures 1 and 2 and in a caudal direction from the rostral end of the neural tube (closure 3). The open regions of neural folds, termed neuropores, gradually shorten, leading to complete closure of the anterior neuropore (between closures 2 and 3) on embryonic day 9 and the hindbrain neuropore (between closures 1 and 2) a few hours later. Cranial NTDs (anencephaly, see sidebar) result from failure of closure 2, or incomplete "zippering" between closures 1 and 2, which closes the midbrain and hindbrain. If fusion does not progress from the anterior end of the neural plate (closure 3), the resultant phenotype is a split face usually accompanied by forebrain anencephaly.

Craniorachischisis: a condition where the neural tube remains open throughout the hindbrain and entire spinal region

Neuropores: open regions of neural folds, which gradually shorten and close, forming the sealed primary neural tube

Neural plate: a thickened region of ectoderm, induced on the dorsal surface of the postgastrulation embryo

ANENCEPHALY

Cranial NTDs result from failed closure of the neural folds in the mid/hindbrain. This leads to exencephaly in which exposed neuroepithelium bulges from the brain. Subsequent degeneration of this tissue results in anencephaly.

Unlike the cranial region where closure proceeds bidirectionally, spinal neurulation is entirely caudally directed as the embryo continues to grow. Primary neurulation completes with final closure of the posterior neuropore on embryonic day 10. Impaired progression of closure, and consequently the presence of a persistently open posterior neuropore, results in spina bifida, and the size of the ensuing lesion relates directly to the axial level at which closure stops.

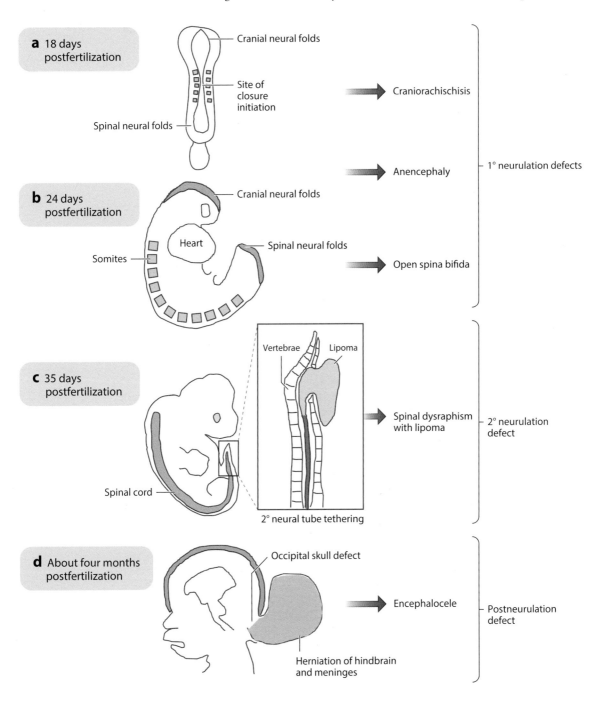

a 18 days postfertilization

Cranial neural folds

Site of closure initiation

Spinal neural folds

Craniorachischisis

Anencephaly

b 24 days postfertilization

Cranial neural folds

Heart

Somites

Spinal neural folds

Open spina bifida

1° neurulation defects

c 35 days postfertilization

Vertebrae   Lipoma

Spinal cord

2° neural tube tethering

Spinal dysraphism with lipoma

2° neurulation defect

d About four months postfertilization

Occipital skull defect

Encephalocele

Herniation of hindbrain and meninges

Postneurulation defect

Primary Neurulation in Humans

Examination of human embryos suggests that initiation of closure is discontinuous, as in the mouse (Nakatsu et al. 2000, O'Rahilly & Müller 2002). Bending of the neural plate begins at approximately 18 days after fertilization, with an event equivalent to closure 1 at approximately 21 days and completion of closure at the posterior neuropore by 26–28 days postfertilization (**Figure 1a,b**). It appears that closure of the forebrain and midbrain in human embryos may be achieved by progression between the site of closure 1 and the rostral end of the neural plate without an intervening initiation site analogous to closure 2 (O'Rahilly & Müller 2002, Sulik et al. 1998).

PCP: planar cell polarity

Secondary Neurulation

In mice and humans, the neural tube caudal to the midsacral region is continuous with the caudal end of the primary neural tube but forms by a distinct process, termed secondary neurulation (Copp & Brook 1989, Schoenwolf 1984). This process involves condensation of a population of tail bud–derived cells to form an epithelial rod that undergoes canalization to form the lumen of the tube in the lower sacral and coccygeal regions. Malformations resulting from disturbance of secondary neurulation are closed (skin covered) and often involve tethering of the spinal cord, with associated ectopic lipomatous material (**Figure 1c**) (Lew & Kothbauer 2007).

MECHANISMS UNDERLYING NEURAL TUBE CLOSURE

Studies of neurulation-stage embryos, both normal and developing NTDs, provide insights into key molecular and cellular pathways underlying the morphological tissue movements of neural tube closure (Copp & Greene 2010). The occurrence of isolated NTDs at cranial or caudal levels in humans and different mouse models suggests the likely involvement of region-specific mechanisms, dependent on different gene products, in addition to ubiquitous requirements that are essential at all levels.

Shaping of the Neural Plate: Convergent Extension Is Required to Initiate Closure

Concomitant with the onset of neural tube closure, the neural plate undergoes narrowing in the mediolateral axis (convergence) and elongation in the rostrocaudal axis (extension), owing to intercalation of cells at the midline (Keller 2002). Convergent extension depends on activity of a noncanonical Wnt signaling pathway, homologous to the planar cell polarity (PCP) pathway first described in *Drosophila* as regulating cell polarity in the plane of epithelia (Goodrich & Strutt 2011).

Figure 1

Diagrammatic representation of the developmental origin of malformations broadly classified as neural tube defects in humans. (*a,b*) Disorders of primary neurulation include craniorachischisis (*a*) in which the neural tube fails to initiate closure, leaving most of the brain and the entire spine open. If closure initiates successfully, then the cranial and/or spinal neural folds may fail to close (*b*), generating exencephaly/anencephaly and open spina bifida (myelomeningocele), respectively. (*c*) Disorders of secondary neurulation comprise failure of the neural tube to separate completely from adjacent tissues, resulting in tethering and diminished mobility. The spinal cord is covered by skin and often associated with fatty tissue accumulation (lipoma) through as-yet-unknown mechanisms. (*d*) Postneurulation defects can arise when the bony structure of the skeleton fails to develop fully. Herniation of the meninges, with or without brain tissue, through a skull defect (shown here as occipital but sometimes parietal or fronto-ethmodial) generates encephalocele, while an analogous defect in the spinal region produces meningocele.

Signaling occurs via Frizzled (Fzd) membrane receptors and cytoplasmic Dishevelled (Dvl) but without stabilization of β-catenin.

Functional disruption of PCP mediators prevents convergent extension, and the neural plate remains broad in *Xenopus* (Wallingford & Harland 2001, 2002) and mouse embryos (Greene et al. 1998, Ybot-Gonzalez et al. 2007b). Hence, closure 1 fails, leading to craniorachischisis in mice homozygous for mutations in core PCP genes including *Vangl2* and *Celsr1*, or double mutants for *Dvl-1* and *-2*, or *Fzd-3* and *-6* (Juriloff & Harris 2012a). Craniorachischisis also results from mutation of the PCP-related genes *Scrb1* (Murdoch et al. 2001) and *Ptk7* (Lu et al. 2004) or genes encoding accessory proteins, such as *Sec24b*, which affects Vangl2 transport (Merte et al. 2010). Ultimately, failure of closure initiation in PCP-mutant embryos is thought to result from insufficient proximity of the neural folds, owing to the broadened midline.

Failure of closure 1 in most of the core PCP mutant embryos precludes analysis of a requirement for convergent extension at later stages of neurulation. However, spina bifida occurs in some *loop-tail* heterozygotes (*Vangl2^{Lp/+}*) (Copp et al. 1994) and in compound heterozygotes of *Vangl2^{Lp/+}* with mutations of *Ptk7*, *Sec24b*, or *Sdc4* (Escobedo et al. 2013, Lu et al. 2004, Merte et al. 2010). Moreover, non-canonical Wnt signaling is compromised in *Lrp6* null embryos that develop spina bifida (Gray et al. 2013). These observations suggest that PCP signaling may continue to be required as spinal neurulation proceeds.

Despite the entirely open spinal neural tube in *Vangl2^{Lp/Lp}* embryos with craniorachischisis, closure does occur in the forebrain and much of the midbrain, implying that PCP-dependent convergent extension is not required throughout the cranial region. Nonetheless, exencephaly is observed in digenic combinations of *Vangl2^{Lp/+}* with some Wnt pathway genes (e.g., *Dvl3^{+/−}*, *Fzd1^{+/−}*, and *Fzd2^{+/−}*) (Etheridge et al. 2008, Yu et al. 2010). Exencephaly also develops in mutants for the PCP effector genes *Fuz* or *Intu*, but the role of these genes in cilium-dependent hedgehog signaling seems more likely to explain their loss-of-function effect on cranial neural tube closure than does a role in regulating convergent extension (Gray et al. 2009, Heydeck & Liu 2011, Zeng et al. 2010) (see Bending of the Neural Folds: Regulation by Shh and BMP Signaling, below). Thus, components of PCP signaling potentially affect neural tube closure via multiple cellular mechanisms.

Bending of the Neural Folds: Regulation by Shh and BMP Signaling

To achieve closure, the neuroepithelium must bend to bring the tips of the neural folds into apposition. Bending occurs in a stereotypical manner at hinge points: a median hinge point (MHP) in the midline and paired dorsolateral hinge points (DLHPs) that arise laterally (Shum & Copp 1996). The morphology varies along the body axis with differing modes in the upper (MHP only), midspine (MHP and DLHPs), and caudal (DLHPs only) regions of the primary spinal neural tube.

The mechanisms underlying neuroepithelial bending are not fully understood, but one notable feature of the MHP is the predominance of wedge-shaped cells (wider basally than apically) compared with nonbending regions (Schoenwolf & Smith 1990). At neural plate stages, the neuroepithelium is a pseudostratified epithelium in which nuclei move to the basal pole during S-phase, owing to interkinetic nuclear migration (see sidebar). Prolongation of S-phase at the MHP provides a possible means by which regulation of the cell cycle may contribute to cell wedging and hence MHP formation (Schoenwolf & Smith 1990).

Bending is regulated by signals emanating from nonneural tissues dorsal and ventral to the neural folds (reviewed by Greene & Copp 2009). The MHP is induced by signals from the notochord, located immediately ventral to the midline of the neuroepithelium (Smith & Schoenwolf 1989, Ybot-Gonzalez et al. 2002). At the molecular level, notochord-derived Shh induces the floor plate of the neural tube at the MHP (Chiang et al. 1996, Placzek & Briscoe 2005). However, this action is

INTERKINETIC NUCLEAR MIGRATION

Interkinetic nuclear migration occurs in the neuroepithelium and describes the apico-basal migration of the nucleus according to the phase of the cell cycle such that S-phase occurs at the basal aspect and mitosis at the apical aspect.

not essential for spinal neural tube closure, which completes in the absence of a floor plate in mouse embryos lacking Shh or Fox A2 (Ang & Rossant 1994, Chiang et al. 1996). Thus, the MHP may be functionally important in floor plate development but is not essential for neural tube closure.

In contrast to the MHP, DLHPs appear essential for the neural tube to close in the low spinal region. For example, *Zic2* mutant embryos, in which DLHPs are absent, develop severe spina bifida (Ybot-Gonzalez et al. 2007a). The formation of DLHPs is actively regulated; the interplay of inhibitory and inductive signals determines their appearance at different axial levels (Copp & Greene 2013). These signals include inhibitory effects of Shh from the notochord and BMP signaling from the surface ectoderm at the dorsal tips of the neural folds. These signals are opposed by the BMP antagonist noggin, whose expression in the dorsal neural folds is sufficient to induce DLHPs (Ybot-Gonzalez et al. 2002, 2007a).

In contrast to the effects of an absence of Shh signaling, NTDs do result from mutations that enhance Shh signaling, for example, through deficient function of inhibitory or cilia-related genes such as *Gli3*, *Rab23*, *Fkbp8*, *Tulp3*, and *Ift40* (Miller et al. 2013, Murdoch & Copp 2010). Mutants involving increased Shh signaling display NTDs at cranial and/or spinal levels. Although spina bifida in some of these models appears to be associated with suppression of dorsolateral bending of the neural folds (Murdoch & Copp 2010), the mechanism underlying cranial NTDs is not clear.

Cranial Neurulation: Additional Complexity and Sensitivity to Disruption

The neural folds in the cranial region bend in the midline and dorsolaterally as in the midspinal region, but the closure process appears morphologically more complex. The folds are initially biconvex, with the tips facing away from the midline, and then switch to a biconcave shape allowing the tips to approach in the midline. The additional complexity of cranial compared with spinal neurulation appears to be reflected in a more extensive genetic underpinning and a greater sensitivity to disruption, at least in rodents. Exencephaly occurs in approximately three times as many knockout mouse models as does spina bifida and is the NTD type most commonly induced by teratogens (Copp et al. 1990, Harris & Juriloff 2010).

Cranial neurulation may rely on specific contributory factors that are not involved in the spinal region such as expansion of the mesenchyme underlying the neural folds (Greene & Copp 2009, Zohn & Sarkar 2012). Moreover, disruption of the actin cytoskeleton prevents closure in the cranial but not the spinal region (Morriss-Kay & Tuckett 1985, Ybot-Gonzalez & Copp 1999). Similarly, exencephaly is observed, but spinal neurulation completes successfully in null mutants for several cytoskeletal components (e.g., *n-cofilin*, *vinculin*) (Gurniak et al. 2005, Xu et al. 1998). Nevertheless, apically located actin microfilaments are present throughout the neuroepithelium (Sadler et al. 1982), and functional disruption of the cytoskeleton-associated proteins *MARCKS-related protein* or *Shroom3* causes both spinal and cranial NTDs (Hildebrand & Soriano 1999, Xu et al. 1998), suggesting that regulation of the actomyosin cytoskeleton plays a role in closure in both regions. Shroom proteins appear to play a key role: Expression of Shroom in *Xenopus* is sufficient to induce apical constriction of epithelial cells, whereas functional disruption inhibits neural fold bending and suppresses closure (Haigo et al. 2003).

Adhesion and Fusion of the Neural Folds

Once the neural folds meet at the dorsal midline, processes of adhesion, fusion, and remodeling give rise to two discrete epithelial layers, with the nascent neural tube overlain by an intact surface ectoderm (Pai et al. 2012). At the closure site, the neural fold tips are composed of neuroepithelium continuous with the nonneural surface ectoderm. The cell type that adheres first may differ at varying axial levels (Geelen & Langman 1979, Ray & Niswander 2012). Nevertheless, at all levels initial contact appears to involve subcellular protrusions, resembling lamellipodia and filopodia, observed by electron microscopy (Geelen & Langman 1979) and in live embryos (Pyrgaki et al. 2010). The molecular basis of adhesion is not well characterized, perhaps owing to functional redundancy among the proteins involved. However, a role for the interaction of cell surface ephrin receptors with Eph ligands is suggested by the occurrence of cranial NTDs in mice lacking ephrin-A5 or EphA7 (Holmberg et al. 2000) and by delayed spinal closure in embryos exposed to peptides that block ephrin-A/EphA interactions (Abdul-Aziz et al. 2009).

Knockout of protease-activated receptors (PAR1 and PAR2) in the surface ectoderm also causes cranial NTDs, implicating a role for signaling via these G protein–coupled receptors in closure (Camerer et al. 2010). Further evidence for the function of the nonneural ectoderm is provided by *Grhl2* null mutants, which fail in closure throughout the cranial region and exhibit spina bifida (Brouns et al. 2011, Rifat et al. 2010, Werth et al. 2010). *Grhl2* is expressed in the surface ectoderm overlying the neural folds and regulates expression of several components of the apical adhesion junction complex, including E-cadherin (Pyrgaki et al. 2011, Werth et al. 2010).

Regulation of Cell Proliferation and Cell Death

During neurulation the embryo grows rapidly. Cell cycle exit and neuronal differentiation begin in the neuroepithelium shortly after closure, and maintenance of adequate proliferation in the neuroepithelium appears crucial for closure, particularly in the cranial region. Thus, in mice, NTDs can be caused by exposure to antimitotic agents (Copp et al. 1990) or mutation of genes encoding proteins associated with cell-cycle progression (e.g., *neurofibromin 1*, *nucleoporin*) or prevention of neuronal differentiation (e.g., Notch pathway genes *Hes1*, *Hes3*, *RBP-Jκ*) (Harris & Juriloff 2007, 2010). Conversely, excessive cell proliferation is also associated with NTDs in several mouse models, such as *Phactr4* mutants (Kim et al. 2007).

Characteristic patterns of apoptotic cell death occur in the neural folds and the midline of the closed neural tube (Geelen & Langman 1979, Massa et al. 2009, Yamaguchi et al. 2011). Increased cell death could inhibit closure by compromising the functional and/or mechanical integrity of the neuroepithelium. It is associated with NTDs in several teratogen-induced and genetic models, although only rarely has a direct causal link been definitively established (Copp & Greene 2013, Fukuda et al. 2011). The occurrence of exencephaly in mice lacking apoptosis-related genes such as *caspase3* or *Apaf1* suggests a requirement for apoptosis in closure (Harris & Juriloff 2010). However, forebrain and spinal closure occurs normally in these models and pharmacological suppression of apoptosis does not cause NTDs, suggesting that it is dispensable to complete closure (Massa et al. 2009).

CLINICAL FEATURES OF NEURAL TUBE DEFECTS

Owing to their multifactorial causation, NTDs represent a group of disorders. However, after failure of neural fold closure has occurred, defects originating from various primary causes may share similar pathogenic features.

Open NTDs and Associated Conditions

Open NTDs can result from failure of closure at a de novo initiation site or incomplete progression of closure following successful initiation (**Figure 1**). Where embryos are available for examination, as in experimental models, NTDs can be recognized during or immediately after neurulation stages owing to the persistently open neural folds. However, at later embryonic and fetal stages, the morphological appearance changes considerably owing to secondary changes and degeneration.

In cranial NTDs, the open neural folds undergo growth and differentiation and typically appear to bulge from the developing brain, termed exencephaly. Inability to form the skull vault over the open region causes the exposed neural tissue to degenerate, leading to the characteristic appearance of anencephaly, observed later in human or rodent pregnancy (Wood & Smith 1984, Seller 1995). Both anencephaly and craniorachischisis (~10% of NTDs) are lethal conditions at or shortly after birth.

Open neural folds in the spinal region prevent the sclerotome-derived vertebral arches from covering the neuroepithelium, the consequent opening in the vertebral column giving rise to the term spina bifida (Copp et al. 2013). The neural tissues may be contained within a meninges-covered sac that protrudes through the open vertebrae (myelomeningocele; spina bifida cystica) or exposed directly to the amniotic fluid (myelocele). Babies born with open spina bifida usually survive with appropriate medical care but suffer neurological impairment, the severity of which depends on the level of the lesion. Associated conditions include hydrocephalus, Chiari malformation type II, and vertebral abnormalities as well as genitourinary and gastrointestinal disorders.

Diagnosis, Treatment, and Maternal-Fetal Surgery

NTDs can be diagnosed prenatally by ultrasound (Cameron & Moran 2009). However, where prenatal diagnosis is not routinely available and/or therapeutic abortion is not an option, many babies with NTDs are born. Postnatal medical care for babies born with open spina bifida usually involves surgery to close and cover the lesion. Multiple subsequent surgeries are commonly required to alleviate tethering of the spinal cord, treat hydrocephalus, and/or address orthopedic and urological problems.

As open NTDs arise early during pregnancy, there is a prolonged period during which secondary neurological damage may occur owing to exposure of nervous tissue to the amniotic fluid environment. These considerations provided impetus for the development of in utero fetal surgery for spina bifida, which may improve neurological outcomes compared with postnatal repair, although with fetal and maternal risks (Adzick et al. 1998, 2011). Experimental models of spina bifida are being used to investigate the possible combination of surgical intervention with additional therapy, intended to remediate neural damage. Examples include the implantation of biodegradable scaffolds to promote neural regeneration and/or neural stem cells to populate the damaged spinal cord (Saadai et al. 2011, 2013).

Disorders of the Closed Neural Tube

This review focuses on open NTDs, characterized by failure of neural tube closure. Various other conditions are also associated with abnormalities of the closed brain or spinal cord and are often categorized as NTDs under a broader definition (**Figure 1**). There is also a less well-defined group of closed spinal NTDs in which the vertebral arches are malformed but covered by skin. These conditions, including spina bifida occulta and spinal dysraphisms, vary widely in clinical presentation. The more severe subtypes are associated with various abnormalities of the spinal cord, lipoma (**Figure 1c**), and/or anorectal abnormalities. The embryonic origin of closed spina bifida is not well defined but is hypothesized to involve abnormalities of secondary neurulation (Copp et al. 2013).

Abnormal development of the skull or vertebrae may also allow herniation of the closed neural tube through the affected bony region, as in encephalocele (**Figure 1d**) or meningocele, respectively.

CAUSES OF NTDs

NTDs are among the most common birth defects worldwide with a prevalence that varies from 0.5 to more than 10 per 1,000 pregnancies. This variance likely reflects differing contributions from risk factors such as nutritional status, prevalence of obesity and diabetes, usage of folic acid supplementation and/or fortification, the presence of environmental toxicants, and differing genetic predisposition among ethnic groups. In most populations, there is also a striking gender bias: Anencephaly is more prevalent among females than males. Many NTD mouse strains also show a female preponderance among cranial NTDs, apparently reflecting a fundamental higher sensitivity of cranial neural tube closure to disturbance in female embryos (Juriloff & Harris 2012b). Overall, although studies have identified numerous risk factors, these may account for less than half of NTDs, suggesting that additional genetic and nongenetic factors remain to be identified (Agopian et al. 2013).

Environment Factors

Various teratogenic agents induce NTDs in rodent models (Copp et al. 1990, Copp & Greene 2010). In humans, teratogens that have been associated with NTDs include the anticonvulsant drug valproic acid (Wlodarczyk et al. 2012) and the fungal product fumonisin (Missmer et al. 2006). Other nongenetic risk factors include maternal fever and excessive use of hot tubs (Moretti et al. 2005), consistent with the induction of NTDs by hypothermia in rodent models.

Maternal obesity and diabetes are well-recognized risk factors for NTDs (Correa et al. 2003). Determining the cause of diabetes-related NTDs is hampered by the complexity of the diabetic milieu, although hyperglycemia alone is sufficient to cause NTDs in cultured rodent embryos. It has been proposed that NTDs may result from increased oxidative stress, altered expression of genes such as *Pax3*, and neuroepithelial cell apoptosis (Fine et al. 1999, Reece 2012). Recent findings suggest that activation of apoptosis signal-regulating kinase 1 (ASK1) in hyperglycemic conditions leads to activation of the apoptosis mediator caspase 8 by stimulating the FoxO3a transcription factor (Yang et al. 2013).

Nutritional factors and folate. The historical link between lower socioeconomic status and higher risk of birth defects led to examination of the possible involvement of nutritional factors in NTDs. Lower blood levels of the B-vitamin folate were observed in mothers of NTD fetuses (Smithells et al. 1976), prompting an intervention trial of a folic acid–containing multivitamin supplement to prevent NTD recurrence (Schorah 2008, Smithells et al. 1981). A multicenter randomized controlled trial confirmed that maternal folic acid supplementation (at 4 mg per day) significantly reduces the recurrence risk (Wald et al. 1991). Additional clinical trials provided evidence for reduction of occurrence risk (Berry et al. 1999, Czeizel et al. 2011, Czeizel & Dudás 1992).

Questions remain concerning the mechanism by which folic acid prevents NTDs (Blom et al. 2006, Copp et al. 2013). Although maternal folate status is a risk factor, in most cases, maternal

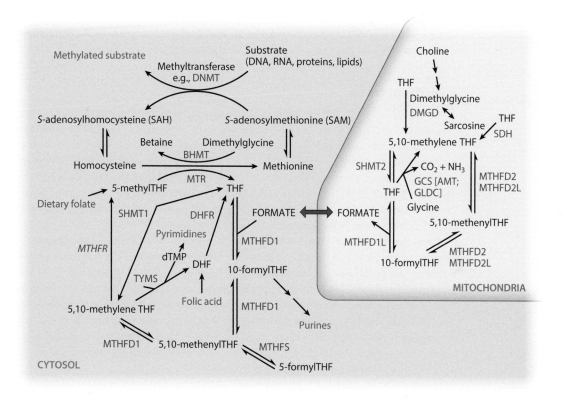

Figure 2

Overview of folate one-carbon metabolism. Folates provide a backbone for the transfer of one-carbon units. Key outputs (*green*) include nucleotide biosynthesis and methylation. Among methylation cycle intermediates, homocysteine may also be converted to cystathionine in the transulfuration pathway and *S*-adenosyl methionine is involved in polyamine biosynthesis. Folate-mediated one-carbon metabolism (FOCM) is compartmentalized: one-carbon units from the mitochondria enter cytoplasmic FOCM as formate while reactions of thymidylate biosynthesis also operate in the nucleus (catalyzed by SHMT1, TYMS, and DHFR). In loss-of-function mouse models, neural tube defects (NTDs) arise in mutants for Mthfd1l and genes encoding the glycine cleavage system (GCS). Shmt1 and Mthfr null mice are viable to birth but may develop NTDs under folate-deficient conditions. Entry points of folate, from dietary sources or folic acid, are indicated in blue. Enzymes that catalyze steps of one-carbon metabolism and methylation reactions are indicated in purple: BHMT, betaine-homocysteine S-methyltransferase; DNMT, DNA methyltransferase; DHFR, dihydrofolate reductase; DMGD, dimethylglycine dehydrogenase; GCS, glycine cleavage system (including AMT, aminomethyltransferase; GLDC, glycine decarboxylase); MTHFD (1, 1L, 2, 2L), methylenetetrahydrofolate dehydrogenase; MTHFR, methylenetetrahydrofolate reductase; MTHFS, methenyltetrahydrofolate synthetase; SDH, sarcosine dehydrogenase; SHMT, serine hydroxymethyltransferase.

folate levels are within the normal range and rarely clinically deficient. Nonetheless, data have shown an inverse relationship between blood folate concentration and risk of an affected pregnancy (Daly et al. 1995). Suboptimal folate levels may contribute to NTD development in individuals who are genetically susceptible. Such a gene-environment interaction has been demonstrated in mice, where folate deficiency does not cause NTDs unless deficiency is present in combination with a mutation of a predisposing gene, such as *Pax3* (Burren et al. 2008).

Folate one-carbon metabolism (**Figure 2**) comprises a complex network of interlinked reactions that mediate transfer of one-carbon groups for several biosynthetic processes (Stover 2009). Among these, attention has focused particularly on the requirement for nucleotide biosynthesis and methylation reactions in neural tube closure. Abnormal thymidylate and purine biosynthesis have been identified in mouse NTD models (Beaudin et al. 2011, Fleming & Copp 1998) and in a proportion of NTD cases (Dunlevy et al. 2007), whereas deficient methylation may also be implicated in NTDs (see Gene-Regulatory Mechanisms and NTDs, below).

Genetics of NTDs

Most NTDs occur sporadically, with a relative scarcity of multigenerational families. Nevertheless, there is strong evidence for a genetic component in the etiology of NTDs, and the pattern of inheritance favors a multifactorial polygenic or oligogenic model, as opposed to an effect of single genes with partial penetrance (Harris & Juriloff 2007). Most studies of NTD genetics have focused on one or more candidate genes (reviewed by Boyles et al. 2005, Greene et al. 2009, Harris & Juriloff 2010). In general, candidates have been (*a*) human orthologs of genes whose mutation causes NTDs in mice, of which there are more than 200 examples; or (*b*) genes related to environmental risk factors, particularly folate metabolism.

Case-control association studies have implicated several genes, whereas mutation screening by sequencing has identified putative pathogenic mutations. However, the definitive assignment of a gene variant as causative is complicated by the apparent multigenic nature of NTDs and by the large number of possible candidate genes, modifier genes, epigenetic factors, and environmental influences. Moreover, where putative mutations have been identified in specific genes, each has been involved in only a small proportion of NTD patients, suggesting that there is considerable heterogeneity underlying the genetic basis of NTDs. Thus, although the morphological and cellular bases of neural tube closure have become increasingly well understood, the genetic basis of NTDs in individual cases remains largely unclear.

Gene-gene interactions and effect of modifier genes. Mouse studies suggest three broad mechanisms by which genetic interactions may result in NTDs. First, in some instances functional redundancy makes it necessary for two orthologous genes to be mutated [e.g., *Dvl1-Dvl2* (Hamblet et al. 2002), *Cdx1-Cdx2* double knockouts (Savory et al. 2011)], in order to reveal a requirement in neural tube closure. Second, additive effects of heterozygous mutations may result in NTDs that resemble those of individual homozygotes [e.g., *Dvl3* with *Vangl2^{Lp}* (Etheridge et al. 2008)]. Third, variation in the penetrance and expressivity of NTD phenotypes between inbred mouse strains is widely reported to reflect variants in modifier genes. For example, the rate of exencephaly resulting from *Cecr1* mutation is strongly affected by strain background (Davidson et al. 2007). Whereas the identity of modifier genes for NTDs has rarely been determined, a variant in *Lmnb1* is present in some mouse strains and significantly increases the frequency of NTDs in *curly tail* (*Grhl3^{ct}*) embryos (De Castro et al. 2012).

Genes implicated through experimental models. In mice, mutation of genes encoding components of the PCP pathway causes NTDs (see Shaping of the Neural Plate, above). Sequencing of PCP genes in humans has identified putative mutations in *CELSR1*, *VANGL1*, *VANGL2*, *FZD6*, *SCRIB1*, and *DVL2* in some patients with craniorachischisis, spina bifida, anencephaly, or closed forms of spina bifida (Chandler et al. 2012; De Marco et al. 2013; Kibar et al. 2007; Lei et al. 2010, 2013; Robinson et al. 2012; and reviewed by Juriloff & Harris 2012a). As in mice, heterozygous human PCP mutations could hypothetically interact with other genetic NTD risk factors in a digenic or polygenic fashion to cause a range of NTD types. This interaction could involve summation of multiple variants in PCP genes. For example, a putative mutation in *DVL2* was identified in a spina bifida patient in combination with a second, previously identified missense variant in *VANGL2* (De Marco et al. 2013).

Among other genes implicated in NTDs from mouse models, association studies have not provided evidence for a major contribution to risk, and few positive results have emerged from sequencing-based mutation screens. As data begin to emerge from large-scale exome sequencing studies of NTD patients, it will become possible to evaluate the contribution of multiple genes in the same patient cohorts and the mutational load associated with individual risk.

Analysis of genes related to environmental risk factors. The identification of environmental factors such as maternal diabetes and folate status as risk factors for NTDs provides impetus for analysis of related genes in affected families. Risk could be associated with maternal genotype if genetic variation alters maternal metabolism and secondarily affects the developing embryo. However, the inheritance of maternal alleles by the embryo complicates interpretation of such effects. Alternatively, a genetically determined abnormality in the embryo itself could influence risk of NTDs, potentially through interaction with a predisposing environmental factor. For example, it may be informative to analyze genetic data on folate-related genes in the context of maternal folate status (Etheredge et al. 2012).

Association with risk of spina bifida has been reported for several genes implicated in diabetes, obesity, glucose metabolism, and oxidative stress. These potential "risk" genes include *GLUT1*, *SOD1*, and *SOD2* (Davidson et al. 2008, Kase et al. 2012). Maternal variants in the obesity-related genes *FTO*, *LEP*, and *TCF7L2* are also associated with NTDs, consistent with maternal obesity as a risk factor (Lupo et al. 2012).

Genes related to folate one-carbon metabolism have been perhaps the most intensively studied group of candidates for NTDs (reviewed by Blom et al. 2006, Greene et al. 2009, Shaw et al. 2009). The C677T polymorphism of *MTHFR*, which encodes an alanine-to-valine substitution, has been associated with NTDs. The TT genotype is found at higher frequency among NTD cases than in controls in some populations (e.g., Irish) but not others (e.g., Hispanics) (Botto & Yang 2000). Several studies indicate positive associations with other folate-related genes, including *MTRR*, although these have generally not been observed in all study populations.

In mice, mutations in folate-metabolizing enzymes (e.g., *Mthfd1*) are sometimes lethal before the stage of neural tube closure (e.g., Christensen et al. 2013, MacFarlane et al. 2009), whereas others do not disrupt closure (e.g., Chen et al. 2001, Di Pietro et al. 2002). Null embryos for the folate receptor, *Folr1*, die preneurulation but develop NTDs when supplemented with sufficient folic acid to prevent early lethality (Piedrahita et al. 1999). NTDs are also observed in *Shmt1* knockouts, under folate-deficient conditions (Beaudin et al. 2011). In contrast, NTDs occur spontaneously in mice carrying loss-of-function alleles of *Amt* (Narisawa et al. 2012) or *Mthfd1L* (Momb et al. 2013), both of which encode enzymes of mitochondrial folate metabolism (see sidebar and **Figure 2**) (Tibbetts & Appling 2010). The homologous genes in humans have also been linked to NTDs. Missense mutations have been identified in NTD patients in *AMT* as well as in *GLDC*, which encodes its partner enzyme in the glycine cleavage system (Narisawa et al. 2012). Genetic associations with NTDs have been reported for *MTHFD1L* (Parle-McDermott et al. 2009) and

MITOCHONDRIAL FOLATE METABOLISM

Mitochondrial folate metabolism comprises a set of reactions of folate one-carbon metabolism that are compartmentalized in mitochondria and generate formate that is transferred to the cytoplasm where it acts as a one-carbon donor. AMT, GLDC (components of the glycine cleavage system), and MTHFD1L are specific to mitochondrial folate metabolism.

SLC23A32 (*MFTC*), encoding a mitochondrial folate transporter (Pangilinan et al. 2012). Altogether, these findings suggest that NTD risk is influenced by the function of mitochondrial folate metabolism, a major source of one-carbon units to the cytoplasm.

Gene-Regulatory Mechanisms and NTDs

Identification of causative genes may be complicated, in addition to the potential multigenic nature of NTDs, by the potential involvement of aberrant gene expression, perhaps resulting from mutations in regulatory elements. For example, mutations resulting in insufficient expression of *Grhl3* or excess expression of *Grhl2* cause NTDs in mice in the absence of coding mutations (Brouns et al. 2011, Gustavsson et al. 2007). Further complexity may be added by the potential for regulation by epigenetic modifications such as DNA methylation, histone modification, or chromatin remodeling, each of which has been associated with NTDs in mice and in some cases in humans (reviewed by Greene et al. 2011, Harris & Juriloff 2010). For example, methylation of LINE-1 genomic elements was lower than normal in DNA of anencephalic but not spina bifida fetuses (Wang et al. 2010).

A simple model predicts a positive correlation between folate status and methylation. However, data from human pregnancy suggest that the relationship is not straightforward (Crider et al. 2012). A recent study found an inverse correlation of LINE-1 methylation with maternal and cord blood folate, whereas different imprinted genes showed positive or negative associations (Haggarty et al. 2013). Somewhat counterintuitively, use of folic acid supplements was associated with reduced LINE-1 methylation.

A requirement for DNA methylation in mouse neural tube closure is suggested by the occurrence of NTDs in knockouts of *Dnmt3b*, encoding a DNA methyltransferase, and in embryos cultured with 5-azacytidine (Matsuda & Yasutomi 1992, Okano et al. 1999). Similarly, inhibition of the methylation cycle reduces DNA methylation and causes NTDs in cultured mouse embryos (Burren et al. 2008, Dunlevy et al. 2006). However, *Mthfr* null embryos do not develop NTDs despite a significant reduction in global DNA methylation (Chen et al. 2001), nor is there an exacerbating effect of *Mthfr* loss-of-function on *Pax3* or *curly tail* mutants, although both show increased rates of NTDs under folate-deficient conditions (Burren et al. 2008, De Castro et al. 2010, Pickell et al. 2009). Thus, questions remain about the relationships among folate status, DNA methylation, and risk of NTDs.

Other epigenetic mechanisms include various modifications of histone proteins, which potentially misregulate genes that influence neurulation. NTDs occur in mice carrying mutations in the histone demethylases *Jarid2* (Takeuchi et al. 1999) and *Fbxl10* (Fukuda et al. 2011). Similarly, histone acetylases and deacetylases, which regulate the equilibrium of histone acetylation, are implicated in NTDs. An acetylase-specific knockin mutation of *Gcn5* causes cranial NTDs (Bu et al. 2007), as does loss-of-function of another histone acetylase, p300 (Yao et al. 1998). Increased acetylation is also associated with NTDs. For example, cranial NTDs occur in mice carrying mutations in the histone deacetylases *Sirt1* or *Hdac4* (Cheng et al. 2003, Vega et al. 2004). The teratogenic effects of valproic acid and trichostin A may also be mediated through their inhibition of histone deacetylases (Finnell et al. 2002).

PRIMARY PREVENTION OF NTDs

Once the neural tube has failed to close, ensuing damage to the exposed neural tissue is irreversible, despite possible palliative benefit of in utero surgery (see Diagnosis, Treatment, and

Maternal-Fetal Surgery). Therefore, primary prevention is the optimal approach for reducing the burden of NTDs.

Folic Acid Supplementation and Fortification

The reduction in risk of NTDs following maternal folic acid supplementation led to public health recommendations that women who may become pregnant should consume 0.4 mg of folic acid daily or 4 mg daily following a previous affected pregnancy (Czeizel et al. 2011). To ensure that additional folate was received, food fortification programs were introduced in many countries. This approach has raised blood folate levels and has been associated with lower NTD frequency (Crider et al. 2011). The magnitude of effect varies, with the greatest reduction found where preexisting rates were highest (Blencowe et al. 2010, Rosenthal et al. 2013). Some countries have delayed a decision on fortification owing to safety concerns (e.g., possible enhancement of bowel cancer), but a recent meta-analysis found no evidence for increased cancer rates following folic acid supplementation (Vollset et al. 2013).

Folate-Resistant NTDs

Folic acid supplementation in clinical trials has not approached 100% NTD prevention, and an estimated one-third of NTDs may be folic acid resistant (Blencowe et al. 2010). A study in the United States, where folate fortification of food is mandatory, found no apparent protective effect of folic acid supplements (Mosley et al. 2009), suggesting that increased dosage would not necessarily provide additional preventive effects.

Given the multifactorial causation of NTDs it seems reasonable to suppose that optimal prevention will require a combination of multiple interventions. Possible approaches may relate to folate one-carbon metabolism. For example, as with folate, evidence shows a graded relationship between lower levels of circulating vitamin B_{12} and increased risk of an NTD-affected pregnancy (Molloy et al. 2009). Perhaps use of B_{12} supplements would further reduce NTD frequency, although this approach remains to be tested.

Another possibility is that folic acid cannot ameliorate some defects that result from abnormal folate metabolism, owing to defects in the intervening enzymes required to transfer one-carbon units to key downstream metabolites. In this case, supplementation with alternative folates, such as 5-methylTHF (Czeizel et al. 2011), or key downstream molecules may be advantageous. For example, supplementation with formate prevented NTDs in *Mthfd1L* null mice (Momb et al. 2013), whereas combinations of thymidine and purine precursors prevented NTDs in *curly tail* mice, in which folic acid is not protective (Leung et al. 2013).

In addition to low levels of folate and vitamin B_{12}, lower maternal levels of other vitamins, including vitamin C, have been reported in NTDs (Smithells et al. 1976). Conversely, intake of several vitamins and maternal diet are associated with lower risk of NTDs, which suggests that nutrients other than folic acid may be beneficial (Chandler et al. 2012, Sotres-Alvarez et al. 2013). Experimental analysis of individual vitamins found that *myo*-inositol deficiency caused NTDs in cultured rodent embryos (Cockroft 1988). Inositol supplementation significantly reduced NTD frequency in *curly tail* mice (Greene & Copp 1997) and in rodent models of diabetes (Reece et al. 1997), and inositol is in clinical testing for prevention of NTD recurrence.

SUMMARY

Experimental models provide systems for analysis of the developmental events of neural tube closure, and fundamental cellular and morphological processes continue to be defined in more detail. In principle, NTDs may result from insufficiency of one or more of the key driving forces (e.g., cellular properties and/or morphogenetic movements) that are necessary to achieve closure, for example, through mutation of a PCP gene. Alternatively, a genetic lesion or environmental insult may disrupt the closure process even where the underlying machinery is intact, for example through induction of aberrant cellular behaviors such as excess apoptosis. Experimental models require careful analysis to disentangle these possibilities. A key challenge will be to understand how the molecular and cellular determinants of neurulation relate to the biomechanical forces required to fold the neuroepithelium to achieve closure.

Advances in exome and whole-genome sequencing may help researchers begin to understand the genetic basis of NTDs in humans. The multifactorial complexity of NTDs means that analysis of data from such studies will present a major challenge. Moreover, investigators will need to integrate genetic data with information on epigenetic and environmental factors to obtain a more complete understanding of the cause of individual NTDs.

Folic acid supplementation provides a means to reduce NTD risk and represents a major public health advance. Nevertheless, the heterogeneity of NTDs suggests that primary prevention may be achieved best by multiple interventions, and use of additional micronutrients alongside folic acid may provide additional opportunities to further reduce risk.

DISCLOSURE STATEMENT

The authors are not aware of any affiliations, memberships, funding, or financial holdings that might be perceived as affecting the objectivity of this review.

ACKNOWLEDGMENTS

Work in the authors' laboratory is funded by the Medical Research Council (J003794), the Newlife Foundation (11-1206), and the Wellcome Trust (087525).

LITERATURE CITED

Abdul-Aziz NM, Turmaine M, Greene ND, Copp AJ. 2009. EphrinA-EphA receptor interactions in mouse spinal neurulation: implications for neural fold fusion. *Int. J. Dev. Biol.* 53:559–68

Adzick NS, Sutton LN, Crombleholme TM, Flake AW. 1998. Successful fetal surgery for spina bifida. *Lancet* 352:1675–76

Adzick NS, Thom EA, Spong CY, Brock JW III, Burrows PK, et al. 2011. A randomized trial of prenatal versus postnatal repair of myelomeningocele. *N. Engl. J. Med.* 364:993–1004

Agopian AJ, Tinker SC, Lupo PJ, Canfield MA, Mitchell LE. 2013. Proportion of neural tube defects attributable to known risk factors. *Birth Defects Res. A Clin. Mol. Teratol.* 97(1):42–46

Ang S-L, Rossant J. 1994. *HNF*-3β is essential for node and notochord formation in mouse development. *Cell* 78:561–74

Beaudin AE, Abarinov EV, Noden DM, Perry CA, Chu S, et al. 2011. *Shmt1* and de novo thymidylate biosynthesis underlie folate-responsive neural tube defects in mice. *Am. J. Clin. Nutr.* 93:789–98

Berry RJ, Li Z, Erickson JD, Li S, Moore CA, et al. 1999. Prevention of neural-tube defects with folic acid in China. *N. Engl. J. Med.* 341(20):1485–90

Blencowe H, Cousens S, Modell B, Lawn J. 2010. Folic acid to reduce neonatal mortality from neural tube disorders. *Int. J. Epidemiol.* 39(Suppl. 1):i110–21

Blom HJ, Shaw GM, den Heijer M, Finnell RH. 2006. Neural tube defects and folate: case far from closed. *Nat. Rev. Neurosci.* 7(9):724–31

Botto LD, Yang Q. 2000. 5,10-Methylenetetrahydrofolate reductase gene variants and congenital anomalies: a HuGE review. *Am. J. Epidemiol.* 151(9):862–77

Boyles AL, Hammock P, Speer MC. 2005. Candidate gene analysis in human neural tube defects. *Am. J. Med. Genet. C. Semin. Med. Genet.* 135(1):9–23

Brouns MR, De Castro SC, Terwindt-Rouwenhorst EA, Massa V, Hekking JW, et al. 2011. Over-expression of *Grhl2* causes spina bifida in the *Axial defects* mutant mouse. *Hum. Mol. Genet.* 20:1536–46

Bu P, Evrard YA, Lozano G, Dent SY. 2007. Loss of Gcn5 acetyltransferase activity leads to neural tube closure defects and exencephaly in mouse embryos. *Mol. Cell. Biol.* 27:3405–16

Burren KA, Savery D, Massa V, Kok RM, Scott JM, et al. 2008. Gene-environment interactions in the causation of neural tube defects: folate deficiency increases susceptibility conferred by loss of *Pax3* function. *Hum. Mol. Genet.* 17:3675–85

Camerer E, Barker A, Duong DN, Ganesan R, Kataoka H, et al. 2010. Local protease signaling contributes to neural tube closure in the mouse embryo. *Dev. Cell* 18:25–38

Cameron M, Moran P. 2009. Prenatal screening and diagnosis of neural tube defects. *Prenat. Diag.* 29:402–11

Chandler AL, Hobbs CA, Mosley BS, Berry RJ, Canfield MA, et al. 2012. Neural tube defects and maternal intake of micronutrients related to one-carbon metabolism or antioxidant activity. *Birth Defects Res. A Clin. Mol. Teratol.* 94(11):864–74

Chen Z, Karaplis AC, Ackerman SL, Pogribny IP, Melnyk S, et al. 2001. Mice deficient in methylenetetrahydrofolate reductase exhibit hyperhomocysteinemia and decreased methylation capacity, with neuropathology and aortic lipid deposition. *Hum. Mol. Genet.* 10(5):433–43

Cheng HL, Mostoslavsky R, Saito S, Manis JP, Gu Y, et al. 2003. Developmental defects and p53 hyperacetylation in Sir2 homolog (SIRT1)-deficient mice. *Proc. Natl. Acad. Sci. USA* 100(19):10794–99

Chiang C, Litingtung Y, Lee E, Young KE, Corden JL, et al. 1996. Cyclopia and defective axial patterning in mice lacking *Sonic hedgehog* gene function. *Nature* 383:407–13

Christensen KE, Deng L, Leung KY, Arning E, Bottiglieri T, et al. 2013. A novel mouse model for genetic variation in 10-formyltetrahydrofolate synthetase exhibits disturbed purine synthesis with impacts on pregnancy and embryonic development. *Hum. Mol. Genet.* 22(18):3705–19

Cockroft DL. 1988. Changes with gestational age in the nutritional requirements of postimplantation rat embryos in culture. *Teratology* 38:281–90

Copp AJ, Brook FA. 1989. Does lumbosacral spina bifida arise by failure of neural folding or by defective canalisation? *J. Med. Genet.* 26:160–66

Copp AJ, Brook FA, Estibeiro JP, Shum ASW, Cockroft DL. 1990. The embryonic development of mammalian neural tube defects. *Prog. Neurobiol.* 35:363–403

Copp AJ, Checiu I, Henson JN. 1994. Developmental basis of severe neural tube defects in the *loop-tail* (*Lp*) mutant mouse: use of microsatellite DNA markers to identify embryonic genotype. *Dev. Biol.* 165:20–29

Copp AJ, Greene NDE. 2010. Genetics and development of neural tube defects. *J. Pathol.* 220:217–30

Copp AJ, Greene NDE. 2013. Neural tube defects—disorders of neurulation and related embryonic processes. *Wiley Interdiscip. Rev. Dev. Biol.* 2(2):213–27

Copp AJ, Greene NDE, Murdoch JN. 2003. The genetic basis of mammalian neurulation. *Nat. Rev. Genet.* 4:784–93

Copp AJ, Stanier P, Greene ND. 2013. Neural tube defects: recent advances, unsolved questions, and controversies. *Lancet Neurol.* 12(8):799–810

Correa A, Botto L, Liu YC, Mulinare J, Erickson JD. 2003. Do multivitamin supplements attenuate the risk for diabetes-associated birth defects? *Pediatrics* 111:1146–51

Crider KS, Bailey LB, Berry RJ. 2011. Folic acid food fortification—its history, effect, concerns, and future directions. *Nutrients* 3(3):370–84

Crider KS, Yang TP, Berry RJ, Bailey LB. 2012. Folate and DNA methylation: a review of molecular mechanisms and the evidence for folate's role. *Adv. Nutr.* 3(1):21–38

Czeizel AE, Dudás I. 1992. Prevention of the first occurrence of neural-tube defects by periconceptional vitamin supplementation. *N. Engl. J. Med.* 327:1832–35

Czeizel AE, Dudás I, Paput L, Bánhidy F. 2011. Prevention of neural-tube defects with periconceptional folic acid, methylfolate, or multivitamins? *Ann. Nutr. Metab.* 58:263–71

Daly LE, Kirke PN, Molloy A, Weir DG, Scott JM. 1995. Folate levels and neural tube defects: implications for prevention. *JAMA* 274(21):1698–702

Davidson CE, Li Q, Churchill GA, Osborne LR, McDermid HE. 2007. Modifier locus for exencephaly in *Cecr2* mutant mice is syntenic to the 10q25.3 region associated with neural tube defects in humans. *Physiol. Genomics* 31:244–51

Davidson CM, Northrup H, King TM, Fletcher JM, Townsend I, et al. 2008. Genes in glucose metabolism and association with spina bifida. *Reprod. Sci.* 15(1):51–58

De Castro SC, Leung KY, Savery D, Burren K, Rozen R, et al. 2010. Neural tube defects induced by folate deficiency in mutant curly tail (Grhl3) embryos are associated with alteration in folate one-carbon metabolism but are unlikely to result from diminished methylation. *Birth Defects Res. A Clin Mol. Teratol.* 88:612–18

De Castro SC, Malhas A, Leung K-Y, Gustavsson P, Vaux DJ, et al. 2012. Lamin B1 polymorphism influences morphology of the nuclear envelope, cell cycle progression, and risk of neural tube defects in mice. *PLoS Genet.* 8(11):e1003059

De Marco P, Merello E, Consales A, Piatelli G, Cama A, et al. 2013. Genetic analysis of *Disheveled 2* and *Disheveled 3* in human neural tube defects. *J. Mol. Neurosci.* 49(3):582–88

Di Pietro E, Sirois J, Tremblay ML, Mackenzie RE. 2002. Mitochondrial NAD-dependent methylenetetrahydrofolate dehydrogenase-methenyltetrahydrofolate cyclohydrolase is essential for embryonic development. *Mol. Cell. Biol.* 22(12):4158–66

Dunlevy LPE, Burren KA, Mills K, Chitty LS, Copp AJ, Greene NDE. 2006. Integrity of the methylation cycle is essential for mammalian neural tube closure. *Birth Defects Res. A Clin. Mol. Teratol.* 76:544–52

Dunlevy LPE, Chitty LS, Burren KA, Doudney K, Stojilkovic-Mikic T, et al. 2007. Abnormal folate metabolism in foetuses affected by neural tube defects. *Brain* 130:1043–49

Escobedo N, Contreras O, Muñoz R, Farías M, Carrasco H, et al. 2013. Syndecan 4 interacts genetically with Vangl2 to regulate neural tube closure and planar cell polarity. *Development* 140(14):3008–17

Etheredge AJ, Finnell RH, Carmichael SL, Lammer EJ, Zhu H, et al. 2012. Maternal and infant gene-folate interactions and the risk of neural tube defects. *Am. J. Med. Genet. A* 158A(10):2439–46

Etheridge SL, Ray S, Li S, Hamblet NS, Lijam N, et al. 2008. Murine Dishevelled 3 functions in redundant pathways with Dishevelled 1 and 2 in normal cardiac outflow tract, cochlea, and neural tube development. *PLoS Genet.* 4:e1000259

Fine EL, Horal M, Chang TI, Fortin G, Loeken MR. 1999. Evidence that elevated glucose causes altered gene expression, apoptosis, and neural tube defects in a mouse model of diabetic pregnancy. *Diabetes* 48(12):2454–62

Finnell RH, Waes JGV, Eudy JD, Rosenquist TH. 2002. Molecular basis of environmentally induced birth defects. *Annu. Rev. Pharmacol. Toxicol.* 42:181–208

Fleming A, Copp AJ. 1998. Embryonic folate metabolism and mouse neural tube defects. *Science* 280:2107–9

Fukuda T, Tokunaga A, Sakamoto R, Yoshida N. 2011. Fbxl10/Kdm2b deficiency accelerates neural progenitor cell death and leads to exencephaly. *Mol. Cell Neurosci.* 46:614–24

Geelen JAG, Langman J. 1979. Ultrastructural observations on closure of the neural tube in the mouse. *Anat. Embryol.* 156:73–88

Goodrich LV, Strutt D. 2011. Principles of planar polarity in animal development. *Development* 138(10):1877–92

Gray JD, Kholmanskikh S, Castaldo BS, Hansler A, Chung H, et al. 2013. LRP6 exerts non-canonical effects on Wnt signaling during neural tube closure. *Hum. Mol. Genet.* 22(21):4267–81

Gray RS, Abitua PB, Wlodarczyk BJ, Szabo-Rogers HL, Blanchard O, et al. 2009. The planar cell polarity effector Fuz is essential for targeted membrane trafficking, ciliogenesis and mouse embryonic development. *Nat. Cell Biol.* 11:1225–32

Greene NDE, Copp AJ. 1997. Inositol prevents folate-resistant neural tube defects in the mouse. *Nat. Med.* 3:60–66

Greene NDE, Copp AJ. 2009. Development of the vertebrate central nervous system: formation of the neural tube. *Prenat. Diag.* 29:303–11

Greene NDE, Gerrelli D, Van Straaten HWM, Copp AJ. 1998. Abnormalities of floor plate, notochord and somite differentiation in the *loop-tail (Lp)* mouse: a model of severe neural tube defects. *Mech. Dev.* 73:59–72

Greene NDE, Stanier P, Copp AJ. 2009. Genetics of human neural tube defects. *Hum. Mol. Genet.* 18:R113–29

Greene NDE, Stanier P, Moore GE. 2011. The emerging role of epigenetic mechanisms in the etiology of neural tube defects. *Epigenetics* 6:875–83

Gurniak CB, Perlas E, Witke W. 2005. The actin depolymerizing factor n-cofilin is essential for neural tube morphogenesis and neural crest cell migration. *Dev. Biol.* 278(1):231–41

Gustavsson P, Greene NDE, Lad D, Pauws E, de Castro SCP, et al. 2007. Increased expression of *Grainyhead-like-3* rescues spina bifida in a folate-resistant mouse model. *Hum. Mol. Genet.* 16(21):2640–46

Haggarty P, Hoad G, Campbell DM, Horgan GW, Piyathilake C, McNeill G. 2013. Folate in pregnancy and imprinted gene and repeat element methylation in the offspring. *Am. J. Clin. Nutr.* 97(1):94–99

Haigo SL, Hildebrand JD, Harland RM, Wallingford JB. 2003. Shroom induces apical constriction and is required for hingepoint formation during neural tube closure. *Curr. Biol.* 13(24):2125–37

Hamblet NS, Lijam N, Ruiz-Lozano P, Wang J, Yang Y, et al. 2002. Dishevelled 2 is essential for cardiac outflow tract development, somite segmentation and neural tube closure. *Development* 129:5827–38

Harris MJ, Juriloff DM. 2007. Mouse mutants with neural tube closure defects and their role in understanding human neural tube defects. *Birth Defects Res. A Clin. Mol. Teratol.* 79(3):187–210

Harris MJ, Juriloff DM. 2010. An update to the list of mouse mutants with neural tube closure defects and advances toward a complete genetic perspective of neural tube closure. *Birth Defects Res. A Clin. Mol. Teratol.* 88:653–69

Heydeck W, Liu A. 2011. PCP effector proteins inturned and fuzzy play nonredundant roles in the patterning but not convergent extension of mammalian neural tube. *Dev. Dyn.* 240:1938–48

Hildebrand JD, Soriano P. 1999. Shroom, a PDZ domain-containing actin-binding protein, is required for neural tube morphogenesis in mice. *Cell* 99(5):485–97

Holmberg J, Clarke DL, Frisén J. 2000. Regulation of repulsion versus adhesion by different splice forms of an Eph receptor. *Nature* 408:203–6

Juriloff DM, Harris MJ. 2012a. A consideration of the evidence that genetic defects in planar cell polarity contribute to the etiology of human neural tube defects. *Birth Defects Res. A Clin. Mol. Teratol.* 94(10):824–40

Juriloff DM, Harris MJ. 2012b. Hypothesis: the female excess in cranial neural tube defects reflects an epigenetic drag of the inactivating X chromosome on the molecular mechanisms of neural fold elevation. *Birth Defects Res. A Clin. Mol. Teratol.* 94(10):849–55

Kase BA, Northrup H, Morrison AC, Davidson CM, Goiffon AM, et al. 2012. Association of copper-zinc superoxide dismutase (SOD1) and manganese superoxide dismutase (SOD2) genes with nonsyndromic myelomeningocele. *Birth Defects Res. A Clin. Mol. Teratol.* 94(10):762–69

Keller R. 2002. Shaping the vertebrate body plan by polarized embryonic cell movements. *Science* 298:1950–54

Kibar Z, Torban E, McDearmid JR, Reynolds A, Berghout J, et al. 2007. Mutations in VANGL1 associated with neural-tube defects. *N. Engl. J. Med.* 356(14):1432–37

Kim TH, Goodman J, Anderson KV, Niswander L. 2007. Phactr4 regulates neural tube and optic fissure closure by controlling PP1-, Rb-, and E2F1-regulated cell-cycle progression. *Dev. Cell* 13(1):87–102

Lei Y, Zhu H, Duhon C, Yang W, Ross ME, et al. 2013. Mutations in planar cell polarity gene *SCRIB* are associated with spina bifida. *PLoS ONE* 8(7):e69262

Lei YP, Zhang T, Li H, Wu BL, Jin L, Wang HY. 2010. VANGL2 mutations in human cranial neural-tube defects. *N. Engl. J. Med.* 362:2232–35

Leung KY, De Castro SC, Savery D, Copp AJ, Greene ND. 2013. Nucleotide precursors prevent folic acid-resistant neural tube defects in the mouse. *Brain* 136(Pt. 9):2836–41

Lew SM, Kothbauer KF. 2007. Tethered cord syndrome: an updated review. *Pediatr. Neurosurg.* 43(3):236–48

Lu X, Borchers AG, Jolicoeur C, Rayburn H, Baker JC, Tessier-Lavigne M. 2004. PTK7/CCK-4 is a novel regulator of planar cell polarity in vertebrates. *Nature* 430:93–98

Lupo PJ, Canfield MA, Chapa C, Lu W, Agopian AJ, et al. 2012. Diabetes and obesity-related genes and the risk of neural tube defects in the national birth defects prevention study. *Am. J. Epidemiol.* 176:1101–9

MacFarlane AJ, Perry CA, Girnary HH, Gao D, Allen RH, et al. 2009. *Mthfd1* is an essential gene in mice and alters biomarkers of impaired one-carbon metabolism. *J. Biol. Chem.* 284(3):1533–39

Massa V, Savery D, Ybot-Gonzalez P, Ferraro E, Rongvaux A, et al. 2009. Apoptosis is not required for mammalian neural tube closure. *Proc. Natl. Acad. Sci. USA* 106:8233–38

Matsuda M, Yasutomi M. 1992. Inhibition of cephalic neural tube closure by 5-azacytidine in neurulating rat embryos in vitro. *Anat. Embryol.* 185:217–23

Merte J, Jensen D, Wright K, Sarsfield S, Wang Y, et al. 2010. Sec24b selectively sorts Vangl2 to regulate planar cell polarity during neural tube closure. *Nat. Cell Biol.* 12:41–46

Miller KA, Ah-Cann CJ, Welfare MF, Tan TY, Pope K, et al. 2013. Cauli: a mouse strain with an ift140 mutation that results in a skeletal ciliopathy modelling Jeune syndrome. *PLoS. Genet.* 9(8):e1003746

Missmer SA, Suarez L, Felkner M, Wang E, Merrill AH Jr, et al. 2006. Exposure to fumonisins and the occurrence of neural tube defects along the Texas-Mexico border. *Environ. Health Perspect.* 114:237–41

Molloy AM, Kirke PN, Troendle JF, Burke H, Sutton M, et al. 2009. Maternal vitamin B12 status and risk of neural tube defects in a population with high neural tube defect prevalence and no folic acid fortification. *Pediatrics* 123:917–23

Momb J, Lewandowski JP, Bryant JD, Fitch R, Surman DR, et al. 2013. Deletion of *Mthfd1l* causes embryonic lethality and neural tube and craniofacial defects in mice. *Proc. Natl. Acad. Sci. USA* 110:549–54

Moretti ME, Bar-Oz B, Fried S, Koren G. 2005. Maternal hyperthermia and the risk for neural tube defects in offspring: systematic review and meta-analysis. *Epidemiology* 16:216–19

Morriss-Kay GM, Tuckett F. 1985. The role of microfilaments in cranial neurulation in rat embryos: effects of short-term exposure to cytochalasin D. *J. Embryol. Exp. Morphol.* 88:333–48

Mosley BS, Cleves MA, Siega-Riz AM, Shaw GM, Canfield MA, et al. 2009. Neural tube defects and maternal folate intake among pregnancies conceived after folic acid fortification in the United States. *Am. J. Epidemiol.* 169:9–17

Murdoch JN, Copp AJ. 2010. The relationship between sonic Hedgehog signalling, cilia and neural tube defects. *Birth Defects Res. A Clin. Mol. Teratol.* 88:633–52

Murdoch JN, Rachel RA, Shah S, Beermann F, Stanier P, et al. 2001. *Circletail*, a new mouse mutant with severe neural tube defects: chromosomal localisation and interaction with the *loop-tail* mutation. *Genomics* 78:55–63

Nakatsu T, Uwabe C, Shiota K. 2000. Neural tube closure in humans initiates at multiple sites: evidence from human embryos and implications for the pathogenesis of neural tube defects. *Anat. Embryol.* 201(6):455–66

Narisawa A, Komatsuzaki S, Kikuchi A, Niihori T, Aoki Y, et al. 2012. Mutations in genes encoding the glycine cleavage system predispose to neural tube defects in mice and humans. *Hum. Mol. Genet.* 21:1496–503

O'Rahilly R, Müller F. 2002. The two sites of fusion of the neural folds and the two neuropores in the human embryo. *Teratology* 65:162–70

Okano M, Bell DW, Haber DA, Li E. 1999. DNA methyltransferases Dnmt3a and Dnmt3b are essential for de novo methylation and mammalian development. *Cell* 99(3):247–57

Pai YJ, Abdullah NL, Mohd-Zin SW, Mohammed RS, Rolo A, et al. 2012. Epithelial fusion during neural tube morphogenesis. *Birth Defects Res. A Clin. Mol. Teratol.* 94:817–23

Pangilinan F, Molloy AM, Mills JL, Troendle JF, Parle-McDermott A, et al. 2012. Evaluation of common genetic variants in 82 candidate genes as risk factors for neural tube defects. *BMC Med. Genet.* 13:62

Parle-McDermott A, Pangilinan F, O'Brien KK, Mills JL, Magee AM, et al. 2009. A common variant in *MTHFD1L* is associated with neural tube defects and mRNA splicing efficiency. *Hum. Mutat.* 30:1650–56

Pickell L, Li D, Brown K, Mikael LG, Wang XL, et al. 2009. Methylenetetrahydrofolate reductase deficiency and low dietary folate increase embryonic delay and placental abnormalities in mice. *Birth Defects Res. A Clin. Mol. Teratol.* 85(6):531–41

Piedrahita JA, Oetama B, Bennett GD, van Waes J, Kamen BA, et al. 1999. Mice lacking the folic acid-binding protein Folbp1 are defective in early embryonic development. *Nat. Genet.* 23(2):228–32

Placzek M, Briscoe J. 2005. The floor plate: multiple cells, multiple signals. *Nat. Rev. Neurosci.* 6:230–40

Pyrgaki C, Liu A, Niswander L. 2011. Grainyhead-like 2 regulates neural tube closure and adhesion molecule expression during neural fold fusion. *Dev. Biol.* 353:38–49

Pyrgaki C, Trainor P, Hadjantonakis AK, Niswander L. 2010. Dynamic imaging of mammalian neural tube closure. *Dev. Biol.* 344:941–47

Ray HJ, Niswander L. 2012. Mechanisms of tissue fusion during development. *Development* 139(10):1701–11

Reece EA. 2012. Diabetes-induced birth defects: what do we know? What can we do? *Curr. Diab. Rep.* 12(1):24–32

Reece EA, Khandelwal M, Wu YK, Borenstein M. 1997. Dietary intake of *myo*-inositol and neural tube defects in offspring of diabetic rats. *Am. J. Obstet. Gynecol.* 176:536–39

Rifat Y, Parekh V, Wilanowski T, Hislop NR, Auden A, et al. 2010. Regional neural tube closure defined by the Grainy head-like transcription factors. *Dev. Biol.* 345:237–45

Robinson A, Escuin S, Doudney K, Vekemans M, Stevenson RE, et al. 2012. Mutations in the planar cell polarity genes *CELSR1* and *SCRIB* are associated with the severe neural tube defect craniorachischisis. *Hum. Mutat.* 33:440–47

Rosenthal J, Casas J, Taren D, Alverson CJ, Flores A, Frias J. 2013. Neural tube defects in Latin America and the impact of fortification: a literature review. *Public Health Nutr.* 17:537–50

Saadai P, Nout YS, Encinas J, Wang A, Downing TL, et al. 2011. Prenatal repair of myelomeningocele with aligned nanofibrous scaffolds—a pilot study in sheep. *J. Pediatr. Surg.* 46(12):2279–83

Saadai P, Wang A, Nout YS, Downing TL, Lofberg K, et al. 2013. Human induced pluripotent stem cell-derived neural crest stem cells integrate into the injured spinal cord in the fetal lamb model of myelomeningocele. *J. Pediatr. Surg.* 48(1):158–63

Sadler TW, Greenberg D, Coughlin P, Lessard JL. 1982. Actin distribution patterns in the mouse neural tube during neurulation. *Science* 215:172–74

Savory JG, Mansfield M, Rijli FM, Lohnes D. 2011. Cdx mediates neural tube closure through transcriptional regulation of the planar cell polarity gene *Ptk7*. *Development* 138:1361–70

Schoenwolf GC. 1984. Histological and ultrastructural studies of secondary neurulation in mouse embryos. *Am. J. Anat.* 169:361–76

Schoenwolf GC, Smith JL. 1990. Epithelial cell wedging: a fundamental cell behavior contributing to hinge point formation during epithelial morphogenesis. In *Control of Morphogenesis*, Vol. 1, ed. RE Keller, D Fristrom, pp. 325–34. London: Saunders

Schorah C. 2008. Dick Smithells, folic acid, and the prevention of neural tube defects. *Birth Defects Res. A Clin. Mol. Teratol.* 85:254–59

Seller MJ. 1995. Sex, neural tube defects, and multisite closure of the human neural tube. *Am. J. Med. Genet.* 58:332–36

Shaw GM, Lu W, Zhu H, Yang W, Briggs FB, et al. 2009. 118 SNPs of folate-related genes and risks of spina bifida and conotruncal heart defects. *BMC Med. Genet.* 10(1):49

Shum ASW, Copp AJ. 1996. Regional differences in morphogenesis of the neuroepithelium suggest multiple mechanisms of spinal neurulation in the mouse. *Anat. Embryol.* 194:65–73

Smith JL, Schoenwolf GC. 1989. Notochordal induction of cell wedging in the chick neural plate and its role in neural tube formation. *J. Exp. Zool.* 250:49–62

Smithells RW, Sheppard S, Schorah CJ. 1976. Vitamin deficiencies and neural tube defects. *Arch. Dis. Child.* 51:944–50

Smithells RW, Sheppard S, Schorah CJ, Seller MJ, Nevin NC, et al. 1981. Apparent prevention of neural tube defects by periconceptional vitamin supplementation. *Arch. Dis. Child.* 56:911–18

Sotres-Alvarez D, Siega-Riz AM, Herring AH, Carmichael SL, Feldkamp ML, et al. 2013. Maternal dietary patterns are associated with risk of neural tube and congenital heart defects. *Am. J. Epidemiol.* 177(11):1279–88

Stover PJ. 2009. One-carbon metabolism-genome interactions in folate-associated pathologies. *J. Nutr.* 139:2402–5

Sulik KK, Zuker RM, Dehart DB, Cessot F, Delezoide AL, et al. 1998. Normal patterns of neural tube closure differ in the human and the mouse. *Proc. Greenwood Genet. Center* 18:129–30

Takeuchi T, Kojima M, Nakajima K, Kondo S. 1999. *Jumonji* gene is essential for the neurulation and cardiac development of mouse embryos with a C3H/He background. *Mech. Dev.* 86(1–2):29–38

Tibbetts AS, Appling DR. 2010. Compartmentalization of mammalian folate-mediated one-carbon metabolism. *Annu. Rev. Nutr.* 30:57–81

Vega RB, Matsuda K, Oh J, Barbosa AC, Yang X, et al. 2004. Histone deacetylase 4 controls chondrocyte hypertrophy during skeletogenesis. *Cell* 119(4):555–66

Vollset SE, Clarke R, Lewington S, Ebbing M, Halsey J, et al. 2013. Effects of folic acid supplementation on overall and site-specific cancer incidence during the randomised trials: meta-analyses of data on 50 000 individuals. *Lancet* 381:1029–36

Wald N, Sneddon J, Densem J, Frost C, Stone R, MRC Vitamin Study Res Group. 1991. Prevention of neural tube defects: results of the Medical Research Council Vitamin Study. *Lancet* 338:131–37

Wallingford JB, Harland RM. 2001. *Xenopus* Dishevelled signaling regulates both neural and mesodermal convergent extension: parallel forces elongating the body axis. *Development* 128(13):2581–92

Wallingford JB, Harland RM. 2002. Neural tube closure requires Dishevelled-dependent convergent extension of the midline. *Development* 129(24):5815–25

Wang L, Wang F, Guan J, Le J, Wu L, et al. 2010. Relation between hypomethylation of long interspersed nucleotide elements and risk of neural tube defects. *Am. J. Clin. Nutr.* 91:1359–67

Werth M, Walentin K, Aue A, Schönheit J, Wuebken A, et al. 2010. The transcription factor grainyhead-like 2 regulates the molecular composition of the epithelial apical junctional complex. *Development* 137(22):3835–45

Wlodarczyk BJ, Palacios AM, George TM, Finnell RH. 2012. Antiepileptic drugs and pregnancy outcomes. *Am. J. Med. Genet. A* 158A(8):2071–90

Wood LR, Smith MT. 1984. Generation of anencephaly: 1. Aberrant neurulation and 2. Conversion of exencephaly to anencephaly. *J. Neuropath. Exp. Neurol.* 43:620–33

Xu WM, Baribault H, Adamson ED. 1998. Vinculin knockout results in heart and brain defects during embryonic development. *Development* 125:327–37

Yamaguchi Y, Shinotsuka N, Nonomura K, Takemoto K, Kuida K, et al. 2011. Live imaging of apoptosis in a novel transgenic mouse highlights its role in neural tube closure. *J. Cell Biol.* 195:1047–60

Yang P, Li X, Xu C, Eckert RL, Reece EA, et al. 2013. Maternal hyperglycemia activates an ASK1-FoxO3a-Caspase 8 pathway that leads to embryonic neural tube defects. *Sci. Signal.* 6(290):ra74

Yao TP, Oh SP, Fuchs M, Zhou ND, Ch'ng LE, et al. 1998. Gene dosage-dependent embryonic development and proliferation defects in mice lacking the transcriptional integrator p300. *Cell* 93:361–72

Ybot-Gonzalez P, Cogram P, Gerrelli D, Copp AJ. 2002. Sonic hedgehog and the molecular regulation of neural tube closure. *Development* 129:2507–17

Ybot-Gonzalez P, Copp AJ. 1999. Bending of the neural plate during mouse spinal neurulation is independent of actin microfilaments. *Dev. Dyn.* 215:273–83

Ybot-Gonzalez P, Gaston-Massuet C, Girdler G, Klingensmith J, Arkell R, et al. 2007a. Neural plate morphogenesis during mouse neurulation is regulated by antagonism of Bmp signalling. *Development* 134:3203–11

Ybot-Gonzalez P, Savery D, Gerrelli D, Signore M, Mitchell CE, et al. 2007b. Convergent extension, planar-cell-polarity signalling and initiation of mouse neural tube closure. *Development* 134:789–99

Yu H, Smallwood PM, Wang Y, Vidaltamayo R, Reed R, Nathans J. 2010. Frizzled 1 and frizzled 2 genes function in palate, ventricular septum and neural tube closure: general implications for tissue fusion processes. *Development* 137:3707–17

Zeng H, Hoover AN, Liu A. 2010. PCP effector gene Inturned is an important regulator of cilia formation and embryonic development in mammals. *Dev. Biol.* 339:418–28

Zohn IE, Sarkar AA. 2012. Does the cranial mesenchyme contribute to neural fold elevation during neurulation? *Birth Defects Res. A Clin. Mol. Teratol.* 94(10):841–48

Functions and Dysfunctions of Adult Hippocampal Neurogenesis

Kimberly M. Christian,[1,2] Hongjun Song,[1,2,3] and Guo-li Ming[1,2,3]

[1]Institute for Cell Engineering, [2]Department of Neurology, [3]The Solomon Snyder Department of Neuroscience, Johns Hopkins University School of Medicine, Baltimore, Maryland 21205; email: shongju1@jhmi.edu, gming1@jhmi.edu

Annu. Rev. Neurosci. 2014. 37:243–62

First published online as a Review in Advance on May 29, 2014

The *Annual Review of Neuroscience* is online at neuro.annualreviews.org

This article's doi: 10.1146/annurev-neuro-071013-014134

Keywords

neural stem cells, development, brain disorders, dentate gyrus

Abstract

Adult neurogenesis, a developmental process of generating functionally integrated neurons, occurs throughout life in the hippocampus of the mammalian brain and showcases the highly plastic nature of the mature central nervous system. Significant progress has been made in recent years to decipher how adult neurogenesis contributes to brain functions. Here we review recent findings that inform our understanding of adult hippocampal neurogenesis processes and special properties of adult-born neurons. We further discuss potential roles of adult-born neurons at the circuitry and behavioral levels in cognitive and affective functions and how their dysfunction may contribute to various brain disorders. We end by considering a general model proposing that adult neurogenesis is not a cell-replacement mechanism, but instead maintains a plastic hippocampal neuronal circuit via the continuous addition of immature, new neurons with unique properties and structural plasticity of mature neurons induced by new-neuron integration.

Contents

INTRODUCTION

One major advance in modern neuroscience is the appreciation of the extent of plasticity in the mature nervous system. Many critical functions of the nervous system depend on its mutability: its ability to process external information, to encode novel associations among events and objects in the world, and to generate adaptive behavior. Plasticity thought to underlie these functions occurs at multiple levels, from epigenetic modifications of gene expression, to neuronal activity-dependent modulation of synaptic strength, to tuning of integrated circuits that carry multimodal sensory information. Perhaps the most striking form of structural plasticity in the adult nervous system is the de novo generation and integration of new neurons into the existing circuitry through a process known as adult neurogenesis (Kempermann & Gage 1999). Originally thought to occur only during embryonic development, active adult neurogenesis has now been shown in almost all mammalian species examined so far (Lledo et al. 2006, Ming & Song 2005). Active neurogenesis occurs in two discrete regions: the subventricular zone of the lateral ventricle, from where newborn neurons migrate to the olfactory bulb or striatum (Ernst et al. 2014) and differentiate mostly into interneurons, and the dentate gyrus of the hippocampus, where newborn granule cells are integrated into the local circuitry (Ming & Song 2011) (**Figure 1***a*). Significant progress has been made in the past decade to understand the generation, development, and integration of adult-born neurons, molecular and regulatory mechanisms, and potential contributions of adult neurogenesis to brain function and dysfunction (Ming & Song 2011, Zhao et al. 2008).

    Adult hippocampal neurogenesis has garnered significant interest because of its potential to influence information processing in the medial temporal lobe, a neural substrate for many forms of learning and memory and a site of pathophysiology associated with various neurological disorders

(Squire 1992). The canonical signaling network of the hippocampus consists of synaptically connected principal neurons located in three major subregions to form the trisynaptic circuit: granule cells in the dentate gyrus and pyramidal neurons in CA1 and CA3 (**Figure 1***a*). Information flows from the entorhinal cortex through medial and lateral perforant pathways to the dentate gyrus, then to CA3 pyramidal cells via mossy fiber axons of granule cells, then to CA1 pyramidal cells via Schaffer collateral projections of CA3 neurons, and finally to the subiculum and back to the entorhinal cortex. This primary hippocampal circuit forms a closed loop wherein sensory information from specific cortical areas converges onto the entorhinal cortex, is processed through the hippocampal circuitry, and returns to the region of origin in the entorhinal cortex. In addition, there are direct projections from the entorhinal cortex to CA3 and CA1 and multiple modulatory inputs from other brain regions to the hippocampus. The dentate gyrus also contains diverse γ-aminobutyric acid (GABA)-ergic inhibitory neurons within the hilus region and the molecular layer, which mediate feedforward and feedback inhibition, and hilar glutamatergic mossy cells, which receive inputs primarily from granule cells and then innervate granule cells and local interneurons.

Within the hippocampus, dentate granule cells are the only neurons to be continuously generated. Young adult rats generate an estimated 9,000 new cells each day in the dentate gyrus, about 6% of the total granule cell population each month (Cameron & McKay 2001), whereas adult humans add 700 new neurons in each hippocampus per day, corresponding to an annual turnover of 1.75% of the renewing neuronal population (Spalding et al. 2013). The significant number of new neurons, together with dynamic regulation of adult neurogenesis by various physiological and pathological stimuli (Ma et al. 2009), suggests that adult neurogenesis may be integral to certain brain functions. Indeed, behavior analyses in animal models support a critical role for dentate newborn neurons in several hippocampus-dependent functions (Aimone et al. 2011). Electrophysiological analyses have identified special properties of immature adult-born neurons (Ge et al. 2007, Schmidt-Hieber et al. 2004, Snyder et al. 2001), providing a mechanistic basis for their unique contributions to neural processes. In addition, many studies have implicated dysfunction of adult hippocampal neurogenesis in an increasing number of brain disorders (Braun & Jessberger 2013, Ming & Song 2009, Sahay & Hen 2007, Winner et al. 2011). Focusing on adult hippocampal neurogenesis in this review, we start with a summary of the recent progress in our understanding of the adult neurogenesis processes and special properties of newborn neurons and follow with a discussion of models for how adult neurogenesis contributes to circuit regulation and behavior under normal and pathological conditions. Interested readers can consult other recent reviews on general topics of adult neural stem cells and neurogenesis (Bonaguidi et al. 2012, Gage & Temple 2013, Göritz & Frisen 2012, Kriegstein & Alvarez-Buylla 2009, Ming & Song 2011) and functions of olfactory bulb adult neurogenesis (Lepousez et al. 2013).

PROCESSES OF ADULT HIPPOCAMPAL NEUROGENESIS

Tremendous progress has been made in recent years, mostly using rodents as experimental models, in understanding the origin of newborn neurons and their development, maturation, and integration into the existing neuronal circuitry in the adult hippocampus. Genetic fate-mapping studies have demonstrated that neural precursors located within the subgranular zone, between the granule cell layer and the hilus, are the source of newborn neurons in the dentate gyrus (Dhaliwal & Lagace 2011). Clonal lineage-tracing analyses have further identified radial glia-like precursors as multipotent neural stem cells, capable of repeated self-renewal and generation of both neurons and astrocytes but not oligodendrocytes (Bonaguidi et al. 2011). During neurogenic cell

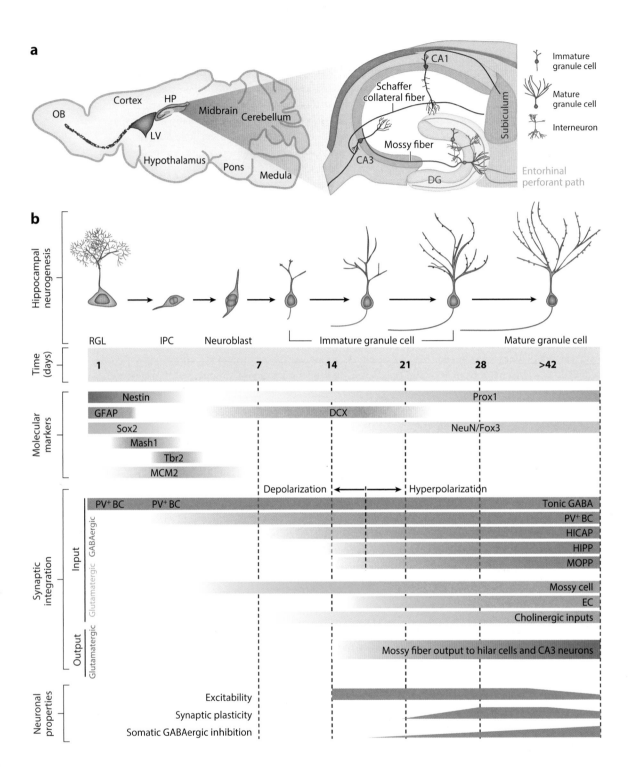

division, these neural stem cells give rise to intermediate progenitor cells, which in turn give rise to proliferating neuroblasts, postmitotic immature neurons, and finally mature dentate granule cells (**Figure 1b**). Nonradial precursors within the subgranular zone can also give rise to newborn neurons, although their identity remains unclear (Bonaguidi et al. 2012). The majority of the newborn cells in rodents die within the first four days of their birth (Sierra et al. 2010) or within one to three weeks after birth (Tashiro et al. 2006). Ultimately, less than 25% of newborn neurons survive to become mature neurons and synaptically integrate under normal conditions. Significant efforts have been devoted to understanding how newborn neurons become integrated into existing circuits.

Synaptic Inputs

Studies using oncoretroviruses for birth-dating and labeling have shown that newborn granule cells in young adult mice develop a single primary dendrite with multiple branches that reaches the molecular layer within 7 days and exhibits rapid growth between 7 and 17 days, followed by modest growth for at least two months (Sun et al. 2013). Functional electrophysiological character-ization of labeled newborn neurons in acute slices has revealed a stereotypic integration process in which GABAergic synapses precede glutamatergic synapse formation (Espósito et al. 2005, Ge et al. 2006, Overstreet-Wadiche et al. 2006b). Although radial glia-like precursors exhibit functional GABA$_A$ receptors and tonic responses to ambient GABA (Song et al. 2012), the first functional synaptic inputs appear to form onto proliferating neuroblasts within four days of birth (Song et al. 2013, Tozuka et al. 2005). Postmitotic newborn neurons continue to exhibit tonic GABA responses while their GABAergic synaptic responses mature (Espósito et al. 2005, Ge et al. 2006). Several recent studies, using paired recording (Markwardt et al. 2011), optogenetics (Song et al. 2013), and rabies virus–based retrograde transsynaptic tracing (Deshpande et al. 2013, Li et al. 2013, Vivar et al. 2012), have identified multiple interneuron subtypes that innervate new-born neurons within weeks of birth (**Figure 1b**), including parvalbumin-expressing basket cells, somatostatin-expressing HIPP (hilar perforant path-associated) cells, HICAP cells (hilar interneu-ron with commissural-associational pathway-associated axon terminals), and MOPP (molecular layer perforant pathway) cells, such as neurogliaform cells/Ivy cells. The sequence of GABAergic synapse formation by various interneuron subtypes remains unknown. Notably, newborn neurons exhibit initial depolarizing responses to GABA, which gradually shift to hyperpolarizing responses within two to three weeks of birth (Ge et al. 2006, Overstreet Wadiche et al. 2005). Depolariz-ing GABAergic signaling promotes survival, maturation, and synapse formation and activation in newborn neurons (Chancey et al. 2013, Ge et al. 2006, Pontes et al. 2013, Song et al. 2013, Tozuka et al. 2005). Initial GABAergic synaptic inputs are not sufficient to elicit action potentials and are therefore unlikely to be directly involved in information processing. Instead, new neuron outputs

---

Figure 1

Summary of basic processes of neurogenesis in the young adult mouse hippocampus. (*a*) A sagittal section view of an adult rodent brain highlighting two restricted regions that exhibit active adult neurogenesis—the hippocampus (HP) and lateral ventricle (LV)—which generate new neurons that mostly migrate into the olfactory bulb. More detailed hippocampal structure is further illustrated with the primary trisynaptic circuit formed by three principal neuronal subtypes. (*b*) Summary of the developmental processes of adult hippocampal neurogenesis, including time course of marker expression, developmental stages, synaptic integration, and special neuronal properties associated with different stages. BC, basket cells; DG, dentate gyrus; EC, entorhinal cortex; HICAP, hilar interneuron with commissural-associational pathway-associated axon terminals; HIPP, hilar perforant path-associated interneurons; IPCs, intermediate progenitor cells; MOPP, molecular-layer perforant pathway cells; OB, olfactory bulb; PV$^+$, parvalbumin-expressing interneurons; RGL, radial glia-like cell.

are controlled by glutamatergic synaptic inputs. Electrophysiological analyses have shown that the first detectable glutamatergic synaptic responses emerge in 11–14-day-old newborn neurons, and these responses mature over the next several weeks, accompanied by increased density of dendritic spines (Chancey et al. 2014, Espósito et al. 2005, Ge et al. 2006). Recent optogenetic and rabies virus–based retrograde transsynaptic tracing suggested that glutamatergic synaptic inputs onto newborn neurons originating from mossy cells form ahead of those by perforant pathway fibers from the entorhinal cortex (**Figure 1b**) (Chancey et al. 2014, Deshpande et al. 2013, Kumamoto et al. 2012). In addition, studies revealed inputs from cholinergic septal neurons at early stages (Deshpande et al. 2013, Vivar et al. 2012). Upon maturation, adult-born granule cells appear to exhibit general properties that are indistinguishable from developmentally born granule cells (Ge et al. 2007; Laplagne et al. 2006, 2007), although differences in some specific characteristics cannot be ruled out.

Synaptic Outputs

Newborn neurons extend a single axon from the base of the cell body that follows a stereotypic pathway through the hilus to reach CA3 within 7 days and establishes mature primary projection patterns within 21 days (Sun et al. 2013). Electron microscopic analyses of retrovirally labeled newborn neurons have shown synaptic structures associated with cells in both hilus and CA3 within 14 days and mossy fiber en passant boutons reaching morphological maturation within 8 weeks (Faulkner et al. 2008, Toni et al. 2008). Optogenetic activation of newborn neurons confirmed functional glutamatergic synaptic outputs onto hilar mossy cells and interneurons and CA3 neurons (Gu et al. 2012, Toni et al. 2007). Two- to four-week-old adult-born neurons synthesize and corelease GABA, in addition to glutamate (Cabezas et al. 2012, 2013). However, this GABA release appears to modulate presynaptic mossy fiber excitability only by activating GABA$_B$ autoreceptors and GABAergic postsynaptic responses have not been detected (Cabezas et al. 2012). In general, we know much less about properties of synaptic outputs of newborn neurons compared with their inputs, information critically needed to better understand adult neurogenesis functions. Future studies using new tools, such as optogenetics, anterograde transsynaptic tracing, and whole-mount imaging, are needed to provide a more complete picture about different targets of adult-born neurons and temporal dynamics of functional synapse formation.

A common feature of pre- and postsynaptic integration of newborn neurons is the apparent competition with mature granule cells for innervation from afferent axons and efferent connections to invade and replace preexisting synapses (Toni et al. 2007, 2008). Therefore, adult neurogenesis not only continuously adds new individual units to the dentate gyrus, but also induces structural plasticity of mature neurons, including mature granule cells and hilar mossy cells and interneurons, presynaptic terminals of entorhinal inputs, and postsynaptic sites on CA3 neurons.

Basic characterization of the adult hippocampal neurogenesis process has provided critical information on when and how newborn neurons could contribute to brain functions. For example, 14-day-old adult-born neurons already exhibit functional glutamatergic synaptic inputs and outputs and can therefore participate in neural processing during immature stages. Developmental and synaptic integration patterns of adult-born neurons are largely consistent with those described for dentate granule cells generated during development (Liu et al. 1996, Overstreet-Wadiche et al. 2006a, Zhao et al. 2006); therefore, it seems that adult-born neurons may participate in the same neuronal circuits as do preexisting ones. A fundamental question follows: How can a small population of adult-born neurons make meaningful contributions to brain functions in the presence of millions of mature neurons of the same type?

SPECIAL PROPERTIES OF ADULT-BORN DENTATE GRANULE NEURONS

One significant advance in the field came from discoveries of special properties of adult-born neurons while they were immature. These distinct cellular and circuit-level properties work together to determine their potential to make a functional contribution.

Distinct Cellular Properties

Electrophysiological analyses showed that, compared with mature neurons, immature adult-born neurons are highly excitable (Dieni et al. 2013, Mongiat et al. 2009). As a result, they are very efficient in generating action potentials, even with weak glutamatergic inputs (Marín-Burgin et al. 2012). Immature newborn neurons also exhibit a lower induction threshold and larger amplitude of associative long-term potentiation (LTP) of perforant path synaptic inputs compared with mature granule cells in acute slices under identical conditions (Ge et al. 2007, Schmidt-Hieber et al. 2004). This enhanced synaptic plasticity is partially due to a lack of strong GABAergic inhibition in immature neurons (Ge et al. 2008). Adult-born neurons exhibit such properties only during a critical period between approximately three and six weeks after birth and depend on developmentally regulated synaptic expression of NR2B-containing N-methyl-D-aspartate (NMDA) receptors (Ge et al. 2007). Similarly, in vivo field recordings showed that four-week-old newborn neurons exhibit enhanced LTP at mossy fiber synaptic outputs onto CA3 neurons (Gu et al. 2012). Therefore, adult-born neurons have distinct cellular properties compared with mature neurons, and the transient nature of such properties may provide a fundamental mechanism allowing adult-born neurons within this critical period to serve as major mediators of experience-induced plasticity.

Distinct Circuitry Properties

One major difference between immature and mature granule neurons is in GABAergic inputs and, in particular, a lack of strong perisomatic inhibition of immature neurons (Ge et al. 2008, Li et al. 2012, Marín-Burgin et al. 2012). As a result, four-week-old newborn neurons exhibit a lower firing threshold owing to an enhanced excitation/inhibition balance involving feedforward inhibitory circuitry. Combined with higher intrinsic excitability, weak afferent activity recruits a substantial proportion of immature neurons while activating few mature granule cells, as shown by calcium imaging in acute slices (Marin-Burgin et al. 2012). These observations suggest a model in which immature neurons with a low activation threshold and input specificity comprise a population of integrators that are broadly tuned during a finite developmental period and may encode most features of the incoming afferent information. However, mature granule cells generated during both development and adult neurogenesis, owing to their high activation thresholds and input specificity, serve as pattern separators. In this model, activity patterns entering the dentate gyrus could undergo differential encoding through immature neuronal cohorts that are highly responsive and integrative and, in parallel, through a large population of mature granule cells with sparse activity and high input specificity.

Long-distance modulatory inputs may also differentially affect immature and mature granule cells and impact information processing. Dentate granule cells are known to receive dopaminergic inputs from the ventral tegmental area (Gasbarri et al. 1997). Dopamine causes a long-lasting attenuation of medial perforant path inputs to newborn neurons through D1-like receptors and decreases their capacity to express LTP, whereas dopamine activation via D2-like receptors suppresses synaptic inputs onto mature granule cells but does not influence their LTP expression

(Mu et al. 2011). Whether other long-projection modulatory inputs differentially regulate newborn and mature granule cells remains to be determined.

These studies, mostly in vitro characterizations in acute slice preparations, have demonstrated differential properties of adult-born neurons during immature stages and provided the framework for how adult-born neurons could make unique contributions to specific brain functions. Indeed, many computational modeling and animal behavior studies support the immature neuron model of how adult neurogenesis may contribute to hippocampal functions (Deng et al. 2010). This model does not rule out potential contributions of mature adult-born neurons because these neurons are also plastic in response to neuronal activity and could be involved in different aspects of learning and memory (Lemaire et al. 2012, Ramirez-Amaya et al. 2006). One critical parameter of this model is the rate of new-neuron maturation or the duration of immature states. The entire neurodevelopmental process takes an estimated eight weeks in young adult mice (Ge et al. 2007, Zhao et al. 2006). This maturation rate is affected by numerous environmental, pathological, and pharmacological factors (Piatti et al. 2011, Zhao et al. 2006) and exhibits significant differences among species (Brus et al. 2013). Notably, neuronal maturation in the dentate gyrus of adult macaque monkeys appears to be longer than six months (Kohler et al. 2011). It is tempting to speculate that the lengthened time course for adult-born neuron maturation in nonhuman primates, and possibly in humans, may help to maintain immature neuronal properties over a longer life span.

Much remains to be learned about basic properties of adult-born neurons. One major roadblock is the lack of effective approaches to directly examine physiological properties of newborn neurons in awake behaving animals. Our current methodology is limited to immediate early gene expression as an indirect readout of neuronal activation, which has produced conflicting results on whether adult-born neurons are preferentially recruited into active networks over preexisting neurons upon specific behavioral stimulation (Ramirez-Amaya et al. 2006, Stone et al. 2011). Future studies of newborn neurons in vivo, such as targeted recording of firing properties (Leutgeb et al. 2007; Neunuebel & Knierim 2012, 2014) or calcium imaging using miniature microscopes (Ziv et al. 2013), will provide essential new information and significantly advance the field.

POTENTIAL MODES OF ADULT-BORN NEURON CONTRIBUTION TO BRAIN FUNCTIONS

Adult-born neurons could impact brain functions directly via two modes: first, as an information-processing unit and, second, as an active modulator of local circuitry to shape mature neuron firing, synchronization, and network oscillations (**Figure 2**). One hallmark of the dentate gyrus is its sparse activation as shown by both in vivo recording of putative granule cells and immediate early gene expression (Neunuebel & Knierim 2012, 2014; Ramírez-Amaya et al. 2005). Although these neurons are small in absolute number, preferential recruitment of excitable immature neurons with enhanced plasticity would allow this population to be a major player in information processing in the trisynaptic circuit.

In the second mode, newborn neurons could actively modulate local circuit activity, for example, maintaining a basal tone of excitation/inhibition or facilitating information encoding by increasing signal to noise in the region and/or by priming circuits to respond (**Figure 2**). Immature newborn neurons target hilar basket cells (Toni et al. 2008), which provide strong inhibition of a large number of mature granule cells and regulate network oscillations (Freund 2003). Indeed, in vivo recordings from the dentate gyrus in anesthetized mice have shown that elimination of adult neurogenesis leads to decreased amplitude of perforant path–evoked responses and a marked increase in both the amplitude of spontaneous γ-frequency bursts in the dentate gyrus and the

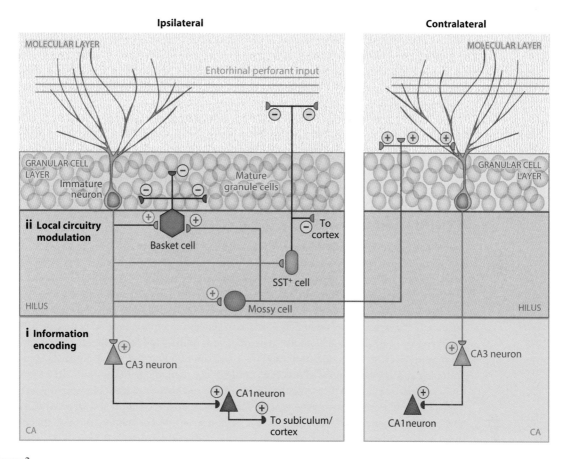

Figure 2

Circuitry properties of newborn neurons in the adult hippocampus. A schematic illustration of connectivity of newborn neurons in the adult dentate gyrus, highlighting the two modes by which newborn neurons can contribute to hippocampal functions, direct information processing via CA3 neurons (*i*, direct pathway) and modulation of local circuitry via hilar interneurons and mossy cells (*ii*, indirect pathway). SST+ neurons, somatostatin-expressing interneurons.

synchronization of dentate neuron firing to these bursts (Lacefield et al. 2012). Immediate early gene expression analysis has also shown increased activation of dentate granule cells in response to specific learning tasks after ablation of adult neurogenesis (Burghardt et al. 2012). In addition to interneurons, newborn neurons also innervate hilar mossy cells (Toni et al. 2008), which are glutamatergic and activate local interneurons and contralateral newborn and mature granule cells. Together, a modulator mode provides an amplification mechanism that allows a small number of newborn neurons to impact the global function of the dentate gyrus across both hemispheres. In addition, hippocampal somatostatin-expressing interneurons, including those in the hilus, send distal projections and directly modulate inhibition in the entorhinal cortex (Melzer et al. 2012). This finding raises the intriguing possibility that newborn neurons shape network properties well beyond the dentate gyrus by regulating these long projection interneurons. More studies are needed to investigate potential roles of new neurons as an active modulator of proximal and distal circuitry, especially in awake behaving animals.

Adult neurogenesis could also contribute indirectly to brain functions through alternations in structure properties of the circuitry, a possibility that has been rarely tested experimentally (**Figure 2**). For example, adult neurogenesis disrupts existing synapses of mature neurons owing to competition during new-neuron synaptic integration (Toni et al. 2007, 2008). The impact of such structural plasticity of mature neurons on brain functions is not well understood. In addition, adult hippocampal neurogenesis generates astrocytes that migrate into the hilus, the granule cell layer, and the molecular layer (Bonaguidi et al. 2011, Encinas et al. 2011). Given the critical role of astrocytes in regulating various brain functions (Clarke & Barres 2013), including adult hippocampal neurogenesis (Song et al. 2002), this is a largely untapped area that warrants future exploration.

FUNCTIONS OF ADULT HIPPOCAMPAL NEUROGENESIS IN COGNITION AND MOOD REGULATION

Immediately after the initial discovery of neurogenesis in the postnatal rat hippocampus, Altman (1967) postulated that newborn neurons are critical for learning and memory. Studies have since shown that hippocampus-dependent learning and memory (Gould et al. 1999), experience, mood, behavioral states, and antidepressants dynamically regulate multiple adult hippocampal neurogenesis processes (Deng et al. 2010, Sahay & Hen 2007). The first experimental evidence in mammals of a casual role of adult neurogenesis in generation or modification of specific behaviors came from a study in which blockade of neurogenesis in the adult mouse by an antimitotic agent disrupts trace eye-blink conditioning and trace fear conditioning, but not contextual feature conditioning and spatial memory (Shors et al. 2001). Since then, the field has gradually transitioned from correlative studies with manipulations that lack specificity to more sophisticated genetic and optogenetic approaches with enhanced temporal and spatial resolution and targeted behavioral protocols. Computational modeling has also been instrumental in framing possible functions of adult hippocampal neurogenesis and its underlying mechanisms (Aimone & Gage 2011). Although the idea is still under debate, the dorsal and ventral hippocampus are likely involved in fine-tuned, spatially discrete memory processes and affective behaviors, respectively, and adult hippocampal neurogenesis has been implicated in both functions (Deng et al. 2010, Kitabatake et al. 2007, Sahay et al. 2011b).

Cognitive Functions

A central dogma regarding information processing through the trisynaptic circuit of the hippocampus has been that the dentate gyrus mediates pattern separation, the ability to distinguish similar stimuli and contexts, whereas CA3 mediates pattern completion, the reinstatement of activity patterns correlated with complete contexts and associations using only partial or degraded information. Building on Marr's description of the information-processing capacity of the hippocampus based on its structure and intrinsic connectivity, an emergent model proposed that the densely packed dentate gyrus can support the orthogonalization of inputs arriving from the entorhinal cortex (pattern separation) (Marr 1971, Treves & Rolls 1994). The dentate gyrus, in turn, directly projects to CA3, a hippocampal subregion with extensive recurrent connections and thus the putative capacity to reactivate stored patterns using partial inputs (pattern completion). The idea that the dentate gyrus could support pattern separation received empirical support following the observation that cells in the dentate gyrus region are sparsely activated and have very low firing rates. Many behavioral assays of the dentate gyrus and CA3 function have been designed to test this hypothesis, and indeed results suggest that each of these regions may play a role in pattern separation and completion, respectively (Hunsaker & Kesner 2013). Recent evidence also supports an essential role for newborn granule neurons in mediating pattern separation, as inferred from behavioral deficits in mnemonic discrimination when adult neurogenesis is impaired (Aimone et al. 2011, Sahay et al. 2011b). For example, ablation of adult neurogenesis via irradiation impairs the ability of the mice to discriminate stimuli with little spatial separation, but not stimuli widely separated in space, in two spatial memory tasks (Clelland et al. 2009). Conversely, increasing the number of adult-born neurons by deleting the proapoptotic gene *Bax* from adult neural precursors and their progeny enhances their ability to differentiate between overlapping contextual representations (Sahay et al. 2011a). In one striking example, mice engineered to block synaptic release via tetanus toxin light-chain expression in mature dentate granule cells, but not in most immature neurons younger than four weeks, exhibit improved discrimination of very similar contexts in a fear-conditioning test, and blockade of adult neurogenesis via irradiation impairs context discrimination, suggesting a predominant role of immature neurons in mediating pattern separation (Nakashiba et al. 2012). However, these mice exhibit defects in the water-maze task and fear-conditioning tasks with partial cues presented, which suggests a critical role for mature granule cells in rapid pattern completion. One must recognize that the computational definition of pattern separation and completion implies a strict input–output relationship at the neural level that is not necessarily congruent with the use of this concept to describe behavior (Santoro 2013). It is still not clear, for example, that an animal that appears to pattern complete by behaving indistinguishably in a full-cue versus partial-cue context is doing so by using attractor dynamics in the CA3 region or by engaging mechanisms that compensate for the activation of only a subset of synapses to recruit the full complement of synapses involved in encoding (Knierim & Zhang 2012). Likewise, it is not clear that similar contexts are encoded by minimally overlapping ensembles of dentate granule cells.

Adult hippocampal neurogenesis has also been implicated in other aspects of contextual and spatial memory (Deng et al. 2010). By varying the timing between adult neurogenesis ablation and behavioral tests, studies have also pinpointed essential roles of adult-born immature neurons at various stages for these functions, especially when the task is difficult (Deng et al. 2009, Denny et al. 2012). These results corroborate findings of special cellular properties associated with immature stages and support the immature neuron model of the contribution of adult neurogenesis to brain functions. Adult neurogenesis has also been implicated in memory consolidation and the reorganization of memory traces to extrahippocampal structures, such as the prefrontal cortex (Kitamura et al. 2009). One study has shown that decreased adult hippocampal neurogenesis is accompanied by a prolonged period of hippocampus-dependent associative fear memory, whereas increased adult neurogenesis is associated with accelerated reorganization of memory traces that rely less on the hippocampus. In another interesting study, longitudinal activity data were collected on a large number of inbred mice, which shared one large enriched environment, to explore the relationship among cognitive challenges, adult hippocampal neurogenesis, and the development of individual behavioral traits. The size of the roaming area explored by an individual mouse was positively correlated with the amount of hippocampal neurogenesis (Freund et al. 2013), suggesting that one function of adult neurogenesis may be to shape the neuronal circuitry according to individual needs and improve adaptability over the life course of the individual.

There are many inconsistencies in the current literature regarding effects of various manipulations of adult neurogenesis levels on behavioral test outcomes, which have been summarized in previous reviews (Deng et al. 2010, W.R. Kim et al. 2012). Many possible factors contribute to these contradictory findings, such as differences in the genetic background, experimental manipulation, and behavioral paradigms. One major limitation of traditional approaches is the chronic nature of manipulations that affect adult neurogenesis throughout multiple phases of the learning process; therefore, it is not always clear whether adult-born neurons contribute to encoding, consolidation, storage, and/or retrieval processes. In addition, compensatory network changes may occur following ablation of adult neurogenesis (Singer et al. 2011). Newly available tools now allow for specific manipulation of the activity of adult-born neurons at distinct stages of maturation and during specific stages of learning and recall (W.R. Kim et al. 2012). One recent study of retrovirally targeted newborn neurons in the adult mouse dentate gyrus showed that optogenetic suppression of four-week-old, but not two- or eight-week-old, newborn neurons during recall trials impairs contextual fear memory and spatial memory retrieval (Gu et al. 2012). Given the increasing availability of sophisticated genetic models to target specific populations of neural progenitor subtypes or newborn neurons at specific maturation stages, combined with optogenetic and pharmacogenetic tools to manipulate neuronal activity with spatial and temporal precision, future studies will be able to directly address questions of how and when newborn neurons contribute to neural function.

Mood Regulation

It is now well-established that stress negatively regulates progenitor proliferation and new-neuron survival (Gould et al. 1992), whereas clinical antidepressant treatments, including electroconvulsive therapy and chemical antidepressants (Malberg et al. 2000), promote proliferation of neural progenitors and maturation of newborn neurons during adult hippocampal neurogenesis (Sahay & Hen 2007, Warner-Schmidt & Duman 2006). Such effects are evolutionarily conserved from rodents to nonhuman primates (Perera et al. 2007) and potentially to humans (Boldrini et al. 2009). Ablation of adult neurogenesis does not appear to alter affective phenotypes at basal levels but abolishes some antidepressant-induced behaviors in rodents (Santarelli et al. 2003) and nonhuman primates (Perera et al. 2011). Emerging evidence suggests a critical role of adult hippocampal neurogenesis in the stress response by suppressing the hypothalamic-pituitary-adrenal (HPA) axis. In mice with adult neurogenesis ablated, mild stress leads to increased levels of stress hormones and greater stress responses in behavioral tests (Schloesser et al. 2009, Snyder et al. 2011). Furthermore, blockade of adult neurogenesis abolishes the antidepressant effect of hippocampal regulation of the HPA axis after chronic stress (Surget et al. 2011). The mechanism by which adult-born neurons regulate the HPA axis under basal conditions and upon antidepressant treatment remains to be determined. Anatomical studies have revealed that whereas the dorsal hippocampus projects primarily to cortical areas that mediate cognitive processes such as learning and memory and navigation and exploration, the ventral hippocampus projects to the limbic system, including the amygdala, the nucleus accumbens, and the hypothalamus (Fanselow & Dong 2010). Future studies are needed to address how adult-born neurons, especially those in ventral regions, influence neural pathways involved in emotional experience and affective states. It will also be interesting to test the immature neuron model of adult neurogenesis function in the context of mood regulation. Development of optogenetic and pharmacogenetic approaches to target newborn neurons at specific maturation stages in the dorsal or ventral dentate gyrus will facilitate these efforts (Kheirbek et al. 2013).

DYSFUNCTION OF ADULT HIPPOCAMPAL NEUROGENESIS IN BRAIN DISORDERS

The dentate gyrus is vulnerable to cell death; but as one of the most labile structures in the brain in terms of population dynamics, it is also subject to the consequences of dysregulated adult neurogenesis. A substantial body of literature addresses changes of adult hippocampal neurogenesis in rodents, and limited reports in humans, in the context of various pathophysiological conditions, including aging, epilepsy, stroke, degenerative neurological disorders, and neuropsychiatric disorders (Kempermann et al. 2008, Parent 2003, Sahay & Hen 2007, Winner et al. 2011). In most cases, whether these changes represent adaptive responses to various pathophysiological conditions, or are part of the pathophysiology that contributes to the condition, is unknown. Examples in animal models now suggest that dysfunction of adult hippocampal neurogenesis may play a causal role in brain disorders. There are two modes by which dysregulated adult hippocampal neurogenesis can contribute to dysfunction of the hippocampus: a loss-of-function mode due to decreased new-neuron production and integration, and a gain-of-function mode due to aberrant development and integration of new neurons.

Dysfunction via Loss-of-Function

Fragile X syndrome, the most common form of inherited intellectual disability, is caused by the functional loss of fragile X mental retardation protein (FMRP). *Fmrp* null mice exhibit deficits in some forms of hippocampus-dependent learning, accompanied by reduced adult hippocampal neurogenesis due to impaired neuronal differentiation and survival (Luo et al. 2010). Deletion of *Fmrp* specifically in adult neural progenitor cells using *nestin-CreER^{T2}* mice recapitulates defects in both adult neurogenesis and hippocampus-dependent learning, and furthermore, restoration of *Fmrp* expression in adult neural progenitors alone is sufficient to rescue learning deficits in *Fmrp* null mice (Guo et al. 2011). These striking results suggest a causal role of adult neurogenesis dysfunction in learning impairments associated with fragile X syndrome, at least in animal models. Whether this is generalizable to other disorders remains to be seen.

One major regulator of adult hippocampal neurogenesis is aging. Over the subject's lifetime, the rate of adult neurogenesis decreases dramatically, from rodents to primates, which may contribute to the dysfunction of hippocampus (Ming & Song 2011). The rate of decline of adult neurogenesis during aging is much more robust in rodents (about tenfold) than in humans (about fourfold) (Spalding et al. 2013).

Dysfunction via Gain-of-Function

Dentate granule cells may play a central role in the pathogenesis of temporal lobe epilepsy, one of the most common human seizure-related disorders (Houser 1992). In animal models of epilepsy, pilocarpine-induced status epilepticus leads to a dramatic and prolonged increase in dentate neural progenitor proliferation (Parent et al. 1997). However, many of these newborn neurons integrate aberrantly, displaying hilar basal dendrites with spines, ectopic hilar localization of the cell body, and mossy fiber sprouting (Jessberger et al. 2007, Kron et al. 2010), similar to what has been observed in postmortem dentate gyri of patients with temporal lobe epilepsy (Houser 1992). Eliminating cohorts of newborn neurons decreases status epilepticus–induced mossy fiber sprouting and ectopic granule cells (Kron et al. 2010) and attenuates spontaneous recurrent seizures in mice (Jung et al. 2004). Separately, deletion of PTEN (phosphatase and tensin homolog deleted on chromosome ten) in a small percentage of dentate granule cells born postnatally is sufficient to cause spontaneous seizure within four weeks, accompanied by aberrant granule cell morphology seen in epilepsy (Pun et al. 2012). Collectively, these studies provide strong evidence that dysfunction of adult hippocampal neurogenesis plays a causal role in epileptogenesis.

In another gain-of-function example, retrovirus-mediated knockdown of *Disrupted in Schizophrenia I* (*DISC1*), a risk gene for major mental illness (Thomson et al. 2013), leads to aberrant integration of newborn dentate granule neurons in the adult mouse hippocampus, including ectopic location of the cell body to the outer granule cell layer and molecular layer, aberrant axonal targeting beyond CA3, hyperexcitability, and aberrant formation of synaptic inputs and outputs, due in part to hyperactivation of the mTOR pathway in newborn neurons (Duan et al. 2007; Faulkner et al. 2008; J.Y. Kim et al. 2009, 2012). Dysregulated adult hippocampal neurogenesis following DISC1 knockdown in one cohort of newborn neurons is sufficient to cause several behavioral phenotypes, including pronounced learning and memory deficits (in the object-place recognition task and the spatial version of the Morris water maze), as well as clear anxiety and depression-like phenotypes (in the forced-swim test and elevated plus maze) (Zhou et al. 2013). Inactivation of these aberrant neurons reverses specific behavioral phenotypes, indicating a causal role of adult neurogenesis dysfunction in behavioral impairments.

Notably, the impact of adult neurogenesis dysfunction due to gain-of-function on animal behavior is generally more pronounced than that seen in loss-of-function conditions. This finding may not be surprising because the complete absence or removal of a system may trigger the recruitment of alternative pathways to compensate (Singer et al. 2011), whereas miswiring of newborn neurons can be more detrimental, especially given their high excitability and unique properties. These findings have significant implications for future cell-replacement therapy in which correct wiring of transplanted neurons could be essential for functional benefits and for avoiding potential side effects.

CONCLUSION

In the past decade we have witnessed rapid advances in the adult neurogenesis field, with significant progress in (*a*) the characterization of this phenomenon in different species, including humans; (*b*) the delineation of neurogenic processes and properties of adult-born neurons; (*c*) exploration of its function at circuitry and behavioral levels; and (*d*) an appreciation of how dysfunction of adult neurogenesis may contribute to brain disorders. Despite these tremendous findings, understanding the function of adult hippocampal neurogenesis remains a central goal in the field. Perhaps one of the most frequently asked questions is why it occurs in the dentate gyrus. The dentate gyrus is one of the two regions with continuous neurogenesis from rodents to humans. Fully addressing these questions will require a multidisciplinary approach and new technologies. First, we need to know more about basic properties of the dentate gyrus and how it processes information and contributes to hippocampal functions, which will provide the framework to delineate the contribution of adult neurogenesis. Second, we need to have a better knowledge of the dentate circuitry, especially synaptic outputs of newborn neurons. Third, recordings of newborn neurons at different maturation stages in awake behaving animals will provide critical information to test current models of how adult-born neurons contribute to brain functions and dysfunctions. Fourth, the field needs to address contradictory results from behavioral analyses using newly available tools with better cell-type specificity and higher temporal and spatial resolution. Fifth, we need to consider how other plasticity associated with adult neurogenesis also contributes to brain functions, such as the generation of new astrocytes and induced structural changes in mature neurons. Sixth, comparative studies of two primary neurogenic regions in different species have proven highly informative, and ultimately, we want to understand the function of adult neurogenesis in humans.

It was originally proposed that adult neurogenesis is not a cell-replacement mechanism in which dying individual neurons are functionally replaced by new neurons, but instead continuously provides new cohorts of immature neurons with properties and information-processing capacities that are distinct from those of existing mature neurons (Ge et al. 2007). This immature neuron model of the adult neurogenesis contribution to brain function has gained significant support over the past few years from additional comparisons of immature and mature neurons, computational modeling, and animal behavioral analyses. Building on this model is the plastic dentate gyrus hypothesis: Adult neurogenesis represents a continuous developmental process that maintains a highly plastic dentate circuitry, collectively with the addition of immature neurons with unique properties and new astrocytes and with the continued structural plasticity of associated mature neurons in broader brain regions. The heterogeneous nature of the dentate gyrus, with a small immature neuronal cohort that is highly plastic and excitable and with a large population of mature granule cells that is sparsely activated with high input specificity, offers unique information-processing power that can adapt to dynamic needs over the lifetime.

Understanding the physiological function of adult neurogenesis not only provides a new prospective on the plasticity of the mature nervous system, but also has significant implications for our understanding of several brain disorders and regenerative medicine. Recent evidence supports a critical contribution of dysfunctional postnatal neurogenesis, via both loss-of-function and gain-of-function modes, to developmental disorders and may be a crucial mechanism that initiates the onset of disorders such as autism and schizophrenia (Ming & Song 2009). Recent animal model studies have suggested that treating molecular deficits underlying neurodevelopmental disorders could result in significant amelioration of associated behavioral phenotypes, even when treatments were initiated in adults (Ehninger et al. 2008). Therefore, targeting adult neurogenesis could be a novel potential therapeutic strategy for these disorders. Basic principles learned from normal and dysregulated adult neuronal development and synaptic integration of newborn neurons will also provide invaluable information for the future development of cell-replacement therapy.

DISCLOSURE STATEMENT

The authors are not aware of any affiliations, memberships, funding, or financial holdings that might be perceived as affecting the objectivity of this review.

ACKNOWLEDGMENTS

We apologize that we could not cite all primary literatures owing to space limitations. The research in the authors' laboratories was supported by National Institutes of Health (R01NS048271, R01HD069184 to G-l.M. & R37NS047344, R21ES021957 to H.S.), the Maryland Stem Cell Research Fund (K.C. & G-l.M.), NARSAD (K.C. & G-l.M.), the Simons Foundation (H.S.), and the Dr. Miriam and Sheldon G. Adelson Medical Research Foundation (G-l.M.).

LITERATURE CITED

Aimone JB, Deng W, Gage FH. 2011. Resolving new memories: a critical look at the dentate gyrus, adult neurogenesis, and pattern separation. *Neuron* 70:589–96

Aimone JB, Gage FH. 2011. Modeling new neuron function: a history of using computational neuroscience to study adult neurogenesis. *Eur. J. Neurosci.* 33:1160–69

Altman J. 1967. *The Neurosciences: Second Study Program*, ed. GC Quarton, T Melnechuck, FO Schmitt, pp. 723–43. New York: Rockefeller Univ. Press

Boldrini M, Underwood MD, Hen R, Rosoklija GB, Dwork AJ, et al. 2009. Antidepressants increase neural progenitor cells in the human hippocampus. *Neuropsychopharmacology* 34:2376–89

Bonaguidi MA, Song J, Ming GL, Song H. 2012. A unifying hypothesis on mammalian neural stem cell properties in the adult hippocampus. *Curr. Opin. Neurobiol.* 22:754–61

Bonaguidi MA, Wheeler MA, Shapiro JS, Stadel RP, Sun GJ, et al. 2011. In vivo clonal analysis reveals self-renewing and multipotent adult neural stem cell characteristics. *Cell* 145:1142–55

Braun SMG, Jessberger S. 2013. Adult neurogenesis in the mammalian brain. *Front. Biol.* 8:295–304

Brus M, Keller M, Lévy F. 2013. Temporal features of adult neurogenesis: differences and similarities across mammalian species. *Front. Neurosci.* 7:135

Burghardt NS, Park EH, Hen R, Fenton AA. 2012. Adult-born hippocampal neurons promote cognitive flexibility in mice. *Hippocampus* 22:1795–808

Cabezas C, Irinopoulou T, Cauli B, Poncer JC. 2013. Molecular and functional characterization of GAD67-expressing, newborn granule cells in mouse dentate gyrus. *Front. Neural Circuits* 7:60

Cabezas C, Irinopoulou T, Gauvain G, Poncer JC. 2012. Presynaptic but not postsynaptic GABA signaling at unitary mossy fiber synapses. *J. Neurosci.* 32:11835–40

Cameron HA, McKay RD. 2001. Adult neurogenesis produces a large pool of new granule cells in the dentate gyrus. *J. Comp. Neurol.* 435:406–17

Chancey JH, Adlaf EW, Sapp MC, Pugh PC, Wadiche JI, Overstreet-Wadiche LS. 2013. GABA depolarization is required for experience-dependent synapse unsilencing in adult-born neurons. *J. Neurosci.* 33:6614–22

Chancey JH, Poulsen DJ, Wadiche JI, Overstreet-Wadiche L. 2014. Hilar mossy cells provide the first glutamatergic synapses to adult-born dentate granule cells. *J. Neurosci.* 34:2349–54

Clarke LE, Barres BA. 2013. Emerging roles of astrocytes in neural circuit development. *Nat. Rev. Neurosci.* 14:311–21

Clelland CD, Choi M, Romberg C, Clemenson GD Jr, Fragniere A, et al. 2009. A functional role for adult hippocampal neurogenesis in spatial pattern separation. *Science* 325:210–13

Deng W, Aimone JB, Gage FH. 2010. New neurons and new memories: How does adult hippocampal neurogenesis affect learning and memory? *Nat. Rev. Neurosci.* 11:339–50

Deng W, Saxe MD, Gallina IS, Gage FH. 2009. Adult-born hippocampal dentate granule cells undergoing maturation modulate learning and memory in the brain. *J. Neurosci.* 29:13532–42

Denny CA, Burghardt NS, Schachter DM, Hen R, Drew MR. 2012. 4- to 6-week-old adult-born hippocampal neurons influence novelty-evoked exploration and contextual fear conditioning. *Hippocampus* 22:1188–201

Deshpande A, Bergami M, Ghanem A, Conzelmann KK, Lepier A, et al. 2013. Retrograde monosynaptic tracing reveals the temporal evolution of inputs onto new neurons in the adult dentate gyrus and olfactory bulb. *Proc. Natl. Acad. Sci. USA* 110:E1152–61

Dhaliwal J, Lagace DC. 2011. Visualization and genetic manipulation of adult neurogenesis using transgenic mice. *Eur. J. Neurosci.* 33:1025–36

Dieni CV, Nietz AK, Panichi R, Wadiche JI, Overstreet-Wadiche L. 2013. Distinct determinants of sparse activation during granule cell maturation. *J. Neurosci.* 33:19131–42

Duan X, Chang JH, Ge S, Faulkner RL, Kim JY, et al. 2007. Disrupted-In-Schizophrenia 1 regulates integration of newly generated neurons in the adult brain. *Cell* 130:1146–58

Ehninger D, Li W, Fox K, Stryker MP, Silva AJ. 2008. Reversing neurodevelopmental disorders in adults. *Neuron* 60:950–60

Encinas JM, Michurina TV, Peunova N, Park J-H, Tordo J, et al. 2011. Division-coupled astrocytic differentiation and age-related depletion of neural stem cells in the adult hippocampus. *Cell Stem Cell* 8:566–79

Ernst A, Alkass K, Bernard S, Salehpour M, Perl S, et al. 2014. Neurogenesis in the striatum of the adult human brain. *Cell* 156:1072–83

Espósito MS, Piatti VC, Laplagne DA, Morgenstern NA, Ferrari CC, et al. 2005. Neuronal differentiation in the adult hippocampus recapitulates embryonic development. *J. Neurosci.* 25:10074–86

Fanselow MS, Dong HW. 2010. Are the dorsal and ventral hippocampus functionally distinct structures? *Neuron* 65:7–19

Faulkner RL, Jang M-H, Liu X-B, Duan X, Sailor KA, et al. 2008. Development of hippocampal mossy fiber synaptic outputs by new neurons in the adult brain. *Proc. Natl. Acad. Sci. USA* 105:14157–62

Freund J, Brandmaier AM, Lewejohann L, Kirste I, Kritzler M, et al. 2013. Emergence of individuality in genetically identical mice. *Science* 340:756–59

Freund TF. 2003. Interneuron Diversity series: Rhythm and mood in perisomatic inhibition. *Trends Neurosci.* 26:489–95

Gage FH, Temple S. 2013. Neural stem cells: generating and regenerating the brain. *Neuron* 80:588–601

Gasbarri A, Sulli A, Packard MG. 1997. The dopaminergic mesencephalic projections to the hippocampal formation in the rat. *Prog. Neuropsychopharmacol. Biol. Psychiatry* 21:1–22

Ge S, Goh EL, Sailor KA, Kitabatake Y, Ming GL, Song H. 2006. GABA regulates synaptic integration of newly generated neurons in the adult brain. *Nature* 439:589–93

Ge S, Sailor KA, Ming GL, Song H. 2008. Synaptic integration and plasticity of new neurons in the adult hippocampus. *J. Physiol.* 586:3759–65

Ge S, Yang CH, Hsu KS, Ming GL, Song H. 2007. A critical period for enhanced synaptic plasticity in newly generated neurons of the adult brain. *Neuron* 54:559–66

Göritz C, Frisén J. 2012. Neural stem cells and neurogenesis in the adult. *Cell Stem Cell* 10:657–59

Gould E, Beylin A, Tanapat P, Reeves A, Shors TJ. 1999. Learning enhances adult neurogenesis in the hippocampal formation. *Nat. Neurosci.* 2:260–65

Gould E, Cameron HA, Daniels DC, Woolley CS, McEwen BS. 1992. Adrenal hormones suppress cell division in the adult rat dentate gyrus. *J. Neurosci.* 12:3642–50

Gu Y, Arruda-Carvalho M, Wang J, Janoschka SR, Josselyn SA, et al. 2012. Optical controlling reveals time-dependent roles for adult-born dentate granule cells. *Nat. Neurosci.* 15:1700–6

Guo W, Allan AM, Zong R, Zhang L, Johnson EB, et al. 2011. Ablation of Fmrp in adult neural stem cells disrupts hippocampus-dependent learning. *Nat. Med.* 17:559–65

Houser CR. 1992. Morphological changes in the dentate gyrus in human temporal lobe epilepsy. *Epilepsy Res. Suppl.* 7:223–34

Hunsaker MR, Kesner RP. 2013. The operation of pattern separation and pattern completion processes associated with different attributes or domains of memory. *Neurosci. Biobehav. Rev.* 37:36–58

Jessberger S, Zhao C, Toni N, Clemenson GD Jr, Li Y, Gage FH. 2007. Seizure-associated, aberrant neurogenesis in adult rats characterized with retrovirus-mediated cell labeling. *J. Neurosci.* 27:9400–7

Jung KH, Chu K, Kim M, Jeong SW, Song YM, et al. 2004. Continuous cytosine-b-D-arabinofuranoside infusion reduces ectopic granule cells in adult rat hippocampus with attenuation of spontaneous recurrent seizures following pilocarpine-induced status epilepticus. *Eur. J. Neurosci.* 19:3219–26

Kempermann G, Gage FH. 1999. New nerve cells for the adult brain. *Sci. Am.* 280:48–53

Kempermann G, Krebs J, Fabel K. 2008. The contribution of failing adult hippocampal neurogenesis to psychiatric disorders. *Curr. Opin. Psychiatry* 21:290–95

Kheirbek MA, Drew LJ, Burghardt NS, Costantini DO, Tannenholz L, et al. 2013. Differential control of learning and anxiety along the dorsoventral axis of the dentate gyrus. *Neuron* 77:955–68

Kim JY, Duan X, Liu CY, Jang M-H, Guo JU, et al. 2009. DISC1 regulates new neuron development in the adult brain via modulation of AKT-mTOR signaling through KIAA1212. *Neuron* 63:761–73

Kim JY, Liu CY, Zhang F, Duan X, Wen Z, et al. 2012. Interplay between DISC1 and GABA signaling regulates neurogenesis in mice and risk for schizophrenia. *Cell* 148:1051–64

Kim WR, Christian K, Ming GL, Song H. 2012. Time-dependent involvement of adult-born dentate granule cells in behavior. *Behav. Brain Res.* 227:470–79

Kitabatake Y, Sailor KA, Ming GL, Song H. 2007. Adult neurogenesis and hippocampal memory function: new cells, more plasticity, new memories? *Neurosurg. Clin. N. Am.* 18:105–13, x

Kitamura T, Saitoh Y, Takashima N, Murayama A, Niibori Y, et al. 2009. Adult neurogenesis modulates the hippocampus-dependent period of associative fear memory. *Cell* 139:814–27

Knierim JJ, Zhang K. 2012. Attractor dynamics of spatially correlated neural activity in the limbic system. *Annu. Rev. Neurosci.* 35:267–85

Kohler SJ, Williams NI, Stanton GB, Cameron JL, Greenough WT. 2011. Maturation time of new granule cells in the dentate gyrus of adult macaque monkeys exceeds six months. *Proc. Natl. Acad. Sci. USA* 108:10326–31

Kriegstein A, Alvarez-Buylla A. 2009. The glial nature of embryonic and adult neural stem cells. *Annu. Rev. Neurosci.* 32:149–84

Kron MM, Zhang H, Parent JM. 2010. The developmental stage of dentate granule cells dictates their contribution to seizure-induced plasticity. *J. Neurosci.* 30:2051–59

Kumamoto N, Gu Y, Wang J, Janoschka S, Takemaru K, et al. 2012. A role for primary cilia in glutamatergic synaptic integration of adult-born neurons. *Nat. Neurosci.* 15:399–405, S1

Lacefield CO, Itskov V, Reardon T, Hen R, Gordon JA. 2012. Effects of adult-generated granule cells on coordinated network activity in the dentate gyrus. *Hippocampus* 22:106–16

Laplagne DA, Espósito MS, Piatti VC, Morgenstern NA, Zhao C, et al. 2006. Functional convergence of neurons generated in the developing and adult hippocampus. *PLoS Biol.* 4:e409

Laplagne DA, Kamienkowski JE, Espósito MS, Piatti VC, Zhao C, et al. 2007. Similar GABAergic inputs in dentate granule cells born during embryonic and adult neurogenesis. *Eur. J. Neurosci.* 25:2973–81

Lemaire V, Tronel S, Montaron MF, Fabre A, Dugast E, Abrous DN. 2012. Long-lasting plasticity of hippocampal adult-born neurons. *J. Neurosci.* 32:3101–8

Lepousez G, Valley MT, Lledo PM. 2013. The impact of adult neurogenesis on olfactory bulb circuits and computations. *Annu. Rev. Physiol.* 75:339–63

Leutgeb JK, Leutgeb S, Moser MB, Moser EI. 2007. Pattern separation in the dentate gyrus and CA3 of the hippocampus. *Science* 315:961–66

Li Y, Aimone JB, Xu X, Callaway EM, Gage FH. 2012. Development of GABAergic inputs controls the contribution of maturing neurons to the adult hippocampal network. *Proc. Natl. Acad. Sci. USA* 109:4290–95

Li Y, Stam FJ, Aimone JB, Goulding M, Callaway EM, Gage FH. 2013. Molecular layer perforant path-associated cells contribute to feed-forward inhibition in the adult dentate gyrus. *Proc. Natl. Acad. Sci. USA* 110:9106–11

Liu YB, Lio PA, Pasternak JF, Trommer BL. 1996. Developmental changes in membrane properties and postsynaptic currents of granule cells in rat dentate gyrus. *J. Neurophysiol.* 76:1074–88

Lledo PM, Alonso M, Grubb MS. 2006. Adult neurogenesis and functional plasticity in neuronal circuits. *Nat. Rev. Neurosci.* 7:179–93

Luo Y, Shan G, Guo W, Smrt RD, Johnson EB, et al. 2010. Fragile X mental retardation protein regulates proliferation and differentiation of adult neural stem/progenitor cells. *PLoS Genet.* 6:e1000898

Ma DK, Kim WR, Ming GL, Song H. 2009. Activity-dependent extrinsic regulation of adult olfactory bulb and hippocampal neurogenesis. *Ann. N. Y. Acad. Sci.* 1170:664–73

Malberg JE, Eisch AJ, Nestler EJ, Duman RS. 2000. Chronic antidepressant treatment increases neurogenesis in adult rat hippocampus. *J. Neurosci.* 20:9104–10

Marín-Burgin A, Mongiat LA, Pardi MB, Schinder AF. 2012. Unique processing during a period of high excitation/inhibition balance in adult-born neurons. *Science* 335:1238–42

Markwardt SJ, Dieni CV, Wadiche JI, Overstreet-Wadiche L. 2011. Ivy/neurogliaform interneurons coordinate activity in the neurogenic niche. *Nat. Neurosci.* 14:1407–9

Marr D. 1971. Simple memory: a theory for archicortex. *Philos. Trans. R. Soc. B* 262:23–81

Melzer S, Michael M, Caputi A, Eliava M, Fuchs EC, et al. 2012. Long-range-projecting GABAergic neurons modulate inhibition in hippocampus and entorhinal cortex. *Science* 335:1506–10

Ming GL, Song H. 2005. Adult neurogenesis in the mammalian central nervous system. *Annu. Rev. Neurosci.* 28:223–50

Ming GL, Song H. 2009. DISC1 partners with GSK3β in neurogenesis. *Cell* 136:990–92

Ming GL, Song H. 2011. Adult neurogenesis in the mammalian brain: significant answers and significant questions. *Neuron* 70:687–702

Mongiat LA, Espósito MS, Lombardi G, Schinder AF. 2009. Reliable activation of immature neurons in the adult hippocampus. *PLoS ONE* 4:e5320

Mu Y, Zhao C, Gage FH. 2011. Dopaminergic modulation of cortical inputs during maturation of adult-born dentate granule cells. *J. Neurosci.* 31:4113–23

Nakashiba T, Cushman JD, Pelkey KA, Renaudineau S, Buhl DL, et al. 2012. Young dentate granule cells mediate pattern separation, whereas old granule cells facilitate pattern completion. *Cell* 149:188–201

Neunuebel JP, Knierim JJ. 2012. Spatial firing correlates of physiologically distinct cell types of the rat dentate gyrus. *J. Neurosci.* 32:3848–58

Neunuebel JP, Knierim JJ. 2014. CA3 retrieves coherent representations from degraded input: direct evidence for CA3 pattern completion and dentate gyrus pattern separation. *Neuron* 81:416–27

Overstreet Wadiche L, Bromberg DA, Bensen AL, Westbrook GL. 2005. GABAergic signaling to newborn neurons in dentate gyrus. *J. Neurophysiol.* 94:4528–32

Overstreet-Wadiche LS, Bensen AL, Westbrook GL. 2006a. Delayed development of adult-generated granule cells in dentate gyrus. *J. Neurosci.* 26:2326–34

Overstreet-Wadiche LS, Bromberg DA, Bensen AL, Westbrook GL. 2006b. Seizures accelerate functional integration of adult-generated granule cells. *J. Neurosci.* 26:4095–103

Parent JM. 2003. Injury-induced neurogenesis in the adult mammalian brain. *Neuroscientist* 9:261–72

Parent JM, Yu TW, Leibowitz RT, Geschwind DH, Sloviter RS, Lowenstein DH. 1997. Dentate granule cell neurogenesis is increased by seizures and contributes to aberrant network reorganization in the adult rat hippocampus. *J. Neurosci.* 17:3727–38

Perera TD, Coplan JD, Lisanby SH, Lipira CM, Arif M, et al. 2007. Antidepressant-induced neurogenesis in the hippocampus of adult nonhuman primates. *J. Neurosci.* 27:4894–901

Perera TD, Dwork AJ, Keegan KA, Thirumangalakudi L, Lipira CM, et al. 2011. Necessity of hippocampal neurogenesis for the therapeutic action of antidepressants in adult nonhuman primates. *PLoS ONE* 6:e17600

Piatti VC, Davies-Sala MG, Espósito MS, Mongiat LA, Trinchero MF, Schinder AF. 2011. The timing for neuronal maturation in the adult hippocampus is modulated by local network activity. *J. Neurosci.* 31:7715–28

Pontes A, Zhang Y, Hu W. 2013. Novel functions of GABA signaling in adult neurogenesis. *Front. Biol.* 8:496–507

Pun RY, Rolle IJ, Lasarge CL, Hosford BE, Rosen JM, et al. 2012. Excessive activation of mTOR in postnatally generated granule cells is sufficient to cause epilepsy. *Neuron* 75:1022–34

Ramirez-Amaya V, Marrone DF, Gage FH, Worley PF, Barnes CA. 2006. Integration of new neurons into functional neural networks. *J. Neurosci.* 26:12237–41

Ramírez-Amaya V, Vazdarjanova A, Mikhael D, Rosi S, Worley PF, Barnes CA. 2005. Spatial exploration-induced Arc mRNA and protein expression: evidence for selective, network-specific reactivation. *J. Neurosci.* 25:1761–68

Sahay A, Hen R. 2007. Adult hippocampal neurogenesis in depression. *Nat. Neurosci.* 10:1110–15

Sahay A, Scobie KN, Hill AS, O'Carroll CM, Kheirbek MA, et al. 2011a. Increasing adult hippocampal neurogenesis is sufficient to improve pattern separation. *Nature* 472:466–70

Sahay A, Wilson DA, Hen R. 2011b. Pattern separation: a common function for new neurons in hippocampus and olfactory bulb. *Neuron* 70:582–88

Santarelli L, Saxe M, Gross C, Surget A, Battaglia F, et al. 2003. Requirement of hippocampal neurogenesis for the behavioral effects of antidepressants. *Science* 301:805–9

Santoro A. 2013. Reassessing pattern separation in the dentate gyrus. *Front. Behav. Neurosci.* 7:96

Schloesser RJ, Manji HK, Martinowich K. 2009. Suppression of adult neurogenesis leads to an increased hypothalamo-pituitary-adrenal axis response. *NeuroReport* 20:553–57

Schmidt-Hieber C, Jonas P, Bischofberger J. 2004. Enhanced synaptic plasticity in newly generated granule cells of the adult hippocampus. *Nature* 429:184–87

Shors TJ, Miesegaes G, Beylin A, Zhao M, Rydel T, Gould E. 2001. Neurogenesis in the adult is involved in the formation of trace memories. *Nature* 410:372–76

Sierra A, Encinas JM, Deudero JJ, Chancey JH, Enikolopov G, et al. 2010. Microglia shape adult hippocampal neurogenesis through apoptosis-coupled phagocytosis. *Cell Stem Cell* 7:483–95

Singer BH, Gamelli AE, Fuller CL, Temme SJ, Parent JM, Murphy GG. 2011. Compensatory network changes in the dentate gyrus restore long-term potentiation following ablation of neurogenesis in young-adult mice. *Proc. Natl. Acad. Sci. USA* 108:5437–42

Snyder JS, Kee N, Wojtowicz JM. 2001. Effects of adult neurogenesis on synaptic plasticity in the rat dentate gyrus. *J. Neurophysiol.* 85:2423–31

Snyder JS, Soumier A, Brewer M, Pickel J, Cameron HA. 2011. Adult hippocampal neurogenesis buffers stress responses and depressive behaviour. *Nature* 476:458–61

Song H, Stevens CF, Gage FH. 2002. Astroglia induce neurogenesis from adult neural stem cells. *Nature* 417:39–44

Song J, Sun J, Moss J, Wen Z, Sun GJ, et al. 2013. Parvalbumin interneurons mediate neuronal circuitry—neurogenesis coupling in the adult hippocampus. *Nat. Neurosci.* 16:1728–30

Song J, Zhong C, Bonaguidi MA, Sun GJ, Hsu D, et al. 2012. Neuronal circuitry mechanism regulating adult quiescent neural stem-cell fate decision. *Nature* 489:150–54

Spalding KL, Bergmann O, Alkass K, Bernard S, Salehpour M, et al. 2013. Dynamics of hippocampal neurogenesis in adult humans. *Cell* 153:1219–27

Squire LR. 1992. Memory and the hippocampus: a synthesis from findings with rats, monkeys, and humans. *Psychol. Rev.* 99:195–231

Stone SS, Teixeira CM, Zaslavsky K, Wheeler AL, Martinez-Canabal A, et al. 2011. Functional convergence of developmentally and adult-generated granule cells in dentate gyrus circuits supporting hippocampus-dependent memory. *Hippocampus* 21:1348–62

Sun GJ, Sailor KA, Mahmood QA, Chavali N, Christian KM, et al. 2013. Seamless reconstruction of intact adult-born neurons by serial end-block imaging reveals complex axonal guidance and development in the adult hippocampus. *J. Neurosci.* 33:11400–11

Surget A, Tanti A, Leonardo ED, Laugeray A, Rainer Q, et al. 2011. Antidepressants recruit new neurons to improve stress response regulation. *Mol. Psychiatry* 16:1177–88

Tashiro A, Sandler VM, Toni N, Zhao C, Gage FH. 2006. NMDA-receptor-mediated, cell-specific integration of new neurons in adult dentate gyrus. *Nature* 442:929–33

Thomson PA, Malavasi EL, Grünewald E, Soares DC, Borkowska M, Millar JK. 2013. DISC1 genetics, biology and psychiatric illness. *Front. Biol.* 8:1–31

Toni N, Laplagne DA, Zhao C, Lombardi G, Ribak CE, et al. 2008. Neurons born in the adult dentate gyrus form functional synapses with target cells. *Nat. Neurosci.* 11:901–7

Toni N, Teng EM, Bushong EA, Aimone JB, Zhao C, et al. 2007. Synapse formation on neurons born in the adult hippocampus. *Nat. Neurosci.* 10:727–34

Tozuka Y, Fukuda S, Namba T, Seki T, Hisatsune T. 2005. GABAergic excitation promotes neuronal differentiation in adult hippocampal progenitor cells. *Neuron* 47:803–15

Treves A, Rolls ET. 1994. Computational analysis of the role of the hippocampus in memory. *Hippocampus* 4:374–91

Vivar C, Potter MC, Choi J, Lee JY, Stringer TP, et al. 2012. Monosynaptic inputs to new neurons in the dentate gyrus. *Nat. Commun.* 3:1107

Warner-Schmidt JL, Duman RS. 2006. Hippocampal neurogenesis: opposing effects of stress and antidepressant treatment. *Hippocampus* 16:239–49

Winner B, Kohl Z, Gage FH. 2011. Neurodegenerative disease and adult neurogenesis. *Eur. J. Neurosci.* 33:1139–51

Zhao C, Deng W, Gage FH. 2008. Mechanisms and functional implications of adult neurogenesis. *Cell* 132:645–60

Zhao C, Teng EM, Summers RG Jr, Ming GL, Gage FH. 2006. Distinct morphological stages of dentate granule neuron maturation in the adult mouse hippocampus. *J. Neurosci.* 26:3–11

Zhou M, Li W, Huang S, Song J, Kim JY, et al. 2013. mTOR Inhibition ameliorates cognitive and affective deficits caused by Disc1 knockdown in adult-born dentate granule neurons. *Neuron* 77:647–54

Ziv Y, Burns LD, Cocker ED, Hamel EO, Ghosh KK, et al. 2013. Long-term dynamics of CA1 hippocampal place codes. *Nat. Neurosci.* 16:264–66

Emotion and Decision Making: Multiple Modulatory Neural Circuits

Elizabeth A. Phelps,[1,2,3] Karolina M. Lempert,[1] and Peter Sokol-Hessner[1,2]

[1]Department of Psychology, [2]Center for Neural Science, New York University, New York, NY 10003;

[3]Nathan Kline Institute, Orangeburg, New York, NY 10963; email: liz.phelps@nyu.edu, karolina.lempert@gmail.com, psh234@nyu.edu

Annu. Rev. Neurosci. 2014. 37:263–87

First published online as a Review in Advance on May 29, 2014

The *Annual Review of Neuroscience* is online at neuro.annualreviews.org

This article's doi: 10.1146/annurev-neuro-071013-014119

Keywords

striatum, orbitofrontal cortex, amygdala, insular cortex, mood, stress

Abstract

Although the prevalent view of emotion and decision making is derived from the notion that there are dual systems of emotion and reason, a modulatory relationship more accurately reflects the current research in affective neuroscience and neuroeconomics. Studies show two potential mechanisms for affect's modulation of the computation of subjective value and decisions. Incidental affective states may carry over to the assessment of subjective value and the decision, and emotional reactions to the choice may be incorporated into the value calculation. In addition, this modulatory relationship is reciprocal: Changing emotion can change choices. This research suggests that the neural mechanisms mediating the relation between affect and choice vary depending on which affective component is engaged and which decision variables are assessed. We suggest that a detailed and nuanced understanding of emotion and decision making requires characterizing the multiple modulatory neural circuits underlying the different means by which emotion and affect can influence choices.

Contents

BEYOND DUAL SYSTEMS IN THE MIND AND BRAIN: A MODULATORY ROLE FOR EMOTION IN DECISION MAKING

The prevalent view of the role of emotion in decision making in economics, psychology, and, more recently, neuroscience is the dual systems approach. In economics, choices have been characterized as relying on either System 1 or System 2, with emotion as one of the factors contributing to the more automatic, less deliberative system 1 (Kahneman 2011). In psychology, the terms "hot" versus "cool" have been used to describe decisions driven by affect or not (e.g., Figner et al. 2009; see sidebar, Dual-Process Theories). In neuroscience, brain-imaging research has been used to argue that the human mind is "vulcanized" such that our highly developed prefrontal cortex can be used to overcome the emotional or limbic responses that may sway us to perform irrationally (Cohen 2005). These modern dual-system accounts of the relation between emotion and decision making have a long history. The idea that opposing forces of emotion and reason compete in the human mind is prevalent in Western thought, highlighted by a range of scholars including philosophers such as Plato and Kant and the father of psychoanalysis, Sigmund Freud (Peters 1970). The intuitive nature of this distinction is also apparent in the everyday language used when reflecting on decisions as being made with the heart or the head.

The notion that there are distinct systems for emotion and cognition was also apparent in early theories of brain anatomy. Building on earlier work by Paul Broca and James Papez, Paul Maclean introduced the term "limbic system" in 1952 to describe the phylogenetically older brain regions that lined the inner border of the cortex that he proposed were responsible for basic emotional responses (Lambert 2003) (see **Figure 1a**). The limbic system quickly became known as the emotional center of the brain, with the neocortex underlying higher cognitive functions, including reason. This early theory was highly influential in its time; however, as basic research into neuroanatomy and structure-function relationships progressed over the past several decades, the limbic system concept did not hold up. For example, a region of the neocortex,

DUAL-PROCESS THEORIES

Dual-process theories are a dominant class of theories of human decision making that argue for the existence of two separate, opposing decision systems. Choice results from competition between these two systems: One is generally emotional, fast, automatic, "hot," and/or subconscious, whereas the other is cognitive, slow, deliberative, "cool," and/or explicit. Some specific theories using this structure include the following.

- System 1/System 2: Emotion is considered part of System 1, which "operates automatically and quickly, with little or no effort and no sense of voluntary control. System 2 allocates attention to the effortful mental activities that demand it, including complex computations. The operations of System 2 are often associated with the subjective experience of agency, choice, and concentration" (Kahneman 2011, pp. 20–21).
- Hot/cool: In this theory, positing a hot emotional system and a cool cognitive system, "risk taking [is] the result of a competition between two neural systems...Affective processing is spontaneous and automatic, operates by principles of similarity and contiguity, and influences behavior by affective impulses.... The cognitive-control system... is the neural basis of deliberative processing, which is effortful, controlled, and operates according to formal rules of logic... [and] it is the neural basis of inhibitory control, a mechanism that can block affective impulses and therefore enables deliberative decision making even in affect-charged situations" (Figner et al. 2009, p. 710).

the orbitofrontal cortex, is important in emotion (Damasio 2005), and the hippocampus, a key component of the limbic system, is critical for the basic cognitive function of memory (Squire 2004). Some researchers have tried to modify the limbic system concept to more accurately reflect the emotion/cognition division (e.g., Cohen 2005, Rolls 2013) (see **Figure 1b**). However, as our understanding of both the complexity of emotion and its underlying neural systems expands, there is clearly no clean delineation between brain regions underlying emotion and cognition. There is no clear evidence for a unified system that drives emotion. Thus affective neuroscientists and neuroanatomists have suggested that the limbic system concept is no longer useful and should be abandoned to facilitate the development of a more complete and detailed understanding of the representation of emotion in the brain (see LeDoux 2000 for a discussion).

Given that affective neuroscientists now generally view the limbic system concept as obsolete, perhaps it is also time to revisit the usefulness of dual system models to characterize the relation between emotion and decision making. Without a clear instantiation of an emotion system in the brain, it is difficult to conceive of a psychological model that relies on such a system. The importance of neural instantiations for psychological theories has become increasingly apparent as the discipline of cognitive neuroscience evolves. When examining other cognitive functions, such as memory, attention, and perception, a more fine-grained analysis of specific brain circuitries underlying the relation between factors indicative of emotion and those of cognition has emerged. This research suggests a modulatory role for emotion's influence on cognition, and vice versa (see Phelps 2006 for a review). Translating this modulatory view to the study of decision making suggests that affective processes may influence a primary factor underlying choice behavior: the computation of subjective value.

In this review, we explore a range of means by which affective factors may influence choices and highlight investigations of the neural circuitry mediating emotion's modulation of decision making. One challenge in approaching this literature is the recognition that emotion is not a unitary construct, but rather a compilation of component affective processes. Although the precise nature of these component processes is a topic of theoretical debate that goes beyond the scope of this

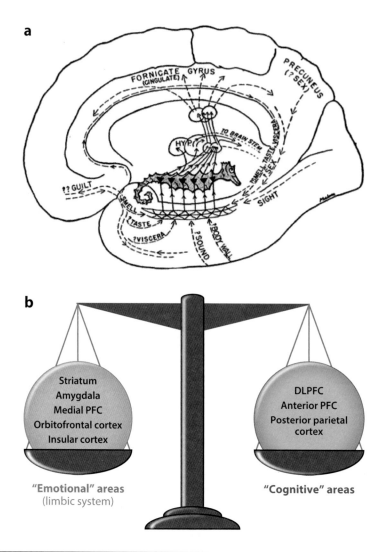

Figure 1

(*a*) The limbic system, centered on the hippocampus, as conceptualized by MacLean (1949). The limbic system concept, as an integrated brain circuit for emotion, has not been supported by more recent neuroanatomical evidence and investigations of brain function (LeDoux 2000). Panel reproduced with permission from Lippincott Williams & Wilkins. (*b*) Dual-process theories of emotion and decision making suggest that choices reflect the outcome of a competition between systems. In this framework, emotional or limbic areas (updated to include other regions; Cohen 2005) are associated with automatic, often irrational choices, whereas cognitive areas are implicated in more deliberative, rational decision making. DLPFC, dorsolateral prefrontal cortex; PFC, prefrontal cortex.

review (e.g., Ekman & Davidson 1994, Scherer 2000, Barrett 2006), a few general distinctions have emerged that will aid in characterizing this literature. The term affect is generally used as the overarching term to describe this collection of component processes, whereas the term emotion refers to a discrete reaction to an internal or external event that can yield a range of synchronized responses, including physiological responses (e.g., flight or fight), facial and/or bodily expressions, subjective feelings, and action tendencies, such as approach or avoid. Although these reactions are synchronized in response to an event, they may not all be present and their intensity can vary independently. The discrete nature of these emotional reactions is their distinct quality relative to other affective components, although how long they last, from relatively transient to much longer, can vary depending on the nature of the eliciting event and intensity. For the purposes of this review, we differentiate a discrete emotional reaction from a stress response, which is characterized by specific physiological and neurohormonal changes that disrupt homeostasis resulting from a real, imagined, or implied threat (Ulrich-Lai & Herman 2009). The impact of these neurohormonal changes lasts beyond the stressor itself (Dickerson & Kemeny 2004) and induces a relatively lasting affective state. Another lasting affective state is mood, which is defined predominantly by subjective feelings that are not necessarily linked to a specific event. Like emotions, moods can elicit action tendencies. Although these affective processes provide a basis for our characterization of the existing literature on the neural basis of affect and decision making, they do not capture the range of affective experience that may be relevant to understanding decision processes more broadly (see Scherer 2000, 2005 for a more in-depth discussion of component process models of emotion).

One could argue that choice itself is indicative of an affective response because it signals an evaluation of preference, motivation, or subjective value assigned to the choice options. The view that value and emotion are inherently intertwined is more common among psychologists and neuroscientists (e.g., Rolls & Grabenhorst 2008) than economists (e.g., Kahneman 2011), but for this review we focus on evidence that independent affective components modulate the assessment of subjective value and the decision. With this aim in mind, we limit our discussion largely to studies that have explicitly measured and/or manipulated a factor commonly linked to emotion or affect. We do not include studies in which emotion is inferred from choices or blood-oxygenation-level-dependent (BOLD) signal because there is limited evidence of unique BOLD patterns linked to specific affective components (Phelps 2009), a problem commonly known as reverse inference (Poldrack 2006). Given that our primary interest is to characterize the current literature examining the neuroscience of emotion's modulation of decision making, we discuss the growing psychological literature on affect and decision making (Lerner et al. 2015) only if there is also some link to the underlying neural circuitry.

Two broad categories of research explore the modulation of decision making by emotion or affect. The first explores how a decision is altered when it occurs during a specific affective state. In this class of studies, the affective state is incidental to the choice itself but nevertheless modulates the decision. The second class of studies examines how the emotional reaction elicited by the choice itself is incorporated into the computation of subjective value. In the final section, we examine how processes that alter emotion can change choices, highlighting the reciprocal, modulatory relationship between emotion and decision making.

AFFECT CARRYOVER: HOW INCIDENTAL AFFECT MODULATES DECISIONS

One means by which emotion can influence choices is through incidental affect. Incidental affect is a baseline affective state that is unrelated to the decision itself. Studies investigating incidental affect trigger an affective state prior to the decision-making task and evaluate its impact on choices. Below we describe two incidental affective states that have been shown to influence decisions.

Stress

Although stress is a term that is widely used to mean many different things, one clear neurobiological indication of a stress reaction is activation of the hypothalamic-pituitary-adrenal (HPA) axis. Stress reactions are also accompanied by sympathetic nervous system arousal, which can be more transient, but HPA axis activation results in a cascade of neuroendocrine changes, most notably glucocorticoid release, that can have a relatively lasting impact. These neurohormonal changes influence function in several brain regions implicated in decision making (see Arnsten 2009, Ulrich-Lai & Herman 2009 for reviews) (see **Figure 2***a*).

Several studies examining stress and decision making highlight the impact of stress on the prefrontal cortex (PFC). Even relatively mild stress can impair performance on PFC-dependent tasks, such as working memory, owing to the negative impact of catecholamines and glucocorticoids on PFC function (Arnsten 2009). The impact of stress on other brain regions varies. For example, although mild stress can enhance hippocampal function, more intense and/or prolonged stress impairs the hippocampus (McEwen 2007). In contrast, performance on striatal-dependent tasks is often enhanced with stress (Packard & Goodman 2012), and dopaminergic neurons in the ventral tegmental area and the striatum show transient and lasting stress-specific responses (Ungless et al. 2010). In addition, amygdala function is generally enhanced with stress, and the amygdala modulates some stress effects on the hippocampus, striatum, and PFC (e.g., Roozendaal et al. 2009). Given the uneven effects of stress on the neural circuits that mediate decision making, the impact of stress may vary depending on the intensity and duration of the stressor, as well as on the specific variables assessed in the decision-making task.

The PFC is proposed to play a role in goal-directed decisions, whereas the striatum is generally linked to choices based on habits (Balleine & O'Doherty 2010). To explore this trade-off between PFC- and striatal-mediated choices under stress, Dias-Ferreira and colleagues (2009) examined how chronic stress affected later performance on a devaluation task in rodents. Devaluation tasks assess whether choices are habitual or directed toward a reinforcement goal by altering the value of the reward. If reducing the value of the reward changes behavior, the task is said to be goal directed, whereas if devaluing the reward does not alter behavior, the task is habitual. After training rats to press a lever to receive a food reward, the rats were fed the food to satiety, thus devaluing subsequent reward presentations. Rats who had not been previously stressed reduced their response rate, reflecting the devalued reward outcome. In contrast, stressed rats failed to modify their behavior following devaluation, consistent with habitual responding. Dias-Ferreira et al. (2009) found that chronic, restraint stress resulted in neuronal atrophy of the medial PFC and dorsal medial striatum, a circuit known to be involved in goal-directed actions. They also observed neuronal hypertrophy of the dorsal lateral striatum, a region linked to habit learning.

In humans, it is not possible to experimentally induce chronic stress and observe its long-term consequences, but several techniques have been used to induce acute, mild stress that reliably results in HPA-axis activation (see Dickerson & Kemeny 2004). These acute, mild stressors have been shown to impair performance on PFC-dependent tasks and reduce PFC BOLD responses (e.g., Qin et al. 2009, Raio et al. 2013). Using a devaluation paradigm similar to that described above, Schwabe & Wolf (2009) found that acute stress yields a similar shift from goal-directed to habitual choices in humans. In a follow-up series of studies, Schwabe and colleagues (2010, 2011) administered drugs targeting glucocorticoid and noradrenergic activity to explore the neurohormonal changes that might underlie this stress-induced shift to habitual actions. They observed that administering both hydrocortisone and an α2-adrenoceptor antagonist (yohimbine), which increases noradrenaline levels in the brain, resulted in the shift to habitual actions typically observed with stress, but neither drug alone was sufficient to do so (Schwabe et al. 2010). Conversely, if stressed participants were administered a beta-adrenergic antagonist (propranolol), they failed to demonstrate the typical shift to habitual actions, despite intact increased cortisol with stress (Schwabe et al. 2011). This observation suggests that both glucocorticoids and noradrenaline are necessary neurohormonal components underlying the shift from goal-directed to habitual actions with stress. These findings are consistent with research in nonhuman animals showing that elevated

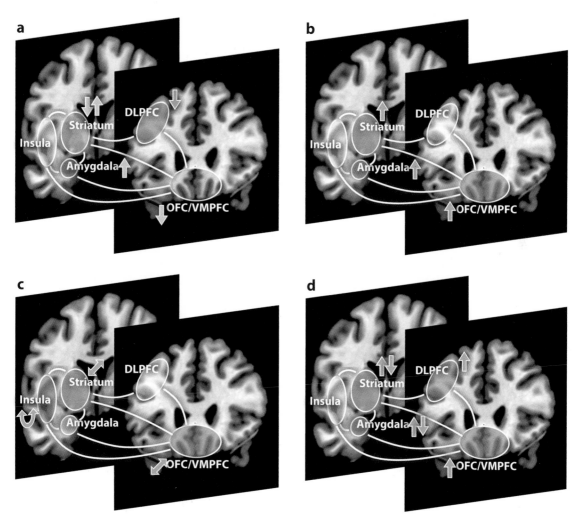

Figure 2

Potential candidates for multiple modulatory neural circuits involved in affect and decision making. (*a*) Decision making under stress. Even mild stress impairs the dorsolateral prefrontal cortex (DLPFC), leading to decreased goal-directed behavior and increased habitual behavior (Schwabe & Wolf 2009, Otto et al. 2013). Stress also impairs the orbitofrontal cortex (OFC)/ventromedial prefrontal cortex (VMPFC) and enhances amygdala function, whereas different subdivisions of the striatum may show increased or decreased reactivity with stress (Arnsten 2009, Roozendaal et al. 2009). (*b*) Emotion contributes to valuation. The amygdala influences the computation of subjective value in the striatum and the OFC/VMPFC and modulates learning from reinforcement (Li et al. 2011, Sokol-Hessner et al. 2012, Rudebeck et al. 2013). (*c*) The relationship between subjective value and BOLD activity. A meta-analysis of fMRI studies (Bartra et al. 2013) shows a linear relationship between subjective value and activity in the OFC/VMPFC and the ventral striatum, whereas the relationship between insula activity and subjective value is U-shaped, suggesting that the insula may contribute to value computation in situations of high arousal or salience. (*d*) Cognitive emotion regulation. The influence of emotion on choice can be altered using cognitive emotion regulation techniques mediated by the DLPFC and VMPFC. Emotional reactions can be increased or decreased with these techniques (Ochsner & Gross 2005), leading to corresponding changes in the amygdala and striatum (Delgado et al. 2008a,b; Sokol-Hessner et al. 2013).

noradrenaline levels during stress alter executive control and PFC function, and glucocorticoids play a role in the exaggeration and persistence of this effect. The impact of these neurohormonal changes on the PFC is mediated through the amygdala (Arnsten 2009).

This balance between PFC and striatal contributions to decision making can also be observed in tasks that tap into model-based and model-free reinforcement learning (Daw et al. 2005). In model-free learning, one learns which choice is beneficial through previous experience with its reinforcing consequences, whereas model-based learning requires a model of the environment that allows one to engage in a series of choices that maximize reward. Theoretical models (Daw et al. 2005) and functional magnetic resonance imaging (fMRI) studies (Gläscher et al. 2010) suggest that although both model-based and model-free decisions engage striatal-based reinforcement learning mechanisms, model-based choices also depend on interactions with the lateral PFC. Using a decision-making task that yields different patterns of choices depending on whether one is using a model-free or model-based strategy, Otto and colleagues (2013) found that stress attenuated model-based, but not model-free, contributions to choice behavior. Relatively high baseline working-memory capacity had a protective effect, attenuating the deleterious effect of stress on choices.

Several studies have examined the impact of stress on tasks of risky decision making, although the nature of these findings varies depending on a number of factors. Porcelli & Delgado (2009) found that stress exacerbated the "reflection effect," which is the tendency to be risk seeking when choosing between possible losses and risk averse when choosing between potential gains. However, other risky decision-making tasks have reported that participants are more risk seeking overall under stress (Starcke et al. 2008) or more risk seeking in the loss domain only (Pabst et al. 2013). In addition, the impact of stress on risky decisions may depend on the level of risk (von Helversen & Reiskamp 2013) and may interact with gender (e.g., Preston et al. 2007, Lighthall et al. 2009). As this series of studies indicates, there are likely several decision and individual difference variables that will need to be disentangled to determine how stress may influence different aspects of risky decisions.

The only brain-imaging study on stress and risky decision making to date (Lighthall et al. 2012) used a task in which participants earn points for inflating virtual balloons but must "cash out" before the balloon explodes or risk losing their points. Consistent with earlier findings (Lighthall et al. 2009), this task demonstrated a gender interaction: Males were more risk seeking, and females less risk seeking, following the stressor. There was also an interaction in BOLD responses: Males in the stress condition showed greater activation in the insula and putamen while making decisions, whereas females showed the opposite pattern. Although the precise roles of these regions in this task is unclear, the insula has been implicated in signaling aversive outcomes and weighing differences in expected value in risky decision making (Clark et al. 2008), whereas the putamen is known to play a role in habitual behavior (Balleine & O'Doherty 2010).

Limited research in other decision-making domains has examined the impact of stress. Studies of intertemporal choice have found that stress exaggerates the tendency to discount future rewards in favor of smaller immediate rewards (Kimura et al. 2013) or that this effect depends on the level of perceived stress (Lempert et al. 2012). Studies of moral decision making find that stress decreases the likelihood of making utilitarian judgments in personal moral decisions (i.e., inflicting harm to maximize good consequences; Youssef et al. 2012) and correlates with egocentric moral decisions (Starcke et al. 2011). Finally, stress results in more prosocial decisions (i.e., more trust and less punishment) but less generosity as well (von Dawans et al. 2012, Vinkers et al. 2013). Most of these studies hypothesize that their findings could be attributed to the impact of stress on executive control and PFC function, although direct evidence of diminished PFC involvement due to stress in these tasks is lacking.

As this emerging research on stress and decision making indicates, a range of processes are tapped in decision-making tasks, and stress has broad, and uneven, effects on brain function. In addition, there are significant individual differences in response to stress and differential effects of chronic, acute, mild, or severe stress. In spite of these caveats, the extensive literature characterizing the impact of stress on brain function can be leveraged to understand one of the means by which an affective state might influence choices. To the extent that we can identify specific neural circuits linked to specific decision variables, such as in goal-directed versus habitual actions and model-based versus model-free decisions, we can start to characterize the distinct impact of stress.

Mood

Although stress responses are often accompanied by negative feelings, mood states are characterized primarily by subjective feelings with little concordant psychophysiological or neurohormonal changes (Scherer 2005). Research on the neural basis of moods is limited by two significant constraints. First, moods are relatively lasting states, which prevents quickly switching from one mood to another—a necessary requirement to detect within-subject differential BOLD responses, which are optimal for fMRI studies. Second, given that the primary measure of a mood state is via subjective report, it is challenging to assess moods in nonhuman animals. However, substantial psychological evidence indicates that moods affect decisions and provides some hints of the neural changes that may mediate these effects.

The influence of mood on the neural systems of decision making has been explored in a social decision-making task, the ultimatum game. In this game, there are two players: a proposer who is given a sum of money to divide with a responder, who can choose whether to accept or reject the proposer's offer. If the responder rejects the offer, both players receive nothing. In theory, the responder should accept any offer because the alternative is nothing at all. However, previous research has shown that offers around 20% of the total sum are rejected approximately half the time, presumably to punish the proposer for an unfair offer (Thaler 1988).

Studies inducing mood states prior to the ultimatum game show that participants in the role of the responder were more likely to reject unfair offers when they were in a sad (Harlé & Sanfey 2007) or disgust mood (Moretti & di Pellegrino 2010). BOLD responses during the ultimatum game were examined in two groups of participants in the role of the responder. One group underwent a sad mood induction procedure prior to scanning, and the other underwent a control, neutral mood induction task (Harlé et al. 2012). As expected, the sad mood group rejected more offers. During the presentation of unfair offers, investigators noted significantly more BOLD activation in the bilateral insula, the ventral striatum, and the anterior cingulate in the sad group relative to the control group. During the presentation of fair offers, there were no group differences in the insula; in the ventral striatum, however, the control group showed greater BOLD reward activity relative to the sad group. BOLD responses in the insula mediated the relationship between self-reported sadness and the tendency to reject unfair offers. In the context of this study, the authors suggested that the insula supports the integration of a negative mood state into the decision process. The findings in the ventral striatum are interpreted as reflecting reduced reward sensitivity when sad. This network of regions is proposed to underlie the infusion of sadness into the choice (Harlé et al. 2012).

Psychological research suggests that the infusion of mood into the computation of subjective value results from the carryover of the general action tendencies elicited by mood states onto the decision process. This proposed carryover effect is known as an appraisal tendency (Lerner et al. 2004). For example, Lerner and colleagues (2004) induced a sad, disgust, or neutral mood and explored its impact on the endowment effect—the phenomenon in which the price one is willing to accept to sell an owned item is greater than the price one would pay to buy the same item. They found that a sad mood reversed the endowment effect (i.e., higher buy prices than sell prices), whereas a disgust mood led to a reduction in both buy and sell prices. It was suggested that sadness is an indication that the current situation is unfavorable, which enhances the appraisal of the subjective value of choice options that change the situation. Disgust, however, is linked to a tendency to move away from or expel what is disgusting, which carries over to a tendency to reduce the subjective value of all items.

Numerous studies have shown that moods also influence risky choices. For example, sad moods can increase preferences toward high-risk options, whereas anxious moods bias preferences toward low-risk options (Raghunathan & Pham 1999). Consistent with this concept, fear results in less risk seeking and anger results in more risk seeking (Lerner & Keltner 2001). Finally, positive moods can exaggerate the tendency to overweigh losses relative to gains (i.e., loss aversion) in risky gambles (Isen et al. 1988), and some of these effects of mood on risky decisions may vary by gender (Fessler et al. 2004). Studies investigating the neural systems of risky decision making have highlighted the roles of the orbitofrontal cortex (OFC) in risk-prediction errors, or in updating assessments of risk (e.g., O'Neill & Schultz 2013), and the insular cortex in the representation of risk (e.g., Knutson & Bossaerts 2007). Both of these regions have also been implicated in the representation of mood states (Lane et al. 1997, Damasio et al. 2000). Although neural evidence has yet to indicate how mood states shift the neural representation of risk assessment, this overlap in the neural circuitry mediating mood and risk provides a starting point for investigations on this topic.

Incidental Affect: Summary and Other Factors

A range of incidental affective states may bias decisions. We have highlighted two such states, stress and mood, which have different effects on choices. Stress results in changes in brain function in several regions that have been implicated in different aspects of the decision process, most notably impaired function of the PFC. To date, we know relatively more concerning stress effects on the brain than we do about how to distinguish different decision variables that engage unique neural circuits. Investigations of mood and decision making, however, are limited by the sparse literature on the neural basis of moods. The best hypothesis at this point is that moods somehow shift neural processing in regions that are involved in the assessment of subjective value, such as the OFC. Our relatively poor understanding of the neural basis of moods is exacerbated by the fact that animal models are intrinsically limited for studying phenomena characterized by subjective states. However, given the extensive psychological research on this topic, and the prevalence of mood states in everyday life, unpacking the neural mechanisms of moods and decisions is critical if we ever hope to achieve a relatively nuanced and rich understanding of human decision making.

The carryover effect of incidental affect on decisions has also been linked to other factors not discussed above because they lacked either the measurement or manipulation of affect or evidence for the underlying neural mechanisms. For instance, studies of Pavlovian-to-instrumental transfer demonstrate that actions occurring in the presence of affective Pavlovian cues are modified consistent with the motivational valence of these cues (i.e., performed with more or less vigor; see Huys et al. 2011). Both the striatum and amygdala are highlighted as regions important in integrating the affective value of the Pavlovian cue with the value of the instrumental action (e.g., Corbit & Balleine 2005, Corbit et al. 2007). A similar line of psychological research, known as affective priming, examines how the presence of an emotional cue, such as an angry or happy face, shifts subsequent choices (Winkielman et al. 2005). These studies suggest that the emotional reaction to the cue carries over to the decision. Although emotional reactions are event driven and discrete and moods are lasting subjective states that do not require an eliciting event, the mechanism by which Pavlovian-to-instrumental transfer and affective priming are proposed to alter choices is similar to the notion of appraisal tendency discussed earlier. That is, emotional reactions, like moods, produce action tendencies that bleed over to the appraisal of the subjective value of concurrent choice options.

Finally, an understanding of the influence of incidental affect on choice behavior would be incomplete without considering how individual variability in baseline affective tendencies may alter decisions. Individuals' affect dispositions vary (Scherer 2005). For instance, some people are generally more anxious and others are more cheery. These traits with an affective flavor can influence choices, much like transient mood states. Just as anxious or fear mood states result in less risk taking, higher trait anxiety is also linked to less risky decisions, perhaps because anxiety results in more negative appraisals of subjective value (see Hartley & Phelps 2012 for a review). Of course, more extreme negative affect dispositions, such as trait anxiety, are linked to psychopathologies that have profound functional consequences, including maladaptive decisions. Accordingly, patients with anxiety disorders are more risk averse than are healthy individuals (Giorgetta et al. 2012) and are also more likely to punish in social decision-making tasks (Grecucci et al. 2013a). Given the clear link between maladaptive decisions and psychopathology, Sharp et al. (2012) proposed that decision science is an important tool to aid in our characterization of a range of psychological disorders.

EMOTION AS VALUE: HOW EMOTIONAL REACTIONS TO THE CHOICE MODULATE DECISIONS

Theories concerning the function of emotions universally highlight the role of emotions in driving actions (e.g., Frijda 2007). A classic example is the fight-or-flight response first characterized by Cannon (1915), in which a potentially threatening event (such as a predator) alters the physiological state to facilitate adaptive action (i.e., quickly escaping). Unlike the influence of incidental stress in biasing concurrent but unrelated choices, described above, in the predator example it is the choice options that evoke the emotional reaction, which in turn drives the choice. In this case, the emotional reaction is an important component of the value computation. Although this example may seem extreme, since we rarely encounter threats to our survival in our everyday lives, the principle applies in more subtle ways in our daily choices. That is, our emotional reactions to choice options or outcomes contribute to the determination of subjective value.

Emotional reactions vary widely in both intensity and quality. The neural circuits mediating the influence of emotion on choices may vary with these emotional qualities, and our understanding of the neural representation of different kinds of emotional reactions is still relatively limited. However, for emotions related to threat, extensive, cross-species research into the underlying neural circuits has been carried out. This literature provides a starting point to explore emotion's modulation of choice (see Phelps & LeDoux 2005 for a review).

The amygdala is a central component of this circuitry and is known to have a critical role in associating aversive, threatening events with neutral cues (i.e., Pavlovian fear conditioning). One of the amygdala's subregions, the lateral nucleus, is the site of synaptic plasticity linking neutral cues and aversive events. The lateral nucleus projects to both the central and basal nuclei. The central nucleus sends signals to the hypothalamus and brain stem nuclei, which mediate the physiological threat response, whereas the basal nucleus projects to the striatum. The striatum helps integrate motivation with action values, and in the presence of a conditioned threat cue, the basal nucleus input is critical for avoidance actions (LeDoux & Gorman 2001). This circuitry, with independent pathways mediating physiological threat reactions and avoidance actions, may also play a broader role in different decision contexts.

In addition to the amygdala's influence on the striatum, the amygdala has reciprocal connections with the OFC, and lesion studies in nonhuman primates have demonstrated that this connectivity with the amygdala contributes to the representation of value in the OFC (Rudebeck et al. 2013). Finally, the amygdala, the OFC, and the striatum share connectivity with the insula, a region also commonly linked to emotion's influence on decisions (e.g., Naqvi et al. 2007).

To determine how and when emotion influences the value computation, it is necessary to both measure and/or manipulate the emotional response and specify the decision variables. How emotion may influence a choice depends not only on the qualities of the emotional reaction, but also on the characteristics of the choice. Below we review studies exploring the neural systems mediating emotion's modulation of subjective value for different types of decision tasks (see **Figure 2b,c**).

Risky Decisions

Risky decisions involve comparing choice options with varying probabilities of losses or gains. One of the first studies that assessed a specific emotion variable in humans and linked it to brain function used a risky decision-making paradigm known as the Iowa gambling task (IGT). The IGT presents participants with four decks of cards: two yielding small gains and losses (safe decks) and two yielding larger gains, but also occasional large losses (risky decks). Participants are asked to select cards sequentially from the different decks to win or lose money. Preferentially choosing from the safe decks results in a more favorable long-term outcome, and healthy control participants learn this through trial and error. In contrast, patients with OFC or amygdala lesions fail to shift their preference to the safe decks over time. Bechara and colleagues (1997, 1999) measured the skin conductance response (SCR), an indication of autonomic nervous system arousal, during choices and found that control participants developed an anticipatory SCR prior to selecting from the risky decks, whereas OFC- and amygdala-lesioned patients did not. The authors proposed that the anticipatory arousal response is a bodily (somatic) signal that steers participants away from less profitable, risky choices, an idea they refer to as the somatic marker hypothesis. Several studies over the years have challenged the primary assumption of the somatic marker hypothesis (Maia & McClelland 2004, Fellows & Farah 2005), a challenge further supported by evidence suggesting that autonomic responses and avoidance actions are driven by separate neural circuits (LeDoux & Gorman 2001). In spite of these caveats, this study was the first to clearly identify some of the neural circuitry mediating the integration of emotion in risky decisions.

Risky decision-making tasks vary widely, and several decision factors that influence choices may come into play. Two decision variables that are often confounded are risk sensitivity and loss aversion. Loss aversion is the tendency to weigh losses more than gains when considering the choice options. Someone who is highly loss averse may also appear to be risk averse, even if she or he is generally risk seeking in choices with minimal potential loss. Using a gambling task that enabled independent assessment of risk sensitivity and loss aversion, Sokol-Hessner and colleagues (2009) found that higher relative SCRs to losses versus gains were linked to greater loss aversion. No relationship was noted between arousal and risk sensitivity. Similarly, greater BOLD signal in the amygdala to losses relative to gains also correlated with loss aversion, but this response was unrelated to risk sensitivity (Sokol-Hessner et al. 2012). Consistent with these imaging results, patients with amygdala lesions show reduced loss aversion overall (DeMartino et al. 2010), and administering a beta-adrenergic blocker (propranolol), which has previously been shown to diminish the amygdala's modulation of memory (Phelps 2006), also reduces loss aversion but does not affect risk sensitivity (Sokol-Hessner et al. 2013). This series of studies provides strong evidence that the amygdala plays a critical role in mediating aversion to losses but that it is not linked to risk tendencies. Given that most risky decision tasks do not independently model loss aversion and risk sensitivity, some observed effects of emotion in risky decision-making tasks may be due to loss aversion and not to risk attitudes per se.

Both the IGT and the task used by Sokol-Hessner and colleagues (2009) engage neural and physiological systems that serve to identify potential negative outcomes, prompting avoidance actions. However, other risky decision tasks have relatively few losses, as illustrated in pay-to-play games similar to slot machines, in which the only loss that occurs is when participants pay to play and do not win. Studies with these tasks have shown that emotional responses, including pleasantness ratings, SCR, and cardiovascular measures, to wins and near misses predict gambling propensity, including probable pathological gambling (Lole et al. 2011, Clark et al. 2012). These findings have been linked to increased BOLD activity in the striatum and insula during near misses (Clark et al. 2008, Chase & Clark 2010). In this decision-making context, emotions may drive people to take risky choices and not avoid them.

Another potentially important factor in risky decision-making tasks is whether the risks are known and static or unknown and changing. In dynamic and uncertain environments, the decision maker must learn the risk involved in different choices, and this risk may change over time. In dynamic, risky decision-making tasks, autonomic arousal, assessed via pupil dilation, was associated with more uncertain, exploratory decisions (Jepma & Niewenhuis 2011) and surprising outcomes (Preuschoff et al. 2011). Arousal, as well as amygdala BOLD responses, has been linked to associability (Li et al. 2011), a learning signal related to the unexpected or surprising nature of the cue, which serves to gate updating values from prediction errors coded in the striatum (see also Roesch et al. 2012). These studies suggest that in dynamic choice contexts the emotional response may be a component of ongoing predictions and evaluation necessary for learning.

Risky decision-making tasks vary widely on several dimensions, including the content of the choice (e.g., gains versus losses) and its context (e.g., static or dynamic). As the above studies demonstrate, such factors matter in part because they may shift the modulatory role of emotion: from avoiding bad outcomes to seeking favorable ones, to weighing and incorporating new information in changing environments. Future research will need to dissociate these possible roles and influences of emotion by carefully identifying the decision variables at play, the shared and separate neural circuitry used, and the underlying computations driving choices.

Social Decisions

In our everyday lives, the stimuli most likely to elicit emotional responses are other people. Social decision-making tasks investigate how choices are influenced by social context. For most of these tasks, the shift in decisions is simply due to the presence of another person, even if that person is anonymous. This observation is apparent in the ultimatum game described above in which responder participants routinely reject potential profit to punish the proposer for unfair offers. Not surprisingly, such rejections of financial gain are not observed if the proposer is a computer (van't Wout et al. 2006).

Studies examining the neural basis of rejection in the ultimatum game report an increased BOLD signal in the insula during unfair offers that is correlated with rejection rate (Sanfey et al. 2003). Arousal, assessed via SCR, was also increased during unfair offers and correlated with rejection rates—a pattern not observed when playing against a computer (van't Wout et al. 2006). In this case, the subjective value of the unfair offers was modulated by the social context of the choice. Increased physiological arousal has also been correlated with choice behavior in a social, moral decision-making task, and patients with damage to the ventromedial PFC (VMPFC), including the OFC, show both reduced physiological arousal and diminished impact of the social context on decisions (Moretto et al. 2010). The insula and OFC are two regions linked to emotion's influence on risky decisions (see above), but in these tasks the emotional reaction is driven by the interpersonal nature of the decision, as opposed to other decision variables.

Although the simple presence of another person can evoke an emotional reaction that may influence choices, who that person is may also matter. The influence of individual characteristics in social decision making has been investigated in cross-race interactions. The impact of race group on decisions was examined using the trust game in which a participant must decide whether to invest money with a partner. Trust decisions correlated with nonconscious, negative evaluative race attitudes for Black versus White, such that participants with stronger negative implicit attitudes invested less with Black compared with White partners (Stanley et al. 2011). BOLD responses during this task showed greater amygdala activation for Black versus White partners, scaled for the size of the investment, whereas striatum activation reflected the race-based discrepancy in trust decisions (Stanley et al. 2012). These findings are consistent with a model in which the amygdala codes race-related evaluative information and the striatum integrates this information with the action value.

The impact of social factors on decisions has been demonstrated with many different decision tasks, and there have been numerous investigations of the neural circuitry and neurochemistry mediating these effects (see Rilling & Sanfey 2011, Kubota et al. 2012 for reviews). However, relatively few studies have examined whether the impact of social context on choice is related to emotional responses. Given the emotional salience of other people, it is possible that emotion mediates the influence of many social factors on decisions. Only by assessing emotional reactions during these decision tasks can we start to delineate the impact of emotion evoked by the social situation from other factors that are uniquely social.

Intertemporal Choice

Intertemporal choice tasks measure preferences between options available at different points in time. In general, people tend to prefer immediate rewards to rewards received after a delay, even when the delayed reward is larger. This phenomenon, known as temporal discounting, has been linked to many maladaptive behaviors, including poor retirement savings, obesity, and drug addiction.

Investigations of the neural systems mediating intertemporal choice have reported conflicting results. One study reported greater BOLD responses in the OFC and the striatum during choices with an immediate reward option and greater BOLD signal in the DLPFC related to choosing the delay option (McClure et al. 2004). This BOLD pattern was interpreted as supporting a theory proposed in economics (Laibson 1997) suggesting that immediate rewards engender a greater emotional response, as reflected in the striatum and OFC BOLD responses, whereas choosing the delayed reward requires cognitive control of this emotional impulse, thus engaging the DLPFC. Consistent with this proposed inhibitory role for the DLPFC, Figner and colleagues (2010) found that disrupting DLPFC function through transcranial magnetic stimulation resulted in greater temporal discounting; however, in contrast to this proposed model, so did lesions of the OFC (Sellitto et al. 2010). Another study found that BOLD signal in the VMPFC and striatum correlated with subjective value of both immediate and delayed rewards (Kable & Glimcher 2010), consistent with the known roles for these regions in the representation and updating of value (see Bartra et al. 2013 for review). The investigators suggested that increased BOLD responses to immediate reward options observed in the earlier study were due to the fact that immediate rewards generally had a greater subjective value than did delayed rewards. However, none of these studies assessed emotional responses.

To determine if emotion plays a role in temporal discounting, Lempert and colleagues (2013) measured arousal, as assessed with pupil dilation, during an intertemporal choice task. Surprisingly, emotional arousal did not reliably correlate with the subjective value of either immediate or delayed rewards, but rather this relationship varied depending on the structure of the choice set. Greater arousal responses were observed when rewards were better than expected, regardless of whether those rewards were immediate or delayed. These findings conflict with the model proposed by McClure and colleagues (2004), which suggests that it is the emotional response to the immediate choice that drives discounting, and more closely align with the study by Kable & Glimcher (2007), which proposed a unified neural representation of subjective value of immediate and delayed rewards; both may be influenced by emotion depending on the task environment.

Further support for the notion that both immediate and delayed rewards elicit emotional responses that influence choices comes from studies investigating how altering the emotional salience of the delayed reward increases patience. For example, manipulating the mental representation of a future reward to make it more concrete can change its emotional intensity and the choice. Benoit and colleagues (2011) gave participants a typical intertemporal choice task but asked them to imagine specific ways they could spend the delayed reward in the future. This manipulation increased subjective ratings of vividness and emotional intensity of the future reward and resulted in less temporal discounting. This effect was associated with increased coupling between the VMPFC and the hippocampus. A study using a similar task replicated these behavioral results and found, consistent with Kable & Glimcher (2007), that subjective value for delayed rewards correlated with BOLD signal in the striatum and the OFC, whereas activation of the dorsal anterior cingulate, and its connectivity with the hippocampus and the amygdala, mediated the change in discount rate (Peters & Buchel 2010). These same neural circuits are known to be involved in the future projection of personal events and their modulation by emotion (Sharot et al. 2007).

Studies of intertemporal choice provide a compelling example of the influence of affective neuroscience on decision science: The predominant theory used to explain the tendency to discount future rewards in economics relies on dual systems, one impulsive (emotion) and one that controls these impulses (cognitive control; Laibson 1997). As the discussion above indicates, to the extent that emotion plays a role in this behavior, emotion's contribution varies depending on the choice environment and the task structure. This variability in the role of emotion provides an opportunity for investigators to manipulate task parameters that alter emotion to influence the tendency of subjects to discount future rewards, a topic we discuss in more detail in the next section.

Emotion as Value: Summary and Related Phenomena

In addition to the decision tasks described above, emotion is also thought to contribute to subjective value computation in drug addiction, although the measurement and quantification of emotion and the understanding of the underlying neural mechanisms lag behind theory. For example, intense cue-driven motivation, termed craving, is central to addiction theory (Skinner & Aubin 2010). Cravings are a major factor that contribute to relapse, but their source is complex; studies have variously connected them to the insula (Naqvi et al. 2007), the striatum (Kober et al. 2010), and the PFC (Rose et al. 2011). Nevertheless, understanding the systems that induce these motivational desires will ultimately lead to significant advances in the treatment of such disorders of choice.

A primary function of emotion is to provide a signal to the organism that a stimulus or event may be relevant for present or future survival or well-being (e.g., Frijda 2007). Thus it is not surprising that emotional reactions modulate a range of cognitive functions, such as memory, attention, and perception (Phelps 2006). It is also not surprising that part of the calculation of the value of decision options should include the nature of the emotional response elicited by those options or potential outcomes. How this occurs, however, varies depending on the decision variables assessed and the specific emotional reaction. As our review of this literature indicates, there is likely a collection of neural circuits underlying emotion's modulation of the value calculation.

Across studies of the neural basis of decision making, the OFC/VMPFC and the striatum are cited as necessary for the coding of subjective value; the striatum is specifically linked to updating values from reinforcement (prediction errors) via dopaminergic projections from the ventral tegmental area (e.g., Bartra et al. 2013). The studies outlined above examining the impact of emotion also implicate, in addition to these regions, the insula and amygdala. The insula is a large region linked to numerous functions relevant to decision making, including the anticipation of pain (Ploghaus et al. 1999) and monetary loss (Knutson & Bossaerts 2007), as well as the representation of disgust (Phillips et al. 1997) and physiological arousal (Critchley et al. 2000). A recent meta-analysis of fMRI studies examining the coding of subjective value found, not surprisingly, that the OFC/VMPFC and the ventral striatum emerged as two regions with BOLD responses that positively correlate with subjective value. In contrast, the insula, along with some other striatal regions and the dorsomedial PFC, showed greater BOLD responses for both more positive or negative subjective value. Bartra et al. (2013) suggest that the insula may integrate emotional salience or arousal linked to the decision variables into the value computation, regardless of its valence. As mentioned above, connectivity between the amygdala and the ventral striatum is critical for enabling avoidance behavior to acquired threats (LeDoux & Gorman 2001), and the amygdala contributes to value coding in the OFC (Rudebeck et al. 2013). The amygdala may play a role in avoidance across a range of decision tasks (e.g., Stanley et al. 2012, Sokol-Hessner et al. 2013), as well as in modulating learning from both positive and negative reinforcement more broadly (e.g., Roesch et al. 2012; Murray & Rudebeck 2013) (see **Figure 2b,c**). The limited research to date on the integration of emotion into value computation is starting to yield a network of regions, but our understanding of precisely how these regions interact in more complex human decision-making tasks is still relatively unclear.

CHANGING AFFECT, CHANGING CHOICES

Clinical interventions for a range of psychopathologies are focused on changing affect. Outside the clinic, the ability to regulate the appropriateness of emotional responses to circumstances is a major component of healthy, adaptive social behavior and well-being. Although we often describe emotions as reactions to environmental stimuli that are beyond our control, affective scientists have long recognized the fluidity of our emotional lives and our ability to alter or determine our emotions. A major focus of basic research in affective neuroscience over the past decade has been to understand how emotions can be modified and how we can utilize this flexibility of emotion to develop more effective clinical interventions or more satisfying and healthy lives (Hartley & Phelps 2010, Davidson & Begley 2012).

To the extent that affect and emotions are incorporated into the assessment of subjective value, changing emotions should also change choices. Although several techniques have been used to change emotion in the laboratory across species (see Hartley & Phelps 2010 for a review), and all these techniques are presumed to influence later choices, only a few have been implemented directly during decision-making tasks to assess how choices are altered. Below we review the research examining one such technique and highlight some other potential mechanisms for future investigation (see **Figure 2d**).

Cognitive Emotion Regulation

The common wisdom that one can see the glass as half full or half empty captures the essence of cognitive emotion regulation. Our emotional reactions are determined, in part, by how we appraise or interpret the circumstance or event (Scherer 2005). Although some individuals may have a general tendency to see the world in a positive or negative light, the ability to shift emotion through changing one's interpretation of an event, known as reappraisal, can also be taught and consciously applied. In a typical reappraisal task, the participant is asked to think about the stimulus differently to reduce its negative emotional consequences.

Many studies have investigated the neural systems that mediate the cognitive regulation of negative emotions as assessed through subjective reports or physiological responses (see Ochsner & Gross 2005). They typically report increased DLPFC BOLD responses during regulation versus attend conditions accompanied by decreased amygdala activation. The DLPFC is proposed to implement the executive control needed to actively reinterpret the stimulus during reappraisal, whereas the amygdala is involved in the expression of the emotional response. There is relatively sparse direct connectivity between the DLPFC and amygdala, so it is unlikely that the DLPFC directly influences amygdala function but rather does so through more ventral PFC regions. The VMPFC is known to have reciprocal connections with the amygdala that inhibit emotional reactions following extinction learning in Pavlovian fear-conditioning tasks, and it is proposed to mediate the influence of the DLPFC on the amygdala (Delgado et al. 2008b); however, other studies have suggested that the ventrolateral PFC (VLPFC) plays this role (Buhle et al. 2014). This DLPFC-VMPFC/VLPFC-amygdala circuitry is thought to underlie the cognitive control of diminishing negative emotional reactions, but it may also play a role in increasing negative affect depending on the reappraisal strategy (Otto et al. 2014). Emotion regulation strategies can also be employed to reduce arousal associated with anticipated monetary reward. These strategies engage overlapping regions of the DLPFC and VMPFC and yield decreased BOLD reward responses in the striatum (Delgado et al. 2008a). A similar circuitry has been implicated in the cognitive control of cravings (Kober et al. 2010).

In a risky decision-making task, a reappraisal strategy altered both arousal and choices. As described earlier, Sokol-Hessner and colleagues (2009) found that the relative SCR response to losses relative to gains correlated selectively with loss aversion but was unrelated to risk sensitivity. A similar pattern was observed for amygdala BOLD signal (Sokol-Hessner et al. 2013). In a variation of this task, participants were instructed to reappraise the significance of the choice by thinking of it as one of many, or to "think like a trader" building a portfolio. Using this strategy reduced the SCR to losses, and this reduction was correlated with diminished loss aversion, with no effect on risk sensitivity (Sokol-Hessner et al. 2009). Mirroring the SCR results, reduced amygdala BOLD responses to losses during regulation also correlated with a reduction in loss aversion. In contrast, baseline BOLD responses in the DLPFC, VMPFC, and striatum increased with regulation (Sokol-Hessner et al. 2013). These findings suggest that using a reappraisal strategy to change emotion and choices engages the same neural circuitry that is observed in more typical emotion regulation tasks. A similar reappraisal strategy that either emphasized or de-emphasized the importance of each individual choice was found to both increase and decrease subjective value in a risky decision task (Braunstein et al. 2014). In addition, in an intertemporal choice study described above, reframing the interpretation of a future reward resulted in more patience (Benoit et al. 2011).

Emotion regulation strategies have also been used to change the tendency to punish in the ultimatum game. Van't Wout and colleagues (2010) asked participants to play the ultimatum game while utilizing a cognitive emotion reappraisal strategy. In the responder role, participants who reappraised the motivations of the proposer in suggesting an unfair offer were less likely to reject it. This cognitive emotion regulation manipulation carried over to future choices. When the participants were subsequently put in the proposer role, they were less likely to propose unfair offers. In a follow-up fMRI study, participants in the responder role were asked to imagine either negative intentions of the proposer or positive intentions. Relative to a baseline condition, these reappraisal strategies resulted in rejecting more or fewer unfair offers, respectively, and subjective emotional responses varied as well. Consistent with previous research, activation of the insula predicted the rejection of unfair offers (Sanfey et al. 2003, Harlé et al. 2012), and the regulation strategies resulted in both increased insula BOLD responses with the negative intention strategy and decreased BOLD signal with the positive intention strategy. As expected, given the general emotion regulation circuitry outlined above, the DLPFC showed increased activation during both reappraisal conditions relative to baseline (Grecucci et al. 2013b).

As these studies indicate, cognitive emotion regulation techniques are flexible strategies that can rapidly change emotional reactions. The reappraisal strategies described above were adapted to the specific decision situation but had the same effect of altering the emotional response and modulating decisions, and they engaged typical cognitive emotion regulation regions. This confluence of evidence provides strong support for the notion that emotion is a critical component of the assessment of subjective value in these tasks because changing emotion also changed the choice.

Changing Affect: Summary and Other Potential Techniques

The notion that changing affect alters decisions was also demonstrated in the incidental affect manipulations described above. In those studies, inducing stress or a mood in the laboratory changed choices. In contrast, cognitive emotion regulation techniques alter choices by changing the appraisal of the decision variables that elicit emotional reactions. These techniques are powerful because they are flexible, can alter emotion in different ways, and can be quickly acquired and utilized without changing the situation. However, they require an effortful application of the strategy, which may not always be ideal. With practice, these strategies may become more automatic and less deliberate. Consistent with this, novice stock traders have demonstrated more physiological arousal to volatility in the stock market, a finding attributed to loss aversion, as compared with more senior stock traders, who show less arousal and better choices (Lo & Repin 2002), perhaps resulting from a broader perspective on market volatility gained from experience.

Although the flexibility of cognitive emotion regulation techniques can be an advantage, there are also some potential disadvantages. Cognitive emotion regulation strategies are less successful in stressful situations (Raio et al. 2013), perhaps owing to their dependence on the DLPFC. In addition, when emotional reactions consistently result in maladaptive choices, it may be useful to have a technique that leads to a more lasting change. Affective neuroscience has identified a few such strategies, but their impact on emotional reactions linked to decision making has not yet been widely investigated. For example, extinction training has been used to reduce acquired affective responses by repeatedly presenting the cue without the associated reinforcement. Although this technique can be effective, it leaves the original association intact; thus the unwanted affective response may return. To induce a more lasting change in learned associations underlying emotional responses and instrumental actions, researchers have recently investigated techniques that change the original associative, affective memory by altering its re-storage after retrieval or reconsolidation. Our understanding of reconsolidation mechanisms and how to target these processes in humans is still in its infancy, but this technique may lead to exciting advances in reducing the impact of maladaptive emotional reactions on choices (see Hartley & Phelps 2010 for a review).

Finally, an interesting twist in investigations examining the relation between emotion and decision making is that choices themselves can alter emotions. For example, animals given the opportunity to learn to avoid shocks show a lasting benefit, exhibiting diminished fear responses and faster and more robust extinction in subsequent tasks in which they do not have control over the shock reinforcer (Maier & Watkins 2010, Hartley et al. 2014). This research suggests that this persistent impact of choice on future threat reactions results from alterations in the brain stem–prefrontal–amygdala circuitry underlying the generation and control of learned threat associations (Maier & Watkins 2010). In humans, the opportunity for choice enhances subjective affective ratings of choice options and concurrently increases BOLD reward responses in the striatum (Leotti & Delgado 2011, 2014). Psychological theories have emphasized the importance of perception of control over one's environment on well-being (e.g., Bandura et al. 2003), as well as the impact of choices on preferences (e.g., Festinger 1957). Studies examining the neural basis of the impact of choice on emotional reactions and preferences (e.g., Sharot et al. 2009) are starting to provide a neurobiological framework for these psychological findings.

To the extent that affect and emotion influence choices, changing affective responses will alter our decisions. The emerging research on techniques to change affect shows that a range of mechanisms to modify emotions can be differentially applied in different decision contexts. Some are flexible and rapid, such as cognitive emotion regulation, and others are more lasting, such as targeting reconsolidation. In addition, choices themselves can change affect, which in theory should change subsequent choices. If we can discover and characterize more effective means to alter emotion, we should be able to harness these techniques to help optimize decisions.

MOVING PAST DUAL SYSTEMS TO MULTIPLE MODULATORY NEURAL CIRCUITS

In this overview of the current neuroscience literature exploring the relationship between affect and decision making, we have attempted to identify the neural circuits that mediate this interaction. What is emerging is clearly incompatible with the notion of two systems. Rather the literature suggests that there are multiple neural circuits underlying the modulation of decision making by emotion or affect. As our breakdown of affective components and decision tasks demonstrates, the specific neural circuits involved vary depending on which affective component is engaged and which decision variables are assessed. Thus, we suggest an alternative approach to understand the relation between emotion and decision making, which entails characterizing and identifying the multiple neural circuits underlying the different means by which emotion and affect influence decisions (see also Sanfey & Chang 2008).

Of course, this multiple modulatory neural circuits approach is not nearly as parsimonious as the dual systems account, and one could argue that dual systems is simply a rough and useful heuristic to characterize the role of emotion in decision making. Although referring to the heart and head may be useful in thinking about some types of decisions outside the laboratory, the reference to dual systems as the primary psychological and neural theory for understanding the relationship between emotion and decisions is still relatively common in the scientific literature (e.g., Cohen 2005, Greene 2007, Figner et al. 2009, Reyna & Brainerd 2011, Paxton et al. 2012). Much as some researchers have suggested that scientists abandon the limbic system concept because its continued use impedes progress in understanding the detailed and complex neural basis of affect (LeDoux 2000), we argue that the repeated reference to dual systems of emotion and reason in research on decision making potentially limits scientific advances by discouraging investigations that capture the detailed and nuanced relationships between unique aspects of affect and choices. Furthermore, it suggests to the layperson and scientists in other disciplines that the intuitive and historical notion of competing forces of emotion and reason is based on scientific fact. Perpetuation of this idea may dampen enthusiasm for efforts to further explore the complex interactions of affect and decision making and may result in the development of potentially misguided or nonoptimal techniques to inhibit emotion in order to promote rational decision making.

Despite its complexity, our proposed conceptualization of the relationship between affect and decision making begins to capture the subtleties involved in understanding their interaction. Both affect and decision making are general terms that describe a collection of factors and processes, only some of which are explored above. Investigating affect and emotion is challenging by itself, both because manipulating and measuring affect in the laboratory is difficult and because there is debate about how best to characterize affective variables. Differentiating the collection of unique variables that influence choice in any given situation is also challenging for decision science. In spite of these caveats, initial attempts to measure or manipulate affective components and to relate them to specific aspects of decision tasks have yielded exciting advances. As the disciplines of affective neuroscience and neuroeconomics advance, we can build on this progress to further characterize the multiple neural circuits that mediate the modulatory relationship between emotion and decision making.

DISCLOSURE STATEMENT

The authors are not aware of any affiliations, memberships, funding, or financial holdings that might be perceived as affecting the objectivity of this review.

ACKNOWLEDGMENTS

The authors thank Catherine Stevenson, Jackie Reitzes, and Catherine Hartley for assistance with manuscript preparation and Sandra Lackovic for assistance with figure preparation. This work was partially funded by a grant from the National Institutes of Health (AG039283) to E.A.P.

LITERATURE CITED

Arnsten AF. 2009. Stress signaling pathways that impair prefrontal cortex structure and function. *Nat. Rev. Neurosci.* 10(6):410–22

Balleine BW, O'Doherty JP. 2010. Human and rodent homologies in action control: cortico-striatal determinants of goal-directed and habitual action. *Neuropsychopharmacology* 35(1):48–69

Bandura A, Caprara GV, Barbaranelli C, Gerbino M, Pastorelli C. 2003. Role of affective self-regulatory efficacy in diverse spheres of psychosocial functioning. *Child Dev.* 74(3):769–82

Barrett LF. 2006. Are emotions natural kinds? *Perspect. Psychol. Sci.* 1(1):28–58

Bartra O, McGuire JT, Kable JW. 2013. The valuation system: a coordinate-based meta-analysis of BOLD fMRI experiments examining neural correlates of subjective value. *NeuroImage* 76:412–27

Bechara A, Damasio H, Damasio AR, Lee GP. 1999. Different contributions of the human amygdala and ventromedial prefrontal cortex to decision-making. *J. Neurosci.* 19(13):5473–81

Bechara A, Damasio H, Tranel D, Damasio AR. 1997. Deciding advantageously before knowing the advantageous strategy. *Science* 275:1293–95

Benoit RG, Gilbert SJ, Burgess PW. 2011. A neural mechanism mediating the impact of episodic prospection on farsighted decisions. *J. Neurosci.* 31:6771–79

Braunstein ML, Herrera SJ, Delgado MR. 2014. Reappraisal and expected value modulate risk taking. *Cogn. Emot.* 28(1):172–81

Buhle JT, Silvers JA, Wager TD, ONyemekwu C, Kober H, et al. 2014. Cognitive reappraisal of emotion: A meta-analysis of human neuroimaging studies. *Cereb. Cortex.* In press

Cannon WB. 1915. *Bodily Changes in Pain, Hunger, Fear and Rage: An Account of Recent Researches into the Function of Emotional Excitement*. New York: D. Appleton

Chase HW, Clark L. 2010. Gambling severity predicts midbrain response to near-miss outcomes. *J. Neurosci.* 30(18):6180–87

Clark L, Bechara A, Damasio H, Aitken MR, Sahakian BJ, Robbins TW. 2008. Differential effects of insular and ventromedial prefrontal cortex lesions on risky decision-making. *Brain* 131:1311–22

Clark L, Crooks B, Clarke K, Aitken MR, Dunn BD. 2012. Physiological responses to near-miss outcomes and personal control during simulated gambling. *J. Gambl. Stud.* 28(1):123–37

Cohen JD. 2005. The vulcanization of the human brain: a neural perspective on interactions between cognition and emotion. *J. Econ. Perspect.* 19:3–24

Corbit LH, Balleine BW. 2005. Double dissociation of basolateral and central amygdala lesions on the general and outcome-specific forms of Pavlovian-instrumental transfer. *J. Neurosci.* 25(4):962–70

Corbit LH, Janak PH, Balleine BW. 2007. General outcome-specific forms of Pavlovian-instrumental transfer: the effect of shifts in motivational state and inactivation of the ventral tegmental area. *Eur. J. Neurosci.* 26:3141–49

Critchley HD, Elliot R, Mathias CJ, Dolan RJ. 2000. Neural activity relating to generation and representation of galvanic skin conductance responses: a functional magnetic resonance imaging study. *J. Neurosci.* 20(8):3033–40

Damasio AR. 2005. *Descartes' Error: Emotion, Reason and the Human Brain*. New York: Penguin

Damasio AR, Grabowski TJ, Bechara A, Damasio H, Ponto LL, et al. 2000. Subcortical and cortical brain activity during the feeling of self-generated emotions. *Nat. Neurosci.* 3(10):1049–56

Davidson RJ, Begley S. 2012. *The Emotional Life of Your Brain: How Its Unique Patterns Affect the Way You Think, Feel, and Live—and How You Can Change Them*. New York: Hudson Street

Daw ND, Niv Y, Dayan P. 2005. Uncertainty-based competition between prefrontal and dorsolateral striatal systems for behavioral control. *Nat. Neurosci.* 8(27):1704–11

De Martino B, Camerer CF, Adolphs R. 2010. Amygdala damage eliminates monetary loss aversion. *Proc. Natl. Acad. Sci. USA* 107(8):3788–92

Delgado MR, Gillis MM, Phelps EA. 2008a. Regulating the expectation of reward via cognitive strategies. *Nat. Neurosci.* 11:880–81

Delgado MR, Nearing KI, LeDoux JE, Phelps EA. 2008b. Neural circuitry underlying the regulation of conditioned fear and its relation to extinction. *Neuron* 59(5):829–38

Dias-Ferreira E, Sousa JC, Melo I, Morgado P, Mesquita AR, et al. 2009. Chronic stress causes frontostriatal reorganization and affects decision-making. *Science* 325:621–25

Dickerson SS, Kemeny ME. 2004. Acute stressors and cortisol responses: a theoretical integration and synthesis of laboratory research. *Psychol. Bull.* 130:355–91

Ekman P, Davidson RJ. 1994. *The Nature of Emotion: Fundamental Questions*. New York: Oxford Univ. Press

Fellows LK, Farah MJ. 2005. Different underlying impairments in decision-making following ventromedial and dorsolateral frontal lobe damage in humans. *Cereb. Cortex* 15:58–63

Fessler DMT, Pillsworth EG, Flamson TJ. 2004. Angry men and disgusted women: an evolutionary approach to the influence of emotion on risk-taking. *Organ. Behav. Hum. Decis. Process.* 95:107–23

Festinger L. 1957. *A Theory of Cognitive Dissonance*. Stanford, CA: Stanford Univ. Press

Figner B, Knoch D, Johnson EJ, Krosch AR, Lisanby SH, et al. 2010. Lateral prefrontal cortex and self-control in intertemporal choice. *Nat. Neurosci.* 13:538–39

Figner B, Mackinley RJ, Wilkening F, Weber EU. 2009. Affective and deliberative processes in risky choice: age differences in risk taking in the Columbia Card Task. *J. Exp. Psychol. Learn. Mem. Cogn.* 35(3):709–30

Frijda NH. 2007. *The Laws of Emotion*. Mahwah, NJ: Lawrence Erlbaum

Giorgetta C, Grecucci A, Zuanon S, Perini L, Balestrieri M, et al. 2012. Reduced risk-taking behavior as a trait feature of anxiety. *Emotion* 12(6):1373–83

Gläscher J, Daw N, Dayan P, O'Doherty JP. 2010. States versus rewards: dissociable neural prediction error signals underlying model-based and model-free reinforcement learning. *Neuron* 66(4):585–95

Greene JD. 2007. Why are VMPFC patients more utilitarian? A dual-process theory of moral judgment explains. *Trends Cogn. Sci.* 11(8):322–23

Grecucci A, Giorgetta C, Brambilla P, Zuanon S, Perini L, et al. 2013a. Anxious ultimatums: how anxiety disorders affect socioeconomic behaviour. *Cogn. Emot.* 27(2):230–44

Grecucci A, Giorgetta C, Van't Wout M, Bonini N, Sanfey AG. 2013b. Reappraising the ultimatum: an fMRI study of emotion regulation and decision making. *Cereb. Cortex* 23(2):399–410

Harlé K, Sanfey AG. 2007. Incidental sadness biases social economic decisions in the ultimatum game. *Emotion* 7:876–81

Harlé KM, Chang LJ, van't Wout M, Sanfey AG. 2012. The neural mechanisms of affect infusion in social economic decision-making: a mediating role of the anterior insula. *NeuroImage* 61:32–40

Hartley CA, Gorun A, Reddan MC, Ramirez F, Phelps EA. 2014. Stressor controllability modulates fear extinction in humans. *Neurobiol. Learn. Mem.* In press

Hartley CA, Phelps EA. 2010. Changing fear: the neurocircuitry of emotion regulation. *Neuropsychopharmacology* 35:136–46

Hartley CA, Phelps EA. 2012. Anxiety and decision-making. *Biol. Psychiatry* 72:113–18

Huys QJM, Cools R, Gölzer M, Friedel E, Heinz A, et al. 2011. Disentangling the roles of approach, activation and valence in instrumental and Pavlovian responding. *PLoS Comput. Biol.* 7(4):e1002028

Isen AM, Nygren TE, Ashby FG. 1988. Influence of positive affect on the subjective utility of gains and losses: It is just not worth the risk. *J. Personal. Soc. Psychol.* 55:710–17

Jepma K, Niewenhuis S. 2011. Pupil diameter predicts changes in the exploration-exploitation tradeoff: evidence for the adaptive gain theory. *J. Cogn. Neurosci.* 23:1587–96

Kable JW, Glimcher PW. 2007. The neural correlates of subjective value during intertemporal choice. *Nat. Neurosci.* 10(12):1625–33

Kable JW, Glimcher PW. 2010. An "as soon as possible" effect in human intertemporal decision making: behavioral evidence and neural mechanisms. *J. Neurophysiol.* 103(5):2513–31

Kahneman D. 2011. *Thinking, Fast and Slow.* New York: Farrar, Strauss, and Giroux

Kimura K, Izawa S, Sugaya N, Ogawa N, Yamada KC, et al. 2013. The biological effects of acute psychosocial stress on delay discounting. *Psychoneuroendocrinology* 38(10):2300–8

Knutson B, Bossaerts P. 2007. Neutral antecedents of financial decisions. *J. Neurosci.* 27(31):8174–77

Kober H, Mende-Siedlecki P, Kross EF, Weber J, Mischel W, et al. 2010. Prefrontal-striatal pathway underlies cognitive regulation of craving. *Proc. Natl. Acad. Sci. USA* 107(33):14811–16

Kubota JT, Banaji MR, Phelps EA. 2012. The neuroscience of race. *Nat. Neurosci.* 15:940–48

Laibson D. 1997. Golden eggs and hyberbolic discounting. *Q. J. Econ.* 112(2):443–78

Lambert KG. 2003. The life and career of Paul Maclean: a journey toward neurobiological and social harmony. *Physiol. Behav.* 79(3):343–49

Lane RD, Reiman EM, Ahern GL, Schwartz GE, Davidson RJ. 1997. Neuroanatomical correlates of happiness, sadness and disgust. *Am. J. Psychiatry* 154(7):926–33

LeDoux JE. 2000. Emotion circuits in the brain. *Annu. Rev. Neurosci.* 23:155–84

LeDoux JE, Gorman JM. 2001. A call to action: overcoming anxiety through active coping. *Am. J. Psychiatry* 158:1953–55

Lempert KM, Glimcher PW, Phelps EA. 2013. *Reference-dependence in intertemporal choice.* Poster presented at Annu. Meet. Soc. Neuroecon., 12th, Lausanne, Switz.

Lempert KM, Porcelli AJ, Delgado MR, Tricomi E. 2012. Individual differences in delay discounting under acute stress: the role of trait perceived stress. *Front. Psychol.* 3:251

Leotti LA, Delgado MR. 2011. The inherent reward of choice. *Psychol. Sci.* 22:1310–18

Leotti LA, Delgado MR. 2014. The value of exercising control over monetary gains and losses. *Psychol. Sci.* 25:596–604

Lerner JS, Keltner D. 2001. Fear, anger, and risk. *J. Pers. Soc. Psychol.* 81:146–59

Lerner JS, Li Y, Valdesolo P, Kassam K. 2015. Emotion and decision making. *Annu. Rev. Psychol.* 66:In press

Lerner JS, Small DA, Loewenstein G. 2004. Heart strings and purse strings: carryover effects of emotions on economic decisions. *Psychol. Sci.* 15:337–41

Li J, Schiller D, Schoenbaum G, Phelps EA, Daw ND. 2011. Differential roles of human striatum and amygdala in associative learning. *Nat. Neurosci.* 14:1250–52

Lighthall NR, Mather M, Gorlick MA. 2009. Acute stress increases sex differences in risk seeking in the balloon analogue risk task. *PLoS ONE* 4:e6002

Lighthall NR, Sakaki M, Vasunilashorn S, Nga L, Somayajula S, et al. 2012. Gender differences in reward-related decision processing under stress. *Soc. Cogn. Affect. Neurosci.* 7:476–84

Lo AW, Repin DV. 2002. The psychophysiology of real-time financial risk processing. *J. Cogn. Neurosci.* 14(3):323–39

Lole L, Gonsalvez CJ, Blaszczynski A, Clarke AR. 2012. Electrodermal activity reliably captures physiological differences between wins and losses during gambling on electronic machines. *Psychophysiology* 49(2):154–63

Maclean PD. 1949. Psychosomatic disease and the visceral brain; recent developments bearing on the Papez theory of emotion. *Psychosom. Med.* 11(6):338–53

Maia TV, McClelland JL. 2004. A reexamination of the evidence for the somatic marker hypothesis: what participants really know in the Iowa gambling task. *Proc. Natl. Acad. Sci. USA* 101:16075–80

Maier SF, Watkins LR. 2010. Role of the medial prefrontal cortex in coping and resilience. *Brain Res.* 1355:52–60

McClure SM, Laibson DI, Loewenstein G, Cohen JD. 2004. Separate neural systems value immediate and delayed monetary rewards. *Science* 306(5695):503–7

McEwen BS. 2007. Physiology and neurobiology of stress and adaptation: central role of the brain. *Physiol. Rev.* 87(3):873–904

Moretti L, di Pellegrino G. 2010. Disgust selectively modulates reciprocal fairness in economic interactions. *Emotion* 10:169–80

Moretto G, Làdavas E, Mattioli F, di Pellegrino G. 2010. A psychophysiological investigation of moral judgment after ventromedial prefrontal damage. *J. Cogn. Neurosci.* 22:1888–99

Murray EA, Rudebeck. 2013. The drive to strive: goal generation based on current needs. *Front. Neurosci.* 7:112

Naqvi NH, Rudrauf D, Damasio H, Bechara A. 2007. Damage to the insula disrupts addiction to cigarette smoking. *Science* 315(5811):531–34

Ochsner KN, Gross JJ. 2005. The cognitive control of emotion. *Trends Cogn. Sci.* 9:242–49

O'Neill M, Schultz W. 2013. Risk prediction error in orbitofrontal neurons. *J. Neurosci.* 33(40):15810–14

Otto AR, Raio CM, Chiang A, Phelps EA, Daw NA. 2013. Working-memory capacity protects model-based learning from stress. *Proc. Natl. Acad. Sci. USA* 110(52):20941–46

Otto B, Misra S, Prasad A, McRae K. 2014. Functional overlap of top-down emotion regulation and generation: an fMRI study identifying common neural substrates between cognitive reappraisal and cognitively generated emotions. *Cogn. Affect Behav. Neurosci.* In press

Pabst S, Brand M, Wolf OT. 2013. Stress effects on framed decisions: There are differences for gains and losses. *Front. Behav. Neurosci.* 7:142

Packard MG, Goodman J. 2012. Emotional arousal and multiple memory systems in the mammalian brain. *Front. Behav. Neurosci.* 6:14

Paxton JM, Ungar L, Greene JD. 2012. Reflection and reasoning in moral judgment. *Cogn. Sci.* 36(1):163–77

Peters J, Büchel C. 2010. Episodic future thinking reduces reward delay discounting through an enhancement of prefrontal-mediotemporal interactions. *Neuron* 66:138–48

Peters RS. 1970. Reason and passion. *R. Inst. Philos. Lect.* 4:132–53

Phelps EA. 2006. Emotion and cognition: insights from studies of the human amygdala. *Annu. Rev. Psychol.* 57:27–53

Phelps EA. 2009. The study of emotion in neuroeconomics. In *Neuroeconomics: Decision Making and the Brain*, ed. PW Glimcher, C Camerer, E Fehr, RA Poldrack, pp. 233–50. London: Elsevier

Phelps EA, LeDoux JE. 2005. Contributions of the amygdala to emotion processing: from animal models to human behavior. *Neuron* 48(2):175–87

Phillips ML, Young AW, Senior C, Brammer M, Andrew C, et al. 1997. A specific neural substrate for perceiving facial expressions of disgust. *Nature* 389(6650):495–98

Ploghaus A, Tracey I, Gati JS, Clare S, Menon RS, et al. 1999. Dissociating pain from its anticipation in the human brain. *Science* 284(5422):1979–81

Poldrack RA. 2006. Can cognitive processes be inferred from neuroimaging data? *Trends Cogn. Sci.* 10:59–63

Porcelli AJ, Delgado MR. 2009. Acute stress modulates risk taking in financial decision making. *Psychol. Sci.* 20:278–83

Preston SD, Buchanan TW, Stansfield RB, Bechara A. 2007. Effects of anticipatory stress on decision making in a gambling task. *Behav. Neurosci.* 121:257–63

Preuschoff K, 't Hart BM, Einhäuser W. 2011. Pupil dilation signals surprise: evidence for noradrenaline's role in decision making. *Front. Neurosci.* 5:115

Qin S, Hermans EJ, van Marle HJ, Luo J, Fernández G. 2009. Acute psychological stress reduces working memory-related activity in the dorsolateral prefrontal cortex. *Biol. Psychiatry* 66(1):25–32

Raghunathan R, Pham MT. 1999. All negative moods are not equal: motivational influences of anxiety and sadness on decision-making. *Organ. Behav. Hum. Decis. Process.* 79:56–77

Raio CM, Orederu TA, Palazzolo L, Shurick AA, Phelps EA. 2013. Cognitive emotion regulation fails stress test. *Proc. Natl. Acad. Sci. USA* 110(37):15139–44

Reyna VF, Brainerd CJ. 2011. Dual processes in decision making and developmental neuroscience: a fuzzy-trace model. *Dev. Rev.* 31:180–206

Rilling JK, Sanfey AG. 2011. The neuroscience of social decision-making. *Annu. Rev. Psychol.* 62:23–48

Roesch MR, Esber GR, Li J, Daw ND, Schoenbaum G. 2012. Surprise! Neural correlates of Pearce-Hall and Rescorla-Wagner coexist within the brain. *Eur. J. Neurosci.* 35:1190–200

Rolls ET. 2013. Limbic systems for emotion and for memory, but no single limbic system. *Cortex.* In press

Rolls ET, Grabenhorst F. 2008. The orbitofrontal cortex and beyond: from affect to decision-making. *Prog. Neurobiol.* 86(3):216–44

Roozendaal B, McEwen BS, Chattarji S. 2009. Stress, memory and the amygdala. *Nat. Rev. Neurosci.* 10:423–33

Rose JE, McClernon FJ, Froeliger B, Behm FM, Preud'homme X, Krystal AD. 2011. Repetitive transcranial magnetic stimulation of the superior frontal gyrus modulates craving for cigarettes. *Biol. Psychiatry* 70(8):794–99

Rudebeck PH, Mitz AR, Chacko RV, Murray EA. 2013. Effects of amygdala lesions on reward-value coding in orbital and medial prefrontal cortex. *Neuron* 80(6):1519–31

Sanfey AG, Chang LJ. 2008. Multiple systems in decision making. *Ann. N.Y. Acad. Sci.* 1128:53–62

Sanfey AG, Rilling JK, Aronson JA, Nystrom LE, Cohen JD. 2003. The neural basis of economic decision-making in the ultimatum game. *Science* 300:1755–58

Scherer KR. 2000. Psychological models of emotion. In *The Neuropsychology of Emotion*, ed. J Borod, pp. 137–62. Oxford, UK: Oxford Univ. Press

Scherer KR. 2005. What are emotions? And how can they be measured? *Soc. Sci. Inf.* 44:695–729

Schwabe L, Höffken O, Tegenthoff M, Wolf OT. 2011. Preventing the stress-induced shift from goal-directed to habit action with a β-adrenergic antagonist. *J. Neurosci.* 31(47):17317–25

Schwabe L, Tegenthoff M, Höffken O, Wolf OT. 2010. Concurrent glucocorticoid and noradrenergic activity shifts instrumental behavior from goal-directed to habitual control. *J. Neurosci.* 30(24):8190–96

Schwabe L, Wolf OT. 2009. Stress prompts habit behavior in humans. *J. Neurosci.* 39(22):7191–98

Sellitto M, Ciaramelli E, di Pellegrino G. 2010. Myopic discounting of future rewards after medial orbitofrontal damage in humans. *J. Neurosci.* 8:16429–36

Sharot T, De Martino B, Dolan RJ. 2009. How choice reveals and shapes expected hedonic outcome. *J. Neurosci.* 29(12):3760–65

Sharot T, Riccardi AM, Raio CM, Phelps EA. 2007. Neural mechanisms mediating optimism bias. *Nature* 450:102–5

Sharp C, Monterosso J, Montague PR. 2012. Neuroeconomics: a bridge for translational research. *Biol. Psychiatry* 72:87–92

Skinner MD, Aubin H-J. 2010. Craving's place in addiction theory: contributions of the major models. *Neurosci. Biobehav. Rev.* 34(4):606–23

Sokol-Hessner P, Camerer CF, Phelps EA. 2012. Emotion regulation reduces loss aversion and decreases amygdala responses to losses. *Soc. Cogn. Affect. Neurosci.* 8(3):341–50

Sokol-Hessner P, Hsu M, Curley NG, Delgado MR, Camerer CF, Phelps EA. 2009. Thinking like a trader selectively reduces individuals' loss aversion. *Proc. Natl. Acad. Sci. USA* 106:5035–40

Sokol-Hessner P, Lackovic SF, Tobe RH, Leventhal BL, Phelps EA. 2013. *The effect of propranolol on loss aversion and decision-making.* Poster presented at Soc. Neurosci. Annu. Meet., 43rd, San Diego, CA. Program no. 99.15

Squire LR. 2004. Memory systems of the brain: a brief history and current perspective. *Neurobiol. Learn. Mem.* 82(3):171–77

Stanley DA, Sokol-Hessner P, Banaji MR, Phelps EA. 2011. Implicit race attitudes predict trustworthiness judgments and economic trust decisions. *Proc. Natl. Acad. Sci. USA* 108(19):7710–15

Stanley DA, Sokol-Hessner P, Fareri DS, Perino MT, Delgado MR, et al. 2012. Race and reputation: Perceived racial group trustworthiness influences the neural correlates of trust decisions. *Philos. Trans. R. Soc. B.* 367(1589):744–53

Starcke K, Polzer C, Wolf OT, Brand M. 2011. Does stress alter everyday moral decision-making? *Psychoneuroendocrinology* 36:210–19

Starcke K, Wolf OT, Markowitsch HJ, Brand M. 2008. Anticipatory stress influences decision making under explicit risk conditions. *Behav. Neurosci.* 122(6):1352–60

Thaler RH. 1988. Anomalies: the ultimatum game. *J. Econ. Perspect.* 2:195–206

Ulrich-Lai YM, Herman JP. 2009. Neural regulation of endocrine and autonomic stress responses. *Nat. Rev. Neurosci.* 10:397–409

Ungless MA, Argilli E, Bonci A. 2010. Effects of stress and aversion on dopamine neurons: implications for addiction. *Neurosci. Biobehav. Rev.* 35:151–56

van't Wout M, Chang LJ, Sanfey AG. 2010. The influence of emotion regulation on social interactive decision-making. *Emotion* 10:815–21

van't Wout M, Kahn R, Sanfey AG, Aleman A. 2006. Affective state and decision-making in the ultimatum game. *Exp. Brain Res.* 169:564–68

Vinkers CH, Zorn JV, Cornelisse S, Koot S, Houtepen LC, et al. 2013. Time-dependent changes in altruistic punishment following stress. *Psychoneuroendocrinology* 38(9):1467–75

von Dawans B, Fischbacher U, Kirschbaum C, Fehr E, Heinrichs M. 2012. The social dimension of stress reactivity: acute stress increases prosocial behavior in humans. *Psychol. Sci.* 23(6):651–60

von Helversen B, Rieskamp J. 2013. Does the influence of stress on financial risk taking depend on the riskiness of the decision? *Proc. Natl. Acad. Sci. USA* 35:1546–51

Winkielman P, Berridge KC, Wilbarger JL. 2005. Unconscious affective reactions to masked happy versus angry faces influence consumption behavior and judgments of value. *Personal. Soc. Psychol. Bull.* 31:121–35

Youssef FF, Dookeeram K, Basdeo V, Francis E, Doman M, et al. 2012. Stress alters personal moral decision making. *Psychoneuroendocrinology* 37:491–98

Basal Ganglia Circuits for Reward Value–Guided Behavior

Okihide Hikosaka,[1] Hyoung F. Kim,[1]
Masaharu Yasuda,[1] and Shinya Yamamoto[1,2]

[1]Laboratory of Sensorimotor Research, National Eye Institute, National Institutes of Health, Bethesda, Maryland 20892; email: oh@lsr.nei.nih.gov

[2]System Neurosciences, Human Technology Research Institute, National Institute of Advanced Industrial Science and Technology (AIST), Tsukuba, 305-8568, Japan

Annu. Rev. Neurosci. 2014. 37:289–306

First published online as a Review in Advance on May 14, 2014

The *Annual Review of Neuroscience* is online at neuro.annualreviews.org

This article's doi: 10.1146/annurev-neuro-071013-013924

Keywords

caudate nucleus, substantia nigra, superior colliculus, flexible value, stable value, visual object

Abstract

The basal ganglia are equipped with inhibitory and disinhibitory mechanisms that enable a subject to choose valuable objects and actions. Notably, a value can be determined flexibly by recent experience or stably by prolonged experience. Recent studies have revealed that the head and tail of the caudate nucleus selectively and differentially process flexible and stable values of visual objects. These signals are sent to the superior colliculus through different parts of the substantia nigra so that the animal looks preferentially at high-valued objects, but in different manners. Thus, relying on short-term value memories, the caudate head circuit allows the subject's gaze to move expectantly to recently valued objects. Relying on long-term value memories, the caudate tail circuit allows the subject's gaze to move automatically to previously valued objects. The basal ganglia also contain an equivalent parallel mechanism for action values. Such flexible–stable parallel mechanisms for object and action values create a highly adaptable system for decision making.

Contents

INTRODUCTION

Disinhibition:
removal of a sustained
inhibition, causing an
increase in excitability

The basal ganglia control behavior by disinhibiting desired actions and inhibiting undesired actions (Hikosaka et al. 2000) using serial and parallel inhibitory circuits (**Figure 1**). A key circuit consists of serial inhibitory connections that remove inhibitions (i.e., disinhibition) on target areas. In parallel with this direct pathway, the indirect and hyperdirect pathways enhance inhibitions on target areas (Nambu et al. 2002). These circuits together select desired actions.

How then is an action determined to be desired or undesired? Behavioral studies have shown that a fundamental determinant is reward value (Rangel et al. 2008): Animals choose actions that

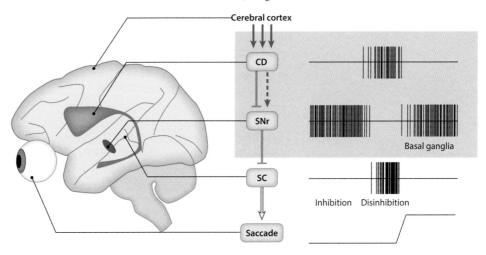

Figure 1

Basal ganglia circuit controlling the initiation of saccadic eye movements. Some neurons in the monkey caudate nucleus (CD) are excited by visual inputs that originate from the cerebral cortices and other areas. The CD neurons inhibit the tonic activity of substantia nigra pars reticulata (SNr) neurons through direct connections or enhance the tonic activity of SNr neurons through indirect connections. Because the SNr connection to the superior colliculus (SC) is inhibitory, the direct signal from the CD leads to a disinhibition of SC neurons (*right*), whereas the indirect signal from the CD leads to an enhanced inhibition of SC neurons. Arrows indicate excitatory connections (or effects). Red lines indicate inhibitory connections. Solid and hatched lines indicate direct and indirect connections, respectively.

will bring rewards and initiate them quickly (Takikawa et al. 2002). This preference is caused, at least in part, by changes in the activity of basal ganglia neurons: They often change their sensory or motor activity depending on the value of the expected reward (Hikosaka et al. 2006).

Studies have suggested that the reward-dependent plasticity of basal ganglia neurons is caused by inputs from dopamine neurons located in the substantia nigra pars compacta (SNc) and the ventral tegmental area (Wickens et al. 2003, Nakamura & Hikosaka 2006). Dopamine neurons typically encode a reward prediction error: excitation when the new outcome is better than expected, and inhibition when the outcome is worse (Schultz 1998). Dopamine-guided plasticity would guide animals to choose actions that lead to better rewards (Schultz et al. 1997). The amount of reward obtained per choice would eventually approach the maximum.

Learning does not stop there, however. The chosen action, as it is repeated, becomes more accurate, quicker, and stereotyped (Sakai et al. 2003) and is eventually carried out automatically (Logan 1985). This automated response is known as habit or skill (Salmon & Butters 1995). This skill-learning process guides animals to obtain maximum rewards per unit time (rather than per choice) (Hikosaka et al. 2013). This adaptation is critical for survival because the chosen action can be shortened, say, from 10 min to 1 min after learning a skill (Newell & Rosenbloom 1981).

How can the brain embrace these two kinds of learning: choice and skill? Studies on procedural learning have suggested that the brain is equipped with parallel mechanisms, one for choice (initial learning) and the other for skill (late learning) (Hikosaka et al. 1999). Underlying neural mechanisms are parallel cortico–basal ganglia and cortico-cerebellar circuits (Middleton & Strick 2000). These data incite a more basic question. How does the brain process two kinds of memories: short-term flexible memories and long-term stable memories? We address this question using data from recent studies on the basal ganglia, particularly studies on how subjects learn to choose more valuable objects.

FLEXIBLE AND STABLE VALUES

Suppose you are in a grocery store trying to buy apples, and you find that there are six kinds of apples. Which would you choose? There are two extreme strategies. The first strategy is to rely on your long-term data: You choose one that you tried before and liked very much. This method is easy but may not be perfect. Your favorite apple might be overripe. You may have no experience with two of the six kinds, and they might be better than your favored one. The second strategy is to ignore the long-term data and rely on short-term data. You may take a sniff of them to estimate their ripeness or touch them to get some idea about their hardness. This short-term data may give you a better determination of which apple is the best. But your onsite estimate may not be perfect. So, neither of the two strategies is perfect.

In everyday life, we often face a similar kind of decision dilemma (Daw et al. 2006, Cohen et al. 2007, Evans 2008). Which strategy is better? The answer depends on the reliability of the information gained. If you have shopped at this particular grocery store for ten years, the long-term data may be more reliable than the short-term data. If you are visiting an unfamiliar grocery store, the short-term data may be more reliable. In many cases, however, both the long-term and short-term data are useful. Such interactions of information processing are formalized in Bayesian decision theory (Körding & Wolpert 2006, McNamara et al. 2006). The scheme shown in **Figure 2** raises the possibility that the brain contains two parallel mechanisms, which selectively process the long-term and short-term data. This article is centered around this hypothesis.

To test the hypothesis, we need to identify brain areas that process information selectively for the long-term or short-term data. Answering this question turns out to be difficult because the three kinds of information covary. In particular, the short-term data affect the decision, which then

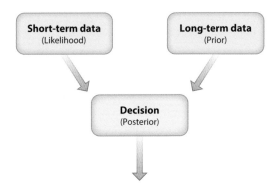

Figure 2

General scheme of decision making. A proper decision should be based on slowly accumulated and to-be-retained data (long-term data) as well as quickly acquired and to-be-erased data (short-term data). One interpretation, following Bayesian theory, may be that long-term data, short-term data, and the decision correspond respectively to prior, likelihood, and posterior.

affects the long-term data (Hikosaka et al. 1999). Nonetheless, we developed two experimental procedures in which the covariation was minimized (Yasuda et al. 2012, Kim & Hikosaka 2013, Yamamoto et al. 2013) (**Figure 3**).

The first procedure is to let the short-term mechanism operate while minimizing the operation of the long-term mechanism (**Figure 3a**). Each object changes its associated reward value between high and low in blocks of trials, creating short-term data. The subject needs to keep track of the

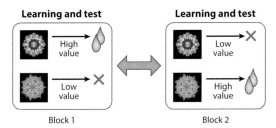

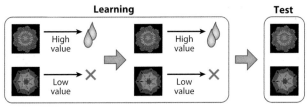

Figure 3

Experimental procedures to selectively activate the flexible and stable value mechanisms. (*a*) Flexible value procedure. Fractal objects change their values across blocks of trials. The values of objects are learned and tested in the same block. (*b*) Stable value procedure. Fractal objects stably retain their values during learning across days. The values of objects are then tested on separate days. During the test session, the objects are not associated with their stable values.

value change so that if more than one object is presented the subject can choose the object that has recently been highly valued. Owing to the frequent reversals of the object values, the long-term mechanism would not accumulate data. We call it a flexible value procedure because the values of individual objects change flexibly.

The second procedure is to let the long-term mechanism express its signals while minimizing the operation of the short-term mechanism (**Figure 3***b*). Each object is associated with a fixed value (i.e., high or low) during long-term training across several days, creating long-term data. Sometime after that, the objects are presented, but the reward is not given or is given in a noncontingent manner. Hence, the short-term mechanism should not operate systematically. We call it a stable value procedure because the values of individual objects remain stable.

Anatomical studies have suggested that the brain consists of parallel distributed networks (Mishkin et al. 1983, Goldman-Rakic 1988). This feature is conspicuously shown for the connections between the cerebral cortex and the basal ganglia (Alexander et al. 1986). On the basis of experiments using the flexible and stable value procedures, we suggest that different regions of the caudate nucleus (CD) in the basal ganglia may serve as the long-term and short-term mechanisms.

CD: caudate nucleus

CDh: head of caudate nucleus

CDb: body of caudate nucleus

CDt: tail of caudate nucleus

DIFFERENTIAL VALUE CODING IN THE CAUDATE NUCLEUS

Many neurons in the monkey basal ganglia respond to visual stimuli (Hikosaka & Wurtz 1983a, Caan et al. 1984, Joseph & Boussaoud 1985, Matsumura et al. 1992, Brown et al. 1995, Sato & Hikosaka 2002). Visual signals are particularly prominent in the CD throughout its rostrocaudal extent: head (CDh)—body (CDb)—tail (CDt) (**Figure 4***a*). Notably, however, the CD visual responses often depend on the behavioral context, especially memory and reward value (Hikosaka et al. 1989, Kawagoe et al. 1998).

Using the flexible–stable value procedures (**Figure 3**), these CD visual responses were categorized into two types (i.e., flexible versus stable value coding) in a regionally distinct manner (Kim & Hikosaka 2013) (**Figure 4**). In the flexible value procedure, CDh neurons generally showed clear value-based modulations, whereas CDt neurons, as a population, showed no value modulations (**Figure 4***b*). In contrast, in the stable value procedure, CDt neurons showed clear value modulations, whereas CDh neurons showed only weak value modulations (**Figure 4***c*). CDb neurons, as a population, showed both flexible and stable value modulations. The proportion of neurons showing significant value modulations were distributed in opposite gradients across the CD subregions (**Figure 4***d,e*).

CDh neurons overall showed value-differential responses in the flexible value procedure but not in the stable value procedure, suggesting that they serve as the short-term mechanism (**Figure 2**). In contrast, CDt neurons showed value-differential responses in the stable value procedure but not in the flexible value procedure (Yamamoto et al. 2013), suggesting that they serve as the long-term mechanism (**Figure 2**).

Note that the CDb, which sits between the CDh and the CDt, as a whole encodes both flexible and stable values (Kim & Hikosaka 2013). The subregional differences in the CD may therefore reflect a gradient of functionality rather than dichotomy. A key factor may be learning speed: CDh neurons learn object values quickly across trials, whereas CDt neurons learn object values slowly across days; CDb neurons may have intermediate learning speeds. If so, the scheme in **Figure 2** may need to be revised: There may be more than two memory mechanisms with different learning speeds (Kording et al. 2007).

These data suggest that different parts in the CD convey different kinds of information that could together guide decision making properly in the flexible value as well as the stable value condition. Obviously, decision making needs to be expressed as motor outputs. When one encounters

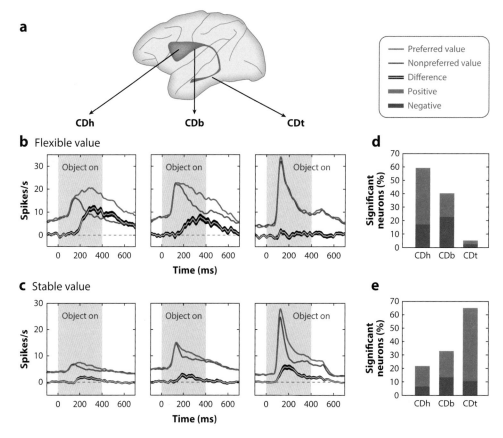

Figure 4

Differential encoding of flexible and stable values in subregions of the caudate nucleus. (*a*) Anatomy of the caudate nucleus and its subregions in the macaque monkey. (*b*) Average responses to the flexibly valued objects of neurons in the CDh, CDb, and CDt. Neuronal responses were averaged for the neurons' preferred (*purple*) and nonpreferred (*green*) values using a cross-validation method. The yellow line indicates the difference between the preferred and nonpreferred responses [mean ± SE (*SE shown in black*)]. (*c*) Average responses to the stably valued objects in the three caudate subregions. (*d*) Proportions of flexible value-coding neurons in the three caudate subregions. (*e*) Proportions of stable value-coding neurons.

a good (high-valued) object, one orients one's gaze by making a saccadic eye movement to the object and probably approaches it (Tatler et al. 2011, Awh et al. 2012). This value-based orienting is universal among animals (Bromberg-Martin et al. 2010) and crucial for survival (Hikosaka et al. 2013). Indeed, gaze (or attention) is strongly attracted by objects with high reward values, and this fundamental decision making is regulated by the CD, as shown in the following sections.

OCULOMOTOR EXPRESSION OF OBJECT VALUES

Primates (humans and monkeys) heavily rely on saccadic eye movements. When faced with a crowded scene, our gaze jumps between objects with saccades (Sheinberg & Logothetis 2001, Henderson 2003, Land 2006), but the saccade pattern varies depending on internal thoughts and goals (Yarbus 1967). When we perform everyday motor activities, our gaze jumps to an object of interest before our hand reaches out to it (Johansson et al. 2001). Macaque monkeys behave similarly (Miyashita et al. 1996). In our daily as well as professional routines, we can easily find objects relevant to our goals (Chun & Nakayama 2000, Rothkopf et al. 2007). Similar skills develop while performing psychophysical tasks (Shiffrin & Schneider 1977, Karni & Sagi 1993, Ahissar & Hochstein 1996, Sigman & Gilbert 2000, Seitz et al. 2009). A common feature among these behaviors is that particular objects (among many others) are assigned high values in each context. In human psychophysical experiments, when a visual object is associated with a reward, attention (Kristjánsson et al. 2010, Anderson et al. 2011, Awh et al. 2012, Chelazzi et al. 2013) and gaze (Anderson & Yantis 2012, Theeuwes & Belopolsky 2012) are attracted if the object appears subsequently. Intrinsically valuable stimuli (e.g., faces) attract gaze with faster saccades (Xu-Wilson et al. 2009).

The gaze of macaque monkeys is also strongly affected by reward values. Suppose the saccade to one position (i.e., left) is followed by a large reward, whereas the saccade to another position (i.e., right) is followed by a small reward. The positional value bias induces a consistent bias in saccades: quicker (i.e., shorter reaction times) and faster (i.e., larger peak velocity) to the high-valued than to the low-valued position (Takikawa et al. 2002). If the position-reward contingency is reversed, the saccade bias reverses within several trials (Lauwereyns et al. 2002). Likewise, in the flexible value procedure shown in **Figure 3a**, saccades are quicker and faster to the high-valued than to the low-valued object (Yasuda et al. 2012, Kim & Hikosaka 2013, Yamamoto et al. 2013). If the two objects are presented simultaneously, the monkey chooses the high-valued object. These experiments examined oculomotor decisions based on flexible values that represent the short-term data (**Figure 2**).

In a real-life situation, however, decisions are sometimes better guided by the long-term data (belief, intuition, or skill) (Bechara et al. 1997, Marewski et al. 2010), which emerges through long-term learning (Crossman 1959, Newell & Rosenbloom 1981, Karni & Sagi 1993). Robust memories are created by associating particular objects or actions (among others) with a consistently high value (Shiffrin & Schneider 1977). Gaze or attention is highly sensitive to such consistent object-value associations (Bichot & Schall 1999, Kyllingsbaek et al. 2001, Peck et al. 2009). This sensitivity is shown more systematically using the stable value procedure (**Figure 3b**). When stably valued objects (some high-valued, the others low-valued) were presented, the monkeys looked at high-valued objects preferentially (Yasuda et al. 2012, Kim & Hikosaka 2013, Yamamoto et al. 2013) (**Supplemental Figure 1**; follow the **Supplemental Material link** from the Annual Reviews home page at **http://www.annualreviews.org**). There are several important features about the value-based gaze bias. First, the gaze bias occurred even though no reward was given after the free viewing. Second, the gaze bias required 4–5 daily learning sessions to fully develop. Third, the gaze bias remained significant long after the learning had stopped (Yasuda et al. 2012).

These findings indicate that the values of objects influence gaze as well as attention, regardless of whether the values are changing flexibly or have been established stably after prolonged experience. This conclusion, in relation to the scheme in **Figure 2** and the findings in **Figure 4**, leads to the following hypothesis: The oculomotor decision is controlled both by the CDh, which represents flexible object values, and by the CDt, which represents stable object values (Kim & Hikosaka 2013).

OCULOMOTOR MECHANISMS IN THE BASAL GANGLIA

We have suggested that biased object values elicit differential responses in CD neurons, which then lead to differential oculomotor responses. Indeed, the CD has a strong influence on oculomotor mechanisms. First, electrical stimulation of the CDh and the CDb (Kitama et al. 1991, Watanabe & Munoz 2011) as well as the CDt (Yamamoto et al. 2012) facilitates the initiation of saccades. CDh–CDb stimulation also suppresses saccades (Watanabe & Munoz 2010). Second, dopamine

GABAergic:
transmitting or
secreting
γ-aminobutyric acid

SNr: substantia nigra
pars reticulata

SC: superior
colliculus

GPe: external
segment of globus
pallidus

depletion in the visual oculomotor region of the CDh–CDb in the monkey induces a severe hemineglect (Kato et al. 1995, Miyashita et al. 1995). Finally, anatomical and physiological studies have shown that neural circuitries originate from the CD to the oculomotor outputs, as described below in detail.

In general, the basal ganglia control body movements by changing the level of inhibitions on target motor structures (Wichmann & DeLong 1996). This finding explains why dysfunctions of the basal ganglia often show up as a too-strong inhibition (e.g., akinesia) or a lack of inhibition (e.g., chorea). This principle is clearly implemented in the oculomotor mechanism in the basal ganglia (**Figure 1**). Direct control over saccade initiation is achieved by the GABAergic connection from the substantia nigra pars reticulata (SNr) to the superior colliculus (SC) (Hikosaka & Wurtz 1983b). SNr neurons fire continuously with high frequencies, thus tonically inhibiting saccadic neurons in the SC.

In normally behaving animals, the SNr–SC tonic inhibition is reduced occasionally and temporarily by the direct GABAergic inhibitory inputs from the striatum (mostly the CD) (Hikosaka et al. 1993) (**Figure 1**), but in particular behavioral contexts (Hikosaka et al. 2000, 2006). This transient causes a disinhibition of SC neurons, allowing a saccade to be generated. In addition to the direct pathway, the CD has indirect pathways to control the SNr, which are mediated by the external segment of the globus pallidus (GPe) and possibly the subthalamic nucleus (STN) (Alexander & Crutcher 1990). Because GPe neurons are GABAergic and inhibitory, the net effect through this indirect pathway is largely facilitatory on SNr neurons (**Figure 1**) and hence suppressive on saccade initiation. Thus, the combination of the direct and indirect pathways seems ideal for selective choice: direct pathway for approach and indirect pathway for avoidance (Mink 1996, Hikosaka et al. 2000).

BASAL GANGLIA GUIDE VALUE-ORIENTED GAZE

The basal ganglia have two output nuclei, the internal segment of the globus pallidus (GPi) and the SNr, but only the SNr projects to the SC (Marín et al. 1998). Moreover, only a subpopulation of SNr neurons project to the SC (Beckstead & Frankfurter 1982, Parent et al. 1983, Francois et al. 1984). Thus, the oculomotor function of the basal ganglia is carried out largely by the particular SNr neurons that project to the SC. In the macaque monkey, a majority of the SC-projecting SNr neurons are localized in the dorsolateral portion of the SNr (Beckstead & Frankfurter 1982, Francois et al. 1984). Notably, the CDt projects focally to the dorsolateral SNr (Saint-Cyr et al. 1990). Indeed, a majority of neurons in the dorsolateral SNr project to the SC and encode stable values of fractal objects similarly to CDt neurons (Yasuda et al. 2012) (**Figure 5**).

The SC-projecting SNr neurons were inhibited by stably high-valued objects and excited by stably low-valued objects (**Figure 5b**). The inhibition may be due to direct inputs from the CDt, whereas the excitation may be due to indirect inputs from the CDt through the GPe (i.e., disinhibition). The categorical responses of SNr neurons can be explained by the following scheme: CDt neurons coding positive values (i.e., stronger responses to high-valued objects; **Figure 4e**, *red*) project to the SNr directly, whereas CDt neurons coding negative values (**Figure 4e**, *blue*) project to the SNr through the GPe. This hypothesis remains to be tested.

Functionally, when a stably high-valued object appears, SNr neurons are largely inhibited, and therefore SC saccadic neurons are disinhibited, thus facilitating saccades to the object (**Figure 5a**). When a stably low-valued object appears, SNr neurons are largely excited, and therefore SC saccadic neurons are further inhibited, thus suppressing saccades to the object. Indeed, the monkeys look at stably high-valued objects intensely while avoiding low-valued objects (Yasuda et al. 2012, Kim & Hikosaka 2013, Yamamoto et al. 2013) (**Supplemental Figure 1**).

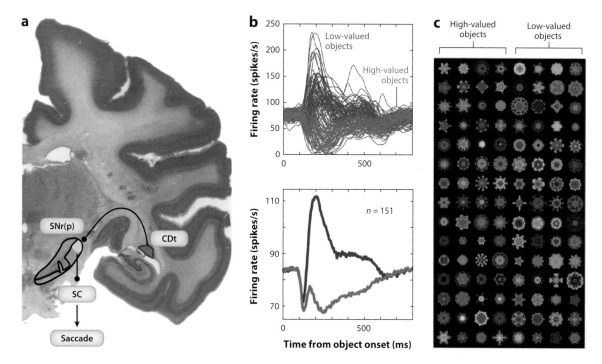

Figure 5

High-capacity memory of stable values encoded by SNr neurons. (*a*) The locations of the CDt and SNr(p) shown on a coronal Nissl-stained section. The CDt (*red*) has a direct inhibitory connection to the dorsolateral SNr(p) (*yellow*), which then inhibits presaccadic neurons in the SC. (*b*, *top*) The responses of an SC-projecting SNr(p) neuron to 120 well-learned objects (*c*). The neuron was inhibited by most high-valued objects (*red*) and excited by most low-valued objects (*blue*). (*b*, *bottom*) The average responses of 151 SNr(p) neurons to high-valued objects (*red*) and low-valued objects (*blue*), which were chosen randomly from ~300 well-learned objects.

The above results suggest that the SNr is the mediator of stable value signals originating from the CDt. However, previous studies have shown that the SNr is also a major mediator of signals from the CDh and CDb (Hikosaka et al. 1993). Then, why do SNr neurons encode stable values but not flexible values? Preliminary observation from our lab suggests that there are in fact flexible value–coding neurons in the SNr, but they are located more anterior-medial-ventrally, largely separate from the stable value–coding neurons (Yasuda & Hikosaka 2013). Thus, the anterior SNr, or SNr(a), would encode flexible values, and the posterior SNr, or SNr(p), would encode stable values. This segregation is consistent with anatomical (Smith & Parent 1986, Saint-Cyr et al. 1990) and electrophysiological (Hikosaka et al. 1993) data. If the result proves true, it would indicate that the flexible value (short-term) mechanism and the stable value (long-term) mechanism are segregated from the input (CD) to the output (SNr) of the basal ganglia (**Figure 6**). One interpretation, following a Bayesian scheme (**Figure 2**), may be that the CDh–SNr(a)–SC circuit serves the likelihood mechanism, whereas the CDt–SNr(p)–SC circuit serves the prior mechanism.

SNr(a): anterior portion of substantia nigra pars reticulata

SNr(p): posterior portion of substantia nigra pars reticulata

MEMORIES REQUIRED FOR THE STABLE VALUE MECHANISM

To make a decision you can rely on short-term data only, long-term data only, or both (**Figure 2**). If you rely only on long-term data, you need to maintain long-term memories about all objects,

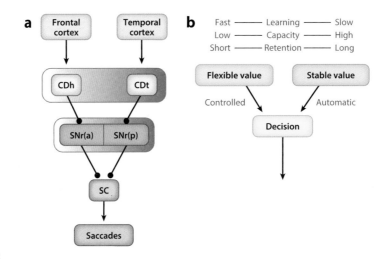

Figure 6

Parallel CD–SNr–SC circuits underlying value-based decision making. (*a*) Anatomical scheme. The CDh receives inputs mainly from the frontal cortical areas (Yeterian & Pandya 1991), whereas the CDt receives inputs mainly from the temporal cortical areas (Saint-Cyr et al. 1990). The CDh and CDt have equivalent downstream mechanisms (as shown in **Figure 1**) but use different neural circuits before reaching the SC. (*b*) Information processing along the parallel CD–SNr–SC circuits, which emulates the general scheme of decision making in **Figure 2**. These circuits have contrasting characteristics in terms of memory and motor output mechanisms.

actions, and events you have ever experienced; thus, the stable value mechanism must have a high-capacity memory that can retain information for a very long time, as suggested previously (Chun & Nakayama 2000) (**Figure 6***b*). Indeed, these memory features are what characterize neurons in the CDt and the SNr(p), which likely serve as the stable value mechanism (**Figure 6**).

As shown in **Figure 5**, SNr(p) neurons virtually categorized many visual objects (up to 300 so far) on the basis of their stable values (high-capacity memory). Some of the stably valued objects were removed from the learning schedule, and after more than 100 days they were shown to the monkey. SNr(p) neurons still categorized them almost perfectly (long-term retention). CDt neurons seem to have similar memory features (Yamamoto et al. 2013). As expected from the SNr(p) neurons' high-capacity memories, the monkey preferentially looked at stably high-valued objects, which were chosen from the ∼300 objects, even after more than 100 days' retention (Yasuda et al. 2012).

The stable value mechanism has two additional, notable features. First, stable values are learned slowly (**Figure 6***b*): The value-differential responses of CDt and SNr(p) neurons as well as the monkey's preferential looking developed gradually across 4–5 daily sessions (Yasuda et al. 2012, Kim & Hikosaka 2013). Second, stable values are signaled quickly. CDt and SNr(p) neurons started signaling the stable value of an object about 100 ms after the object appeared, which allowed the monkey to look at or look away from the object (Yasuda et al. 2012, Kim & Hikosaka 2013, Yamamoto et al. 2013). It seems unlikely that such a quick response is based on conscious thought which is characterized by sequential processing and is hence time-consuming (Sternberg 1969, Treisman & Gelade 1980). It rather suggests that the stable value–based responses occur subconsciously and automatically (**Figure 6***b*). Indeed, categorization or identification of visual objects that requires long-term learning is performed subconsciously (Shiffrin & Schneider 1977, Chun & Jiang 1998, Sigman & Gilbert 2000). Last, the quick and automatic response based on stable values is an essential skill to survive in the competitive world (Hikosaka et al. 2013).

Is the stable value mechanism essential, and if so how? Let us think about two situations: (*a*) choose among novel objects, and (*b*) choose among familiar objects (**Supplemental Figure 2**) (Kim & Hikosaka 2013). On each trial, 2 objects are presented among a set of 8 objects, and the monkey has to choose one. When all were novel objects (**Supplemental Figure 2*a***), the monkey gradually learned to choose the high-valued object, whereas CDh neurons gradually differentiated objects by their flexible values. But CDt neurons showed no differential responses even after 100 trials. Thus, only the flexible value mechanism is necessary in a novel context. When all objects were familiar (previously well learned) (**Supplemental Figure 2*b***), from the first trial, the monkey chose the high-valued object, whereas CDt as well as SNr(p) neurons differentiated all the objects by their stable values. But CDh neurons initially showed no differential responses. Therefore, the stable value mechanism is essential when previously well-learned objects are unexpectedly presented.

CONTROLLED VERSUS AUTOMATIC SACCADES

The schemes in **Figure 6** suggest that the CDh guides saccades on the basis of the flexible values of target objects, whereas the CDt guides saccades on the basis of the stable values of target objects. This hypothesis was supported by inactivating each of the CDh and CDt by injecting a GABA$_A$ agonist (muscimol) (Kim & Hikosaka 2013). During inactivation of the CDh, the flexible value information largely lost its ability to affect saccades (**Figure 7*b***, *center*). During inactivation of the CDt, the stable value information lost its ability to affect saccades (**Figure 7*c***, *bottom*). In contrast, saccades remained affected by the stable value information during the CDh inactivation (**Figure 7*c***, *center*) and by the flexible value information during the CDt inactivation (**Figure 7*b***, *bottom*).

From the motor output viewpoint, saccades have different natures depending on whether they are guided by the CDh or the CDt. As discussed in part in the preceding section, the saccades guided by the CDh can be characterized as controlled (**Figure 6*b***), because they are sensitive to the immediate reward outcome (**Figure 7*b***). In contrast, the saccades guided by the CDt can be characterized as automatic (**Figure 6*b***) because they occur even though no reward was given after the saccades (**Figure 7*c***).

These results are largely consistent with previous studies suggesting that different parts of the basal ganglia support goal-directed behavior and habits (Yin & Knowlton 2006, Redgrave et al. 2010). In particular, the role of the CDt in automatic saccades supports the seminal research by Mishkin and colleagues, which suggests that the connection from the temporal cortex to the CDt serves visual habits (Mishkin et al. 1984, Fernandez-Ruiz et al. 2001).

OBJECT VALUE VERSUS ACTION VALUE

Goal-directed behavior consists of at least two stages: (*a*) find good objects, and (*b*) manipulate the good objects to reach a goal (i.e., reward) (Hikosaka et al. 2013). Each stage is driven by value: stimulus (or object) value and action value (Rudebeck et al. 2008). We have discussed that object values can be flexible or stable, and these two kinds of object values are processed separately in the basal ganglia. Previous studies have suggested that a similar functional organization is present for action values (Hikosaka et al. 1999). Neurons in the anterior striatum (mostly CDh) were more active when the monkey learned a new sequential action (i.e., when action value was flexible), whereas neurons in the posterior striatum (mostly putamen) tended to be more active when the monkey performed well-learned sequences (i.e., when action value was stable) (Miyachi et al. 2002). Inactivation of the anterior striatum disrupted new learning, whereas inactivation of the posterior striatum disrupted the well-learned performance (Miyachi et al. 1997).

Muscimol: an alkaloid extracted from the poison mushroom *Amanita muscaria*, which selectively stimulates GABA$_A$ receptors

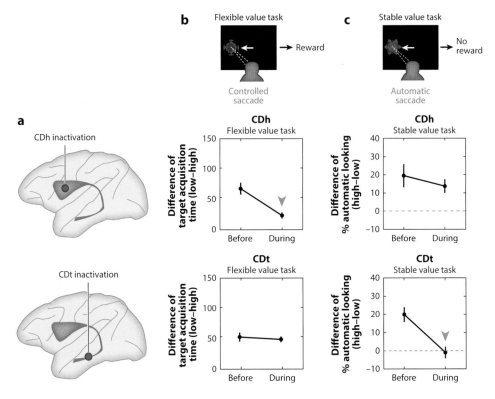

Figure 7

Differential impairments of controlled and automatic saccades by CDh and CDt inactivations. (*a*) Injection sites of muscimol in the caudate nucleus: CDh (*top*) and CDt (*bottom*). (*b*) Effects on the controlled saccades that were predictively influenced by reward feedback in the flexible value procedure (**Figure 3a**). The differences in the target acquisition time between high- and low-valued objects are plotted before and during inactivation (mean ± SE). Data are shown for CDh inactivation (*center*) and CDt inactivation (*bottom*). (*c*) Effects on the automatic saccades that occurred without reward feedback but were influenced by the stable value of the presented object (**Figure 3b**). The differences in the probability of automatic looking between high- and low-valued objects are plotted before and during inactivation (mean ± SE) (same format as in panel *b*). The effects are shown only for contralateral saccades.

These results together suggest that, for both object and action values, the anterior striatum serves as the flexible value mechanism, whereas the posterior striatum serves as the stable value mechanism (**Figure 8**). A key factor underlying this difference would be the speed of learning and forgetting. The combination of quick- and slow-learning mechanisms would enable more robust decision making, as suggested in studies on human motor learning (Kording et al. 2007).

BASAL GANGLIA DYSFUNCTIONS

The results discussed above are relevant to the clinical observation that humans with basal ganglia dysfunctions show impairments in controlled behavior and/or automatic behavior (Brown et al. 1997, Lieberman 2000). Impairments in cognitive tasks, such as set-shifting and value reversal, are observed in patients with Parkinson's disease (Lees & Smith 1983, Cools et al. 1984, Brown & Marsden 1990, Kehagia et al. 2010), Huntington's disease (Lawrence et al. 1996), and focal lesions of the basal ganglia (Caplan et al. 1990, Bhatia & Marsden 1994). In contrast, patients

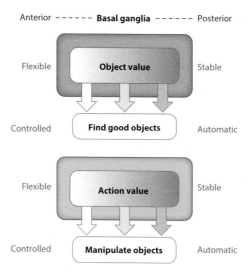

Anterior - - - - - - **Basal ganglia** - - - - - Posterior

Flexible   **Object value**   Stable

Controlled   **Find good objects**   Automatic

Flexible   **Action value**   Stable

Controlled   **Manipulate objects**   Automatic

Figure 8

Hypothetical parallel mechanisms controlling behavior on the basis of object values and action values. The first mechanism aims to find good objects, whereas the second mechanism aims to manipulate the objects. These mechanisms together enable animals and humans to gain access to rewards efficiently. For both mechanisms, the anterior part of the basal ganglia processes flexible values and guides controlled behavior, whereas the posterior part of the basal ganglia processes stable values and guides automatic behavior. However, the object and action mechanisms may not share the same neural circuits within the basal ganglia. Other brain structures may also constitute the mechanisms, especially those connected with the basal ganglia including the cerebral cortex (not shown).

with Parkinson's disease may lose the ability to perform behaviors automatically (Marsden 1982, Redgrave et al. 2010). They may have difficulty in carrying out two motor actions simultaneously (Schwab et al. 1954, Benecke et al. 1986). They are also impaired in performing tasks that require visual skills (Knowlton et al. 1996, Ashby & Maddox 2005). These results are difficult to interpret if we assume that the basal ganglia have a single function. Instead, these findings may reflect parallel functions discussed in this article: Different parts of the basal ganglia process flexible and stable values of objects as well as actions relying on short- and long-term memories, and they guide behavior in a controlled and automatic manner.

CONCLUSIONS

To make a good decision, we need to consider both recently acquired data and previously accumulated data. This process is an important and demanding task for the brain, but underlying mechanisms are unclear. In this article we have shown recent evidence that the CDh and CDt process information selectively and separately for the recently acquired data (flexible values of visual objects) and for the previously acquired data (stable values of visual objects). Such a parallel mechanism is imperative because flexible value information must be erased quickly, whereas stable value information must be retained for a long time. Indeed, the two kinds of information are processed in separate circuits downstream to the CDh and CDt, through different parts of the SNr, until they reach the SC, influencing the initiation of saccades. The CDh-flexible and CDt-stable circuits would work in a complementary manner: A proper estimate of object values can be provided only by the CDh-flexible circuit if the objects are novel, only by the CDt-stable circuit if the objects are familiar but have not been experienced for some time, or otherwise by a compromise between these circuits.

These findings raise many questions. First, do the flexible and stable value signals interact with each other, and if so, how? According to the scheme described above, the SC is the only place where these value signals are integrated. This suggestion seems unlikely, however. Stable values would serve as the prior information from which any neurons involving decision making would benefit. This might be achieved by the outputs of SNr neurons to the thalamus and then to many cortical and subcortical areas (Lynch et al. 1994, Middleton & Strick 1996, Tanibuchi et al. 2009). Second, how do CDh and CDt circuits selectively and differentially acquire flexible and stable values? Somewhere along each circuit, information on each visual object must be modified by the associated reward values. Dopamine may play a key role in this process because neurons along the CDh–CDt circuits receive massive inputs from midbrain dopamine neurons, which carry reward value signals (Schultz 1998). How then can dopamine neurons support short-term learning in the CDh circuit and long-term learning in the CDt circuit? Or, do different neuromodulatory mechanisms work for the CDh–CDt circuits? These questions represent outstanding issues in neuroscience.

DISCLOSURE STATEMENT

The authors are not aware of any affiliations, memberships, funding, or financial holdings that might be perceived as affecting the objectivity of this review.

ACKNOWLEDGMENTS

We thank Reza Shadmehr, Ali Ghazizadeh, and Ilya Monosov for helpful comments and discussions. The work was supported by the Intramural Research Program in the National Eye Institute, part of the National Institutes of Health.

LITERATURE CITED

Ahissar M, Hochstein S. 1996. Learning pop-out detection: specificities to stimulus characteristics. *Vis. Res.* 36:3487–500

Alexander GE, Crutcher MD. 1990. Functional architecture of basal ganglia circuits: neural substrates of parallel processing. *Trends Neurosci.* 13:266–71

Alexander GE, DeLong MR, Strick PL. 1986. Parallel organization of functionally segregated circuits linking basal ganglia and cortex. *Annu. Rev. Neurosci.* 9:357–81

Anderson BA, Laurent PA, Yantis S. 2011. Value-driven attentional capture. *Proc. Natl. Acad. Sci. USA* 108:10367–71

Anderson BA, Yantis S. 2012. Value-driven attentional and oculomotor capture during goal-directed, unconstrained viewing. *Atten. Percept. Psychophys.* 74:1644–53

Ashby FG, Maddox WT. 2005. Human category learning. *Annu. Rev. Psychol.* 56:149–78

Awh E, Belopolsky AV, Theeuwes J. 2012. Top-down versus bottom-up attentional control: a failed theoretical dichotomy. *Trends Cogn. Sci.* 16:437–43

Bechara A, Damasio H, Tranel D, Damasio AR. 1997. Deciding advantageously before knowing the advantageous strategy. *Science* 275:1293–95

Beckstead RM, Frankfurter A. 1982. The distribution and some morphological features of substantia nigra neurons that project to the thalamus, superior colliculus and pedunculopontine nucleus in the monkey. *Neuroscience* 7:2377–88

Benecke R, Rothwell JC, Dick JP, Day BL, Marsden CD. 1986. Performance of simultaneous movements in patients with Parkinson's disease. *Brain* 109:739–57

Bhatia KP, Marsden CD. 1994. The behavioural and motor consequences of focal lesions of the basal ganglia in man. *Brain* 117:859–76

Bichot NP, Schall JD. 1999. Effects of similarity and history on neural mechanisms of visual selection. *Nat. Neurosci.* 2:549–54

Bromberg-Martin ES, Matsumoto M, Hikosaka O. 2010. Dopamine in motivational control: rewarding, aversive, and alerting. *Neuron* 68:815–34

Brown LL, Schneider JS, Lidsky TI. 1997. Sensory and cognitive functions of the basal ganglia. *Curr. Opin. Neurobiol.* 7:157–63

Brown RG, Marsden CD. 1990. Cognitive function in Parkinson's disease: from description to theory. *Trends Neurosci.* 13:21–29

Brown VJ, Desimone R, Mishkin M. 1995. Responses of cells in the tail of the caudate nucleus during visual discrimination learning. *J. Neurophysiol.* 74:1083–94

Caan W, Perrett DI, Rolls ET. 1984. Responses of striatal neurons in the behaving monkey. 2. Visual processing in the caudal neostriatum. *Brain Res.* 290:53–65

Caplan LR, Schmahmann JD, Kase CS, Feldmann E, Baquis G, et al. 1990. Caudate infarcts. *Arch. Neurol.* 47:133–43

Chelazzi L, Perlato A, Santandrea E, Della Libera C. 2013. Rewards teach visual selective attention. *Vis. Res.* 85:58–72

Chun MM, Jiang Y. 1998. Contextual cueing: implicit learning and memory of visual context guides spatial attention. *Cogn. Psychol.* 36:28–71

Chun MM, Nakayama K. 2000. On the functional role of implicit visual memory for the adaptive deployment of attention across scenes. *Vis. Cogn.* 7:65–81

Cohen JD, McClure SM, Yu AJ. 2007. Should I stay or should I go? How the human brain manages the trade-off between exploitation and exploration. *Philos. Trans. R. Soc. B* 362:933–42

Cools AR, van den Bercken JH, Horstink MW, van Spaendonck KP, Berger HJ. 1984. Cognitive and motor shifting aptitude disorder in Parkinson's disease. *J. Neurol. Neurosurg. Psychiatry* 47:443–53

Crossman ERFW. 1959. A theory of the acquisition of speed-skill. *Ergonomics* 2:153–66

Daw ND, O'Doherty JP, Dayan P, Seymour B, Dolan RJ. 2006. Cortical substrates for exploratory decisions in humans. *Nature* 441:876–79

Evans JSBT. 2008. Dual-processing accounts of reasoning, judgment, and social cognition. *Annu. Rev. Psychol.* 59:255–78

Fernandez-Ruiz J, Wang J, Aigner TG, Mishkin M. 2001. Visual habit formation in monkeys with neurotoxic lesions of the ventrocaudal neostriatum. *Proc. Natl. Acad. Sci. USA* 98:4196–201

Francois C, Percheron G, Yelnik J. 1984. Localization of nigrostriatal, nigrothalamic and nigrotectal neurons in ventricular coordinates in macaques. *Neuroscience* 13:61–76

Goldman-Rakic PS. 1988. Topography of cognition: parallel distributed networks in primate association cortex. *Annu. Rev. Neurosci.* 11:137–56

Henderson JM. 2003. Human gaze control during real-world scene perception. *Trends Cogn. Sci.* 7:498–504

Hikosaka O, Nakahara H, Rand MK, Sakai K, Lu X, et al. 1999. Parallel neural networks for learning sequential procedures. *Trends Neurosci.* 22:464–71

Hikosaka O, Nakamura K, Nakahara H. 2006. Basal ganglia orient eyes to reward. *J. Neurophysiol.* 95:567–84

Hikosaka O, Sakamoto M, Miyashita N. 1993. Effects of caudate nucleus stimulation on substantia nigra cell activity in monkey. *Exp. Brain Res.* 95:457–72

Hikosaka O, Sakamoto M, Usui S. 1989. Functional properties of monkey caudate neurons. II. Visual and auditory responses. *J. Neurophysiol.* 61:799–813

Hikosaka O, Takikawa Y, Kawagoe R. 2000. Role of the basal ganglia in the control of purposive saccadic eye movements. *Physiol. Rev.* 80:953–78

Hikosaka O, Wurtz RH. 1983a. Visual and oculomotor functions of monkey substantia nigra pars reticulata. I. Relation of visual and auditory responses to saccades. *J. Neurophysiol.* 49:1230–53

Hikosaka O, Wurtz RH. 1983b. Visual and oculomotor functions of monkey substantia nigra pars reticulata. IV. Relation of substantia nigra to superior colliculus. *J. Neurophysiol.* 49:1285–301

Hikosaka O, Yamamoto S, Yasuda M, Kim HF. 2013. Why skill matters. *Trends Cogn. Sci.* 17:434–41

Johansson RS, Westling G, Bäckström A, Flanagan JR. 2001. Eye-hand coordination in object manipulation. *J. Neurosci.* 21:6917–32

Joseph JP, Boussaoud D. 1985. Role of the cat substantia nigra pars reticulata in eye and head movements. I. Neural activity. *Exp. Brain Res.* 57:286–96

Karni A, Sagi D. 1993. The time course of learning a visual skill. *Nature* 365:250–52

Kato M, Miyashita N, Hikosaka O, Matsumura M, Usui S, Kori A. 1995. Eye movements in monkeys with local dopamine depletion in the caudate nucleus. I. Deficits in spontaneous saccades. *J. Neurosci.* 15:912–27

Kawagoe R, Takikawa Y, Hikosaka O. 1998. Expectation of reward modulates cognitive signals in the basal ganglia. *Nat. Neurosci.* 1:411–16

Kehagia AA, Murray GK, Robbins TW. 2010. Learning and cognitive flexibility: frontostriatal function and monoaminergic modulation. *Curr. Opin. Neurobiol.* 20:199–204

Kim HF, Hikosaka O. 2013. Distinct basal ganglia circuits controlling behaviors guided by flexible and stable values. *Neuron* 79:1001–10

Kitama T, Ohno T, Tanaka M, Tsubokawa H, Yoshida K. 1991. Stimulation of the caudate nucleus induces contraversive saccadic eye movements as well as head turning in the cat. *Neurosci. Res.* 12:287–92

Knowlton BJ, Mangels JA, Squire LR. 1996. A neostriatal habit learning system in humans. *Science* 273:1399–402

Kording KP, Tenenbaum JB, Shadmehr R. 2007. The dynamics of memory as a consequence of optimal adaptation to a changing body. *Nat. Neurosci.* 10:779–86

Körding KP, Wolpert DM. 2006. Bayesian decision theory in sensorimotor control. *Trends Cogn. Sci.* 10:319–26

Kristjánsson A, Sigurjónsdóttir O, Driver J. 2010. Fortune and reversals of fortune in visual search: reward contingencies for pop-out targets affect search efficiency and target repetition effects. *Atten. Percept. Psychophys.* 72:1229–36

Kyllingsbaek S, Schneider WX, Bundesen C. 2001. Automatic attraction of attention to former targets in visual displays of letters. *Percept. Psychophys.* 63:85–98

Land MF. 2006. Eye movements and the control of actions in everyday life. *Prog. Retin. Eye Res.* 25:296–324

Lauwereyns J, Watanabe K, Coe B, Hikosaka O. 2002. A neural correlate of response bias in monkey caudate nucleus. *Nature* 418:413–17

Lawrence AD, Sahakian BJ, Hodges JR, Rosser AE, Lange KW, Robbins TW. 1996. Executive and mnemonic functions in early Huntington's disease. *Brain* 119:1633–45

Lees AJ, Smith E. 1983. Cognitive deficits in the early stages of Parkinson's disease. *Brain* 106:257–70

Lieberman MD. 2000. Intuition: a social cognitive neuroscience approach. *Psychol. Bull.* 126:109–37

Logan GD. 1985. Skill and automaticity: relations, implications, and future directions. *Can. J. Psychol.* 39:367–86

Lynch JC, Hoover JE, Strick PL. 1994. Input to the primate frontal eye field from the substantia nigra, superior colliculus, and dentate nucleus demonstrated by transneuronal transport. *Exp. Brain Res.* 100:181–86

Marewski JN, Gaissmaier W, Gigerenzer G. 2010. Good judgments do not require complex cognition. *Cogn. Process.* 11:103–21

Marín O, Smeets WJ, González A. 1998. Evolution of the basal ganglia in tetrapods: a new perspective based on recent studies in amphibians. *Trends Neurosci.* 21:487–94

Marsden CD. 1982. The mysterious motor function of the basal ganglia: the Robert Wartenberg lecture. *Neurology* 32:514–39

Matsumura M, Kojima J, Gardiner TW, Hikosaka O. 1992. Visual and oculomotor functions of monkey subthalamic nucleus. *J. Neurophysiol.* 67:1615–32

McNamara JM, Green RF, Olsson O. 2006. Bayes' theorem and its applications in animal behaviour. *OIKOS* 112:243–51

Middleton FA, Strick PL. 1996. The temporal lobe is a target of output from the basal ganglia. *Proc. Natl. Acad. Sci. USA* 93:8683–87

Middleton FA, Strick PL. 2000. Basal ganglia and cerebellar loops: motor and cognitive circuits. *Brain Res. Brain Res. Rev.* 31:236–50

Mink JW. 1996. The basal ganglia: focused selection and inhibition of competing motor programs. *Prog. Neurobiol.* 50:381–425

Mishkin M, Malamut B, Bachevalier J. 1984. Memories and habits: two neural systems. In *Neurobiology of Learning and Memory*, ed. G Lynch, JL McGaugh, NM Weinberger, pp. 65–77. New York: Guilford

Mishkin M, Ungerleider LG, Macko KA. 1983. Object vision and spatial vision: two cortical pathways. *Trends Neurosci.* 6:414–17

Miyachi S, Hikosaka O, Lu X. 2002. Differential activation of monkey striatal neurons in the early and late stages of procedural learning. *Exp. Brain Res.* 146:122–26

Miyachi S, Hikosaka O, Miyashita K, Kárádi Z, Rand MK. 1997. Differential roles of monkey striatum in learning of sequential hand movement. *Exp. Brain Res.* 115:1–5

Miyashita K, Rand MK, Miyachi S, Hikosaka O. 1996. Anticipatory saccades in sequential procedural learning in monkeys. *J. Neurophysiol.* 76:1361–66

Miyashita N, Hikosaka O, Kato M. 1995. Visual hemineglect induced by unilateral striatal dopamine deficiency in monkeys. *NeuroReport* 6:1257–60

Nakamura K, Hikosaka O. 2006. Role of dopamine in the primate caudate nucleus in reward modulation of saccades. *J. Neurosci.* 26:5360–69

Nambu A, Tokuno H, Takada M. 2002. Functional significance of the cortico-subthalamo-pallidal 'hyperdirect' pathway. *Neurosci. Res.* 43:111–17

Newell A, Rosenbloom PS. 1981. Mechanisms of skill acquisition and the law of practice. In *Cognitive Skills and Their Acquisition*, ed. JR Anderson, pp. 1–55. Hillsdale, NJ: Erlbaum

Parent A, Mackey A, Smith Y, Boucher R. 1983. The output organization of the substantia nigra in primate as revealed by a retrograde double labeling method. *Brain Res. Bull.* 10:529–37

Peck CJ, Jangraw DC, Suzuki M, Efem R, Gottlieb J. 2009. Reward modulates attention independently of action value in posterior parietal cortex. *J. Neurosci.* 29:11182–91

Rangel A, Camerer C, Montague PR. 2008. A framework for studying the neurobiology of value-based decision making. *Nat. Rev. Neurosci.* 9:545–56

Redgrave P, Rodriguez M, Smith Y, Rodriguez-Oroz MC, Lehericy S, et al. 2010. Goal-directed and habitual control in the basal ganglia: implications for Parkinson's disease. *Nat. Rev. Neurosci.* 11:760–72

Rothkopf CA, Ballard DH, Hayhoe MM. 2007. Task and context determine where you look. *J. Vis.* 7:1–20

Rudebeck PH, Behrens TE, Kennerley SW, Baxter MG, Buckley MJ, et al. 2008. Frontal cortex subregions play distinct roles in choices between actions and stimuli. *J. Neurosci.* 28:13775–85

Saint-Cyr JA, Ungerleider LG, Desimone R. 1990. Organization of visual cortical inputs to the striatum and subsequent outputs to the pallido-nigral complex in the monkey. *J. Comp. Neurol.* 298:129–56

Sakai K, Kitaguchi K, Hikosaka O. 2003. Chunking during human visuomotor sequence learning. *Exp. Brain Res.* 152:229–42

Salmon DP, Butters N. 1995. Neurobiology of skill and habit learning. *Curr. Opin. Neurobiol.* 5:184–90

Sato M, Hikosaka O. 2002. Role of primate substantia nigra pars reticulata in reward-oriented saccadic eye movement. *J. Neurosci.* 22:2363–73

Schultz W. 1998. Predictive reward signal of dopamine neurons. *J. Neurophysiol.* 80:1–27

Schultz W, Dayan P, Montague PR. 1997. A neural substrate of prediction and reward. *Science* 275:1593–99

Schwab RS, Chafetz ME, Walker S. 1954. Control of two simultaneous voluntary motor acts in normals and in parkinsonism. *AMA Arch. Neurol. Psychiatry* 72:591–98

Seitz AR, Kim D, Watanabe T. 2009. Rewards evoke learning of unconsciously processed visual stimuli in adult humans. *Neuron* 61:700–7

Sheinberg DL, Logothetis NK. 2001. Noticing familiar objects in real world scenes: the role of temporal cortical neurons in natural vision. *J. Neurosci.* 21:1340–50

Shiffrin RM, Schneider W. 1977. Controlled and automatic human information processing: II. Perceptual learning, automatic attending, and a general theory. *Psychol. Rev.* 84:127–90

Sigman M, Gilbert CD. 2000. Learning to find a shape. *Nat. Neurosci.* 3:264–69

Smith Y, Parent A. 1986. Differential connections of caudate nucleus and putamen in the squirrel monkey (*Saimiri sciureus*). *Neuroscience* 18:347–71

Sternberg S. 1969. Memory-scanning: mental processes revealed by reaction-time experiments. *Am. Sci.* 57:421–57

Takikawa Y, Kawagoe R, Itoh H, Nakahara H, Hikosaka O. 2002. Modulation of saccadic eye movements by predicted reward outcome. *Exp. Brain Res.* 142:284–91

Tanibuchi I, Kitano H, Jinnai K. 2009. Substantia nigra output to prefrontal cortex via thalamus in monkeys. I. Electrophysiological identification of thalamic relay neurons. *J. Neurophysiol.* 102:2933–45

Tatler BW, Hayhoe MM, Land MF, Ballard DH. 2011. Eye guidance in natural vision: reinterpreting salience. *J. Vis.* 11:5

Theeuwes J, Belopolsky AV. 2012. Reward grabs the eye: oculomotor capture by rewarding stimuli. *Vis. Res.* 74:80–85

Treisman AM, Gelade G. 1980. A feature-integration theory of attention. *Cogn. Psychol.* 12:97–136

Watanabe M, Munoz DP. 2010. Saccade suppression by electrical microstimulation in monkey caudate nucleus. *J. Neurosci.* 30:2700–9

Watanabe M, Munoz DP. 2011. Saccade reaction times are influenced by caudate microstimulation following and prior to visual stimulus appearance. *J. Cogn. Neurosci.* 23:1794–807

Wichmann T, DeLong MR. 1996. Functional and pathophysiological models of the basal ganglia. *Curr. Opin. Neurobiol.* 6:751–58

Wickens JR, Reynolds JN, Hyland BI. 2003. Neural mechanisms of reward-related motor learning. *Curr. Opin. Neurobiol.* 13:685–90

Xu-Wilson M, Zee DS, Shadmehr R. 2009. The intrinsic value of visual information affects saccade velocities. *Exp. Brain Res.* 196:475–81

Yamamoto S, Kim HF, Hikosaka O. 2013. Reward value-contingent changes of visual responses in the primate caudate tail associated with a visuomotor skill. *J. Neurosci.* 33:11227–38

Yamamoto S, Monosov IE, Yasuda M, Hikosaka O. 2012. What and where information in the caudate tail guides saccades to visual objects. *J. Neurosci.* 32:11005–16

Yarbus AL. 1967. *Eye Movements and Vision*. New York: Plenum

Yasuda M, Hikosaka O. 2013. Functional territories in the primate substantia nigra separately signaling flexible and stable values. *Soc. Neuroscience Meet. Plann.* Abstr. 291.05

Yasuda M, Yamamoto S, Hikosaka O. 2012. Robust representation of stable object values in the oculomotor basal ganglia. *J. Neurosci.* 32:16917–32

Yeterian EH, Pandya DN. 1991. Prefrontostriatal connections in relation to cortical architectonic organization in rhesus monkeys. *J. Comp. Neurol.* 312:43–67

Yin HH, Knowlton BJ. 2006. The role of the basal ganglia in habit formation. *Nat. Rev. Neurosci.* 7:464–76

Motion-Detecting Circuits in Flies: Coming into View

Marion Silies, Daryl M. Gohl, and Thomas R. Clandinin

Department of Neurobiology, Stanford University, Stanford, California 94305;
email: msilies@stanford.edu, daryl.gohl@gmail.com, trc@stanford.edu

Annu. Rev. Neurosci. 2014. 37:307–27

The *Annual Review of Neuroscience* is online at
neuro.annualreviews.org

This article's doi:
10.1146/annurev-neuro-071013-013931

Keywords

vision, motion computation, behavior, neurogenetics, evolution, *Drosophila*

Abstract

Visual motion cues provide animals with critical information about their environment and guide a diverse array of behaviors. The neural circuits that carry out motion estimation provide a well-constrained model system for studying the logic of neural computation. Through a confluence of behavioral, physiological, and anatomical experiments, taking advantage of the powerful genetic tools available in the fruit fly *Drosophila melanogaster*, an outline of the neural pathways that compute visual motion has emerged. Here we describe these pathways, the evidence supporting them, and the challenges that remain in understanding the circuits and computations that link sensory inputs to behavior. Studies in flies and vertebrates have revealed a number of functional similarities between motion-processing pathways in different animals, despite profound differences in circuit anatomy and structure. The fact that different circuit mechanisms are used to achieve convergent computational outcomes sheds light on the evolution of the nervous system.

Contents

INTRODUCTION

How do neural circuits capture and process sensory information to shape an organism's perception and guide its behavior? Each of the steps linking sensory input to behavioral output reflects specific neural computations. To dissect these steps, one must understand how animals respond to specific sensory cues, identify the neurons that carry out each computation, characterize their functional properties and connectivity, and use this information to determine how circuit properties relate to behavior. To achieve this synthesis requires quantitative measurements of behavior, detailed anatomical descriptions, physiological characterization of individual neurons, and targeted manipulation of neural activity (**Figure 1**). Computational models can then be used to put this wealth of information together. Finally, by examining evolutionarily conserved or convergent properties of the circuit, one can understand both the genetic constraints and optimal strategies underlying neural computation. Although progress has been made across different levels of analysis in many sensory systems, we do not comprehensively understand how even seemingly simple computations are implemented at the circuit level and are used to guide behavior.

The processing of visual motion cues provides an excellent paradigm for interrogating the functional properties of neural circuits and linking these properties to the animal's behavioral goals. The neural and behavioral mechanisms of motion detection have been studied in a diverse array of organisms (Clifford & Ibbotson 2002). Recently, however, a unique combination of genetic, physiological, and anatomical tools available in the fruit fly (Venken et al. 2011) have provided new mechanistic insights into the properties of motion-detecting circuits. In this review, we focus on these studies, synthesizing work that began in other insects (and has been reviewed previously; see Borst et al. 2010, Borst & Haag 2002) and relating these findings to parallel studies in vertebrates.

To navigate the world, many animals interpret patterns of optic flow, the apparent displacement of the visual scene across the retina caused by self-motion (Egelhaaf et al. 2002). Animals also encounter motion cues that are independent of self-motion and can be associated with predators,

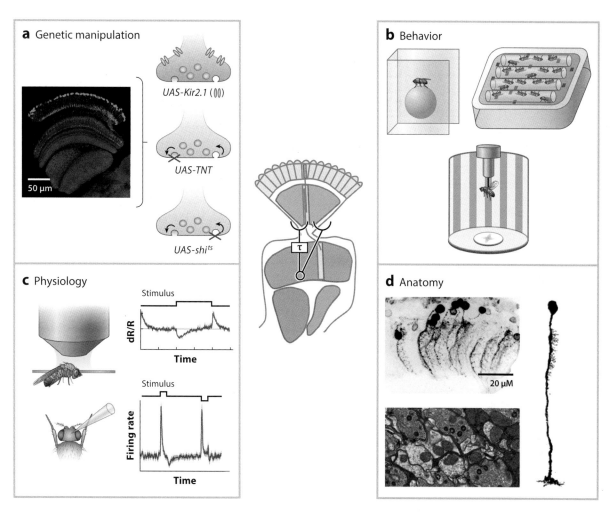

Figure 1

The computations underlying motion processing can be analyzed at many levels. (*a*) Studying motion vision in *Drosophila* allows genetic access to defined expression patterns in the visual system. Genetic manipulations can be used to inactivate defined cell types through expression of different effector molecules that lead to a hyperpolarized membrane potential (*Kir2.1*, indicated by different background color of synapse), blocking of vesicle fusion (*TNT*), or disruption of endocytosis (*shibire^{ts}*). (*b*) Flies display extensive behavioral responses to motion, which can be studied in single flies or populations in walking or flying flies. (*c*) Physiological responses to visual motion cues can be measured in in vivo calcium imaging experiments or electrophysiologically. Examples shown are the lamina neuron L2 responses to light flashes (*top*) and electrophysiological recordings of loom-sensitive lobula complex neurons (modified after de Vries & Clandinin 2012). (*d*) The anatomy of the fly visual system has been studied in exquisite detail through a combination of light microscopy illustrated by a confocal image of a clone of L3 cells (*top left*), an illustration of an L3 cell based on Golgi impregnations (*right*), and electron microscopic studies (*bottom left*). Ultimately the data derived from all these approaches will need to be synthesized and supplemented with algorithmic models to understand the logic of motion computation (*central panel*).

prey, mates, and other objects in the environment (Hildreth & Koch 1987). Thus, by decomposing the spatial pattern of motion signals across the visual field, the brain can derive a rich, highly salient and ethologically relevant set of environmental cues. At the most basic level, all these motion signals are detected by comparing luminance changes in space and time, with comparisons between exactly two points in space, at two points in time, constituting minimal motion signals (Hildreth & Koch

1987). These local signals allow neurons to become direction-selective, responding preferentially to motion in specific directions. Pooling these local motion signatures allows the brain to assemble a representation of the overall pattern of motion in the environment. This information can then be used to guide an appropriate response. Thus, motion detection represents a well-constrained neural computation that can direct a number of distinct, ethologically relevant behaviors.

DROSOPHILA AS A MODEL OF MOTION PROCESSING

Insects have been used as models of motion detection for more than 80 years; early studies examined fruit fly mutants affecting retinal function (Hecht & Wald 1934, Kalmus 1943). Subsequent studies developed computational models based on experiments performed originally in the beetle *Chlorophanus* and extended through a wealth of electrophysiological experiments and computational modeling based on large Diptera (Bishop & Keehn 1966, Borst et al. 2010, Eckert 1981, Franceschini et al. 1989, Hassenstein & Reichardt 1956, Hausen 1982, Marmarelis & McCann 1973, Reichardt & Poggio 1975). However, given its experimental tractability and the genetic advantages, the fruit fly *Drosophila melanogaster* is especially attractive for the dissection of motion-detecting circuits; quantitative measurements of its behavioral responses to motion began ~50 years ago (Buchner 1976, Götz 1964, Heisenberg & Wolf 1984, Tammero et al. 2004, Tammero & Dickinson 2002) (**Figure 1**). In addition, the relative numerical simplicity of the fruit fly's brain provides another critical advantage. Whereas the human optic nerve alone contains the axons of roughly one million neurons, the entire fly visual system contains ~100,000 neurons, ~6 orders of magnitude fewer neurons than that of the human brain (Morante & Desplan 2004). This relative anatomical simplicity, together with an extremely regular columnar organization, has been leveraged to construct a nearly complete connectome for the peripheral visual system (Fischbach & Dittrich 1989; Meinertzhagen & O'Neil 1991; Rivera-Alba et al. 2011; Takemura et al. 2008, 2013). In addition, *Drosophila* exhibits robust orienting responses to visual motion, so-called optomotor responses, which provide quantitative readouts of motion circuitry (Götz 1964, Heisenberg & Wolf 1984, Tammero et al. 2004, Tammero & Dickinson 2002). Finally, all these advantages can be paired with extremely sophisticated genetic tools that allow the activities of individual neurons or neuronal classes to be monitored or manipulated (Venken et al. 2011). Although many challenges remain in identifying and characterizing the neural circuits that process visual motion, the past several years have seen rapid progress. We are approaching the point where we can link specific neural computations to particular cells and circuits and define how the outputs of these circuits control behavior (**Figure 1**).

THE GROSS ANATOMY OF THE *DROSOPHILA* VISUAL SYSTEM

Drosophila, like most Diptera, devotes a large fraction of its brain to the processing of visual information (Morante & Desplan 2004). The *Drosophila* visual system consists of the retina and three distinct optic ganglia: the lamina, medulla, and lobula complex, composed of the lobula and lobula plate. The retina is composed of ~800 ommatidia, each of which contains 8 photoreceptor cells. The two inner photoreceptors, R7 and R8, are involved primarily in spectral discrimination and the detection of polarized light, whereas the outer photoreceptors, R1–R6, express a broad-spectrum opsin (*rh1*) and provide the major inputs to motion-detection circuits (Heisenberg & Buchner 1977, Morante et al. 2007, Wardill et al. 2012). *Drosophila* has a neural superposition eye, in which R1–R6 axons project into the first optic neuropil, the lamina, where a retinotopic map is formed, each point in visual space being represented by a single lamina column (Clandinin & Zipursky 2002, Meinertzhagen & O'Neil 1991, Rivera-Alba et al. 2011). The medulla, the largest optic ganglion, contains a plethora of cell types (Fischbach & Dittrich 1989, Takemura et al. 2008). These are also arranged largely in columnar fashion and form a diverse array of feedforward and feedback connections within the medulla, all other optic ganglia, the central brain, and the contralateral optic lobe (Fischbach & Dittrich 1989). Visual signals are further processed in the lobula and lobula plate, where in the blow fly a population of ~60 lobula plate tangential cells (LPTCs) can be identified that are tuned to specific features of the visual stimulus (such as object size and direction of motion) (Borst & Haag 2002). LPTCs with similar properties, as well as neurons that respond to looming signals, have also been described in *Drosophila* (de Vries & Clandinin 2012, Joesch et al. 2008, Schnell et al. 2010). Finally, the visual system is connected to the central brain by ~2,000 fibers, connecting the two visual hemispheres with one another, with motor centers, and with regions of the brain thought to play roles in sensory integration and memory formation (Fischbach & Dittrich 1989, Heisenberg & Wolf 1984).

MODELS OF MOTION DETECTION IN INSECTS

A strong theoretical framework for understanding motion vision was established through the pioneering work of Hassenstein and Reichardt, who used quantitative behavioral studies of the beetle *Chlorophanus* to develop an algorithmic model for elementary motion detection (Hassenstein & Reichardt 1956). The Hassenstein-Reichardt correlator (HRC) model relies on comparing the changes in contrast between two points in space at two points in time. This model provided a critical framework for a quantitative analysis of behavioral and neuronal responses to motion and made a number of predictions that have been verified experimentally (Hassenstein & Reichardt 1956, Reichardt 1961). These predictions included a quadratic relationship between the magnitude of the stimulus contrast and the size of the motion response, as well as the existence of a class of so-called "reverse phi" visual illusions (Buchner 1976, Clark et al. 2011, Götz 1964, Haag et al. 2004, Katsov & Clandinin 2008, Tuthill et al. 2011). To achieve direction selectivity, the HRC proposes an operation in which the luminance signal from one photoreceptor is compared, after a characteristic time delay, with the signal from an adjacent photoreceptor through a multiplication step (**Figure 2a,b**). This circuit responds strongly only when presented with a motion signal in one direction. By then using arithmetic subtraction to compare the signals produced by one such detector with those of a mirror-symmetric circuit, a signal is generated whose magnitude corresponds to how closely the stimulus matches the filter corresponding to the time delay and whose sign indicates the direction of motion. In spite of the original HRC's explanatory power, researchers have long thought that the sign-correct multiplication implemented by the HRC is biologically implausible. Thus investigators have constructed elaborated models in which contrast changes in different directions, lightening and darkening, are processed separately and then combined (Clark et al. 2011, Eichner et al. 2011, Franceschini et al. 1989, Reichardt 1961). To do this, visual inputs are "half-wave rectified," meaning that each channel transmits only information about either brightening or darkening. Because any point in space can get either brighter or darker over time (corresponding to a moving bright or dark edge), there are four possible combinations of contrast changes at two points in space. The first elaborated model, the so-called four-quadrant multiplier, therefore had four separate channels, one for each combination of contrast change at two points in space, the outputs of which were then summed to generate a signal indistinguishable from that of the original HRC (Reichardt 1961). Moreover, fruit flies can respond to all four possible combinations of contrast (Clark et al. 2011, Eichner et al. 2011, Joesch et al. 2013). A long-standing observation, however, has been that flies can detect apparent motion signals associated with sequential presentations of pairs of bars, even when the time delay between the presentation of the first bar and that of the second is very long (Clark et al. 2011, Egelhaaf & Borst 1992, Eichner et al. 2011). This observation

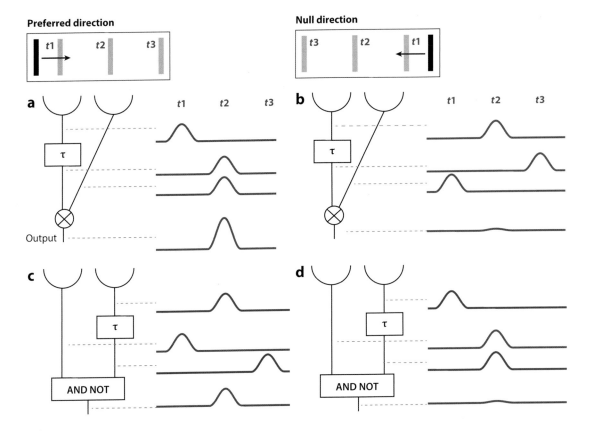

Figure 2

Computational models of elementary motion detection. These schematics illustrate proposed algorithms for elementary motion detection for the Hassenstein-Reichardt correlator (HRC) model (*a,b*) and the Barlow-Levick model (*c,d*). The top panel shows a bar moving in either the preferred direction (*left panels*) or the null direction (*right panels*) for a given elementary motion detector module, with the bar crossing the three gray areas at three sequential times (*t1, t2,* and *t3*). (*a*) For the HRC model, when a bar moving in the preferred direction passes the first photoreceptor, it generates a signal, which is then delayed such that it arrives at a downstream multiplier element coincident with the signal from the bar passing the second photoreceptor; these signals are then multiplied to generate a motion signal. (*b*) When the bar is moving in the null direction, the two signals of the HRC do not arrive at the same time, and the multiplication step does not generate a motion signal. (*c*) In the Barlow-Levick model, motion in the preferred direction generates an excitatory signal followed by a delayed inhibitory signal. Because the excitatory signal arrives at a downstream neuron prior to the arrival of the inhibitory signal, a motion signal is generated. (*d*) For the Barlow-Levick model, motion in the null direction generates a delayed inhibitory signal before the excitatory signal, and thus both signals arrive at the downstream neuron simultaneously, causing the excitation to be nullified by the coincident inhibitory signal.

is puzzling because conventional HRC models take changes in luminance as their inputs rather than the luminance values themselves. Because there is no change in luminance after a static bar has been presented for a long time, HRC models that incorporate physiologically plausible filtering characteristics do not predict significant responses to such stimuli. As a result, recent modeling efforts have incorporated "DC components" (DC stands for direct current, an electronics term that indicates a tonic signal) that mix a small amount of information about absolute luminance values into the contrast signal, creating hybrid HRC-like models that can account for many published observations (Eichner et al. 2011). How the circuit might capture these DC signals remains incompletely understood, and it is unclear whether these models accurately predict the relative strength of responses to apparent motion signals at both short and long timescales. Moreover, neither classical HRC models nor more recent hybrid HRC-like models accurately predict the response of flies to additional types of apparent motion stimuli (Clark et al. 2014). Finally, although elaborated versions of the HRC can explain psychophysical responses to motion in humans (van Santen & Sperling 1984), an algorithmically different model, the motion energy model, is thought to underlie direction selectivity in the vertebrate cortex (Adelson & Bergen 1985, Emerson et al. 1992). In addition, a model with some similarities to an HRC but with very different neural implementation, the Barlow-Levick model, is thought to account for the emergence of direction selectivity in the vertebrate retina (Barlow & Levick 1965, Briggman et al. 2011, Yonehara et al. 2013). Here, luminance signals from two photoreceptors are combined through a logical AND-NOT gate utilizing a temporally delayed, inhibitory mechanism (**Figure 2c,d**). As studies identify the neural circuits that carry out motion detection, a major challenge will be to link identified circuits to an existing or novel computational model. However, it remains unknown where, how, and, in fact, whether the precise neural computations implied by these models are implemented in the fly brain.

INSIGHTS FROM ANATOMY

An important prerequisite to the identification of motion-detecting circuits is a detailed understanding of visual system architecture. Work using both classical Golgi stains as well as more modern single-cell labeling techniques has described almost 100 morphologically distinct cell types in the optic lobe (Fischbach & Dittrich 1989, Morante & Desplan 2008, Raghu & Borst 2011, Tuthill et al. 2013). The axonal and dendritic structures in the visual system are highly stereotyped both within and between animals. These neurons are of diverse forms, with columnar projection neurons in the lamina, the medulla, and the lobula [including, for example, the lamina neurons L1–L5 and the transmedullary (Tm) and transmedullary Y (TmY) cells], as well as a host of feedback neurons between all optic ganglia. In addition, the medulla houses medulla intrinsic (Mi) neurons, distal medulla (Dm) and proximal medulla (Pm) wide-field cells, or medulla tangential (Mt) cells. Further cell types were described in downstream neuropils, most prominently T4 and T5 cells, which provide input to the lobula plate and send dendrites into the medulla and lobula. The lobula plate also houses many LPTCs, such as the horizontal system (HS) and the vertical system (VS) neurons (Borst & Euler 2011, Fischbach & Dittrich 1989).

Pioneering efforts to infer functional organization from anatomy relied on densitometric measurements of dendritic and axonal projections in distinct layers of the visual system and identified three proposed pathways: the L1, the L2, and the L3/R8 pathway (Bausenwein et al. 1992). The L1 and L2 pathways were proposed relays between lamina neuron inputs and the T4 and T5 cells, respectively. T4 and T5 were of particular interest because these cells each have four distinct subtypes that project specifically to one of the four layers in the lobula plate. Functional labeling studies demonstrated that these layers responded preferentially when the flies were shown different directions of motion (Bausenwein & Fischbach 1992, Buchner 1984). Several medulla neurons were proposed to connect L1 and L2 to T4 and T5. In particular, Mi1 was a candidate relay from L1 to T4, and Tm1 represented a possible connection between L2 and T5 (**Figure 3**). These two pathways were proposed to subserve distinct functions, a claim that was only confirmed 25 years later (see below; Bausenwein et al. 1992).

Finally, whereas the L1 and L2 pathways were predicted to function in motion detection, the L3/R8 pathway was implicated in form and color vision (Bausenwein et al. 1992). Although we focus here on studies in *Drosophila*, we also note that both anatomical and physiological studies in other Diptera identified similar cell types and suggested that these neuron types are involved in an evolutionarily conserved task (Strausfeld 1989).

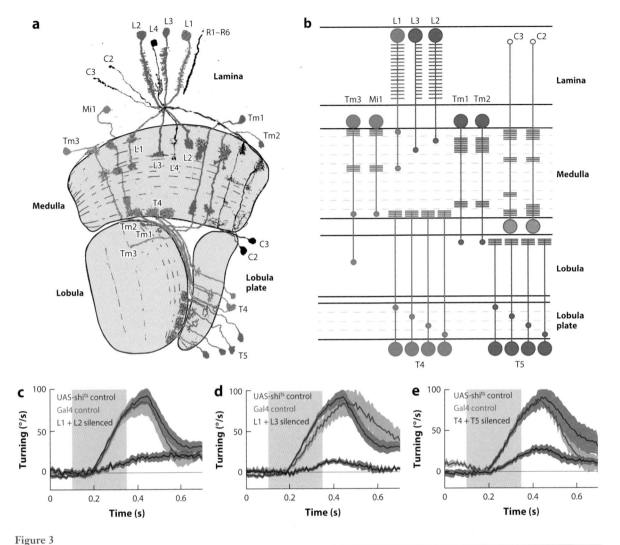

Figure 3

(*a*) Composite of camera lucida drawings based on Golgi stainings. Shown are various cell types in the optic lobe: neurons and candidate neurons of the light edge–detecting pathway (*green*), the dark edge–detecting pathway (*magenta*), and other neurons discussed in this review (modified after Fischbach & Dittrich 1989). (*b*) Schematic representation of neurons and candidate neurons of the light (*green*) and dark edge–detecting pathways (*magenta*) as well as the C2 and C3 feedback neurons. Dendrites are indicated as blocks of horizontal lines and output synapses indicated by small circles. Dashed lines indicate medulla and lobula plate layers. (*c–e*) Turning behavior of walking flies in response to rotating square wave gratings at 100% contrast, presented for 250 ms indicated by the gray patch. Shading denotes +/−1 SEM. Panels *c* and *d* are modified after Silies et al. 2013.

Because investigators could not directly visualize synapses, the accuracy of these anatomical predictions was uncertain. A higher-resolution view of the visual system neuroanatomy was provided by serial electron microscopic (EM) reconstructions, focusing first on the lamina then extending to the medulla (Meinertzhagen & O'Neil 1991; Rivera-Alba et al. 2011; Takemura et al. 2008, 2011, 2013). In the lamina, R cells make strong synaptic connections with three projection neurons, L1–L3 (Meinertzhagen & O'Neil 1991). These studies also confirmed the predicted synaptic connections between L2 and Tm1 and identified Tm2 as well as another lamina projection neuron

L4 as prominent connections downstream of L2 (Meinertzhagen & O'Neil 1991, Takemura et al. 2011). This potential pathway was further supported by EM reconstructions of an entire medulla column and its direct neighbors (Takemura et al. 2013). Synaptic connections were also identified in the predicted L1-Mi1-T4 pathway, and the medulla interneuron Tm3 was identified as a second prominent postsynaptic target of L1. These reconstructions also demonstrated that Mi1 and Tm3 provide the main inputs onto T4. In addition, these EM data used the relative spatial distribution of synapses to construct anatomical receptive fields. These anatomical relationships between synaptic inputs revealed that Mi1 and Tm3 inputs are spatially displaced as seen from the perspective of a T4 neuron, suggesting a mechanism by which motion-detecting circuits could achieve correlation between two spatially separated inputs, a central feature of all motion-detector models. In addition, this anatomical displacement onto T4 dendrites covaried with the directional preference of the lobula plate layer to which a given T4 cell projected, tightening the link between these projections and the emergence of direction selectivity (Takemura et al. 2013).

The pathways identified by these anatomical studies provide an important foundation for identifying the circuits that process visual motion. However, these proposed circuits simplify the circuitry of the lamina and medulla. L1, for example, receives synaptic inputs from L5, C2, and C3 cells that are approximately as numerous as those L1 makes onto Mi1 and Tm3, which suggests that feedback synapses and lateral interactions constitute major circuit elements. Indeed, the lamina is a highly interconnected network where about one-third of all possible connections between neurons are realized (Rivera-Alba et al. 2011). In addition, some discrepancies exist between EM studies (Takemura et al. 2008, 2011, 2013), possibly reflecting variability between individual animals. Finally, although the L3 pathway is omitted from the proposed motion-detection pathways, it also provides input to motion-detecting circuits (Silies et al. 2013, Tuthill et al. 2013), and L3 is connected to both Mi1 and C3 (Takemura et al. 2013). Thus, cross talk between anatomically identified pathways likely shapes responses to visual motion.

Functional validation of these predicted circuits will reveal the extent to which they account for the implementation of motion detection. Several efforts are under way to create tools that allow genetic access to any cell type of interest (Gohl et al. 2011; Lai & Lee 2006; Lee & Luo 1999; Luan et al. 2006; Pfeiffer et al. 2008, 2010). Such tools, combined with the wealth of genetic effectors that can be used to manipulate and monitor neuronal activity, enable detailed functional characterization of motion-detection circuits.

PHYSIOLOGICAL CHARACTERIZATION OF MOTION CIRCUITS

Both behavioral and physiological measures can be used to examine the function of specific neurons in motion detection. Whereas behavioral experiments are ultimately required to understand the link between specific computations and their functional output, physiological experiments are necessary to understand how individual circuit elements contribute to information processing. One key feature of motion vision is the extraction of directional signals from the patterns of luminance changes on the retina. Identifying the cell types in which direction selectivity first arises and learning the mechanisms by which this occurs would represent significant progress in understanding motion computation.

Physiological recordings in larger Diptera demonstrated that the lobula plate contains neurons with direction-selective responses to wide-field visual motion (Borst & Haag 2002). In *Drosophila*, the HS and VS LPTCs respond to motion along the four cardinal axes (Joesch et al. 2008, Schnell et al. 2010). These neurons project their dendrites into specific layers of the lobula plate and send their axons into the central brain, where they interact with descending motor neurons (Borst & Haag 2002, Haag et al. 2007, Wertz et al. 2008). Metabolic labeling experiments in which flies were chronically exposed to motion showed that each layer of the four layers of the lobula plate responds preferentially to one axial direction of motion (Bausenwein & Fischbach 1992, Buchner et al. 1979). These layers also contain the axonal projections of T4 and T5 neurons. Although the prevailing view for many years was that direction selectivity arises for the first time in LPTCs, recordings from blow flies suggested that direction-selective signals might exist at least one synapse earlier in visual pathways, in T4 (Borst & Haag 2002, Douglass & Strausfeld 1996). In vivo calcium imaging experiments using genetically encoded indicators recently demonstrated that T4 and T5 neurons are direction selective (Maisak et al. 2013). Remarkably, each of the T4 and T5 subtypes responds selectively to one of the four cardinal directions of motion, a tuning pattern consistent with their pattern of layer-specific targeting in the lobula plate. Genetic silencing of these cells demonstrated that HS and VS cells are downstream of T4 and T5 and that T4 and T5 preferentially pass information about moving light and dark edges, respectively (Maisak et al. 2013, Schnell et al. 2012).

Although physiological measurements from medulla neurons have not yet been described in fruit flies, the response properties of the lamina neurons that serve as inputs to motion-detecting circuitry have been characterized in detail. Using genetic silencing techniques, the lamina neurons L1 and L2 were initially demonstrated to be redundantly required for optomotor responses (Rister et al. 2007). Subsequent electrophysiological recordings in LPTCs as well as behavioral experiments showed that silencing L1 or L2 individually specifically affected the response to moving light or moving dark edges, respectively, providing direct evidence that they feed into two distinct motion-processing circuits (Clark et al. 2011, Joesch et al. 2010). This finding led to the appealing hypothesis that L1 and L2 might themselves rectify incoming signals to segregate responses to lightening and darkening (Reiff et al. 2010). However, both neurons responded strongly, and approximately linearly, to both contrast increments and decrements (Clark et al. 2011). Detailed characterization of the L2 receptive field revealed an antagonistic center-surround organization that is shaped by γ-aminobutyric acid (GABA)-ergic lateral interactions. These interactions linearize responses to contrast and facilitate selective responses to small moving dark objects (Freifeld et al. 2013). Furthermore, examining the physiological responses of another input, the lamina neuron L3, to moving dark edge detection reveals that this cell is rectified, displaying a preference for darkening, in agreement with the original hypothesis that there might be physiological specialization early in visual processing (Franceschini et al. 1989, Silies et al. 2013).

BEHAVIORAL APPROACHES TO CHARACTERIZING MOTION-DETECTING CIRCUITRY

Behavioral and physiological assays have coevolved with increasingly sophisticated stimulus designs, data analysis, and genetic tools. Behavioral experiments have been a powerful tool for analyzing the effects of specific mutations or neural manipulations on visual processing. In this section, we summarize what these manipulations have revealed about motion-detecting circuits, starting in the outermost layers of the visual system and progressing inward.

Photoreceptor Inputs

One of the initial draws of using fruit flies as models of motion detection lay in the appeal of using genetic screens to identify loci involved in motion perception (Heisenberg et al. 1978, Heisenberg & Wolf 1984). However, with a few notable exceptions, these initial studies identified predominantly mutations affecting photoreceptor function. A central finding was that mutants lacking the outer photoreceptors (R1–R6) showed defective optomotor responses, providing functional evidence that these cells are required for motion detection (Heisenberg & Buchner 1977). This work also identified an unusual allele in *optomotor blind* (*omb*)/*bifid*, a developmental transcription factor (Heisenberg et al. 1978, Pflugfelder et al. 1992). Remarkably, homozygous *omb^H31* mutant flies lacked a lobula plate but retained the medulla and displayed fundamental defects in motion detection (Heisenberg et al. 1978). However, further forward genetic approaches using chemical mutagenesis lacked the desired specificity, and it rapidly became clear that single genes are rarely linked to specific circuit elements or single cell types (Choe & Clandinin 2005). Thus, the broad approach of using genetic techniques to explore motion-detecting circuits lay dormant for many years until new techniques evolved to allow cell-type-specific manipulations of neuronal activity.

Lamina Neurons

The emergence of genetic silencing techniques enabled examination of circuitry downstream of photoreceptors. Several effector molecules were discovered that when expressed in neurons block neurotransmitter release (*TNT*), or block endocytosis, and thus deplete the vesicle pool (*shibire^ts*) or hyperpolarize the cell (*Kir2.1*) (Johns et al. 1999, Kitamoto 2001, Sweeney et al. 1995) (**Figure 1**). Flies in which L1 and L2 were both genetically silenced could not respond to visual motion cues, providing the first causal evidence that these cells serve as inputs to motion detectors (Rister et al. 2007; see also **Figure 3**). These studies used grating stimuli containing both light and dark edges and suggested that L1 and L2 function redundantly under many conditions, with a specialized function for the L2 pathway apparent only at very low contrast. A subsequent study identified a phenotype associated with inactivating L2 in freely moving flies exposed to complex motion stimuli and argued for a specialized input from L2 into motion-detecting pathways (Katsov & Clandinin 2008). Subsequent behavioral studies designed more specific stimuli and corroborated the functional segregation of motion-detection circuits, observed using physiological approaches, into two distinct pathways, one selective for moving light edges and the other selective for moving dark edges (Clark et al. 2011, Joesch et al. 2010).

Although a strict separation of light edge– and dark edge–detecting pathways represents an attractive model, the actual implementation of this edge selectivity appears more complex. First, L1 and L2 are not themselves physiologically specialized to detect specific edge types. Thus, the edge selectivity associated with either L1 or L2 silencing must arise in downstream circuitry. Second, blocking L2 function did not abolish all behavioral responses to moving dark edges, suggesting additional inputs to dark edge detection (Clark et al. 2011, Silies et al. 2013, Tuthill et al. 2013). Indeed, a forward genetic screen based on neuronal silencing identified L3 as an unexpected input to motion detection (Silies et al. 2013). Combinatorial silencing experiments demonstrated that both L3 and L1 genetically interact with L2 and make critical contributions to dark edge detection (**Figure 3**). This work expanded the previous model describing the input architecture to motion-detecting circuits. Consistent with this new view, early Golgi-labeling studies noted that cells in the "L2 pathway" often share projections between medulla layers M2 and M3, and sometimes also M1, the layers targeted by L2, L3, and L1, respectively (Bausenwein et al. 1992). Thus, cross talk between these pathways may occur in the distal medulla. Finally, this more complex view, incorporating inputs from L1–L3, is still oversimplified. Recent work addressed the full repertoire of cell types in the lamina by specifically manipulating the activity of each of the 12 cell types in this ganglion (Tuthill et al. 2013). Genetic silencing and activation of these neurons, combined with a large battery of stimuli, demonstrated that many lamina neurons, including many neurons that have feedback projections from the medulla to the lamina, displayed defects in motion-evoked behavioral responses under at least some conditions. Thus, motion-processing circuits are likely to be highly distributed, with lamina neurons exerting complex effects in shaping responses to visual stimuli.

This distributed processing network suggests that motion detection is not a monolithic computation, but rather a series of processing steps that can be flexibly tuned and combined to guide different motion-evoked behaviors. Although most behavioral studies have focused on optomotor turning responses evoked by rotational motion, other visual motion cues exist and can modulate distinct behaviors. For example, translational motion cues can modulate the forward walking speed of flies (Götz & Wenking 1973, Katsov & Clandinin 2008, Silies et al. 2013). Only L2 and L3 silencing affect this behavior, demonstrating an early behavioral specialization in motion-processing circuits (Katsov & Clandinin 2008, Silies et al. 2013). Recent characterization of the movements of freely moving flies has revealed considerable behavioral complexity, so it will be interesting to determine whether other specific motion cues can influence this rich behavioral repertoire and reveal additional specialization in motion-detecting circuits (Branson et al. 2009; Fry et al. 2009; Straw et al. 2010, 2011).

Lamina and Medulla Interneurons

Because motion vision requires comparing signals from two neighboring points in space, lateral connections between the columnar inputs defined by L1–L3 are required. Although anatomical studies identified one lateral connection between columns taking place at the level of T4 dendrites (Takemura et al. 2013), many other intercolumnar cells that could mediate such comparisons exist. One of these, the second-order interneuron L4, which is located in the lamina, receives its main inputs from L2 cells rather than from photoreceptors (Meinertzhagen & O'Neil 1991, Strausfeld & Campos-Ortega 1973). Anatomical studies of L4 suggested a critical function in motion-detecting circuits (Braitenberg 1970, Rister et al. 2007, Takemura et al. 2011). However, behavioral data examining the function of L4 in motion detection remain conflicting (Silies et al. 2013, Tuthill et al. 2013, Zhu et al. 2009). One study reported that specific inactivation of L4 caused defects in behavioral responses to many motion stimuli (Tuthill et al. 2013), whereas another study could not isolate a requirement for L4 in motion detection, even when L4 was silenced in combination with L1–L3 (Silies et al. 2013). Furthermore, the physiological responses of L4 to light were not affected by silencing its anatomically defined presynaptic input L2 (Silies et al. 2013, Takemura et al. 2011). These studies examined flies in different behavioral states (walking versus flying) and used different effector molecules to achieve silencing, perhaps accounting for these discrepancies. Silencing L4 had strong effects on the ability of flies to fixate a moving stripe and disrupted a visual startle response (Silies et al. 2013, Tuthill et al. 2013). Fixation responses utilize neural pathways that are distinct from those in motion detection (Bahl et al. 2013), which suggests that L4 provides a critical input, an idea that can be easily tested.

In the medulla, as discussed above, anatomical predictions suggest that the medulla interneurons Tm1, Tm2, Mi1, and Tm3 are central to motion detection. However, none of these neurons have yet been characterized physiologically nor have they been genetically silenced. Thus, their precise role in motion-detecting circuitry remains unclear. Behavioral tests in combination with specific genetic silencing of these neurons will be valuable to test anatomical predictions.

Two multicolumnar interneurons C2 and C3 with extensive arborization in the medulla and GABAergic feedback projections onto L1 and L2 in the lamina have been characterized at this level (Fei et al. 2010, Kolodziejczyk et al. 2008, Meinertzhagen & O'Neil 1991, Takemura et al. 2008, Tuthill et al. 2013; **Figure 3**). Silencing C2 alone had only very limited behavioral consequences, whereas silencing C3 caused defects in responses to motion moving from back to front across the eye, a phenotype that was enhanced by simultaneously silencing both cells. Some of the medulla arbors of C3 are oriented along the anteroposterior axis, which suggests that these cells relay directional inhibition back to earlier stages of visual processing (Tuthill et al. 2013).

Thus, behavioral studies of interneurons have shown that lateral interactions shape responses to visual motion. At the same time, anatomical studies have focused on the numerically strongest synaptic connections to derive potential pathways but have also revealed many other connections. Testing these connections requires both specific genetic tools to manipulate neuron types individually and in combination as well as carefully constructed stimuli and behavioral assays that are matched to particular aspects of motion computation.

Connecting the Medulla to the Lobula Plate

Although the medulla, with its ~60 neuron types, remains a central challenge in motion-processing circuitry, a clearer picture emerges at the level of the medulla outputs. T4 and T5 neurons are strong candidates for being involved in motion vision based on anatomy, but flies that lack both T4 and T5 outputs also show strong deficits in optomotor behaviors (Bahl et al. 2013) (**Figure 3**). T4 appears to relay information about moving light edges, whereas T5 preferentially relays information about moving dark edges (Maisak et al. 2013). However, the stimulus used to assign these neurons to specific pathways behaviorally detects imbalances between responses to light edge motion and those to dark edge motion, rather than detecting the absolute strength of the behavioral response to each edge type (Clark et al. 2011). Thus, it is unclear whether T4 and T5 represent the sole outputs of the light and dark edge pathways to downstream motion-processing circuits. Given the high level of interconnectedness, the numerous feedback interactions in the fly optic lobe, and the number of morphologically defined output neurons from the medulla, it would be surprising if this hypothesis was true.

T4 and T5 make synapses in the layers of the lobula plate, which house the dendrites of the wide-field LPTCs of the HS and VS systems. Behavioral experiments using microdissection studies and stimulation in larger Diptera as well as the omb^{H31} mutant that lacks these cells argue that the output of LPTCs guides optomotor behavior (Heisenberg et al. 1978). Verifying this hypothesis using genetic silencing techniques is complicated by the fact that these cells are strongly connected through gap junctions and thus may withstand at least some manipulations to silence their synaptic outputs (Haag & Borst 2004, Schnell et al. 2010). However, as we discuss further below, optogenetic stimulation of LPTCs triggers head-turning responses, demonstrating that these cells are sufficient to drive a part of the behavioral motion-response repertoire (Haikala et al. 2013).

WHAT DO THE OPTIC LOBES TELL THE BRAIN ABOUT VISUAL MOTION?

The optic lobes of the fly comprise at least half of all the neurons in the CNS, containing more than 50,000 neurons per hemisphere. This extensive network is connected to the central brain by a comparatively small number of axon fascicles, passing in both directions (Fischbach & Dittrich 1989). As a result of this processing bottleneck, the information that passes through these connections likely represents highly derived visual features. Recent studies have directly examined the activity patterns seen in output neurons of the visual system, cells whose axons project into various parts of the central brain. First, a small population of neurons that have dendritic arbors in the lobula and lobula plate are exquisitely tuned to the motion of looming objects, stimuli that have long been associated with escape responses in both flies and other insects (Card & Dickinson 2008; de Vries & Clandinin 2012; Fotowat et al. 2009; Joesch et al. 2008, 2010; Schnell et al. 2010). Using a combination of genetic silencing in sighted flies and optogenetic stimulation in blind flies, these loom-detecting neurons were shown to be both necessary for normal escape responses and sufficient to trigger escape behavior absent any visual cue (de Vries & Clandinin 2012). In a related vein, detailed electrophysiological studies of lobula columnar neurons (LCNs) revealed that these cells respond to "visual primitives," features of a scene such as specific motion directions that could be used to guide particular behavioral responses (Mu et al. 2012). These studies also suggested possible downstream processing mechanisms because individual examples of LCNs appeared to display only subtle tuning for specific stimuli. Neurons of the same class project to optic glomeruli, structures in which inputs from LCN neurons with similar tuning properties converge. By analogy with olfactory glomeruli, then, these studies proposed that optic glomeruli contain neurons that integrate the responses of LCN neurons with overlapping tuning to enhance the robustness of the response (Mu et al. 2012). Finally, optogenetic stimulation of a subset of LPTCs, the HS cells, revealed that activation of these cells in one of the two optic lobes using a so-called step function opsin with prolonged inactivation kinetics can drive both head and body turns in the yaw plane (Haikala et al. 2013). Because these cells provide direct inputs to descending motor neurons in the blow fly (Haag et al. 2010, Huston & Krapp 2008, Strausfeld & Gronenberg 1990), each optic lobe likely controls the activation of a motor program that can promote turning in one direction.

Taken together, these studies argue that output neurons in the optic lobe signal the presence of particular visual features that are behaviorally salient and have access to specific subsets of the behavioral repertoire. Nonetheless, it seems improbable that a given motion signal will have invariant access to a specific behavioral response because the meaning of any particular visual feature may depend on the animal's behavioral state. For example, a stationary animal may construe the presence of a looming object as a close approach that should be avoided, whereas an animal that is flying may detect a looming signal as indicative of a perch. Thus, how these visual outputs are gated by information about the animal's behavioral state and goals remains an open question.

In addition to guiding relatively simple behaviors, visual motion signals can also be integrated with other sensory modalities to subserve more complex navigational tasks. For example, optic flow patterns can be combined with olfactory cues to guide search strategies in both freely moving and tethered flies (Duistermars & Frye 2008, Frye & Dickinson 2004). In addition, these visual signals are also gated by wind, causing profound shifts in flies' responses to visual motion signals (Budick et al. 2007). Finally, visual place learning exists in both flies and other insects (Mizunami et al. 1998, Ofstad et al. 2011), and even though flies match visual panoramas to remembered patterns, visual motion cues created by self-motion through these scenes may also guide the trajectories chosen. Identifying and characterizing the brain regions that integrate visual motion signals with other sensory modalities, or that use motion cues to construct neural representations of visual scenes, remain challenges for future work.

WHAT DOES THE REST OF THE FLY'S BRAIN TELL THE OPTIC LOBE ABOUT MOTION?

Visual motion signals are shaped by the statistics of the animal's environment, by the movements of the animal itself, and by the interaction between these two. On evolutionary timescales, information about both of these contrast distributions can be programmed into visual circuitry, as exemplified by the distinct tuning properties of motion-sensitive neurons in different insect species that move at different speeds (O'Carroll et al. 1996). Recent work has demonstrated that both fly and human visual systems exploit the statistics of natural scenes to construct motion detectors that are differentially tuned to the motion of light and dark edges (Clark et al. 2014). In addition, changes in visual experience, such as dramatic alterations in light exposure over hours to days, can change the inputs to motion-processing pathways by altering the size and number of synaptic connections in the lamina (Barth et al. 1997, 2010). Furthermore, over timescales spanning seconds to hours, motion-detecting circuits are also regulated by feedback inputs from the central brain. In particular, the tuning properties of motion-sensitive neurons are altered by the animal's behavioral state. Both calcium-imaging studies as well as electrophysiological recordings of awake-behaving flies have revealed that motion-sensitive neurons increase their gain when animals are behaviorally active and shift their tuning properties to respond more strongly to higher contrast frequencies while flying (Chiappe et al. 2010, Maimon et al. 2010). Building on previous pharmacological studies in larger insects, octopaminergic inputs to the visual system shape these adjustments (Longden & Krapp 2010, Sombati & Hoyle 1984, Suver et al. 2012). Indeed, globally exposing the brain to octopamine had very similar effects to inducing flight, arguing that this shift in behavioral state is signaled by increasing octopamine levels. Although it is unclear whether octopamine acts directly on the direction-selective neurons that change their tuning properties as a function of behavioral state, or acts via some more complex circuit that may impinge on multiple levels of visual processing, these studies demonstrate that at least some motion-processing circuits have access to information about ongoing behavior.

Given this evidence for direct inputs from the central brain acting on motion-detecting circuits in the optic lobes, one might consider the level of detail the central brain provides about behavior. In largely stationary environments, most of the optic flow pattern will be determined by self-motion. As a result, relaying efference copy information about the intended motor program to the visual system would enable very powerful strategies for detecting salient, unexpected motion cues. Information about the animal's intended trajectory could construct a model of the expected motion pattern, which could be used to eliminate anticipated motion signals. Thus, the remaining discrepant motion signals could identify variation in motor outputs or environmental perturbations or could be attributed to the motion of other objects. However, although the utility of such a model is appealing, we do not know whether the optic lobe has access to such detailed information about behavior.

HOW DO MOTION-PROCESSING CIRCUITS EVOLVE?

The ability to detect visual motion is likely universal among animals with image-forming eyes, reflecting the central utility of motion detection for navigation, prey capture, predator avoidance, and the pursuit of conspecifics (Nakayama 1985). In primates, motion-processing circuits have been studied extensively in the cortex, whereas in other vertebrates, including mice and salamanders, direction-selective responses to motion are prominent in retinal ganglion cells and the output channels of the retina and have also been studied in cortex (Born & Bradley 2005, Gollisch & Meister 2010, Masland 2012). In several of these systems, the circuit mechanisms that induce direction selectivity have been characterized in detail. Although we do not know the extent to which the fly achieves direction selectivity using similar circuit mechanisms, investigators have identified a number of parallels in processing strategies. Motion processing is fundamentally constrained by the statistics of the environment (Fitzgerald et al. 2011), and recent work argues that these statistics have imposed particular algorithmic structures on neural circuitry spanning the evolutionary tree (Clark et al. 2014). Morphological parallels between the vertebrate retina and the fly optic lobe have long been noted (Ramón y Cajal & Sanchez 1915, Sanes & Zipursky 2010). The molecular underpinnings of eye development are evolutionarily widespread, and there are functional parallels between regions of the fly central brain and the vertebrate visual cortex (Erclik et al. 2009, Seelig & Jayaraman 2013). Recent work has extended these parallels to the functional level, comparing the properties of lamina neurons in the fly to their anatomical analogs in the vertebrate retina, the bipolar cells. Like bipolar cells, the lamina neuron L2 displays an antagonistic center surround receptive field, with response properties that are well captured by a model that was previously used to describe the responses of a fast OFF bipolar cell type (Baccus et al. 2008, Freifeld et al. 2013). It is unclear whether this functional parallel extends to behavior. Although the L2 neuron provides an important input to neural pathways involved in detecting moving dark edges, the behavioral functions of fast-OFF bipolar cells are unknown. The circuit mechanisms by which these cells acquire their tuning properties are very different; the center-surround organization of L2 is strongly dependent on GABAergic circuitry providing presynaptic inputs onto photoreceptor cells, circuits that are not found in the vertebrate retina (Freifeld et al. 2013). Thus, it is tempting to speculate that evolution has shaped optimal tuning properties to relay information about contrast decrements to downstream circuits. The fact that the circuits that construct these properties in these two very evolutionarily distant systems are themselves different argues for convergent evolutionary processes. Therefore, evolution may have selected for a particular computational algorithm rather than for a specific circuit implementation.

Focusing selective pressure on circuit function has implications for the evolution of circuit structure. Peripheral visual systems of both the fly and the mouse display an incredibly dense network of synaptic connections (Briggman et al. 2011; Helmstaedter et al. 2013; Meinertzhagen & O'Neil 1991; Rivera-Alba et al. 2011; Takemura et al. 2008, 2013). Such dense networks may be advantageous for robustness and wiring economy (Chklovskii et al. 2002, Rivera-Alba et al. 2011). Alternately, such dense connectivity may be caused by evolutionary drift. A dense network of synaptic connections could allow for different circuit implementations that are selectively neutral because they achieve the same algorithmic properties. Thus, advancing our understanding of the neural circuits that implement motion detection in flies and comparing these circuits across functionally similar circuits in other animals will shed new light on the evolution of computation.

DISCLOSURE STATEMENT

The authors are not aware of any affiliations, memberships, funding, or financial holdings that might be perceived as affecting the objectivity of this review.

NOTE ADDED IN PROOF

After this manuscript was accepted for publication, two new studies addressed the circuit components of the dark edge detecting pathway. In an elegant serial EM reconstruction study, Shinomiya et al. (2014) identify the Tm9 neuron as a novel circuit component, linking L3 inputs to T5 outputs. Meier et al. (2014) perform physiological characterizations of L4 and Tm2 interneurons and provide the first functional evidence that a transmedullary neuron (Tm2) is required for direction-selective responses in the lobula plate.

LITERATURE CITED

Adelson EH, Bergen JR. 1985. Spatiotemporal energy models for the perception of motion. *J. Opt. Soc. Am. A* 2:284–99

Baccus SA, Ölveczky BP, Manu M, Meister M. 2008. A retinal circuit that computes object motion. *J. Neurosci.* 28:6807–17

Bahl A, Ammer G, Schilling T, Borst A. 2013. Object tracking in motion-blind flies. *Nat. Neurosci.* 16:730–38

Barlow HB, Levick WR. 1965. The mechanism of directionally selective units in rabbit's retina. *J. Physiol.* 178:477–504

Barth M, Hirsch HV, Meinertzhagen IA, Heisenberg M. 1997. Experience-dependent developmental plasticity in the optic lobe of *Drosophila melanogaster*. *J. Neurosci.* 17:1493–504

Barth M, Schultze M, Schuster CM, Strauss R. 2010. Circadian plasticity in photoreceptor cells controls visual coding efficiency in *Drosophila melanogaster*. *PLoS ONE* 5:e9217

Bausenwein B, Dittrich AP, Fischbach KF. 1992. The optic lobe of *Drosophila melanogaster*. II. Sorting of retinotopic pathways in the medulla. *Cell Tissue Res.* 267:17–28

Bausenwein B, Fischbach KF. 1992. Activity labeling patterns in the medulla of *Drosophila melanogaster* caused by motion stimuli. *Cell Tissue Res.* 270:25–35

Bishop LG, Keehn DG. 1966. Two types of neurones sensitive to motion in the optic lobe of the fly. *Nature* 212:1374–76

Born RT, Bradley DC. 2005. Structure and function of visual area MT. *Annu. Rev. Neurosci.* 28:157–89

Borst A, Euler T. 2011. Seeing things in motion: models, circuits, and mechanisms. *Neuron* 71:974–94

Borst A, Haag J. 2002. Neural networks in the cockpit of the fly. *J. Comp. Physiol. A Neuroethol. Sens. Neural Behav. Physiol.* 188:419–37

Borst A, Haag J, Reiff DF. 2010. Fly motion vision. *Annu. Rev. Neurosci.* 33:49–70

Braitenberg V. 1970. Order and orientation of elements in the visual system of the fly. *Kybernetik* 7:235–42

Branson K, Robie AA, Bender J, Perona P, Dickinson MH. 2009. High-throughput ethomics in large groups of *Drosophila*. *Nat. Methods* 6:451–57

Briggman KL, Helmstaedter M, Denk W. 2011. Wiring specificity in the direction-selectivity circuit of the retina. *Nature* 471:183–88

Buchner E. 1976. Elementary movement detectors in an insect visual system. *Biol. Cybernetics* 24:85–101

Buchner E. 1984. Behavioural analysis of spatial vision in insects. In *Photoreception and Vision in Invertebrates*, ed. MA Ali, pp. 561–621. New York: Plenum

Buchner E, Buchner S, Hengstenberg R. 1979. 2-Deoxy-D-glucose maps movement-specific nervous activity in the second visual ganglion of *Drosophila*. *Science* 205:687–88

Budick SA, Reiser MB, Dickinson MH. 2007. The role of visual and mechanosensory cues in structuring forward flight in *Drosophila melanogaster*. *J. Exp. Biol.* 210:4092–103

Card G, Dickinson MH. 2008. Visually mediated motor planning in the escape response of *Drosophila*. *Curr. Biol.* 18:1300–7

Chiappe ME, Seelig JD, Reiser MB, Jayaraman V. 2010. Walking modulates speed sensitivity in *Drosophila* motion vision. *Curr. Biol.* 20:1470–75

Chklovskii DB, Schikorski T, Stevens CF. 2002. Wiring optimization in cortical circuits. *Neuron* 34:341–47

Choe KM, Clandinin TR. 2005. Thinking about visual behavior; learning about photoreceptor function. *Curr. Top. Dev. Biol.* 69:187–213

Clandinin TR, Zipursky SL. 2002. Making connections in the fly visual system. *Neuron* 35:827–41

Clark DA, Bursztyn L, Horowitz MA, Schnitzer MJ, Clandinin TR. 2011. Defining the computational structure of the motion detector in *Drosophila*. *Neuron* 70:1165–77

Clark DA, Fitzgerald JE, Ales JM, Gohl DM, Silies MA, et al. 2014. Flies and humans share a motion estimation strategy that exploits natural scene statistics. *Nat. Neurosci.* 17:296–303

Clifford CW, Ibbotson MR. 2002. Fundamental mechanisms of visual motion detection: models, cells and functions. *Prog. Neurobiol.* 68:409–37

de Vries SE, Clandinin TR. 2012. Loom-sensitive neurons link computation to action in the *Drosophila* visual system. *Curr. Biol.* 22:353–62

Douglass JK, Strausfeld NJ. 1996. Visual motion-detection circuits in flies: parallel direction- and non-direction-sensitive pathways between the medulla and lobula plate. *J. Neurosci.* 16:4551–62

Duistermars BJ, Frye MA. 2008. Crossmodal visual input for odor tracking during fly flight. *Curr. Biol.* 18:270–75

Eckert H. 1981. The horizontal cells in the lobula plate of the blowfly, *Phaenicia sericata*. *J. Comp. Physiol.* 143:511–26

Egelhaaf M, Borst A. 1992. Are there separate ON and OFF channels in fly motion vision? *Vis. Neurosci.* 8:151–64

Egelhaaf M, Kern R, Krapp HG, Kretzberg J, Kurtz R, Warzecha AK. 2002. Neural encoding of behaviourally relevant visual-motion information in the fly. *Trends Neurosci.* 25:96–102

Eichner H, Joesch M, Schnell B, Reiff DF, Borst A. 2011. Internal structure of the fly elementary motion detector. *Neuron* 70:1155–64

Emerson RC, Bergen JR, Adelson EH. 1992. Directionally selective complex cells and the computation of motion energy in cat visual cortex. *Vision Res.* 32:203–18

Erclik T, Hartenstein V, McInnes RR, Lipshitz HD. 2009. Eye evolution at high resolution: the neuron as a unit of homology. *Dev. Biol.* 332:70–79

Fei H, Chow DM, Chen A, Romero-Calderón R, Ong WS, et al. 2010. Mutation of the *Drosophila* vesicular GABA transporter disrupts visual figure detection. *J. Exp. Biol.* 213:1717–30

Fischbach K-F, Dittrich APM. 1989. The optic lobe of *Drosophila melanogaster*. I. A Golgi analysis of wild-type structure. *Cell Tissue Res.* 258:441–75

Fitzgerald JE, Katsov AY, Clandinin TR, Schnitzer MJ. 2011. Symmetries in stimulus statistics shape the form of visual motion estimators. *Proc. Natl. Acad. Sci. USA* 108:12909–14

Fotowat H, Fayyazuddin A, Bellen HJ, Gabbiani F. 2009. A novel neuronal pathway for visually guided escape in *Drosophila melanogaster*. *J. Neurophysiol.* 102:875–85

Franceschini N, Riehle A, Le Nestour A. 1989. Directionally selective motion detection by insect neurons. See Stavenga & Hardie 1989, pp. 360–90

Freifeld L, Clark DA, Schnitzer MJ, Horowitz MA, Clandinin TR. 2013. GABAergic lateral interactions tune the early stages of visual processing in *Drosophila*. *Neuron* 78:1075–89

Fry SN, Rohrseitz N, Straw AD, Dickinson MH. 2009. Visual control of flight speed in *Drosophila melanogaster*. *J. Exp. Biol.* 212:1120–30

Frye MA, Dickinson MH. 2004. Motor output reflects the linear superposition of visual and olfactory inputs in *Drosophila*. *J. Exp. Biol.* 207:123–31

Gohl DM, Silies MA, Gao XJ, Bhalerao S, Luongo FJ, et al. 2011. A versatile in vivo system for directed dissection of gene expression patterns. *Nat. Methods* 8:231–37

Gollisch T, Meister M. 2010. Eye smarter than scientists believed: neural computations in circuits of the retina. *Neuron* 65:150–64

Götz KG. 1964. Optomotorische untersuchung des visuellen systems einiger augenmutanten der fruchtfliege *Drosophila*. *Kybernetik* 2:77–92

Götz KG, Wenking H. 1973. Visual control of locomotion in the walking fruitfly *Drosophila*. *J. Comp. Physiol. A* 85:235–66

Haag J, Borst A. 2004. Neural mechanism underlying complex receptive field properties of motion-sensitive interneurons. *Nat. Neurosci.* 7:628–34

Haag J, Denk W, Borst A. 2004. Fly motion vision is based on Reichardt detectors regardless of the signal-to-noise ratio. *Proc. Natl. Acad. Sci. USA* 101:16333–38

Haag J, Wertz A, Borst A. 2007. Integration of lobula plate output signals by DNOVS1, an identified premotor descending neuron. *J. Neurosci.* 27:1992–2000

Haag J, Wertz A, Borst A. 2010. Central gating of fly optomotor response. *Proc. Natl. Acad. Sci. USA* 107:20104–9

Haikala V, Joesch M, Borst A, Mauss AS. 2013. Optogenetic control of fly optomotor responses. *J. Neurosci.* 33:13927–34

Hassenstein B, Reichardt W. 1956. Systemtheoretische analyse der zeit-, reihenfolgen- und vorzeichenauswertung bei der bewegungsperzeption des rüsselkäfers chlorophanus. *Z. Naturforschung B* 11:513–24

Hausen K. 1982. Motion sensitive interneurons in the optomotor system of the fly. I. The horizontal cells: structure and signals. *Biol. Cybern.* 45:143–56

Hecht S, Wald G. 1934. The visual acuity and intensity discrimination of *Drosophila*. *J. Gen. Physiol.* 17:517–47

Heisenberg M, Buchner E. 1977. The role of retinula cell types in visual behavior of *Drosophila melanogaster*. *J. Comp. Physiol. A* 187:127–62

Heisenberg M, Wolf R. 1984. *Vision in Drosophila: Genetics of Microbehavior*. Berlin: Springer-Verlag

Heisenberg M, Wonneberger R, Wolf R. 1978. Optomotor-blind[H31]—a *Drosophila* mutant of the lobula plate giant neurons. *J. Comp. Physiol. A* 124:287–96

Helmstaedter M, Briggman KL, Turaga SC, Jain V, Seung HS, Denk W. 2013. Connectomic reconstruction of the inner plexiform layer in the mouse retina. *Nature* 500:168–74

Hildreth EC, Koch C. 1987. The analysis of visual motion: from computational theory to neuronal mechanisms. *Annu. Rev. Neurosci.* 10:477–533

Huston SJ, Krapp HG. 2008. Visuomotor transformation in the fly gaze stabilization system. *PLoS Biol.* 6:e173

Joesch M, Plett J, Borst A, Reiff DF. 2008. Response properties of motion-sensitive visual interneurons in the lobula plate of *Drosophila melanogaster*. *Curr. Biol.* 18:368–74

Joesch M, Schnell B, Raghu SV, Reiff DF, Borst A. 2010. ON and OFF pathways in *Drosophila* motion vision. *Nature* 468:300–4

Joesch M, Weber F, Eichner H, Borst A. 2013. Functional specialization of parallel motion detection circuits in the fly. *J. Neurosci.* 33:902–5

Johns DC, Marx R, Mains RE, O'Rourke B, Marbán E. 1999. Inducible genetic suppression of neuronal excitability. *J. Neurosci.* 19:1691–97

Kalmus H. 1943. The optomotor responses of some eye mutants of *Drosophila*. *J. Genet.* 45:206–13

Katsov AY, Clandinin TR. 2008. Motion processing streams in *Drosophila* are behaviorally specialized. *Neuron* 59:322–35

Kitamoto T. 2001. Conditional modification of behavior in *Drosophila* by targeted expression of a temperature-sensitive shibire allele in defined neurons. *J. Neurobiol.* 47:81–92

Kolodziejczyk A, Sun X, Meinertzhagen IA, Nässel DR. 2008. Glutamate, GABA and acetylcholine signaling components in the lamina of the *Drosophila* visual system. *PLoS ONE* 3:e2110

Lai S-L, Lee T. 2006. Genetic mosaic with dual binary transcriptional systems in *Drosophila*. *Nat. Neurosci.* 9:703–9

Lee T, Luo L. 1999. Mosaic analysis with a repressible cell marker for studies of gene function in neuronal morphogenesis. *Neuron* 22:451–61

Longden KD, Krapp HG. 2010. Octopaminergic modulation of temporal frequency coding in an identified optic flow-processing interneuron. *Front. Syst. Neurosci.* 4:153

Luan H, Peabody NC, Vinson CR, White BH. 2006. Refined spatial manipulation of neuronal function by combinatorial restriction of transgene expression. *Neuron* 52:425–36

Maimon G, Straw AD, Dickinson MH. 2010. Active flight increases the gain of visual motion processing in *Drosophila*. *Nat. Neurosci.* 13:393–99

Maisak MS, Haag J, Ammer G, Serbe E, Meier M, et al. 2013. A directional tuning map of *Drosophila* elementary motion detectors. *Nature* 500:212–16

Marmarelis PZ, McCann GD. 1973. Development and application of white-noise modeling techniques for studies of insect visual nervous system. *Kybernetik* 12:74–89

Masland RH. 2012. The neuronal organization of the retina. *Neuron* 76:266–80

Meier M, Serbe E, Maisak MS, Haag J, Dickson BJ, Borst A. 2014. Neural circuit components of the *Drosophila* OFF motion vision pathway. *Curr. Biol.* 24:385–92

Meinertzhagen IA, O'Neil SD. 1991. Synaptic organization of columnar elements in the lamina of the wild type in *Drosophila melanogaster*. *J. Comp. Neurol.* 305:232–63

Mizunami M, Weibrecht JM, Strausfeld NJ. 1998. Mushroom bodies of the cockroach: their participation in place memory. *J. Comp. Neurol.* 402:520–37

Morante J, Desplan C. 2004. Building a projection map for photoreceptor neurons in the *Drosophila* optic lobes. *Semin. Cell Dev. Biol.* 15:137–43

Morante J, Desplan C. 2008. The color-vision circuit in the medulla of *Drosophila*. *Curr. Biol.* 18:553–65

Morante J, Desplan C, Celik A. 2007. Generating patterned arrays of photoreceptors. *Curr. Opin. Genet. Dev.* 17:314–19

Mu L, Ito K, Bacon JP, Strausfeld NJ. 2012. Optic glomeruli and their inputs in *Drosophila* share an organizational ground pattern with the antennal lobes. *J. Neurosci.* 32:6061–71

Nakayama K. 1985. Biological image motion processing: a review. *Vision Res.* 25:625–60

O'Carroll DC, Bidwell NJ, Laughlin SB, Warrant EJ. 1996. Insect motion detectors matched to visual ecology. *Nature* 382:63–66

Ofstad TA, Zuker CS, Reiser MB. 2011. Visual place learning in *Drosophila melanogaster*. *Nature* 474:204–7

Pfeiffer BD, Jenett A, Hammonds AS, Ngo TT, Misra S, et al. 2008. Tools for neuroanatomy and neurogenetics in *Drosophila*. *Proc. Natl. Acad. Sci. USA* 105:9715–20

Pfeiffer BD, Ngo TT, Hibbard KL, Murphy C, Jenett A, et al. 2010. Refinement of tools for targeted gene expression in *Drosophila*. *Genetics* 186:735–55

Pflugfelder GO, Roth H, Poeck B, Kerscher S, Schwarz H, et al. 1992. The lethal(1)optomotor-blind gene of *Drosophila melanogaster* is a major organizer of optic lobe development: isolation and characterization of the gene. *Proc. Natl. Acad. Sci. USA* 89:1199–203

Raghu SV, Borst A. 2011. Candidate glutamatergic neurons in the visual system of *Drosophila*. *PLoS ONE* 6:e19472

Ramón y Cajal S, Sanchez D. 1915. *Contribución al Conocimiento de los Centros Nerviosos de los Insectos*. Madrid: Imprenta de Hijos de Nicholas Moja

Reichardt W. 1961. Autocorrelation, a principle for the evaluation of sensory information by the central nervous system. In *Sensory Communication*, ed. WA Rosenblith, pp. 303–17. New York/London: MIT Press/Wiley

Reichardt W, Poggio T. 1975. Theory of pattern induced flight orientation of fly *Musca domestica* 0.2. *Biol. Cybernet.* 18:69–80

Reiff DF, Plett J, Mank M, Griesbeck O, Borst A. 2010. Visualizing retinotopic half-wave rectified input to the motion detection circuitry of *Drosophila*. *Nat. Neurosci.* 13:973–78

Rister J, Pauls D, Schnell B, Ting CY, Lee CH, et al. 2007. Dissection of the peripheral motion channel in the visual system of *Drosophila melanogaster*. *Neuron* 56:155–70

Rivera-Alba M, Vitaladevuni SN, Mischenko Y, Lu Z, Takemura SY, et al. 2011. Wiring economy and volume exclusion determine neuronal placement in the *Drosophila* brain. *Curr. Biol.* 21:2000–5

Sanes JR, Zipursky SL. 2010. Design principles of insect and vertebrate visual systems. *Neuron* 66:15–36

Schnell B, Joesch M, Forstner F, Raghu SV, Otsuna H, et al. 2010. Processing of horizontal optic flow in three visual interneurons of the *Drosophila* brain. *J. Neurophysiol.* 103:1646–57

Schnell B, Raghu SV, Nern A, Borst A. 2012. Columnar cells necessary for motion responses of wide-field visual interneurons in *Drosophila*. *J. Comp. Physiol. A Neuroethol. Sens. Neural Behav. Physiol.* 198:389–95

Seelig JD, Jayaraman V. 2013. Feature detection and orientation tuning in the *Drosophila* central complex. *Nature* 503:262–66

Shinomiya K, Karuppudurai T, Lin T-Y, Lu Z, Lee C-H, et al. 2014. Candidate neural substrates for off-edge motion detection in *Drosophila*. *Curr. Biol.* 24:1062–70

Silies M, Gohl DM, Fisher YE, Freifeld L, Clark DA, Clandinin TR. 2013. Modular use of peripheral input channels tunes motion-detecting circuitry. *Neuron* 79:111–27

Sombati S, Hoyle G. 1984. Generation of specific behaviors in a locust by local release into neuropil of the natural neuromodulator octopamine. *J. Neurobiol.* 15:481–506

Stavenga DG, Hardie RC, eds. 1989. *Facets of Vision*. Berlin: Springer-Verlag

Strausfeld NJ. 1989. Beneath the compound eye: neuroanatomical analysis and physiological correlates in the study of insect vision. See Stavenga & Hardie 1989, pp. 317–59

Strausfeld NJ, Campos-Ortega JA. 1973. The L4 monopolar neurone: a substrate for lateral interaction in the visual system of the fly *Musca domestica* (L.). *Brain Res.* 59:97–117

Strausfeld NJ, Gronenberg W. 1990. Descending neurons supplying the neck and flight motor of Diptera: organization and neuroanatomical relationships with visual pathways. *J. Comp. Neurol.* 302:954–72

Straw AD, Branson K, Neumann TR, Dickinson MH. 2011. Multi-camera real-time three-dimensional tracking of multiple flying animals. *J. R. Soc. Interface* 8:395–409

Straw AD, Lee S, Dickinson MH. 2010. Visual control of altitude in flying *Drosophila*. *Curr. Biol.* 20:1550–56

Suver MP, Mamiya A, Dickinson MH. 2012. Octopamine neurons mediate flight-induced modulation of visual processing in *Drosophila*. *Curr. Biol.* 22:2294–302

Sweeney ST, Broadie K, Keane J, Niemann H, O'Kane CJ. 1995. Targeted expression of tetanus toxin light chain in *Drosophila* specifically eliminates synaptic transmission and causes behavioral defects. *Neuron* 14:341–51

Takemura S-Y, Bharioke A, Lu Z, Nern A, Vitaladevuni S, et al. 2013. A visual motion detection circuit suggested by *Drosophila* connectomics. *Nature* 500:175–81

Takemura S-Y, Karuppudurai T, Ting CY, Lu Z, Lee CH, Meinertzhagen IA. 2011. Cholinergic circuits integrate neighboring visual signals in a *Drosophila* motion detection pathway. *Curr. Biol.* 21:2077–84

Takemura S-Y, Lu Z, Meinertzhagen IA. 2008. Synaptic circuits of the *Drosophila* optic lobe: the input terminals to the medulla. *J. Comp. Neurol.* 509:493–513

Tammero LF, Dickinson MH. 2002. The influence of visual landscape on the free flight behavior of the fruit fly *Drosophila melanogaster*. *J. Exp. Biol.* 205:327–43

Tammero LF, Frye MA, Dickinson MH. 2004. Spatial organization of visuomotor reflexes in *Drosophila*. *J. Exp. Biol.* 207:113–22

Tuthill JC, Chiappe ME, Reiser MB. 2011. Neural correlates of illusory motion perception in *Drosophila*. *Proc. Natl. Acad. Sci. USA* 108:9685–90

Tuthill JC, Nern A, Holtz SL, Rubin GM, Reiser MB. 2013. Contributions of the 12 neuron classes in the fly lamina to motion vision. *Neuron* 79:128–40

van Santen JP, Sperling G. 1984. Temporal covariance model of human motion perception. *J. Opt. Soc. Am. A* 1:451–73

Venken KJT, Simpson JH, Bellen HJ. 2011. Genetic manipulation of genes and cells in the nervous system of the fruit fly. *Neuron* 72:202–30

Wardill TJ, List O, Li X, Dongre S, McCulloch M, et al. 2012. Multiple spectral inputs improve motion discrimination in the *Drosophila* visual system. *Science* 336:925–31

Wertz A, Borst A, Haag J. 2008. Nonlinear integration of binocular optic flow by DNOVS2, a descending neuron of the fly. *J. Neurosci.* 28:3131–40

Yonehara K, Farrow K, Ghanem A, Hillier D, Balint K, et al. 2013. The first stage of cardinal direction selectivity is localized to the dendrites of retinal ganglion cells. *Neuron* 79:1078–85

Zhu Y, Nern A, Zipursky SL, Frye MA. 2009. Peripheral visual circuits functionally segregate motion and phototaxis behaviors in the fly. *Curr. Biol.* 19:613–19

Neuromodulation of Circuits with Variable Parameters: Single Neurons and Small Circuits Reveal Principles of State-Dependent and Robust Neuromodulation

Eve Marder, Timothy O'Leary, and Sonal Shruti

Volen Center and Biology Department, Brandeis University, Waltham, Massachusetts 02454;
email: marder@brandeis.edu, toleary@brandeis.edu, sshruti@brandeis.edu

Annu. Rev. Neurosci. 2014. 37:329–46

The *Annual Review of Neuroscience* is online at
neuro.annualreviews.org

This article's doi:
10.1146/annurev-neuro-071013-013958

Keywords

stomatogastric nervous system, central pattern generators, neuropeptides, biogenic amines, intrinsic excitability, synaptic strength

Abstract

Neuromodulation underlies many behavioral states and has been extensively studied in small circuits. This has allowed the systematic exploration of how neuromodulatory substances and the neurons that release them can influence circuit function. The physiological state of a network and its level of activity can have profound effects on how the modulators act, a phenomenon known as state dependence. We provide insights from experiments and computational work that show how state dependence can arise and the consequences it can have for cellular and circuit function. These observations pose a general unsolved question that is relevant to all nervous systems: How is robust modulation achieved in spite of animal-to-animal variability and degenerate, nonlinear mechanisms for the production of neuronal and network activity?

Contents

INTRODUCTION

All nervous systems, large and small, are modulated by numerous amines, neuropeptides, gases, and other molecules (Bargmann 2012, Brezina 2010, Harris-Warrick & Johnson 2010, Levitan 1988, Marder 2012, Stein 2009, Taghert & Nitabach 2012). At the extreme, alterations in modulatory tone are likely to underlie changes in arousal, mood, and other global states of brain networks that dramatically influence an animal's behavior (Alekseyenko et al. 2010, 2013; Beverly et al. 2011; Lee & Dan 2012; Liu et al. 2012). Neuromodulators also have more focused actions that may influence the properties of single neurons, synapses, or local circuits (Marder 2012). Despite the ubiquitous nature of neuromodulatory control systems and their clear relevance for understanding circuit dynamics and behavior (Bargmann 2012, Blitz & Nusbaum 2011, Brezina 2010, Nusbaum & Blitz 2012, Taghert & Nitabach 2012), several fundamental questions relevant to neuromodulation remain mysterious. In this review, we draw on a large literature on the effects of neuromodulators on well-characterized small circuits found in invertebrates to illustrate and elucidate some of the fundamental puzzles and conundrums posed by neuromodulation. Although these issues are easily illustrated in small nervous systems, the general principles we articulate are as relevant to neuromodulation in the human brain as they are in *Caenorhabditis elegans*, crustaceans, or mollusks. We regret that space limitations preclude our doing justice to most of the elegant work on neuromodulation in small invertebrate circuits.

We first discuss numerous issues relevant to the modulation of single neurons, which almost certainly are general features of all nervous systems. We then discuss some computational and experimental studies of modulation of small circuit dynamics. Although we illustrate principles using known small circuits, admittedly with their specific idiosyncrasies, the reader who steps back will see general principles that rise above the system specifics. For example, the challenges in understanding the action of a peptide in the crustacean nervous system frame many of the issues relevant to understanding how modulatory substances such as dopamine influence behavior in mammals. Finally, we pose a significant challenge for future studies: to understand how neuromodulation can be tuned in concert with other circuit properties so that individuals can respond reliably and appropriately to neuromodulatory inputs.

MODULATION OF SINGLE NEURONS AND SYNAPSES

Modulators exert their action by altering the strength of synapses and the conductances and other properties of intrinsic membrane channels (Harris-Warrick & Marder 1991). Therefore the distribution and expression level of different ion channels and receptors are pivotal in determining a cell or network response to a neuromodulator (Marder 2012, Marder & Thirumalai 2002).

Neuromodulation of Single Neurons with Similar Behavior Can Depend Critically on Underlying Parameters

A large number of computational studies have demonstrated that similar neuronal properties can result from vastly different sets of membrane conductances (Goldman et al. 2001; Golowasch et al. 2002; Sobie 2009; Taylor et al. 2006, 2009), and a growing number of experimental studies have shown that identified neurons can have widely varying conductance densities of ionic currents and synaptic strengths (Goaillard et al. 2009; Ransdell et al. 2013; Rinberg et al. 2012; Roffman et al. 2012; Schulz et al. 2006, 2007; Swensen & Bean 2005; Tobin et al. 2009). **Figure 1a** shows an example of model neurons with substantially different sets of conductance densities but with very similar membrane potential waveforms. **Figure 1b** shows data from an experimental study in which voltage-clamp measurements of conductance densities were made from isolated Purkinje neurons with very similar waveforms. Again, it can be seen that very similar membrane potential trajectories can arise from quite different sets of conductances.

These data raise three related questions. First, can neurons with variable underlying conductances give robust or reliable responses to modulation (Goldman et al. 2001, Szücs & Selverston 2006)? Second, can neurons with similar sets of conductances nonetheless respond differently

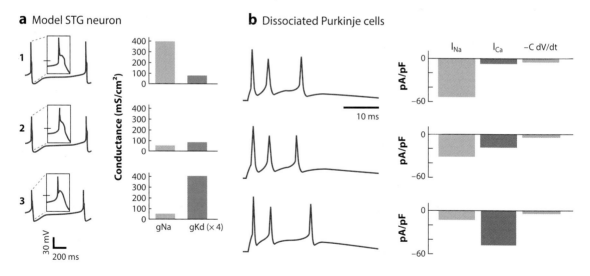

Figure 1

Neurons produce stereotyped physiological behavior with variable underlying membrane conductances. (*a*) Example membrane potential traces from three different single-spike bursting model neurons, adapted from **Figure 2** in Golowasch et al. (2002). Values of two of the six conductance densities in each model are shown to the right of each trace. The time base of each inset is 50 ms; the horizontal tick in the insets corresponds to −30 mV. (*b*) Current-clamp recordings of three dissociated Purkinje neurons showing similar burst-firing behavior, adapted from Swensen & Bean (2005), figure 2A. The voltage-dependent contributions of fast sodium (I_{Na}), calcium (I_{Ca}) and net ionic membrane currents (−C dv/dt) measured in each of the neurons are shown to the right. Abbreviation: STG, stomatogastric ganglion.

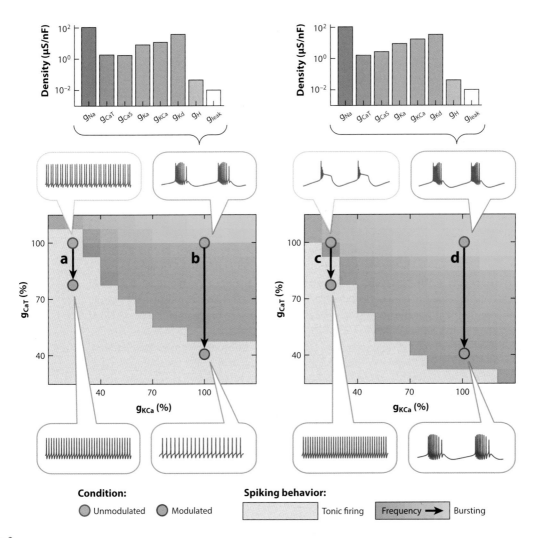

Figure 2

Consequences of precise neuromodulation in populations of neurons with variable underlying conductances. Parameter maps showing conductance densities and corresponding behavior of a population of conductance-based model neurons. The middle heat maps show the regions of conductance values that produce two distinct types of membrane potential behavior, tonic spiking (*gray regions*) and burst firing (*colored regions, shaded according to burst frequency*), as a function of two of the seven voltage-dependent conductances in the model (g_{CaT}, a transient calcium conductance, and g_{KCa}, a calcium-dependent potassium conductance). The left and right plots show the same regions of conductance space for models in which the remaining five conductances have different underlying values. The membrane potential behavior of four base-case neurons is shown (*a–d, top insets*). The underlying conductances for cases *b* and *d*, which have similar membrane potential activity, are displayed at the top of the figure. Modulation causing a 20% (cases *a* and *c*) and 60% (cases *b* and *d*) change in g_{CaT} is indicated by the vertical arrows and results in the membrane potential behavior in the bottom insets. The time base for each trace is 200 ms.

to modulation or perturbation (Goldman et al. 2001)? And third, should we expect qualitatively different responses to neuromodulation or perturbation from neurons of the same cell type?

Figure 2 explores these issues using a computational model with seven voltage- and time-dependent conductances as well as a leak conductance. The behavior of the models as a function of their low-threshold Ca^{2+} conductance (g_{CaT}) and Ca^{2+}-activated K^+ conductance (g_{Kca}) is shown as regions of the parameter space that produce bursting or tonic firing. Note that these regions are similar in form but are not identical because of the differences in other conductances in the two models. The membrane-potential waveforms of two models (points b and d) are quite similar, and the conductance densities, although not identical, are also relatively similar. But, at much lower values of g_{KCa}, the neuron at point a fires doublets of spikes, whereas the neuron at the same location in the other map (point c) fires short bursts of action potentials with long plateaus.

Several important points relevant to state-dependent neuromodulation can be seen in **Figure 2**. A large change in one conductance (in this case g_{CaT}) can produce relatively little change in behavior (the neuron shown in point d maintains its bursting behavior despite a large modulatory change in conductance). The same change in conductance in the neuron shown at point b produces a qualitative change in activity: It brings the neuron across the boundaries between tonic firing and bursting behavior (Goldman et al. 2001). Even a small neuromodulatory change acting on the neurons at points a and c will produce qualitative changes in behavior. Nonetheless, at different starting values of g_{KCa}, much larger changes in g_{CaT} have little or no obvious effect on the neuron's firing properties. In summary, the effect of a modulatory substance that alters one or more of the voltage-dependent conductances of a neuron depends not only on how strong its action is on its target conductance, but also on the values of all the other conductances in the neuron (**Figure 2**).

This example illustrates complications that are inherent even when a single membrane conductance is modulated. It is important to remember that the same neuromodulator can act on numerous voltage-dependent currents in the same neuron and that this can be mediated by one or more receptors (Boyle et al. 1984, Braha et al. 1993, Harris-Warrick et al. 1995, Kiehn & Harris-Warrick 1992, Spitzer et al. 2008, Thompson & Calabrese 1992, Tobin & Calabrese 2005).

Degenerate Neuromodulatory Actions Can Hide Differences in Underlying Mechanisms

Neurons and circuits routinely respond to many neuromodulatory substances, some of which may elicit responses that closely resemble each other (Marder & Bucher 2007, Marder & Eisen 1984, Swensen & Marder 2000). One example is shown in the response of the isolated anterior burster (AB) neuron from the lobster stomatogastric ganglion. This neuron responds to the application of cholinergic agonists, amines, and many neuropeptides with robust bursts of attenuated action potentials riding on top of a large depolarizing slow wave (Ayali & Harris-Warrick 1999, Flamm & Harris-Warrick 1986, Harris-Warrick & Flamm 1987, Hooper & Marder 1987, Marder & Eisen 1984). **Figure 3a** shows an example of AB neurons that were silent prior to the application of dopamine, serotonin, or octopamine. In each case, the amines elicited strong bursting. Nonetheless, the underlying mechanisms of the bursts evoked by dopamine and serotonin differ significantly. The dopamine-elicited bursts depend strongly on underlying Ca^{2+} currents, and the slow wave persists in the presence of TTX and in reduced Na^+ saline (**Figure 3b**) (Harris-Warrick & Flamm 1987). In contrast, TTX entirely suppresses the serotonin- and octopamine-elicited bursts, but these persist in low Ca^{2+} concentrations (**Figure 3b**). These data are consistent with the different neuromodulators activating bursts that initially look the same but, by acting on different currents, result in bursts with different underlying mechanisms (Epstein & Marder 1990, Harris-Warrick & Flamm 1987).

Activity Can Alter the Effects of Neuromodulators

The R15 neuron of *Aplysia* fires in bursts that are strongly modulated by serotonin and ELH (egg-laying hormone) (Adams & Benson 1985) via mechanisms that depend on intracellular $[Ca^{2+}]$

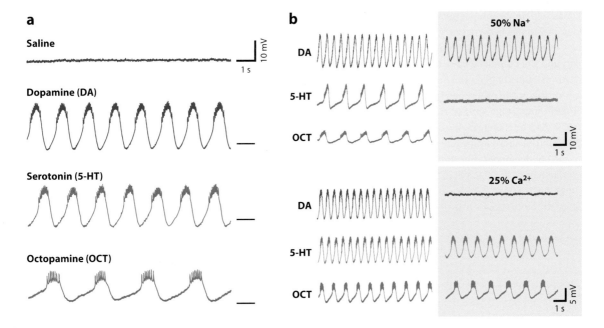

Figure 3

Convergence of distinct modulatory mechanisms on single-neuron behavior. (*a*) Neuromodulator-dependent induction of rhythmic bursting a pharmacologically isolated anterior burster (AB) cell in the stomatogastric ganglion in *Panulirus interruptus*. Shown are a saline (control) recording of a silent AB cell and bursting induced by bath application dopamine (100 μM), serotonin (10 μM), and octopamine (100 μM). Horizontal lines indicate the resting potential of the control recording. Figure adapted from Harris-Warrick & Flamm (1987, figure 1). (*b*) Manipulating extracellular cation concentration reveals distinct mechanisms of modulator-induced bursting. (*Top*) Left traces show recordings from the same isolated neuron in dopamine (100 μM), serotonin (10 μM), and octopamine (100 μM). The neuron was silent in the absence of a neuromodulator (not shown). Right traces show the same neuron in each modulatory condition, 5–7 min after Na$^+$ concentration was reduced to 50% by equimolar replacement with Tris. (*Bottom*) A separate experiment on a different isolated AB cell showing responses to dopamine (100 μM), serotonin (10 μM), and octopamine (100 μM) in control saline (*left traces*) and with Ca^{2+} replaced with equimolar Mg^{2+} (*right traces*). Figures adapted from Harris-Warrick & Flamm (1987, figures 4, 6).

and cAMP (cyclic adenosine monophosphate) (Kramer & Levitan 1990). Consequently, and as illustrated in one study (Kramer & Levitan 1990), changes in neuronal activity that influence intracellular [Ca^{2+}] also influence the modulation of R15 by serotonin and ELH. The general principle of this study is that the effects of any modulator that acts via second-messenger signal transduction pathways will be altered by any other synaptically or intrinsically driven processes that alter the states of those second-messenger systems (Yu et al. 2004). Because most neuro-modulators act via second-messenger systems, researchers expect that modulator action will be constantly influenced by the ongoing activity patterns of the target neuron or muscle, which also modify the intracellular second-messenger pathways of the neuron, in a sense producing cross talk between modulator actions (Antonov et al. 2010, Braha et al. 1993, Yu et al. 2004). As a result, the same modulator may produce disparate responses in the same neuronal target (Spitzer et al. 2008).

When neuromodulators converge onto the same intracellular second-messenger pathways or onto the same membrane current, they may interact in interesting and nonlinear fashions (Brezina 2010). For example, in crustaceans, many peptide neuromodulators bind to different classes of receptors but eventually activate the same membrane current, I$_{MI}$ (Golowasch & Marder 1992,

334   *Marder • O'Leary • Shruti*

Swensen & Marder 2000). In this case, the effects of the modulators can saturate and occlude each other. In other cases, low concentrations of one modulator may enhance the actions of another because of the amplification steps engaged in signal transduction pathways.

MODULATION OF CIRCUITS

We know that modulators alter circuit function. In principle, this alteration could occur if a neuromodulator influenced only one or a few members of the circuit and if neurons which themselves are direct targets for modulation influenced the activity of other circuit neurons. Alternatively, a modulator may act on many, or most, of the neurons of a circuit in a more widespread fashion. Relatively few biological systems have been examined extensively enough to assess this uncertainty, but in the stomatogastric ganglion some modulators clearly act on many circuit neurons while others have more restricted cellular targets (Swensen & Marder 2001).

Modulation of Circuit Dynamics Can Be State Dependent

There are numerous examples in the literature of neuromodulatory actions that depend on the physiological state and initial level of activity in a network (Nadim et al. 2008, Nusbaum & Marder 1989, Weimann et al. 1997, Williams et al. 2013) and, presumably, on the circuit mechanisms that are responsible for those differences in activity. For example, the neuropeptide proctolin strongly increases the frequency of the pyloric rhythm of the crustacean stomatogastric ganglion (STG) when the initial frequency is low, but it has little or no effect on frequency when the starting frequency is high (Hooper & Marder 1987, Nusbaum & Marder 1989). In contrast, the inhibitory allatostatin peptides elicit little effect on robust pyloric rhythms but show much more dramatic effects when the starting frequency is lower (Skiebe & Schneider 1994, Szabo et al. 2011).

A recent modeling study at the circuit level (Gutierrez et al. 2013) makes many of the same points that we have made previously at the single-neuron level. This study examines the behavior of a five-neuron network (**Figure 4a**), loosely inspired by the connectivity of the crab STG, and examines the circuit mechanisms that control the relative coordination of fast and slow oscillators connected through a hub neuron via electrical synapses (**Figure 4a**) (Gutierrez et al. 2013). In the model circuit, the f1 and f2 neurons are fast oscillators and reciprocally inhibit each other. The s1 and s2 neurons are slow oscillators and also reciprocally inhibit each other. The f2 and s2 neurons are electrically coupled to a hub neuron (hn), which is also inhibited by the f1 and s1 neurons (**Figure 4a**). This network can show many different behaviors, depending on the strengths of the electrical and chemical synapses in the network (**Figure 4b**). Four of these behaviors are shown in **Figure 4b**. In case 1, the f1, f2, hn, and s2 neurons are firing at the same frequency, whereas s1 is not following but instead firing every other cycle. In case 2, all five neurons are firing at the same frequency. In case 3, f1 is alone firing rapidly, but the other 4 neurons are firing slowly. In case 4, f1 and f2 are firing rapidly, but hn is firing with s1 and s2 in a slow rhythm.

The parameterscape (**Figure 4c**) is a visualization tool that displays the relative behaviors possible for this five cell circuit as the strength of the electrical synapse (g_{el}) and chemical synapse from f1 and s1 to the hn (g_{synA}) is varied. The locations in parameter space of the four examples in **Figure 4b** are labeled as points in the parameterscape, and the distinct frequency relationships can be read off. This visualization demonstrates clearly how different forms of network coordination emerge as a function of changes in parameters.

State dependence of modulation is easily seen using the arrows in **Figure 4c**. Modulator A increases the strength of g_{synA}. At some starting values of parameters, even a small change in g_{synA} can bring the network across boundaries of qualitative network behavior. For example, if

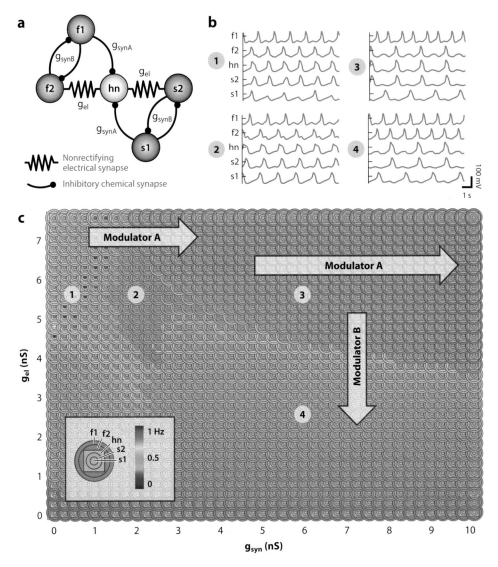

Figure 4

A simple conductance-based model rhythm-generating network captures general principles of neuromodulation in small circuits. (*a*) Model circuit diagram from Gutierrez et al. (2013). Each neuron is a conductance-based oscillating Morris–Lecar model neuron. Synaptic connections are graded, inhibitory chemical synapses (*black blobs*, g_{syna}, g_{synb}) or nonrectifying electrical synapses (resistor symbols, g_{el}). The intrinsic properties of the neurons are tuned to produce two identical fast-bursting cells (f1, f2), two slow cells (s1, s2), and an intermediate-frequency hub neuron (hn); see reference for full simulation details. (*b*) Example circuit activity for different combinations of synaptic parameters g_{syna} and g_{el} as indicated in the parameter map (panel *c*). (*c*) Parameter map, or parameterscape, encoding circuit output as a function of two of the synaptic conductance strengths, g_{syna} and g_{el}. Each symbol in the parameterscape shows the bursting frequency of the five neurons in the circuit in a color-coded ring, as shown in the legend. The hub neuron is represented as a concentric square. Arrows indicate putative modulatory changes in circuit parameters.

we start in region 1, in which all neurons but s1 are active together and s1 is left behind at a much lower frequency, a relatively modest increase in g_{synA} brings the network into an entirely different coordination mode in which all neurons but f1 are slow. But, a much larger change in g_{synA} elicited by modulator A does not alter the network coordination (starting in region 3). In contrast, a relatively small change in g_{el} elicited by modulator B crosses the boundaries between areas of different network coordination (**Figure 4c**).

This kind of state dependence of modulator action on network coordination can be seen in biological systems. **Figure 5a** shows the connectivity for the ~27 neurons of the STG of the crab, *Cancer borealis*. Two distinct rhythms coexist in the STG: the pyloric rhythm and the slower gastric rhythm. Some cells in the STG are capable of participating in both rhythms. When the neuropeptide CCAP was applied to two different preparations (with different starting levels of activity), the IC neuron fired in time with either the fast pyloric rhythm (preparation 1) or the gastric mill rhythm (preparation 2) (Weimann et al. 1997).

Degenerate Modulation of Circuits

Modulators can act in numerous ways on neuronal circuits to produce changes in circuit dynamics that resemble each other but arise from quite different circuit mechanisms (Beverly et al. 2011, Gutierrez et al. 2013, Kintos et al. 2008, Nadim et al. 2008, Rodriguez et al. 2013, Saideman et al. 2007a). For example, in *C. elegans*, robust thermosensory behaviors depend on different circuit configurations at different temperatures, and neuromodulation plays a crucial role in setting the temperature range of these circuits and thus their behavior (Beverly et al. 2011). Similarly, octanol avoidance can be generated by different sets of sensory neurons (Chao et al. 2004) depending on modulatory status (Harris et al. 2010, Mills et al. 2012, Wragg et al. 2007), which in turn depends on whether the animal is starved or fed.

In the stomatogastric nervous system, very similar gastric mill rhythms are elicited by two different manipulations: activation of the specific modulatory projection neuron MCN1, or bath application of the neuropeptide CabPK (**Figure 6**). The network mechanisms engaged by these two treatments are shown schematically. In the MCN1 activated rhythm, DG (the dorsal gastric cell) is not necessarily activated, and the AB inhibition of Int1 is not required (**Figure 6a**). In contrast, CabPK application activates DG, and the AB inhibition of Int1 is necessary for this rhythm (**Figure 6b**) (Kintos et al. 2008; Rodriguez et al. 2013; Saideman et al. 2006, 2007a,b). Although the previous work shows that similar gastric rhythms can be generated by different circuit mechanisms, different forms of the gastric rhythm can also be activated by stimulation of various sensory inputs, and several of these employ the same core circuitry (White & Nusbaum 2011).

Behavioral Selection by Neuromodulation

Understanding how animals use internal state and sensory inputs to produce one behavior or another or different forms of related behaviors is a fundamental problem in neuroscience (Briggman & Kristan 2008, Flavell et al. 2013, Kristan 2008, Palmer & Kristan 2011, Taghert & Nitabach 2012). In many instances, behavior choice depends on changes in internal states that are produced by activating specific sets of neuromodulatory inputs (Friesen & Kristan 2007; Jing et al. 2007, 2008; Wagenaar et al. 2010; Wu et al. 2010). The previous history of activation of some network elements can produce what has been termed latent modulation, a change in the way the network responds to a subsequent stimulus (Dacks & Weiss 2013), as shown in the *Aplysia* feeding network. Buccal motor outputs can be either ingestive or egestive (Jing & Weiss 2001), depending on which sets of higher-level interneurons are activated (Morgan et al. 2000, 2002); however, prestimulation of the input CBI-2 can enhance both subsequently evoked ingestive and egestive behaviors, although these behaviors are mutually exclusive (Dacks & Weiss 2013). This observation is similar to that seen in *C. elegans*, where a G protein–coupled receptor, npr-1, can affect two behaviors elicited by the same sensory neuron (Bargmann 2012) and neuropeptides can alter the responses of olfactory neurons (Chalasani et al. 2010).

a The STG connectome

b Preparation 1   Control   LP   10 μM CCAP

c Preparation 2   Control   LP   PY   PD   10 μM CCAP

OPEN QUESTION: DOES NEUROMODULATION POSE A SPECIAL PROBLEM FOR HOMEOSTATIC REGULATION OF EXCITABILITY AND SYNAPTIC STRENGTH?

A large and growing body of both computational and experimental work argues that the number and distribution of ion channels and receptors expressed by neurons are regulated so that neurons maintain an activity level appropriate to their network function (Baines et al. 2001; Davis 2006; LeMasson et al. 1993; Liu et al. 1998; O'Leary et al. 2010, 2013; O'Leary & Wyllie 2011; Pratt & Aizenman 2007; Pratt et al. 2003; Turrigiano 2008, 2011). This raises a potential problem: Can neuronal excitability or network parameters be tuned so that multiple different neuromodulatory substances, which can be thought of as a variety of different perturbations, alter network performance in the desired direction? Possible answers to this question are illustrated in **Figure 7**, which shows a cartoon of how circuit output depends on two network parameters such as synaptic or intrinsic conductances.

The hypothetical circuit produces two distinct, behaviorally important motor patterns: a motor pattern associated with a basal unmodulated state, A, and a distinct "modulated" state, B. In keeping with experimental observations, the circuit parameters across a population of individuals show variability (Goaillard et al. 2009), and the parameters in each case have been tuned by some activity-dependent process to maintain the unmodulated circuit in the basal state. But what is needed to ensure that a modulator with specific cellular actions will produce a reliable change in behavior across individuals? The problem is posed in **Figure 7a**. Here, the same modulatory action fails to produce similar responses to all the individuals with similar starting behavior. Note that in some cases the modulator brings the network across its state boundaries, but in other cases it fails to do so.

This problem may be solved in two distinct ways. One possibility is shown in **Figure 7b**. In this case, the modulatory effect is not specifically tuned in each individual (arrows are all identical), but the circuit parameters are restricted to portions of the parameter space that will allow the modulator to produce its appropriate action. This case requires the regulatory mechanism to somehow sense how far each circuit is from the transition and tune the circuit parameters such that the modulator reliably pushes the circuit through a transition. Alternatively, both the circuit parameters and the modulator effect could be coordinately tuned (**Figure 7c**). This possibility would predict differences in how individual circuit parameters respond to a modulator because the length and direction of the arrows need to differ. Such a model is consistent with the variable effects observed in modulator-induced membrane currents across different individual preparations (Goaillard et al. 2009). However, we do not know how the cellular mechanisms in the circuit sense

←

Figure 5

Identical neuromodulatory manipulations in the same circuit can produce distinct physiological changes in different animals. (*a*) The connectivity diagram, or connectome, of the stomatogastric ganglion (STG) in *Cancer borealis*. On the left are the neurons that routinely participate in the faster pyloric rhythm, and on the right are neurons that routinely participate in the slower gastric mill rhythm. The five neurons shown in color are in the configuration modeled in **Figure 4**, with the inferior cardiac (IC) neuron in the hn position. (*b*) Extracellular recordings of the dorsal ventricular nerve (dvn, *red*), the median ventricular nerve (mvn, *black*), and the lateral gastric nerve (lgn, *blue*). Cells in panel *a* are colored according to their respective output nerve; bursts of action potentials from each of the cells are indicated. Bath application (10 μM) of the endogenous neuropeptide CCAP (Stangier et al. 1987) induces a version of the slow gastric rhythm evident in the bursts of action potentials in the lgn and a modulating envelope of activity in the mvn (*right*). Adapted from Weimann et al. (1997, figure 2). (*c*) Similar to panel *b* but in a different preparation. A distinct version of the gastric rhythm is induced by CCAP application under identical recording conditions.

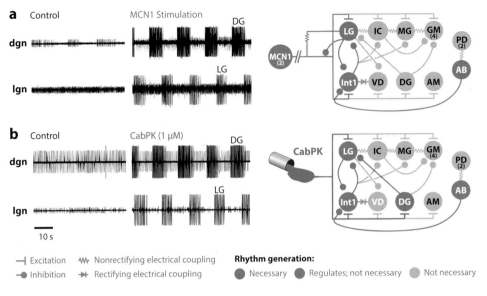

a Control   MCN1 Stimulation

dgn

lgn

b Control   CabPK (1 μM)

dgn

lgn

10 s

⊣ Excitation   ⋀⋀ Nonrectifying electrical coupling   **Rhythm generation:**
●─ Inhibition   ➤ Rectifying electrical coupling   ● Necessary   ● Regulates; not necessary   ● Not necessary

Figure 6

Distinct modulatory mechanisms can converge on the same transition in circuit behavior. (*a*) The gastric mill rhythm can be induced by stimulating the modulatory projection neuron MCN1, as seen in the slow rhythm induced in the extracellular recordings of the lateral and dorsal gastric nerves (lgn and dgn, respectively). (*Right*) A schematic of the gastric mill circuit configured by MCN1 stimulation. In this case, modulatory input by MCN1 excites LG and Int1, inducing bursting. Circuit symbols: excitation (*t-bars*); inhibition (*filled circles*); nonrectifying electrical coupling (*resistors*); rectifying electrical coupling (*diode*). Parallel lines crossing the MCN1 axon represent additional anatomical distance between the MCN1 soma and its axon terminals in the STG. Numbers in parentheses indicate the cell copy number per nervous system for each neuron type when there is more than one neuron. (*b*) Bath application of the neuropeptide CabPK (10 μM) elicits the same gastric mill motor pattern as does MCN1 stimulation but via a distinct circuit configuration (*right*) in a different preparation. In this case, CabPK excites LG, Int1, and DG, and AB activity is necessary for a gastric rhythm. Adapted from Saideman et al. (2007a) and Rodriguez et al. (2013).

deviations from desired modulator effects and which plasticity mechanisms or learning rules exist to shape the modulatory response itself.

A single desired output from a neuron or network can result from multiple sets of conductance parameters (Goldman et al. 2001, Golowasch et al. 2002, Prinz et al. 2004, Taylor et al. 2009), but not all these solutions will respond to a given neuromodulatory influence in the same way (Szücs & Selverston 2006). The space of reliable responses may still be large (**Figure 4**), and achieving reliable modulation may be mitigated by restricting the solutions that biological circuits tend to find during development (Grashow et al. 2009). Nonetheless, the requirement to respond appropriately to neuromodulation adds a constraint to the set of solutions that homeostatic tuning processes should find. Another possibility is that the neuromodulatory receptors and their downstream signaling pathways are themselves tuned to the particular cells and circuits in which they function. If the latter is true, we must then ask how this tuning signal is encoded and what mechanisms underlie the regulation of the modulator receptors and signaling components.

Some neuromodulatory influences, such as those that occur daily or every several hours during feeding or sleep/wake transitions (Lee & Dan 2012, van den Pol 2012), may produce changes in activity on timescales that are relatively rapid compared with slower homeostatic tuning processes. In this case, the changes in activity produced by modulation would contribute directly to the signals

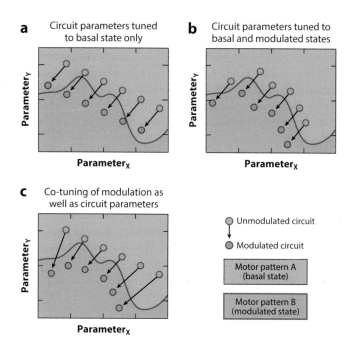

a Circuit parameters tuned to basal state only

b Circuit parameters tuned to basal and modulated states

c Co-tuning of modulation as well as circuit parameters

Parameter_Y

Parameter_X

○ Unmodulated circuit

● Modulated circuit

Motor pattern A (basal state)

Motor pattern B (modulated state)

Figure 7

How are neuromodulator actions tuned to produce reliable behavioral switches in variable circuits whose components are homeostatically tuned? Shown are cartoons of parameter space for a motor circuit in which two behaviorally important activity patterns (A and B) can be produced by modulating two parameters, X and Y (e.g., synaptic or intrinsic conductances). Green points depict the unmodulated parameters, which have been homeostatically tuned to produce a basal output (pattern A). Arrows indicate a quantitative change in both parameters that is produced by a neuromodulator or group of modulators that are responsible for switching activity to motor pattern B. (*a*) A scenario in which unmodulated circuit parameters are tuned, but the magnitude and direction of the modulatory change are not. (*b*) An alternative scenario in which circuit parameters are tuned to give basal output, but in a way that guarantees reliable transition to the modulated state without tuning the effect of the modulator. (*c*) A final scenario in which parameters as well as the modulatory effects are tuned to reliably switch each instance of the network from pattern A to pattern B.

used by neurons as they regulate their conductance densities, and one can imagine that modulator receptors together with their downstream targets can be coregulated. The problem becomes much more challenging when a network must respond appropriately to a modulatory state that is rare or intermittent. In that case, one must suppose that evolutionary pressures constrain the set of solutions available to the homeostatic tuning processes to those that will allow the modulatory processes to move circuit performance in the desired direction. In the absence of such constraints, one imagines that brain disorders such as mental illness and seizures might be far more common than they are.

CONCLUDING REMARKS

We know that neuromodulators are responsible for producing the circuit configurations that evoke specific behaviors. But we are only starting to understand some of the mechanisms that control the neuromodulatory complement of individual neurons (Birren & Marder 2013, Dulcis et al. 2013) and how the release of modulatory substances is controlled (van den Pol 2012). These processes need to be well matched with the processes that control the properties of the target circuits. A remaining challenge is to understand how this matching occurs and thus how networks can function reliably in response to their rich neuromodulatory control systems.

In the examples we have discussed, a common set of themes has emerged. First, modulators act by altering parameters of single neurons and synapses and, consequently, circuits. Second, synaptic and intrinsic membrane parameters themselves vary by organism. Third, how membrane currents and synapses interact means that modulation can exhibit state dependence and nonlinearities. These principles are likely common to all nervous systems, from small to large. A major set of challenges is to understand how nervous systems can operate in basal conditions and respond appropriately to many neuromodulatory perturbations in spite of having intrinsically variable components.

DISCLOSURE STATEMENT

The authors are not aware of any affiliations, memberships, funding, or financial holdings that might be perceived as affecting the objectivity of this review.

ACKNOWLEDGMENTS

This work is supported by NIH grants NS 17813 and MH 46742 and a Charles A. King Trust Fellowship (T.O.).

LITERATURE CITED

Adams WB, Benson JA. 1985. The generation and modulation of endogenous rhythmicity in the *Aplysia* bursting pacemaker neurone R15. *Prog. Biophys. Mol. Biol.* 46:1–49

Alekseyenko OV, Chan YB, Li R, Kravitz EA. 2013. Single dopaminergic neurons that modulate aggression in *Drosophila*. *Proc. Natl. Acad. Sci. USA* 110:6151–56

Alekseyenko OV, Lee C, Kravitz EA. 2010. Targeted manipulation of serotonergic neurotransmission affects the escalation of aggression in adult male *Drosophila melanogaster*. *PLoS ONE* 5:e10806

Antonov I, Kandel ER, Hawkins RD. 2010. Presynaptic and postsynaptic mechanisms of synaptic plasticity and metaplasticity during intermediate-term memory formation in *Aplysia*. *J. Neurosci.* 30:5781–91

Ayali A, Harris-Warrick RM. 1999. Monoamine control of the pacemaker kernel and cycle frequency in the lobster pyloric network. *J. Neurosci.* 19:6712–22

Baines RA, Uhler JP, Thompson A, Sweeney ST, Bate M. 2001. Altered electrical properties in *Drosophila* neurons developing without synaptic transmission. *J. Neurosci.* 21:1523–31

Bargmann CI. 2012. Beyond the connectome: how neuromodulators shape neural circuits. *Bioessays* 34:458–65

Beverly M, Anbil S, Sengupta P. 2011. Degeneracy and neuromodulation among thermosensory neurons contribute to robust thermosensory behaviors in *Caenorhabditis elegans*. *J. Neurosci.* 31:11718–27

Birren SJ, Marder E. 2013. Neuroscience. Plasticity in the neurotransmitter repertoire. *Science* 340:436–37

Blitz DM, Nusbaum MP. 2011. Neural circuit flexibility in a small sensorimotor system. *Curr. Opin. Neurobiol.* 21:544–52

Boyle MB, Klein M, Smith SJ, Kandel ER. 1984. Serotonin increases intracellular Ca^{2+} transients in voltage-clamped sensory neurons of *Aplysia californica*. *Proc. Natl. Acad. Sci. USA* 81:7642–46

Braha O, Edmonds B, Sacktor T, Kandel ER, Klein M. 1993. The contributions of protein kinase A and protein kinase C to the actions of 5-HT on the L-type Ca^{2+} current of the sensory neurons in *Aplysia*. *J. Neurosci.* 13:1839–51

Brezina V. 2010. Beyond the wiring diagram: signalling through complex neuromodulator networks. *Philos. Trans. R. Soc. B* 365:2363–74

Briggman KL, Kristan WB. 2008. Multifunctional pattern-generating circuits. *Annu. Rev. Neurosci.* 31:271–94

Chalasani SH, Kato S, Albrecht DR, Nakagawa T, Abbott LF, Bargmann CI. 2010. Neuropeptide feedback modifies odor-evoked dynamics in *Caenorhabditis elegans* olfactory neurons. *Nat. Neurosci.* 13:615–21

Chao MY, Komatsu H, Fukuto HS, Dionne HM, Hart AC. 2004. Feeding status and serotonin rapidly and reversibly modulate a *Caenorhabditis elegans* chemosensory circuit. *Proc. Natl. Acad. Sci. USA* 101:15512–17

Dacks AM, Weiss KR. 2013. Latent modulation: a basis for non-disruptive promotion of two incompatible behaviors by a single network state. *J. Neurosci.* 33:3786–98

Davis GW. 2006. Homeostatic control of neural activity: from phenomenology to molecular design. *Annu. Rev. Neurosci.* 29:307–23

Dulcis D, Jamshidi P, Leutgeb S, Spitzer NC. 2013. Neurotransmitter switching in the adult brain regulates behavior. *Science* 340:449–53

Epstein IR, Marder E. 1990. Multiple modes of a conditional neural oscillator. *Biol. Cybern.* 63:25–34

Flamm RE, Harris-Warrick RM. 1986. Aminergic modulation in lobster stomatogastric ganglion. II. Target neurons of dopamine, octopamine, and serotonin within the pyloric circuit. *J. Neurophysiol.* 55:866–81

Flavell SW, Pokala N, Macosko EZ, Albrecht DR, Larsch J, Bargmann CI. 2013. Serotonin and the neuropeptide PDF initiate and extend opposing behavioral states in *C. elegans*. *Cell* 154:1023–35

Friesen WO, Kristan WB. 2007. Leech locomotion: swimming, crawling, and decisions. *Curr. Opin. Neurobiol.* 17:704–11

Goaillard J-M, Taylor AL, Schulz DJ, Marder E. 2009. Functional consequences of animal-to-animal variation in circuit parameters. *Nat. Neurosci.* 12:1424–30

Goldman MS, Golowasch J, Marder E, Abbott LF. 2001. Global structure, robustness, and modulation of neuronal models. *J. Neurosci.* 21:5229–38

Golowasch J, Goldman MS, Abbott LF, Marder E. 2002. Failure of averaging in the construction of a conductance-based neuron model. *J. Neurophysiol.* 87:1129–31

Golowasch J, Marder E. 1992. Proctolin activates an inward current whose voltage dependence is modified by extracellular Ca^{2+}. *J. Neurosci.* 12:810–17

Grashow R, Brookings T, Marder E. 2009. Reliable neuromodulation from circuits with variable underlying structure. *Proc. Natl. Acad. Sci. USA* 106:11742–46

Gutierrez GJ, O'Leary T, Marder E. 2013. Multiple mechanisms switch an electrically coupled, synaptically inhibited neuron between competing rhythmic oscillators. *Neuron* 77:845–58

Harris G, Mills H, Wragg R, Hapiak V, Castelletto M, et al. 2010. The monoaminergic modulation of sensory-mediated aversive responses in *Caenorhabditis elegans* requires glutamatergic/peptidergic cotransmission. *J. Neurosci.* 30:7889–99

Harris-Warrick RM, Coniglio LM, Levini RM, Gueron S, Guckenheimer J. 1995. Dopamine modulation of two subthreshold currents produces phase shifts in activity of an identified motoneuron. *J. Neurophysiol.* 74:1404–20

Harris-Warrick RM, Flamm RE. 1987. Multiple mechanisms of bursting in a conditional bursting neuron. *J. Neurosci.* 7:2113–28

Harris-Warrick RM, Johnson BR. 2010. Checks and balances in neuromodulation. *Front. Behav. Neurosci.* 4:47

Harris-Warrick RM, Marder E. 1991. Modulation of neural networks for behavior. *Annu. Rev. Neurosci.* 14:39–57

Hooper SL, Marder E. 1987. Modulation of the lobster pyloric rhythm by the peptide proctolin. *J. Neurosci.* 7:2097–112

Jing J, Vilim FS, Cropper EC, Weiss KR. 2008. Neural analog of arousal: persistent conditional activation of a feeding modulator by serotonergic initiators of locomotion. *J. Neurosci.* 28:12349–61

Jing J, Vilim FS, Horn CC, Alexeeva V, Hatcher NG, et al. 2007. From hunger to satiety: reconfiguration of a feeding network by *Aplysia* neuropeptide Y. *J. Neurosci.* 27:3490–502

Jing J, Weiss KR. 2001. Neural mechanisms of motor program switching in *Aplysia*. *J. Neurosci.* 21:7349–62

Kiehn O, Harris-Warrick RM. 1992. 5-HT modulation of hyperpolarization-activated inward current and calcium-dependent outward current in a crustacean motor neuron. *J. Neurophysiol.* 68:496–508

Kintos N, Nusbaum MP, Nadim F. 2008. A modeling comparison of projection neuron- and neuromodulator-elicited oscillations in a central pattern generating network. *J. Comput. Neurosci.* 24:374–97

Kramer RH, Levitan IB. 1990. Activity-dependent neuromodulation in *Aplysia* neuron R15: intracellular calcium antagonizes neurotransmitter responses mediated by cAMP. *J. Neurophysiol.* 63:1075–88

Kristan WB. 2008. Neuronal decision-making circuits. *Curr. Biol.* 18:R928–32

Lee S-H, Dan Y. 2012. Neuromodulation of brain states. *Neuron* 76:209–22

LeMasson G, Marder E, Abbott LF. 1993. Activity-dependent regulation of conductances in model neurons. *Science* 259:1915–17

Levitan IB. 1988. Modulation of ion channels in neurons and other cells. *Annu. Rev. Neurosci.* 11:119–36

Liu Q, Liu S, Kodama L, Driscoll MR, Wu MN. 2012. Two dopaminergic neurons signal to the dorsal fan-shaped body to promote wakefulness in *Drosophila*. *Curr. Biol.* 22:2114–23

Liu Z, Golowasch J, Marder E, Abbott LF. 1998. A model neuron with activity-dependent conductances regulated by multiple calcium sensors. *J. Neurosci.* 18:2309–20

Marder E. 2012. Neuromodulation of neuronal circuits: back to the future. *Neuron* 76:1–11

Marder E, Bucher D. 2007. Understanding circuit dynamics using the stomatogastric nervous system of lobsters and crabs. *Annu. Rev. Physiol.* 69:291–316

Marder E, Eisen JS. 1984. Electrically coupled pacemaker neurons respond differently to same physiological inputs and neurotransmitters. *J. Neurophysiol.* 51:1362–74

Marder E, Thirumalai V. 2002. Cellular, synaptic and network effects of neuromodulation. *Neural Netw.* 15:479–93

Mills H, Wragg R, Hapiak V, Castelletto M, Zahratka J, et al. 2012. Monoamines and neuropeptides interact to inhibit aversive behaviour in *Caenorhabditis elegans*. *EMBO J.* 31:667–78

Morgan PT, Jing J, Vilim FS, Weiss KR. 2002. Interneuronal and peptidergic control of motor pattern switching in *Aplysia*. *J. Neurophysiol.* 87:49–61

Morgan PT, Perrins R, Lloyd PE, Weiss KR. 2000. Intrinsic and extrinsic modulation of a single central pattern generating circuit. *J. Neurophysiol.* 84:1186–93

Nadim F, Brezina V, Destexhe A, Linster C. 2008. State dependence of network output: modeling and experiments. *J. Neurosci.* 28:11806–13

Nusbaum MP, Blitz DM. 2012. Neuropeptide modulation of microcircuits. *Curr. Opin. Neurobiol.* 22:592–601

Nusbaum MP, Marder E. 1989. A modulatory proctolin-containing neuron (MPN). II. State-dependent modulation of rhythmic motor activity. *J. Neurosci.* 9:1600–7

O'Leary T, van Rossum MC, Wyllie DJ. 2010. Homeostasis of intrinsic excitability in hippocampal neurones: dynamics and mechanism of the response to chronic depolarization. *J. Physiol.* 588:157–70

O'Leary T, Williams AH, Caplan JS, Marder E. 2013. Correlations in ion channel expression emerge from homeostatic tuning rules. *Proc. Natl. Acad. Sci. USA* 110:E2645–54

O'Leary T, Wyllie DJ. 2011. Neuronal homeostasis: time for a change? *J. Physiol.* 589:4811–26

Palmer CR, Kristan WB Jr. 2011. Contextual modulation of behavioral choice. *Curr. Opin. Neurobiol.* 21:520–26

Pratt KG, Aizenman CD. 2007. Homeostatic regulation of intrinsic excitability and synaptic transmission in a developing visual circuit. *J. Neurosci.* 27:8268–77

Pratt KG, Watt AJ, Griffith LC, Nelson SB, Turrigiano GG. 2003. Activity-dependent remodeling of presynaptic inputs by postsynaptic expression of activated CaMKII. *Neuron* 39:269–81

Prinz AA, Bucher D, Marder E. 2004. Similar network activity from disparate circuit parameters. *Nat. Neurosci.* 7:1345–52

Ransdell JL, Nair SS, Schulz DJ. 2013. Neurons within the same network independently achieve conserved output by differentially balancing variable conductance magnitudes. *J. Neurosci.* 33:9950–56

Rinberg A, Taylor AL, Marder E. 2013. The effects of temperature on the stability of a neuronal oscillator. *PLOS Comput. Biol.* 9:e1002857

Rodriguez JC, Blitz DM, Nusbaum MP. 2013. Convergent rhythm generation from divergent cellular mechanisms. *J. Neurosci.* 33:18047–64

Roffman RC, Norris BJ, Calabrese RL. 2012. Animal-to-animal variability of connection strength in the leech heartbeat central pattern generator. *J. Neurophysiol.* 107:1681–93

Saideman SR, Blitz DM, Nusbaum MP. 2007a. Convergent motor patterns from divergent circuits. *J. Neurosci.* 27:6664–74

Saideman SR, Christie AE, Torfs P, Huybrechts J, Schoofs L, Nusbaum MP. 2006. Actions of kinin peptides in the stomatogastric ganglion of the crab *Cancer borealis*. *J. Exp. Biol.* 209:3664–76

Saideman SR, Ma M, Kutz-Naber KK, Cook A, Torfs P, et al. 2007b. Modulation of rhythmic motor activity by pyrokinin peptides. *J. Neurophysiol.* 97:579–95

Schulz DJ, Goaillard JM, Marder E. 2006. Variable channel expression in identified single and electrically coupled neurons in different animals. *Nat. Neurosci.* 9:356–62

Schulz DJ, Goaillard JM, Marder EE. 2007. Quantitative expression profiling of identified neurons reveals cell-specific constraints on highly variable levels of gene expression. *Proc. Natl. Acad. Sci. USA* 104:13187–91

Skiebe P, Schneider H. 1994. Allatostatin peptides in the crab stomatogastric nervous system: inhibition of the pyloric motor pattern and distribution of allatostatin-like immunoreactivity. *J. Exp. Biol.* 194:195–208

Sobie EA. 2009. Parameter sensitivity analysis in electrophysiological models using multivariable regression. *Biophys. J.* 96:1264–74

Spitzer N, Cymbalyuk G, Zhang H, Edwards DH, Baro DJ. 2008. Serotonin transduction cascades mediate variable changes in pyloric network cycle frequency in response to the same modulatory challenge. *J. Neurophysiol.* 99:2844–63

Stangier J, Hilbich C, Beyreuther K, Keller R. 1987. Unusual cardioactive peptide (CCAP) from pericardial organs of the shore crab *Carcinus maenas*. *Proc. Natl. Acad. Sci. USA* 84:575–79

Stein W. 2009. Modulation of stomatogastric rhythms. *J. Comp. Physiol. A Neuroethol. Sens. Neural. Behav. Physiol.* 195:989–1009

Swensen AM, Bean BP. 2005. Robustness of burst firing in dissociated Purkinje neurons with acute or long-term reductions in sodium conductance. *J. Neurosci.* 25:3509–20

Swensen AM, Marder E. 2000. Multiple peptides converge to activate the same voltage-dependent current in a central pattern-generating circuit. *J. Neurosci.* 20:6752–59

Swensen AM, Marder E. 2001. Modulators with convergent cellular actions elicit distinct circuit outputs. *J. Neurosci.* 21:4050–58

Szabo TM, Chen R, Goeritz ML, Maloney RT, Tang LS, et al. 2011. Distribution and physiological effects of B-type allatostatins (myoinhibitory peptides, MIPs) in the stomatogastric nervous system of the crab, *Cancer borealis*. *J. Comp. Neurol.* 519:2658–76

Szücs A, Selverston AI. 2006. Consistent dynamics suggests tight regulation of biophysical parameters in a small network of bursting neurons. *J. Neurobiol.* 66:1584–601

Taghert PH, Nitabach MN. 2012. Peptide neuromodulation in invertebrate model systems. *Neuron* 76:82–97

Taylor AL, Goaillard JM, Marder E. 2009. How multiple conductances determine electrophysiological properties in a multicompartment model. *J. Neurosci.* 29:5573–86

Taylor AL, Hickey TJ, Prinz AA, Marder E. 2006. Structure and visualization of high-dimensional conductance spaces. *J. Neurophysiol.* 96:891–905

Thompson KJ, Calabrese RL. 1992. FMRFamide effects on membrane properties of heart cells isolated from the leech, *Hirudo medicinalis*. *J. Neurophysiol.* 67:280–91

Tobin AE, Calabrese RL. 2005. Myomodulin increases I_h and inhibits the NA/K pump to modulate bursting in leech heart interneurons. *J. Neurophysiol.* 94:3938–50

Tobin AE, Cruz-Bermúdez ND, Marder E, Schulz DJ. 2009. Correlations in ion channel mRNA in rhythmically active neurons. *PLoS ONE* 4:e6742

Turrigiano G. 2011. Too many cooks? Intrinsic and synaptic homeostatic mechanisms in cortical circuit refinement. *Annu. Rev. Neurosci.* 34:89–103

Turrigiano GG. 2008. The self-tuning neuron: synaptic scaling of excitatory synapses. *Cell* 135:422–35

van den Pol AN. 2012. Neuropeptide transmission in brain circuits. *Neuron* 76:98–115

Wagenaar DA, Hamilton MS, Huang T, Kristan WB, French KA. 2010. A hormone-activated central pattern generator for courtship. *Curr. Biol.* 20:487–95

Weimann JM, Skiebe P, Heinzel HG, Soto C, Kopell N, et al. 1997. Modulation of oscillator interactions in the crab stomatogastric ganglion by crustacean cardioactive peptide. *J. Neurosci.* 17:1748–60

White RS, Nusbaum MP. 2011. The same core rhythm generator underlies different rhythmic motor patterns. *J. Neurosci.* 31:11484–94

Williams AH, Calkins A, O'Leary T, Symonds R, Marder E, Dickinson PS. 2013. The neuromuscular transform of the lobster cardiac system explains the opposing effects of a neuromodulator on muscle output. *J. Neurosci.* 33:16565–75

Wragg RT, Hapiak V, Miller SB, Harris GP, Gray J, et al. 2007. Tyramine and octopamine independently inhibit serotonin-stimulated aversive behaviors in *Caenorhabditis elegans* through two novel amine receptors. *J. Neurosci.* 27:13402–12

Wu JS, Vilim FS, Hatcher NG, Due MR, Sweedler JV, et al. 2010. Composite modulatory feedforward loop contributes to the establishment of a network state. *J. Neurophysiol.* 103:2174–84

Yu X, Byrne JH, Baxter DA. 2004. Modeling interactions between electrical activity and second-messenger cascades in *Aplysia* neuron R15. *J. Neurophysiol.* 91:2297–311

The Neurobiology of Language Beyond Single Words

Peter Hagoort[1,2] and Peter Indefrey[2,3]

[1]Max Planck Institute for Psycholinguistics, 6525 XD Nijmegen, The Netherlands;
email: peter.hagoort@donders.ru.nl

[2]Donders Institute for Brain, Cognition and Behaviour, Radboud University Nijmegen,
6525 EN Nijmegen, The Netherlands

[3]Heinrich Heine University, 40225 Düsseldorf, Germany; email: indefrey@phil.hhu.de

Annu. Rev. Neurosci. 2014. 37:347–62

The *Annual Review of Neuroscience* is online at
neuro.annualreviews.org

This article's doi:
10.1146/annurev-neuro-071013-013847

Keywords

syntactic processing, semantic processing, neuropragmatics, Broca's area,
Wernicke's area, meta-analysis

Abstract

A hallmark of human language is that we combine lexical building blocks retrieved from memory in endless new ways. This combinatorial aspect of language is referred to as unification. Here we focus on the neurobiological infrastructure for syntactic and semantic unification. Unification is characterized by a high-speed temporal profile including both prediction and integration of retrieved lexical elements. A meta-analysis of numerous neuroimaging studies reveals a clear dorsal/ventral gradient in both left inferior frontal cortex and left posterior temporal cortex, with dorsal foci for syntactic processing and ventral foci for semantic processing. In addition to core areas for unification, further networks need to be recruited to realize language-driven communication to its full extent. One example is the theory of mind network, which allows listeners and readers to infer the intended message (speaker meaning) from the coded meaning of the linguistic utterance. This indicates that sensorimotor simulation cannot handle all of language processing.

Contents

INTRODUCTION

Language processing is more than memory retrieval and more than the simple concatenation of retrieved lexical items. The expressive power of human language is based on the ability to combine elements from memory in novel ways. This notion led the linguist Wilhelm von Humboldt (1829) to his famous claim that language "makes infinite use of finite means." This central feature of the human language system is, however, not represented in the classical Wernicke-Lichtheim-Geschwind (WLG) model of the neural architecture of language. This model did not go beyond the single word level (Hagoort 2013, Levelt 2013). Even today a large proportion of neuroimaging research on language focuses on processing at or below this level. For example, one of the most influential recent models of the neurobiology of language, that of Hickok & Poeppel (2007), focuses mainly on speech processing at lexical and sublexical levels. Our focus here is instead on processes beyond the retrieval of lexical information. Clearly, retrieval of the lexical building blocks for language (e.g., phonological, morphological, syntactic building blocks) is a crucial component of human language processing (Hagoort 2005, 2013). But equally important is the process of deriving new and complex meaning from the lexical building blocks. This process is referred to in different ways. Here we use the term unification to indicate the real-time assembly of the pieces retrieved from memory into larger structures, with contributions from context. Classic psycholinguistic studies of unification had a strong focus on syntactic analysis (parsing). Unification operations take place at the syntactic processing level, but not only there (cf. Jackendoff 2002, 2007). At the semantic and phonological levels, too, conceptual and lexical elements are combined and integrated into larger structures. It is therefore useful to distinguish between syntactic, semantic, and phonological unification (Hagoort 2005, 2013). Before discussing the neurobiological infrastructure for unification in more detail, we discuss the temporal dynamics of the operations supported by this infrastructure.

Unification and Prediction

One of the most remarkable characteristics of speaking and listening is the speed at which it occurs. Speakers produce easily between 2 and 5 words per second, information that must be decoded by the listener within roughly the same time frame. Given the rate of typical speech we can deduce that word recognition is extremely fast and efficient, taking no more than 200–300 ms. Language processing is highly incremental, with a continuous interaction between the preceding context and the ongoing lexical retrieval operations (Zwitserlood 1989). Here prediction and postlexical integration might work together (cf. Kutas et al. 2015). Electrophysiological data from event-related brain potential (ERP) studies on language provide relevant insights about the time course of the incremental updating of the input representation.

Considering that the acoustic duration of many words is on the order of a few hundred milliseconds, the immediacy of the electrophysiological language-related effects is remarkable. For instance, the early left anterior negativity (ELAN), a syntax-related effect (Friederici et al. 2003), starts at ~100–150 ms after onset of the acoustic word (but see Steinhauer & Drury 2012 for a critical discussion of the ELAN). The onset of the N400 is at ~250 ms, and another language-relevant ERP, the so-called P600, usually starts at ~500 ms. Thus some of these effects occur well before the end of a spoken word. Classifying visual input (e.g., a picture) as depicting an animate or inanimate entity takes the brain ~150 ms (Thorpe et al. 1996). Roughly the same amount of time is needed to classify orthographic input as a letter (Grainger et al. 2008). If we take this time estimate as our reference point, the early appearance of electrophysiological responses to a spoken word is a salient feature of word processing. In physiological terms, the earliness of some of these brain responses suggests that they are not influenced by long-range recurrent feedback to the primary and secondary auditory cortices involved in first-pass acoustic and phonological analysis. Recent modeling work suggests that early ERP effects are best explained by a model with feedforward connections only. Backward connections become essential only after 220 ms (Garrido et al. 2007). The effects of backward connections are, therefore, not manifest in the latency range of early ERP effects such as the ELAN because not enough time has passed for return activity from higher levels to auditory cortex. In the case of speech, the N400 follows the word recognition points closely in time, which suggests that what occurs on-line (that is, in real time) in language comprehension is presumably based partly on predictive processing. Under many circumstances, there is simply not enough time for top-down feedback to exert control over a preceding bottom-up analysis. Very likely, lexical, semantic, and syntactic cues conspire to predict sometimes very detailed characteristics of the next anticipated word, including its syntactic and semantic makeup. A mismatch between contextual prediction and the output of bottom-up analysis results in an immediate brain response recruiting additional processing resources to salvage the on-line interpretation process. Recent ERP studies have provided evidence that context can indeed result in predictions about the syntactic features (i.e., grammatical gender; Van Berkum et al. 2005) and word form (DeLong et al. 2005) of the next word. Lau et al. (2006) showed that the ELAN elicited by a word category violation was modulated by the strength of the expectation for a particular word category in the relevant syntactic slot. More general event knowledge or semantic domains seem to assist with prediction of upcoming information (see Metusalem et al. 2012, Szewczyk & Schriefers 2013, Kutas et al. 2015 for a review of the relevant literature). Whereas the evidence for prediction is strong in the case of language comprehension, the role of prediction in language production (speaking) is less clear. The listener can, at best, guess (predict) what the speaker will say next, but the speaker herself knows where her speech is headed. Hence, in this case, planning an utterance seems to coincide with prediction (Meyer & Hagoort 2013).

Unification refers to the ongoing combinatorial operations in language, which include both context-driven predictions and integration of the lexical information into a representation that spans the whole utterance. The difference is that predictions work forward in time, whereas integration works backward in time; in the latter case, information is integrated into a context that preceded the currently processed lexical item.

Unification is not restricted to linguistic building blocks. ERP studies have shown that non-linguistic information also contributes to unification. Here one can think of information about the speaker (Van Berkum et al. 2008), co-speech gestures (Willems et al. 2007), or world knowledge (Hagoort et al. 2004). These studies found the same effects for nonlinguistic compared with linguistic information on the amplitude and latency of ERPs such as the N400. These data suggest that both linguistic and nonlinguistic contexts have an immediate impact on the interpretation of an utterance (Hagoort & van Berkum 2007).

Shared Circuitry for Comprehension and Production

The WLG model strictly divided between areas involved in language production (Broca's area) and language comprehension (Wernicke's area). It has become clear that this division is incorrect. Lesions in Broca's area impair not only language production but also language comprehension (Caramazza & Zurif 1976), whereas lesions in Wernicke's area also affect language production. More recent neuroimaging studies provided further evidence that the classical view on the role of these regions is no longer tenable. Central aspects of language production and comprehension are subserved by shared neural circuitry (Menenti et al. 2011, Segaert et al. 2012). For instance, Segaert et al. (2012) investigated whether the neurobiological substrate for coding and processing syntactic representations is shared between speaking and listening. To ensure that not just the same brain regions but in fact the same neuronal populations within these regions are involved in both modalities, these investigators used a repetition suppression design (Grill-Spector et al. 2006). For repeated syntactic structures (e.g., actives, passives) within and between modalities (i.e., speaking and listening), they found the same suppression effects for within- and for between-modality repetitions. These effects were localized in the left middle temporal gyrus (MTG), the left inferior frontal gyrus (IFG), and the left supplementary motor area (see **Supplemental Figure 1**. Follow the **Supplemental Material link** from the Annual Reviews home page at **http://www.annualreviews.org**).

Given that speaking and listening have different processing requirements, the same areas may not be recruited to the same degree in these two cases (Indefrey et al. 2004). This difference between speaking and listening, however, does not detract from the finding that when needed, the same areas subserve unification operations in language production and language comprehension.

Even less clear is the degree of separation between networks for syntactic and semantic unification. Separation of these networks is nontrivial because syntactic differences are not without semantic consequences. For instance, sentences containing an object-relative clause (e.g., "The reporter who the senator attacked admitted the error") are found to be more difficult to understand than sentences with a subject-relative clause (e.g., "The reporter who attacked the senator admitted the error"). This increased difficulty is usually attributed to the difference in the syntactic structure of object and subject relative clauses. Chen and coworkers (2006), however, showed that increases in activation of Broca's area for object compared with subject relative clauses are observed only when the subject of the relative clause is inanimate and the subject of the main clause is animate (e.g., "The golfer that the lightning struck") and not for the syntactically identical sentences with animate relative clause subjects (e.g., "The wood that the man chopped heated the cabin"). Chen et al. argue that this finding rules out hypotheses based on syntactic structure complexity and rather suggests that activation of Broca's area reflects the relative difficulty of thematic role assignment; inanimate referents were dispreferred agents of an action and animate referents dispreferred undergoers of action (see also Kuperberg et al. 2008). A difference in syntactic structure clearly has immediate consequences for mapping grammatical roles (subject, object) onto thematic roles (agent, undergoer) and hence for the event structure depicted in these sentences. The event structure is part of the conceptual/semantic representation of the sentence. Despite the intertwined nature of syntactic and semantic unification, for this review we attempt to segregate these two types of unification as much as possible using data from a new meta-analysis of numerous functional imaging studies.

SYNTACTIC VERSUS SEMANTIC UNIFICATION: A META-ANALYSIS

So far, the available information is most clear for the areas involved in syntactic analysis. In a recent meta-analysis of more than 80 neuroimaging studies on sentence processing, Indefrey (2012) identified consistent activations in a comparison of sentences that were syntactically more demanding and control sentences that were syntactically less demanding (e.g., object-relative versus subject-relative sentences). Two regions were systematically more strongly activated when participants were reading or listening to syntactically demanding sentences. The one region is Broca's area in the left inferior frontal gyrus (LIFG), more specifically the pars opercularis [Brodmann area (BA) 44] and the pars triangularis (BA 45). The other region includes the posterior parts of the left superior and middle temporal gyri. These results are remarkably consistent with those from the syntactic adaptation study by Segaert et al. (2012), with the outcome of a representational similarity analysis of magnetoencephalography (MEG) data (Tyler et al. 2013), and with a recent analysis of lesion data in patients with a syntactic comprehension deficit (Tyler et al. 2011). Although the specific contributions of these areas to syntactic analysis is a matter of debate, the involvement of the LIFG and left posterior MTG cannot be denied.

Hereafter we present a new meta-analysis of neuroimaging studies on sentence processing to establish, among other things, the degree to which the network for semantic unification can be segregated from the network for syntactic unification. We analyzed 151 hemodynamic studies on sentence processing that reported activated brain regions with Talairach coordinates or Montreal Neurological Institute (MNI) coordinates (see **Supplemental Table 1**). A subset of 85 studies contrasting sentence-level processing with lower-level control conditions, or syntactically more demanding sentences with less demanding sentences, were taken from previous meta-analyses, such as the one discussed above (Indefrey 2011, 2012). To identify additional studies on sentence-level semantic processing, we conducted a search in Thomson Reuters Web of Knowledge with the search term "fMRI OR functional magnetic resonance imaging OR PET OR Positron Emission Tomography AND semantic AND sentence." This search yielded 409 studies, of which 66 reported contrasts between semantically more demanding and less demanding sentences, contrasts between syntactically more-demanding and less-demanding sentences, or contrasts between sentences and nonsentential control conditions. All these studies tested healthy participants. The activation foci and the spatial extent of 198 contrasts were coded in an anatomical reference system of 112 regions on the basis of the stereotaxic atlas of Talairach & Tournoux (1988) (for details, see caption to **Supplemental Table 2** and Indefrey & Levelt 2000, 2004).

For any particular region the reliability of its activation was assessed using the following estimate: The average number of activated regions reported per experiment divided by the number of regions (112) corresponds to the probability for any particular region to be reported in an experiment, if reports were randomly distributed over regions. Assuming this probability, the chance level for a region to be reported as activated in a certain number of experiments is given by a binomial distribution. The possibility that the agreement of reports about a certain region was coincidental was rejected if the chance level was below 5% (uncorrected for the number of regions). Regions with a chance level below 0.0004 survived a Bonferroni correction for 112 regions and are reported as 0.05 (corrected). This estimate takes into account that not all studies covered the whole brain owing to the heterogeneity of techniques and analysis procedures (for example, analyzing only regions of interest). The procedure also controls for the fact that the average number of activated regions per study differs between contrasts. In contrasts comparing sentences to low-level control conditions, the number of activated regions is typically higher than in contrasts comparing syntactically demanding to less demanding sentences; thus the chances of coincidental agreements between studies are also higher. **Supplemental Tables 2, 3**, and **4** show the summary data for all 112 regions and all comparisons reported here.

Results

The contrasts we analyzed fall into two main categories: About one-third of the contrasts compared sentences to nonsentential stimuli, ranging from word lists to cross-hair fixation or rest conditions. The resulting brain activations can be expected to include whichever brain regions are involved in sentence-level syntactic and semantic processing. However, many other regions subserving lower-level processes, such as lexical retrieval, may also show up. About two-thirds of the contrasts compared syntactically or semantically demanding sentences to less demanding sentences. The latter studies controlled much more tightly for lower-level (e.g., lexical) differences between stimuli so that the resulting activations could be considered specific to syntactic or semantic unification. Note, however, that these studies not only may have missed neural correlates of sentence-level processing that were shared between demanding and less demanding sentences, but also may have induced processes related to higher general cognitive demands, such as attention or error-related processes.

The main manipulations for increasing syntactic demands are the use of sentences containing syntactic violations or word-class ambiguities (e.g., "watch" as noun or verb) and the use of structurally more complex sentences, such as those containing center-embedded object-relative clauses. Manipulations for increasing semantic demands are semantic violations ("The trains were very sour . . . ") and lexical-semantic ambiguities (e.g., "bank") that did not affect the syntactic structure in comparing with the correct control sentences. In addition, as instances of higher semantic demands, we also considered experimental manipulations that complicated the listener's ability to assign an overall meaning without inducing a syntactic difference. These instances include sentences with a metaphoric meaning, sentences inducing semantic operations such as coercion, metonymy, and sentences making connections to the previous discourse context. These instances also include sentences requiring listeners to assess speakers' intentions (irony, indirect replies, or requests; see also the section on neuropragmatics).

Sentences compared with control conditions below sentence level. Compared with control conditions below sentence level, the comprehension of sentences reliably activates the temporal lobes and the posterior IFG bilaterally (**Figure 1a**, **Supplemental Figure 2**, **Supplemental Table 2**), albeit with a clear left hemisphere dominance. Not activated in any of the 53 studies were large parts of the right parietal and right inferior temporal cortices. These areas are thus reliably not involved during sentence comprehension. As shown in **Supplemental Figure 3**, some differences can be seen among the regions involved in processing written and spoken sentences. Some right hemisphere temporal regions are not reliably found in reading, and posterior frontal regions are less frequently found in listening. Of particular interest are the results for a subset of studies that used the most natural kind of sentence processing: The participants just listened or read

---

Figure 1

Schematic representation of the brain showing regions with reliably reported activations for sentences compared with nonsentential stimuli (*a*) and sentences with high syntactic or semantic processing demands compared with simpler sentences (*b,c*). The left posterior inferior frontal gyrus is further subdivided into Brodmann areas (BA) 44 (*above black line*), BA 45 (*below black line, above AC–PC line*) and BA 47 (*below AC–PC line*). Green regions indicate a reliable number of reports. Pink regions indicate no reports in 53 studies. For details, see **Supplemental Tables 2, 3**, and **4** (follow the **Supplemental Material link** from the Annual Reviews home page at **http://www.annualreviews.org**). Abbreviations: AC, anterior commissure; PC, posterior commissure.

a Sentences compared with control conditions below sentence level

All (53 studies)

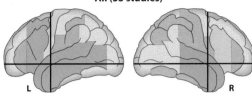

Passive reading (13 studies)

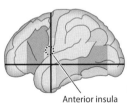

Passive listening (20 studies)

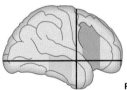

b Sentences with higher compared with sentences with lower processing demands

Higher syntactic demands (57 studies)

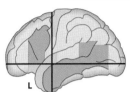
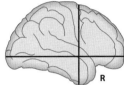
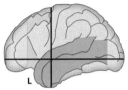
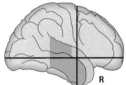

Higher semantic demands (51 studies)

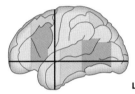
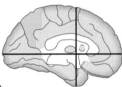
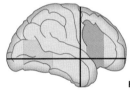
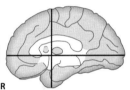

Anterior insula

c Direct comparisons between sentences with high syntactic and high semantic demands

Syntactic–semantic (6 studies)

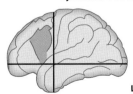
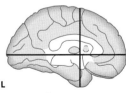

Semantic–syntactic (10 studies)

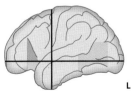
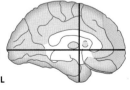

● Reliably activated *p* < 0.05, corrected　　● Reliably activated *p* < 0.05, uncorrected　　● Reliably not activated *p* < 0.05, uncorrected

for comprehension without performing any additional tasks (**Figure 1a**, **Supplemental Figure 4**). Irrespective of the input modality, passive sentence comprehension activates the left MTG, the superior temporal gyrus (STG), and the orbital (BA 47) and triangular (BA 45) parts of LIFG. The most dorsal part of the IFG (pars opercularis, BA 44), however, is not reliably activated during passive listening to (simple) sentences. The same holds when sentence processing is compared with the processing of word lists (see **Supplemental Figure 2**). This result suggests that either BA 44 is not involved in any sentence-level unification process or this unification process is not necessarily active for passive listening, for instance because a good-enough processing strategy (Ferreira et al. 2002) does not require a full compositional analysis.

Sentences with higher demands on syntactic or semantic processing. So far, we have identified a set of candidate regions that may be involved in sentence-level processing. However, the contrast of sentences compared with nonsentential stimuli neither excludes the involvement of these regions in lexical processes nor allows for a distinction between semantic and syntactic unification. This limitation is not present for studies comparing syntactically or semantically more demanding sentences with simpler sentences. In these studies, the words used in both conditions are typically well-matched. A contribution of word-level differences to the resulting brain activation can hence be excluded. Despite that syntactic complexity may always have repercussions for the construction of sentence meaning, semantic and syntactic processes can, in principle, be independently taxed in this type of design. **Figure 1b** shows the results of 57 studies comparing syntactically demanding sentences with less-demanding control sentences and 51 studies comparing semantically demanding sentences with less-demanding control sentences. Higher syntactic processing demands most reliably activate the more dorsal parts of posterior LIFG (BA 44/45), the right posterior IFG and the left posterior STG and MTG. In addition, the left precuneus, the left inferior parietal lobule, and the right posterior MTG all activated reliably. Higher semantic processing demands most reliably activate all parts of posterior LIFG (but BA 45/47 are reported twice as often as is BA 44; see **Supplemental Figure 5**, **Supplemental Table 4**), the right posterior IFG, and the left middle and posterior MTG. In addition, the data indicate a reliable activation of the medial prefrontal cortex that is not seen for higher syntactic processing demands and demonstrate activations of the left anterior insula, angular gyrus, and the posterior ITG.

The results of 16 studies directly comparing sentences with high syntactic and high semantic processing demands (**Figure 1c**, **Supplemental Figure 6**) confirm that the medial prefrontal cortex is important for processing sentences with high semantic processing demands. Direct comparisons also demonstrate a syntactic/semantic gradient in the LIFG: a reliably stronger activation of BA 44 is seen for syntactically, compared with semantically, demanding sentences; a reliably stronger activation of BA 45/47 is observed for semantically, compared with syntactically, demanding sentences.

In sum, compared with studies contrasting relatively undemanding sentences with nonsentential control conditions, studies contrasting higher with lower demands on syntactic or semantic sentence-level processing have found that IFG and some additional regions activate reliably. However, they have found no reliable activation of the anterior temporal lobe, despite some claims that it might be involved in semantic and syntactic combinatorial processing (Hickok & Poeppel 2007, Friederici 2012). We now consider the different kinds of experimental manipulations that induced increased processing demands to determine in which way they contributed to the observed overall results.

Violations, ambiguity, and complexity. Studies comparing sentences with syntactic violations (mostly agreement violations and phrase-structure/word-category violations) with correct sentences have most reliably found BA 44/45 activation (**Supplemental Figure 7**). Studies comparing sentences containing semantic violations with correct sentences have most reliably found activation of all parts of the left posterior IFG; activation of BA 45/47 was again reported more often than was activation of BA 44. Both semantic and syntactic violations generally activate the posterior temporal cortex less frequently than do demanding sentences. Within semantic violations, we also compared violations of semantic selection restrictions (e.g., "Dutch trains are sour"/"My mother ironed a kiss") with violations of world knowledge (e.g., "Dutch trains are white"/"The woman painted the insect"; examples taken from Hagoort et al. 2004 and Kuperberg et al. 2000). Both types of semantic violation activate BA 45/47 (**Supplemental Figure 8**). World knowledge violations also seem to activate widespread medial brain structures.

Supplemental Figure 9 shows the reliably activated regions for sentences containing local syntactic ambiguities (mostly word-class ambiguities, e.g., "He noticed that landing planes frightens some new pilots") or semantic ambiguities ("The reporter commented that modern compounds react unpredictably"; examples from Rodd et al. 2010). Both types of ambiguities activate the posterior IFG bilaterally and the left posterior MTG. For syntactic ambiguities, activation in the left posterior IFG is confined to BA 44. Semantic ambiguities activate the left posterior inferior medial temporal lobe.

Sixty-three studies increased the degree of syntactic or semantic complexity using means other than violations or ambiguities. The majority of studies on syntactic complexity compared sentences containing complex relative clauses with simpler relative clauses. The main manipulation in the remaining studies was the use of noncanonical word order. Our criterion for studies inducing semantic complexity was a comparison of a condition in which understanding the meaning of the sentence required some additional effort compared with that required for syntactically identical control sentences. In most of the studies this goal was achieved by comparing sentences containing a metaphoric meaning (e.g., "A sailboat is a floating leaf"; example from Diaz & Hogstrom 2011) with sentences containing a literal meaning. Whereas a metaphor can be understood without knowing the context of the utterance or the speaker, the same is not true for a different kind of nonliteral sentence. To understand that ironic/sarcastic sentences or indirect replies/requests go beyond a literal interpretation of the sentence requires that a listener assesses the situation and the speaker's intention (see below).

Both syntactic and semantic complexity reliably induce stronger activation of the posterior IFG bilaterally and the left mid and posterior MTG (see **Supplemental Figure 10**). Left posterior IFG activation again shows a gradient with activation of BA 44 for syntactic but not semantic complexity and shows activation of BA 47 for semantic but not syntactic complexity. The posterior STG may selectively be recruited for syntactic complexity. Conversely, semantic complexity induces medial prefrontal activations that are not reliably seen for syntactic complexity manipulations.

Separate analyses of the two main kinds of syntactic complexity studied by investigators (**Supplemental Figure 11**) yielded results that were similar to the overall activation patterns induced by syntactic complexity; therefore, the mechanism that drives these activations seems to be shared by noncanonical word orders and relative clause complexity. Separate analyses of different kinds of semantic complexity, however, yielded differential activation patterns (**Supplemental Figure 12**). Sentences with metaphoric meaning contributed most to the overall activation of BA 45/47 and left posterior MTG. This result replicates the findings of a recent voxel-based meta-analysis on metaphor processing (Bohrn et al. 2012; see contrast "metaphor > literal" in their table 3). A few studies also reported activations of BA 47 and left MTG by comparing literal sentences inducing additional semantic operations (metonymy, coercion, establishment of a causal relationship with preceding context) with literal sentences without such operations. By contrast, sentences that required the listener to assess the speaker's intentions did not reliably activate BA 45 or the left posterior temporal lobe. These kinds of sentences most frequently activated the medial prefrontal cortex (also reliably reported for metaphoric sentences but in a relatively smaller number of studies) and also the right temporoparietal cortex (mainly observed in studies using indirect utterances).

Summary and Discussion

Our meta-analysis comparing reliable activations for the processing of various syntactically and semantically demanding sentences yielded several important results. The most robust result is a distinctive activation pattern in the posterior LIFG; syntactic demands activated more dorsal parts (BA 44/45) and semantic demands activated more ventral parts (BA 45/47) across all kinds of increased processing demands (violations, ambiguity, complexity) and in studies performing direct comparisons of high syntactic and semantic processing demands (see **Figure 2a**). This result indicates that syntactic unification cannot simply be reduced to semantic unification. In particular, BA 44 activation is clearly driven more strongly by syntactic than by semantic demands, suggesting that this region contains neuronal populations involved in syntactic operations as such or that the semantic consequences of syntactic demands (difficulty of thematic role assignment) are processed by neuronal populations that differ from those processing other kinds of semantic unification.

A novel observation is that the dorsal/ventral gradient observed in the left posterior IFG seems to be mirrored in the left posterior temporal lobe. Higher syntactic demands reliably activate the STG and MTG, and higher semantic demands reliably activate the MTG and ITG.

To confirm these gradients, we conducted a spatially more fine-grained analysis on a subset of 28 studies reporting Talairach or MNI coordinates of activation foci in the left posterior temporal cortex. Whereas the mean locations of the activations reported for syntactically (mean $y = -45.0$, $SD = 7.7$) and semantically (mean $y = -46.5$, $SD = 5.9$) demanding sentences did not differ significantly in the rostral/caudal dimension, the mean location of activation foci for semantically demanding sentences was significantly more ventral (middle temporal gyrus, mean $z = 0.9$, $SD = 8.6$) than the mean location of activation foci for syntactically demanding sentences (superior temporal sulcus, mean $z = 10.2$, $SD = 5.7$; see **Figure 2b** and **Supplemental Figure 13**). Activations in the left posterior inferior frontal gyrus reported for the same contrasts in this set of studies confirmed a significantly more ventral and rostral mean location of activation foci for semantically demanding sentences (BA 45, mean $y = 22.2$, $SD y = 6.3$, mean $z = 4.2$, $SD z = 10.8$) as compared with syntactically demanding sentences (BA 44, mean $y = 14.8$, $SD y = 8.0$, mean $z = 15.0$, $SD z = 5.6$).

These corresponding gradients in posterior frontal and temporal regions are remarkably consistent with a functional connectivity pattern found by Xiang et al. (2010), which links seed regions in BA 44, BA 45, and BA 45/47 to left posterior STG, MTG, and ITG, respectively. This finding

---

→

Figure 2

(*a*) Summary of activation patterns for sentences with high syntactic or semantic processing demands compared with simpler sentences. (*b*) Syntactic/semantic gradients in left inferior frontal and posterior temporal cortex based on 28 studies reporting posterior temporal cortex activation for syntactically demanding or semantically demanding sentences compared with less demanding sentences (see **Supplemental Figure 13** for details). The centers represent the mean coordinates of the local maxima, and the radii represent the standard deviations of the distance between the local maxima and their means. Abbreviations: GFm, GFi, middle and inferior frontal gyri; BA, Brodmann area; GTs, GTm, GTi, superior, middle, and inferior temporal gyri; STS, ITS, superior and inferior temporal sulci; Gsm, supramarginal gyrus.

clearly supports the idea that sentence-level unification relies on the coactivation of neuronal populations in a network of posterior frontal and temporal regions, with a similar functional gradient in both parts of the brain.

Another important observation is the degree to which posterior temporal lobe activation differs between violations and other kinds of higher processing demands. Syntactic violations do

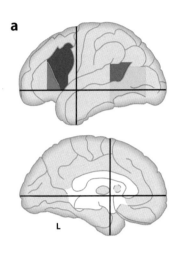

a

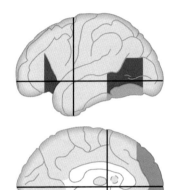

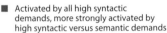

L

L

■ Activated by all high syntactic demands, more strongly activated by high syntactic versus semantic demands

■ Activated by all high syntactic demands (except ambiguity), less strongly activated by high syntactic versus semantic demands

▫ Activated by all high syntactic demands (except violation), less strongly activated by high syntactic versus semantic demands

■ Activated by syntactically complex sentences but not ambiguity and violation

■ Activated by all high semantic demands (except irony) and more strongly activated by high semantic versus syntactic demands

■ Activated by all nonliteral sentences (in particular speaker meaning) and semantic violations, more strongly activated by high semantic versus syntactic demands

■ Activated by semantic ambiguity

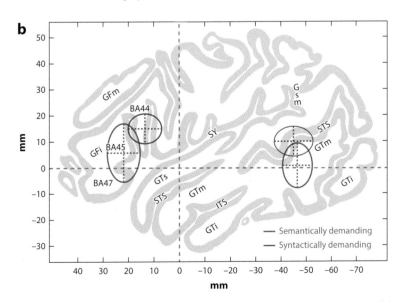

b

not seem to elicit posterior temporal lobe activations reliably, and reports about such activations are relatively infrequent for semantic violations. A plausible account for this dissociation may be based on a distinction between Broca's area subserving sentence-level compositional processes and the posterior temporal lobe subserving the retrieval of lexical syntactic and semantic information (Hagoort 2005). Violation paradigms may tap relatively specifically into a unification stage, whereas ambiguities and complexity manipulations typically also require lexical syntactic and semantic specifications to be retrieved. In sentences with unambiguous violations ("The blouse was on ironed"/"The thunderstorm was ironed"; examples from Friederici et al. 2003), no alternative lexical specifications can be constructed to help remedy the problem; however, not only for ambiguities ("he noticed that landing planes frightens some new pilots"/"the reporter commented that modern compounds react unpredictably"), but also for many kinds of syntactic and semantic complexities, alternative lexical specifications may solve the comprehension problem.

Finally, our results suggest that the neuronal substrate underlying the processing of ironic and indirect utterances differs from that activated in other kinds of high-demand sentences. Such utterances do not seem to induce reliable activation of Broca's area (BA 44/45) and Wernicke's area but, instead, show the most reliable activation in the medial prefrontal cortex and the right temporoparietal cortex. The difficulty in processing these sentences does not seem to be in decoding the meaning of the sentence but in a different aspect of comprehension to which we now turn in more detail.

BEYOND CORE AREAS FOR LANGUAGE: THE CASE OF NEUROPRAGMATICS

In many instances, linguistic expressions are underdetermined with respect to the meaning they convey. What is said and what is understood are often not the same. Communication goes further than the exchange of explicit propositions. In essence, the goal of the speaker is either to change the listener's mind or to commit the addressee to the execution of certain actions, such as closing the window in reply to the statement, "It is cold here." A theory of speech acts is needed to understand how we move from coded meaning to inferred speaker meaning (cf. Grice 1989, Levinson 2013, Hagoort & Levinson 2015). The steps required to progress from coded meaning to speaker meaning involve additional neuronal systems, as recent studies have shown.

In a study on conversational implicatures and indirect requests, for example, Bašnáková and coworkers (2013) contrasted direct and indirect replies, two classes of utterances whose speaker meanings are either very similar to, or markedly different from, their coded meaning. In this study, participants listened to natural spoken dialogue in which the final utterance, e.g., "It is hard to give a good presentation," had different meanings depending on the dialogue context and the immediately preceding question. This final utterance either served as a direct reply (to the question "How hard is it to give a good presentation?") or as an indirect reply (to "Did you like my presentation?"). In the indirect reply condition, activations were observed in the medial prefrontal cortex (mPFC) extending into the right anterior part of the supplementary motor area (SMA), and in the right temporoparietal junction (TPJ). The data thus demonstrated a pattern typical for tasks that involve mentalizing based on a theory of mind (ToM) (Amodio & Frith 2006, Mitchell et al. 2006, Saxe et al. 2006). Although the exact role of all the individual ToM regions is not yet clearly established, both mPFC and right TPJ constitute core regions in ToM tasks (Carrington & Bailey 2009). The most specific hypothesis about the role of the posterior part of right TPJ (Mars et al. 2012) in the mentalizing network is that it is implicated in mental state reasoning, i.e., thinking about other people's beliefs, emotions, and desires (Saxe 2010).

mPFC activation in the Bašnáková et al. study (2013) was found in parts of the mPFC associated with complex sociocognitive processes such as mentalizing or thinking about the intentions of others (such as communicative intentions, right anterior mPFC) or about oneself (right posterior mPFC). The involvement of these regions is also consistently observed in discourse comprehension (e.g., Mason & Just 2009, Mar 2011). This finding might come as no surprise because the motivations, goals, and desires of fictional characters are likely accessed in a manner similar to that of real-life protagonists (Mar & Oatley 2008). In fact, an influential model from the discourse-processing literature (Mason & Just 2009) ascribes to the dorsomedial part of the frontal cortex and the right TPJ a functional role as a protagonist perspective network, which generates expectations about how the protagonists of stories will act on the basis of understanding their intentions.

Although the literature on the neuropragmatics of language is still limited, studies have, with remarkable consistency, demonstrated that understanding the communicative intent of an utterance requires mentalizing. Because the linguistic code underdetermines speaker meaning, the ToM network needs to be invoked to progress from coded meaning to speaker meaning. Despite the popular view that the mirror neuron system (MNS) is sufficient for action understanding (Rizzolatti & Sinigaglia 2010), the MNS does not provide the crucial neural infrastructure for inferring speaker meaning. In addition to core areas for retrieving lexical information from memory and unifying the lexical building blocks to produce and understand multiword utterances, other brain networks are needed to realize language-driven communication to its full extent (for a more extended review of studies on neuropragmatics, see Hagoort & Levinson 2015).

CONCLUSION

We have outlined the contours of the neurobiological infrastructure that supports language processing beyond the comprehension or production of single words. In this context, research has established substantial deviations of the classical WLG model, with its focus on single-word processing. Three major deviations are worth highlighting: (a) Although not discussed in detail here (for more extended discussions, see Hagoort 2013, Amunts & Catani 2015) the connectivity of the language cortex in left perisylvian regions is much more extended than was proposed in the classical model and is certainly not restricted to the arcuate fasciculus; (b) the distribution of labor between the core regions in left perisylvian cortex is fundamentally different than that proposed in the classical model (in contrast to the claims of the WGL model, there is strong evidence for shared circuitry for core aspects of sentence-level language production and comprehension); and (c) the operation of language to its full extent requires a much more extended network than what the classical model assumed. The basic principle of brain organization for higher cognitive functions proposes that these functions are based on the interaction between numerous neuronal circuits and brain regions that support the various contributing functional components. These circuits are not necessarily specialized for language but nevertheless need to be recruited for the sake of successful language processing. One example is the ToM network, which seems critical for speakers to construct utterances with knowledge of the listener in mind and for listeners to progress from coded meaning to speaker meaning. As described above in the section on the temporal profile of unification, the network of language areas in the brain is highly dynamic despite the static nature suggested by pictures of the neuronal infrastructure for language. The specific contribution that any area makes to information processing depends on the input it receives at a certain time step, which itself depends on the computational environment in which it is embedded. Moreover, a large meta-analysis has at last confirmed a clear division of labor in both frontal and temporal areas for syntactic and semantic unification.

DISCLOSURE STATEMENT

The authors are not aware of any affiliations, memberships, funding, or financial holdings that might be perceived as affecting the objectivity of this review.

ACKNOWLEDGMENTS

This research was in part supported by a Spinoza Award to P.H. and by support to P.I. from the Collaborative Research Center 991 of the Deutsche Forschungsgemeinschaft (DFG). We thank Frauke Hellwig for her assistance with the artwork.

LITERATURE CITED

Amodio DM, Frith CD. 2006. Meeting of minds: the medial frontal cortex and social cognition. *Nat. Rev. Neurosci.* 7:268–77

Amunts K, Catani M. 2015. Cytoarchitectonics, receptor architectonics and network topology of language. See Gazzaniga 2015. In press

Bašnáková J, Weber K, Petersson KM, van Berkum J, Hagoort P. 2013. Beyond the language given: the neural correlates of inferring speaker meaning. *Cereb. Cortex.* In press

Bohrn IC, Altmann U, Jacobs AM. 2012. Looking at the brains behind figurative language—a quantitative meta-analysis of neuroimaging studies on metaphor, idiom, and irony processing. *Neuropsychologia* 50:2669–83

Caramazza A, Zurif EB. 1976. Dissociation of algorithmic and heuristic processes in language comprehension: evidence from aphasia. *Brain Lang.* 3:572–82

Carrington SJ, Bailey AJ. 2009. Are there theory of mind regions in the brain? A review of the neuroimaging literature. *Hum. Brain Mapp.* 30:2313–35

Chen E, West WC, Waters G, Caplan D. 2006. Determinants of bold signal correlates of processing object-extracted relative clauses. *Cortex* 42:591–604

DeLong KA, Urbach TP, Kutas M. 2005. Probabilistic word pre-activation during language comprehension inferred from electrical brain activity. *Nat. Neurosci.* 8:1117–21

Diaz MT, Hogstrom LJ. 2011. The influence of context on hemispheric recruitment during metaphor processing. *J. Cogn. Neurosci.* 23:3586–97

Ferreira F, Bailey KGD, Ferraro V. 2002. Good-enough representations in language comprehension. *Curr. Dir. Psychol. Sci.* 11:11–15

Friederici AD. 2012. The cortical language circuit: from auditory perception to sentence comprehension. *Trends Cogn. Sci.* 16:262–68

Friederici AD, Rüschemeyer SA, Hahne A, Fiebach CJ. 2003. The role of left inferior frontal and superior temporal cortex in sentence comprehension: localizing syntactic and semantic processes. *Cereb. Cortex* 13:170–77

Garrido MI, Kilner JM, Kiebel SJ, Friston KJ. 2007. Evoked brain responses are generated by feedback loops. *Proc. Natl. Acad. Sci. USA* 104:20961–66

Gazzaniga MS, ed. 2015. *The Cognitive Neurosciences*. Cambridge, MA: MIT Press. 5th ed. In press

Grainger J, Rey A, Dufau S. 2008. Letter perception: from pixels to pandemonium. *Trends Cogn. Sci.* 12:381–87

Grice P. 1989. *Studies in the Way of Words*. Cambridge, MA: Harvard Univ. Press

Grill-Spector K, Henson R, Martin A. 2006. Repetition and the brain: neural models of stimulus-specific effects. *Trends Cogn. Sci.* 10:14–23

Hagoort P. 2005. On Broca, brain, and binding: a new framework. *Trends Cogn. Sci.* 9:416–23

Hagoort P. 2013. MUC (memory, unification, control) and beyond. *Front. Psychol.* 4:416

Hagoort P, Hald L, Bastiaansen M, Petersson KM. 2004. Integration of word meaning and world knowledge in language comprehension. *Science* 304:438–41

Hagoort P, Levinson SL. 2015. Neuropragmatics. See Gazzaniga 2015. In press

Hagoort P, van Berkum J. 2007. Beyond the sentence given. *Philos. Trans. R. Soc. B* 362:801–11

Hickok G, Poeppel D. 2007. The cortical organization of speech processing. *Nat. Rev. Neurosci.* 8:393–402

Indefrey P. 2011. Neurobiology of syntax. In *The Cambridge Encyclopedia of the Language Sciences*, ed. PC Hogan, pp. 835–38. Cambridge, UK/New York: Cambridge Univ. Press

Indefrey P. 2012. Hemodynamic studies of syntactic processing. In *The Handbook of the Neuropsychology of Language*, ed. M Faust, pp. 209–28. Malden, MA: Blackwell

Indefrey P, Hellwig F, Herzog H, Seitz RJ, Hagoort P. 2004. Neural responses to the production and comprehension of syntax in identical utterances. *Brain Lang.* 89:312–19

Indefrey P, Levelt WJM. 2000. The neural correlates of language production. In *The New Cognitive Neurosciences*, ed. MS Gazzaniga, pp. 845–65. Cambridge, MA: MIT Press

Indefrey P, Levelt WJM. 2004. The spatial and temporal signatures of word production components. *Cognition* 92:101–44

Jackendoff R. 2002. *Foundations of Language: Brain, Meaning, Grammar, Evolution*. Oxford, UK: Oxford Univ. Press

Jackendoff R. 2007. A parallel architecture perspective on language processing. *Brain Res.* 1146:2–22

Kuperberg GR, McGuire PK, Bullmore ET, Brammer MJ, Rabe-Hesketh S, et al. 2000. Common and distinct neural substrates for pragmatic, semantic, and syntactic processing of spoken sentences: an fMRI study. *J. Cogn. Neurosci.* 12:321–41

Kuperberg GR, Sitnikova T, Lakshmanan BM. 2008. Neuroanatomical distinctions within the semantic system during sentence comprehension: evidence from functional magnetic resonance imaging. *NeuroImage* 40:367–88

Kutas M, Federmeier KD, Urbach ThP. 2015. The "negatives" and "positives" of prediction in language. See Gazzaniga 2015. In press

Lau E, Stroud C, Plesch S, Phillips C. 2006. The role of structural prediction in rapid syntactic analysis. *Brain Lang.* 98:74–88

Levelt WJM. 2013. *A History of Psycholinguistics: The Pre-Chomskyan Era*. Oxford, UK: Oxford Univ. Press

Levinson SC. 2013. Action formation and ascription. In *The Handbook of Conversation Analysis*, ed. J Sidnell, T Stivers, pp. 103–30. Malden, MA: Wiley-Blackwell

Mar RA. 2011. The neural bases of social cognition and story comprehension. *Annu. Rev. Psychol.* 62:103–34

Mar RA, Oatley K. 2008. The function of fiction is the abstraction and simulation of social experience. *Perspect. Psychol. Sci.* 3:173–92

Mars RB, Sallet J, Schüffelgen U, Jbabdi S, Toni I, Rushworth MF. 2012. Connectivity-based subdivisions of the human right "temporoparietal junction area": evidence for different areas participating in different cortical networks. *Cereb. Cortex* 22:1894–903

Mason RA, Just MA. 2009. The role of the theory-of-mind cortical network in the comprehension of narratives. *Lang. Linguist. Compass* 3:157–74

Menenti L, Gierhan SME, Segaert K, Hagoort P. 2011. Shared language: overlap and segregation of the neuronal infrastructure for speaking and listening revealed by functional MRI. *Psychol. Sci.* 22:1173–82

Metusalem R, Kutas M, Urbach TP, Hare M, McRae K, Elman JL. 2012. Generalized event knowledge activation during online sentence comprehension. *J. Mem. Lang.* 66:545–67

Meyer AS, Hagoort P. 2013. What does it mean to predict one's own utterances? *Behav. Brain Sci.* 36:367–68

Mitchell JP, Macrae CN, Banaji MR. 2006. Dissociable medial prefrontal contributions to judgments of similar and dissimilar others. *Neuron* 50:655–63

Rizzolatti G, Sinigaglia C. 2010. The functional role of the parieto-frontal mirror circuit: interpretations and misinterpretations. *Nat. Rev. Neurosci.* 11:264–74

Rodd JM, Longe OA, Randall B, Tyler LK. 2010. The functional organisation of the fronto-temporal language system: evidence from syntactic and semantic ambiguity. *Neuropsychologia* 48:1324–35

Saxe R, Moran JM, Scholz J, Gabrieli J. 2006. Overlapping and non-overlapping brain regions for theory of mind and self reflection in individual subjects. *Soc. Cogn. Affect. Neurosci.* 1:229–34

Saxe, R. 2010. The right temporo-parietal junction: a specific brain region for thinking about thoughts. In *Handbook of Theory of Mind*, ed. A Leslie, T German, pp. 1–35. Hove, UK: Psychology Press

Segaert K, Menenti L, Weber K, Petersson KM, Hagoort P. 2012. Shared syntax in language production and language comprehension—an FMRI study. *Cereb. Cortex* 22:1662–70

Steinhauer K, Drury JE. 2012. On the early left-anterior negativity (ELAN) in syntax studies. *Brain Lang.* 120:135–62

Szewczyk JM, Schriefers H. 2013. Prediction in language comprehension beyond specific words: an ERP study on sentence comprehension in Polish. *J. Mem. Lang.* 68:297–314

Talairach J, Tournoux P. 1988. *Co-Planar Stereotaxic Atlas of the Human Brain: 3-D Proportional System: An Approach to Cerebral Imaging.* Stuttgart: Thieme

Thorpe S, Fize D, Marlot C. 1996. Speed of processing in the human visual system. *Nature* 381:520–22

Tyler LK, Cheung TP, Devereux BJ, Clarke A. 2013. Syntactic computations in the language network: characterizing dynamic network properties using representational similarity analysis. *Front. Psychol.* 4:271

Tyler LK, Marslen-Wilson WD, Randall B, Wright P, Devereux BJ, et al. 2011. Left inferior frontal cortex and syntax: function, structure and behaviour in patients with left hemisphere damage. *Brain* 134:415–31

Van Berkum JJ, Brown CM, Zwitserlood P, Kooijman V, Hagoort P. 2005. Anticipating upcoming words in discourse: evidence from ERPs and reading times. *J. Exp. Psychol. Learn. Mem. Cogn.* 31:443–67

Van Berkum JJA, van den Brink D, Tesink C, Kos M, Hagoort P. 2008. The neural integration of speaker and message. *J. Cogn. Neurosci.* 20:580–91

Von Humboldt W. 1829. On the dual. *Nouv. Rev. Ger.* 1:378–81

Willems RM, Özyürek A, Hagoort P. 2007. When language meets action: the neural integration of gesture and speech. *Cereb. Cortex* 17:2322–33

Xiang H-D, Fonteijn HM, Norris DG, Hagoort P. 2010. Topographical functional connectivity pattern in the Perisylvian language networks. *Cereb. Cortex* 20:549–60

Zwitserlood P. 1989. The locus of the effects of sentential-semantic context in spoken-word processing. *Cognition* 32:25–64

Coding and Transformations in the Olfactory System

Naoshige Uchida,[1] Cindy Poo,[2] and Rafi Haddad[1,3]

[1]Department of Molecular and Cellular Biology, Center for Brain Science, Harvard University, Cambridge, Massachusetts 02138; email: uchida@mcb.harvard.edu

[2]Champalimaud Neuroscience Programme, Champalimaud Centre for the Unknown, 1400-038 Lisbon, Portugal; email: cindy.poo@neuro.fchampalimaud.org

[3]Gonda Brain Research Center, Bar Ilan University, Ramat Gan, 52900 Israel; email: rafihaddad@gmail.com

Annu. Rev. Neurosci. 2014. 37:363–85

The *Annual Review of Neuroscience* is online at neuro.annualreviews.org

This article's doi:
10.1146/annurev-neuro-071013-013941

Keywords

neural coding, olfaction, pattern recognition

Abstract

How is sensory information represented in the brain? A long-standing debate in neural coding is whether and how timing of spikes conveys information to downstream neurons. Although we know that neurons in the olfactory bulb (OB) exhibit rich temporal dynamics, the functional relevance of temporal coding remains hotly debated. Recent recording experiments in awake behaving animals have elucidated highly organized temporal structures of activity in the OB. In addition, the analysis of neural circuits in the piriform cortex (PC) demonstrated the importance of not only OB afferent inputs but also intrinsic PC neural circuits in shaping odor responses. Furthermore, new experiments involving stimulation of the OB with specific temporal patterns allowed for testing the relevance of temporal codes. Together, these studies suggest that the relative timing of neuronal activity in the OB conveys odor information and that neural circuits in the PC possess various mechanisms to decode temporal patterns of OB input.

Contents

INTRODUCTION

Cracking the Code

Neurons transmit information using sequences of discrete events: action potentials (spikes). A central question in neuroscience is how trains of action potentials represent specific information. That is, what is a neural code in the brain? Understanding a neural code requires tackling four issues. First, to identify potential neural codes for sensory systems, one must observe which aspects of stimuli cause discernible changes in neural activity (e.g., by determining tuning curves). One critical question is which aspects of the neural activity (e.g., firing rates, spike timing) carry reliable information about the stimulus (representations)? However, understanding the first issue is not sufficient to establish a neural code (or representation) (deCharms & Zador 2000). Second, one needs to understand the mechanisms by which relevant information is transformed into such activity patterns (encoding). Third, for a neural representation (or a neural code) to be functional, it must be read out by downstream neurons (decoding). Ultimately, a given neural code should be read out by animals to guide behavior. Finally, one should ask why a given code is used by the system. That is, what are the computational advantages of a particular coding method (computational merits)? Different brain regions may benefit from different coding methods depending on their evolutionary, structural, and functional constraints.

To understand the fourth issue discussed above, we need to understand the goals of the system as well as the complexity and challenges that the system faces in the real world. For instance, imagine a rat in the wild trying to scavenge for food. The rat must detect and recognize a particular scent by comparing incoming sensory input with the memorized scents of several previously learned food odors (odor memory and recognition). To locate the food source successfully, the rat must also recognize the food odor across a wide concentration range (concentration-invariant recognition) and isolate (or segment) the target odor from complex background odors. These computations must be performed in the presence of various other perturbations (e.g., sensory and neural noise). In addition, the rat might be required to make a very fast decision. Which coding method may be the most advantageous, considering the constraints of the olfactory system and the complexity of natural odor scenes? Furthermore, from an evolutionary perspective, Barlow (1961) proposed that the goal of sensory systems is to encode the maximum amount of relevant information with a small number of spikes (the efficient coding hypothesis). Ultimately, one should be able to explain why a given neural code is suited to fulfill all these requirements over other codes.

Recent experiments as well as computational approaches have begun to address these questions in the olfactory system. Here we review these recent studies and synthesize emerging ideas, focusing mainly on odor coding in the mammalian olfactory system. Note, however, that studies using other animals including insects and fish have contributed significantly to developing the ideas discussed below. Although we discuss some of these studies, readers are encouraged to refer to previous reviews (Friedrich 2013, Laurent 2002, Wilson 2013) for further information.

Olfaction as a Pattern Recognition Problem Across ~1,000 Input Channels

Identification of odorant receptors and elucidation of the remarkable wiring patterns of early olfactory circuits have laid out the basic logic of odor information processing in the brain (Axel 1995, Mori et al. 1999) (**Figure 1a**). Almost all volatile chemicals, even newly synthesized molecules, can be perceived and discriminated from other molecules. To handle such diverse molecular inputs, the olfactory system has developed a unique strategy. It uses a large array of odorant receptors (ORs) (Buck & Axel 1991) whose tuning specificity varies: Some ORs are broadly tuned, whereas others are strongly activated by only a very specific set of molecules (Hallem & Carlson 2006, Nara et al. 2011, Saito et al. 2009). The number of functional ORs varies across species and has been estimated to be ~1,000 in rodents, ~300 in humans, and ~60–350 in insects (Go & Niimura 2008, Olender et al. 2013, Touhara & Vosshall 2009).

Each olfactory sensory neuron (OSN) expresses only one type of OR, and axons of OSNs expressing a given type of OR converge onto typically two small (~50–100 μm) spherical structures in the OB called glomeruli, where axons relay information about the activation level of each receptor type to neurons in the OB. As such, each odor activates an odor-specific pattern of glomeruli (Friedrich & Korsching 1997, Meister & Bonhoeffer 2001, Rubin & Katz 1999, Uchida et al. 2000) (**Figure 1b**). Thus, odor recognition or discrimination can be seen as the process of deciphering activity patterns across the two-dimensional sheet of glomeruli in the OB. Odor recognition is a pattern-recognition problem defined by ~1,000 input channels (odorant receptors or glomeruli in rodents).

Whereas our understanding of peripheral olfactory processing has shown great progress, our understanding of how the brain deciphers patterns of glomerular activation remains elusive and a matter of hot debate. One salient issue has been which features of neural activity (firing rates, spike timing, etc.) are important in conveying odor information. It has long been known that odors elicit distinct temporal spike patterns that are not directly related to the dynamics of the olfactory stimulus. These observations have led to the idea that the exact timing of spikes can be a

Figure 1

(*a*) Olfactory system. Abbreviations: OE, olfactory epithelium; OSN, olfactory sensory neuron; OB, olfactory bulb; OC, olfactory cortex; PG, periglomerular cell; SA, short axon cell; M, mitral cell; T, tufted cell; G, granule cell; P, pyramidal cell; FF, feedforward inhibitory interneuron; FB, feedback inhibitory interneuron (after Shepherd 2001). (*b*) Odor-evoked activity in the rat OB measured using intrinsic signal imaging (Uchida et al. 2000). (*c*) Odor-evoked spiking activity of mitral cells in zebrafish OB (Friedrich & Laurent 2001). (*d–f*) The time course of glomerular activation in the rat OB measured using voltage-sensitive dyes (Spors et al. 2006). The changes in the fluorescent signals in three glomeruli marked in circles in panel *d* are shown in panel *e*.

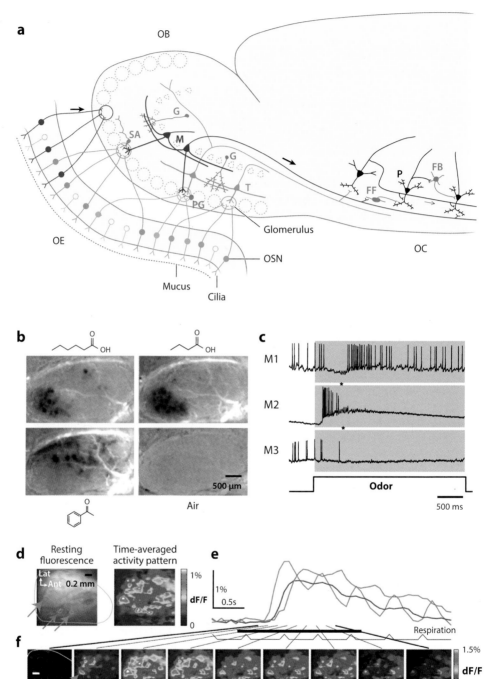

neural code for nontemporal features of stimuli such as odor identity or concentration. However, the functional importance of various temporal coding mechanisms remains to be established.

ODOR REPRESENTATIONS IN THE OLFACTORY BULB

Theories of Temporal Coding in Olfaction

Odor stimulation evokes odor- and cell-specific temporal patterns of activity in the OB. These include modulation of firing rates across various timescales and synchronous oscillations at different frequencies (Bathellier et al. 2010, Wilson & Mainen 2006) (**Figure 1c–e**). Two models have sought to describe the role of temporal coding in olfaction.

Hopfield model: latency coding. Hopfield (1995) proposed a theoretical model for pattern recognition. The model assumes common oscillatory inputs to a population of neurons. When these neurons receive different levels of constant input, those receiving stronger input fire early in each oscillation cycle compared with those receiving weaker input (**Figure 2a–c**). Using this simple principle, the model converts input strength into spike timing (latency). This model has been particularly attractive to the field of olfaction for several reasons. First, as mentioned above, the olfactory system can be seen as a pattern-recognition mechanism using ~1,000 input channels. The Hopfield model provides a mechanism by which the strengths of 1,000 glomerular inputs can be transformed into a pattern of spike times across postsynaptic neurons [mitral (M) and tufted (T) cells]. Second, oscillations are prominent in the olfactory system. Third, this model, in principle, allows rapid encoding of complex input patterns in just one oscillation cycle using one spike per neuron. Fourth, when odor concentration changes, the timing of each neuron's spiking shifts together and the relative timing across neurons remains unchanged, providing a mechanism for concentration-invariant odor recognition.

Brody & Hopfield (2003) extended the Hopfield model, discussed above, to include a mechanism that reads out a specific pattern of inputs (Brody & Hopfield 2003). The model assumed varying strengths of biases that make each M/T cell easy or difficult to fire (**Figure 2d**). The spike timing of each M/T cell is determined by the sum of OSN inputs and a bias term. When a specific set of M/T cells receives the same amount of the summed input (OSN and bias inputs), these M/T cells fire synchronously, and neurons receiving inputs from these M/T cells are effectively activated (**Figure 2e**). This model also achieved concentration-invariant recognition as well as odor segmentation.

Experimental results have provided some support for the Hopfield model. In rodents, investigators have shown that the latency of glomerular activations shows stimulus-specific patterns (Spors et al. 2006, Spors & Grinvald 2002), and M/T cell spike timing with respect to respiration cycle at the theta frequency range (7–12 Hz) is advanced with increased odor concentrations (Cang & Isaacson 2003, Fukunaga et al. 2012, Margrie & Schaefer 2003). In larva, *Xenopus laevis*, which do not sniff, latency from stimulus onset across M/T cells (on the order of tens to hundreds of milliseconds) reliably conveyed stimulus information about odor identity and concentration (Junek et al. 2010). These studies suggest that spike latency with respect to slow oscillations (e.g., respiration cycle) or latency from odor onset on the order of hundreds of milliseconds can convey reliable odor information. However, in many of these cases, shorter spike latency is accompanied by increased spike counts. Furthermore, one study in locusts showed that spike timing with respect to fast oscillations (15–30 Hz, or beta frequency) does not appear to support coding of odor concentration (Stopfer et al. 2003). Which of these two (latency or spike counts) sends more reliable information, or is used by the system, remains to be examined.

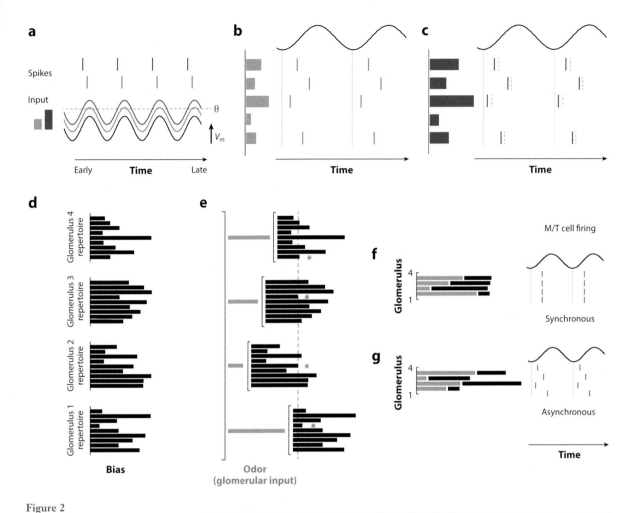

Figure 2

(*a–c*) Hopfield model (Hopfield 1995). A pattern of inputs (*gray bars*) is transformed into the timing of spikes in an oscillation cycle (panel *b*). (*a*) The neuron membrane potential has an internal rhythm (*black*). Two stimuli (*gray and blue*) increase membrane potential. The neuron spikes when the membrane potential crosses a certain threshold (θ). Higher-intensity stimulation shifts the timing of spikes toward early time points while the overall pattern is preserved (panel *c*). (*d–g*) Brody & Hopfield model (Brody & Hopfield 2003). Each black bar represents bias received by each mitral (M)/tufted (T) cell (panel *d*). Upon odor stimulation, each M/T cell receives the sum of odor-evoked inputs (panel *e, gray*) and bias input (panel *e, black*). When a set of M/T cells (*red asterisk* in panel *e*) receives the same amount of total inputs, these M/T cells fire synchronously (panel *f*). When these M/T cells are stimulated with a different odor, the total inputs do not match, and the neurons thus fire asynchronously (panel *g*).

Laurent model: slow-evolving temporal coding, decorrelations. The Hopfield model provided one way to look at activity patterns across a population of neurons using spike latency. But temporal dynamics of neural responses in the olfactory system goes beyond the first spikes. Work by Laurent and colleagues in locusts has provided a different way to look at more complex temporal patterns (Laurent 2002, Laurent et al. 2001).

Principal neurons in insect antennal lobes (ALs) [called projection neurons (PNs)] (Laurent et al. 1996, Wehr & Laurent 1996) or M/T cells in the zebrafish OB (Friedrich & Laurent 2001) respond to odors with complex, odor- and cell-specific temporal patterns over hundreds of milliseconds or a few seconds. These responses consist of successive periods of up- and downmodulations of firing rates. Populations of neurons exhibit synchronized oscillatory activity, but each neuron only transiently participates in this population activity. The identities of neurons that participate in the oscillatory ensemble change over time.

Does the temporally evolving ensemble provide any computational merit? The data in zebrafish OB demonstrate that ensemble activity patterns become gradually decorrelated: The patterns of odor-evoked activity across neurons initially reflect the similarity (or classes) of odors (thus odor representations for similar odors are correlated); however, over the time course of hundreds of milliseconds to seconds, these correlations are gradually reduced and the ensemble activity patterns become less similar (Brown et al. 2005, Friedrich & Laurent 2001, Mazor & Laurent 2005). In addition to examining the similarity (correlation) of the trial-averaged responses, the authors also quantified the reliability of the neural response at each time point by estimating how well an unbiased observer can classify neural responses on a trial-by-trial basis using a simple linear decoder (decoding analysis). Using this analysis, the rate of successfully classifying odors indeed improved over a similarly slow timescale. PNs in locusts also have similar slow temporal dynamics; however, classification success is highest 200–300 ms after odor onset (Stopfer et al. 2003). In contrast, *Drosophila* PNs have weak temporal dynamics and the ensemble activity patterns do not show decorrelations (Olsen et al. 2010). The significance of slow temporal decorrelation remains to be clarified in the future (Friedrich 2013).

Studying Odor Representations in Behaving Animals

Humans' sense of smell relies on inhalation of odor molecules into the nose. In awake animals, sniffing patterns change dramatically depending on behavioral contexts. During active exploration, rodents sniff at the theta frequency range (6–10 Hz). Behavioral experiments in rodents have shown that fine odor discrimination can be made with just one sniff or 250 ms (Abraham et al. 2004, Rinberg et al. 2006b, Uchida & Mainen 2003, Zariwala et al. 2013). How is odor information encoded in such a short time period? Do the temporal dynamics of neural responses evolve that quickly? Furthermore, investigators have suggested that during rapid sniffing, OB neurons lose their phase locking to respiration (Carey et al. 2009, Kay & Laurent 1999). With these constraints, can either latency coding or slow temporal coding work at the level of M/T cells in awake rodents? Recent studies have begun to address these questions (Cury & Uchida 2010, Shusterman et al. 2011). Studies in anesthetized animals using new techniques (e.g., optogenetics and in vivo whole-cell recording) have also provided insights into mechanisms of temporal dynamics of OB neurons. In the following sections, we review these studies.

Respiration Coupling of Spontaneous Activity in the Mammalian OB

In awake rodents, spontaneous firing of M/T cells is generally high (on average 10–25 spikes per second) compared with those under anesthesia (Rinberg et al. 2006a). A recent study in behaving rats compared spontaneous firing patterns across different respiration frequencies (Cury & Uchida 2010) (**Figure 3**). This study found that spontaneous firing of many M/T cells is locked to the respiration cycle. Each M/T cell fired maximally at a particular latency from inhalation onset, and the timing was conserved across various respiration frequencies. Latency to spike varied among M/T cells, such that they tiled the entire cycle of rapid respiration (sniffing) (**Figure 3**). In anesthetized mice, M/T cells that innervate the same glomeruli (sister M/T cells) tend to fire with similar latencies when compared with M/T cells that were connected to different glomeruli (Dhawale et al. 2010). Using whole-cell recording in anesthetized mice, a recent study found that T cells consistently fired earlier in the respiratory cycle relative to inhalation than did M cells

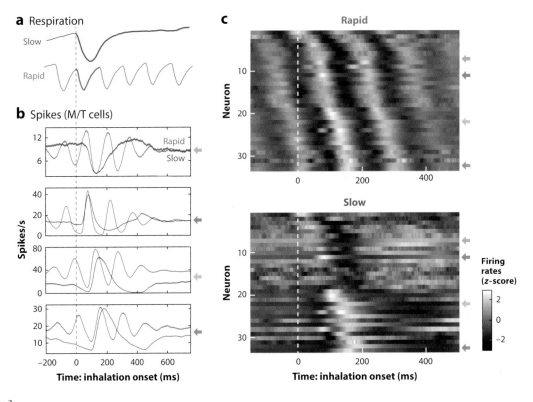

Figure 3

Spontaneous activity of mitral/tufted (M/T) cells (Cury & Uchida 2010). (*a*) Example respiration patterns during slow (*blue*) versus rapid (*red*) breathing; downward change, inhalation; upward change, exhalation. (*b*) Firing patterns of 4 M/T cells during slow (*blue*) and rapid (*red*) respiration. The traces are aligned by inhalation onset. (*c*) Firing patterns of 33 M/T cells. Firing rates are normalized using *z*-scores, calculated over rapid sniffing events, and plotted using the color scale at right. In both graphs, neurons are sorted from bottom to top by increasing latency to the peak firing rate observed in rapid sniffing peri-event time histograms (PETHs) (0–160 ms) following inhalation onset. The four neurons in panel *b* are indicated by colored arrows.

(Fukunaga et al. 2012). Together, these studies revealed a highly organized temporal pattern of spontaneous firing of M/T cells at the population level, which can serve as a temporal frame of specific coding mechanisms. For instance, respiration-coupled sequential activation of M/T cells can provide a reference time frame for odor-evoked responses.

Odor-Evoked Responses in the Mammalian Olfactory Bulb

Odor-evoked responses have often been detected as a gross firing rate change over time. Studies found that fewer M/T cells changed their average firing rate in response to odors in awake animals compared with anesthetized animals (Davison & Katz 2007, Rinberg et al. 2006a), suggesting that odor responses are sparse in awake animals. However, neurons can also change their spike timing in addition to the overall frequency. Indeed, recent studies in awake animals show that many neurons change their spike timing in an odor-specific manner (Cury & Uchida 2010, Gschwend et al. 2012, Shusterman et al. 2011) (**Figure 4***b*). These responses consisted of transient changes in firing rate within a respiratory cycle, which is often obscured in analyses examining average firing rates. Sister M/T cells that fire at similar respiratory phases prior to odor stimulation become

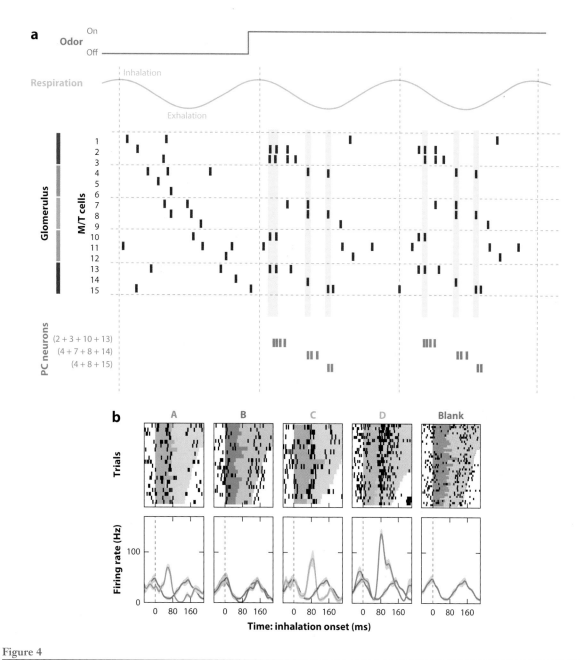

Figure 4

(*a*) A model of respiration-coupled, spike-timing-based odor coding. Modified from Wilson & Mainen (2006). The PC neurons act as coincidence detectors of specific M/T cell firing patterns. (*b*) Odor-evoked activity of an example mitral/tufted (M/T) cell in the OB to four odors and blank airstream [also overlaid on all odor responses (*gray*)] (Cury & Uchida 2010).

coupled to different respiratory phases with odor stimulation (Dhawale et al. 2010). Taking into account these temporal changes in M/T cell spikes, odor-evoked responses are not sparse in the OB of awake animals (15–60% of M/T cells responded to a given odor) (Cury & Uchida 2010, Gschwend et al. 2012, Shusterman et al. 2011).

Do fine temporal structures within a respiratory cycle convey odor information? A recent study addressed this question by comparing the ability to identify an odor from the neural response when taking into account subsniff temporal structures versus only the firing rate over entire sniff cycles (Cury & Uchida 2010). The authors found that odor-decoding performance improved significantly when subsniff temporal structures were considered. Furthermore, most of the information was conveyed within the first 100 ms after inhalation onset. These initial fine-scale temporal patterns were found to be well preserved between slow and rapid modes of odor sampling, whereas patterns based on the overall firing rates in the entire sniff cycle were poorly conserved (Cury & Uchida 2010). Decoding using the first spike latency from inhalation onset performed poorly compared with decoding using entire subsniff temporal patterns. This result is in part because high spontaneous firing rates prevent reliable detections of the onset of odor-evoked activity. In this study, activity from all recorded M/T cells was included for decoding. As discussed above, Fukunaga et al. (2012) suggests that the firing of M cells occurs later in the respiration cycles. A specific subset of OB neurons may still be able to encode reliable odor information by latency alone. Nevertheless, these studies demonstrate that subsniff temporal patterns, in particular initial portions of the dynamic response, can reliably convey odor information.

Odor Representations by Dynamic Neural Assemblies (Trajectories)

As discussed above, odors trigger evolving dynamics in an ensemble of neurons. Rather than looking at single neurons one by one, analyzing the behavior of the ensemble as a whole can provide useful insights. The pattern of time-varying activity of N neurons can be described as a trajectory in N-dimensional activity space. Using this multidimensional space, investigators can quantify the dynamics of neuron ensembles (e.g., velocity, acceleration). Furthermore, using appropriate dimension reduction methods (e.g., principal component analysis), investigators can visualize the dynamics of a neuron ensemble (Stopfer et al. 2003). With prolonged odor stimulation, an ensemble activity of PNs in locust AL changed immediately after odor onset and, after about 1 s, reached a steady attractor-like state (Mazor & Laurent 2005). When an odor stimulus was terminated, the ensemble activity moved quickly again and returned to its steady-state baseline activity. Remarkably, decoding analyses showed that discriminability of odors was highest at the time of maximum velocity rather than during an attractor-like, steady state (Mazor & Laurent 2005). A similar analysis applied to OSNs produced a contrasting result: The highest level of odor discriminability was achieved during the steady state, and odor termination caused much less movement in the ensemble trajectory than at odor onset (Fdez Galán et al. 2004). Thus, compared with OSNs, PNs emphasize more sharp changes in odor stimuli in odor stimulation. Similar observations were made in the rat OB during both rapid sniffing and slow breathing, although a steady state was not achieved during the timescale of a single inhalation period, and the speed at which ensemble patterns changed was much faster (in ~35 ms, ensemble patterns became substantially different) (Cury & Uchida 2010). These results suggest that the ensemble activity evolves not necessarily to gradually reach a more informative steady state but to establish that temporal dynamics itself (or its trajectory) is informative or can be a carrier of information.

Why would transient dynamics be more suitable than steady-state neural responses in representing information? First, computational analysis suggests that information can be coded by a change in the ensemble activity or by the time derivative of ensemble trajectories (Moazzezi &

Dayan 2008, 2010). Changes evoked by odor stimulation can be largely invariant to changes in the initial set point of a neuronal ensemble caused by background odors or a slow context-dependent change in baseline activities. Second, in locust, investigators showed that ensemble trajectories evoked by varying concentrations of the same odor change continuously and fall onto a specific manifold (or low-dimensional space) despite the fact that individual neurons often respond nonlinearly across changing concentrations (Stopfer et al. 2003). This emerging property of ensemble dynamics may provide a basis for concentration-invariant odor recognition. Furthermore, reaching a steady state might require a relatively long constant stimulus (Fdez Galán et al. 2004, Mazor & Laurent 2005). Such a slow process may not be adaptive given the various time pressures in natural environments.

ENCODING: HOW ARE TEMPORAL PATTERNS OF MITRAL/TUFTED CELLS GENERATED?

So far, our discussion has focused on which features of neural activity contain reliable stimulus information and are computationally attractive. We now turn to the neural mechanisms that generate these activities (encoding).

Temporal Dynamics During Spontaneous Activity

The spontaneous spiking of M/T cells (or PNs in AL) in response to clean, nonodorous air is driven primarily by inputs from OSNs. Closing the nostril in mammals or cutting antennae in insects greatly reduces spontaneous activity of OB (or AL) neurons (Joseph et al. 2012, Onoda & Mori 1980, Sobel & Tank 1993). OSNs can likely be activated by pressure or mechanical stimulation, and this type of activation mechanism shares the same signaling pathway as that used for odor-evoked activation (Grosmaitre et al. 2007). Knocking out the cyclic nucleotide–gated channel CNGA2 eliminated respiration-coupled spontaneous activity of M/T cells. However, it is unclear how ordered, yet diverse, respiratory-phase preferences of M/T cells are generated. Different phase preferences among nonsister M/T cells can be explained by different levels of activations among glomeruli as observed in various imaging experiments. M/T cells have different levels of hyperpolarization-activated current (I_h or "sag" potential) that makes M/T cells difficult to fire. These I_h currents are more similar among sister M/T cells than among nonsister pairs (Angelo et al. 2012). In addition, sister M/T cells are connected with gap junctions (Schoppa & Urban 2003). These mechanisms can explain the similar phase preferences in sister M/T cells and the diversity seen in nonsister cells. Further diversities in respiratory-phase preferences can be attributed to differences in intrinsic properties of M/T cells (Padmanabhan & Urban 2010) as well as neural circuits within the OB. For example, recent studies have shown that inhibitory inputs play a role in delaying M cell firing compared with T cells (Fukunaga et al. 2012), in shaping their odor tuning (Kikuta et al. 2013), or in making their responses more transient (Olsen et al. 2010).

Odor-Evoked Temporal Dynamics

Odors increase OSN activity and add more complex temporal dynamics on top of spontaneous activity. First, odor molecules pass through a rather complex environment (or filter) of mucosae, which contain various odor-binding proteins and enzymes that modify odor molecules. Odor molecules then bind to ORs with different affinities, resulting in various latencies and temporal dynamics of OSN firing and glomerular activations (Spors et al. 2006, Spors & Grinvald 2002). OB circuits that are engaged with increased OSN input also add more complex temporal dynamics.

For example, some evidence indicates that inhibition determines respiratory-phase preferences of M versus T cells (Fukunaga et al. 2012), and deep short axon cells provide feedforward inhibition onto granule cells that limit the time window of granule cell spikes (Boyd et al. 2012).

In the visual system, relatively simple models have been successfully used to describe the relationship between sensory inputs and spiking outputs (Meister & Berry 1999). One successful model consists of linear filtering of incoming inputs followed by a nonlinear spike-generating mechanism [linear-nonlinear (L-N) model]. Geffen et al. (2009) presented locust ALs with flickering odor stimuli. Although recorded neurons showed diverse response dynamics, a simple L-N model with just three parameters could predict odor-evoked responses. The linear filters obtained had diverse time-varying waveforms but could be described as a superposition of two filters. These two waveforms were thought to represent direct excitatory inputs (ON-filter) from OSNs and inhibitory inputs (OFF-filter) from inhibitory interneurons. The shape and the time course of the ON-filter can be determined in part by odorant-receptor interactions, whereas the OFF-filter reflects complex, indirect inputs from inhibitory interneurons. It remains to be seen how these filters change with different odor concentrations or with odor mixtures. In rodents, Khan et al. (2008) showed that M/T cell responses to odor mixtures could be predicted by simple linear combinations of filters extracted from responses to component odorants. These approaches are very useful in quantifying relative contributions of different factors that generate seemingly complex temporal patterns of M/T cell activities.

DECODING: HOW DO DOWNSTREAM NEURONS READ OUT INPUTS FROM THE OB?

To establish which features of the observed neural activity in the OB actually constitute a neural code, it is essential to understand how downstream neurons read out these activity patterns and how they are used by the animal. We now discuss odor coding and decoding in downstream neurons.

Odor Representations in the Olfactory Cortex

M/T cells in the mammalian OB project to several areas, including the anterior olfactory nucleus (AON), the piriform cortex (PC), the tenia tecta, the olfactory tubercle, the cortical amygdala, and the entorhinal cortex (Haberly 2001). These areas are traditionally called the olfactory cortex (OC), although some of these areas do not contain pyramidal neurons that form distinct layers (and thus are not cortical). In insects, PNs in the AL, analogous to M/T cells in the OB, project to the mushroom body (MB) and the lateral horn.

In insects, neural activity in the MB, which is thought to be involved in associative learning, is quite distinct from neural activity observed in the AL (Perez-Orive et al. 2002). First, spontaneous activity of MB neurons (called Kenyon cells) is nearly nonexistent [mean: 0.052 spikes per second, median: 0.011 spikes per second (Jortner et al. 2007)]. Second, an odor activates many fewer neurons in the MB than it does in the AL. Third, odor-evoked responses consist of a burst of a few spikes (often a single spike). Fourth, MB neurons respond to odors in a more concentration-invariant manner. These results thus showed that odor representations in insects are drastically transformed from dynamic and dense representations in the AL to sparse and simple representations in the MB. Sparse representations in the MB provide more explicit information about odor identity, which appears to be suitable for their roles in the formation, storage, and recall of associative memories (Perez-Orive et al. 2002).

Fewer studies have recorded neural activity in the mammalian olfactory cortices, particularly in awake animals. However, recent studies have begun to document basic characteristics of both spontaneous and odor-evoked activities in olfactory cortices. For example, the anterior part of the PC, the most studied among the mammalian olfactory cortices, has relatively low, but not quite silent, spontaneous activity [in awake animals, 6.15 ± 9.01 spikes per second, mean $\pm$ standard deviation (Miura et al. 2012)]. Neurons that respond to an odor are broadly distributed in space without apparent spatial organization (Illig & Haberly 2003, Miura et al. 2012, Poo & Isaacson 2009, Rennaker et al. 2007, Stettler & Axel 2009). A single odor activates about 10–30% of neurons (Miura et al. 2012, Poo & Isaacson 2009, Stettler & Axel 2009). The breadth of odor tuning varies among neurons; although most neurons select to a small number of odors, some neurons are broadly activated by many (Miura et al. 2012, Poo & Isaacson 2009, Zhan & Luo 2010). Recording in awake rats showed that temporal responses of anterior PC neurons are simple, consisting mainly of transient burst spiking that is tightly locked to sniff onset (Miura et al. 2012), and the spike counts in this sniff-locked neural activity convey reliable odor information (Gire et al. 2013b, Miura et al. 2012). In contrast, spike timing conveyed little additional information about odor identity compared with information provided by total spike counts over the entire sniff cycle (Miura et al. 2012). Furthermore, the overall firing rates of ensemble neurons correlated with the trial-by-trial accuracy of perceptual decisions in a psychophysical odor-discrimination task. These results suggested that there is a profound transformation in the way odors are represented in the OB and the anterior PC: In the PC, odor representations are distributed and do not show odor-specific spatial patterns as seen in the vertebrate OB (Friedrich & Korsching 1997, Meister & Bonhoeffer 2001, Rubin & Katz 1999, Uchida et al. 2000). Furthermore, the importance of spike timing–based coding is much reduced in the anterior PC compared with the OB.

Convergence and Divergence of Olfactory Bulb Projections to Piriform Cortex

Anatomical tracing studies have shown that neighboring M/T cells, including M/T cells belonging to the same glomerulus, project broadly across the PC without apparent spatial order (Ghosh et al. 2011, Igarashi et al. 2012, Nagayama et al. 2010, Sosulski et al. 2011). Conversely, a small cortical region receives input from M/T cells belonging to a distributed set of glomeruli (Miyamichi et al. 2011). A recent study in *Drosophila* demonstrated quantitatively that individual Kenyon cells receive information from a nearly random set of glomeruli (Caron et al. 2013). These results show that spatially segregated channels of odor information become integrated in the PC (or the MB). In brain slice experiments, investigators found that coincident inputs from multiple M/T cells are required to activate PC neurons (Apicella et al. 2010). Measuring the response of the PC population to odor mixtures revealed interactions between odors, exhibiting cross-odor suppression as well as supralinear excitation (Stettler & Axel 2009, Wilson & Sullivan 2011, Yoshida & Mori 2007). Additionally, recent studies in intact animals used glutamate uncaging and optogenetic photostimulation of glomeruli and recording in anterior PC to show that anterior PC neurons respond to activation of a specific and dispersed set of glomeruli (Davison & Ehlers 2011, Haddad et al. 2013). These experiments are consistent with the long-held view that PC neurons integrate information across dispersed glomeruli and act as combination detectors (**Figure 4a**). However, a recent study using optogenetic stimulation in behaving mice showed that mice can detect the activation of a single glomerulus using light (Smear et al. 2013). These results suggest that even though naturalistic glomerular activity that is spread across the OB is integrated to activate anterior PC neurons, the animal could also detect strong activation of a single glomerulus.

Can Neurons in the Piriform Cortex Read Out Temporal Patterns in the Olfactory Bulb?

Are PC neurons also sensitive to the temporal patterns of inputs? As discussed above, a neural recording experiment suggested that odor representations are transformed from spike timing–based to firing rate–based representations between the OB and anterior PC. A recent study tested this idea directly by recording from neurons in anterior and posterior parts of the PC while varying the timing of optogenetic stimulation of two foci of glomeruli or M/T cells in anesthetized mice (Haddad et al. 2013). This study showed that firing rate responses of PC neurons depended on the order and the lag of input activations. Information conveyed by the firing rate increased in more central brain regions, and conversely, information conveyed by temporal patterns decreased. PC neurons' sensitivity to relative timing of activation did not depend on the time at which photostimulation was given with respect to the respiratory cycle. These results demonstrate that neurons in the PC can read out relative timing of activation across glomeruli and M/T cells. This study, however, used a relatively long duration of stimulation (~80 ms) to activate each spot, and the relative timing was varied on the order of tens of milliseconds. Whereas this stimulation protocol was aimed to mimic the time course of glomerular activation, M/T cells exhibit much faster dynamics. Furthermore, this study examined interactions between only two foci. It remains to be examined whether PC neurons are also sensitive to finer and more complex temporal and spatial patterns.

In contrast, a recent study in zebrafish showed that responses of neurons in the posterior zone of the dorsal telencephalon (Dp) did not depend on whether a set of M/T cells is activated synchronously or asynchronously, suggesting that Dp neurons effectively discard information about synchrony in the OB (Blumhagen et al. 2011). Although it is difficult to compare the Haddad and Blumhagen studies directly because each study used different stimulation protocols, it is possible that the difference in temporal sensitivity is the result of the difference in these neural circuits. For instance, feedforward inhibition in zebrafish Dp is relatively weak, whereas that in rodents plays a major role in shaping responses of PC neurons. Furthermore, temporal modulation in zebrafish OB is much slower (seconds) than that in the mouse OB (tens of milliseconds). Therefore, temporal sensitivity of neural circuits may be different between these species.

Can Animals Read Out Temporal Patterns in the OB?

The mere presence of the neural activity that contains reliable stimulus information does not prove its functional importance. The encoded stimulus information needs to be read out by downstream brain regions and ultimately utilized by the animal to discriminate sensory stimuli successfully. Recording neuronal activity in animals performing a psychophysical task is therefore critical for understanding the functional importance of neural codes. For example, discrimination of sensory stimuli in a behaving animal should cofluctuate with information content of the neural code used on a trial-by-trial basis. Such correlations can provide additional evidence that a given neural code is used by the animal (Britten et al. 1996, Luna et al. 2005, O'Connor et al. 2013). Cury & Uchida (2010) observed that the firing rates of subsniff, short epochs, but not the firing rates over the entire sniff cycle, cofluctuated with an animal's decisions (whether the animal decided to make a choice after one sniff or to take another sniff).

The importance of neural activation timing at the behavioral level was studied more directly by activating the olfactory circuit electrically or optogenetically in behaving animals (Monod et al. 1989; Smear et al. 2011, 2013). An earlier study showed that rats could not discriminate electrical stimulation of the OB delivered during inhalation versus exhalation (Monod et al. 1989). However, it was later shown that mice can discriminate optogenetic stimulation of OSNs given at certain times in the respiration cycle (sniff phase) with a precision of ~10 ms (Smear et al. 2011). This study, however, could not examine the role of relative timing because the same OSN population was activated at different times relative to inhalation onset without changing the relative timing across neurons. This limitation raises the question of whether perceived differences are due merely to the timing of stimulus onset with respect to the respiration cycle or to changes in perceived quality or intensity of the stimuli. It remains to be examined whether the animal can discriminate more complex stimulus patterns (e.g., relative time code). Nevertheless, these experiments demonstrate that optogenetic activation (or electrical stimulation) can provide a powerful means to directly test the functional relevance of specific coding schemes.

The Role of Inhibitory and Recurrent Circuits in Decoding Olfactory Bulb Inputs

Which mechanisms generate the activity patterns of PC neurons? How are spatiotemporal activity patterns in the OB recognized by the neural circuits in the PC? Principal excitatory neurons in the PC receive inputs directly from M/T cells. However, PC principal cells also receive prominent inhibitory connections as well as excitatory recurrent connections, which shape their responses (**Figure 5***a*). The functional properties of cortical circuits play a critical role in the representations of sensory stimuli. The time course and strength of synaptic inputs primarily determine the timing and selectivity of spike output of PC neurons. These synaptic circuits have been studied using anatomical methods and brain slice in vitro electrophysiology.

Inhibition in the PC is provided by a local network of interneurons. Feedforward inhibitory inputs onto PC principal neurons are mediated by superficial layer 1 (L1) GABAergic interneurons that are directly activated by M/T cells. They selectively target the apical dendritic compartments of principal neurons in the PC and provide inhibition with a short latency (<10 ms) relative to the onset of M/T cell excitation (Stokes & Isaacson 2010, Suzuki & Bekkers 2012). Complementary to feedforward inhibition, feedback inhibition is mediated by deep L2/3 interneurons that target the soma and basal dendrites of PC principal neurons. Recruitment of these deep L2/3 interneurons requires the activation of PC neurons and therefore provides inhibition at a later onset (tens of milliseconds) (Stokes & Isaacson 2010, Suzuki & Bekkers 2012). The relative time course of cortical inhibition and OB input can restrict integration of specific synaptic inputs to a limited time window. Thus cortical inhibition can play a role in increasing PC neurons' sensitivity to synchronous synaptic inputs as well as in generating order sensitivity (**Figure 5***b,c*).

PC principal neurons also make excitatory connections with each other; these recurrent inputs are restricted to L2/3 and basal dendritic compartments (Haberly & Price 1978, Johnson et al. 2000). Recurrent connections may play an important role in shaping odor-evoked responses (Haberly 2001; Haberly & Presto 1986; Hasselmo & Bower 1990; Johnson et al. 2000; Ketchum & Haberly 1993a,b; Luskin & Price 1983a,b). Indeed, studies using in vivo whole-cell recordings as well as optogenetics demonstrated that excitatory recurrent activity functionally influences whether PC principal neurons are recruited by OB M/T cell inputs (Franks et al. 2011, Poo & Isaacson 2011). Temporally, recurrent excitation comes at a delay (10–50 ms) relative to direct M/T cell excitation (**Figure 5***c*). Together with cortical inhibition, these cortical circuits in the PC provide a mechanism for PC neurons to detect temporally patterned OB inputs.

In addition to synaptic delays, other properties of the cortical circuit such as short-term synaptic plasticity and location of the targeted dendritic compartment also contribute to the activation of neurons embedded in a cortical circuit (Behabadi et al. 2012, Branco et al. 2010, Gabernet et al. 2005, Higley & Contreras 2006, Larkum et al. 2009). In PC, feedforward inhibition decays rapidly

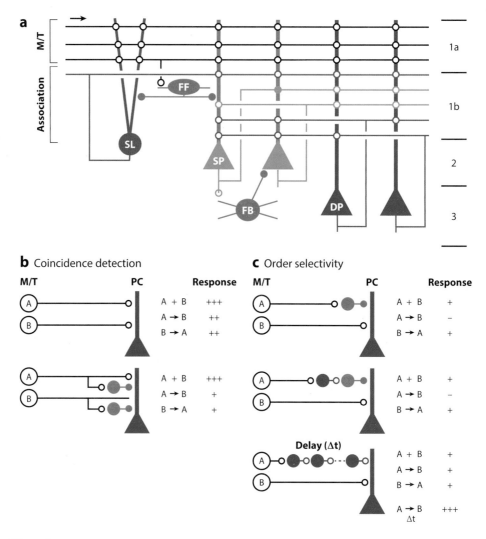

Figure 5

(*a*) Schematic diagram of neural circuits in the piriform cortex. Abbreviations: DP, deep pyramidal neuron; FB, feedback inhibitory interneuron; FF, feedforward inhibitory interneuron; M/T, mitral/tufted; SL, semilunar cell; SP, superficial pyramidal neuron. (*b*) Neural circuits for coincidence detection. Red circles represent inhibitory interneurons providing feedforward inhibition. This piriform cortex (PC) neuron responds strongly when both inputs A and B arrive at the same time (i.e., act as a coincident detector). *Top*: a small lag between stimulus A and B will result in a small decrease in the response due to a relatively large temporal integration window. *Bottom*: Feedforward inhibition narrows the window for temporal integration of the two inputs. (*c*) Neural circuits for order selectivity. *Top*, feedforward inhibition. *Middle*, feedback inhibition; *bottom*, recurrent excitation. Red circles represent inhibitory interneurons, and blue circles represent excitatory neurons.

throughout a train of OB input, whereas feedback inhibition grows as a result of more PC principal neurons being recruited. Recruitment of these two types of inhibition is separated in time, but it is also segregated in space: Feedfoward inhibition targets distal apical dendrites of principal neurons, whereas feedback inhibition is restricted to the soma and basal dendrites. Excitatory OB projections and recurrent input within the PC are also similarly separated in their temporal recruitment and target location. Although the specific manner in which these synaptic properties are engaged in vivo still remains elusive, they provide strong clues about how PC circuits could serve in the discrimination of temporal patterns of OB inputs.

The idea that the OC provides suitable neural circuits that recognize specific spatiotemporal patterns of inputs dates back to the 1980s: Structural features of seemingly random and combinatorial inputs from the OB and recurrent neural circuits in PC resemble those used in theoretical models of autoassociative memory or neural circuits of the hippocampus and entorhinal cortex known to play important roles in learning and memory (Haberly 2001, Lynch et al. 1986). These theories suggested that PC neural circuits may play a role in recognizing not only specific combinations of inputs but also their temporal patterns (Kleinfeld 1986, Sompolinsky & Kanter 1986, Tank & Hopfield 1987). These studies have also suggested that delayed input to PC neurons, which is an essential part of recognizing temporal patterns, can be generated by recurrent excitation and that recurrent inhibitory circuits may provide gain control mechanisms that stabilize and accelerate computations by otherwise-unstable recurrent networks (Chance & Abbott 2000, Jin & Seung 2002). Recent experiments have begun to test the idea that the PC performs odor recognition based on partial inputs (pattern completion) (Barnes et al. 2008, Chapuis & Wilson 2011). Interplay between experiments and theory will be increasingly important to advance our understanding of how the OC recognizes spatiotemporal input patterns and performs its functions at the systems level.

COMPUTATIONAL MERITS: DIFFERENT CODING SCHEMES IN THE OLFACTORY BULB AND THE PIRIFORM CORTEX

As discussed above, accumulating evidence suggests that temporal patterns of activity in the OB play a role in conveying odor information, whereas a firing rate–based code becomes more important in the PC. Why, then, do the OB and the PC use different coding schemes?

Temporal coding might be efficient in terms of firing economy (Gire et al. 2013a). Indeed, temporal coding can provide mechanisms that do not require an increase in firing rates or that require only one spike per neuron (Hopfield 1995). However, according to the efficiency principle, one would not expect to see high spontaneous firing rates as observed in M/T cells. One possible explanation for the high spontaneous firing rate in M/T cells is that cyclic activations of OSNs through airflow facilitate the detection of odorants at low concentrations. When cyclic activations are high enough to cause OSNs to fire spontaneously, the presence of even low concentrations of odorant can be detected as a shift in the timing of spikes in each cycle. This principle is akin to stochastic resonance, a phenomenon whereby a signal that is normally too weak to be detected by a sensor can be boosted by adding white noise to the signal, which contains a wide spectrum of frequencies. Relatively high firing rates and dependency on temporal coding might be, at least in part, explained by this mechanism. Furthermore, the temporal dynamics observed in the OB may reflect the fact that olfaction is a chemical sense utilizing various mechanisms that generate temporally rich neural responses (e.g., interactions between odorants with mucosa and odorant receptors, as discussed above). In addition, although the vast number of inhibitory neurons in the OB may have evolved for different purposes, such as gain control, they have also become a source of temporal responses. The olfactory system may have evolved to exploit these inherent properties of the system by enabling downstream neurons to read the OB temporal code and convert the temporal code into a rate code.

There are substantial anatomical differences between the OB and the PC: Whereas a relatively small number of neurons (20–50 M cells) transmit odor information that converges through each input channel (glomerulus) in the OB, PC contains at least 2 orders of magnitude more neurons (Shepherd 2003). Whereas efficiency, in terms of the amount of information transmitted per neuron and per unit time, is crucial in the OB, the PC can distribute the information to a large number of neurons. One advantage of a rate-based code over a temporal code is that downstream areas can more readily read out such a code or combine it with other kinds of information encoded in rates. This ability might then facilitate proposed functions of the PC such as the formation of associative memories (Franks et al. 2011, Haberly 2001).

CONCLUDING REMARKS

The challenge of understanding neural coding of sensory information lies in the ability to address comprehensively the key issues of representation, encoding, decoding, and computational merit. In recent years, our understanding of the neural coding in the olfactory system has been advanced by new experimental results and the formulation of theories. In particular, studies that aim to reconcile computational frameworks, constraints of neural circuit dynamics, and neural recording during behavior have played important roles. New techniques such as optogenetics have also allowed researchers to test specific hypotheses regarding neural coding. Here, we review recent evidence for how odor information encoded in the OB is subsequently transformed and read out (decoded) by neurons in PC. One limiting factor of current results is that most of the studies using behavior have focused on the relatively simple task of odor discrimination. As we discussed earlier, specific neural coding may become more critical for computationally challenging and ethologically important tasks such as concentration-invariant recognition (including odor identification at an extremely low concentration) (Uchida & Mainen 2007) and odor segmentation. To address these issues, it will be very important to develop more complex behavioral paradigms both in humans and in experimental animals to elucidate neural coding and computations performed in the olfactory system.

DISCLOSURE STATEMENT

The authors are not aware of any affiliations, memberships, funding, or financial holdings that might be perceived as affecting the objectivity of this review.

ACKNOWLEDGMENTS

We thank Zachary Mainen, Jeffry Isaacson, Jeremiah Cohen and David Gire for their critical comments on the manuscript.

LITERATURE CITED

Abraham NM, Spors H, Carleton A, Margrie TW, Kuner T, Schaefer AT. 2004. Maintaining accuracy at the expense of speed: Stimulus similarity defines odor discrimination time in mice. *Neuron* 44:865–76

Angelo K, Rancz EA, Pimentel D, Hundahl C, Hannibal J, et al. 2012. A biophysical signature of network affiliation and sensory processing in mitral cells. *Nature* 488:375–78

Apicella A, Yuan Q, Scanziani M, Isaacson JS. 2010. Pyramidal cells in piriform cortex receive convergent input from distinct olfactory bulb glomeruli. *J. Neurosci.* 30:14255–60

Axel R. 1995. The molecular logic of smell. *Sci. Am.* 273:154–59

Barlow H. 1961. Possible principles underlying the transformations of sensory messages. In *Sensory Communication*, ed. W Rosenblith, pp. 217–34. Cambridge, MA: MIT Press

Barnes DC, Hofacer RD, Zaman AR, Rennaker RL, Wilson DA. 2008. Olfactory perceptual stability and discrimination. *Nat. Neurosci.* 11:1378–80

Bathellier B, Gschwend O, Carleton A. 2010. Temporal coding in olfaction. In *The Neurobiology of Olfaction*, ed. A Menini, pp. 329–51. Boca Raton, FL: CRC Press

Behabadi BF, Polsky A, Jadi M, Schiller J, Mel BW. 2012. Location-dependent excitatory synaptic interactions in pyramidal neuron dendrites. *PLoS Comput. Biol.* 8:e1002599

Blumhagen F, Zhu P, Shum J, Schärer Y-P, Yaksi E, et al. 2011. Neuronal filtering of multiplexed odour representations. *Nature* 479:493–98

Boyd AM, Sturgill JF, Poo C, Isaacson JS. 2012. Cortical feedback control of olfactory bulb circuits. *Neuron* 76:1161–74

Branco T, Clark BA, Häusser M. 2010. Dendritic discrimination of temporal input sequences in cortical neurons. *Science* 329:1671–75

Britten KH, Newsome WT, Shadlen MN, Celebrini S, Movshon JA. 1996. A relationship between behavioral choice and the visual responses of neurons in macaque MT. *Vis. Neurosci.* 13:87–100

Brody CD, Hopfield JJ. 2003. Simple networks for spike-timing-based computation, with application to olfactory processing. *Neuron* 37:843–52

Brown SL, Joseph J, Stopfer M. 2005. Encoding a temporally structured stimulus with a temporally structured neural representation. *Nat. Neurosci.* 8:1568–76

Buck L, Axel R. 1991. A novel multigene family may encode odorant receptors: a molecular basis for odor recognition. *Cell* 65:175–87

Cang J, Isaacson JS. 2003. *In vivo* whole-cell recording of odor-evoked synaptic transmission in the rat olfactory bulb. *J. Neurosci.* 23:4108–16

Carey RM, Verhagen JV, Wesson DW, Pírez N, Wachowiak M. 2009. Temporal structure of receptor neuron input to the olfactory bulb imaged in behaving rats. *J. Neurophysiol.* 101:1073–88

Caron SJC, Ruta V, Abbott LF, Axel R. 2013. Random convergence of olfactory inputs in the *Drosophila* mushroom body. *Nature* 497:113–17

Chance FS, Abbott LF. 2000. Divisive inhibition in recurrent networks. *Network* 11:119–29

Chapuis J, Wilson DA. 2011. Bidirectional plasticity of cortical pattern recognition and behavioral sensory acuity. *Nat. Neurosci.* 15:155–61

Cury KM, Uchida N. 2010. Robust odor coding via inhalation-coupled transient activity in the mammalian olfactory bulb. *Neuron* 68:570–85

Davison IG, Ehlers MD. 2011. Neural circuit mechanisms for pattern detection and feature combination in olfactory cortex. *Neuron* 70:82–94

Davison IG, Katz LC. 2007. Sparse and selective odor coding by mitral/tufted neurons in the main olfactory bulb. *J. Neurosci.* 27:2091–101

deCharms RC, Zador A. 2000. Neural representation and the cortical code. *Annu. Rev. Neurosci.* 23:613–47

Dhawale AK, Hagiwara A, Bhalla US, Murthy VN, Albeanu DF. 2010. Non-redundant odor coding by sister mitral cells revealed by light addressable glomeruli in the mouse. *Nat. Neurosci.* 13:1404–12

Fdez Galán R, Sachse S, Galizia CG, Herz AVM. 2004. Odor-driven attractor dynamics in the antennal lobe allow for simple and rapid olfactory pattern classification. *Neural Comput.* 16:999–1012

Franks KM, Russo MJ, Sosulski DL, Mulligan AA, Siegelbaum SA, Axel R. 2011. Recurrent circuitry dynamically shapes the activation of piriform cortex. *Neuron* 72:49–56

Friedrich RW. 2013. Neuronal computations in the olfactory system of zebrafish. *Annu. Rev. Neurosci.* 36:383–402

Friedrich RW, Korsching SI. 1997. Combinatorial and chemotopic odorant coding in the zebrafish olfactory bulb visualized by optical imaging. *Neuron* 18:737–52

Friedrich RW, Laurent G. 2001. Dynamic optimization of odor representations by slow temporal patterning of mitral cell activity. *Science* 291:889–94

Fukunaga I, Berning M, Kollo M, Schmaltz A, Schaefer AT. 2012. Two distinct channels of olfactory bulb output. *Neuron* 75:320–29

Gabernet L, Jadhav SP, Feldman DE, Carandini M, Scanziani M. 2005. Somatosensory integration controlled by dynamic thalamocortical feed-forward inhibition. *Neuron* 48:315–27

Geffen MN, Broome BM, Laurent G, Meister M. 2009. Neural encoding of rapidly fluctuating odors. *Neuron* 61:570–86

Ghosh S, Larson SD, Hefzi H, Marnoy Z, Cutforth T, et al. 2011. Sensory maps in the olfactory cortex defined by long-range viral tracing of single neurons. *Nature* 472:217–20

Gire DH, Restrepo D, Sejnowski TJ, Greer C, De Carlos JA, et al. 2013a. Temporal processing in the olfactory system: Can we see a smell? *Neuron* 78:416–32

Gire DH, Whitesell JD, Doucette W, Restrepo D. 2013b. Information for decision-making and stimulus identification is multiplexed in sensory cortex. *Nat. Neurosci.* 16:991–93

Go Y, Niimura Y. 2008. Similar numbers but different repertoires of olfactory receptor genes in humans and chimpanzees. *Mol. Biol. Evol.* 25:1897–907

Grosmaitre X, Santarelli LC, Tan J, Luo M, Ma M. 2007. Dual functions of mammalian olfactory sensory neurons as odor detectors and mechanical sensors. *Nat. Neurosci.* 10:348–54

Gschwend O, Beroud J, Carleton A. 2012. Encoding odorant identity by spiking packets of rate-invariant neurons in awake mice. *PLoS One* 7:e30155

Haberly LB. 2001. Parallel-distributed processing in olfactory cortex: new insights from morphological and physiological analysis of neuronal circuitry. *Chem. Senses* 26:551–76

Haberly LB, Presto S. 1986. Ultrastructural analysis of synaptic relationships of intracellularly stained pyramidal cell axons in piriform cortex. *J. Comp. Neurol.* 248:464–74

Haberly LB, Price JL. 1978. Association and commissural fiber systems of the olfactory cortex of the rat. *J. Comp. Neurol.* 178:711–40

Haddad R, Lanjuin A, Madisen L, Zeng H, Murthy VN, Uchida N. 2013. Olfactory cortical neurons read out a relative time code in the olfactory bulb. *Nat. Neurosci.* 16:949–57

Hallem EA, Carlson JR. 2006. Coding of odors by a receptor repertoire. *Cell* 125:143–60

Hasselmo ME, Bower JM. 1990. Afferent and association fiber differences in short-term potentiation in piriform (olfactory) cortex of the rat. *J. Neurophysiol.* 64:179–90

Higley MJ, Contreras D. 2006. Balanced excitation and inhibition determine spike timing during frequency adaptation. *J. Neurosci.* 26:448–57

Hopfield JJ. 1995. Pattern recognition computation using action potential timing for stimulus representation. *Nature* 376:33–36

Igarashi KM, Ieki N, An M, Yamaguchi Y, Nagayama S, et al. 2012. Parallel mitral and tufted cell pathways route distinct odor information to different targets in the olfactory cortex. *J. Neurosci.* 32:7970–85

Illig KR, Haberly LB. 2003. Odor-evoked activity is spatially distributed in piriform cortex. *J. Comp. Neurol.* 457:361–73

Jin DZ, Seung HS. 2002. Fast computation with spikes in a recurrent neural network. *Phys. Rev. E Stat. Nonlin. Soft Matter Phys.* 65:051922

Johnson DM, Illig KR, Behan M, Haberly LB. 2000. New features of connectivity in piriform cortex visualized by intracellular injection of pyramidal cells suggest that "primary" olfactory cortex functions like "association" cortex in other sensory systems. *J. Neurosci.* 20:6974–82

Jortner RA, Farivar SS, Laurent G. 2007. A simple connectivity scheme for sparse coding in an olfactory system. *J. Neurosci.* 27:1659–69

Joseph J, Dunn FA, Stopfer M. 2012. Spontaneous olfactory receptor neuron activity determines follower cell response properties. *J. Neurosci.* 32:2900–10

Junek S, Kludt E, Wolf F, Schild D. 2010. Olfactory coding with patterns of response latencies. *Neuron* 67:872–84

Kay LM, Laurent G. 1999. Odor- and context-dependent modulation of mitral cell activity in behaving rats. *Nat. Neurosci.* 2:1003–9

Ketchum KL, Haberly LB. 1993a. Membrane currents evoked by afferent fiber stimulation in rat piriform cortex. I. Current source-density analysis. *J. Neurophysiol.* 69:248–60

Ketchum KL, Haberly LB. 1993b. Membrane currents evoked by afferent fiber stimulation in rat piriform cortex. II. Analysis with a system model. *J. Neurophysiol.* 69:261–81

Khan AG, Thattai M, Bhalla US. 2008. Odor representations in the rat olfactory bulb change smoothly with morphing stimuli. *Neuron* 57:571–85

Kikuta S, Fletcher ML, Homma R, Yamasoba T, Nagayama S. 2013. Odorant response properties of individual neurons in an olfactory glomerular module. *Neuron* 77:1122–35

Kleinfeld D. 1986. Sequential state generation by model neural networks. *Proc. Natl. Acad. Sci. USA* 83:9469–73

Larkum ME, Nevian T, Sandler M, Polsky A, Schiller J. 2009. Synaptic integration in tuft dendrites of layer 5 pyramidal neurons: a new unifying principle. *Science* 325:756–60

Laurent G. 2002. Olfactory network dynamics and the coding of multidimensional signals. *Nat. Rev. Neurosci.* 3:884–95

Laurent G, Stopfer M, Friedrich RW, Rabinovich MI, Volkovskii A, Abarbanel HDI. 2001. Odor encoding as an active, dynamical process: experiments, computation, and theory. *Annu. Rev. Neurosci.* 24:263–97

Laurent G, Wehr M, Davidowitz H. 1996. Temporal representations of odors in an olfactory network. *J. Neurosci.* 16:3837–47

Luna R, Hernández A, Brody CD, Romo R. 2005. Neural codes for perceptual discrimination in primary somatosensory cortex. *Nat. Neurosci.* 8:1210–19

Luskin MB, Price JL. 1983a. The laminar distribution of intracortical fibers originating in the olfactory cortex of the rat. *J. Comp. Neurol.* 216:292–302

Luskin MB, Price JL. 1983b. The topographic organization of associational fibers of the olfactory system in the rat, including centrifugal fibers to the olfactory bulb. *J. Comp. Neurol.* 216:264–91

Lynch G, Shepherd GM, Black Killackey HP. 1986. *Synapses, Circuits, and the Beginnings of Memory.* Cambridge, MA: MIT Press

Margrie TW, Schaefer AT. 2003. Theta oscillation coupled spike latencies yield computational vigour in a mammalian sensory system. *J. Physiol.* 546:363–74

Mazor O, Laurent G. 2005. Transient dynamics versus fixed points in odor representations by locust antennal lobe projection neurons. *Neuron* 48:661–73

Meister M, Berry MJ 2nd. 1999. The neural code of the retina. *Neuron* 22:435–50

Meister M, Bonhoeffer T. 2001. Tuning and topography in an odor map on the rat olfactory bulb. *J. Neurosci.* 21:1351–60

Miura K, Mainen ZF, Uchida N. 2012. Odor representations in olfactory cortex: distributed rate coding and decorrelated population activity. *Neuron* 74:1087–98

Miyamichi K, Amat F, Moussavi F, Wang C, Wickersham I, et al. 2011. Cortical representations of olfactory input by trans-synaptic tracing. *Nature* 472:191–96

Moazzezi R, Dayan P. 2008. Change-based inference for invariant discrimination. *Network* 19:236–52

Moazzezi R, Dayan P. 2010. Change-based inference in attractor nets: linear analysis. *Neural Comput.* 22:3036–61

Monod B, Mouly AM, Vigouroux M, Holley A. 1989. An investigation of some temporal aspects of olfactory coding with the model of multi-site electrical stimulation of the olfactory bulb in the rat. *Behav. Brain Res.* 33:51–63

Mori K, Nagao H, Yoshihara Y. 1999. The olfactory bulb: coding and processing of odor molecule information. *Science* 286:711–15

Nagayama S, Enerva A, Fletcher ML, Masurkar AV, Igarashi KM, et al. 2010. Differential axonal projection of mitral and tufted cells in the mouse main olfactory system. *Front. Neural Circuits* 4:120

Nara K, Saraiva LR, Ye X, Buck LB. 2011. A large-scale analysis of odor coding in the olfactory epithelium. *J. Neurosci.* 31:9179–91

O'Connor DH, Hires SA, Guo ZV, Li N, Yu J, et al. 2013. Neural coding during active somatosensation revealed using illusory touch. *Nat. Neurosci.* 16:958–65

Olender T, Nativ N, Lancet D. 2013. HORDE: comprehensive resource for olfactory receptor genomics. *Methods Mol. Biol.* 1003:23–38

Olsen SR, Bhandawat V, Wilson RI. 2010. Divisive normalization in olfactory population codes. *Neuron* 66:287–99

Onoda N, Mori K. 1980. Depth distribution of temporal firing patterns in olfactory bulb related to air-intake cycles. *J. Neurophysiol.* 44:29–39

Padmanabhan K, Urban NN. 2010. Intrinsic biophysical diversity decorrelates neuronal firing while increasing information content. *Nat. Neurosci.* 13:1276–82

Perez-Orive J, Mazor O, Turner GC, Cassenaer S, Wilson RI, Laurent G. 2002. Oscillations and sparsening of odor representations in the mushroom body. *Science* 297:359–65

Poo C, Isaacson JS. 2009. Odor representations in olfactory cortex: "sparse" coding, global inhibition, and oscillations. *Neuron* 62:850–61

Poo C, Isaacson JS. 2011. A major role for intracortical circuits in the strength and tuning of odor-evoked excitation in olfactory cortex. *Neuron* 72:41–48

Rennaker RL, Chen C-FF, Ruyle AM, Sloan AM, Wilson DA. 2007. Spatial and temporal distribution of odorant-evoked activity in the piriform cortex. *J. Neurosci.* 27:1534–42

Rinberg D, Koulakov A, Gelperin A. 2006a. Sparse odor coding in awake behaving mice. *J. Neurosci.* 26:8857–65

Rinberg D, Koulakov A, Gelperin A. 2006b. Speed-accuracy tradeoff in olfaction. *Neuron* 51:351–58

Rubin BD, Katz LC. 1999. Optical imaging of odorant representations in the mammalian olfactory bulb. *Neuron* 23:499–511

Saito H, Chi Q, Zhuang H, Matsunami H, Mainland JD. 2009. Odor coding by a mammalian receptor repertoire. *Sci. Signal.* 2:ra9

Schoppa NE, Urban NN. 2003. Dendritic processing within olfactory bulb circuits. *Trends Neurosci.* 26:501–6

Shepherd GM. 2001. Computational structure of the olfactory system. In *Olfaction: a Model System for Computational Neuroscience*, ed. JL Davis, H Eichenbaum, pp. 3–42. Cambridge, MA: MIT Press

Shepherd GM. 2003. *The Synaptic Organization of the Brain.* New York: Oxford Univ. Press

Shusterman R, Smear MC, Koulakov AA, Rinberg D. 2011. Precise olfactory responses tile the sniff cycle. *Nat. Neurosci.* 14:1039–44

Smear M, Resulaj A, Zhang J, Bozza T, Rinberg D. 2013. Multiple perceptible signals from a single olfactory glomerulus. *Nat. Neurosci.* 16:1687–91

Smear M, Shusterman R, O'Connor R, Bozza T, Rinberg D. 2011. Perception of sniff phase in mouse olfaction. *Nature* 479:397–400

Sobel EC, Tank DW. 1993. Timing of odor stimulation does not alter patterning of olfactory bulb unit activity in freely breathing rats. *J. Neurophysiol.* 69:1331–37

Sompolinsky H, Kanter I. 1986. Temporal association in asymmetric neural networks. *Phys. Rev. Lett.* 57:2861–64

Sosulski DL, Bloom ML, Cutforth T, Axel R, Datta SR. 2011. Distinct representations of olfactory information in different cortical centres. *Nature* 472:213–16

Spors H, Grinvald A. 2002. Spatio-temporal dynamics of odor representations in the mammalian olfactory bulb. *Neuron* 34:301–15

Spors H, Wachowiak M, Cohen LB, Friedrich RW. 2006. Temporal dynamics and latency patterns of receptor neuron input to the olfactory bulb. *J. Neurosci.* 26:1247–59

Stettler DD, Axel R. 2009. Representations of odor in the piriform cortex. *Neuron* 63:854–64

Stokes CCA, Isaacson JS. 2010. From dendrite to soma: dynamic routing of inhibition by complementary interneuron microcircuits in olfactory cortex. *Neuron* 67:452–65

Stopfer M, Jayaraman V, Laurent G. 2003. Intensity versus identity coding in an olfactory system. *Neuron* 39:991–1004

Suzuki N, Bekkers JM. 2012. Microcircuits mediating feedforward and feedback synaptic inhibition in the piriform cortex. *J. Neurosci.* 32:919–31

Tank DW, Hopfield JJ. 1987. Neural computation by concentrating information in time. *Proc. Natl. Acad. Sci. USA* 84:1896–900

Touhara K, Vosshall LB. 2009. Sensing odorants and pheromones with chemosensory receptors. *Annu. Rev. Physiol.* 71:307–32

Uchida N, Mainen ZF. 2003. Speed and accuracy of olfactory discrimination in the rat. *Nat. Neurosci.* 6:1224–29

Uchida N, Mainen ZF. 2007. Odor concentration invariance by chemical ratio coding. *Front. Syst. Neurosci.* 1:3

Uchida N, Takahashi YK, Tanifuji M, Mori K. 2000. Odor maps in the mammalian olfactory bulb: domain organization and odorant structural features. *Nat. Neurosci.* 3:1035–43

Wehr M, Laurent G. 1996. Odour encoding by temporal sequences of firing in oscillating neural assemblies. *Nature* 384:162–66

Wilson DA, Sullivan RM. 2011. Cortical processing of odor objects. *Neuron* 72:506–19

Wilson RI. 2013. Early olfactory processing in *Drosophila*: mechanisms and principles. *Annu. Rev. Neurosci.* 36:217–41

Wilson RI, Mainen ZF. 2006. Early events in olfactory processing. *Annu. Rev. Neurosci.* 29:163–201

Yoshida I, Mori K. 2007. Odorant category profile selectivity of olfactory cortex neurons. *J. Neurosci.* 27:9105–14

Zariwala HA, Kepecs A, Uchida N, Hirokawa J, Mainen ZF. 2013. The limits of deliberation in a perceptual decision task. *Neuron* 78:339–51

Zhan C, Luo M. 2010. Diverse patterns of odor representation by neurons in the anterior piriform cortex of awake mice. *J. Neurosci.* 30:16662–72

Chemogenetic Tools to Interrogate Brain Functions

Scott M. Sternson[1] and Bryan L. Roth[2]

[1]Janelia Farm Research Campus, Howard Hughes Medical Institute, Ashburn, Virginia 20147;
email: sternsons@janelia.hhmi.org

[2]Department of Pharmacology and Division of Chemical Biology and Medicinal Chemistry,
University of North Carolina Chapel Hill School of Medicine, Chapel Hill,
North Carolina 27599; email: bryan_roth@med.unc.edu

Annu. Rev. Neurosci. 2014. 37:387–407

First published online as a Review in Advance on
June 16, 2014

The *Annual Review of Neuroscience* is online at
neuro.annualreviews.org

This article's doi:
10.1146/annurev-neuro-071013-014048

Keywords

DREADDs, designer receptors exclusively activated by designer drugs,
PSAM, PSEM

Abstract

Elucidating the roles of neuronal cell types for physiology and behavior is
essential for understanding brain functions. Perturbation of neuron electrical activity can be used to probe the causal relationship between neuronal
cell types and behavior. New genetically encoded neuron perturbation tools
have been developed for remotely controlling neuron function using small
molecules that activate engineered receptors that can be targeted to cell types
using genetic methods. Here we describe recent progress for approaches
using genetically engineered receptors that selectively interact with small
molecules. Called "chemogenetics," receptors with diverse cellular functions
have been developed that facilitate the selective pharmacological control over
a diverse range of cell-signaling processes, including electrical activity, for
molecularly defined cell types. These tools have revealed remarkably specific behavioral physiological influences for molecularly defined cell types
that are often intermingled with populations having different or even opposite functions.

Contents

INTRODUCTION

In an amazingly prescient article published in 1979, Francis Crick (1979) engaged in a "wish list" of items that would be essential for understanding brain function:

> For example, a method that would make it possible to inject one neuron with a substance that would then clearly stain all the neurons connected to it, and no others, would be invaluable. . . . [Similarly,] a method by which all neurons of just one type could be inactivated, leaving the others more or less unaltered [is also needed].

Today, in 2014, we have tools both to identify (Wall et al. 2010) and to visualize (Chung et al. 2013, Hama et al. 2011, Ke et al. 2013) neuronal connectivity in intact brains. Efficient technologies are also readily available to selectively silence genetically identified neurons using small molecules (Armbruster et al. 2007, Lechner et al. 2002, Magnus et al. 2011) and photons (Li et al. 2005, Zhang et al. 2007).

In a later article, Crick also presaged the need for methods to "turn neurons on" (Crick 1999), and now optical (Boyden et al. 2005, Li et al. 2005), small-molecule (Alexander et al. 2009, Armbruster et al. 2007, Magnus et al. 2011), and photochemical (Callaway & Katz 1993, Kokel et al. 2013, Zemelman et al. 2002) technologies are also available. This review focuses on methods that utilize small molecules to activate (Alexander et al. 2009, Armbruster et al. 2007, Magnus et al. 2011) and inhibit (Armbruster et al. 2007, Lechner et al. 2002, Magnus et al. 2011) neuronal firing. Over the years, a number of terms have been used to describe small-molecule-mediated activation of engineered proteins including pharmacogenetics (Sasaki et al. 2011), pharmacosynthetics (Farrell & Roth 2012), and chemogenetics (Strobel 1998). Here we use the term chemogenetics because it was used first to describe this approach, whereas the term pharmacogenetics is not appropriate given its connotations in pharmacology and genetics (Farrell & Roth 2012). We also highlight new tools that allow the precise modulation of signaling (Armbruster et al. 2007, Farrell et al. 2013, Guettier et al. 2009b, Nakajima & Wess 2012) in genetically defined neurons, glia, and other cell types.

EARLY CHEMOGENETIC TECHNOLOGIES BASED ON G PROTEIN–COUPLED RECEPTORS

G protein–coupled receptors (GPCRs) represent the largest class of neuronal signal-transducing molecules (Allen & Roth 2011). Depending on the specific downstream effector system initiated,

GPCRs can excite, inhibit, or otherwise modulate neuronal firing (Farrell & Roth 2013). Initial attempts at modulating cellular signaling using chemogenetic approaches utilized GPCRs that were engineered by site-directed mutagenesis to bind nonnatural ligands. In a pioneering study, Strader et al. (1991) designed a mutant β2-adrenergic receptor that was unable to bind the native ligand adrenaline but could be activated by 1-(3′,4′-dihydroxyphenyl)-3-methyl-L-butanone (L-185,870) (**Figure 1**). Although L-185,870 had relatively low potency for the engineered receptor ($EC_{50} \sim 40$ uM), the results represented an essential proof of concept for this general approach. Further modifications led to a highly engineered β2-adrenergic receptor with even higher potency for L-185,870 (~7 uM) (Small et al. 2001) and nonresponsiveness to native ligands.

The next advance occurred with the creation of a family of engineered receptors dubbed RASSLs (receptor activated solely by synthetic ligand). The initial RASSL was an engineered k-opioid receptor (KOR) that was insensitive to native peptide ligands but could be activated potently by the synthetic KOR agonist spiradoline (Coward et al. 1998). This KOR RASSL (**Figure 1**) was subsequently used in the first chemogenetic study from which remote control of cardiac activity was achieved (Redfern et al. 1999). For these experiments, the KOR RASSL was conditionally and reversibly expressed in cardiac myocytes using tetracycline-inducible expression driven by a myosin heavy chain promoter (αMHC–tTA). Subsequently, RASSL technology was used to unravel the code for sweet and bitter taste (Mueller et al. 2005, Zhao et al. 2003). Several other RASSLs have also been generated (for a review, see Conklin et al. 2008), although their utility

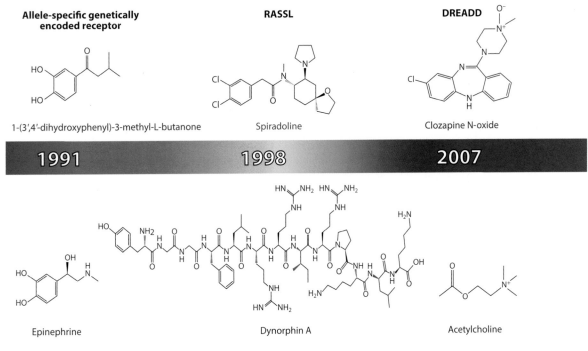

Figure 1

Evolutionary timeline of GPCR-based chemogenetic approaches listing the main corresponding tools starting with allele-specific engineered β-adrenergic receptors, RASSLs, and DREADDs. Relevant structures shown include (*top*) engineered ligands and (*bottom*) endogenous ligands. Green text indicates no pharmacologic activity at the native target; red indicates activity. Abbreviations: DREADD, designer receptor exclusively activated by designer drug; GPCR, G protein–coupled receptor; RASSL, receptor activated solely by synthetic ligand.

in the neurosciences has been hampered owing to the pharmacological activities of the cognate ligands (e.g., spiradoline is a potent KOR agonist) and to the fact that some, but not all (Chang et al. 2007), RASSLs have high levels of constitutive activity (Hsiao et al. 2008, 2011; Sweger et al. 2007).

Since then, researchers have engineered several other receptor-ligand pairs based on 5-HT$_{2A}$ serotonin (Kristiansen et al. 2000, Westkaemper et al. 1999) and adenosine (Gao et al. 2006; Jacobson et al. 2001, 2005) receptors. Optically activated chimeric opsins that can activate canonical GPCR signaling cascades (Airan et al. 2009) have also been created. In each case, and in distinction to RASSLs, the orthologous ligands (with the exception of the opsins) showed greatly attenuated activity at the native receptor and greatly enhanced activity at the engineered receptor. Additionally, the affinities and potencies for the native orthologous ligands were greatly reduced. The chemogenetic and optogenetic tools are important because they demonstrate the potential that many GPCRs could be engineered to bind relatively inactive cognate ligands. However, given the relatively weak potency of synthetic ligands (Kristiansen et al. 2000, Westkaemper et al. 1999) and adenosine (Gao et al. 2006; Jacobson et al. 2001, 2005) or given modest signaling (Airan et al. 2009), they have not been broadly adopted as tools for remotely controlling neuronal signaling.

CHEMOGENETIC CONTROL OF NEURONAL AND NON-NEURONAL SIGNALING USING DREADD TECHNOLOGY

Fundamental problems associated with these early attempts to control GPCR signaling, as stated above, were that the ligands were not particularly well suited for in vivo studies because of the effects on cognate and noncognate receptors and that the engineered receptors occasionally had high levels of constitutive activity. To overcome these problems, Armbruster & Roth (2005) developed a platform they termed DREADD (designer receptor exclusively activated by designer drug) in which directed molecular evolution in yeast was used to activate GPCRs via pharmacologically inert, drug-like small molecules (Alexander et al. 2009, Armbruster et al. 2007). As initially described (Armbruster et al. 2007, Dong et al. 2010, Rogan & Roth 2011), an engineered human M3-muscarinic receptor was subjected to random mutagenesis, expressed in genetically engineered yeast (Schmidt et al. 2003), and grown in media containing clozapine N-oxide (CNO) (**Figure 1**). CNO was chosen because of its excellent ability to penetrate the central nervous system (Bender et al. 1994), favorable pharmacokinetics in mice (Bender et al. 1994) and humans (Jann et al. 1994), and inert pharmacology (Armbruster et al. 2007).

Under screening conditions, only yeast that express a mutant M3-muscarinic receptor that can be activated by CNO survive. After several cycles of selection and mutagenesis as well as comprehensive bioinformatics and pharmacological characterization, researchers selected an M3-muscarinic receptor with two mutations (Y149C, A239G) that fulfilled the following criteria:

- Nanomolecular potency for activation by CNO
- Relative insensitivity to acetylcholine (the native ligand)
- No detectible constitutive activity.

The resulting Y149C, A239G M3-muscarinic receptor was the first DREADD and is now known as hM3Dq to indicate its selectivity for Gαq-mediated signaling pathways (**Figure 2**). Because these two residues (e.g., Y149C and A239G) are conserved among all muscarinic receptors throughout evolution from *Drosophila* to humans, they can create an entire family of DREADD-based muscarinic receptors (vis hM1Dq, hM2Di, hM3Dq, hM4Di, hM5Dq), all of which are potently activated by CNO, insensitive to acetylcholine, and devoid of constitutive activity (Armbruster et al. 2007). M1-, M3-, and M5-DREADDs all couple to Gαq, whereas M2- and M4-DREADDs couple to Gαi-G proteins (**Figure 2**). Subsequently, Guettier et al. (2009)

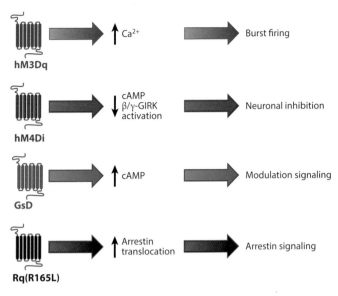

Figure 2

Main DREADD-based tools currently available as well as their typical uses in neuroscience.

created a chimeric muscarinic-adrenergic receptor DREADD (GsD) that selectively activates Gαs and activates neuronal cAMP-mediated signaling (Farrell et al. 2013).

In the initial study Armbruster et al. (2007) reported that hM4Di-DREADD could also induce neuronal silencing via Gαi-mediated activation of G protein inwardly rectifying potassium channels in hippocampal neurons in vitro. Armbruster et al. (2007) also predicted that hM4Di would also be useful for silencing neuronal activity in vivo. Subsequently, many groups have independently reported the successful attenuation of neuronal firing by CNO-mediated activation of hM4Di (Atasoy et al. 2012; Brancaccio et al. 2013; Ferguson et al. 2011; Kozorovitskiy et al. 2012; Krashes et al. 2011; Parnaudeau et al. 2013; Ray et al. 2011, 2012; Sasaki et al. 2011). In every instance, the attenuation of neuronal firing was accompanied by striking behavioral and/or physiological consequences (**Table 1**) and the imputation of distinct populations of genetically identified neurons as mediators of behavior and/or physiology. hM4Di has also been used to deconstruct signaling pathways involved in the migration of tumor cells (Yagi et al. 2011) and to interrogate orthosteric (Alvarez-Curto et al. 2011a,b) and allosteric (Abdul-Ridha et al. 2013, Nawaratne et al. 2008) signaling modes of Gαi-mediated receptors.

With regard to hM3Dq and other Gαq-DREADDs, Alexander et al. (2009) discovered that activating genetically encoded hM3Dq in hippocampal principal cells by CNO induced slow depolarization and burst firing. Since then, many groups have independently reported successful activation of neuronal firing by CNO-mediated activation of hM3Dq in a variety of contexts (Atasoy et al. 2012, Brancaccio et al. 2013, Garner et al. 2012, Kong et al. 2012, Krashes et al. 2011, Sasaki et al. 2011, Vrontou et al. 2013). Additionally, hM3Dq has been used to interrogate the consequences of acute and chronic activation of Gαq-mediated signaling in pancreatic β-cells (Guettier et al. 2009, Jain et al. 2013). In every reported instance, activation of Gαq signaling led to striking behavioral and/or physiological consequences (**Table 1**).

Gαs-DREADD (GsD) was initially used in studies of pancreatic β-cells in vitro and in vivo to deconstruct the signaling pathways essential for insulin secretion (Guettier et al. 2009). Subsequent studies demonstrated that CNO-mediated activation of GsD potently and efficaciously

Table 1 Representative recent publications using DREADD technology

DREADD	Experiment	Result	Reference(s)
hM3Dq +/− hM4Di	Remote control of feeding	Identification of neurons that encode hunger	Atasoy et al. (2012), Kong et al. (2012), Krashes et al. (2011)
hM3Dq	Generation of a synthetic memory trace	Memory encoded sparsely	Garner et al. (2012)
hM4Di	Alteration in neuronal plasticity	Altered striatal connectivity	Kozorovitskiy et al. (2012)
hM4Di	5-HT neuron silencing	Behavior and physiological consequences	Ray et al. (2011)
hM3Dq	Identification of neurons responsible for pleasurable sensation	DRG neurons identified as target of MGPR4 orphan receptor	Vrontou et al. (2013)
GsD	Modulation of cAMP	Modulates circadian clock; regulates insulin secretion	Brancaccio et al. (2013), Guettier et al. (2009)

Abbreviations: DREADD, designer receptor exclusively activated by designer drug; DRG, dorsal root ganglion; GsD, Gαs-DREADD.

augments cAMP-mediated signaling in a variety of neuronal contexts (Brancaccio et al. 2013, Farrell et al. 2013, Ferguson et al. 2011). Given the central role of cAMP-mediated signaling in reward (Carlezon et al. 1998), memory (Kida et al. 2002), and psychoactive drug actions (Carlezon et al. 1998, Pliakas et al. 2001), GsD will likely prove useful for deconstructing the role of cAMP-mediated signaling pathways in genetically defined neurons in vivo (for a recent example, see Ferguson et al. 2011).

GPCRs signal not only via G proteins but also by activating β-arrestin-mediated signaling pathways (Luttrell et al. 1999; for a review, see Allen & Roth 2011). Arrestin-mediated signaling has been implicated in the actions of many psychoactive drugs (Beaulieu et al. 2009) including opioids (Bohn et al. 1999), lithium (Beaulieu et al. 2008), and antipsychotics (Allen et al. 2011). Thus, the creation of a DREADD that specifically activates β-arrestin signaling would be valuable for delineating the role(s) of arrestin-ergic signaling in a variety of cellular contexts (Allen & Roth 2011). Recently, Nakajima & Wess (2012) reported that a mutant M3-muscarinic receptor designated Rq(R165L) could selectively activate arrestin signaling without perturbing G protein–mediated pathways (**Figure 2**). Although the potency of CNO is likely too low to be of great utility for studies in vivo, Rq(R165L) serves as a nice proof of concept for the selective activation of arrestin-ergic signaling by DREADDs.

Importantly, in every reported instance, the expression of either hM4Di or hM3Dq DREADDs has no apparent effect on baseline behaviors, neuronal function, or morphology (see Alexander et al. 2009). Additionally, CNO administration in the absence of DREADD expression has no measurable effect on any of the many monitored behavioral and physiological outcomes (see Alexander et al. 2009; Ray et al. 2011, 2012). In Guettier et al. (2009), a small effect on pancreatic β-cell activity was noted when GsD was overexpressed, although other studies have reported no effect of baseline GsD expression in a variety of neuronal contexts (Brancaccio et al. 2013, Farrell et al. 2013, Ferguson et al. 2011).

Loffler et al. (2012) have raised a concern that some of the effects of CNO could be mediated by the relatively inefficient conversion of CNO to clozapine. A formal pharmacokinetic study in mice, however, disclosed no apparent conversion of CNO to clozapine—at least following acute administration (Guettier et al. 2009). In addition, and as summarized above, in every experiment reported to date, CNO has no effect on any observed phenomenon in either mice or rats when administered in the absence of DREADD expression. However, a small fraction of CNO is interconverted to clozapine (~10% by mass) in guinea pigs and humans (Jann et al. 1994). Thus, investigators contemplating the use of CNO in humans (or other primates or guinea pigs) will need to design experiments so that the dose of CNO is kept relatively low and that appropriate controls are performed (e.g., CNO administration in the absence of DREADD expression).

To summarize, various DREADDs allow for chemogenetic activation of various canonical (e.g., Gαs, Gαq, Gαi) and noncanonical (e.g., β-arrestin) signaling pathways in essentially any cellular context (**Figure 2**). In all neurons examined to date, activation of these pathways with CNO leads to burst firing with hM3Dq and attenuation/silencing of neuronal firing with hM4Di. Activation of GsD and arrestin-biased DREADD in neurons likely modulates neuronal activity contexts (Brancaccio et al. 2013, Farrell et al. 2013, Ferguson et al. 2011).

The advantages of DREADDs over other approaches such as optogenetics are as follows:

- CNO can be administered orally and noninvasively (e.g., via drinking water).
- CNO kinetics predict a relatively prolonged duration of neuronal activation, inhibition, or modulation (e.g., minutes to hours).
- CNO-mediated activation of DREADDs requires no specialized equipment.
- CNO is readily available.
- CNO diffuses widely following administration, allowing for the modulation of signaling and activity in distributed neuronal populations.
- CNO has been administered to humans and is a known metabolite of widely prescribed medication.

The main disadvantage of the DREADD system is the lack of precise temporal control as is achieved with light-mediated systems such as optogenetics and optopharmacology. This disadvantage could soon be overcome with photocaged CNO (B.L. Roth, manuscript in preparation). Another useful tool that may soon be available involves having additional GPCR-ligand pairs available to allow for multiplexing control over signaling (E. Vardy and B.L. Roth, submitted manuscript). Finally, it may also be useful to obtain more potent control over arrestin signaling than is currently afforded by arrestin-biased DREADD; such technology is also in development (Y. Gotoh and B.L. Roth, manuscript in preparation).

CHEMOGENETIC CONTROL OF IONIC CONDUCTANCE

Ion channels are especially well suited for manipulating neuronal activity because they directly control the electrical properties of cells. Early work in manipulating the electrical properties of neurons involved genetically targeted expression of ion channels that chronically alter neuronal conductance. For example, neuronal overexpression of inward rectifier potassium channels resulted in suppression of neuronal excitability, but prolonged expression led to toxicity and could also result in compensatory responses (Ehrengruber et al. 1997, Nadeau et al. 2000). By allowing rapid, remote control over different ion conductances, ligand-gated ion channels (LGICs) are better suited for temporal control over neuronal activity. LGICs have been widely exploited for neuronal stimulation or silencing to examine causal relationships between electrical activity and animal behavior, primarily by intracranial administration of agonists for glutamate (Stanley et al. 1993) and GABA (Hikosaka & Wurtz 1985) receptors. However, to perturb a localized subset of neurons in the brain, small molecules must be locally targeted, typically through a cannula that destroys overlying neural tissue. A greater drawback is that these perturbations are not cell type specific owing both to the widespread expression of glutamate and GABA receptors on neurons and to the absence of pharmacologically distinct LGICs on most cell types.

More recently, several LGICs that are optimized primarily for use in mammals have been developed for cell type–specific pharmacological control of neuron electrical activity. LGICs suitable as ectopically expressed tools for neuronal activity perturbation also require a selective ligand that does not activate endogenous ion channels. Three categories of LGIC tools have been developed for cell type–selective neuron perturbation: (*a*) invertebrate LGICs, (*b*) ectopic expression of endogenous mammalian LGICs in the context of a global knockout background for that channel, and (*c*) engineered ligand and ion channel systems.

Invertebrate LGICs

Invertebrate LGICs with pharmacological properties distinct from mammalian ion channels have been exploited to perturb electrical activity in genetically targeted neuron populations via transgenic expression in the mammalian brain. Glutamate-gated chloride (GluCl) channels from the roundworm *Caenorhabditis elegans* have been developed as selective neuronal silencers (Slimko et al. 2002). GluCl channels are high conductance chloride channels formed as heteromers of GluClα and GluClβ subunits, both of which must be expressed to produce functional channels (Slimko et al. 2002). GluCl conductance can be activated by the antiparasite drug ivermectin (IVM), which is a high potency allosteric agonist. IVM is commonly administered at low doses to mammals as an antiparasite medication without obvious neurological side effects, implying selective action on the GluCl channels of parasites over endogenous mammalian LGICs. In neurons, IVM gating of GluCl channels results in the suppression of neuronal activity (**Figure 3**), primarily through a large drop in input resistance across the neuronal membrane (Slimko et al. 2002). The effect on neuronal activity can be understood using Ohm's law ($V = IR$), where the voltage change (V) is directly proportional to the current (I) across the membrane as a function of the membrane resistance (R) of the cell. Reduction of neuronal input resistance by opening the GluCl channels acts as an electrical shunt, thereby reducing the influence of an injected or synaptic current on the membrane potential and, hence, decreases neuronal excitability.

An additional consideration for the use of GluCl channels in mammals regards their affinity for glutamate, which is an abundant neurotransmitter in mammalian brains. Activation of exogenous GluCl by endogenous glutamate has been minimized by a single-point mutation (Y182F) in the glutamate binding pocket of GluClβ, which substantially reduces the potency of glutamate activation of GluCl (Li et al. 2002). IVM sensitivity is only weakly changed, likely owing to the distinct GluCl binding sites for IVM and glutamate (Hibbs & Gouaux 2011).

This modified GluCl channel is useful for silencing neurons in behaving mice in conjunction with minimally invasive intraperitoneal IVM administration. Targeted delivery of GluClα/β subunits to defined neuron populations by viral vectors and transgenic overexpression has been used to suppress neuronal function in the striatum, amygdala, and hypothalamus. Inhibition of striatal neurons on one side of the brain during amphetamine administration resulted in rotation during locomotion (Lerchner et al. 2007). Given this behavioral readout, dosing and pharmacokinetic properties have indicated that IVM has a slow onset of action, requiring dosing one day prior to behavioral test, and that IVM clearance requires multiple days. This is likely due to the high lipophilicity of IVM, which presumably aids brain penetration, but can also result in accumulation in fat depots in the body (McKellar et al. 1992), reducing brain access and acting as a reservoir for IVM after the initial dose. Using these dosing parameters, researchers have also used the IVM/GluCl system to suppress PKCδ-expressing neurons in the central nucleus of the amygdala,

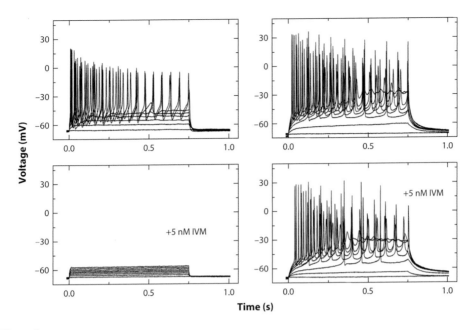

Figure 3

Neuronal silencing with invertebrate ligand-gated ion channels. (*Left*) Coexpression of GluClα and GluClβ (subunits of glutamate-gated chloride channels) in neurons does not reduce cellular excitability, but electrical activity is strongly suppressed in the presence of ivermectin (IVM). (*Right*) Untransfected cells are not silenced by IVM. Figure modified from Slimko et al. (2002).

which increases fear responses (freezing) to a conditioned stimulus previously paired with a foot shock (Haubensak et al. 2010). In addition, GluCl/IVM-mediated suppression of neurons in the ventromedial hypothalamic nucleus, a region that can induce aggressive behavior when activated, reduces aggression of male mice toward male intruders (Lin et al. 2011).

Further modifications to GluCl channels that improved protein trafficking and IVM sensitivity have also been reported (Frazier et al. 2013), and these may prove useful for reducing potential toxicity or off-target effects associated with high IVM doses that are required for neuron silencing in the brain. As an alternative method, the mammalian glycine receptor, another chloride channel, has been recently engineered for neuronal perturbation via the introduction of mutations that render the channel sensitive to IVM allosteric activation and that reduce glycine sensitivity (Lynagh & Lynch 2010). These channels require only a single subunit to be delivered to neurons to achieve neuronal silencing. In principle, chemical modifications to IVM could also provide variants with faster pharmacokinetic properties, which would facilitate more acute perturbations.

Mammalian LGICs

Tools for selective perturbation of neuronal activity have also been developed using mammalian LGICs, which enables use of an extensive range of selective small-molecule ligands for these channels. By ectopically targeting the LGIC to the cell type of interest, researchers use these tools to adapt nonessential mammalian LGICs for selective neuronal activation or silencing. Because these LGICs are also expressed endogenously, selective channel expression is performed on a global knockout background for the endogenous LGIC gene to avoid activation of endogenous channels.

This strategy has been used to demonstrate cell type–selective chemical activation of neurons via targeted expression of the TRPV1 ion channel. TRPV1 is a nonselective cation channel that is gated by the small molecule capsaicin (the molecule in chili peppers that renders them spicy), resulting in neuronal depolarization (Arenkiel et al. 2007, Zemelman et al. 2003). Because capsaicin and other TRPV1 agonists can act on endogenous channels, TRPV1 must be targeted to specific cell types in mice in which endogenous *Trpv1* has been genetically inactivated (**Figure 4a**). This has been carried out by ectopically targeting TRPV1 to dopamine neurons in *Trpv1*$^{-/-}$ mice. In these mice, capsaicin results in robust activation of dopamine neurons, elevated release of dopamine in the striatum, and increased locomotor activity (Guler et al. 2012). Although the use of this chemogenetic strategy for neuronal activation requires extensive mouse breeding, it also offers the convenience of no surgical procedures to selectively express TRPV1 or to deliver its agonist capsaicin.

A related strategy has been developed for neuron inhibition using cell type–selective activation of GABA$_A$ receptors by the allosteric agonist zolpidem. *Gabrg2* encodes the GABA$_A$ receptor γ2 subunit, which is essential for GABA$_A$ receptor function and is critical for zolpidem sensitivity. However, zolpidem sensitivity in GABA$_A$ receptors is eliminated in mice with a targeted mutation of the γ2 subunit at amino acid position 77 from isoleucine to phenylalanine (I77F), while retaining GABA responsiveness. To utilize zolpidem as a selective neuronal silencer, mice were engineered to be zolpidem insensitive in every neuron except the cell type that was targeted for silencing (**Figure 4b**). This was accomplished by replacing exon 4 of *Gabrg2*, which encodes I77 of the γ2 subunit, with an I77F mutation. The modified exon 4 was flanked by loxP sites so that it could be removed cell type selectively by crossing to a Cre recombinase–expressing mouse line (Wulff et al. 2007). Then, mice carrying a transgene encoding the native *Gabrg2* cDNA sequence under a cell type–specific promoter could be bred onto this background to add zolpidem-sensitive γ2 subunits only to the cell populations associated with Cre recombinase removal of the I77F-encoding exon. The result was zolpidem sensitivity only in the Cre-expressing cell type of interest. Applying this strategy to cerebellar Purkinje cells resulted in pronounced ataxia in the presence of zolpidem that was not apparent in I77F mutant mice (Wulff et al. 2007). A key aspect of this silencing approach is that neurons are not directly inhibited by zolpidem, but instead the efficacy of endogenous GABA-releasing synaptic input is potentiated in the zolpidem-sensitive neuron populations.

Engineered LGICs and Ligands

A newer approach to chemogenetic manipulation of neuron electrical activity is based on chimeric ion channels that were developed using concerted genetic and chemical engineering of selective interactions between ion channels and small-molecule agonists (Magnus et al. 2011).

Figure 4

Electrical activity perturbation strategies using mammalian ligand-gated ion channels. (*a*) Scheme for neuronal activation that involves cell type–selective Cre recombinase–dependent expression of TRPV1 on a global *Trpv1* knockout background. *Slc6a3*$^{+/Cre}$ restricts ectopic TRPV1 expression to dopamine neurons. Intraperitoneal delivery of the TRPV1 agonist capsaicin results in dopamine neuronal activation and increases locomotion in mice. Panel modified from Guler et al. (2012). (*b*) Scheme for silencing of neuronal activity using the GABA$_A$ receptor allosteric agonist zolpidem. The endogenous zolpidem sensitivity (mediated by the GABA$_A$ receptor γ2 subunit at position F77) is eliminated with global knockin of a loxP-flanked exon encoding F77I. In a cell type of interest (in this case, Purkinje neurons targeted selectively using the L7 promoter), Cre recombinase is expressed to remove the loxP-flanked exon. The same cell type–specific promoter is also used to transgenically express the zolpidem-sensitive γ2 subunit. Neurons in mice with all these components can be selectively silenced with zolpidem. Via this system, Purkinje neuron silencing was shown to affect motor behavior. Abbreviation: GFP, green fluorescent protein. Panel modified from Wulff et al. (2007).

These engineered LGICs overcome restrictions of earlier LGIC-based tools such as limited characterization of invertebrate channels, the need to knock out endogenous mammalian ion channel genes, and the generally limited capability to optimize either channel properties or the pharmacokinetic properties of ligands.

This chemical and genetic engineering strategy for cell type–specific control over ion conductance is based on classic experiments demonstrating that the extracellular ligand binding domain (LBD) of the α7 nicotinic acetylcholine receptor behaves as an independent actuator module that can be transplanted onto the ion pore domains (IPDs) of other members of the large Cys-loop

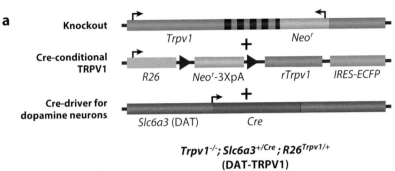

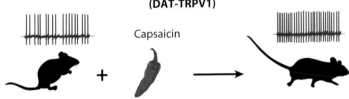

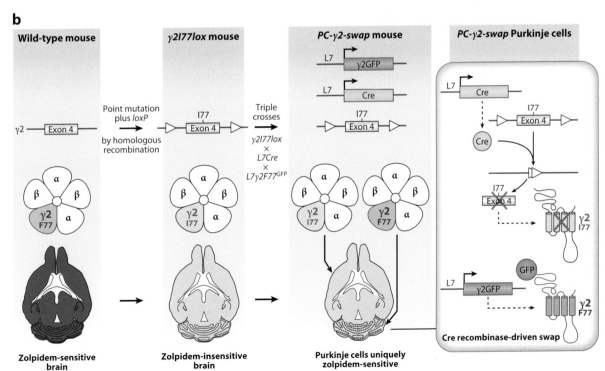

receptor ion channel family. Thus, splicing the α7 nAChR LBD to the IPD of the serotonin receptor (5HT3a) produces a channel (α7-5HT3) with α7 nAChR pharmacology and 5HT3a conductance properties (Eisele et al. 1993). An analogous engineered channel has been developed by fusing the α7 nAChR LBD to the chloride-selective glycine receptor (GlyR) IPD, which renders an acetylcholine-responsive chloride channel (α7-GlyR) (Grutter et al. 2005). This modular property is a strong foundation for optimizing functional characteristics. Moreover, because α7 nAChR LBDs assemble into homomeric pentamers, chimeric LGICs based on this motif self-assemble without needing to express additional cofactors, which facilitates their use as tools targeted to molecularly defined neuronal populations.

The major challenge to use these chimeric ion channels and their ligands as cell type–selective perturbation tools is that α7 nAChR is endogenously expressed in many neuron populations and α7 nAChR agonists can perturb these other cell groups. As described above, this problem has been typically addressed by eliminating the endogenous allele, which usually requires expensive and time-intensive mouse breeding approaches. For chimeric channels using the extracellular LBD of the α7 nAChR, an alternative solution was used. The ligand recognition properties of the α7 nAChR LBD were engineered using a "bump-hole" strategy (Bishop et al. 2000, Hwang & Miller 1987, Lin et al. 2001, Westkaemper et al. 1999) in which LBD mutations generate "holes" that allow binding of bulky ("bumped") chemical analogs of ligands that would not otherwise bind the endogenous LBD. An α7 nAChR agonist, quinuclidinyl benzamide PNU-282987, was used as a starting point for agonist design because it crosses the blood-brain barrier (Walker et al. 2006); is highly selective for α7 nAChR over other isoforms; and is highly selective against a broad panel of vertebrate ion channels, GPCRs, and transporters (Bodnar et al. 2005). A library of mutated ion channels was tested in an activity-based screen against a library of "bumped" quinuclidinyl benzamides. From this screen, multiple mutated ion channel and complementary agonist combinations were identified with ligands that did not activate the unmodified receptor. Furthermore, several combinations of agonist/mutated LBDs were engineered to be orthogonal to each other, allowing their use in concert. These mutated LBDs were called pharmacologically selective actuator modules (PSAM; pronounced "sam"), and distinct PSAMs are represented by the specific mutation that renders their selectivity, e.g., PSAM[L141F]. The cognate agonists were called pharmacologically selective effector molecules (PSEM; pronounced "sem") and are referred to with specific numbers, e.g., PSEM[89S] (**Figure 5a**).

A variety of PSAM/PSEM combinations allowed for the generation of pharmacologically selective ion channels that have distinct ion conductance properties and that can be gated without

Figure 5

Pharmacologically selective actuators and effectors for control of ion conductance. (*a*) Design scheme for chimeric LGICs composed of LBD and IPD modules. LBD mutations yield a PSAM that selectively binds PSEMs (*red*) but not the endogenous ligand (ACh) (*yellow*). PSEMs do not bind the unmodified LBD. (*b*) Combinatorial generation of pharmacologically selective LGICs with diverse conductance properties by joining PSAM and IPD modules. (*c*) (*Left*) Chimeric LGICs for neuronal activation channels built from a nonspecific cation-selective IPD (5HT3 HC, a high-conductance variant of the serotonin 3 receptor) and two pharmacologically distinct PSAMs. Application of the appropriate PSEM leads to sustained neuronal activation (PSEM application for 120 s; traces are cell-attached recordings). (*Right*) The pharmacological selectivity of one of the chimeric channels is shown in a neuron expressing PSAM[L141F,Y115F]-5HT3 HC, which leads to neuronal activation in the presence only of PSEM[89S] but not of PSEM[22S]. (*d*) Using the same PSAM as in panel *c* with the IPD from GlyR results in a pharmacologically selective chloride channel. This channel suppresses neuronal activity even in response to high current injection (*left*) owing to the shunting properties of the open channel that strongly reduces neuronal input resistance (R_{in}). Figure modified from Magnus et al. (2011). Abbreviations: ACh, acetylcholine; GlyR, glycine receptor; IPD, ion pore domain; LBD, ligand binding domain; LGIC, ligand-gated ion channel; PSAM, pharmacologically selective actuator module; PSEM, pharmacologically selective effector molecule; WASH, washout of ligand with artificial cerebrospinal fluid.

activating either the endogenous α7 nAChR or other PSAM-containing channels (**Figure 5b**). PSAMs were fused to IPDs from several members of the Cys-loop LGIC family: serotonin, glycine, GABA C, and nicotinic acetylcholine receptors. Because the IPD determines the ionic conductance properties, PSAM-IPD chimeric channels activated with the corresponding PSEMs provided pharmacological control of ion conductance for either nonspecific cations, chloride or calcium (Magnus et al. 2011).

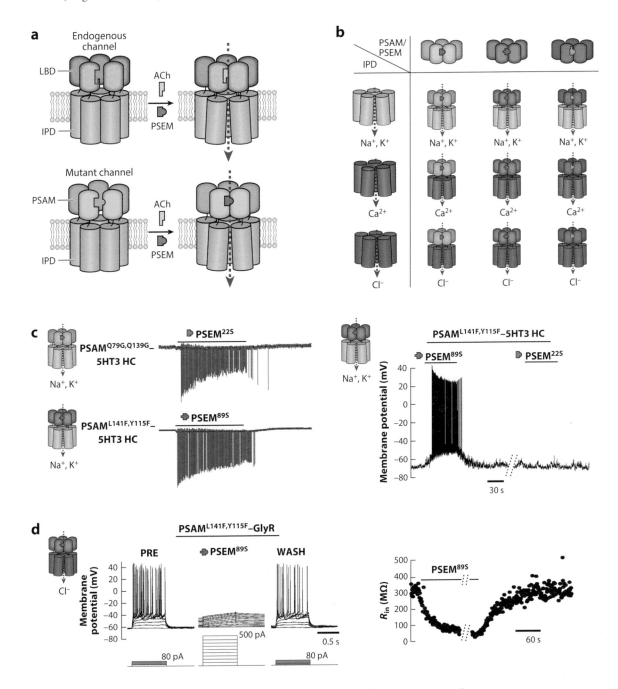

Multiple pharmacologically selective cation channels were generated by fusing different PSAMs to the 5HT3 IPD (**Figure 5c**). Neurons expressing these channels depolarized and fired action potentials for minutes during PSEM application. Action potential activity ceased shortly after PSEM removal. The same PSEM molecules could also be used to activate chimeric chloride channels by fusion of the cognate PSAM LBDs with GlyR or GABA C IPDs (**Figure 5d**). These channels sharply reduced the input resistance of neurons and strongly inhibited neuronal excitability in the presence of the appropriate PSEM (**Figure 5d**). PSAM-GlyR-expressing neurons were electrically shunted by PSEM application and could not be activated even with injection of hundreds of picoamps of depolarizing current, but washout of the PSEM restored neuron excitability within minutes.

Because of the strong shunting properties of PSAM-GlyR silencing, these channels are especially useful for suppressing neuronal activity, even during strong, concerted excitatory synaptic input. This was strikingly demonstrated in experiments dissecting the role of two interneuron cell types in CA1 hippocampal circuits. Strong synaptic drive into CA1 by patterned optical stimulation of Schaeffer collateral axon projections activates a local circuit involving multiple interneurons that shapes the output of principal CA1 pyramidal neurons. $PSAM^{L141F,Y115F}$-GlyR channels selectively silenced molecularly defined subpopulations of CA1 interneurons during strong synaptic stimulation. This allowed precise dissection of the relative contribution of these two interneuron circuit components to the input-output properties of hippocampal principal cells (Lovett-Barron et al. 2012).

The in vivo efficacy of PSAM-GlyR silencing with a PSEM was demonstrated under especially challenging conditions designed to test silencing capability during concerted depolarizing input from activation of the light-activated ion channel channelrhodopsin-2 (ChR2). ChR2 was expressed in hypothalamic neurons that induce feeding behavior when optically stimulated along with $PSAM^{L141F,Y115F}$-GlyR. During optical stimulation of hypothalamic neurons with an implanted optical fiber and in the absence of the cognate PSEM ligand, mice consumed food rapidly within minutes of photoactivation. Intraperitoneal delivery of $PSEM^{89S}$ led to suppression of ChR2-evoked feeding. This effect was completely reversed the following day when ChR2 activation once again was sufficient to evoke feeding (Magnus et al. 2011). These experiments demonstrated that this selective ligand and LGIC system can serve as a powerful neuronal silencer in vivo. Subsequent studies have confirmed the efficacy of $PSAM^{L141F,Y115F}$-GlyR and a related channel, $PSAM^{L141F}$-GlyR, for neuronal silencing in vivo by suppression of contextual fear learning (Lovett-Barron et al. 2014) and inhibition of neurons that are critically important in a skilled reaching task (Esposito et al. 2014). Furthermore, an additional study applied $PSAM^{L141F,Y115F}$-GlyR for neuronal silencing and $PSAM^{L141F,Y115F}$-5HT3 HC for neuronal activation to bidirectionally control hippocampal interneuron activity to reduce or enhance foreign object recognition learning, respectively (Donato et al. 2013). These experiments highlight the flexibility of PSAM-IPD ion channels for gain-of-function and loss-of-function neuronal activity manipulations and also illustrate the in vivo efficacy of their cognate PSEM ligands.

Further extension of the chemogenetic toolbox can be based on PSAM/PSEM selectivity modules that are transferable to functionally diverse IPDs, which provide access to a combinatorial array of LGICs based on these components (**Figure 5b**). The combinations of PSAM/PSEMs and IPDs enable production of additional cell type–selective tools for pharmacological control over LGICs with multiple ionic conductances. These chimeric ion channels can also be further elaborated by applying extensive prior work on IPD structure-function relationships within the Cys-loop receptor superfamily, including mutations that modify channel conductance (Kelley et al. 2003), ion selectivity (Bertrand et al. 1993, Galzi et al. 1992, Gunthorpe & Lummis 2001,

Keramidas et al. 2000), intracellular interactions (Jansen et al. 2008, Temburni et al. 2000, Xu et al. 2006), and desensitization (Breitinger et al. 2001, Galzi et al. 1992, Revah et al. 1991).

OTHER CHEMOGENETIC APPROACHES FOR NEUROBIOLOGY

Chemogenetic methods have also been developed to control other aspects of neuronal function. Cell type–selective pharmacological control of synaptic release has been demonstrated with ligand-induced dimerization technologies. After fusion of the small protein FKBP to the synaptic vesicle–associated protein synaptobrevin, a dimeric FKBP-binding molecule (AP20187) could reversibly oligomerize these critical proteins for synaptic vesicle fusion and neurotransmitter release. In mice engineered to express Synaptobrevin-FKBP in Purkinje neurons, delivery of the ligand directly to the brain results in reversible motor deficits (Karpova et al. 2005). This pioneering work on molecular inhibitors of synaptic transmission highlights the value of considering approaches to selectively suppress synaptic release, which may be especially useful in applications involving suppression of specific axon projections that define a subset of the circuit interactions of a particular cell type.

Kinases are critical for many cell signaling pathways in neurons, but most lack highly selective inhibitors to block activity. To selectively inhibit kinase activity, kinases have been engineered to bind to modified ATP binding-site inhibitors that do not bind to endogenous kinase (Bishop et al. 2000). Selective inhibition of kinases in the brain using pharmacologically selective alleles has provided insight into a number of areas including neurotrophin signaling (Chen et al. 2005), dendritic spine development (Ultanir et al. 2012), and epilepsy (Liu et al. 2013). Targeting pharmacologically selective kinase alleles to specific neuron populations affords cell type–specific modulation of these signaling pathways with a high degree of temporal control.

Another promising area for further development is cell type–specific enzymatic targeting of small molecules (**Figure 6**). In this approach, a small-molecule fluorophore or drug-like molecule is chemically derivatized with a "masking" group that renders it inactive. The masking group is selected to be inert to endogenous enzymatic degradation pathways, but it is labile to an exogenous enzyme that can be targeted as a transgene to specific cell populations. This approach targets fluorophores masked with an ester that is inert to endogenous neuronal esterases. Only specific subsets of neurons expressing a transgene for the enzyme porcine liver esterase showed accumulation of the small-molecule dye (Tian et al. 2012). The effectiveness of this method has also been

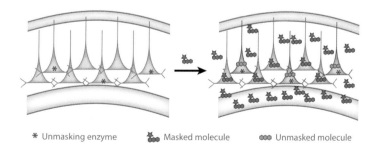

* Unmasking enzyme   🐾 Masked molecule   ⊙⊙⊙ Unmasked molecule

Figure 6

Scheme for cell type–selective targeting of small molecules to a specific subpopulation of neurons within a brain area via targeted unmasking enzyme expression. A subset of neurons expresses an unmasking enzyme (*red asterisk*). A masked molecule (*gray*) that is pharmacologically inert is applied to the entire region and diffuses into every cell. Only the neurons that express the unmasking enzyme liberate the active form of the molecule (*orange*).

demonstrated for a pharmacological inhibitor targeted to specific cells. In addition, a similar approach has been applied in vivo in a zebrafish model (Pisharath 2007, Pisharath et al. 2007). These methods have the potential to enable cell type–specific pharmacology for intracellular signaling pathways in the context of complex heterogeneous tissues such as the brain.

PERSPECTIVES FOR INTEGRATED USE OF VARIOUS TECHNOLOGIES

The growing diversity of pharmacologically orthogonal ion channels, DREADDs, and optogenetic tools can also be used for multiple cell type–specific perturbations in the same animal (Krashes et al. 2014, Magnus et al. 2011). Moreover, because these tools have been rationally developed by concerted chemical and genetic engineering approaches, new variants of these tools can be tailored for specific experimental requirements. One key consideration for applying chemogenetic and optogenetic methods either individually or in concert is the timescales for control of neuronal activity. For example, optogenetic tools provide millisecond precision but require considerable levels of energy to be delivered to the brain for longer timescale perturbations; with PSAM-IPD systems, the pharmacokinetics of PSEMs show rapid onset and brain clearance in one hour (Magnus et al. 2011); and, for DREADDs, the ligand CNO persists for several hours and can be applied for days (Krashes et al. 2011). Therefore, selection of the most suitable tools can now be based on both the mechanism of action for perturbing neuronal function (e.g., ion conductance or G protein signaling) as well as the temporal dynamics of the circuit functions that are under investigation. In the future, we envision that these various technologies will be increasingly used in combination to identify how multiple types of neuronal and nonneuronal perturbations can result in distinctive alterations in circuit dynamics and animal behavior.

DISCLOSURE STATEMENT

The authors are not aware of any affiliations, memberships, funding, or financial holdings that might be perceived as affecting the objectivity of this review.

LITERATURE CITED

Abdul-Ridha A, Lane JR, Sexton PM, Canals M, Christopoulos A. 2013. Allosteric modulation of a chemogenetically modified G protein–coupled receptor. *Mol. Pharmacol.* 83:521–30

Airan RD, Thompson KR, Fenno LE, Bernstein H, Deisseroth K. 2009. Temporally precise in vivo control of intracellular signalling. *Nature* 458:1025–29

Alexander GM, Rogan SC, Abbas AI, Armbruster BN, Pei Y, et al. 2009. Remote control of neuronal activity in transgenic mice expressing evolved G protein-coupled receptors. *Neuron* 63:27–39

Allen JA, Roth BL. 2011. Strategies to discover unexpected targets for drugs active at G protein–coupled receptors. *Annu. Rev. Pharmacol. Toxicol.* 51:117–44

Allen JA, Yost JM, Setola V, Chen X, Sassano MF, et al. 2011. Discovery of β-arrestin-biased dopamine D2 ligands for probing signal transduction pathways essential for antipsychotic efficacy. *Proc. Natl. Acad. Sci. USA* 108:18488–93

Alvarez-Curto E, Prihandoko R, Tautermann CS, Zwier JM, Pediani JD, et al. 2011a. Developing chemical genetic approaches to explore G protein-coupled receptor function: validation of the use of a receptor activated solely by synthetic ligand (RASSL). *Mol. Pharmacol.* 80:1033–46

Alvarez-Curto E, Ward RJ, Pediani JD, Milligan G. 2011b. Ligand regulation of the quaternary organization of cell surface M3 muscarinic acetylcholine receptors analyzed by fluorescence resonance energy transfer (FRET) imaging and homogeneous time-resolved FRET. *J. Biol. Chem.* 285:23318–30

Arenkiel BR, Peca J, Davison IG, Feliciano C, Deisseroth K, et al. 2007. In vivo light-induced activation of neural circuitry in transgenic mice expressing channelrhodopsin-2. *Neuron* 54:205–18

Armbruster B, Roth B. 2005. Creation of designer biogenic amine receptors via directed molecular evolution. *Neuropsychopharmacology* 30:S265

Armbruster BN, Li X, Pausch MH, Herlitze S, Roth BL. 2007. Evolving the lock to fit the key to create a family of G protein-coupled receptors potently activated by an inert ligand. *Proc. Natl. Acad. Sci. USA* 104:5163–68

Atasoy D, Betley JN, Su HH, Sternson SM. 2012. Deconstruction of a neural circuit for hunger. *Nature* 488:172–77

Beaulieu JM, Gainetdinov RR, Caron MG. 2009. Akt/GSK3 signaling in the action of psychotropic drugs. *Annu. Rev. Pharmacol. Toxicol.* 49:327–47

Beaulieu JM, Marion S, Rodriguiz RM, Medvedev IO, Sotnikova TD, et al. 2008. A β-arrestin 2 signaling complex mediates lithium action on behavior. *Cell* 132:125–36

Bender D, Holschbach M, Stocklin G. 1994. Synthesis of n.c.a. carbon-11 labelled clozapine and its major metabolite clozapine-*N*-oxide and comparison of their biodistribution in mice. *Nucl. Med. Biol.* 21:921–25

Bertrand D, Galzi JL, Devillers-Thiery A, Bertrand S, Changeux JP. 1993. Mutations at two distinct sites within the channel domain M2 alter calcium permeability of neuronal α7 nicotinic receptor. *Proc. Natl. Acad. Sci. USA* 90:6971–75

Bishop AC, Ubersax JA, Petsch DT, Matheos DP, Gray NS, et al. 2000. A chemical switch for inhibitor-sensitive alleles of any protein kinase. *Nature* 407:395–401

Bodnar AL, Cortes-Burgos LA, Cook KK, Dinh DM, Groppi VE, et al. 2005. Discovery and structure-activity relationship of quinuclidine benzamides as agonists of α7 nicotinic acetylcholine receptors. *J. Med. Chem.* 48:905–8

Bohn LM, Lefkowitz RJ, Gainetdinov RR, Peppel K, Caron MG, Lin FT. 1999. Enhanced morphine analgesia in mice lacking β-arrestin 2. *Science* 286:2495–98

Boyden ES, Zhang F, Bamberg E, Nagel G, Deisseroth K. 2005. Millisecond-timescale, genetically targeted optical control of neural activity. *Nat. Neurosci.* 8:1263–68

Brancaccio M, Maywood ES, Chesham JE, Loudon AS, Hastings MH. 2013. A Gq-Ca^{2+} axis controls circuit-level encoding of circadian time in the suprachiasmatic nucleus. *Neuron* 78:714–28

Breitinger HG, Villmann C, Becker K, Becker CM. 2001. Opposing effects of molecular volume and charge at the hyperekplexia site α1(P250) govern glycine receptor activation and desensitization. *J. Biol. Chem.* 276:29657–63

Callaway EM, Katz LC. 1993. Photostimulation using caged glutamate reveals functional circuitry in living brain slices. *Proc. Natl. Acad. Sci. USA* 90:7661–65

Carlezon WA Jr, Thome J, Olson VG, Lane-Ladd SB, Brodkin ES, et al. 1998. Regulation of cocaine reward by CREB. *Science* 282:2272–75

Chang WC, Ng JK, Nguyen T, Pellissier L, Claeysen S, et al. 2007. Modifying ligand-induced and constitutive signaling of the human 5-HT4 receptor. *PLoS ONE* 2:e1317

Chen X, Ye H, Kuruvilla R, Ramanan N, Scangos KW, et al. 2005. A chemical-genetic approach to studying neurotrophin signaling. *Neuron* 46:13–21

Chung K, Wallace J, Kim SY, Kalyanasundaram S, Andalman AS, et al. 2013. Structural and molecular interrogation of intact biological systems. *Nature* 497:332–37

Conklin BR, Hsiao EC, Claeysen S, Dumuis A, Srinivasan S, et al. 2008. Engineering GPCR signaling pathways with RASSLs. *Nat. Methods* 5:673–78

Coward P, Wada HG, Falk MS, Chan SD, Meng F, et al. 1998. Controlling signaling with a specifically designed Gi-coupled receptor. *Proc. Natl. Acad. Sci. USA* 95:352–57

Crick F. 1999. The impact of molecular biology on neuroscience. *Philos. Trans. R. Soc. B* 354:2021–25

Crick FH. 1979. Thinking about the brain. *Sci. Am.* 241:219–32

Donato F, Rompani SB, Caroni P. 2013. Parvalbumin-expressing basket-cell network plasticity induced by experience regulates adult learning. *Nature* 504:272–76

Dong S, Rogan SC, Roth BL. 2010. Directed molecular evolution of DREADDs: a generic approach to creating next-generation RASSLs. *Nat. Protoc.* 5:561–73

Ehrengruber MU, Doupnik CA, Xu Y, Garvey J, Jasek MC, et al. 1997. Activation of heteromeric G protein-gated inward rectifier K$^+$ channels overexpressed by adenovirus gene transfer inhibits the excitability of hippocampal neurons. *Proc. Natl. Acad. Sci. USA* 94:7070–75

Eisele JL, Bertrand S, Galzi JL, Devillers-Thiery A, Changeux JP, Bertrand D. 1993. Chimaeric nicotinic-serotonergic receptor combines distinct ligand binding and channel specificities. *Nature* 366:479–83

Esposito MS, Capelli P, Arber S. 2014. Brainstem nucleus MdV mediates skilled forelimb motor tasks. *Nature* 508:351–56

Farrell MS, Pei Y, Wan Y, Yadav PN, Daigle TL, et al. 2013. A Gαs DREADD mouse for selective modulation of cAMP production in striatopallidal neurons. *Neuropsychopharmacology* 38:854–62

Farrell MS, Roth BL. 2013. Pharmacosynthetics: reimagining the pharmacogenetic approach. *Brain Res.* 1511:6–20

Ferguson SM, Eskenazi D, Ishikawa M, Wanat MJ, Phillips PE, et al. 2011. Transient neuronal inhibition reveals opposing roles of indirect and direct pathways in sensitization. *Nat. Neurosci.* 14:22–24

Frazier SJ, Cohen BN, Lester HA. 2013. An engineered glutamate-gated chloride (GluCl) channel for sensitive, consistent neuronal silencing by ivermectin. *J. Biol. Chem.* 288:21029–42

Galzi JL, Devillers-Thiery A, Hussy N, Bertrand S, Changeux JP, Bertrand D. 1992. Mutations in the channel domain of a neuronal nicotinic receptor convert ion selectivity from cationic to anionic. *Nature* 359:500–5

Gao ZG, Duong HT, Sonina T, Kim SK, Van Rompaey P, et al. 2006. Orthogonal activation of the reengineered A3 adenosine receptor (neoceptor) using tailored nucleoside agonists. *J. Med. Chem.* 49:2689–702

Garner AR, Rowland DC, Hwang SY, Baumgaertel K, Roth BL, et al. 2012. Generation of a synthetic memory trace. *Science* 335:1513–16

Grutter T, de Carvalho LP, Dufresne V, Taly A, Edelstein SJ, Changeux JP. 2005. Molecular tuning of fast gating in pentameric ligand-gated ion channels. *Proc. Natl. Acad. Sci. USA* 102:18207–12

Guettier JM, Gautam D, Scarselli M, de Azua IR, Li JH, et al. 2009. A chemical-genetic approach to study G protein regulation of β cell function in vivo. *Proc. Natl. Acad. Sci. USA* 106:19197–202

Guler AD, Rainwater A, Parker JG, Jones GL, Argilli E, et al. 2012. Transient activation of specific neurons in mice by selective expression of the capsaicin receptor. *Nat. Commun.* 3:746

Gunthorpe MJ, Lummis SC. 2001. Conversion of the ion selectivity of the 5-HT$_{3A}$ receptor from cationic to anionic reveals a conserved feature of the ligand-gated ion channel superfamily. *J. Biol. Chem.* 276:10977–83

Hama H, Kurokawa H, Kawano H, Ando R, Shimogori T, et al. 2011. Scale: a chemical approach for fluorescence imaging and reconstruction of transparent mouse brain. *Nat. Neurosci.* 14:1481–88

Haubensak W, Kunwar PS, Cai H, Ciocchi S, Wall NR, et al. 2010. Genetic dissection of an amygdala microcircuit that gates conditioned fear. *Nature* 468:270–76

Hibbs RE, Gouaux E. 2011. Principles of activation and permeation in an anion-selective Cys-loop receptor. *Nature* 474:54–60

Hikosaka O, Wurtz RH. 1985. Modification of saccadic eye movements by GABA-related substances. I. Effect of muscimol and bicuculline in monkey superior colliculus. *J. Neurophysiol.* 53:266–91

Hsiao EC, Boudignon BM, Chang WC, Bencsik M, Peng J, et al. 2008. Osteoblast expression of an engineered G$_s$-coupled receptor dramatically increases bone mass. *Proc. Natl. Acad. Sci. USA* 105:1209–14

Hsiao EC, Nguyen TD, Ng JK, Scott MJ, Chang WC, et al. 2011. Constitutive G$_s$ activation using a single-construct tetracycline-inducible expression system in embryonic stem cells and mice. *Stem Cell Res. Ther.* 2:11

Hwang YW, Miller DL. 1987. A mutation that alters the nucleotide specificity of elongation factor Tu, a GTP regulatory protein. *J. Biol. Chem.* 262:13081–85

Jacobson KA, Gao ZG, Chen A, Barak D, Kim SA, et al. 2001. Neoceptor concept based on molecular complementarity in GPCRs: a mutant adenosine A$_3$ receptor with selectively enhanced affinity for amine-modified nucleosides. *J. Med. Chem.* 44:4125–36

Jacobson KA, Ohno M, Duong HT, Kim SK, Tchilibon S, et al. 2005. A neoceptor approach to unraveling microscopic interactions between the human A$_{2A}$ adenosine receptor and its agonists. *Chem. Biol.* 12:237–47

Jain S, Ruiz de Azua I, Lu H, White MF, Guettier JM, Wess J. 2013. Chronic activation of a designer G$_q$-coupled receptor improves β cell function. *J. Clin. Investig.* 123:1750–62

Jann MW, Lam YW, Chang WH. 1994. Rapid formation of clozapine in guinea-pigs and man following clozapine-*N*-oxide administration. *Arch. Int. Pharmacodyn. Ther.* 328:243–50

Jansen M, Bali M, Akabas MH. 2008. Modular design of Cys-loop ligand-gated ion channels: functional 5-HT3 and GABA Rho1 receptors lacking the large cytoplasmic M3M4 loop. *J. Gen. Physiol.* 131:137–46

Karpova AY, Tervo DG, Gray NW, Svoboda K. 2005. Rapid and reversible chemical inactivation of synaptic transmission in genetically targeted neurons. *Neuron* 48:727–35

Ke MT, Fujimoto S, Imai T. 2013. SeeDB: a simple and morphology-preserving optical clearing agent for neuronal circuit reconstruction. *Nat. Neurosci.* 16:1154–61

Kelley SP, Dunlop JI, Kirkness EF, Lambert JJ, Peters JA. 2003. A cytoplasmic region determines single-channel conductance in 5-HT$_3$ receptors. *Nature* 424:321–24

Keramidas A, Moorhouse AJ, French CR, Schofield PR, Barry PH. 2000. M2 pore mutations convert the glycine receptor channel from being anion- to cation-selective. *Biophys. J.* 79:247–59

Kida S, Josselyn SA, Pena de Ortiz S, Kogan JH, Chevere I, et al. 2002. CREB required for the stability of new and reactivated fear memories. *Nat. Neurosci.* 5:348–55

Kokel D, Cheung CY, Mills R, Coutinho-Budd J, Huang L, et al. 2013. Photochemical activation of TRPA1 channels in neurons and animals. *Nat. Chem. Biol.* 9:257–63

Kong D, Tong Q, Ye C, Koda S, Fuller PM, et al. 2012. GABAergic RIP-Cre neurons in the arcuate nucleus selectively regulate energy expenditure. *Cell* 151:645–57

Kozorovitskiy Y, Saunders A, Johnson CA, Lowell BB, Sabatini BL. 2012. Recurrent network activity drives striatal synaptogenesis. *Nature* 485:646–50

Krashes M, Koda S, Ye CP, Rogan SC, Adams A, et al. 2011. Rapid, reversible activation of AgRP neurons drives feeding behavior. *J. Clin. Investig.* 121:1424–28

Krashes MJ, Shah BP, Madara JC, Olson DP, Strochlic DE, et al. 2014. An excitatory paraventricular nucleus to AgRP neuron circuit that drives hunger. *Nature* 507:238–42

Kristiansen K, Kroeze W, Willins D, Gelber E, Savage J, et al. 2000. A highly conserved aspartic acid (D155) anchors the terminal amine moiety of tryptamines and is involved in membrane targeting of the 5-HT$_{2A}$ serotonin receptor but does not participate in activation via a "salt-bridge disruption" mechanism. *J. Pharmacol. Exp. Ther.* 293:735–46

Lechner HA, Lein ES, Callaway EM. 2002. A genetic method for selective and quickly reversible silencing of mammalian neurons. *J. Neurosci.* 22:5287–90

Lerchner W, Xiao C, Nashmi R, Slimko EM, van Trigt L, et al. 2007. Reversible silencing of neuronal excitability in behaving mice by a genetically targeted, ivermectin-gated Cl- channel. *Neuron* 54:35–49

Li P, Slimko EM, Lester HA. 2002. Selective elimination of glutamate activation and introduction of fluorescent proteins into a *Caenorhabditis elegans* chloride channel. *FEBS Lett.* 528:77–82

Li X, Gutierrez DV, Hanson MG, Han J, Mark MD, et al. 2005. Fast noninvasive activation and inhibition of neural and network activity by vertebrate rhodopsin and green algae channelrhodopsin. *Proc. Natl. Acad. Sci. USA* 102:17816–21

Lin D, Boyle MP, Dollar P, Lee H, Lein ES, et al. 2011. Functional identification of an aggression locus in the mouse hypothalamus. *Nature* 470:221–26

Lin Q, Jiang F, Schultz PG, Gray NS. 2001. Design of allele-specific protein methyltransferase inhibitors. *J. Am. Chem. Soc.* 123:11608–13

Liu G, Gu B, He X-P, Joshi RB, Wackerle HD, et al. 2013. Transient inhibition of TrkB kinase after status epilepticus prevents development of temporal lobe epilepsy. *Neuron* 79:31–38

Loffler S, Korber J, Nubbemeyer U, Fehsel K. 2012. Comment on "Impaired respiratory and body temperature control upon acute serotonergic neuron inhibition." *Science* 337:646; author reply 646

Lovett-Barron M, Kaifosh P, Kheirbek MA, Danielson N, Zaremba JD, et al. 2014. Dendritic inhibition in the hippocampus supports fear learning. *Science* 343:857–63

Lovett-Barron M, Turi GF, Kaifosh P, Lee PH, Bolze F, et al. 2012. Regulation of neuronal input transformations by tunable dendritic inhibition. *Nat. Neurosci.* 15:423–30, S1–S3

Luttrell LM, Ferguson SS, Daaka Y, Miller WE, Maudsley S, et al. 1999. β-arrestin-dependent formation of β2 adrenergic receptor-Src protein kinase complexes. *Science* 283:655–61

Lynagh T, Lynch JW. 2010. An improved ivermectin-activated chloride channel receptor for inhibiting electrical activity in defined neuronal populations. *J. Biol. Chem.* 285:14890–97

Magnus CJ, Lee PH, Atasoy D, Su HH, Looger LL, Sternson SM. 2011. Chemical and genetic engineering of selective ion channel-ligand interactions. *Science* 333:1292–96

McKellar QA, Midgley DM, Galbraith EA, Scott EW, Bradley A. 1992. Clinical and pharmacological properties of ivermectin in rabbits and guinea pigs. *Vet. Rec.* 130:71–73

Mueller KL, Hoon MA, Erlenbach I, Chandrashekar J, Zuker CS, Ryba NJ. 2005. The receptors and coding logic for bitter taste. *Nature* 434:225–29

Nadeau H, McKinney S, Anderson DJ, Lester HA. 2000. ROMK1 (Kir1.1) causes apoptosis and chronic silencing of hippocampal neurons. *J. Neurophysiol.* 84:1062–75

Nakajima K, Wess J. 2012. Design and functional characterization of a novel, arrestin-biased designer G protein-coupled receptor. *Mol. Pharmacol.* 82:575–82

Nawaratne V, Leach K, Suratman N, Loiacono RE, Felder CC, et al. 2008. New insights into the function of M4 muscarinic acetylcholine receptors gained using a novel allosteric modulator and a DREADD (designer receptor exclusively activated by a designer drug). *Mol. Pharmacol.* 74:1119–31

Parnaudeau S, O'Neill PK, Bolkan SS, Ward RD, Abbas AI, et al. 2013. Inhibition of mediodorsal thalamus disrupts thalamofrontal connectivity and cognition. *Neuron* 77:1151–62

Pisharath H. 2007. Validation of nitroreductase, a prodrug-activating enzyme, mediated cell death in embryonic zebrafish (*Danio rerio*). *Comp. Med.* 57:241–46

Pisharath H, Rhee JM, Swanson MA, Leach SD, Parsons MJ. 2007. Targeted ablation of β cells in the embryonic zebrafish pancreas using *E. coli* nitroreductase. *Mech. Dev.* 124:218–29

Pliakas AM, Carlson RR, Neve RL, Konradi C, Nestler EJ, Carlezon WA Jr. 2001. Altered responsiveness to cocaine and increased immobility in the forced swim test associated with elevated cAMP response element-binding protein expression in nucleus accumbens. *J. Neurosci.* 21:7397–403

Ray RS, Corcoran AE, Brust RD, Kim JC, Richerson GB, et al. 2011. Impaired respiratory and body temperature control upon acute serotonergic neuron inhibition. *Science* 333:637–42

Ray RS, Corcoran AE, Brust RD, Soriano LP, Nattie EE, Dymecki SM. 2012. Egr2-neurons control the adult respiratory response to hypercapnia. *Brain Res.* 1511:115–25

Redfern CH, Coward P, Degtyarev MY, Lee EK, Kwa AT, et al. 1999. Conditional expression and signaling of a specifically designed G_i-coupled receptor in transgenic mice. *Nat. Biotechnol.* 17:165–69

Revah F, Bertrand D, Galzi JL, Devillers-Thiery A, Mulle C, et al. 1991. Mutations in the channel domain alter desensitization of a neuronal nicotinic receptor. *Nature* 353:846–49

Rogan SC, Roth BL. 2011. Remote control of neuronal signaling. *Pharmacol. Rev.* 63:291–315

Sasaki K, Suzuki M, Mieda M, Tsujino N, Roth B, Sakurai T. 2011. Pharmacogenetic modulation of orexin neurons alters sleep/wakefulness states in mice. *PLoS ONE* 6:e20360

Schmidt C, Li B, Bloodworth L, Erlenbach I, Zeng FY, Wess J. 2003. Random mutagenesis of the M3 muscarinic acetylcholine receptor expressed in yeast. Identification of point mutations that "silence" a constitutively active mutant M3 receptor and greatly impair receptor/G protein coupling. *J. Biol. Chem.* 278:30248–60

Slimko EM, McKinney S, Anderson DJ, Davidson N, Lester HA. 2002. Selective electrical silencing of mammalian neurons in vitro by the use of invertebrate ligand-gated chloride channels. *J. Neurosci.* 22:7373–79

Small KM, Brown KM, Forbes SL, Liggett SB. 2001. Modification of the β 2-adrenergic receptor to engineer a receptor-effector complex for gene therapy. *J. Biol. Chem.* 276:31596–601

Stanley BG, Ha LH, Spears LC, Dee MG 2nd. 1993. Lateral hypothalamic injections of glutamate, kainic acid, D,L-α-amino-3-hydroxy-5-methyl-isoxazole propionic acid or N-methyl-D-aspartic acid rapidly elicit intense transient eating in rats. *Brain Res.* 613:88–95

Strader CD, Gaffney T, Sugg EE, Candelore MR, Keys R, et al. 1991. Allele-specific activation of genetically engineered receptors. *J. Biol. Chem.* 266:5–8

Strobel SA. 1998. Ribozyme chemogenetics. *Biopolymers* 48:65–81

Sweger EJ, Casper KB, Scearce-Levie K, Conklin BR, McCarthy KD. 2007. Development of hydrocephalus in mice expressing the G_i-coupled GPCR Ro1 RASSL receptor in astrocytes. *J. Neurosci.* 27:2309–17

Temburni MK, Blitzblau RC, Jacob MH. 2000. Receptor targeting and heterogeneity at interneuronal nicotinic cholinergic synapses in vivo. *J. Physiol.* 525(Pt. 1):21–29

Tian L, Yang Y, Wysocki LM, Arnold AC, Hu A, et al. 2012. Selective esterase-ester pair for targeting small molecules with cellular specificity. *Proc. Natl. Acad. Sci. USA* 109:4756–61

Ultanir SK, Hertz NT, Li G, Ge WP, Burlingame AL, et al. 2012. Chemical genetic identification of NDR1/2 kinase substrates AAK1 and Rabin8 uncovers their roles in dendrite arborization and spine development. *Neuron* 73:1127–42

Vrontou S, Wong AM, Rau KK, Koerber HR, Anderson DJ. 2013. Genetic identification of C fibres that detect massage-like stroking of hairy skin in vivo. *Nature* 493:669–73

Walker DP, Wishka DG, Piotrowski DW, Jia S, Reitz SC, et al. 2006. Design, synthesis, structure-activity relationship, and in vivo activity of azabicyclic aryl amides as α7 nicotinic acetylcholine receptor agonists. *Bioorg. Med. Chem.* 14:8219–48

Wall NR, Wickersham IR, Cetin A, De La Parra M, Callaway EM. 2010. Monosynaptic circuit tracing in vivo through Cre-dependent targeting and complementation of modified rabies virus. *Proc. Natl. Acad. Sci. USA* 107:21848–53

Westkaemper R, Glennon R, Hyde E, Choudhary M, Khan N, Roth B. 1999. Engineering in a region of bulk tolerance into the 5-HT$_{2A}$ receptor. *Eur. J. Med. Chem.* 34:441–47

Wulff P, Goetz T, Leppa E, Linden AM, Renzi M, et al. 2007. From synapse to behavior: rapid modulation of defined neuronal types with engineered GABAA receptors. *Nat. Neurosci.* 10:923–29

Xu J, Zhu Y, Heinemann SF. 2006. Identification of sequence motifs that target neuronal nicotinic receptors to dendrites and axons. *J. Neurosci.* 26:9780–93

Yagi H, Tan W, Dillenburg-Pilla P, Armando S, Amornphimoltham P, et al. 2011. A synthetic biology approach reveals a CXCR4-G13-Rho signaling axis driving transendothelial migration of metastatic breast cancer cells. *Sci. Signal.* 4:ra60

Zemelman BV, Lee GA, Ng M, Miesenbock G. 2002. Selective photostimulation of genetically chARGed neurons. *Neuron* 33:15–22

Zemelman BV, Nesnas N, Lee GA, Miesenbock G. 2003. Photochemical gating of heterologous ion channels: remote control over genetically designated populations of neurons. *Proc. Natl. Acad. Sci. USA* 100:1352–57

Zhang F, Wang LP, Brauner M, Liewald JF, Kay K, et al. 2007. Multimodal fast optical interrogation of neural circuitry. *Nature* 446:633–39

Zhao GQ, Zhang Y, Hoon MA, Chandrashekar J, Erlenbach I, et al. 2003. The receptors for mammalian sweet and umami taste. *Cell* 115:255–66

Meta-Analysis in Human Neuroimaging: Computational Modeling of Large-Scale Databases

Peter T. Fox,[1,2,3,4] Jack L. Lancaster,[1,2] Angela R. Laird,[5] and Simon B. Eickhoff[6]

[1]Research Imaging Institute and [2]Department of Radiology, University of Texas Health Science Center at San Antonio, San Antonio, Texas 78229; email: Fox@uthscsa.edu, jlancaster@uthscsa.edu

[3]South Texas Veterans Health Care System, San Antonio, Texas 78229

[4]State Key Lab for Brain and Cognitive Sciences, University of Hong Kong, Pokfulam, Hong Kong

[5]Department of Physics, Florida International University, Miami, Florida 33199; email: angie.laird@fiu.edu

[6]Institute of Clinical Neuroscience and Medical Psychology, Heinrich Heine University of Düsseldorf, 40225 Düsseldorf, Germany; email: simon.b.eickhoff@gmail.com

Annu. Rev. Neurosci. 2014. 37:409–34

The *Annual Review of Neuroscience* is online at neuro.annualreviews.org

This article's doi: 10.1146/annurev-neuro-062012-170320

Keywords

human brain mapping, activation likelihood estimation, ALE, magnetic resonance imaging, MRI, fMRI

Abstract

Spatial normalization—applying standardized coordinates as anatomical addresses within a reference space—was introduced to human neuroimaging research nearly 30 years ago. Over these three decades, an impressive series of methodological advances have adopted, extended, and popularized this standard. Collectively, this work has generated a methodologically coherent literature of unprecedented rigor, size, and scope. Large-scale online databases have compiled these observations and their associated meta-data, stimulating the development of meta-analytic methods to exploit this expanding corpus. Coordinate-based meta-analytic methods have emerged and evolved in rigor and utility. Early methods computed cross-study consensus, in a manner roughly comparable to traditional (nonimaging) meta-analysis. Recent advances now compute coactivation-based connectivity, connectivity-based functional parcellation, and complex network models powered from data sets representing tens of thousands of subjects. Meta-analyses of human neuroimaging data in large-scale databases now stand at the forefront of computational neurobiology.

Contents

Meta-analysis:
retrospective combination of previously reported results to better estimate the reliability of those results

Coordinate-based meta-analysis (CBMA): meta-analysis method(s) developed specifically for use with functional and structural neuroimaging data reported within a standardized coordinate space

INTRODUCTION

Meta-analysis is most generally defined as the post hoc combination of numerical results from prior, independent studies. The original (and still most common) use of meta-analysis was to pool subsignificant effects from several small studies to determine which effects would achieve significance in larger samples (Pearson 1904). In particular, the method was developed to predict which adverse events were rare but real drug side effects and which were unrelated, random events. Neuroimaging meta-analysis, in contrast, pools statistically significant results to further improve predictive power, to build analytic tools and models, and to detect emergent properties of neural systems through large-scale data mining and computational modeling.

Coordinate-based meta-analysis (CBMA) methods collectively comprise a fairly recent, extremely powerful, rapidly evolving family of methods for mining and synthesizing the human neuroscience imaging literature. These methods rely on the widespread adoption by the neuroimaging research community of whole-brain analysis methods that reference a coordinate space, a unique and important accomplishment in its own right. In the earliest applications of CBMA, the primary objective was to report consensus locations and spatial probability distributions for specific functional areas and for widely used task-activation paradigms. The field has rapidly moved beyond this objective, now creating meta-analytic, synthetic images; performing experiments by

SPATIAL NORMALIZATION

Spatial normalization is the process of transforming a brain image from its natural size and shape (native space) into a standardized form that references a 3-D template image and coordinate space (template space). This allows brain images to be averaged across individuals and compared across groups.

contrasting these synthetic images in new ways; modeling the network properties of meta-analytic images; modeling interregional functional connectivity from very-large-scale meta-analyses; extracting highly plausible cortical parcellation schemes based on coactivation spatial probabilities; and, most recently, developing consensus-based functional attributions for regions and networks. In a very literal sense, meta-analyses are providing a steadily expanding, progressively enriched, potentially limitless consensus statement regarding the brain's structural and functional organization. That is, meta-analysis is creating the "collective mind." A truly remarkable feature of these studies is the emergent properties they disclose. Studies that reported solely activation locations for a limited number of specific paradigms are being mined to model interregional connectivity and connectivity-based regional parcellation. The purpose of this review is to introduce the reader to this high-impact, rapidly evolving, highly exciting area of research.

COMMUNITY STANDARDS IN NEUROIMAGING

The neuroimaging community enjoys the enviable status of having developed analytic and reporting standards that not only provide excellent per-study sensitivity, but also enable a growing repertoire of spatial meta-analytic methods. The core analytic standards of the field are spatial normalization (coordinate-based anatomy) (see sidebar, Spatial Normalization), statistical parametric imaging, and local-maxima extraction.

Spatial Normalization

Spatial normalization is the most fundamental analytic standard underlying neuroimaging meta-analysis. The first human atlas that referenced a standardized, stereotaxic coordinate space was created by Jean Talairach, a French neurosurgeon and pioneer of quantitative human brain mapping. Talairach published a series of stereotactic atlases, the first of which (Talairach et al. 1967) reported functional locations (from electrical cortical stimulation) and structural boundaries (from pneumoencephalography) in standardized coordinates. Talairach was also the first person to describe the human brain in terms of functional and structural "probability distributions" (Talairach et al. 1967). The most widely used template is the Talairach & Tournoux (1988) atlas, produced specifically for image-based registration. The origin of Talairach's x-y-z coordinate space is the anterior commissure through which three orthogonal planes are oriented. The principal axis is the line connecting the anterior and posterior commissures (AC–PC line: the y axis). The remaining two axes are the x (right–left) and the z (superior–inferior).

The first algorithm for spatial normalization of tomographic brain images was a nine-parameter affine transformation published by Fox et al. (1985), with the goal of "facilitating direct comparison of experimental results from different laboratories" (p. 149), i.e., in anticipation of CBMA. Shortly thereafter, Fox and colleagues introduced functional-image averaging, which applied spatial standardization to combine images across subjects, decreasing image noise and improving the signal-to-noise ratio (Fox et al. 1988, Fox & Mintun 1989). Since that time, methods have

Coordinate space: any of a number of reference spaces defined by an anatomical template and used to analyze and report neuroimaging data, the two most widely used being Talairach Space and MNI Space

Stereotactic coordinates: three-dimension anatomical addresses (x, y, z) defined relative to a reference space and used to analyze and report functional and structural brain-image-derived observations

steadily improved, expanding to nonaffine transformations including deformation field methods, which compute a unique warping vector for each image voxel (Toga 1998). Averaging functional images in standardized space is now the norm in human neuroscience imaging; tens of thousands of studies have been reported according to this convention (Fox 1995b).

Statistical Parametric Maps

Statistical parametric maps (SPMs) or statistical parametric images (SPIs) are 3-D arrays of group-wise statistical parameters computed from primary ("raw"), per-subject neuroimaging data. The original and still most widely reported type of SPM is computed from functional imaging data. SPMs can be generated from various types of functional imaging data, including $H_2^{15}O$ positron emission tomography (PET) and functional magnetic resonance imaging (fMRI). PET, the original SPM data source, is now rarely used, having been replaced by fMRI.

To generate functional SPMs, functional images are acquired under contrasting behavioral conditions that induce different brain activity patterns. Raw images are converted point-by-point (voxel-by-voxel) into images of statistical parameters that express the strength and consistency of the task-induced changes relative to an error term. Prior to the introduction of SPIs, analysis of functional images relied on regions of interest (ROIs), which sampled the data space in a predefined manner. Brain activations, however, can be quite discrete. Small errors in ROI placement can make large differences in the magnitude and statistical significance of observed effects. SPIs process the entire data matrix, allowing all task-induced changes to be detected. Thus, SPIs increase both the power and the objectivity of functional image analysis. Statistical parametric imaging was introduced to human brain mapping by Fox and colleagues (Fox et al. 1988, Fox & Mintun 1989). Friston and colleagues (1991, 1995) extended this image-analysis strategy to include multi-condition contrasts and computation of a wide range of statistical parameters, coining the term statistical parametric mapping. Correction for multiple comparisons (i.e., for the number of voxels tested) using Gaussian random field theory was a crucial additional advance. The SPM software packages, distributed by the Wellcome Trust Center for NeuroImaging (**http://www.fil.ion.ucl.ac.uk**), have been enormously influential and are among the most widely used worldwide.

Voxel-based morphometry (VBM) is the application of this same basic voxel-wise SPM strategy to high-resolution structural images to detect between-group anatomical differences (Ashburner & Friston 2000). Because the contrast computed (subtraction performed) is between-subjects, it does not have the advantage of within-subject task-control contrast (to eliminate background structure) permitted by functional SPM/SPI analyses. Without this within-subject control (used to eliminate background signal), VBM is intrinsically noisy and generally requires larger sample sizes to obtain significant effects and has a high false-positive rate. These limitations, however, make VBM even more likely to derive benefit from meta-analysis.

Local Maxima

As detected by SPM/SPI analysis, task-induced changes in brain activity or between-group differences in brain anatomy most often take the form of foci that are strongest (biggest effect size and most significant) at the center and fall off in an approximately Gaussian manner. A concise means of describing the location of an activated volume was necessary for analysis and publication. Fox and colleagues (1986, 1987) proposed that an area of activation could be viewed as a local maximum, the centroid of which could be estimated using a 3-D center-of-mass (COM) algorithm. Although it was new to brain imaging, this strategy had been known to astronomers for centuries as "vernier acuity" and to vision scientists for decades as "hyperacuity" (Fox et al. 1986). As applied to PET images with a spatial resolution (full-width at half maximum) of 1.8 cm, the

---

Voxel: a VO-lume pi-XEL, or data point within a 3-D image array. Statistical analyses of 3-D image volumes can be performed at corresponding voxels across data sets, i.e., in a voxel-wise fashion

Statistical parametric map/image (SPM/SPI): transforming brain images from raw, individual (per-subject) data sets in native space to 3-D images of statistical parameters (Z score, T scores, F-values, R-values, p-values, etc.) in a standard coordinate space

Voxel-based morphometry (VBM): a voxel-wise analysis method for detecting between-group (e.g., patients versus controls) differences in brain anatomy within a standardized coordinate space

spatial precision of hyperacute response localization was shown to be submillimeter (Mintun et al. 1989). This somewhat startling spatial precision argued strongly that response coordinates are a very valuable parameter to report and, subsequently, to compile in databases and to meta-analyze.

Current convention is to publish the locations of activation sites as the x-y-z coordinates of the COM or (less accurately) the peak voxel of each local maximum (or minimum). Additional data provided typically include the peak value of the statistical parameter forming the cluster (e.g., peak z value), the p-value of the peak voxel, the extent of the cluster (mm^3) when thresholded to some p or z level, and various anatomical (e.g., hemisphere, lobe, gyrus) or functional (e.g., primary motor cortex) descriptors. The most widely used anatomical descriptors are hierarchical, volumetric labels that reference the 1988 Talairach atlas (Lancaster et al. 2000). An absolute requirement for inclusion of data in CBMA is that results are reported as COM (or peak voxel) addresses of local maxima detected in spatially standardized SPM/SPIs.

Data Volume

Papers following the above-described standards began appearing in the mid-1980s. Since then, the rate of publication has steadily risen. In 2005, we estimated the standards-compliant functional imaging literature to be no fewer than 4,000 articles ($\sim$16,000 experiments) with $\sim$500 new articles (2,000 experiments) published per year (Fox et al. 2005a). As of 2007, Derrfuss & Mar (2009) estimated the conforming functional literature to be $\sim$5,800 papers, with $\sim$1,000 new conforming papers being published each year. The conforming literature now appears to be growing at a rate of more than 2,000 papers per year and likely exceeds 20,000 papers. With such a large volume of well-standardized data, the field of human neuroimaging provides uniquely fertile ground for meta-analysis.

Template Troubles

CBMA relies on the comparability of coordinates across studies. Satisfying this requirement does not mean that all studies must use the same brain template or the same normalization algorithm. Rather, it means that transforms capable of accurately converting between templates must be available, ideally validated prior to release of a new template. To correct some shortcomings of the Talairach & Tournoux (1988) atlas as a warping template, investigators at the Montreal Neurologic Institute (MNI) released a series of structural-MRI-derived templates created as averages of multiple subjects, seeking to create a template representative of a group rather than of an individual (reviewed in Evans et al. 2012). Unfortunately, the process used to create the template altered the origin and orientation (relative to the AC–PC standard) and expanded the brain size beyond the normal range, deviations which were not corrected prior to release. Compounding the problem, the discrepancies were not reported when the templates were released. Consequently, coordinates from studies using MNI templates (in MNI Space) did not correspond to coordinates or anatomical labels referenced to the Talairach space, although users (and reviewers and editors) were largely unaware of this discrepancy. When those in the field became aware of the problem, unvalidated (and ineffective) corrective transforms were made available online and variably applied, worsening the field's collective state of coordinate confusion. For a field at the forefront of computational biology, these were extremely unfortunate missteps. Eventually, Lancaster et al. (2007) quantified and corrected the discrepancies. The impact of these discrepancies on meta-analysis was quantified by Laird et al. (2010), specifically endorsing the use of Lancaster's transforms to move from MNI Space to Talairach Space or vice versa. Despite these efforts, the negative impact of coordinate confusion on database curation and meta-analysis is not fully resolved.

DATABASE DEVELOPMENT

Neuroimaging databases can be classified by the level of processing the data sets have received and the number of subjects per data set. Primary data set repositories contain per-subject images that typically have been processed to remove artifacts but not to compute statistical effects. At the other extreme are coordinate-data repositories, which share reduced data (tables of coordinates) extracted from SPMs computed from groups of subjects. An intermediate option is to provide group-wise SPMs as image volumes, i.e., without reduction to COM coordinates. The advantages and disadvantages of managing and utilizing each data type fall beyond the scope of this review, which is focused on meta-analyses of reported coordinates.

Coordinate Data Databases

BrainMap® (Fox et al. 1994; Fox & Lancaster 1994, 2002) was the first online database of neuroimaging results. BrainMap was initially created as a personal compilation (a spreadsheet) of brain-activation results in standardized coordinates from published and pilot (unpublished) studies carried out by Fox and colleagues using $H_2{}^{15}O$ PET in the laboratory of Marcus Raichle and Michel Ter-Pogossian. Funded in 1988 by the J.S. McDonnell Foundation, BrainMap was unveiled in November 1992, at the first of seven BrainMap workshops (Gibbons 1992). From the outset, the BrainMap strategy aimed to provide coordinate-based results linked to experimental meta-data—emphasizing experimental designs and behavioral conditions—to enable coordinate-based meta-analysis (Fox & Lancaster 1994). One of the core conceptual developments was the BrainMap coding scheme, a taxonomy of experimental design that provides meta-data descriptors intended not simply for retrieval of like studies but also for data-driven inferences concerning the functional properties of specific brain regions and networks (Fox et al. 1994). The BrainMap taxonomy has continued to evolve as the field has matured, with various conceptual validations (Fox et al. 2005b). The BrainMap taxonomy has also helped inform comparable taxonomies for primary, raw data repositories (Turner & Laird 2011).

At BrainMap's inception, the developers' intent was for it to be a community-curated database. It quickly became clear that community curation was antithetical to quality control. Thus, the model evolved to encourage submission of data sets coded in BrainMap terms using software available at **http://www.BrainMap.org**, but all papers would be reviewed and edited by the BrainMap development team. Community members most active in submitting papers for entry are those in the process of performing a meta-analysis, which is symbiotic because it results in a peer-reviewed publication citing the resource (Laird et al. 2005b). At the time of this writing, BrainMap contained 11,103 functional-imaging experiments (88,000 coordinates) from 2,336 peer-reviewed publications and contained 2,444 structural-imaging experiments (16,311 coordinates) from 796 publications.

Derrfuss & Mar (2009) argued that a comprehensive database of the neuroimaging literature was a highly desirable objective and opined that the best way to achieve this objective was in a commercial, subscription-based publication format. Although both points are likely correct, neither has been realized. As an alternative, Neurosynth was created as an online, open-source, uncurated resource that automatically extracts coordinates from neuroimaging articles for meta-analysis (Yarkoni et al. 2011). Rather than use manual coding (à lá BrainMap), Neurosynth uses text-mining and machine-learning techniques to provide frequency-based weightings for behavioral and cognitive terms appearing in the coordinate-containing articles. These weightings are used to drive meta-analyses that can be performed directly from the Neurosynth web-interface, an offline computation in the BrainMap model. At time of writing, Neurosynth contained 5,809 studies (see Related Resources).

Primary Data Databases

Progress toward open sharing of primary (per-subject) experimental neuroimaging data (a community objective dating back to the late 1980s) has been slow, limited by the variability among sites in instrumentation and data-acquisition parameters, by the sheer size of the data sets, by patient confidentiality issues, and by the desire of investigators to protect their invested effort (and grant money) by maintaining access control. Despite these barriers, large, primary data sets are becoming steadily more accessible. The International Consortium for Brain Mapping (ICBM) was a pioneer in this effort (Mazziotta et al. 1995); ICBM data are still being actively downloaded. The Alzheimer's Disease Neuroimaging Initiative (ADNI; Butcher 2007) is a similarly managed, disease-specific initiative. A common data-sharing model is to provide open access to online descriptions of available data, with comprehensive access being managed by an oversight committee.

COORDINATE-BASED META-ANALYSIS

As indicated above, the traditional use of meta-analysis was to achieve statistical significance across studies for effects that failed to achieve significance in individual studies. In neuroimaging, however, the primary use of meta-analysis has been to synthesize the published literature (of significant results) to compute consensus effects and, thereby, place constraints on the interpretation, design, and analysis of subsequent studies (Fox et al. 1998). The first neuroimaging meta-analyses were reported in the context of primary data. That is, coordinates from extant reports were tabulated and plotted to constrain interpretations of new, primary data (Frith et al. 1991). Shortly thereafter, stand-alone neuroimaging meta-analyses began to appear (Tulving et al. 1994, Fox 1995a, Picard & Strick 1996), serving as quantitative reviews and for hypothesis generation. Although the first neuroimaging meta-analyses were statistically informal, this soon changed.

Functional Volumes Modeling

Anticipating the long-term impact of quantitative meta-analysis, Fox, Lancaster, and colleagues launched the BrainMap database in advance of (but hoping to stimulate) the invention of suitable meta-analytic algorithms. The first step in the evolution of quantitative CBMA tools was taken by Paus (1996), who computed and interpreted means and standard deviations of the x-y-z addresses in a review of studies of the frontal eye fields. This initiative was extended by correcting raw estimates of spatial location and variance for sample size to create scalable models of location probabilities [functional volumes models (FVM)] and to suggest uses of such models for data analysis (Fox et al. 1997, 1999, 2001). A specific limitation of FVM was that users needed to select responses that should be grouped, based on expert knowledge of the tasks performed and the region being studied. Although the models produced could be used by nonexperts, creating the models required considerable expertise.

Activation Likelihood Estimation

Activation likelihood estimation (ALE) (Turkeltaub et al. 2002) (**Figure 1**) and related algorithms (Chein et al. 2002, Wager et al. 2003) moved CBMA a quantum leap forward. ALE input data are activation-location coordinates from conceptually related studies (e.g., all Stroop tasks). ALE models the uncertainty in localization of activation foci using Gaussian probability density distributions. The voxel-wise union of these distributions yields the ALE value, a voxel-wise estimate of the likelihood of activation, given the input data. As with FVM, a great advantage of ALE

ALE: activation likelihood estimation (for functional meta-analyses) or anatomical likelihood estimation (for anatomical meta-analyses)

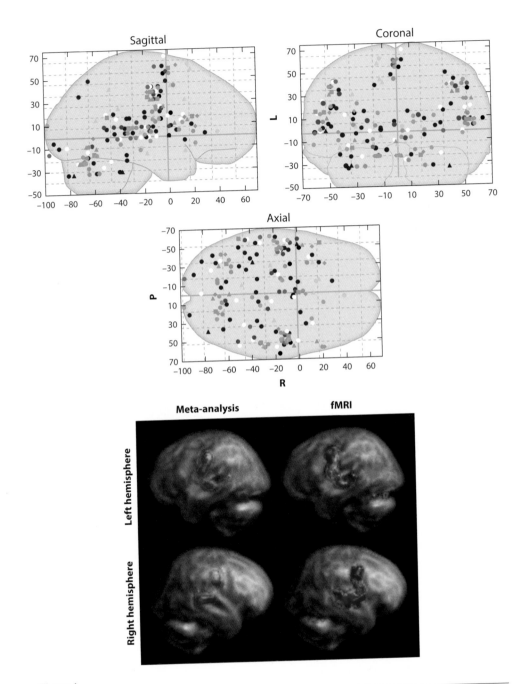

Figure 1

ALE meta-analysis. Activation likelihood estimation (ALE) meta-analysis is illustrated. The top panels show spatial coordinates for single-word reading tasks reported by 11 PET studies (16 experiments; 117 subjects). The bottom panel shows a meta-analytic image computed using the ALE algorithm (*left column*) and an fMRI study performed using the same task (single-word reading). Images are reproduced with permission from Turkeltaub and colleagues (2002), the original ALE publication.

is that the tables of coordinates routinely reported by neuroimaging studies are its input data: Raw data are not required. Unlike FVM, however, ALE requires no user selection of comparable coordinates for modeling; rather, once a set of experiments (e.g., a group of experiments using a similar paradigm) is selected for meta-analysis, the entire set of reported coordinates is used, thereby greatly increasing the reproducibility and objectivity of the analysis.

In the original implementation of ALE, the investigators acknowledged several limitations. For example, while applying the false discovery rate (FDR) method to compute voxel-wise significance, Turkeltaub used a fixed-effects analysis that did not correct for multiple comparisons; the size of the modeled Gaussian distribution was rationalized on the basis of the spatial resolution of the input images, rather than on a formal estimate of spatial uncertainty; a method for comparing two ALE maps was lacking; and there was no correction for the variable number of activations reported per experiment or the number of experiments per paper. Many of these limitations have subsequently been addressed by Turkeltaub and others. Laird et al. (2005a) provided an FDR correction for multiple comparisons and a method for ALE–ALE statistical contrasts. Eickhoff et al. (2009) introduced empirical estimates of between-subject and between-template spatial variability (a modification of the FVM spatial probability model) in place of user-selected Gaussian filtering. In addition, Eickhoff et al. (2009) modified the permutation test for above-chance clustering between experiments in an anatomically constrained space (gray-matter only), a transition from fixed-effects to random-effects inference. Turkeltaub et al. (2012) added corrections for the variable numbers of foci per experiment and experiments per paper to prevent undue weighting of ALE maps by individual experiments (e.g., with large numbers of foci) or individual papers (e.g., with multiple similar experiments). Eickhoff et al. (2012) provided an explicit solution for determining statistical significance rather than relying on FDR. Each of these improvements increased the statistical rigor and specificity of ALE.

Since its introduction, ALE has been applied to many aspects of normal brain function (Decety & Lamm 2007, Costafreda et al. 2008, Spreng et al. 2009). It has also been applied to functional and structural data in numerous disorders, including autism spectrum disorders (Duerden et al. 2012, Nickl-Jockschat et al. 2012), schizophrenia (Glahn et al. 2005, 2008; Ellison-Wright et al. 2008; Ellison-Wright & Bullmore 2009; Minzenberg et al. 2009; Ragland et al. 2009), epilepsy (Barron et al. 2012), Huntington's disease (Dogan et al. 2012, Lambrecq et al. 2013), obsessive-compulsive disorder (Menzies et al. 2008), depression (Fitzgerald et al. 2008), and developmental stuttering (Brown et al. 2005). The most interesting ALE applications do not merely merge previous results, but instead include highly previously ignored regions, resolve conflicting views, validate new paradigms, and generate new hypotheses for experimental testing. A more extensive list of ALE studies and algorithms is available at **http://www.brainmap.org/pubs**.

Activation likelihood estimation: a widely used family of algorithms for CBMA of functional and structural neuroimaging observations

WITHIN-PARADIGM NETWORKS: CLIQUE ANALYSIS

Graph theory network modeling constructs—framing models as nodes connected by edges—have proven extremely well adapted for primary neuroimaging data sets of various types (Lohmann & Bohn 2002, Bullmore & Sporns 2009; see also sidebar, Graph Theory and Neuroimaging). For meta-analytic data sets, they have proven no less apt. In an early foray in this domain, Neumann applied replicator dynamics—a network discovery technique from theoretical biology based on principles of natural selection—to model the network properties of the Stroop task (Neumann et al. 2005). Noting that Neumann's method was limited to a single, dominant clique, Lancaster extended the algorithm to multiple subnets, adapting the Jaccard similarity measure for meta-analytic use, and quantitatively contrasted the two approaches (Lancaster et al. 2005). Note that for both these within-paradigm network-modeling approaches and the more advanced methods that followed (discussed below), graph "edges" are emergent properties, computed as coactivation patterns across studies with no comparable parameter being reported by the included studies.

BETWEEN-PARADIGM NETWORKS: META-ANALYTIC CONNECTIVITY MODELING

Meta-analytic approaches to assessing interregional connectivity are conceptually similar to functional non-meta-analytic approaches (e.g., resting state fMRI; Biswal et al. 1995) because they use temporal covariations in regional activation to detect connectivity. Whereas in a typical fMRI study the unit of time is the second, in the meta-analytic approach the unit of time is the study. Regions in which activations co-occur across studies (i.e., regions that are mutually predictive) are functionally connected; regions that do not co-occur are not connected. Higher probability of co-occurrence should reflect greater strength of functional connectivity.

Region to Region

The concept of a coactivation-based, meta-analytic connectivity mapping (MACM) was introduced by Koski & Paus (2000). To identify frontal lobe projections to the anterior cingulate gyrus, they manually collected and examined data from 107 studies, reporting differential connection patterns within different subregions of the anterior cingulate. Although they regarded their new approach as intrinsically plausible, Koski & Paus acknowledged that it lacked formal validation. In view of this shortcoming, they recommended replications using larger data sets, the development of statistically more sophisticated approaches, and validation of the approach against alternative connectivity measures, all of which eventually came to pass. Note that this application was region-to-region because it limited its scope to connections between frontal lobe and anterior cingulate gyrus.

Region to Whole Brain

The first region-to-whole-brain coactivation meta-analysis was reported by Postuma & Dagher (2006). Having identified 126 peer-reviewed, whole-brain studies with activations in caudate or putamen, the authors computed the first whole-brain, meta-analytic functional connectivity images. In these images, observed coactivation patterns were "consistent with the concept of spatially segregated cortico-striatal connections as predicted by previous anatomical labeling studies in nonhuman primates" (p. 1513). As with Koski & Paus (2000), no validation other than plausibility was offered.

The region-to-whole-brain analysis strategy was adopted and extended by Robinson et al. (2010). Using the Harvard/Oxford atlas to define the amygdala, the entire BrainMap database as the data source, and ALE to compute co-occurrence spatial probabilities, Robinson mapped the coactivation profile of the left and right amygdala. At that time, the BrainMap database contained 170 and 156 experiments for the left and right amygdala, respectively. To validate the approach she termed "meta-analytic connectivity modeling" (MACM), Robinson compared the amygdala MACM results with those obtained by various tract-tracing methods in rhesus monkeys, as reported in the CoCoMac database, finding startlingly good correspondence. Robinson et al. (2012) applied a similar strategy to the caudate nuclei, adding functional filtering using the BrainMap behavioral domain meta-data (**Figure 2**). Projection patterns were confirmed by diffusion tensor imaging (DTI) probabilistic tractography. Following Robinson's validation of MACM by comparing it with primate connectivity and DTI tractography, multiple other validations have been published, of which three are briefly presented here.

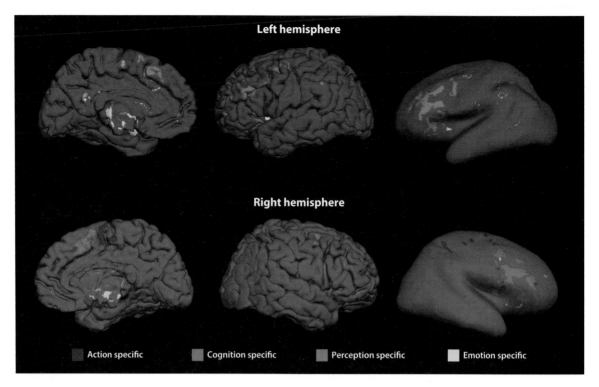

Left hemisphere

Right hemisphere

Action specific   Cognition specific   Perception specific   Emotion specific

Figure 2

Meta-analytic connectivity modeling (MACM) with behavioral domain filtering. The images illustrate the regional and behavioral specificity of connections of the caudate nucleus. Connectivity was computed as coactivation frequency with a seed region (caudate nucleus), sampling across the entire BrainMap database. Activation likelihood estimation (ALE) was used to compute statistical significance of co-occurrences. Behavioral filtering used the top tier of the BrainMap behavioral domain hierarchy. The projection patterns closely matched those established in the primate literature and were confirmed by diffusion tensor imaging (DTI) tractography. Images are reproduced from Robinson and colleagues (2012), the original report of behaviorally filtered MACM connectivity mapping, with permission.

MACM-derived connectivity patterns were compared with those obtained using DTI tractography by Eickhoff and colleagues (2010). Investigators compared connectivity patterns for two subdivisions of the human parietal operculum previously established using postmortem cytoarchitectonics (OP1 and OP4; Eickhoff et al. 2010). For MACM, the opercular regions of interest jointly extracted 245 experiments from the BrainMap database. For DTI, 18 healthy volunteers were studied. Comparison between techniques showed close (but not perfect) correspondence (**Figure 3**). Also, DTI tractography will provide connectivity limited to first-order (direct) connections, whereas MACM—showing all co-occurrences—would be expected to yield both direct and indirect connections. Furthermore, DTI will be intrinsically biased toward heavily myelinated connections, whereas MACM should preclude this bias.

For the nucleus accumbens, Cauda et al. (2011) compared MACM-derived connectivity with the resting-state fMRI connectivity. For the nucleus accumbens region of interest, BrainMap provided 57 experiments, a relatively small input data set. For resting-state fMRI, 17 healthy subjects were studied. Despite the limited amount of BrainMap data utilized, the MACM proved robust (**Figure 3**), as did the resting-state connectivity map. Overall, the two techniques converged, with resting-state connectivity showing somewhat greater sensitivity than did MACM. In this context, it is important to note that the sensitivity of MACM is strongly influenced by the size of the seed region and the volume of data in the BrainMap database. As the database becomes more densely populated, MACM will become more sensitive and allow progressively finer anatomical connectivity parcellations.

Narayana et al. (2012) compared MACM-derived connectivity with cortical-stimulation-based connectivity, using concurrent transcranial magnetic stimulation (TMS) and $H_2{}^{15}O$ PET to map remote projections of the supplementary motor area (SMA). For the SMA seed region, BrainMap provided 266 experiments from 187 papers. As with prior validation, the two techniques converged nicely (**Figure 3**). It should be noted that Robinson, Eickhoff, Cauda, and Narayana all used BrainMap's behavioral domain and/or paradigm class meta-data (discussed further below) to interpret the functional roles of both the seed regions and their connections.

Coactivation-Based Parcellation

Coactivation-based parcellation (CBP) is an extremely exciting extension of the region-to-whole-brain meta-analytic connectivity-mapping approach. Connectivity-based parcellation of structural (DTI) data has provided close correspondence between structurally and functionally defined borders, using the boundary between SMA and pre-SMA as a demonstration case (Johannes-Berg et al. 2004). Eickhoff and colleagues (2011) applied connectivity-based parcellation to BrainMap data for the same brain regions (SMA and pre-SMA) and obtained virtually identical borders (Eickhoff et al. 2011). Providing further validation, Bzdok et al. (2012) applied connectivity-based parcellation to the amygdala, demonstrating close correspondence to previously defined cytoarchitectonic borders (Amunts et al. 2005) (**Figure 4**). This technique has subsequently been applied to differentiate "what" and "where" pathways in parietal cortex (Rottschy et al. 2012) and to classify subregions of dorsolateral prefrontal cortex (DLPFC; Cieslik et al. 2013), the medial prefrontal cortex (Bzdok et al. 2013), the cingulate cortex (Torta et al. 2013), and Broca's area (Brodmann area 44; Clos et al. 2013). In each of these applications, concurrent resting-state functional connectivity (using fMRI) corroborated the BrainMap-derived connectivity patterns.

INDEPENDENT COMPONENTS ANALYSIS

In the two preceding sections, we have reviewed methods that drew connectivity inferences between two seed regions (less complex) and between a seed region and the whole brain (more

---

Figure 3

Meta-analytic connectivity mapping (MACM) validations. The images illustrate three independent validations of MACM-derived connectivity patterns as forming one two-panel row. In each row, the left panel compares the strength of an MACM-derived projection with that of an alternative connectivity-mapping method. The right panels show the MACMs derived applying three different seed regions to the BrainMap database. The top row used DTI tractography as the validation methodology, seeding two regions within the parietal operculum (OP1, OP4; Eickhoff et al. 2010). The middle row used resting-state fMRI as the validation methodology, seeding nucleus accumbens (Cauda et al. 2011). The bottom row used concurrent transcranial magnetic stimulation (TMS)/PET as the validation methodology, seeding the supplementary motor area (SMA) (Narayana et al. 2012). Other abbreviations: aIPC, anterior inferior parietal cortex; aIPS, anterior intraparietal sulcus; aSPC, anterior superior parietal cortex; Broca, Broca's area; L-M1, left primary motor cortex; L-PCG/BA2/3, left postcentral gyrus, Brodmann areas 2 and 3; L-PMD, left dorsal premotor cortex; L-SPL/PreC, left superior parietal lobule/precuneus; M1, primary motor cortex; PCG, postcentral gyrus; pIPC, posterior inferior parietal cortex; PMC, premotor cortex; pSPC, posterior superior parietal cortex; R-M1, right primary motor cortex; R-PCG, right postcentral gyrus; R-PMD, right dorsal premotor cortex; R-STG (BA 4), right superior temporal gyrus, Brodmann area 41; VL/VA, ventrolateral nuclei/ventrolateral anterior nuclei; VPL/VPI, ventroposterior lateral and inferior nuclei. Images are reproduced with permission from each manuscript.

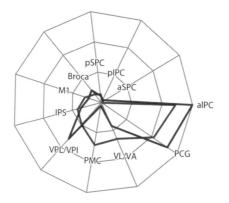

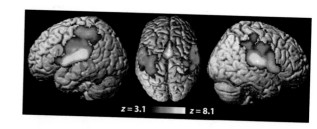

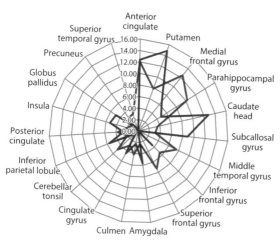

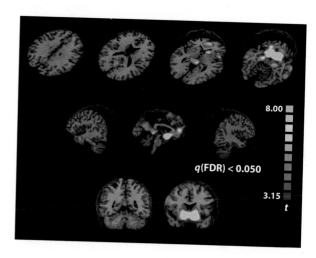

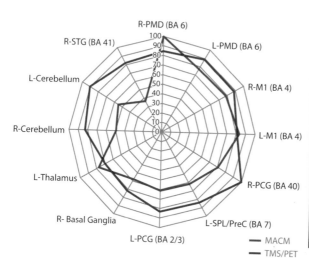

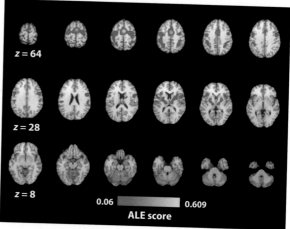

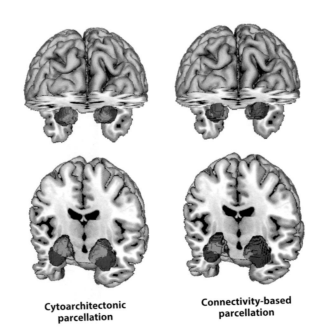

**Cytoarchitectonic
parcellation**

**Connectivity-based
parcellation**

Figure 4

Co-activation-based parcellation. Connectivity-based parcellation of the amygdala as derived using the BrainMap database (*right*) proved highly similar to that observed (in a separate postmortem sample) using cytoarchitecture (*left*). Laterobasal nuclei group (*blue*); centromedial nuclei group (*red*); superficial nuclei group (*green*). Images were reproduced with permission from Bzdok et al. (2012).

complex). In the next two sections we review methods that draw connectivity inferences by comparing coactivations between every brain voxel with every other voxel (all-to-all). The two most general classes of network-discovery methods addressed are independent components analysis (ICA) and graph theoretical modeling. Both approaches have been applied to the entire BrainMap database (or large subsets thereof), i.e., across behaviorally inhomogeneous paradigms.

Toro et al. (2008) used the BrainMap database to generate a comprehensive atlas of the brain's functional connectivity. At the time, BrainMap included 3,402 experiments (conditional contrasts) reporting a total of 27,909 activated locations. For each experiment, a binary, per-study activation volume was generated. From these, the co-occurrence pattern likelihood was computed between all voxels using likelihood ratios. This generated 45,000 unique coactivation maps (one for each 4 mm^3 voxel in the brain). Reproducibility of the coactivation map was assessed by estimating the similarity between pairs of partial coactivation maps that used disjoint random subsamples of experiments for group sizes of 500, 1,000, 1,500, 2,000, 2,500, and 3,000 experiments. The correlation between maps was significant and increased asymptotically with the number of experiments and was strong with even the 500-experiment group. Thus, the coactivation maps did not depend on a particular choice of experiments, and a robust structure in the meta-analytic functional connectivity can be recovered even with a moderate number of studies.

Intrigued by Toro's observations, Smith et al. (2009) took this strategy a step further and applied ICA to the entire BrainMap data volume. ICA has been widely used to demonstrate intrinsic connectivity networks in the resting brain using fMRI [i.e., resting-state networks (RSNs)]. Although observed at rest, Fox & Raichle (2007) proposed that RSNs represent basic organizational units of the brain and that they are "functional networks" used during task performance. Smith

ICA: independent components analysis

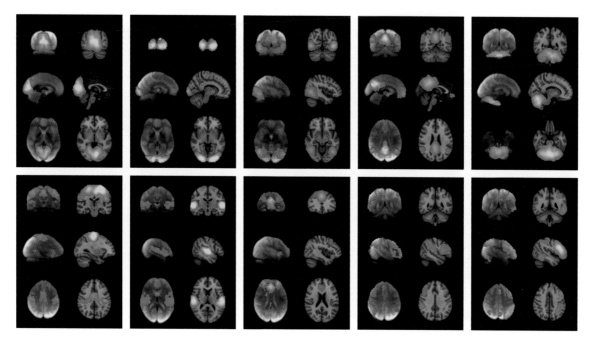

Figure 5

The comparability of independent components derived from meta-analysis and resting-state networks is illustrated. Each of the ten panels shows one well-matched pair of networks from two, 20-component independent components analyses (ICA). In each panel, the left-side images are derived from a meta-analysis of the BrainMap database (~30,000 subjects); the right-side images are from a 36-subject resting-state fMRI database. Images are reproduced with permission from Smith and colleagues (2009).

et al. (2009) tested this hypothesis by comparing ICA decompositions of resting-state fMRI to those derived from the BrainMap data. At the time of this data extraction (Fox & Raichle 2007), BrainMap contained 7,432 experiments, representing imaging studies from 29,671 human subjects. In parallel, ICA was performed using resting-state fMRI data from 36 healthy volunteers. Decompositions were performed into both 20 and 70 components.

Of the 20 components generated separately from the two data sets, ten maps from each set were unambiguously paired between data sets, with a minimum correlation $r = 0.25$ ($p < 10^{-5}$, corrected for multiple comparisons and for spatial smoothness). These ten well-matched pairs of networks are shown in **Figure 5**. With an ICA dimensionality of 70, the primary networks split into subnets in similar (but not identical) ways, continuing to show close correspondence between BrainMap and RSN components. This finding argues that the full repertoire of functional networks utilized by the brain in action (coded in BrainMap) is continuously and dynamically active even when at rest and, vice versa, that RSNs represent an intrinsic functional architecture of the brain that is drawn on to support task performance.

GRAPH THEORY MODELING APPROACHES

As noted above, graph theory network modeling constructs have proven to be well suited for modeling both primary and reduced neuroimaging data. In recent work, these graph theoretical constructs have been applied meta-analytically for voxel-wise, all-to-all (whole-brain to whole-brain) network discovery (see sidebar, Graph Theory and Neuroimaging).

Bayesian Network Discovery

In 2005, Neumann et al. introduced the use of graph theory modeling techniques to within-paradigm meta-analysis (above) using the Stroop paradigm. In 2008, Neumann et al. introduced the use of hierarchical Gaussian analysis to ALE output data (again using the Stroop paradigm). In 2010, Neumann et al. introduced the use of Bayesian network graphs to represent statistical dependencies using all-to-all voxel-wise analyses as a starting point. Bayesian network graphs are probabilistic models that represent a set of random variables and their probabilistic interdependencies. More formally, a Bayesian network is a directed acyclic graph (DAG) that comprises a set of nodes (vertices) and directed links (edges) connecting these nodes. Bayesian networks were chosen for three reasons. First, they belong to the class of directed graphical models, which permits investigation of directed interdependencies between the activation(s) of multiple brain regions. Second, the structure of Bayesian networks can be inferred from observed data. Thus, statistical interdependencies between the brain regions can be inferred from observations across multiple imaging experiments. Third, the theory for learning Bayesian networks from data is well established.

In application to neuroimaging meta-analysis, Neumann's approach used coactivation patterns of brain regions across imaging studies to learn the structure of the underlying DAGs. This was done first by computing an ALE map of a large subset (2,505 experiments) of the BrainMap data. This map was then restricted to the 49 most commonly occurring regions and further restricted to the 13 most commonly co-occurring regions using three separate applications of the replicator dynamics process, each of which identified subsets of regions. The regions included part of the posterior medial frontal cortex primarily covering SMAs and pre-SMAs, the anterior cingulate cortex, posterior parts of the lateral prefrontal cortex bilaterally, the dorsal premotor cortex bilaterally, the left and right anterior insula, the left and right thalamus, the left and right anterior intraparietal sulcus, and the left cerebellum. For these regions, DAGs were computed for groupings provided by each run of replicator dynamics and for the collection of all regions. Although this approach began with the entire BrainMap corpus, the analytic strategy promoted serial, data-driven reductions in the scope of the analysis to specific brain regions. This is distinctly unlike ICA (above), which groups voxels into components without ever reducing the total volume under consideration from the whole brain.

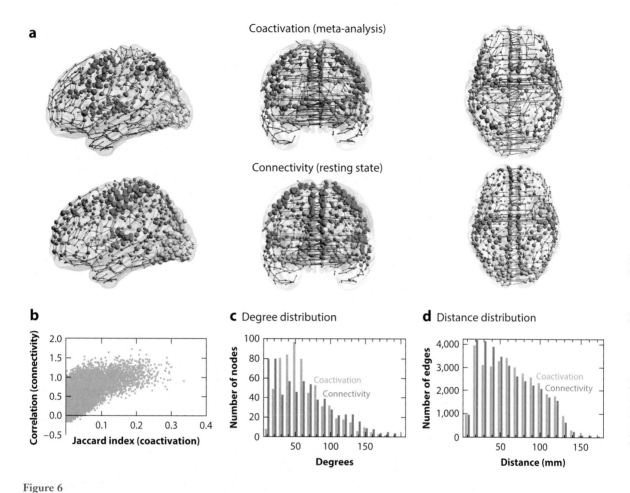

a

Coactivation (meta-analysis)

Connectivity (resting state)

b

c Degree distribution

d Distance distribution

Figure 6

Topological analysis. The comparability of topological networks derived from meta-analysis of the BrainMap database (*top row*) and resting-state fMRI in 27 subjects (*middle row*) is illustrated. In anatomical space (*a, upper two rows*) the size of the nodes is proportional to their weighted degree (strength), and their color corresponds to module membership. The relationship between the coactivation metric (Jaccard index; *b*) and the connectivity metric for every pair of regions is graphed (*b*). The degree and distance distributions for both networks are plotted (*c* and *d*, respectively). Images are reproduced with permission from Crossley et al. (2013).

Large-Scale Topological Modeling

Expanding the all-to-all graph theory approach to neuroimaging meta-analysis, Crossley et al. (2013) estimated the relative frequency at which each pair of regions in standard space was coactivated by multiple tasks reported in the primary literature. Nodes that frequently coactivated were connected by an edge. The resulting functional coactivation graph (**Figure 6**) had complex topological properties such as modules, hubs, and a rich club. The community structure of the meta-analytic network could be linked to the tasks in the primary literature, demonstrating that the modules were functionally specialized (for perception, action, emotion, etc.), whereas the rich club was more diversely coactivated by tasks requiring both action and cognition. It was also notable that the modules defined by this graph theoretical analysis of the BrainMap database were almost exactly supersets of the independent components identified by ICA-based meta-analysis

(Smith et al. 2009), which indicates that different methods of network meta-analysis can generate convergent rather than contradictory insights. This also demonstrates the important role that open-access databases can play in supporting comparative evaluation of alternative methodologies.

FUNCTIONAL ONTOLOGIES

For systems modeling, meta-analysis has the substantial advantage of being able to filter its findings with the behavioral meta-data associated with each experiment in the BrainMap database. Behavioral filtering has been widely used when selecting papers for inclusion in a meta-analysis (quantified and discussed in Fox et al. 2005b). A more recently developed use of behavioral meta-data is to characterize the behavioral properties of specific networks (Robinson et al. 2010, Cauda et al. 2011). Statistical methods to test for between-region differences in behavioral domain profiles have been developed (Lancaster et al. 2012) and are currently being extended to paradigm class data. Using this approach, given sufficient numbers of experiments and well-developed behavioral meta-data, unique behavioral characterization of individual brain regions may be a viable possibility. Characterization will be done, however, using complex behavioral profiles rather than using a one- or two-word term ("put," "get," "move," "selection for action") assigning a mental operation to each brain region, as Posner et al. (1988) had suggested. Our approach is more concordant with the views of Price & Friston (2005), who argued that the mapping between mental operations and brain regions is a many-to-many mapping in which a single region can be involved in many cognitive processes and a single elementary process engages multiple regions. It is also concordant with Poldrack's (2006) argument that the cognitive "reverse inference" (i.e., that a specific mental operation is necessarily engaged if a particular brain region is activated) is intrinsically weak, owing in part to participation of individual regions in multiple cognitive operations. An extension of the behavioral domain profile approach is to extract profiles for multiple regions jointly, i.e., to characterize a functional network. This strategy was employed by Laird et al. (2009) in work that behaviorally categorized the default mode network (DMN), examining behavioral domain profiles of individual areas and of groups of areas (i.e., subnetworks).

Another strategy for meta-analytic structure-function inference was pioneered by Smith and colleagues (2009), in the context of applying ICA to the BrainMap database. **Figure 7** (*left side*) is a heat map showing the respective contributions of BrainMap behavioral domains to individual components in the ICA shown in **Figure 5**. Close inspection reveals that some components have very high behavioral specificity, whereas other components have contributions from a wide range of behavioral domains. The ICA-based strategy of Smith and colleagues has been extended by Laird et al. (2011b) both by enriching the meta-data included in the analysis and by applying hierarchical clustering analysis to sort components into functionally related groupings (**Figure 7**). Even though this approach provides a much more refined association of components with behaviors, some components still show limited behavioral specificity. The most likely explanations for this lack of behavioral specificity in some networks are twofold: First, the behavioral specificity of some

Figure 7

BrainMap meta-data behavioral interpretations. Mapping of BrainMap meta-data onto ICA components is shown. Note that behavioral meta-data form discrete groupings, which functionally characterize the spatial groupings provided by ICA. (*a*) Twenty behavioral domain categorizations were correlated with the ten ICA-derived components shown in **Figure 5**. (*b*) The meta-data analysis has been extended to include 50 behavioral domain categories and 75 paradigm class categories. Hierarchical clustering was used to group the ICA into spatially and behaviorally related clusters for all 20 ICA components. Figures are reproduced with permission from Smith et al. (2009) (*panel a*) and Laird et al. (2011b) (*panel b*).

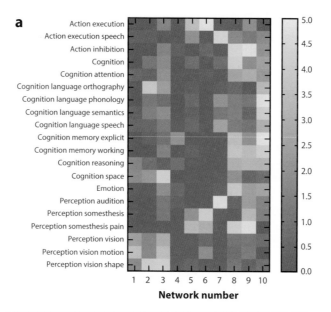

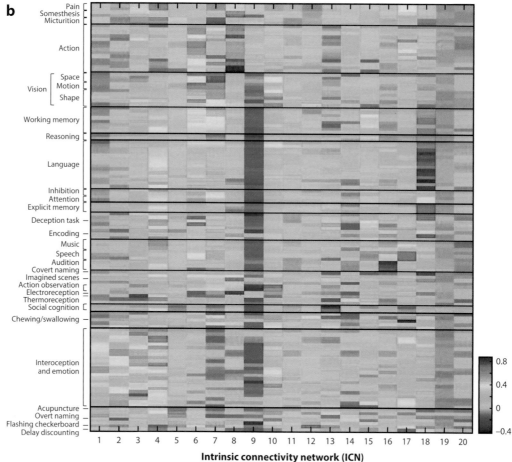

regions and networks ("hubs" in the terminology of topological modeling) is almost certainly low. Hub regions are engaged in a wide variety of tasks and will defy precise behavioral characterization. Second, a more evolved functional ontology is needed, as other studies have argued (Price & Friston 2005, Poldrack 2006). Relative to the second cause, we suggest that the approaches illustrated here provide the tools for ontology development to proceed programmatically. Such development can be determined by targeting networks that show limited behavioral domain specificity and enriching the meta-data, e.g., by adding levels to the coding hierarchy. This work is ongoing (Fox et al. 2005b, Laird et al. 2011a). Ultimately, behavior categorizations that are reflected in the network properties of the brain will have superior intrinsic validity and utility as compared with those based solely on cognitive theory.

A closing point of some importance is that meta-analysis offers the most versatile, most powerful extant approach for discovering the behavioral significance of networks mapped using DTI tractography, cortical thickness covariances, or resting-state fMRI. DTI and cortical-thickness covariances, being anatomical techniques, contain no behavioral information. Resting-state fMRI, being performed at rest, is not under experimental control, leaving the behavior unspecified. Both DTI and resting-state fMRI provide very similar connectivity maps to MACM. Consequently, behavioral characterizations provided for MACM-defined pathways should be reasonably applied to pathways defined by the other techniques.

META-ANALYSES AS TOOLS

The family of CBMA methods described above may appear to be conceptually discrete methods. In practice, however, they tend to be applied serially, with simpler forms of meta-analysis providing input for more advanced forms. For example, clique analyses (described above) take an ALE volume as input and compute a paradigm-specific system model. Similarly, Neumann et al. (2010) used ALE to identify nodes as preparations for doing Bayesian network discovery. The MACM approach of Robinson et al. (2010) used ALE images as priors. Although these are examples of CBMAs providing priors for CBMAs, the strategy is more general. Karlsgodt et al. (2010), for example, used ALE to select ROIs for an analysis of brain-behavior pleiotropy (a one-to-many mapping) of visual working memory. Perhaps the most advanced and impactful use of CBMA to provide priors is in the domain of graphical modeling.

Many system-level modeling approaches commonly applied to functional neuroimaging data (e.g., structural equation modeling and dynamic causal modeling) are confirmatory methods that require strong a priori hypotheses about the regions involved (nodes) and their interdependencies (edges). Well-chosen priors improve model fit (Stephan et al. 2009). Given the ability of the several approaches described above to provide fairly complete, data-driven models, their use as priors for graphical modeling seems quite promising. Perhaps the first application of this strategy was reported by Laird and colleagues (2008), who used an ALE meta-analysis of TMS/PET studies of the primary motor cortex to inform a structural equations model (SEM) analysis of a TMS/PET data set. The goodness-of-fit of the ALE-based model to the data was quite striking, endorsing the value of this strategy. A subsequent application of the strategy used previously published ALE meta-analyses of stuttered and nonstuttered speech (Brown et al. 2005) as priors for fitting PET data during cued speech in persons with and without stuttering (Price et al. 2009). Again, the goodness-of-fit of the ALE-based models to data sets was striking. Furthermore, this strategy allowed excellent between-group (stuttering versus nonstuttering) discrimination with group sizes as small as 15 (power > 0.8). This finding strongly suggests a role for this analysis and modeling approach to treatment trials using graphical models to characterize the brain mechanisms of action of treatments in patient groups.

SUMMARY POINTS

1. Spatial normalization (use of standardized coordinates) has produced a large (tens of thousands of peer-reviewed papers), well-standardized neuroimaging literature reporting hundreds of thousands of functional and structural experimental observations. BrainMap® and other online databases make these data readily available.

2. Sophisticated meta-analytic methods have been developed specifically for this data type or adapted to it from other neuroimaging applications.

3. The most basic meta-analytic method, ALE, demonstrates cross-study reliability of regional observations, filtering out nonreplicating findings within a group of similar experiments. This approach works equally well on functional (task-activation) and structural (between-group anatomical differences) data.

4. Between-experiment coactivation patterns are a reliable index of functional and structural connectivity, with multiple cross-methodology validations in the literature.

5. Coactivation patterns differ between subfields of larger cortical and nuclear structures, allowing connectivity-based parcellation to be computed meta-analytically.

6. Network modeling approaches originally developed for primary-image data sets (e.g., ICA, graph-analytic modeling) are proving to be well suited for analysis of large-scale, reduced-image (standardized coordinates) data sets, with coactivation being the driving observation. Similarity of observations between primary-data and reduced-data (meta-analytic) analyses is the rule and provides additional validation of this strategy.

7. Behavioral meta-data linked per-experiment to coordinate data in online databases provide a rich interpretive framework not otherwise available for network modeling.

8. Meta-analytically vetted regional effects and meta-analytically derived network models can be used as tools (a priori constraints) to analyze primary data sets, decreasing (or eliminating) the need for corrections for multiple comparisons and increasing the likelihood of finding significant effects.

FUTURE ISSUES

1. Advanced meta-analytic network modeling methods (e.g., high-dimensionality ICA) would benefit from substantially larger data sets than are currently available.

2. More high-quality data are available in the literature than can be effectively curated. Greater efficiency of data entry is needed but without sacrificing quality control of data and meta-data.

3. As the scope and sophistication of neuroimaging experimental designs progress, the meta-data taxonomies used by BrainMap and other online databases will need to expand accordingly.

4. Greater care needs to be taken by software providers when releasing new templates and methods to ensure comparability of coordinates (same coordinate = same brain location) with prior literature.

5. Data-driven network analyses conjointly using coactivation patterns and behavioral meta-data need to be developed, advancing beyond using coactivations for network extraction and meta-data for network interpretation.

6. Use of meta-analysis-derived products (ROIs, spatial templates, network models) as priors for data analysis and modeling should be expanded.

DISCLOSURE STATEMENT

The authors are not aware of any affiliations, memberships, funding, or financial holdings that might be perceived as affecting the objectivity of this review.

ACKNOWLEDGMENTS

This work was supported by awards from the National Institutes of Health (MH74457, RR024387, MH084812, NS062254, AA019691, EB015314) and the Congressionally Directed Medical Research Program (W81XWH0820112). The sidebar on graph theory was contributed by Ed Bullmore and Nicolas Crossley. Portions of this review were adapted from Fox & Friston (2012).

LITERATURE CITED

Achard S, Salvador R, Whitcher B, Suckling J, Bullmore E. 2006. A resilient, low-frequency, small-world human brain functional network with highly connected association cortical hubs. *J. Neurosci.* 26:63–72

Alexander-Bloch A, Giedd JN, Bullmore E. 2013. Imaging structural co-variance between human brain regions. *Nat. Rev. Neurosci.* 14:322–36

Amunts K, Kedo O, Kindler M, Pieperhoff P, Mohlberg H, et al. 2005. Cytoarchitectonic mapping of the human amygdala, hippocampal region and entorhinal cortex: intersubject variability and probability maps. *Anat. Embryol.* 210:343–52

Ashburner J, Friston KJ. 2000. Voxel-based morphometry—the methods. *NeuroImage* 11(6):805–21

Barron DS, Fox PM, Laird AR, Robinson JL, Fox PT. 2012. Thalamic medial dorsal nucleus atrophy in medial temporal lobe epilepsy: a VBM meta-analysis. *NeuroImage Clin.* 2:25–32

Biswal B, Yetkin FZ, Haughton VM, Hyde JS. 1995. Functional connectivity in the motor cortex of resting human brain using echo-planar MRI. *Magn. Reson. Med.* 34(4):537–41

Brown S, Ingham RJ, Ingham JC, Laird AR, Fox PT. 2005. Stuttered and fluent speech production: an ALE meta-analysis of functional neuroimaging studies. *Hum. Brain Mapp.* 25:105–17

Bullmore ET, Bassett DA. 2011. Brain graphs: graphical models of the human brain connectome. *Annu. Rev. Clin. Psychol.* 7:113–40

Bullmore ET, Sporns O. 2009. Complex brain networks: graph theoretical analysis of structural and functional systems. *Nat. Rev. Neurosci.* 10(3):186–98

Butcher J. 2007. Alzheimer's researchers open the doors to data sharing. *Lancet Neurol.* 6:480–81

Bzdok D, Laird AR, Zilles K, Fox PT, Eickhoff SB. 2013. An investigation of the structural, connectional, and functional subspecialization in the human amygdala. *Hum. Brain Mapp.* 34:3247–66

Bzdok D, Langner R, Schilback L, Engemann DA, Laird AR, et al. 2013. Segregation of the human medial prefrontal cortex in social cognition. *Front. Hum. Neurosci.* 7:232

Cauda F, Cavanna AE, D'agata F, Sacco K, Duca S, Geminiani GC. 2011. Functional connectivity and coactivation of the nucleus accumbens: a combined functional connectivity and structure-based meta-analysis. *J. Cogn. Neurosci.* 23:2864–77

Cieslik EC, Zilles K, Caspers S, Roski C, Kellermann TS, et al. 2013. Is there "one" DLPFC in cognitive action control? Evidence for heterogeneity from co-activation-based parcellation. *Cereb. Cortex* 23:2677–89

Chein JM, Fissell K, Jacobs S, Fiez JA. 2002. Functional heterogeneity within Broca's area during verbal working memory. *Physiol. Behav.* 77:635–39

Clos M, Amunts K, Laird AR, Fox PT, Eickhoff SB. 2013. Tackling the multifunctional nature of Broca's region meta-analytically: co-activation-based parcellation of area 44. *NeuroImage* 83:174–88

Costafreda SG, Brammer MJ, David AS, Fu CH. 2008. Predictors of amygdala activation during the processing of emotional stimuli: a meta-analysis of 385 PET and fMRI studies. *Brain Res. Rev.* 58:57–70

Crossley NA, Mechelli A, Vértes PE, Winton-Brown TT, Patel AX, et al. 2013. Cognitive relevance of the community structure of the human brain functional coactivation network. *Proc. Natl. Acad. Sci. USA* 110:11583–88

Decety J, Lamm C. 2007. The role of the right temporoparietal junction in social interaction: how low-level computational processes contribute to meta-cognition. *Neuroscientist* 13:580–93

Derrfuss J, Mar RA. 2009. Lost in localization: the need for a universal coordinate database. *NeuroImage* 48:1–7

Dogan I, Eickhoff SB, Schulz JB, Shah JN, Laird AR, et al. 2013. Consistent neurodegeneration and its association with clinical progression in Huntington's disease: a coordinate-based meta-analysis. *Neurodegener. Dis.* 12(1):23–35

Duerden EG, Mak-Fan KM, Taylor MJ, Roberts SW. 2012. Regional differences in grey and white matter in children and adults with autism spectrum disorders: an activation likelihood estimate (ALE) meta-analysis. *Autism Res.* 5:49–66

Eickhoff SB, Bzdok D, Laird AR, Kurth F, Fox PT. 2012. Activation likelihood estimation meta-analysis revisited. *NeuroImage* 59:2349–61

Eickhoff SB, Bzdok D, Laird AR, Roski C, Caspers S, et al. 2011. Co-activation patterns distinguish cortical modules, their connectivity and functional differentiation. *NeuroImage* 57(3):938–49

Eickhoff SB, Jbabdi S, Caspers S, Laird AR, Fox PT, et al. 2010. Anatomical and functional connectivity of cytoarchitectonic areas within the human parietal operculum. *J. Neurosci.* 30:6409–21

Eickhoff SB, Laird AR, Grefkes C, Wang LE, Zilles K, Fox PT. 2009. Coordinate-based activation likelihood estimation meta-analysis of neuroimaging data: a random-effects approach based on empirical estimates of spatial uncertainty. *Hum. Brain Mapp.* 30:2907–26

Ellison-Wright I, Bullmore E. 2009. Meta-analysis of diffusion tensor imaging studies in schizophrenia. *Schizophr. Res.* 108:3–10

Ellison-Wright I, Glahn DC, Laird AR, Thelen SM, Bullmore E. 2008. The anatomy of first-episode and chronic schizophrenia: an anatomical likelihood estimation meta-analysis. *Am. J. Psychiatry* 165:1015–23

Evans AC, Janke AL, Collins DL, Bailet S. 2012. Brain templates and atlases. *NeuroImage* 62(2):911–22

Fitzgerald PB, Laird AR, Maller J, Daskalakis ZJ. 2008. A meta-analytic study of changes in brain activation in depression. *Hum. Brain Mapp.* 29:683–95

Fox MD, Raichle ME. 2007. Spontaneous fluctuations in brain activity observed with functional magnetic resonance imaging. *Nat. Rev. Neurosci.* 8(9):700–11

Fox PT. 1995a. Broca's area: motor encoding in somatic space. *Behav. Brain Sci.* 18:344–45

Fox PT. 1995b. Spatial normalization origins: objectives, applications, and alternatives. *Hum. Brain Mapp.* 3:161–64

Fox PT, Friston KJ. 2012. Distributed processing; distributed functions? *NeuroImage* 61:407–26

Fox PT, Huang AY, Parsons LM, Xiong J-H, Rainey L, Lancaster JL. 1999. Functional volumes modeling: scaling for group size in averaged images. *Hum. Brain Mapp.* 8:143–50

Fox PT, Huang A, Parsons LM, Xiong J-H, Zamarippa F, et al. 2001. Location-probability profiles for the mouth region of human primary motor–sensory cortex: model and validation. *NeuroImage* 13:196–209

Fox PT, Laird AR, Fox SP, Fox PM, Uecker AM, et al. 2005a. BrainMap taxonomy of experimental design: description and evaluation. *Hum. Brain Mapp.* 25:185–98

Fox PT, Laird AR, Lancaster JL. 2005b. Coordinate-based voxel-wise meta-analysis: dividends of spatial normalization. Report of a virtual workshop. *Hum. Brain Mapp.* 5:1–5

Fox PT, Lancaster JL. 1994. Neuroscience on the net. *Science* 266:994–96

Fox PT, Lancaster JL. 2002. Mapping context and content: the BrainMap model. *Nat. Rev. Neurosci.* 3:319–21

Fox PT, Lancaster JL, Parsons LM, Xiong JH, Zamarripa F. 1997. Functional volumes modeling: theory and preliminary assessment. *Hum. Brain Mapp.* 5:306–11

Presented the first large-scale graph-analytic meta-analysis; derived similar networks for BrainMap and resting-state fMRI data sets.

Introduced coactivation-based parcellation as a meta-analytic method.

Described and validated the meta-data taxonomy of the BrainMap database.

Announced the concept, purpose, structure, and online status of the BrainMap database.

Fox PT, Miezin FM, Allman JM, Van Essen DC, Raichle ME. 1987. Retinotopic organization of human visual cortex mapped with positron-emission tomography. *J. Neurosci.* 7(3):913–22

Fox PT, Mikiten S, Davis G, Lancaster JL. 1994. BrainMap: a database of human functional brain mapping. In *Functional Neuroimaging: Technical Foundations*, ed. RW Thatcher, M Hallet, T Zeffiro, ER John, M Huerta. San Diego: Academic

Fox PT, Mintun MA. 1989. Noninvasive functional brain mapping by change-distribution analysis of average PET images of $H_2^{15}O$ tissue activity. *J. Nucl. Med.* 30:141–49

Fox PT, Mintun MA, Raichle ME, Miezin FM, Allman JM, Van Essen DC. 1986. Mapping human visual cortex with position emission tomography. *Nature* 323:806–9

Fox PT, Mintun MA, Reiman EM, Raichle ME. 1988. Enhanced detection of focal brain responses using intersubject averaging and change-distribution analysis of subtracted PET images. *J. Cereb. Blood Flow Metab.* 8:642–53

Fox PT, Parsons LM, Lancaster JL. 1998. Beyond the single study: function/location meta-analysis in cognitive neuroimaging. *Curr. Opin. Neurobiol.* 8:178–87

Fox PT, Perlmutter JS, Raichle ME. 1985. A stereotactic method of anatomical localization for positron emission tomography. *J. Comput. Assist. Tomogr.* 9:141–53

Friston KJ, Ashburner J, Frith CD, Poline J-B, Heather JD, Frackowiak RSJ. 1995. Spatial registration and normalization of images. *Hum. Brain Mapp.* 3:165–89

Friston KJ, Frith CD, Liddle PF, Frackowiak RSJ. 1991. Comparing functional (PET) images: the assessment of significant change. *J. Cereb. Blood Flow Metab.* 11:690–99

Frith CD, Friston KJ, Liddle PF, Frackowiak RS. 1991. Willed action and the prefrontal cortex in man: a study with PET. *Proc. R. Soc. B* 244:241–46

Gibbons A. 1992. Databasing the brain. *Science* 258:1872–73

Glahn DC, Laird AR, Ellison-Wright I, Thelen SM, Robinson JL, et al. 2008. Meta-analysis of gray matter anomalies in schizophrenia: application of anatomic likelihood estimation and network analysis. *Biol. Psychiatry* 64:774–81

Glahn DC, Ragland JD, Abramoff A, Barrett J, Laird AR, et al. 2005. Beyond hypofrontality: a quantitative meta-analysis of functional neuroimaging studies of working memory in schizophrenia. *Hum. Brain Mapp.* 25:60–69

Hagmann P, Kurant M, Gigandet X, Thiran P, Wedeen VJ, et al. 2007. Mapping human whole-brain structural networks with diffusion MRI. *PLoS ONE* 2:e597

Johansen-Berg H, Behrens TEJ, Robson MD, Drobnjak I, Rushworth MFS, et al. 2004. Changes in connectivity profiles define functionally distinct regions in human medial frontal cortex. *Proc. Natl. Acad. Sci. USA* 101(36):13335–40

Karlsgodt KH, Kochunov P, Winkler AM, Laird AR, Almasy L, et al. 2010. A multimodal assessment of the genetic control over working memory. *J. Neurosci.* 30:8197–202

Koski L, Paus T. 2000. Functional connectivity of the anterior cingulate cortex within the human frontal lobe: a brain-mapping meta-analysis. *Exp. Brain Res.* 133:55–65

Laird AR, Eickhoff SB, Fox PM, Uecker AM, Ray KL, et al. 2011a. The BrainMap strategy for standardization, sharing and meta-analysis of neuroimaging data. *BMC Res. Notes* 4:349

Laird AR, Eickhoff SB, Li K, Robin DA, Glahn DC, Fox PT. 2009. Investigating the functional heterogeneity of the default mode network using coordinate-based meta-analytic modeling. *J. Neurosci.* 29:14496–505

Laird AR, Fox PM, Eickhoff SB, Turner JA, Ray KL, et al. 2011b. Behavioral interpretations of intrinsic connectivity networks. *J. Cogn. Neurosci.* 23:4022–37

Laird AR, Robinson JL, McMillan KM, Tordesillas-Gutiérrez D, Moran ST, et al. 2010. Comparison of the disparity between Talairach and MNI coordinates in functional neuroimaging data: validation of the Lancaster transform. *NeuroImage* 51:677–83

Laird AR, Robbins JM, Li K, Price LR, Cykowski MD, et al. 2008. Modeling motor connectivity using TMS/PET and structural equation modeling. *NeuroImage* 41:424–36

Laird AR, Fox PM, Price CJ, Glahn DC, Uecker AM, et al. 2005a. ALE meta-analysis: controlling the false discovery rate and performing statistical contrasts. *Hum. Brain Mapp.* 25:155–64

Laird AR, Lancaster JL, Fox PT. 2005b. BrainMap: the social evolution of a human brain mapping database. *Neuroinformatics* 3:65–78

Introduced the use of spatial normalization (standardized coordinates) to human brain mapping.

Introduced the concept that meta-analysis of coactivations was an index of functional connectivity.

Lambrecq V, Langbour N, Guehl D, Biolac B, Burbaud P, Rotge JY. 2013. Evolution of gray matter loss in Huntington's disease: a meta-analysis. *Eur. J. Neurol.* 20:315–21

Lancaster JL, Laird AR, Eickhoff SB, Martinez MJ, Fox MP, Fox PT. 2012. Automated regional behavioral analysis for human brain images. *Front. Neuroinform.* 6:23

Lancaster JL, Laird AR, Fox M, Glahn DE, Fox PT. 2005. Automated analysis of meta-analysis networks. *Hum. Brain Mapp.* 25:174–84

Lancaster JL, Tordesillas-Gutiérrez D, Martinez M, Salinas F, Evans A, et al. 2007. Bias between MNI and Talairach coordinates analyzed using the ICBM-152 brain template. *Hum. Brain Mapp.* 28:1194–205

Lancaster JL, Woldorff MG, Parsons LM, Liotti M, Freitas CS, et al. 2000. Automated Talairach atlas labels for functional brain mapping. *Hum. Brain Mapp.* 10(3):120–31

Lohmann G, Bohn S. 2002. Using replicator dynamics for analyzing fMRI data of the human brain. *IEEE Trans. Med. Imaging* 21(5):485–92

Mazziotta JC, Toga TW, Evans A, Fox P, Lancaster J. 1995. A probabilistic atlas of the human brain: theory and rationale for its development. The International Consortium for Brain Mapping (ICBM). *NeuroImage* 2(2A):89–101

Menzies L, Chamberlain SR, Laird AR, Thelen SM, Sahakian BJ, Bullmore ET. 2008. Integrating evidence from neuroimaging and neuropsychological studies of obsessive-compulsive disorder: the orbitofronto-striatal model revisited. *Neurosci. Biobehav. Rev.* 32:525–49

Meunier D, Achard S, Morcom A, Bullmore E. 2009. Age-related changes in modular organization of human brain functional networks. *NeuroImage* 44:715–23

Mintun MA, Fox PT, Raichle ME. 1989. A highly accurate method of localizing regions of neuronal activation in the human brain with positron emission tomography. *J. Cereb. Blood Flow Metab.* 9(1):96–103

Minzenberg MJ, Laird AR, Thelen SM, Carter CS, Glahn DC. 2009. Meta-analysis of 41 functional neuroimaging studies of executive function in schizophrenia. *Arch. Gen. Psychiatry* 66:811–22

Narayana S, Laird AR, Tandon N, Franklin C, Lancaster JL, Fox PT. 2012. Electrophysiological and functional connectivity of the human supplementary motor area. *NeuroImage* 62:250–65

Neumann J, Fox PT, Turner R, Lohmann G. 2010. Learning partially directed functional networks from meta-analysis imaging data. *NeuroImage* 49:1372–84

Neumann J, Lohmann G, Derrfuss J, Yves von Cramon D. 2005. The meta-analysis of functional imaging data using replicator dynamics. *Hum. Brain Mapp.* 25:165–73

Neumann J, Yves von Cramon D, Lohmann G. 2008. Model-based clustering of meta-analytic functional imaging data. *Hum. Brain Mapp.* 29:177–92

Nickl-Jockschat T, Habel U, Michel TM, Manning J, Laird AR, et al. 2012. Brain structure anomalies in autism spectrum disorder (ASD)—a meta-analysis of VBM studies using anatomic likelihood estimation (ALE). *Hum. Brain Mapp.* 33:1470–89

Paus T. 1996. Location and function of the human frontal eye-field: a selective review. *Neuropsychologia* 34:475–83

Pearson K. 1904. Report on certain enteric fever inoculation statistics. *Br. Med. J.* 3:1243–46

Picard N, Strick PL. 1996. Motor areas of the medial wall: a review of their location and functional activation. *Cereb. Cortex* 6:342–53

Poldrack RA. 2006. Can cognitive processes be inferred from neuroimaging data? *Trends Cogn. Sci.* 10:59–63

Posner M, Petersen S, Fox PT, Raichle M. 1988. Localization of cognitive operations in the human brain. *Science* 240:1627–31

Postuma RB, Dagher A. 2006. Basal ganglia functional connectivity based on a meta-analysis of 126 positron emission tomography and functional magnetic resonance imaging publications. *Cereb. Cortex* 16:1508–21

Price CJ, Friston KJ. 2005. Functional ontologies for cognition: the systematic definition of structure and function. *Cogn. Neuropsychol.* 22:262–75

Price LR, Laird AR, Fox PT. 2009. Modeling dynamic functional neuroimaging data using structural equation modeling. *Struct. Equ. Modeling* 16:147–62

Ragland JD, Laird AR, Ranganath C, Blumenfeld RS, Gonzales SM, Glahn DC. 2009. Prefrontal activation deficits during episodic memory in schizophrenia. *Am. J. Psychiatry* 166:863–74

Robinson JL, Laird AR, Glahn DC, Blangero J, Sanghera MK, et al. 2012. The functional connectivity of the human caudate: an application of meta-analytic connectivity modeling with behavioral filtering. *NeuroImage* **60:117–29**

Robinson JL, Laird AR, Glahn DC, Lovallo WR, Fox PT. 2010. Metaanalytic connectivity modeling: delineating the functional connectivity of the human amygdala. *Hum. Brain Mapp.* 31:173–84

Rottschy C, Caspers S, Roski C, Reetz K, Dogan I, et al. 2012. Differentiated parietal connectivity of frontal regions for "what" and "where" memory. *Brain Struct. Funct.* 218:1551–67

Smith SM, Fox PT, Miller KL, Glahn DC, Fox PM, et al. 2009. Correspondence of the brain's functional architecture during activation and rest. *Proc. Natl. Acad. Sci. USA* **106:13040–45**

Sporns O, Honey CJ, Kötter R. 2007. Identification and classification of hubs in brain networks. *PLoS ONE* 2:e1049

Spreng RN, Mar RA, Kim AS. 2009. The common neural basis of autobiographical memory, prospection, navigation, theory of mind and the default mode: a quantitative meta-analysis. *J. Cogn. Neurosci.* 21:489–510

Stephan KI, Tittgemeyer M, Knosche TR, Moran RJ, Friston KJ. 2009. Tractography based priors for dynamic causal models. *NeuroImage* 47:1628–38

Talairach J, Szikla G, Tournoux P, Prosalentis A, Bordas-Ferrier M. 1967. *Atlas d'Anatomie Stéréotaxique du Télencéphale*. Paris: Masson

Talairach J, Tournoux P. 1988. *Co-Planar Stereotaxic Atlas of the Human Brain*. New York: Thieme

Toga AW. 1998. *Brain Warping*. San Diego: Academic

Toro R, Fox PT, Paus T. 2008. Functional coactivation map of the human brain. *Cereb. Cortex* **18:2553–59**

Torta DM, Costa T, Duca S, Fox PT, Cauda F. 2013. Parcellation of the cingulate cortex at rest and during tasks: a meta-analytic clustering and experimental study. *Front. Hum. Neurosci.* 7:275

Tulving E, Kapur S, Craik FI, Moscovitch M, Huole S. 1994. Hemispheric encoding/retrieval asymmetry in episodic memory: positron emission tomography findings. *Proc. Natl. Acad. Sci. USA* 91:2016–20

Turkeltaub PE, Eden GF, Jones KM, Zeffiro TA. 2002. Meta-analysis of the functional neuroanatomy of single-word reading: method and validation. *NeuroImage* **16:765–80**

Turkeltaub PE, Eickhoff SB, Laird AR, Fox M, Wiener M, Fox PT. 2012. Minimizing within-experiment and within-group effects in activation likelihood estimation meta-analyses. *Hum. Brain Mapp.* 33:1–13

Turner JA, Laird AR. 2011. The cognitive paradigm ontology: design and application. *Neuroinformatics* 10:57–66

Wager TD, Phan KL, Liberzon I, Taylor SF. 2003. Valence, gender, and lateralization of functional brain anatomy in emotion: a meta-analysis of findings in neuroimaging. *NeuroImage* 19:513–31

Yarkoni T, Poldrack RA, Nichols TE, Van Essen DC, Wager TD. 2011. Large-scale automated synthesis of human functional neuroimaging data. *Nat. Methods* 8(8):665–70

Decoding Neural Representational Spaces Using Multivariate Pattern Analysis

James V. Haxby,[1,2] Andrew C. Connolly,[1] and J. Swaroop Guntupalli[1]

[1]Department of Psychological and Brain Sciences, Center for Cognitive Neuroscience, Dartmouth College, Hanover, New Hampshire 03755; email: james.v.haxby@dartmouth.edu, andrew.c.connolly@dartmouth.edu, swaroopgj@gmail.com

[2]Center for Mind/Brain Sciences (CIMeC), University of Trento, Rovereto, Trentino 38068, Italy

Annu. Rev. Neurosci. 2014. 37:435–56

First published online as a Review in Advance on June 25, 2014

The *Annual Review of Neuroscience* is online at neuro.annualreviews.org

This article's doi: 10.1146/annurev-neuro-062012-170325

Keywords

neural decoding, MVPA, RSA, hyperalignment, population response, fMRI

Abstract

A major challenge for systems neuroscience is to break the neural code. Computational algorithms for encoding information into neural activity and extracting information from measured activity afford understanding of how percepts, memories, thought, and knowledge are represented in patterns of brain activity. The past decade and a half has seen significant advances in the development of methods for decoding human neural activity, such as multivariate pattern classification, representational similarity analysis, hyperalignment, and stimulus-model-based encoding and decoding. This article reviews these advances and integrates neural decoding methods into a common framework organized around the concept of high-dimensional representational spaces.

Contents

INTRODUCTION

Information is encoded in patterns of neural activity. This information can come from our experience of the world or can be generated by thinking. One of the great challenges for systems neuroscience is to break this code. Developing algorithms for decoding neural activity involves many modalities of measurement—including single-unit recording, electrocorticography (ECoG), electro- and magnetoencephalography (EEG and MEG), and functional magnetic resonance imaging (fMRI)—in various species. All decoding methods are multivariate analyses of brain activity patterns that are distributed across neurons or cortical regions. These methods are referred to generally as multivariate pattern analysis (MVPA). This review focuses on the progress made in the past decade and a half in the development of methods for decoding human neural activity as measured with fMRI. We make occasional references to decoding analyses of single-unit recording data in monkeys and of ECoG and MEG data in humans to illustrate the general utility of decoding methods and to indicate the potential for multimodal decoding.

Prior to the discovery that within-area patterns of response in fMRI carried information that afforded decoding of stimulus distinctions (Haxby et al. 2001, Cox & Savoy 2003, Haxby 2012), it was generally believed that the spatial resolution of fMRI allowed investigators to ask only which task or stimulus activated a region globally. Thus, fMRI studies focused on associating brain regions with functions. A region's function was identified by determining which task activated it most strongly. The introduction of decoding using MVPA has revolutionized fMRI research by changing the questions that are asked. Instead of asking what a region's function is, in terms of a single brain state associated with global activity, fMRI investigators can now ask what information is represented in a region, in terms of brain states associated with distinct patterns of activity, and how that information is encoded and organized.

Multivariate pattern (MVP) classification distinguishes patterns of neural activity associated with different stimuli or cognitive states. The first demonstrations of MVP classification showed that different high-level visual stimulus categories (faces, animals, and objects) were associated with distinct patterns of brain activity in the ventral object vision pathway (Haxby et al. 2001, Cox & Savoy 2003). Subsequent work has shown that MVP classification can also distinguish many other brain states, for example low-level visual features in the early visual cortex (Haynes & Rees 2005, Kamitani & Tong 2005) and auditory stimuli in the auditory cortex (Formisano et al. 2008,

MEG: magnetoen-
cephalography

fMRI: functional
magnetic resonance
imaging

**Multivariate pattern
analysis (MVPA):**
analysis of brain
activity patterns with
methods such as
pattern classification,
RSA, hyperalignment,
or stimulus-model-
based encoding and
decoding

REPRESENTATIONAL SPACE

Representational space is a high-dimensional space in which each neural response or stimulus is expressed as a vector with different values for each dimension. In a neural representational space, each pattern feature is a measure of local activity, such as a voxel or a single neuron. In a stimulus representational space, each feature is a stimulus attribute, such as a physical attribute or semantic label.

Staeren et al. 2009), as well as more abstract brain states such as intentions (Haynes et al. 2007, Soon et al. 2008) and the contents of working memory (Harrison & Tong 2009).

Whereas MVP classification simply demonstrates reliable distinctions among brain states, more recently introduced methods characterize how these brain states are organized. Representational similarity analysis (RSA) (Kriegeskorte et al. 2008a) analyzes the geometry of representations in terms of the similarities among brain states. RSA can show that the representations of the same set of stimuli in two brain regions have a different structure (Kriegeskorte et al. 2008a,b; Connolly et al. 2012a,b), whereas MVP classification may find that the classification accuracy is equivalent in those regions. Stimulus-model-based encoding and decoding algorithms show that brain activity patterns can be related to the constituent features of stimuli or cognitive states. This innovation affords predictions of patterns of brain response to novel stimuli based on their features (Kay et al. 2008, Mitchell et al. 2008). It also affords reconstruction of stimuli from brain activity patterns based on predictions of the stimulus features (Miyawaki et al. 2008, Naselaris et al. 2009, Nishimoto et al. 2011, Horikawa et al. 2013).

Several excellent reviews have focused on MVP classification (Norman et al. 2006, Haynes & Rees 2006, O'Toole et al. 2007, Pereira et al. 2009, Tong & Pratte 2012), RSA (Kriegeskorte & Kievet 2013), or stimulus-model-based encoding and decoding (Naselaris et al. 2011). Here we integrate neural decoding methods into a common framework organized around the concept of high-dimensional representational spaces (see sidebar). In all these methods, brain activity patterns are analyzed as vectors in high-dimensional representational spaces. Neural decoding then analyzes these spaces in terms of (*a*) reliably distinctive locations of pattern response vectors (MVP classification), (*b*) the proximity of these vectors to each other (RSA), or (*c*) mapping of vectors from one representational space to another—from one subject's neural representational space to a model space that is common across subjects (hyperalignment) or from stimulus feature spaces to neural spaces (stimulus-model-based encoding).

CORE CONCEPT: REPRESENTATIONAL SPACES

The core concept that underlies neural decoding and encoding analyses is that of high-dimensional representational vector spaces. Neural responses—brain activity patterns—are analyzed as vectors in a neural representational space. Brain activity patterns are distributed in space and time. The elements, or features, of these patterns are local measures of activity, and each of these local measures is a dimension in the representational space. Thus, if neural responses measured with fMRI have 1,000 voxels, the representational space is 1,000-dimensional. If a population response has spike rates for 600 cells, the representational space is 600-dimensional. If fMRI responses with 1,000 voxels include six time points, the response vectors are analyzed in a 6,000-dimensional space.

For fMRI, measures of local activity are usually voxels (volume elements in brain images), but there are numerous alternatives, such as nodes on the cortical surface, the average signal for an area, a principal or independent component, or a measure of functional connectivity between a pair

Representational similarity analysis (RSA): analysis of the pattern of similarities among response vectors

Response vector: a brain activity pattern expressed as the response strengths for features of that pattern, e.g., voxels, single neurons, or model dimensions

Hyperalignment: transformation of individual representational spaces into a model representational space in which each dimension has a common tuning function

of locations. For single-unit recording, local measures can be single-neuron spike rates, multiple-unit spike rates, or local field potentials, among other possibilities. Similarly, EEG and MEG responses can be analyzed as time-varying patterns of activity distributed over sensors or sources, with numerous possibilities for converting activity into frequencies, principal or independent components, or measures of synchrony between sources.

The computational advantages of representational vector spaces extend beyond neural representational spaces to representational spaces for stimuli or cognitive states. For example, a visual stimulus can be modeled as a set of features based on response properties of neurons in V1, as higher-order visual features, or as a set of semantic labels. Sounds—voices and music—can be modeled as sets of acoustic features, words can be modeled as sets of semantic features, actions as sets of movement and goal features, etc. Once the description of the stimulus is in a representational space, various computational manipulations can be applied for relating stimulus representational spaces to neural representational spaces.

All the major varieties of neural decoding and encoding analyses follow from this conversion of patterns of brain activity or stimuli to single points in high-dimensional representational vector spaces. MVP classification uses machine learning methods to define decision boundaries in a neural representational space that best distinguish a set of response vectors for one brain state from others. RSA analyzes the similarity between response vectors as distances in the representational space. Stimulus-model-based encoding predicts the location of the neural response vector for a new stimulus on the basis of the coordinates of that stimulus in a stimulus feature space. Stimulus-model-based decoding tests whether the encoding-based prediction allows correct classification of neural response vectors to new stimuli. Building a model of a neural representational space that is common across brains requires hyperalignment to rotate the coordinate axes of individual representational spaces to minimize the difference in the locations of response vectors for the same stimuli. Thus, stimuli and other cognitive events are represented as vectors in neural representational spaces as well as in stimulus representational spaces, and the computational task for understanding representation becomes one of characterizing the geometries within spaces and relating the geometries of these spaces to each other.

Numerically, a set of response vectors in a representational space is a matrix in which each column is a local pattern feature (e.g., voxel) and each row is a response vector (**Figure 1**). The values in each column reflect the differential responses of that pattern feature to conditions or stimuli. This profile of differential responses is called the tuning function. All the neural decoding and encoding methods can be understood in terms of analyzing or manipulating the geometry of the response vectors in a high-dimensional space. Computationally, these analyses and manipulations are performed using linear algebra. Here we illustrate the concepts related to high-dimensional representational spaces in two-dimensional figures by showing only two dimensions at a time— the equivalent of a two-voxel brain or a two-neuron population. The linear algebra for these two-dimensional toy examples is the same as the linear algebra for representational spaces with many more dimensions and larger matrices. Most of the algorithms that we discuss here can be implemented using PyMVPA (**http://www.pymvpa.org**; Hanke et al. 2009), a Python-based software platform that includes tutorials and sample data sets.

The geometries in a neural representational space, as defined here, are distinctly different from the geometries of cortical anatomy and of cortical topographies. A cortical topography can be thought of as a two-dimensional manifold. Neural encoding can be thought of as the problem of projecting high-dimensional representations into this low-dimensional topography. Decoding is the problem of projecting a neural response in a two-dimensional topography into a high-dimensional representational space. The methods that we discuss here are computational algorithms that attempt to model these transformations.

Machine learning: a branch of artificial intelligence that builds and evaluates induction algorithms that learn data patterns and associate them with labels

Pattern feature: a single element in a distributed pattern, such as a voxel, a single neuron, or a model space dimension

Tuning function: the profile of differential responses to stimuli or brain states for a single pattern feature

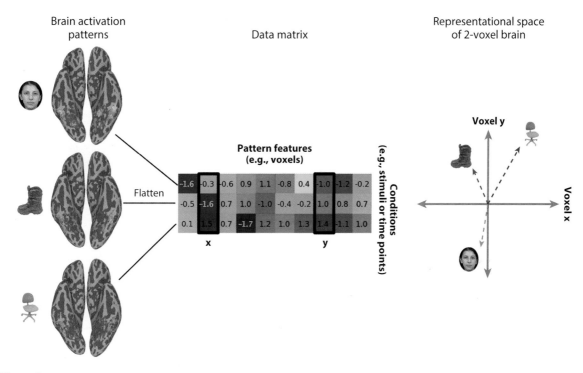

Figure 1

Multivariate pattern analysis (MVPA) is a family of methods that treats the measured fMRI signal as a set of pattern vectors stored in an N × M matrix with N observations (e.g., stimulus conditions, time points) and M features (e.g., voxels, cortical surface nodes) define an M-dimensional vector space. The goal of MVPA analyses is to analyze the structure of these high-dimensional representational spaces.

In a neural representational space, brain responses are vectorized, thus discarding the spatial relationships among cortical locations and the temporal relationships among time points. Thus, the approaches that we present here do not attempt to model the spatial structure of cortical topographies or how high-dimensional functional representations are packed into these topographies. An approach to modeling cortical topographies based on the principle of spatial continuity of function can be found in work by Aflalo & Graziano (2006, 2011) and Graziano & Aflalo (2007a,b). Anatomy and, in particular, cortical topography are important aspects of neural representation, of course. Although decoding methods discard anatomical and topographic information when brain responses are analyzed in high-dimensional representational spaces, the anatomical location of a representational space can be investigated using searchlight analyses (Kriegeskorte et al. 2006, Chen et al. 2011, Oosterhof et al. 2011), and the topographic organization of that representation can be recovered by projecting response vectors and linear discriminants from a common model representational space into individual subjects' topographies (Haxby et al. 2011).

MULTIVARIATE PATTERN CLASSIFICATION

MVP classification uses machine learning algorithms to classify response patterns, associating each neural response with an experimental condition. Pattern classification involves defining sectors in the neural representational space in which all response vectors represent the same class of information, such as a stimulus category (e.g., Haxby et al. 2001, Cox & Savoy 2003), an attended stimulus (e.g., Kamitani & Tong 2005), or a cognitive state (e.g., Haynes et al. 2007).

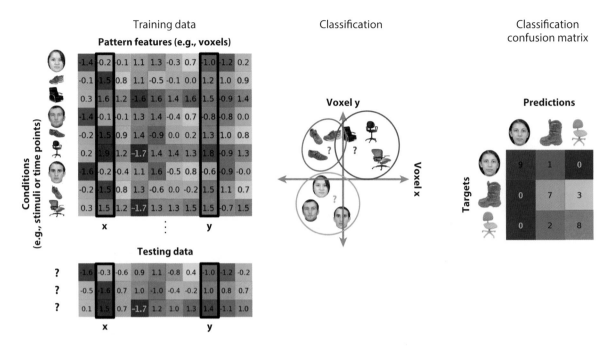

Training data

Pattern features (e.g., voxels)

Classification

Classification confusion matrix

Voxel y

Voxel x

Predictions

Targets

Conditions (e.g., stimuli or time points)

Testing data

Figure 2

MVP classification analyses involve partitioning data matrices into different sets for training and testing a pattern classifier. A classifier is trained to designate sectors of the vector space to the labels provided for the samples in the training set. Test samples are then classified as belonging to the labeled class associated with the sector in which they reside. Classification accuracy is measured as the proportion of predicted labels that match the actual label (target) for each test item. A confusion matrix provides information about the patterns of correct classifications (on the diagonal) and misclassifications (off diagonal).

An MVP classification analysis begins with dividing the data into independent training and test data sets (**Figure 2**). The decision rules that determine the confines of each class of neural response vectors are developed on training data. The border between sectors for different conditions is called a decision surface. The validity of the classifier is then tested on the independent test data. For valid generalization testing, the test data must play no role in the development of the classifier, including data preprocessing (Kriegeskorte et al. 2009). Each test data response vector is then classified as another exemplar of the condition associated with the sector in which it is located.

Classifier accuracy is the percentage of test vectors that are correctly classified. A more revealing assessment of classifier performance is afforded by examining the confusion matrix. A confusion matrix presents the frequencies for all classifications of each experimental condition, including the details about misclassifications. Examination of misclassifications adds information about which conditions are most distinct and which are more similar. This information is analyzed using additional methods in RSA (see next section). Examination of classification accuracy for each condition separately can alert the investigator to whether average accuracy is really dependent on a small number of conditions, rather than an accurate reflection of performance across all or most conditions. Thus, average classification accuracy is a useful metric but discards information that can be discovered by examining the classification confusion matrix. Confusion matrices are shown in **Figure 3** for two category perception experiments (Haxby et al. 2011; Connolly et al. 2012a,b). From the first experiment, on the perception of faces and objects, the confusion matrix reveals that if misclassified, faces are classified as other faces and objects as other objects. Moreover, the classifier cannot distinguish female from male faces. From the second experiment, on the

Test data: the data set used to test the validity of a decision rule that was derived on training data

Training data: the portion of a data set that is used to derive the decision rule for pattern classification

Decision surface: a surface that defines the boundary between sectors in a representational space and is used to classify vectors

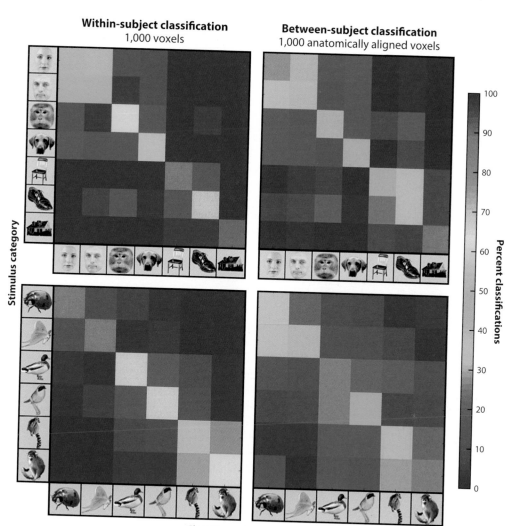

Figure 3

Confusion matrices for two experiments that measured responses to visual objects in human ventral temporal (VT) cortex. The patterns of misclassifications show that when items are misclassified they are more likely to be confused with items from the same superordinate category: faces and small objects (*top*); and primates, birds, and bugs (*bottom*). Classification performed on data matrices from the same subject (i.e., the same set of features) produces higher overall accuracies (*a*) than does between-subject classification (BSC) (*b*) where the features (voxels) have been aligned on the basis of a standard anatomical template.

perception of animals, the confusion matrix reveals that misclassifications are usually within animal class (primates, birds, and insects) and rarely across classes.

MVP classification uses machine learning classifiers to develop the decision rules. In general, different classifiers produce similar results, and some classifiers tend to perform better than others. A seminal MVP classification study (Haxby et al. 2001) used a one-nearest-neighbor classifier that classified a test response vector as the category for the training data vector that was closest in the neural representational space. Distance between vectors was measured using correlation, which is the cosine of the angle between mean-centered vectors. Because a single vector was used for each class in the training data—the mean pattern for each class in half the data—the decision surfaces were linear hyperplanes separating each pair of unique classes. Nearest-neighbor methods are fast and conceptually clear, but most have found that other classifiers provide slightly higher accuracies. Cox & Savoy (2003) were the first to use support vector machine (SVM) (Cortez & Vapnik 1995) classifiers for fMRI. SVM classifiers fine-tune the position of the decision surface on the basis of the vectors that are closest to the surface, i.e., the support vectors, by maximizing the distances from the surface to these borderline cases. Other regression-based methods, such as linear discriminant analysis (LDA) (e.g., Carlson et al. 2003, O'Toole et al. 2005), are also effective and can include regularization methods for selecting features, such as sparse multinomial logistic regression (SMLR) (Yamashita et al. 2008). Most MVP classification analyses have used linear classifiers—meaning that the decision surface is planar—some for theoretically driven reasons (Kamitani & Tong 2005) but mostly for simplicity and to avoid overfitting the noise in the training data, which leads to larger performance decrements in generalization testing.

Until recently, almost all MVP classification analyses had built a new classifier for each individual brain. Cox & Savoy (2003) showed that classifier performance dropped drastically if based on other subjects' data. The performance decrement for between-subject classification (BSC) relative to within-subject classification (WSC) shows that the structure of activity patterns differs across subjects. This variance could be due to the inadequacy of methods for aligning cortical topographies based on anatomical features. Some of the more successful BSC analyses are of large-scale patterns that involve many, widely distributed areas (Shinkareva et al. 2008, 2011), suggesting that larger-scale topographies may be aligned adequately based on anatomy and that poor BSC performance occurs when distinctions are found in finer-scale topographies. Low BSC accuracies could also be due to idiosyncratic neural codes. **Figure 3** shows the confusion matrices for both WSC and BSC of two category-perception experiments, illustrating the severity of the problem (Haxby et al. 2011). Accuracies dropped from 63% to 45% for 7 categories of faces and objects and from 69% to 37% for 6 animal species. A recently developed method for aligning the neural representational spaces across brains—hyperalignment—affords accuracies for BSC that are equal to, and sometimes higher than, the accuracies for WSC (Haxby et al. 2011), suggesting that the neural codes for different individuals are common rather than idiosyncratic. Use of hyperalignment to build a model of a common neural representational space is reviewed in a later section.

REPRESENTATIONAL SIMILARITY ANALYSIS

RSA examines the structure of representations within a representational space in terms of distances between response vectors (**Figure 4**). The complete set of distances among all pairs of response vectors is known as the dissimilarity matrix (DSM) (**Figure 5b,c**). Whereas MVP classification analyzes whether the vectors for different conditions are clearly distinct, RSA analyzes how they are related to each other. This approach confers several advantages. First, RSA can reveal that representations in different brain areas differ even if MVP classification is equivalent in those areas (Kriegeskorte et al. 2008a,b; Connolly et al. 2012a,b). Second, by converting the locations of response vectors from a set of feature coordinates to a set of distances between vectors, the geometry of the representational space is now in a format that is not dependent on the feature coordinate axes. This conversion allows comparison to DSMs for the same conditions in other spaces that have different feature coordinate axes, such as the voxels of another subject's brain or of another brain region (Kriegeskorte et al. 2008a,b; Connolly et al. 2012a,b). It even affords comparison of representational spaces based on stimulus feature models or on other types of brain activity measurement, such as single-unit recordings or MEG.

Representational similarity analysis

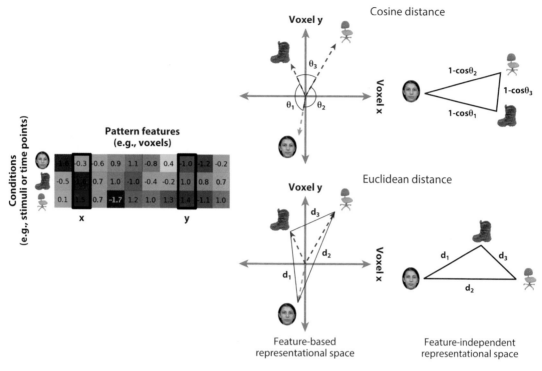

Figure 4

Representational similarity analysis examines the patterns of distances between vectors in the high-dimensional vector space. Measures of angular similarity such as cosine and Pearson product-moment correlation are standard measures that are most sensitive to the relative contributions of feature dimensions. These similarity measures are transformed into dissimilarities by subtracting them from 1. Another standard measure of the distance between vectors is Euclidean distance, which is more sensitive to overall differences in vector length or magnitude.

Investigation of the similarity of neural representations from fMRI data dates back to an early paper by Edelman et al. (1998), which was the first to use multidimensional scaling to visualize the representational space for visual objects. Two groups reanalyzed data from an early MVP classification study on the distributed representation of faces and objects (Haxby et al. 2001; reanalyzed by Hanson et al. 2004; O'Toole et al. 2005, 2007) and found similar similarity structures using different distance measures, one based on intermediate-layer weights in a neural network classifier and the other based on misclassifications. Both found that the strongest dissimilarity was between the animate (human faces and cats) and inanimate (houses, chairs, shoes, bottles, and scissors) categories and, within the inanimate domain, a strong dissimilarity between houses and all the smaller objects. These basic elements of the structure of face, animal, and object representations were corroborated and greatly amplified by subsequent studies in monkeys (Kiani et al. 2007) and humans (Kriegeskorte et al. 2008b). Kiani et al. (2007) measured the responses of single neurons in the monkey inferior temporal (IT) cortex to a large variety of faces, animals, and objects and calculated correlations among response vectors as indices of the similarity of the population response vectors. The results revealed the major distinctions between animate and inanimate stimuli, with a clear distinction between faces and bodies, but went deeper to show a similarity structure for the

Figure 5

Three examples of representational similarity analysis (RSA). (*a*) Dendrogram derived from multiple single-unit recordings in macaque inferior temporal (IT) cortex (from Kiani et al. 2007) shows hierarchical category structure with remarkable detail to the level of different classes of animal body type. (*b*) An example of cross-modal and cross-species RSA analysis for a common set of stimuli (from Kriegeskorte et al. 2008b) shows a high degree of common structure in the representational spaces between humans and monkeys, however, with less definition of subordinate categories within humans. (*c*) A targeted study of subordinate class structure for animal categories (from Connolly et al. 2012b) shows detailed structure for animal classes in human VT cortex.

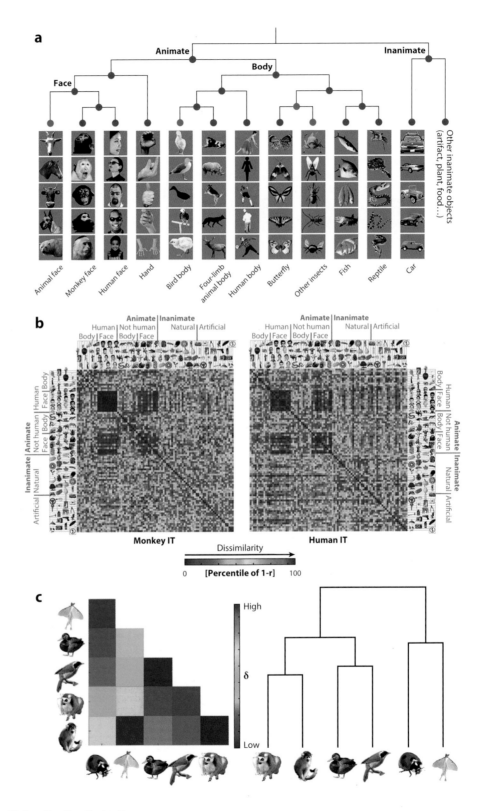

representations of animals that appears to reflect knowledge about similarities among species (see **Figure 5a**). Kriegeskorte et al. (2008a,b) used a subset of the stimuli from the Kiani et al. study in an fMRI study in humans and showed that the basic similarity structure of representations in human ventral temporal cortex is quite similar to that for representations in the monkey IT cortex (**Figure 5b**). Although Kriegeskorte et al. did not show the detailed structure among animal species that was evident in the monkey data, subsequent targeted studies by Connolly et al. (2012b) show that this structure is also evident in the human ventral temporal (VT) cortex (**Figure 5c**).

RSA can be applied in many different ways to discover the structure of neural representational geometries. These approaches include data-driven analyses and model-driven analyses. Data-driven RSA discovers and describes the similarity structures that exist in different cortical fields. Model-driven RSA searches for cortical fields whose similarity structures are predicted using stimulus or cognitive models, including behavioral ratings of perceived similarity.

Cortical fields that have representational spaces with similarity structures of interest can be identified in numerous ways. Cortical fields that show significant MVP classification can be further analyzed with RSA to describe the similarity structure of the conditions that can be distinguished (**Figure 6a**). Cortical fields can also be identified by virtue of having similarity structures that

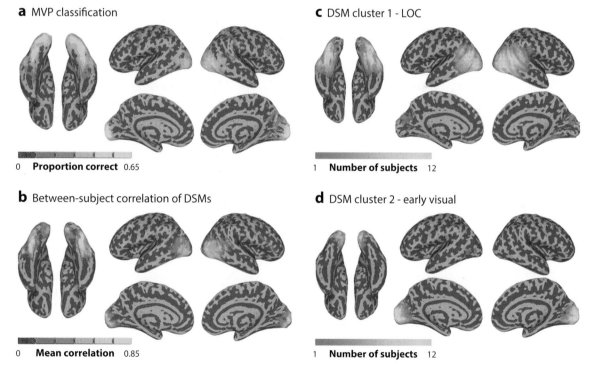

a MVP classification

0   **Proportion correct**   0.65

b Between-subject correlation of DSMs

0   **Mean correlation**   0.85

c DSM cluster 1 - LOC

1   **Number of subjects**   12

d DSM cluster 2 - early visual

1   **Number of subjects**   12

Figure 6

MVPA searchlight analyses (Kriegeskorte et al. 2006) for identifying cortical fields of interest (from Connolly et al. 2012a). MVP classification accuracies (*a*) and consistency in local similarity structures across subjects (*b*) identify similarly large swaths of the visually responsive cortex. Clustering of voxels based on similarities between locally defined searchlight dissimilarity matrix (DSMs) provides a means to identify cortical fields with unique shared structure such as the lateral occipital complex (LOC) (*c*) and the early visual cortex (*d*).

are consistent across subjects (**Figure 6b**). The similarity structures in different cortical fields identified in these ways, however, may differ. Connolly et al. (2012b) developed a clustering method for finding different similarity structures in different cortical locations that were shared across subjects (**Figure 6c,d**).

Understanding the similarity structure in a cortical field requires examining that structure. Often, the DSM itself is too complicated and high-dimensional to see the structure clearly. The full DSM, therefore, is often distilled into a lower-dimensional illustration to facilitate examination. The most common methods for reducing the dimensionality of a DSM are hierarchical clustering to produce a dendrogram (e.g., **Figure 6a,c**), multidimensional scaling (MDS), and related methods such as DISTATIS (Abdi et al. 2009, 2012a). The dendrogram in **Figure 5c** (Connolly et al. 2012b), for example, reveals that animal species from the same class are most similar to each other and that vertebrate classes (primates and birds) are more similar to each other than they are to the invertebrate class (insects). MDS often reveals that a low-dimensional subspace can account for a large portion of a similarity structure. For example, in Connolly et al. (2012b), the similarities among animal classes were captured by a single dimension—the animacy continuum—that ranged from primates to birds to insects and was associated with a distinctive coarse scale topography in human VT cortex that had previously been attributed to the distinction between animate and inanimate stimuli. Finer within-class distinctions (e.g., moths versus ladybugs), however, were based on other dimensions and finer-scale topographies.

The meaning of the similarity structure in a representational space can also be investigated by comparing it to a DSM generated by a model of the experimental conditions based on stimulus features or behavioral ratings (Kriegeskorte et al. 2008a). For example, Connolly et al. (2012b) showed that the DSM in the first cluster (**Figure 6b**) correlated highly with a DSM based on behavioral ratings of similarities among animals, whereas the second cluster correlated highly with a DSM based on visual features from a model of V1 neuron responses. Carlin et al. (2011) used RSA to investigate the representation of the direction of another's eye gaze that is independent of head angle. They constructed a model similarity structure of gaze direction with invariance across head angle. In addition, they constructed models of similarity structure due to confounding factors, such as image similarity, and searched for areas with a similarity structure that correlated with their DSM of interest after partialling out any shared variance due to confounding factors—illustrating how RSA may be used to test a well-controlled model.

One of the great advantages of RSA is that it strips a cluster of response vectors out of a feature-based representational space into a representational space based on relative distances among vectors. This format allows comparison of representational geometries across subjects, across brain regions, across measurement modalities, and even across species. The second-order isomorphism across these spaces is afforded by the feature-independent format of DSMs. For example, between-subject similarity of DSMs has been exploited to afford between-subject MVP classification (Abdi et al. 2012b, Raizada & Connolly 2012).

The feature-independent second-order isomorphism, however, does have some cost. Stripping representational spaces of features makes it impossible to compare population codes in terms of the constituent tuning functions of those features. Thus, one cannot investigate whether the spaces in different subjects share the same feature tuning functions or how these tuning function codes differ for different brain regions. One cannot predict the response to a new stimulus in a subject on the basis of the responses to that stimulus in other subjects. One cannot predict the tuning function for individual neural features in terms of stimulus features, precluding investigators from predicting the response pattern vector for a new stimulus on the basis of its features. The next two sections review methods that do afford these predictions using hyperalignment and stimulus-model-based encoding and decoding.

BUILDING A COMMON MODEL OF A NEURAL REPRESENTATIONAL SPACE

MVP classification usually builds a new classifier for each individual brain [within-subject classification (WSC)]. Except for distinctions that are carried by large, coarse-scale topographies, BSC based on other subjects' anatomically aligned response vectors yields accuracies that are much lower than those for WSC (see **Figure 3**). Basing decoding on classifiers that are tailored to individual representational spaces leaves open the question of whether the population responses in different brains use the same or idiosyncratic codes, in terms of the tuning functions for individual features within those responses.

Procrustes transformation: aligns two patterns of vectors by finding an orthogonal transformation that minimizes distances between paired vectors in two matrices

Building a common model of a representational space requires an algorithm for aligning the representational spaces of individual subjects' brains into that common space. Anatomical alignment, using affine transformations of the brain volume and rubber-sheet warping of the cortical manifold, does not afford BSC accuracies that approach WSC accuracies (Cox & Savoy 2003, Haxby et al. 2011, Conroy et al. 2013). Algorithms for function-based, rubber-sheet alignment of the cortical manifold, based on either the tuning functions or the functional connectivity of cortical nodes, improve BSC but still do not afford BSC accuracies that are equivalent to WSC accuracies (Sabuncu et al. 2010, Conroy et al. 2013).

A recently developed algorithm for aligning individual neural representational spaces into a common model space, hyperalignment, does afford BSC accuracies that are equivalent to, and sometimes exceed, WSC accuracies (Haxby et al. 2011). The algorithm revolves around a transformation matrix that is calculated for each individual subject that rotates that subject's representational space into the common model space (**Figure 7**). Valid parameters with broad general validity for high-level visual representations can be calculated on the basis of brain responses measured while subjects watch a complex, dynamic stimulus. This method was demonstrated using data collected while subjects watched the full-length action movie *Raiders of the Lost Ark*, reasoning that the subjects' visual cortices represent the same visual information while they watch the movie. The response vectors in different subjects' brains, however, are not aligned because voxels in the same anatomical locations do not have the same tuning functions. Hyperalignment uses the Procrustes transformation (Schönemann 1966) to rotate the coordinate axes of an individual's representational space to bring that subject's response vectors into optimal alignment with another subject's vectors. Iterative alignments of individual representational spaces to each other produced a single common representational space for the VT cortex. Each individual representational space could then be rotated into that common model space. The dimensionality of the common model space was reduced using principal components analysis. Optimal BSC accuracy for validation testing across experiments required more than 30 dimensions. Thus, the common model of the representational space in the VT cortex has 35 dimensions, meaning that the transformation matrix is an orthogonal matrix with 35 columns for the 35 common model dimensions and the same number of rows as the number of voxels in an individual subject's VT cortex (**Figure 7**).

The hyperalignment transformation matrix derived from responses to the movie provides the keys that unlock an individual's neural code. The parameters derived from the movie have general validity across a wide range of visual stimuli and can be applied to data from any experiment, making it possible to decode a subject's response vectors for a wide variety of stimuli based on other subjects' brain responses. BSC of data from two category perception experiments, after transformation into the common model dimensions using parameters derived from movie viewing, was equivalent to WSC. In subsequent work, we have found that the algorithm produces valid common models of representational spaces in early visual cortex, in lateral occipital and lateral temporal visual cortices, in auditory cortices, and in motor cortices.

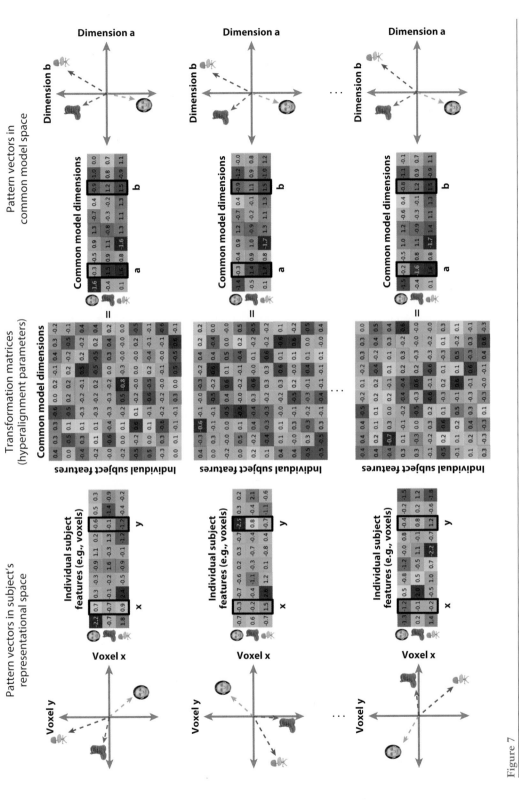

Figure 7

Intersubject hyperalignment of neural representational spaces. Hyperalignment aligns individual subjects' representational spaces into a common model representational space using high-dimensional rotation characterized by an orthogonal transformation matrix for each subject. A dimension in this common model space is a weighted sum of voxels in each individual subject (a column of the transformation matrix), which is functionally equivalent across subjects as reflected by the alignment of pattern vectors in the common model space.

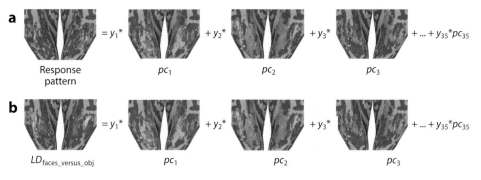

Figure 8

Modeling response patterns as weighted sums of common model dimensions. (*a*) Any response pattern in a subject can be modeled as a weighted sum of patterns representing common model dimensions. (*b*) Category-selective regions defined by contrasts, such as faces-versus-objects for the fusiform face area, can be modeled in the same way.

The transformation matrix provides a set of basis functions that can model any pattern of brain response. Each column in the matrix, corresponding to one common model dimension, is a set of weights distributed across voxels. A pattern of brain response in those voxels is a weighted sum of these patterns of weights (**Figure 8a**). Models of VT cortex based on single dimensions, such as contrasts that define category-selective regions, are modeled well in the 35-dimensional model space (**Figure 8b**), but these single dimensions account for only a small portion of the variance in responses to a dynamic and varied natural stimulus such as the movie. For example, the contrast between responses to faces and responses to objects, which defines the fusiform face area (FFA) (Kanwisher et al. 1997), accounts for only 12% of the variance that is accounted for by the 35-dimensional model. This result indicates that models based on simple, univariate contrasts are insufficient as models of neural representational spaces.

The use of a complex, dynamic stimulus is essential for deriving transformation matrix parameters that afford general validity across a wide range of stimuli. Transformation matrices can also be calculated on the basis of responses to more controlled experiments, such as the category perception experiments. These transformation matrices are valid for modeling the response vectors for stimuli in that experiment but, when applied to data from other experiments, do not afford BSC of new stimuli (Haxby et al. 2011). This result indicates that data from a limited sampling of brain states, such as those sampled in a standard category perception experiment, do not provide a sufficient basis for building a common model of a neural representational space.

STIMULUS-MODEL-BASED ENCODING AND DECODING

For MVP classification, RSA, and hyperalignment, a response vector to be decoded is compared with response vectors for that same stimulus measured in the same subject or in other subjects. These methods cannot predict the response pattern for a novel stimulus or experimental condition. Stimulus-model-based methods extend neural decoding to novel stimuli by predicting the response to stimulus features rather than to whole stimuli.

The stimuli used to produce training data for stimulus-model-based decoding are analyzed into constituent features. Feature sets used for this type of analysis include models of V1 neuron response profiles, namely oriented Gabor filters (Kay et al. 2008, Naselaris et al. 2009), visual motion energy filters (Nishimoto et al. 2011), semantic features (Mitchell et al. 2008, Naselaris

ECoG:
electrocorticography

et al. 2009), and acoustic features of music (Casey et al. 2012). These features can be continuous or binary. Thus, each stimulus is characterized as a vector in the high-dimensional stimulus feature space (**Figure 9**). The response for each feature in a neural representational space (e.g., voxel) is then modeled as a weighted sum of the stimulus features. For example, a V1 voxel may have the strongest weights for Gabor filters of a certain orientation and spatial frequency in a particular location, reflecting the orientation selectivity and retinotopic receptive field for that voxel. The prediction equation for each voxel is calculated on the basis of regularized regression analysis of responses to the stimuli in the training data. The result is a linear transformation matrix in which each column is a neural pattern feature and each row is a stimulus feature (**Figure 9**). The values in each column are the regression weights for predicting the response of that neural feature given a set of stimulus feature values.

Applying the encoding parameter transformation matrix to the stimulus feature values for a new stimulus thus predicts the response in each voxel (**Figure 9**). This new stimulus can be any new stimulus in the same domain that can be described with the same stimulus features. For example, using Gabor filters, investigators can predict the response to any natural still image (Kay et al. 2008, Naselaris et al. 2009). Using motion energy filters, the response to any video can be predicted (Nishimoto et al. 2011). Using acoustic features, the response to any clip of music can be predicted (Casey et al. 2012). The validity of the transformation can then be tested using either MVP classification or Bayesian reconstruction of the stimulus. Each type of validation testing involves analysis of response vectors for new stimuli. For MVP classification, neural response vectors are predicted for stimuli in the validation testing set using the linear transformation matrix estimated from an independent set of training stimuli. The classifier then tests whether the measured response vector is more similar to the predicted response vector for that stimulus than to predicted response vectors for other stimuli in the testing set. For Bayesian reconstruction, the algorithm identifies predicted response vectors generated from a large set of stimuli (priors) that are most similar to the measured response vector. The reconstructed stimulus is then produced from those matching stimuli. For example, Nishimoto et al. (2011) found the 30 videos that generated predicted response vectors that most closely matched a measured response vector. They then produced a video by averaging those 30 video priors. The reconstructed videos bear an unmistakable resemblance to the viewed video, albeit lacking detail, providing a convincing proof of concept.

Related methods have been used to reconstruct 10×10 contrast images (Miyawaki et al. 2008) and decode the contents of dream imagery (Horikawa et al. 2013). A study of brain activity measured over the auditory language cortex using electrocorticography (ECoG) in surgery patients used related methods to reconstruct the spectrogram for spoken words, producing auditory stimuli that closely resembled the original words (Pasley et al. 2012).

In general, stimulus-model-based decoding is limited to Bayesian reconstruction based on similarities to a set of priors. Simply generating a stimulus based on predicted features is infeasible because the dimensionality of the neural representational space, given current brain measurement techniques, is much lower than that of stimulus feature spaces that are complete enough to construct a stimulus. Some researchers have speculated, however, that perception also involves a process of Bayesian reconstruction based on similarity to prior perceptual experiences, a process that could operate effectively with incomplete specification of stimulus features (Friston & Kiebel 2009).

MULTIPLEXED TOPOGRAPHIES FOR POPULATION RESPONSES

The spatial resolution of fMRI used in most neural decoding studies is 2–3 mm. Consequently, each fMRI voxel contains more than 100,000 neurons and roughly 10–100 cortical columns.

Stimulus-model-based encoding and decoding

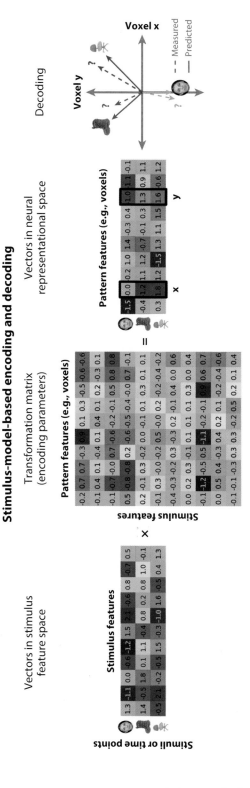

Figure 9

Stimulus-model-based encoding and decoding involve deriving a transformation matrix that affords prediction of responses of neural features from stimulus-model features. Each column in the encoding transformation matrix corresponds to a neural feature and provides the weights for stimulus features to estimate that neural feature's response. Decoding responses to novel stimuli involves comparing measured response patterns to predicted patterns for stimulus priors.

The fact that neural activity resampled into this coarse spatial grid can be effectively decoded suggests that the topographies for distinct neural representations of the decoded information include lower spatial frequency patterns. In fact, decoding of different distinctions appears to be based on topographies of different spatial scales. Coarse distinctions, such as the difference between animate and inanimate stimuli (Martin 2007, Mahon et al. 2009) or animal species at different levels on the animacy continuum (Connolly et al. 2012b), are found in a coarse topography from lateral to medial VT cortex. Finer distinctions, such as those between old and young human faces (Op de Beeck et al. 2010, Brants et al. 2011) or between two types of birds (Connolly et al. 2012b), are carried by finer-scale topographies. The finer-scale topographies for subordinate distinctions, such as old versus young faces, are not restricted to the areas of maximal activity for the superordinate category (Op de Beeck et al. 2010, Brants et al. 2011, Haxby et al. 2011). Thus, the topographies for different distinctions appear to be overlapping and exist at multiple scales.

The topographic organization of the cortex reflects the problem of projecting a high-dimensional representational space, composed of features with complex, interrelated tuning functions, into a two-dimensional manifold. Kohonen (1982, 2001) proposed that cortical maps self-organize to locate neurons with related tuning functions close to each other. Aflalo & Graziano (2006, 2011; Graziano & Aflalo 2007a,b) have used this principle of spatial continuity of function to account for the topographic organization of the motor cortex (Aflalo & Graziano 2006) and the coarse-scale topographic organization of the extrastriate visual cortex (Aflalo & Graziano 2011). Others have used this principle to account for multiplexed functional topographies in the primary visual (Durbin & Mitchison 1990) and auditory (Schreiner 1995) cortices. Accounting for seemingly disordered multiplexed functional topographies requires an adequate model of the high-dimensional representational space. Haxby et al. (2011) showed that the topographies in VT cortex that support a wide range of stimulus distinctions, including distinctions among responses to complex video segments, can be modeled with 35 basis functions. These pattern bases are of different spatial scales and define gradients in different locations. Low-dimensional models of neural representation, such as those exemplified by category-selective regions (Kanwisher 2010, Weiner & Grill-Spector 2011), however, are not sufficient to model complex, multiplexed functional topographies that support these distinctions.

FUTURE DIRECTIONS

Recent advances in computational methods for neural decoding have revealed that the information provided from measurement of human brain activity is far more detailed and specific than was previously thought possible. This review of the current state of the art shows that these methods can be integrated in a framework organized around the concept of high-dimensional representational spaces. The development of algorithms for neural encoding and decoding, however, has only begun. Although the power of neural decoding is limited by brain activity measurement methods—and further technological breakthroughs will bring greater power and sensitivity to neural decoding projects—new computational methods can direct investigators to additional important topics. Three areas for future investigation are addressed below.

Individual and Group Differences

Most neural decoding work to date has focused on the commonality of neural representational spaces across subjects. The methods for aligning representational spaces across subjects, namely RSA and hyperalignment, however, can also be adapted to investigate how an individual's representational space differs from others' or how groups differ. Developing methods for examining individual and group differences would facilitate studies of how factors such as development, education, genetics, and clinical disorders influence neural representation.

Between-Area Transformations in a Processing Pathway

RSA affords one way to draw distinctions among how representations are structured in different parts of a processing pathway (Kriegeskorte et al. 2008b, Connolly et al. 2012b). Modeling the transformation of representations from one cortical field to another, however, would help elucidate how information is processed within a pathway, leading to the construction of representations laden with meaning from representations of low-level physical stimulus properties. For example, the manifold of response vectors that correspond to different views of the same face in early visual cortex is complex and does not afford easy separation of the responses to one individual face from the responses to another, whereas the manifold in the anterior temporal cortex may untangle these manifolds, affording viewpoint-invariant identity recognition (DiCarlo & Cox 2007, Freiwald & Tsao 2010). Determining the structure of these transformations and the role of input from multiple regions is a major challenge for future work.

Multimodality Decoding

fMRI has very coarse temporal resolution. Neural decoding studies with other measurement modalities, such as single-unit recording (e.g., Hung et al. 2005, Freiwald & Tsao 2010), ECoG (Pasley et al. 2012), and MEG (Carlson et al. 2011, Sudre et al. 2012), have shown how population codes for different types of information emerge over time as measured in tens of milliseconds. Similar tracking of population codes has been demonstrated with fMRI but is severely limited by the temporal characteristics of the hemodynamic response (Kohler et al. 2013). Using multiple modalities, representational spaces could be modeled in which some dimensions reflect different time points for the same spatial feature and other dimensions reflect spatiotemporal gradients or wavelets. The potential of multimodal neural decoding is largely unexplored.

DISCLOSURE STATEMENT

The authors are not aware of any affiliations, memberships, funding, or financial holdings that might be perceived as affecting the objectivity of this review.

LITERATURE CITED

Abdi H, Dunlop JP, Williams LJ. 2009. How to compute reliability estimates and display confidence and tolerance intervals for pattern classifiers using the Bootstrap and 3-way multidimensional scaling (DISTATIS). *NeuroImage* 45:89–95

Abdi H, Williams LJ, Conolly AC, Gobbini MI, Dunlop JP, Haxby JV. 2012a. Multiple Subject Barycentric Discriminant Analysis (MUSUBADA): how to assign scans to categories without using spatial normalization. *Comp. Math. Methods Med.* 2012:634165

Abdi H, Williams LJ, Valentin D, Bennani-Dosse M. 2012b. STATIS and DISTATIS: optimum multitable principal component analysis and three way metric multidimensional scaling. *Wiley Interdiscip. Rev. Comput. Stat.* 4:124–67

Aflalo TN, Graziano MSA. 2006. Possible origins of the complex topographic organization of motor cortex: reduction of a multidimensional space onto a two-dimensional array. *J. Neurosci.* 26:6288–97

Aflalo TN, Graziano MSA. 2011. Organization of the macaque extrastriate cortex re-examined using the principle of spatial continuity of function. *J. Neurophysiol.* 105:305–20

Brants M, Baeck A, Wagemans J, Op de Beeck H. 2011. Multiple scales of organization for object selectivity in ventral visual cortex. *NeuroImage* 56:1372–81

Carlin JD, Calder AJ, Kriegeskorte N, Nili H, Rowe JB. 2011. A head view-invariant representation of gaze direction in anterior superior temporal cortex. *Curr. Biol.* 21:1817–21

Carlson TA, Hogendoorn H, Kanai R, Mesik J, Turret J. 2011. High temporal resolution decoding of object position and category. *J. Vis.* 11:1–17

Carlson TA, Schrater P, He S. 2003. Patterns of activity in the categorical representations of objects. *J. Cogn. Neurosci.* 15:704–17

Casey M, Thompson J, Kang O, Raizada R, Wheatley T. 2012. Population codes representing musical timbre for high-level fMRI categorization of music genres. In *Machine Learning and Interpretation in Neuroimaging*, Ser. Vol. 7263, ed. G Langs, I Rish, M Grosse-Wentrup, B Murphy, pp. 36–41. Berlin/Heidelberg: Springer-Verlag

Chen Y, Namburi P, Elliott LT, Heinzle J, Soon CS, et al. 2011. Cortical surface-based searchlight decoding. *NeuroImage* 56:582–92

Connolly AC, Gobbini MI, Haxby JV. 2012a. Three virtues of similarity-based multi-voxel pattern analysis: an example from the human object vision pathway. In *Understanding Visual Population Codes (UVPC): Toward A Common Multivariate Framework for Cell Recording and Functional Imaging*, ed. N Kriegeskorte, G Kreiman, pp. 335–55. Cambridge, MA: MIT Press

Connolly AC, Guntupalli JS, Gors J, Hanke M, Halchenko YO, et al. 2012b. The representation of biological classes in the human brain. *J. Neurosci.* 32:2608–18

Conroy BR, Singer BD, Guntupalli JS, Ramadge PJ, Haxby JV. 2013. Inter-subject alignment of human cortical anatomy using functional connectivity. *NeuroImage* 81:400–11

Cortes C, Vapnik V. 1995. Support-vector networks. *Mach. Learn.* 20:273–97

Cox DD, Savoy RL. 2003. Functional magnetic resonance imaging (fMRI) "brain reading": detecting and classifying distributed patterns of fMRI activity in human visual cortex. *NeuroImage* 19:261–70

DiCarlo JJ, Cox DD. 2007. Untangling invariant object recognition. *Trends Cogn. Sci.* 11:333–41

Durbin R, Mitchison G. 1990. A dimension reduction framework for understanding cortical maps. *Nature* 343:644–47

Edelman S, Grill-Spector K, Kushnir T, Malach R. 1998. Toward direct visualization of the internal shape space by fMRI. *Psychobiology* 26:309–21

Formisano I, De Martino F, Bonte M, Goebel R. 2008. "Who" is saying "what"? Brain-based decoding of human voice and speech. *Science* 322:970–73

Freiwald WA, Tsao DY. 2010. Functional compartmentalization and viewpoint generalization within the macaque face-processing system. *Science* 330:845–51

Friston K, Kiebel S. 2009. Predictive coding under the free-energy principle. *Phil. Trans. R. Soc. B.* 364:1211–21

Graziano MSA, Aflalo TN. 2007a. Mapping behavioral repertoire onto the cortex. *Neuron* 56:239–51

Graziano MSA, Aflalo TN. 2007b. Rethinking cortical organization: moving away from discrete areas arranged in hierarchies. *Neuroscientist* 13:138–47

Hanke M, Halchenko YO, Sederberg PB, Hanson SJ, Haxby JV, Pollman S. 2009. PyMVPA: A Python toolbox for multivariate pattern analysis of fMRI data. *Neuroinformatics* 7:37–53

Hanson SJ, Toshihiko M, Haxby JV. 2004. Combinatorial codes in ventral temporal lobe for object recognition: Haxby (2001) revisited: is there a "face" area? *NeuroImage* 23:156–67

Harrison SA, Tong F. 2009. Decoding reveals the contents of visual working memory in early visual areas. *Nature* 458:632–35

Haxby JV. 2012. Multivariate pattern analysis of fMRI: the early beginnings. *NeuroImage* 62:852–55

Haxby JV, Gobbini MI, Furey ML, Ishai A, Schouten JL, Pietrini P. 2001. Distributed and overlapping representations of faces and objects in ventral temporal cortex. *Science* 293:2425–30

Haxby JV, Guntupalli JS, Connolly AC, Halchenko YO, Conroy BR, et al. 2011. A common, high-dimensional model of the representational space in human ventral temporal cortex. *Neuron* 72:404–16

Haynes JD, Rees G. 2005. Predicting the orientation of invisible stimuli from activity in human primary visual cortex. *Nat. Neurosci.* 8:686–91

An integrated software system based on Python for performing neural decoding.

Initial paper on multivariate pattern classification of fMRI.

Introduces hyperalignment for building a common high-dimensional model of a neural representational space.

Haynes JD, Rees G. 2006. Decoding mental states from brain activity in humans. *Nat. Rev. Neurosci.* 7:523–34

Haynes JD, Sakai K, Rees G, Gilbert S, Frith C, Passingham RE. 2007. Reading hidden intentions in the human brain. *Curr. Biol.* 17:323–28

Horikawa T, Tamaki M, Miyawaki Y, Kamitani Y. 2013. Neural decoding of visual imagery during sleep. *Science* 340:639–42

Hung CP, Kreiman G, Poggio T, DiCarlo JJ. 2005. Fast readout of object identity from macaque inferior temporal cortex. *Science* 310:863–66

Kamitani Y, Tong F. 2005. Decoding the visual and subjective contents of the human brain. *Nat. Neurosci.* 8:679–85

Kanwisher N. 2010. Functional specificity in the human brain: a window into the functional architecture of the mind. *Proc. Natl. Acad. Sci. USA* 107:11163–70

Kanwisher N, McDermott J, Chun MM. 1997. The fusiform face area: a module in human extrastriate cortex specialized for face perception. *J. Neurosci.* 17:4302–11

Kay KN, Naselaris T, Prenger RJ, Gallant JL. 2008. Identifying natural images from human brain activity. *Nature* 452:352–55

Kiani R, Esteky H, Mirpour K, Tanaka K. 2007. Object category structure in response patterns of neuronal population in monkey inferior temporal cortex. *J. Neurophysiol.* 97:4296–309

Kohler PJ, Fogelson SV, Reavis EA, Meng M, Guntupalli JS, et al. 2013. Pattern classification precedes region-average hemodynamic response in early visual cortex. *NeuroImage* 78:249–60

Kohonen T. 1982. Self-organizing formation of topologically correct feature maps. *Biol. Cybern.* 43:59–69

Kohonen T. 2001. *Self-Organizing Maps.* Berlin: Springer

Kriegeskorte N, Goebel R, Bandettini P. 2006. Information-based functional brain mapping. *Proc. Natl. Acad. Sci. USA* 103:3863–68

Kriegeskorte N, Kievet RA. 2013. Representational geometry: integrating cognition, computation, and the brain. *Trends Cogn. Sci.* 17:401–12

Kriegeskorte N, Mur M, Bandettini P. 2008a. Representational similarity analysis—connecting the branches of systems neuroscience. *Front. Syst. Neurosci.* 2:4

Kriegeskorte N, Mur M, Ruff DA, Kiani R, Bodurka J, et al. 2008b. Matching categorical object representations in inferior temporal cortex of man and monkey. *Neuron* 60:1126–41

Kriegeskorte N, Simmons WK, Bellgowan PSF, Baker CI. 2009. Circular analysis is systems neuroscience: the dangers of double dipping. *Nat. Neurosci.* 12:535–40

Mahon BZ, Anzellotti S, Schwarzbach J, Zampini M, Caramazza A. 2009. Category-specific organization in the human brain does not require visual experience. *Neuron* 63:397–405

Martin A. 2007. The representation of object concepts in the brain. *Annu. Rev. Psychol* 58:25–45

Mitchell TM, Shinkareva SV, Carlson A, Chang K-M, Malave VL, et al. 2008. Predicting human brain activity associated with the meanings of nouns. *Science* 320:1191–95

Miyawaki Y, Uchida H, Yamashita O, Sato M, Morito Y, et al. 2008. Visual image reconstruction from human brain activity using a combination of multiscale local image decoders. *Neuron* 60:915–29

Naselaris T, Kay KN, Nishimoto S, Gallant JL. 2011. Encoding and decoding in fMRI. *NeuroImage* 56:400–10

Naselaris T, Prenger RJ, Kay KN, Oliver M, Gallant JL. 2009. Bayesian reconstruction of natural images from human brain activity. *Neuron* 63:902–15

Nishimoto S, Vu AT, Naselaris T, Bejamini Y, Yu B, Gallant JL. 2011. Reconstructing visual experience from brain activity evoked by natural movies. *Curr. Biol.* 21:1641–46

Norman KA, Polyn SM, Detre GJ, Haxby JV. 2006. Beyond mind-reading: multi-voxel pattern analysis of fMRI data. *Trends Cogn. Sci.* 10:424–30

Oosterhof NN, Wiestler T, Downing PE, Diedrichsen J. 2011. A comparison of volume-based and surface-based multi-voxel pattern analysis. *NeuroImage* 56:593–600

Op de Beeck H, Brants M, Baeck A, Wagemans J. 2010. Distributed subordinate specificity for bodies, faces, and buildings in human ventral visual cortex. *NeuroImage* 49:3414–25

O'Toole AJ, Jiang F, Abdi H, Haxby JV. 2005. Partially distributed representations of objects and faces in ventral temporal cortex. *J. Cogn. Neurosci.* 17:580–90

First paper to show that MVPA can decode an early visual feature, namely edge orientation.

Introduced stimulus-model-based encoding and decoding of natural images.

Introduces RSA as a common format for the geometry of representational spaces.

First demonstration that RSA affords comparison of representational spaces across species.

This paper introduced decoding of words and concepts based on semantic feature models.

Reviews methods for stimulus-model-based encoding and decoding.

O'Toole AJ, Jiang F, Abdi H, Pénard N, Dunlop JP, Parent MA. 2007. Theoretical, statistical, and practical perspectives on pattern-based classification approaches to the analysis of functional neuroimaging data. *J. Cogn. Neurosci.* 191:735–52

Pasley BN, David SV, Mesgarani N, Flinker A, Shamma SA, et al. 2012. Reconstructing speech from human auditory cortex. *PLoS Biol.* 10:e1001251

Pereira F, Mitchell T, Botvinick M. 2009. Machine learning classifiers and fMRI: a tutorial overview. *NeuroImage* 45(Suppl. 1):S199–209

Raizada RDS, Connolly AC. 2012. What makes different people's representations alike: neural similarity space solves the problem of across-subject fMRI decoding. *J. Cogn. Neurosci.* 24:868–77

Sabuncu M, Singer BD, Conroy B, Bryan RE, Ramadge PJ, Haxby JV. 2010. Function-based intersubject alignment of human cortical anatomy. *Cereb. Cortex* 20:130–40

Schönemann PH. 1966. A generalized solution of the orthogonal procrustes problem. *Psychometrika* 31:1–10

Schreiner CE. 1995. Order and disorder in auditory cortical maps. *Curr. Opin. Neurobiol.* 5:489–96

Shinkareva SV, Malave VL, Mason RA, Mitchell TM, Just MA. 2011. Commonality of neural representations of words and pictures. *NeuroImage* 54:2418–25

Shinkareva SV, Mason RA, Malave VL, Wang W, Mitchell TM, Just MA. 2008. Using fMRI brain activation to identify cognitive states associated with perception of tools and dwellings. *PLoS ONE* 1:e1394

Soon CS, Brass M, Heinze HJ, Haynes JD. 2008. Unconscious determinants of free decisions in the human brain. *Nat. Neurosci.* 5:543–45

Staeren N, Renvall H, De Martino F, Goebel R, Formisano E. 2009. Sound categories are represented as distributed patterns in the human auditory cortex. *Curr. Biol.* 19:498–502

Sudre G, Pomerleau D, Palatucci M, Wehbe L, Fyshe A, et al. 2012. Tracking neural coding of perceptual and semantic features of concrete nouns. *NeuroImage* 62:451–63

Tong F, Pratte MS. 2012. Decoding patterns of human brain activity. *Annu. Rev. Psychol.* 63:483–509

Weiner K, Grill-Spector K. 2011. The improbable simplicity of the fusiform face area. *Trends Cogn. Sci.* 16:251–54

Yamashita O, Sato M-A, Yoshioka T, Tong F, Kamitani Y. 2008. Sparse estimation automatically selects voxels relevant for the decoding of fMRI activity patterns. *NeuroImage* 42:1414–29

Measuring Consciousness in Severely Damaged Brains

Olivia Gosseries,[1,2,3] Haibo Di,[1,4] Steven Laureys,[1] and Mélanie Boly[1,2,5]

[1]Coma Science Group, Cyclotron Research Center and Neurology Department, University of Liege, and University Hospital of Liege, 4000 Liege, Belgium; email: ogosseries@ulg.ac.be, dihaibo@yahoo.com.cn, steven.laureys@ulg.ac.be, mboly@ulg.ac.be

[2]Center for Sleep and Consciousness, Department of Psychiatry, [3]Postle Laboratory, Department of Psychology and Psychiatry, University of Wisconsin, Madison, Wisconsin 53719

[4]International Vegetative State and Consciousness Science Institute, Hangzhou Normal University, Hangzhou, China

[5]Department of Neurology, University of Wisconsin, Madison, Wisconsin 53792

Annu. Rev. Neurosci. 2014. 37:457–78

First published online as a Review in Advance on June 23, 2014

The *Annual Review of Neuroscience* is online at neuro.annualreviews.org

This article's doi: 10.1146/annurev-neuro-062012-170339

Keywords

vegetative state, minimally conscious state, clinical assessment, neuroimaging, neural correlates of consciousness

Abstract

Significant advances have been made in the behavioral assessment and clinical management of disorders of consciousness (DOC). In addition, functional neuroimaging paradigms are now available to help assess consciousness levels in this challenging patient population. The success of these neuroimaging approaches as diagnostic markers is, however, intrinsically linked to understanding the relationships between consciousness and the brain. In this context, a combined theoretical approach to neuroimaging studies is needed. The promise of such theoretically based markers is illustrated by recent findings that used a perturbational approach to assess the levels of consciousness. Further research on the contents of consciousness in DOC is also needed.

Contents

INTRODUCTION

Vegetative state (VS)/unresponsive wakefulness syndrome (UWS): patients who are aroused but not aware of themselves and their surroundings

Minimally conscious state (MCS): patients who are aroused and show fluctuating signs of awareness without being able to functionally communicate

Disorders of consciousness (DOC): refers to patients with severe acquired brain injuries in an altered state of consciousness; includes coma, VS/UWS, and MCS

EMCS: emergence of the minimally conscious state (i.e., functional communication or object use)

Clinical and neuroimaging studies have made significant progress in the differential diagnosis, treatment, and ethical management of patients in a coma, in a vegetative state/unresponsive wakefulness syndrome (VS/UWS), and in a minimally conscious state (MCS) (Giacino et al. 2014). In this review, we discuss the state of the science for clinical assessment of disorders of consciousness (DOC) and the potential use of neuroimaging to diagnose consciousness.

Following severe damage to the brain, caused by trauma, stroke, or anoxia, patients can fall into a coma. Coma is a transient state characterized by a complete absence of wakefulness and awareness (Plum & Posner 1983). The recovery of wakefulness without signs of awareness heralds a transition to VS/UWS (Laureys et al. 2010, Multi-Society Task Force on PVS 1994a). In contrast, patients in MCS show reproducible nonreflexive behaviors but remain unable to communicate (Giacino et al. 2002). The MCS entity has been divided into MCS+ and MCS−, depending on the complexity of behavioral responses (i.e., presence or absence of language functions, respectively) (Bruno et al. 2012). Emergence of MCS (EMCS) occurs when patients regain accurate communication and/or functional use of objects. Finally, locked-in syndrome (LIS) patients can be misdiagnosed as DOC despite preserved awareness because of a complete paralysis of voluntary muscles, except vertical eye movements (Bauer et al. 1979). **Table 1** summarizes diagnostic criteria for DOC and related states.

CLINICAL ASSESSMENT OF CONSCIOUSNESS

The clinical assessment of the level of consciousness is based primarily on observation of spontaneous and stimulus-evoked behaviors. Arousal is measured by eye-opening, whereas awareness is assessed by patient's command-following or the assessor's search for other nonreflexive behaviors. Misdiagnosis of unawareness is very frequent (up to 40%) when diagnosis is based solely on clinical consensus, without use of appropriate behavioral scales (Schnakers et al. 2009). The most sensitive scale to differentiate MCS from VS/UWS is, to date, the revised version of the Coma Recovery Scale (CRS-R) (Giacino et al. 2004, Seel et al. 2010). In the intensive care unit, a routine use of the Full Outline of Unresponsiveness scale, which is faster to administer, is also recommended

Table 1 Diagnostic criteria for patients with severe acquired brain injuries

Clinical entities	DOC	Definition
Coma (Plum & Posner 1983)	Yes	No wakefulness
		No awareness of self or environment
Vegetative state/unresponsive wakefulness syndrome (Laureys et al. 2010, Multi-Society Task Force on PVS 1994a)	Yes	Wakefulness
		No awareness of self or environment
		No sustained, reproducible, purposeful behavioral responses to external stimuli
		No language comprehension or expression
		Relatively preserved hypothalamic and brain stem autonomic functions
		Bowel and bladder incontinence
		Variably preserved cranial-nerve and spinal reflexes
Minimally conscious state (Bruno et al. 2011b, Giacino et al. 2002)	Yes	Wakefulness
		Fluctuating awareness with reproducible, purposeful behavioral responses to external stimuli
Minimally conscious state minus	Yes	Visual pursuit
		Reaching for objects
		Orientation to noxious stimulation
		Contingent behavior
Minimally conscious state plus	Yes	Following commands
		Intentional communication
		Intelligible verbalization
Emergence from minimally conscious state (Giacino et al. 2002)	No	Functional communication
		Functional object use
Locked-in syndrome (American Congress of Rehabilitation Medicine 1995)	No	Wakefulness
		Awareness
		Aphonia or hypophonia
		Quadriplegia or quadriparesis
		Presence of communication through the eyes
		Preserved cognitive abilities

DOC, disorders of consciousness.

(Wijdicks et al. 2005). Specific assessment material should also be employed to increase sensitivity (see sidebar, Clinical Assessment). On the patient side, some factors potentially causing decreased responsiveness should be noted: motor impairment, aphasia, agnosia, blindness or deafness, fluctuation of vigilance, and the presence of pain (Schnakers 2012). Other medical complications (e.g., infections) and sedating medications may also complicate the assessment of DOC (Whyte et al. 2013). These elements should be investigated. The sidebar Clinical Assessment provides our recommendations concerning clinical assessment of DOC. The sidebar Clinical Management describes how recent advances in clinical diagnosis have affected treatment, prognosis, and ethical issues in DOC.

Even if the border zone between patients in VS/UWS and MCS is, at present, well delimited, bedside assessment of consciousness is intrinsically gated by behavioral responsiveness. It is now increasingly more recognized that the absence of observed purposeful behaviors at the bedside cannot be taken as definitive proof of the absence of consciousness. If persistent doubts concerning a patient's consciousness level exist, neuroimaging techniques such as positron emission tomography (PET), functional magnetic resonance imaging (fMRI), and electroencephalography (EEG) can be useful to complement behavioral diagnosis.

Locked-in syndrome (LIS): patients who are aroused and aware but who cannot move except to make eye movements

Coma Recovery Scale-Revised (CRS-R): behavioral scale developed to assess the levels of consciousness in patients recovering from coma, and especially to differentiate conscious from unconscious patients

ACTIVE NEUROIMAGING PARADIGMS

As previously mentioned, there is a significant risk that decreased behavioral responsiveness in brain-damaged patients may be due at least partially to motor impairment. In this context, neuroimaging paradigms that identify nonreflexive brain activation patterns in response to commands, while bypassing motor output, may be helpful. A positive response to these paradigms could, in principle, be considered reasonable evidence for the presence of consciousness in a given patient.

CLINICAL ASSESSMENT

1. **What to know before starting?**
   - The **terminology** of DOC (see **Table 1**)
   - The **signs of MCS**: reproducible responses to command, visual pursuit, automatic motor response (e.g., scratching, grabbing objects), adapted emotional behavior, localization to noxious stimulation, intelligible verbalization, object recognition and localization, nonfunctional communication, resistance to eye-opening (Giacino et al. 2002, van Ommen et al. 2013)
   - The **signs of EMCS**: functional communication and object use (Giacino et al. 2002)
   - **Reflex behaviors:** auditory startle, blinking to threat, flexion withdrawal/stereotyped to pain, yawning, oral reflexes (Giacino et al. 2002)
   - **Debated behavior:** visual fixation (Bruno et al. 2010), localization to sound (Cheng et al. 2013)
2. **What to do before starting?**
   - Collect patient's past and current **medical history:** sensory deficits, cause of coma, time since onset, localized pain, sedative medication
   - Always consider the patient **conscious** even if apparently unresponsive. Explain the aim of the exam and the need for full collaboration
   - Place the patient in **sitting position**
   - All **limbs** must be **visible**
   - Ensure enough light and quiet **environment** with a period of rest before starting
   - Apply **arousal protocol** if needed (Giacino et al. 2004)
   - Perform a few minutes of **observation** of spontaneous behavior
3. **What to do during the assessment?**
   - Assess **all modalities:** audition, vision, motricity/tactile stimulation, oromotor behavior, communication, arousal
   - Use the **Coma Recovery Scale-Revised**
   - **Use specific tools:** mirror for visual pursuit (Vanhaudenhuyse et al. 2008), own name for auditory localization (Cheng et al. 2013), oral and written commands, colorful objects, meaningful/emotional stimuli
   - **Way to assess:** assess the most reactive part of the body (from medical history, spontaneous behavior), ask several command-following questions based on spontaneous behaviors, use finger for blinking to threat, evaluate visual pursuit in horizontal and vertical planes
   - Give **encouragement** to the patient
   - If signs of **fatigue:** break and/or arousal protocol
4. **Other recommendations**
   - **Repeat assessments** combining morning and afternoon evaluations, minimum 5 times total for a final diagnosis
   - **Extended evaluation time** (20–60 min) needed
   - **Qualified and trained assessor**

CLINICAL MANAGEMENT

Advances in the understanding of brain function in noncommunicative severely brain-damaged patients go hand in hand within their clinical management. There is currently no standard of care to guide clinical management of patients with DOC. Once signs of consciousness are detected at the bedside (Seel et al. 2010) or via neuroimaging (Stender et al. 2014), the next step is to find a way for these patients to communicate. Standardized protocols searching for reliable responses to commands can be used to develop a binary code (Whyte et al. 1999). Communication-enabling brain computer interfaces can also be used via active paradigms in EEG and fMRI (Chatelle et al. 2012a, Lulé et al. 2013), or even by measuring changes in pupil size (Stoll et al. 2013).

Pharmacological treatments such as amantadine (Giacino et al. 2012) and zolpidem (Thonnard et al. 2014, Whyte et al. 2014) should be systematically tried in DOC patients because they can potentially improve patients' levels of awareness (Gosseries et al. 2013). Amantadine has been correlated with an increased metabolism in the frontoparietal network in an MCS patient (Schnakers et al. 2008a), whereas Zolpidem decreased low-frequency EEG activity in several patients with DOC (Williams et al. 2013). If signs of discomfort are observed, using for instance the Nociception Coma Scale-Revised (Chatelle et al. 2012b), pain medication should be given (Schnakers & Zasler 2007). This scale has been shown to selectively capture residual activity in pain matrix regions (e.g., anterior cingulated cortex) in severely brain-damaged patients (Chatelle et al. 2014). In some cases, trials of therapeutic interventions including invasive thalamic brain stimulation (Schiff et al. 2007), spinal cord stimulation (Yamamoto et al. 2013), and noninvasive transcranial direct current stimulation are indicated (Thibaut et al. 2014).

Patients in MCS have more chance of recovery than do patients in VS/UWS (Luauté et al. 2010, Noé et al. 2012). Other prognostic factors are the CRS-R total score on admission (i.e., >6) (Estraneo et al. 2013), a young age (Howell et al. 2013), a traumatic etiology (Multi-Society Task Force on PVS 1994b), an early time since onset (Whyte et al. 2009), the presence of pupillary light reflexes (Fischer et al. 2006), the absence of medical complications (Whyte et al. 2013), and specialized early treatment (Seel et al. 2013). VS/UWS patients who show preserved fMRI activation of associative cortices also have higher chances to recover (Di et al. 2008, Vogel et al. 2013). Finally, the presence of long-latency event-related potential components in response to stimuli (Estraneo et al. 2013, Fischer et al. 2006, Steppacher et al. 2013, Xu et al. 2012) or preserved default mode network (DMN) connectivity (Norton et al. 2012) are also indicative of a better recovery.

Advances in clinical diagnosis and detection of residual cognitive function in patients with DOC also raise new ethical questions about withdrawal of nutrition and hydration in this patient population (Fernández-Espejo & Owen 2013, Kitzinger & Kitzinger 2014). Legal precedence in several countries has established the right of the medical team to withdraw artificial nutrition and hydration from patients in VS/UWS, but not those in MCS (Ferreira 2007, Manning 2012). Opinions on these end-of-life decisions vary, however, depending not only on the diagnosis of the patient, but also on the profession and the cultural background of the clinicians (Demertzi et al. 2011). Moreover, caregivers who consider that VS/UWS patients likely feel pain are more often opposed to withdrawal of life-sustaining therapy (Demertzi et al. 2009, 2013). Another ethical concern is the quality of life in chronic DOC patients. This question is difficult to address in the absence of communication with the patient. In this context, it is striking to note, however, that most LIS patients report subjective near-to-normal quality of life (Bruno et al. 2011a).

To be able to draw such strong inferences, however, these active paradigms must select only positive responses in nonreflexive brain activation patterns following task instruction. Indeed, if a reflex, involuntary brain activation led to a positive response in these paradigms, they would lose their value as a diagnostic tool for willful response to command and, hence, for the presence of consciousness in noncommunicative brain-damaged patients. Thus, validation studies should be performed to ensure that the passive listening of the instruction to perform a task cannot elicit a brain activity pattern similar to the one from a voluntary response. The most effective control

Positron emission tomography (PET): invasive neuroimaging technique that measures brain metabolism energy turnover

Functional magnetic
resonance imaging
(fMRI): noninvasive
neuroimaging
technique that
measures neuronal
activation based on
blood-oxygen-level-
dependent (BOLD)
changes

Electroencephalo-
graphy (EEG):
noninvasive technique
that allows
practitioners to record
electrical activity in
the brain through
electrodes placed on
the scalp

Active paradigm:
procedure that
requires the subject to
perform a specific task
on request

would be to ask subjects to listen to the task instruction while being told beforehand not to perform the task. Ideally, two different commands should also be tested and different reproducible responses should be obtained for each.

An appropriately controlled diagnostic test is the tennis imagery paradigm (Boly et al. 2007, Monti et al. 2010, Owen et al. 2006) and its variants (Bardin et al. 2011). In this fMRI paradigm, patients are instructed to repetitively alternate 30 s of motor imagery (i.e., playing tennis) or spatial navigation mental imagery (i.e., walking in your house) with 30 s of rest. To obtain a brain response to command, fMRI data are analyzed by detecting task-specific motor or spatial navigation neural activation during the periods in which the patient was instructed to perform the task, as compared with periods of rest. The 30-s imagery task duration ensures that the response assessed is not simply due to passive processing of verbal instruction. Validation studies have also been performed to verify that no activation is seen when an assessor instructs the patient not to perform the task. Moreover, comparing brain activation patterns in response to the instruction to imagine spatial navigation assesses specificity. In another recent properly controlled fMRI task, investigators used an increase in brain activation during attention to the words "yes" or "no" presented in a stream of numbers as a patient's response to a command (Naci & Owen 2013). In a separate experiment, this task was controlled for the absence of reflexive activation and, thus, for its specificity to detect only conscious responses (Naci et al. 2013). In addition, the search for a differential response to attention to "yes" or "no" ensures that brain activity patterns are specific to the question asked, which further corroborates the nonreflexivity of the response.

Some properly designed EEG paradigms are currently available to clinicians who seek command-following without motor output in brain-damaged patients. A paradigm designed by Schnakers et al. (2008c) uses differential EEG responses during attention to names as a response to command. In this paradigm, sequences of names containing the patient's own name are presented, in both passive and active conditions. In the active condition, the patients are instructed to count her or his own name or to count another target name. The search for a difference between active and passive conditions as well as between runs with attention to the patient's own name and runs with attention to another name offers a control for both the presence of nonreflexive responses and for specificity. Finally, Cruse et al. (2011) designed an EEG paradigm to detect oscillatory changes after the instruction to imagine squeezing one's hand or moving one's feet. Here again a control experiment shows no response when the subjects are instructed not to do the task. In addition, the comparison of the EEG activity differences for the imagery of moving the hand versus that of moving the foot ensures specificity.

In all the previously cited active paradigms, a positive response can be considered as a reasonable surrogate for the presence of consciousness in brain-damaged patients. Thus, these tasks may be used as additional diagnostic tools in the clinical assessment of consciousness. In fact, these paradigms have already allowed investigators to identify behaviorally VS/UWS answering to command using brain activity (Cruse et al. 2011, Monti et al. 2010, Naci & Owen 2013, Owen et al. 2006) (see also **Figure 1**). Once identified, these patients are not to be considered unconscious anymore but should switch to a diagnostic category of functional MCS (Vogel et al. 2013) or MCS* (Gosseries et al. 2014, Stender et al. 2014).

The main limitations of the active paradigm are that negative findings occur often in DOC and that they are uninterpretable. Recent cohort studies have indeed shown that only a minority, about 20%, of DOC patients can positively respond to this approach (Monti et al. 2010, Stender et al. 2014). Negative results obtained with command-following approaches could be due not to patient unconsciousness, but to other reasons such as aphasia, apraxia, fluctuating vigilance, or simply the patient's unwillingness to collaborate. Thus, negative findings in the active paradigm can never exclude the possibility that the patient has retained awareness.

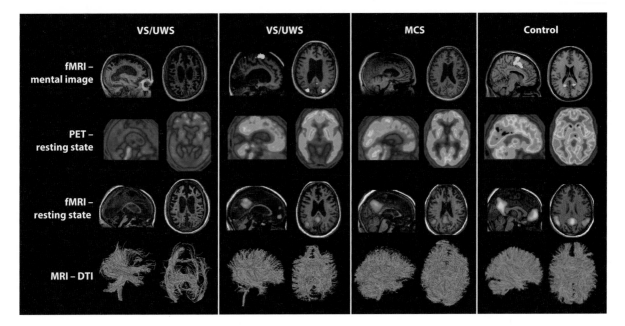

Figure 1

Multimodal diagnosis assessment in disorders of consciousness. Illustrative neuroimaging results in two vegetative state/unresponsive wakefulness syndrome (VS/UWS) patients, one minimally conscious state (MCS) patient, and one healthy control showing possible dissociations between active and passive paradigms and how they usefully complement each other in the evaluation of patients. This figure demonstrates, for example, that fMRI mental imagery tasks (motor imagery on the *left*, navigation imagery on the *right*) show positive results in the control subject and in the second VS/UWS patient. PET and fMRI resting-state results typically show a strong decrease in brain activity and anatomy [here, diffusion tensor imaging (DTI)] in the first VS/UWS patient and show partially preserved brain activity in the second VS/UWS patient as in the MCS patient. Negative responses to active paradigms in MCS patients frequently occur. Figure adapted from Gosseries et al. (2014).

Neuroimaging assessment of DOC should encompass not only active paradigm but also general measures of brain function (the so-called passive approaches). A global assessment of brain function is generally useful and can be especially helpful in the presence of negative results in active paradigms. In the next section, we review potential uses of these passive neuroimaging assessment studies for consciousness diagnosis in DOC.

NEURAL CORRELATE OF CONSCIOUSNESS

In the past few years, numerous studies identified distinct patterns of brain activity in VS/UWS as compared with MCS (Laureys & Schiff 2012). These state-of-the-art studies held to the following safeguards to ensure an accurate clinical diagnosis as well as an appropriate design to draw inferences about group-level differences in a given population study. First, clinical diagnosis should be performed using repeated CRS-R testing by trained assessors (Giacino et al. 2004, Seel et al. 2010). Second, a sufficient number of patients should be studied to obtain a representative sample of each population. It is indeed common that about 20% of patients in VS/UWS present an atypical brain activity pattern. To increase sensitivity, quantitative statistical group analyses can also be used. We now review general patterns of brain function demonstrated in recent studies of VS/UWS and MCS patient populations.

Passive paradigm:
procedure without any specific instruction where the subject does not do anything in particular

Spontaneous Brain Activity

There are three common ways to measure spontaneous regional brain activity using neuroimaging. PET measures regional brain metabolism, whereas fMRI and EEG quantify oscillations at the second and millisecond scales, respectively. Early PET studies identified decreased metabolism in frontoparietal cortices in VS/UWS patients as compared with controls (Beuthien-Baumann et al. 2003, Laureys et al. 1999a), resuming to normal after recovery of consciousness (Laureys et al. 1999b). In MCS patients, lateral frontoparietal area metabolism is preserved (**Figure 2a**) (Thibaut et al. 2012). In addition, MCS+ patients show preserved metabolism in language and sensorimotor areas (Bruno et al. 2012).

EEG studies reported higher delta power in VS/UWS (Lehembre et al. 2012) and more frequent high delta power microstates in VS/UWS as compared with MCS patients (**Figure 2c**) (Fingelkurts et al. 2012b). These results are in line with other studies that show lower bispectral index values (Schnakers et al. 2008b) and decreased spectral entropy in VS/UWS (Gosseries et al. 2011). Moreover, in contrast with MCS, VS/UWS patients do not present with preserved EEG sleep-wake patterns (Landsness et al. 2011). Finally, the amplitude of low-frequency fluctuations of resting-state fMRI signals in the precuneus is higher in MCS as compared with VS/UWS (**Figure 2b**) (Huang et al. 2013).

Response to Stimuli

For regional spontaneous activity, brain reactivity to sensory stimuli can be evaluated with PET, fMRI, or EEG. PET studies suggest that VS/UWS patients typically activate only primary sensory cortices in response to noxious or auditory stimuli (Laureys et al. 2000a, 2002). In contrast, MCS patients show preserved higher-order areas of activation, encompassing the frontoparietal cortices (**Figure 2d**) (Boly et al. 2005, 2004). Likewise, most VS/UWS patients display fMRI activation of only low-level cortices in response to sensory stimuli (Coleman et al. 2009, Di et al. 2007). In contrast, MCS patients typically recruit a more widespread set of associative sensory cortices. Default mode network (DMN) activation in response to self-referential stimuli is also stronger in MCS as compared with VS/UWS patients (**Figure 2e**) (Huang et al. 2013, Qin et al. 2010). Finally, DMN deactivation is also preserved in MCS patients but is virtually absent in VS/UWS patients (Crone et al. 2011).

The mismatch negativity (MMN), an early negative waveform elicited by a deviant tone in a repetitive series, has been one of the most widely studied EEG components in patients with DOC. MMN, as with other long latency components, is found more often in individual MCS patients than in VS/UWS patients (Fischer et al. 2010, Höller et al. 2011, Qin et al. 2008). Another long-latency positive component, the P3, is also found more consistently in MCS (Bekinschtein et al. 2009, Faugeras et al. 2012), although it can be detected in some VS/UWS patients (Perrin et al. 2006). Likewise, statistical group analyses suggested that MMN and P3 amplitude are higher in MCS (Boly et al. 2011, Faugeras et al. 2012). The higher amplitude of long latency components in MCS patients as compared with VS/UWS patients could be linked to preserved function in cerebral backward connections (**Figure 2f**) (Boly et al. 2011).

Functional Connectivity

Functional connectivity studies assess how different brain areas interact with each other. These studies have been performed with numerous conditions in healthy subjects and patient populations. They have now been successfully applied in several ways to differentiate MCS patients from VS/UWS patient populations. These studies assume that if brain areas causally

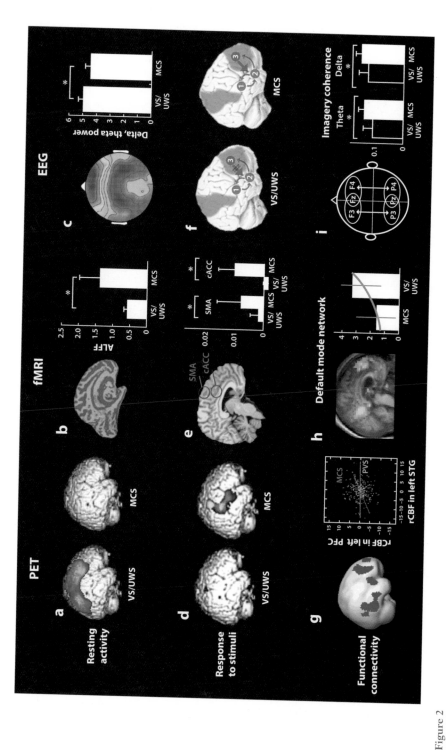

Figure 2

Neural correlates of consciousness in severely damaged brains. PET, fMRI, and EEG results using measures of spontaneous brain activity, response to stimuli, and functional connectivity in vegetative state/unresponsive wakefulness syndrome (VS/UWS) patients and minimally conscious state (MCS) patients. For example, panel *f* shows that, during auditory stimulations, patients in VS/UWS lack backward connections between inferior frontal gyrus (3) and superior temporal gyrus (2) with preserved connections of the primary auditory cortex (1) as compared to patients in MCS. Other abbreviations: ALFF, amplitude of low-frequency fluctuations; SMA, supplementary motor area; cACC, caudal anterior cingulate cortex; rCBF: regional cerebral blood flow; STG, superior temporal gyrus; PFC, prefrontal cortex; PVS, persistent vegetative state; F, frontal; P, parietal; z, central; even number (here, 4) refers to electrode position on the right hemisphere whereas odd number (here, 3) refers to those on the left hemisphere. Asterisks (*) indicate $p < 0.05$. Figure adapted from Boly et al. (2004, 2008, 2011); Fingelkurts et al. (2012b); Huang et al. (2013); Lehembre et al. (2012); Qin et al. (2010); Thibaut et al. (2012); Vanhaudenhuyse et al. (2010).

interact, the time course of their activity should be correlated. This claim usually but not always rests on the assumption of direct anatomical connectivity between the regions studied (Greicius et al. 2009). PET functional connectivity studies assess the correlation in metabolic activity between different brain areas during rest or during sensory stimulation. These studies revealed impaired frontoparietal cortico-cortical and thalamo-cortical connectivity in VS/UWS patients as compared with healthy volunteers (Laureys et al. 1999a, 2000b). As compared with VS/UWS patients, MCS patients show preserved PET functional connectivity in frontoparietal cortices (**Figure 2g**) (Boly et al. 2004). Functional MRI resting-state connectivity studies assess correlations in blood-oxygen-level-dependent (BOLD) signal magnitude among brain regions over the course of a single task-free acquisition session. These resting-state fMRI studies identified preserved connectivity in both lateral and medial frontoparietal areas in MCS patients as compared with VS/UWS patients (**Figure 2h**) (Huang et al. 2013; Kotchoubey et al. 2013; Ovadia-Caro et al. 2012; Soddu et al. 2011a,b; Vanhaudenhuyse et al. 2010). Finally, EEG functional connectivity studies assess similarities in signal amplitude or oscillatory phase (in given frequency bands) between scalp electrodes or between brain regions if performed in source space. Coherence and cross-approximate entropy EEG studies confirmed stronger frontoparietal connectivity in MCS patients as compared with VS/UWS patients (**Figure 2i**) (Lehembre et al. 2012, Wu et al. 2011). The organization of oscillatory brain connectivity in interacting modules is also preserved in MCS patients as compared with VS/UWS patients (Fingelkurts et al. 2013), especially in the DMN (Fingelkurts et al. 2012a). Overall, functional connectivity studies suggest a link between preserved cerebral functional interactions and higher consciousness level (e.g., arousal and/or cognitive functions) in MCS patients as compared with VS/UWS patients.

Individual Results Analysis

As illustrated above, virtually any available neuroimaging technique can reveal different group patterns of brain function in VS/UWS and MCS patients. Even if group separation is clear, at the individual level outliers exist. The interpretation of outliers can be problematic. Combining different techniques may be helpful to better document a patient's general brain function (see **Figure 1**); however, even multimodal assessments may not provide an ultimate solution.

Let us consider this concept in more detail using an example. Suppose we use PET to assess 10 patients unambiguously diagnosed at the bedside as VS/UWS. In our experience, out of these 10 patients, 7 will show a classical frontoparietal hypometabolic PET pattern, and 3 will have preserved metabolism of PET. Among the 3 latter patients, typically only 1 will show a positive response to fMRI or EEG active paradigms. Two out of these 3 will not. What do we do then? What can we infer if the patient does not respond to the active paradigm but has a relatively normal PET? Is high PET metabolism always a definitive marker of the presence of consciousness? If a given neuroimaging measure was a definitive marker of consciousness, it should be consistent in other states of unconsciousness, such as sleep, anesthesia, or seizures. And we know that during epileptic seizures, PET metabolism can be normal, or even increased, even though subjects are unconscious (Engel et al. 1982). Preserved brain metabolism at PET is thus not necessarily definitive proof of the presence of consciousness. **Table 2** illustrates that, to date, none of the classical neuroimaging techniques mentioned above are sufficient to diagnose consciousness. To identify a definitive brain signature of consciousness, developing a theoretical framework to define the mechanisms that link consciousness and the brain is a necessary step (see sidebar, On the Nature of Consciousness, and **Figure 3**). We describe the concrete application of such a theoretical framework to the neuroimaging-based diagnosis of consciousness in the next section.

Table 2   Comparison of neuroimaging findings in different states of unconsciousness

Techniques	VS/UWS > MCS	Alike in other states	Different in other states
PET metabolism	Decrease (FP)	Propofol anesthesia (Fiset et al. 1999), sleep (Braun et al. 1997, Maquet et al. 1990)	Epilepsy (Engel et al. 1982), K complex (Picchioni et al. 2009)
fMRI: oscillation (ALFF)	Decrease (precuneus)	Isoflurane anesthesia (Wang et al. 2011)	Sleep, midazolam anesthesia (Kiviniemi et al. 2005)
EEG: oscillations (delta)	Increase	Sleep (Mascetti et al. 2011)	Epilepsy (Blumenfeld 2005)
PET: response to stimuli	Decrease	Propofol anesthesia (Bonhomme et al. 2001)	TBD
fMRI: response to stimuli	Decrease	Propofol anesthesia (Gosseries et al. 2012, Vanhaudenhuyse et al. 2012)	K complex (Dang-Vu et al. 2011)
EEG: response to stimuli	Decrease	Propofol anesthesia (Heinke et al. 2004)	Burst suppression anesthesia (Kroeger & Amzica 2007)
PET: functional connectivity	Decrease (FP)	Isoflurane, halothane anesthesia (White & Alkire 2003)	TBD
fMRI: functional connectivity	Decrease (FP)	Propofol (Boveroux et al. 2010), sevoflurane anesthesia (Martuzzi et al. 2011)	Sleep (Boly et al. 2012b, Horovitz et al. 2008)
EEG: functional connectivity	Decrease	Propofol, sevoflurane, ketamine anesthesia (Boly et al. 2012a, Lee et al. 2013)	Sleep (Langheim et al. 2011), propofol anesthesia (Barrett et al. 2012, Murphy et al. 2011)

Abbreviations: ALFF, amplitude of low-frequency fluctuations; EEG, electroencephalography; fMRI, functional magnetic resonsance imaging; FP, frontoparietal cortices; MCS, minimally conscious state; PET, positron emission tomography; TBD, to be determined; VS/UWS, vegetative state/unresponsive wakefulness syndrome.

Brain island

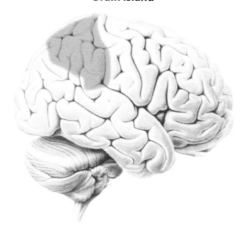

Figure 3

Brain island. See sidebar, On the Nature of Consciousness, for references.

ON THE NATURE OF CONSCIOUSNESS

To develop a mechanistic account of the relationship between consciousness and the brain, forging a comprehensive theory of consciousness is a necessary step. Developing a theory of consciousness is not only useful at a conceptual level, but would also have direct practical implications for assessing patients with DOC. A thoroughly validated theory of consciousness is ultimately the only way to make strong inferences about the presence or absence of consciousness in unresponsive brain-damaged patients where all the other approaches fail.

Let us consider a hypothetical example of an unresponsive brain-damaged patient, whose PET scan shows an island of preserved activity in the right posterior parietal cortex (**Figure 3**). The patient shows only reflexive spontaneous behavior, no behavioral response to command, and no ability to communicate. He also does not follow commands on active paradigms. Moreover, afferent pathways are damaged, impairing the recruitment of cortical areas in response to sensory stimulation. Strikingly, however, brain anatomy, resting metabolism, and fast EEG activity are well preserved in the right posterior parietal cortex.

What can we infer about the presence or absence of consciousness in such a patient? Is anybody home? Is the presence of a well-functioning parietal cortex alone enough for some amount of consciousness (even though, of course, it would be lacking some attributes)? And if so, what could we infer about the contents of consciousness? Would there be any visual, auditory, or verbal content? Would he feel any pain? Would he have any degree of self-awareness? Answering such questions exclusively on the basis of empirical data would clearly not be possible because one cannot directly ask an isolated parietal cortex if it is conscious. Instead, one needs a theory of consciousness that starts from the fundamental features of consciousness itself, provides general principles concerning the necessary and sufficient conditions for consciousness, leads to measures of consciousness that are generally applicable, and provides some guidance about how the quality of experience is determined by the neuroanatomical and neurophysiological organization of brain structures. Thus, in our view, the science of coma and the science of consciousness go hand in hand.

FROM EXPLORATORY TO EXPLANATORY NEURAL CORRELATES OF CONSCIOUSNESS

In the past two decades, several neuroscientific theories hypothesized about the relationships between the brain and consciousness (Block 2011, Dehaene & Changeux 2011, Lamme 2006, Lau & Rosenthal 2011, Tononi 2008, Tononi & Edelman 1998). Such theories can help identify brain markers of the presence or absence of consciousness using neuroimaging. We illustrate this point using the integrated information theory of consciousness (IITC) (Tononi 2012).

IITC states that consciousness is related to a system's capacity for information integration (Tononi 2008, 2012). In the case of the brain, the theory predicts that consciousness-supporting networks should present an optimal balance between functional integration and differentiation (Boly et al. 2009). This hypothesis has recently been tested using transcranial magnetic stimulation (TMS) coupled with high-density EEG. This technique allows investigators to directly measure effective connectivity responses (i.e., TMS-induced causal interactions between distant brain areas) with EEG (Massimini et al. 2009). Our group, in collaboration with Massimini (from the University of Milan) and Tononi (from the University of Wisconsin-Madison), has applied TMS-EEG to assess brain function during sleep, under anesthesia, and in brain-damaged patients. Results of these studies show clear-cut differences in TMS-EEG responses between conscious and unconscious subjects in all conditions. During non–rapid eye movement sleep (NREM), under general anesthesia (e.g., midazolam), and in VS/UWS patients, TMS typically triggers a stereo-typical slow wave that stays local, which indicates a breakdown of effective connectivity (Ferrarelli et al. 2010, Massimini et al. 2005, Rosanova et al. 2012). In contrast, during normal wakefulness,

Transcranial magnetic stimulation (TMS): technique that allows investigators to stimulate the brain noninvasively, which induces neuronal depolarization and discharge of action potentials

NREM: non–rapid eye movement sleep

REM: rapid eye movement sleep

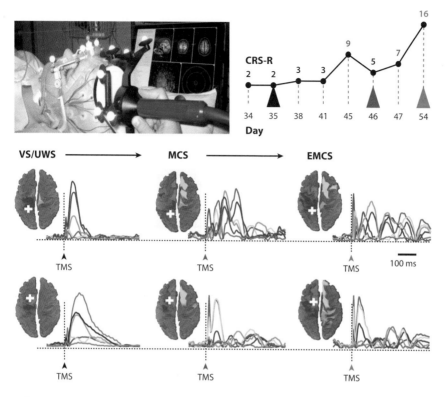

Figure 4

TMS-EEG responses during recovery from coma. TMS-EEG measurements in a patient evolving from vegetative/unresponsive wakefulness syndrome (VS/UWS, *black arrow*) to a minimally conscious state (MCS, *blue arrow*), then to emergence of MCS (EMCS, *red arrow*). The figure illustrates both the spreading and time courses of cortical currents evoked by TMS when stimulating parietal (*top*) and frontal (*bottom*) cortices (*white crosses*). In VS/UWS patients, the response stays local and stereotyped and becomes widespread and differentiated in MCS and EMCS patients. Other abbreviations: CRS-R, Coma Recovery Scale-Revised; EEG, electroencephalography; TMS, transcranial magnetic stimulation. Figure adapted from Rosanova et al. (2012).

in MCS, EMCS, and LIS patients, or during rapid eye movement (REM) sleep, brain activation patterns to TMS are always complex, i.e., widespread and differentiated (**Figure 4**) (Massimini et al. 2005, 2010; Rosanova et al. 2012).

We recently designed a new empirical measure known as the perturbational complexity index (PCI) to quantify in one number the difference in TMS-EEG responses present between states of consciousness and states of unconsciousness (Casali et al. 2013). PCI estimates both the information content and the integration of brain activations through the computation of the normalized Lempel-Ziv complexity (Lempel & Ziv 1976) of the significant EEG spatiotemporal responses to TMS. According to our current results, PCI is remarkably reliable to differentiate consciousness from unconsciousness within and across subjects and conditions: It is always high (i.e., above 0.31) in healthy awake subjects, in MCS, EMCS and LIS patients, as well as during REM sleep, but is invariably low (i.e., below 0.31) during NREM sleep, in patients in VS/UWS and under anesthesia-induced unconsciousness (using midazolam, propofol, or xenon) (**Figure 5**). PCI also allows a clear-cut differentiation between patients in VS/UWS and those who recovered

PCI: perturbational complexity index

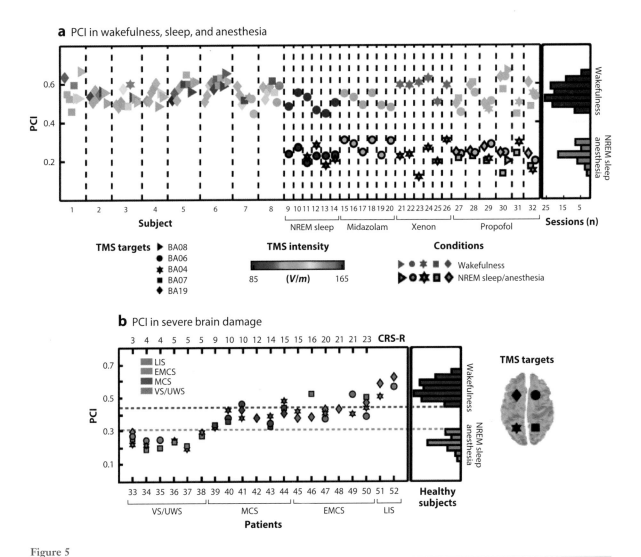

a PCI in wakefulness, sleep, and anesthesia

Figure 5

Perturbational complexity index (PCI) as a marker of consciousness. (*a*) PCI in wakefulness, sleep, and anesthesia. PCI calculated during wakefulness ranges between 0.44 and 0.67, whereas PCI calculated during unconsciousness [i.e., non-rapid eye movement (NREM) sleep and midazolam, xenon, or propofol anesthesia] ranges between 0.12 and 0.31. The histograms display the distributions of PCI across subjects during conscious (*dark gray bars*) and unconscious (*light gray bars*) conditions. (*b*) PCI in severe brain damage. PCI follows the level of consciousness assessed with the Coma Recovery Scale-Revised (CRS-R). It progressively increases from vegetative state/unresponsive wakefulness syndrome (VS/UWS) to minimally conscious state (MCS) and emergence of the MCS (EMCS). VS/UWS values are in the same range as those observed during NREM sleep and general anesthesia. PCI for EMCS and locked-in (LIS) patients are in the same range as healthy awake subjects. Patients in MCS show intermediate PCI values but never below the threshold of unconsciousness (*gray dashed line*, PCI = 0.31). Other abbreviation: TMS, transcranial magnetic stimulation. Figure adapted from Casali et al. (2013).

consciousness (i.e., MCS, EMCS and LIS) at the single-subject level. Further studies on larger samples should confirm these inaugural results. In sum, the highly promising aspect of this theoretically based index of consciousness levels motivates interest in a theoretical framework to help design clinically applicable diagnostic tools for consciousness.

CONTENTS OF CONSCIOUSNESS: WHAT IS IT LIKE TO BE IN AN MCS?

Previous sections discuss progress concerning the diagnosis of the level of consciousness in DOC. However, another outstanding question remains essentially unaddressed: What is the content of consciousness in MCS or in behaviorally VS/UWS patients reclassified by neuroimaging as MCS*? What is it like to be in an MCS? Contents of consciousness are usually assessed by obtaining subjects' reports. In MCS patients, no report can be obtained because no accurate communication is possible. Generalizing neural correlates of conscious content observed in healthy volunteers to interpret MCS brain findings is also problematic because of the presence of the brain lesions and the possible ensuing reorganization. Studies of cognition in MCS using EEG and fMRI active paradigms could help address this question, at least in part. Making inferences about the content of consciousness in noncommunicative patients is a question that can only be addressed fully if empirical studies are complemented by a general theoretical framework (see sidebar, On the Nature of Consciousness, above).

CONCLUSIONS

Recent years witnessed numerous advances in the diagnosis and understanding of brain function in DOC. Research combining clinical, neuroimaging, and theoretical approaches will likely lead to continued fruitful advances in the diagnosis and treatment of these patients.

We offer a few take-home messages:

1. Consciousness is tricky to diagnose clinically; consider the patient as conscious until all evidence is collected.
2. Active paradigms, when properly designed, can successfully probe evidence of the presence of consciousness in unresponsive patients; caution in interpreting negative results is needed, however.
3. Neuroimaging and electrophysiological studies have identified consistent group differences in brain activity patterns in MCS patients as compared with VS/UWS patients. Single-subject level interpretation of these results is nevertheless often limited.
4. Theoretically based neuroimaging approaches (such as PCI) are highly promising to identify reliable single-subject level markers of consciousness. Larger population studies of PCI as a consciousness meter are ongoing.
5. More research on the contents of consciousness in DOC patients is needed.

DISCLOSURE STATEMENT

The authors are not aware of any affiliations, memberships, funding, or financial holdings that might be perceived as affecting the objectivity of this review.

ACKNOWLEDGMENTS

This article was funded by the Belgian National Funds for Scientific Research (FNRS), Fonds Léon Fredericq, James S. McDonnell Foundation, Mind Science Foundation, European

Commission, Concerted Research Action, Public Utility Foundation "Université Européenne du Travail," "Fondazione Europea di Ricerca Biomedica," the National Natural Science Foundation of China (30870861), the Belgian American Educational Foundation (BAEF), the funding of Science and Technology Department of Zhejiang Province (2008C14098), and Hangzhou Normal University (HNUEYT). O.G. received support from NIH grants MH064498 and MH095984 to Bradley R. Postle and Giulio Tononi. O.G. is a postdoctoral researcher, and S.L. is research director at FNRS. We also thank Giulio Tononi for constructive discussions and Aurore Thibaut, Lizette Heine, Francesco Gomez, and Carol Di Perri for providing neuroimaging images.

LITERATURE CITED

Am. Congr. Rehabil. Med. 1995. Recommendations for use of uniform nomenclature pertinent to patients with severe alterations of consciousness. *Arch. Phys. Med. Rehabil.* 76:205–9

Bardin JC, Fins JJ, Katz DI, Hersh J, Heier LA, et al. 2011. Dissociations between behavioural and functional magnetic resonance imaging-based evaluations of cognitive function after brain injury. *Brain* 134:769–82

Barrett AB, Murphy M, Bruno MA, Noirhomme Q, Boly M, et al. 2012. Granger causality analysis of steady-state electroencephalographic signals during propofol-induced anaesthesia. *PLoS ONE* 7:e29072

Bauer G, Gerstenbrand F, Rumpl E. 1979. Varieties of the locked-in syndrome. *J. Neurol.* 221:77–91

Bekinschtein TA, Dehaene S, Rohaut B, Tadel F, Cohen L, Naccache L. 2009. Neural signature of the conscious processing of auditory regularities. *Proc. Natl. Acad. Sci. USA* 106:1672–77

Beuthien-Baumann B, Handrick W, Schmidt T, Burchert W, Oehme L, et al. 2003. Persistent vegetative state: evaluation of brain metabolism and brain perfusion with PET and SPECT. *Nucl. Med. Commun.* 24:643–49

Block N. 2011. Perceptual consciousness overflows cognitive access. *Trends Cogn. Sci.* 15:567–75

Blumenfeld H. 2005. Consciousness and epilepsy: Why are patients with absence seizures absent? *Prog. Brain Res.* 150:271–86

Boly M, Coleman MR, Davis MH, Hampshire A, Bor D, et al. 2007. When thoughts become action: an fMRI paradigm to study volitional brain activity in non-communicative brain injured patients. *NeuroImage* 36:979–92

Boly M, Faymonville M, Peigneux P, Lambermont B, Damas F, et al. 2005. Cerebral processing of auditory and noxious stimuli in severely brain injured patients: differences between VS and MCS. *Neuropsychol. Rehabil.* 15:283–89

Boly M, Faymonville ME, Schnakers C, Peigneux P, Lambermont B, et al. 2008. Perception of pain in the minimally conscious state with PET activation: an observational study. *Lancet Neurol.* 7:1013–20

Boly M, Garrido MI, Gosseries O, Bruno MA, Boveroux P, et al. 2011. Preserved feedforward but impaired top-down processes in the vegetative state. *Science* 332:858–62

Boly M, Massimini M, Tononi G. 2009. Theoretical approaches to the diagnosis of altered states of consciousness. *Prog. Brain Res.* 177:383–98

Boly M, Moran R, Murphy M, Boveroux P, Bruno MA, et al. 2012a. Connectivity changes underlying spectral EEG changes during propofol-induced loss of consciousness. *J. Neurosci.* 32:7082–90

Boly M, Perlbarg V, Marrelec G, Schabus M, Laureys S, et al. 2012b. Hierarchical clustering of brain activity during human nonrapid eye movement sleep. *Proc. Natl. Acad. Sci. USA* 109:5856–61

Bonhomme V, Fiset P, Meuret P, Backman S, Plourde G, et al. 2001. Propofol anesthesia and cerebral blood flow changes elicited by vibrotactile stimulation: a positron emission tomography study. *J. Neurophysiol.* 85:1299–308

Boveroux P, Vanhaudenhuyse A, Bruno MA, Noirhomme Q, Lauwick S, et al. 2010. Breakdown of within- and between-network resting state functional magnetic resonance imaging connectivity during propofol-induced loss of consciousness. *Anesthesiology* 113:1038–53

Braun AR, Balkin TJ, Wesenten NJ, Carson RE, Varga M, et al. 1997. Regional cerebral blood flow throughout the sleep-wake cycle. An $H_2{}^{15}O$ PET study. *Brain* 120(Pt. 7):1173–97

Bruno M-A, Bernheim JL, Ledoux D, Pellas F, Demertzi A, Laureys S. 2011a. A survey on self-assessed well-being in a cohort of chronic locked-in syndrome patients: happy majority, miserable minority. *BMJ Open* 1:e000039

Bruno M-A, Majerus S, Boly M, Vanhaudenhuyse A, Schnakers C, et al. 2012. Functional neuroanatomy underlying the clinical subcategorization of minimally conscious state patients. *J. Neurol.* 259:1087–98

Bruno M-A, Vanhaudenhuyse A, Schnakers C, Boly M, Gosseries O, et al. 2010. Visual fixation in the vegetative state: an observational case series PET study. *BMC Neurol.* 10:35

Bruno M-A, Vanhaudenhuyse A, Thibaut A, Moonen G, Laureys S. 2011b. From unresponsive wakefulness to minimally conscious PLUS and functional locked-in syndromes: recent advances in our understanding of disorders of consciousness. *J. Neurol.* 258:1373–84

Casali AG, Gosseries O, Rosanova M, Boly M, Sarasso S, et al. 2013. A theoretically based index of consciousness independent of sensory processing and behavior. *Sci. Transl. Med.* 5:198ra05

Chatelle C, Chennu S, Noirhomme Q, Cruse D, Owen AM, Laureys S. 2012a. Brain-computer interfacing in disorders of consciousness. *Brain Inj.* 26:1510–22

Chatelle C, Majerus S, Whyte J, Laureys S, Schnakers C. 2012b. A sensitive scale to assess nociceptive pain in patients with disorders of consciousness. *J. Neurol. Neurosurg. Psychiatry* 83:1233–37

Chatelle C, Thibaut A, Bruno MA, Boly M, Bernard C, et al. 2014. Nociception coma scale-revised scores correlate with metabolism in the anterior cingulate cortex. *Neurorehabil. Neural Repair* 28:149–52

Cheng L, Gosseries O, Ying L, Hu X, Yu D, et al. 2013. Assessment of localisation to auditory stimulation in post-comatose states: use the patient's own name. *BMC Neurol.* 13:27

Coleman MR, Davis MH, Rodd JM, Robson T, Ali A, et al. 2009. Towards the routine use of brain imaging to aid the clinical diagnosis of disorders of consciousness. *Brain* 132:2541–52

Crone JS, Ladurner G, Höller Y, Golaszewski S, Trinka E, Kronbichler M. 2011. Deactivation of the default mode network as a marker of impaired consciousness: an fMRI study. *PLoS ONE* 6:e26373

Cruse D, Chennu S, Chatelle C, Bekinschtein TA, Fernández-Espejo D, et al. 2011. Bedside detection of awareness in the vegetative state: a cohort study. *Lancet* 378:2088–94

Dang-Vu TT, Bonjean M, Schabus M, Boly M, Darsaud A, et al. 2011. Interplay between spontaneous and induced brain activity during human non-rapid eye movement sleep. *Proc. Natl. Acad. Sci. USA* 108:15438–43

Dehaene S, Changeux JP. 2011. Experimental and theoretical approaches to conscious processing. *Neuron* 70:200–27

Demertzi A, Ledoux D, Bruno MA, Vanhaudenhuyse A, Gosseries O, et al. 2011. Attitudes towards end-of-life issues in disorders of consciousness: a European survey. *J. Neurol.* 258:1058–65

Demertzi A, Racine E, Bruno M, Ledoux D, Gosseries O, et al. 2013. Pain perception in disorders of consciousness: neuroscience, clinical care, and ethics in dialogue. *Neuroethics* 6:37–50

Demertzi A, Schnakers C, Ledoux D, Chatelle C, Bruno MA, et al. 2009. Different beliefs about pain perception in the vegetative and minimally conscious states: a European survey of medical and paramedical professionals. *Prog. Brain Res.* 177:329–38

Di H, Boly M, Weng X, Ledoux D, Laureys S. 2008. Neuroimaging activation studies in the vegetative state: predictors of recovery? *Clin. Med.* 8:502–7

Di H, Yu SM, Weng XC, Laureys S, Yu D, et al. 2007. Cerebral response to patient's own name in the vegetative and minimally conscious states. *Neurology* 68:895–99

Engel J Jr, Kuhl DE, Phelps ME. 1982. Patterns of human local cerebral glucose metabolism during epileptic seizures. *Science* 218:64–66

Estraneo A, Moretta P, Loreto V, Lanzillo B, Cozzolino A, et al. 2013. Predictors of recovery of responsiveness in prolonged anoxic vegetative state. *Neurology* 80:464–70

Faugeras F, Rohaut B, Weiss N, Bekinschtein T, Galanaud D, et al. 2012. Event related potentials elicited by violations of auditory regularities in patients with impaired consciousness. *Neuropsychologia* 50:403–18

Fernández-Espejo D, Owen AM. 2013. Detecting awareness after severe brain injury. *Nat. Rev. Neurosci.* 14:801–9

Ferrarelli F, Massimini M, Sarasso S, Casali A, Riedner B, et al. 2010. Breakdown in cortical effective connectivity during midazolam-induced loss of consciousness. *Proc. Natl. Acad. Sci. USA* 107:2681–86

Ferreira N. 2007. Latest legal and social developments in the euthanasia debate: bad moral consciences and political unrest. *Med. Law* 26:387–407

Fingelkurts AA, Fingelkurts AA, Bagnato S, Boccagni C, Galardi G. 2012a. DMN operational synchrony relates to self-consciousness: evidence from patients in vegetative and minimally conscious states. *Open Neuroimag. J.* 6:55–68

Fingelkurts AA, Fingelkurts AA, Bagnato S, Boccagni C, Galardi G. 2012b. EEG oscillatory states as neuro-phenomenology of consciousness as revealed from patients in vegetative and minimally conscious states. *Conscious Cogn.* 21:149–69

Fingelkurts AA, Fingelkurts AA, Bagnato S, Boccagni C, Galardi G. 2013. Dissociation of vegetative and minimally conscious patients based on brain operational architectonics: factor of etiology. *Clin. EEG Neurosci.* 44:209–20

Fischer C, Luaute J, Morlet D. 2010. Event-related potentials (MMN and novelty P3) in permanent vegetative or minimally conscious states. *Clin. Neurophysiol.* 121:1032–42

Fischer C, Luauté J, Némoz C, Morlet D, Kirkorian G, Mauguière F. 2006. Improved prediction of awakening or nonawakening from severe anoxic coma using tree-based classification analysis. *Crit. Care Med.* 34:1520–24

Fiset P, Paus T, Daloze T, Plourde G, Meuret P, et al. 1999. Brain mechanisms of propofol-induced loss of consciousness in humans: a positron emission tomographic study. *J. Neurosci.* 19:5506–13

Giacino JT, Ashwal S, Childs N, Cranford R, Jennett B, et al. 2002. The minimally conscious state: definition and diagnostic criteria. *Neurology* 58:349–53

Giacino JT, Fins JJ, Laureys S, Schiff ND. 2014. Disorders of consciousness after acquired brain injury: the state of the science. *Nat. Rev. Neurol.* 10:99–114

Giacino JT, Kalmar K, Whyte J. 2004. The JFK Coma Recovery Scale-Revised: measurement characteristics and diagnostic utility. *Arch. Phys. Med. Rehabil.* 85:2020–29

Giacino JT, Whyte J, Bagiella E, Kalmar K, Childs N, et al. 2012. Placebo-controlled trial of amantadine for severe traumatic brain injury. *N. Engl. J. Med.* 366:819–26

Gosseries O, Boly M, Vanhaudenhuyse A, Bruno M, Phan-Ba R, et al. 2012. *Interaction between spontaneous fluctuation and auditory evoked activity during wakefulness and loss of consciousness*. Presented at Eur. Neurol. Soc. Annu. Meet., Prague, Czech Repub.

Gosseries O, Charland-Verville V, Thonnard M, Bodart O, Laureys S, Demertzi A. 2013. Amantadine, apomorphine and zolpidem in the treatment of disorders of consciousness. *Curr. Pharm. Des.* In press

Gosseries O, Schnakers C, Ledoux D, Vanhaudenhuyse A, Bruno MA, et al. 2011. Automated EEG entropy measurements in coma, vegetative state/unresponsive wakefulness syndrome and minimally conscious state. *Funct. Neurol.* 26:25–30

Gosseries O, Zasler N, Laureys O. 2014. Recent advances in disorders of consciousness: focus on the diagnosis. *Brain Inj.* In press

Greicius MD, Supekar K, Menon V, Dougherty RF. 2009. Resting-state functional connectivity reflects structural connectivity in the default mode network. *Cereb. Cortex* 19:72–78

Heinke W, Kenntner R, Gunter TC, Sammler D, Olthoff D, Koelsch S. 2004. Sequential effects of increasing propofol sedation on frontal and temporal cortices as indexed by auditory event-related potentials. *Anesthesiology* 100:617–25

Höller Y, Bergmann J, Kronbichler M, Crone JS, Schmid EV, et al. 2011. Preserved oscillatory response but lack of mismatch negativity in patients with disorders of consciousness. *Clin. Neurophysiol.* 122:1744–54

Horovitz SG, Fukunaga M, de Zwart JA, van Gelderen P, Fulton SC, et al. 2008. Low frequency BOLD fluctuations during resting wakefulness and light sleep: a simultaneous EEG-fMRI study. *Hum. Brain Mapp.* 29:671–82

Howell K, Grill E, Klein AM, Straube A, Bender A. 2013. Rehabilitation outcome of anoxic-ischaemic encephalopathy survivors with prolonged disorders of consciousness. *Resuscitation* 84:1409–15

Huang Z, Dai R, Wu X, Yang Z, Liu D, et al. 2013. The self and its resting state in consciousness: an investigation of the vegetative state. *Hum. Brain Mapp.* 35:1997–2008

Kitzinger C, Kitzinger J. 2014. Withdrawing artificial nutrition and hydration from minimally conscious and vegetative patients: family perspectives. *J. Med. Ethics.* In press

Kiviniemi VJ, Haanpää H, Kantola JH, Jauhiainen J, Vainionpää V, et al. 2005. Midazolam sedation increases fluctuation and synchrony of the resting brain BOLD signal. *Magn. Reson. Imaging* 23:531–37

Kotchoubey B, Merz S, Lang S, Markl A, Müller F, et al. 2013. Global functional connectivity reveals highly significant differences between the vegetative and the minimally conscious state. *J. Neurol.* 260:975–83

Kroeger D, Amzica F. 2007. Hypersensitivity of the anesthesia-induced comatose brain. *J. Neurosci.* 27:10597–607

Lamme VA. 2006. Towards a true neural stance on consciousness. *Trends Cogn. Sci.* 10:494–501

Landsness E, Bruno M-A, Noirhomme Q, Riedner B, Gosseries O, et al. 2011. Electrophysiological correlates of behavioural changes in vigilance in vegetative state and minimally conscious state. *Brain* 134:2222–32

Langheim FJ, Murphy M, Riedner BA, Tononi G. 2011. Functional connectivity in slow-wave sleep: identification of synchronous cortical activity during wakefulness and sleep using time series analysis of electroencephalographic data. *J. Sleep Res.* 20:496–505

Lau H, Rosenthal D. 2011. Empirical support for higher-order theories of conscious awareness. *Trends Cogn. Sci.* 15:365–73

Laureys S, Celesia GG, Cohadon F, Lavrijsen J, Léon-Carrión J, et al. 2010. Unresponsive wakefulness syndrome: a new name for the vegetative state or apallic syndrome. *BMC Med.* 8:68

Laureys S, Faymonville ME, Degueldre C, Fiore GD, Damas P, et al. 2000a. Auditory processing in the vegetative state. *Brain* 123(Pt. 8):1589–601

Laureys S, Faymonville ME, Luxen A, Lamy M, Franck G, Maquet P. 2000b. Restoration of thalamocortical connectivity after recovery from persistent vegetative state. *Lancet* 355:1790–91

Laureys S, Faymonville ME, Peigneux P, Damas P, Lambermont B, et al. 2002. Cortical processing of noxious somatosensory stimuli in the persistent vegetative state. *NeuroImage* 17:732–41

Laureys S, Goldman S, Phillips C, Van Bogaert P, Aerts J, et al. 1999a. Impaired effective cortical connectivity in vegetative state: preliminary investigation using PET. *NeuroImage* 9:377–82

Laureys S, Lemaire C, Maquet P, Phillips C, Franck G. 1999b. Cerebral metabolism during vegetative state and after recovery to consciousness. *J. Neurol. Neurosurg. Psychiatry* 67:121

Laureys S, Schiff ND. 2012. Coma and consciousness: paradigms (re)framed by neuroimaging. *NeuroImage* 61:478–91

Lee U, Ku S, Noh G, Baek S, Choi B, Mashour GA. 2013. Disruption of frontal-parietal communication by ketamine, propofol, and sevoflurane. *Anesthesiology* 118:1264–75

Lehembre R, Bruno M-A, Vanhaudenhuyse A, Chatelle C, Cologan V, et al. 2012. Resting state EEG study of comatose patients: a connectivity and frequency analysis to find differences between vegetative and minimally conscious states. *Funct. Neurol.* 27:41–47

Lempel A, Ziv J. 1976. On the complexity of finite sequences. *IEEE Trans. Inf. Theory* 22:75–81

Luauté J, Maucort-Boulch D, Tell L, Quelard F, Sarraf T, et al. 2010. Long-term outcomes of chronic minimally conscious and vegetative states. *Neurology* 75:246–52

Lulé D, Noirhomme Q, Kleih SC, Chatelle C, Halder S, et al. 2013. Probing command following in patients with disorders of consciousness using a brain-computer interface. *Clin. Neurophysiol.* 124:101–6

Manning J. 2012. Withdrawal of life-sustaining treatment from a patient in a minimally conscious state. *J. Law Med.* 19:430–35

Maquet P, Dive D, Salmon E, Sadzot B, Franco G, et al. 1990. Cerebral glucose utilization during sleep-wake cycle in man determined by positron emission tomography and [^{18}F]2-fluoro-2-deoxy-D-glucose method. *Brain Res.* 513:136–43

Martuzzi R, Ramani R, Qiu M, Shen X, Papademetris X, Constable RT. 2011. A whole-brain voxel based measure of intrinsic connectivity contrast reveals local changes in tissue connectivity with anesthetic without a priori assumptions on thresholds or regions of interest. *NeuroImage* 58:1044–50

Mascetti L, Foret A, Bourdiec AS, Muto V, Kussé C, et al. 2011. Spontaneous neural activity during human non-rapid eye movement sleep. *Prog. Brain Res.* 193:111–18

Massimini M, Boly M, Casali A, Rosanova M, Tononi G. 2009. A perturbational approach for evaluating the brain's capacity for consciousness. *Prog. Brain Res.* 177:201–14

Massimini M, Ferrarelli F, Huber R, Esser SK, Singh H, Tononi G. 2005. Breakdown of cortical effective connectivity during sleep. *Science* 309:2228–32

Massimini M, Ferrarelli F, Murphy M, Huber R, Riedner B, et al. 2010. Cortical reactivity and effective connectivity during REM sleep in humans. *Cogn. Neurosci.* 1:176–83

Monti MM, Vanhaudenhuyse A, Coleman MR, Boly M, Pickard JD, et al. 2010. Willful modulation of brain activity in disorders of consciousness. *N. Engl. J. Med.* 362:579–89

Multi-Society Task Force on PVS. 1994a. Medical aspects of the persistent vegetative state (1). *N. Engl. J. Med.* 330:1499–508

Multi-Society Task Force on PVS. 1994b. Medical aspects of the persistent vegetative state (2). *N. Engl. J. Med.* 330:1572–79

Murphy M, Bruno MA, Riedner BA, Boveroux P, Noirhomme Q, et al. 2011. Propofol anesthesia and sleep: a high-density EEG study. *Sleep* 34:283–91A

Naci L, Cusack R, Jia VZ, Owen AM. 2013. The brain's silent messenger: using selective attention to decode human thought for brain-based communication. *J. Neurosci.* 33:9385–93

Naci L, Owen AM. 2013. Making every word count for nonresponsive patients. *JAMA Neurol.* 70:1235–41

Noé E, Olaya J, Navarro MD, Noguera P, Colomer C, et al. 2012. Behavioral recovery in disorders of consciousness: a prospective study with the Spanish version of the Coma Recovery Scale-Revised. *Arch. Phys. Med. Rehabil.* 93:428–33

Norton L, Hutchison RM, Young GB, Lee DH, Sharpe MD, Mirsattari SM. 2012. Disruptions of functional connectivity in the default mode network of comatose patients. *Neurology* 78:175–81

Ovadia-Caro S, Nir Y, Soddu A, Ramot M, Hesselmann G, et al. 2012. Reduction in inter-hemispheric connectivity in disorders of consciousness. *PLoS ONE* 7:e37238

Owen AM, Coleman MR, Boly M, Davis MH, Laureys S, Pickard JD. 2006. Detecting awareness in the vegetative state. *Science* 313:1402

Perrin F, Schnakers C, Schabus M, Degueldre C, Goldman S, et al. 2006. Brain response to one's own name in vegetative state, minimally conscious state, and locked-in syndrome. *Arch. Neurol.* 63:562–69

Picchioni D, Killgore WD, Balkin TJ, Braun AR. 2009. Positron emission tomography correlates of visually-scored electroencephalographic waveforms during non-rapid eye movement sleep. *Int. J. Neurosci.* 119:2074–99

Plum F, Posner JB. 1983. *The Diagnosis of Stupor and Coma*. Philadelphia, PA: Davis

Qin P, Di H, Liu Y, Yu S, Gong Q, et al. 2010. Anterior cingulate activity and the self in disorders of consciousness. *Hum. Brain Mapp.* 31:1993–2002

Qin P, Di H, Yan X, Yu S, Yu D, et al. 2008. Mismatch negativity to the patient's own name in chronic disorders of consciousness. *Neurosci. Lett.* 448:24–28

Rosanova M, Gosseries O, Casarotto S, Boly M, Casali AG, et al. 2012. Recovery of cortical effective connectivity and recovery of consciousness in vegetative patients. *Brain* 135:1308–20

Schiff ND, Giacino JT, Kalmar K, Victor JD, Baker K, et al. 2007. Behavioural improvements with thalamic stimulation after severe traumatic brain injury. *Nature* 448:600–3

Schnakers C. 2012. Clinical assessment of patients with disorders of consciousness. *Arch. Ital. Biol.* 150:36–43

Schnakers C, Hustinx R, Vandewalle G, Majerus S, Moonen G, et al. 2008a. Measuring the effect of amantadine in chronic anoxic minimally conscious state. *J. Neurol. Neurosurg. Psychiatry* 79:225–27

Schnakers C, Ledoux D, Majerus S, Damas P, Damas F, et al. 2008b. Diagnostic and prognostic use of bispectral index in coma, vegetative state and related disorders. *Brain Inj.* 22:926–31

Schnakers C, Perrin F, Schabus M, Majerus S, Ledoux D, et al. 2008c. Voluntary brain processing in disorders of consciousness. *Neurology* 71:1614–20

Schnakers C, Vanhaudenhuyse A, Giacino JT, Ventura M, Boly M, et al. 2009. Diagnostic accuracy of the vegetative and minimally conscious state: clinical consensus versus standardized neurobehavioral assessment. *BMC Neurol.* 9:35

Schnakers C, Zasler ND. 2007. Pain assessment and management in disorders of consciousness. *Curr. Opin. Neurol.* 20:620–26

Seel RT, Douglas J, Dennison AC, Heaner S, Farris K, Rogers C. 2013. Specialized early treatment for persons with disorders of consciousness: program components and outcomes. *Arch. Phys. Med. Rehabil.* 94:1908–23

Seel RT, Sherer M, Whyte J, Katz DI, Giacino JT, et al. 2010. Assessment scales for disorders of consciousness: evidence-based recommendations for clinical practice and research. *Arch. Phys. Med. Rehabil.* 91:1795–813

Soddu A, Vanhaudenhuyse A, Bahri M, Bruno MA, Boly M, et al. 2011a. Identifying the default-mode component in spatial IC analyses of patients with disorders of consciousness. *Hum. Brain Mapp.* 33:778–96

Soddu A, Vanhaudenhuyse A, Demertzi A, Bruno MA, Tshibanda L, et al. 2011b. Resting state activity in patients with disorders of consciousness. *Funct. Neurol.* 26:37–43

Stender J, Gosseries O, Bruno M, Charland-Verville V, Vanhaudenhuyse A, et al. 2014. Diagnostic precision of multimodal neuroimaging methods in disorders of consciousness—a clinical validation study. *Lancet.* In press

Steppacher I, Eickhoff S, Jordanov T, Kaps M, Witzke W, Kissler J. 2013. N400 predicts recovery from disorders of consciousness. *Ann. Neurol.* 73:594–602

Stoll J, Chatelle C, Carter O, Koch C, Laureys S, Einhäuser W. 2013. Pupil responses allow communication in locked-in syndrome patients. *Curr. Biol.* 23:R647–48

Thibaut A, Bruno MA, Chatelle C, Gosseries O, Vanhaudenhuyse A, et al. 2012. Metabolic activity in external and internal awareness networks in severely brain-damaged patients. *J. Rehab. Med.* 44:487–94

Thibaut A, Bruno MA, Ledoux D, Demertzi A, Laureys S. 2014. tDCS in patients with disorders of consciousness: Sham-controlled randomized double-blind study. *Neurology.* 82:1112–18

Thonnard M, Gosseries O, Demertzi A, Lugo Z, Vanhaudenhuyse A, et al. 2014. Effect of zolpidem in chronic disorders of consciousness: a prospective open-label study. *Funct. Neurol.* 11:1–6

Tononi G. 2008. Consciousness as integrated information: a provisional manifesto. *Biol. Bull.* 215:216–42

Tononi G. 2012. Integrated information theory of consciousness: an updated account. *Arch. Ital. Biol.* 150:56–90

Tononi G, Edelman GM. 1998. Consciousness and complexity. *Science* 282:1846–51

van Ommen J, Gosseries O, Bruno MA, Vanhaudenhuyse A, Thibaut A, et al. 2013. *Resistance to eye opening in patients with disorders of consciousness: reflex or voluntary?* Presented at Eur. Neurol. Soc. Annu. Meet., Barcelona

Vanhaudenhuyse A, Boveroux P, Bruno M, Gosseries O, Noirhomme Q, et al. 2012. *Does self-referential stimuli perception decrease with diminished level of consciousness?* Presented at Eur. Neurol. Soc. Annu. Meet., Prague, Czech Repub.

Vanhaudenhuyse A, Noirhomme Q, Tshibanda LJ, Bruno MA, Boveroux P, et al. 2010. Default network connectivity reflects the level of consciousness in non-communicative brain-damaged patients. *Brain* 133:161–71

Vanhaudenhuyse A, Schnakers C, Brédart S, Laureys S. 2008. Assessment of visual pursuit in post-comatose states: use a mirror. *J. Neurol. Neurosurg. Psychiatry* 79:223

Vogel D, Markl A, Yu T, Kotchoubey B, Lang S, Müller F. 2013. Can mental imagery functional magnetic resonance imaging predict recovery in patients with disorders of consciousness? *Arch. Phys. Med. Rehabil.* 94:1891–98

Wang K, van Meer MP, van der Marel K, van der Toorn A, Xu L, et al. 2011. Temporal scaling properties and spatial synchronization of spontaneous blood oxygenation level-dependent (BOLD) signal fluctuations in rat sensorimotor network at different levels of isoflurane anesthesia. *NMR Biomed.* 24:61–67

White NS, Alkire MT. 2003. Impaired thalamocortical connectivity in humans during general-anesthetic-induced unconsciousness. *NeuroImage* 19:402–11

Whyte J, DiPasquale M, Vaccaro M. 1999. Assessment of command-following in minimally conscious brain injured patients. *Arch. Phys. Med. Rehabil.* 80:653–60

Whyte J, Gosseries O, Chervoneva I, DiPasquale MC, Giacino J, et al. 2009. Predictors of short-term outcome in brain-injured patients with disorders of consciousness. *Prog. Brain Res.* 177:63–72

Whyte J, Nordenbo AM, Kalmar K, Merges B, Bagiella E, et al. 2013. Medical complications during inpatient rehabilitation among patients with traumatic disorders of consciousness. *Arch. Phys. Med. Rehabil.* 94:1877–83

Whyte J, Rajan R, Rosenbaum A, Katz D, Kalmar K, et al. 2014. Zolpidem and restoration of consciousness. *Am. J. Phys. Med. Rehabil.* 93:101–13

Wijdicks EF, Bamlet WR, Maramattom BV, Manno EM, McClelland RL. 2005. Validation of a new coma scale: the FOUR score. *Ann. Neurol.* 58:585–93

Williams ST, Conte MM, Goldfine AM, Noirhomme Q, Gosseries O, et al. 2013. Common resting brain dynamics indicate a possible mechanism underlying zolpidem response in severe brain injury. *eLife* 2:e01157

Wu DY, Cai G, Zorowitz RD, Yuan Y, Wang J, Song WQ. 2011. Measuring interconnection of the residual cortical functional islands in persistent vegetative state and minimal conscious state with EEG nonlinear analysis. *Clin. Neurophysiol.* 122:1956–66

Xu W, Jiang G, Chen Y, Wang X, Jiang X. 2012. Prediction of minimally conscious state with somatosensory evoked potentials in long-term unconscious patients after traumatic brain injury. *J. Trauma Acute Care Surg.* 72:1024–29

Yamamoto T, Katayama Y, Obuchi T, Kobayashi K, Oshima H, Fukaya C. 2013. Deep brain stimulation and spinal cord stimulation for vegetative state and minimally conscious state. *World Neurosurg.* 80:S30.e1–9

Generating Human Neurons In Vitro and Using Them to Understand Neuropsychiatric Disease

Sergiu P. Paşca,[1] Georgia Panagiotakos,[2] and Ricardo E. Dolmetsch[3]

[1]Department of Psychiatry and Behavioral Sciences, Stanford University School of Medicine, Stanford, California 94305; email: spasca@stanford.edu

[2]Doctoral Program in Neurosciences, Stanford University School of Medicine, Stanford, California 94305; email: panagiog@stanford.edu

[3]Novartis Institutes for Biomedical Research, Cambridge, Massachusetts 02139; email: ricardo.dolmetsch@novartis.com

Annu. Rev. Neurosci. 2014. 37:479–501

First published online as a Review in Advance on June 23, 2014

The *Annual Review of Neuroscience* is online at neuro.annualreviews.org

This article's doi: 10.1146/annurev-neuro-062012-170328

Keywords

human neurons, pluripotent stem cells, evolution, neural development, neuropsychiatric disorders

Abstract

Recent advances in cell reprogramming enable investigators to generate pluripotent stem cells from somatic cells. These induced pluripotent cells can subsequently be differentiated into any cell type, making it possible for the first time to obtain functional human neurons in the lab from control subjects and patients with psychiatric disorders. In this review, we survey the progress made in generating various neuronal subtypes in vitro, with special emphasis on the characterization of these neurons and the identification of unique features of human brain development in a dish. We also discuss efforts to uncover neuronal phenotypes from patients with psychiatric disease and prospects for the use of this platform for drug development.

Contents

INTRODUCTION

The adult brain consists of various neuronal subtypes that connect with remarkable precision to give rise to the functional circuits that underlie behavior. This complexity emerges from a sequence of elaborately coordinated developmental processes that result in the generation of distinct neuronal cell types at specific times and locations (Farkas & Huttner 2008, Okano & Temple 2009, Tiberi et al. 2012, Wonders & Anderson 2006). A long-standing objective in the field of neural stem cell biology has been to recapitulate these developmental events in vitro to obtain a supply of specific neural derivatives. These cells could then be used to study normal neuronal function, to understand how different cell types are affected in disorders of the brain, and to develop potential cell-replacement strategies. Numerous approaches have been undertaken to drive the differentiation of pluripotent human embryonic stem (hES) cells toward specific neural fates using an exogenous application of mitogens and patterning molecules (Amoroso et al. 2013; Aubry et al. 2008; Carri et al. 2013; Chambers et al. 2012; Eiraku et al. 2008; Elkabetz et al. 2008; Erceg et al. 2012; Espuny-Camacho et al. 2013; Fasano et al. 2010; Kriks et al. 2011; Lee et al. 2007a,b; Maroof et al. 2013; Nicholas et al. 2013; Perrier et al. 2004; Shi et al. 2012b). By specifying temporal and positional identity in undifferentiated cells, these methods have been used successfully to model several aspects of neural development in the lab.

The advent of induced pluripotent stem cell (iPSC) technology has provided a unique opportunity to examine the development and function of normal and diseased human neurons in vitro. By inducing the expression of a set of transcription factors required to maintain hES cells, several research groups have demonstrated that human somatic cells can be reprogrammed to become pluripotent, capable of generating derivatives of all three germ layers (Park et al. 2008, Takahashi et al. 2007, Yu et al. 2007). Since their original application, iPSC methods have been refined to utilize nonintegrating vectors, RNA, or small molecules and to enhance reprogramming efficiency (Anokye-Danso et al. 2011, Onder et al. 2012, Rais et al. 2013, Warren et al. 2010, Yoshioka et al. 2013). Alternate paradigms have also been developed to enable direct conversion of somatic cells into induced neurons (iNs) (Caiazzo et al. 2011, Pang et al. 2011, Son et al. 2011, Yoo et al. 2011, Zhang et al. 2013). These approaches allow for an unprecedented interrogation of cellular and molecular features of neural cells from human subjects with a specific genetic background, particularly subjects with mutations giving rise to neuropsychiatric disorders. By applying differentiation strategies developed for hES cells to patient-derived iPSCs, stem cell biologists can now overcome the unavailability of human neurons from early developmental stages or from diseased brains and can now explore human brain development and disease-relevant defects using this platform.

Although we are still far from approaching in vitro the level of neuron diversity in the human brain, we now have populations of neurons that we can generate reliably using directed differentiation methods rooted in developmental neurobiology. These subtypes include midbrain dopamine neurons and cortical excitatory and inhibitory neurons (Espuny-Camacho et al. 2013, Fasano et al. 2010, Kriks et al. 2011, Maroof et al. 2013, Nicholas et al. 2013, Paşca et al. 2011, Perrier et al. 2004, Roy et al. 2006, Shi et al. 2012b). In this review, we outline major components of neuronal identity to be considered when examining in vitro–derived human neurons, with an eye toward identifying specific cellular defects associated with neuropsychiatric disorders. We also review approaches currently available for generating human neurons, with an emphasis on human-specific features that can be modeled in vitro.

CRITERIA FOR DEFINING NEURONAL IDENTITY

There is a growing awareness in neuroscience that developing methods to define neuronal subtypes will be essential for understanding how neural circuits are built and how they are disrupted in disease. This awareness is reflected in the recent Brain Initiative interim report (Brain Work. Group 2013), which includes a census of cell types as a high-priority research area. In the context of disease modeling in vitro, developing a set of criteria to define specific classes of neurons is crucial to reliably identify cellular defects.

Neuronal identity encompasses a number of individual multidimensional phenotypic features, including the expression of lineage-specific transcription factors, patterns of axonal projections, dendritic structure, and electrophysiological properties. Many studies have emphasized the use of individual characteristics to classify neurons, and more recent efforts have attempted to combine phenotypic properties to examine specific types of neurons. Morphological features such as dendritic branching patterns, coupled with the repertoire of ion channels decorating dendrites and the resulting integration and firing properties of individual neurons, have been suggested as one set of axes along which the identity of specific excitatory cortical neurons could be defined (Migliore & Shepherd 2005, Spruston 2008). Similarly, combinatorial patterns of gene expression have been proposed as another axis that defines morphological and functional neuronal diversity (Hobert 2011). In the mouse cerebral cortex, for example, networks of lineage-specific transcription factors have been used to classify populations of excitatory neurons on the basis of their axonal projection patterns or dendritic morphology (Alcamo et al. 2008; Arlotta et al. 2005; Chen et al. 2005, 2008; Cubelos et al. 2010; Greig et al. 2013; Hevner et al. 2001; Lai et al. 2008; Leone et al. 2008; Molyneaux et al. 2005, 2009; Srinivasan et al. 2012). Current efforts are increasingly moving large-scale gene expression and quantitative biochemical analysis into the single-cell domain, building on earlier work correlating single-cell reverse transcription-polymerase chain reaction (RT-PCR) gene expression data with neuronal electrical properties (Eberwine et al. 2012, Mackler et al. 1992, Qiu et al. 2012, Toledo-Rodriguez et al. 2004).

A particularly extensive and systematic classification scheme has been proposed in the study of GABA (γ-aminobutyric acid)-ergic interneurons of the cerebral cortex, in which subtype diversity is comprehensively examined on the basis of the integration of anatomical, physiological, biochemical, and molecular features (Ascoli et al. 2008, DeFelipe et al. 2013). Applying a similar schema to in vitro differentiation paradigms (**Figure 1**), complete characterization of human neurons derived from pluripotent stem cells should integrate individual characteristics comprising cellular morphology, combinatorial single-cell gene expression, neurotransmitter identity, electrical properties, and ultimately in vivo hodology and integration following transplantation. As

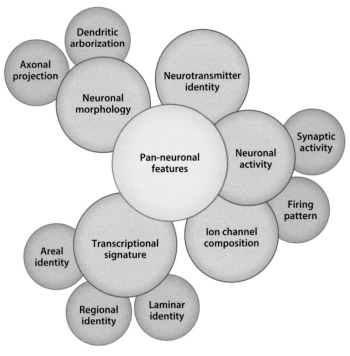

Figure 1

Diagram illustrating the multidimensional features underlying neuronal identity.

we gather more data, the boundaries between cell types are becoming less clear, and significant cell-cell variability can be observed. Nonetheless, quantifying this variability, especially during development, is essential when asking disease-related questions, and comprehensive characterization of human neurons relying on multiple parameters including functional properties will become an important consideration for stem cell–based approaches.

APPROACHES FOR THE GENERATION OF HUMAN NEURONS IN VITRO

Numerous methods have been developed for the directed differentiation of human neurons from pluripotent stem cells. These approaches use defined conditions to recapitulate developmental events that give rise to the diverse subtypes of neurons in the brain. During normal development, neural induction begins immediately after gastrulation when the neural plate forms. In vitro neural induction methods mimic this process, progressively restricting the differentiation potential of pluripotent cells to obtain specified progenitors (Chambers et al. 2009). In the developing embryo, upon formation of the neural tube, neural progenitor cells (NPCs) proliferate extensively in neurogenic zones lining the ventricles termed ventricular (VZ) and subventricular zones (SVZ) (Lu et al. 2000). These proliferative events give way to the differentiation of NPCs into intermediate progenitors (IPCs) and neurons (Farkas & Huttner 2008). Through the actions of morphogens, cells acquire positional identity (Altmann & Brivanlou 2001, Falk & Sommer 2009, Le Dreau & Marti 2012). Along the dorsal-ventral axis, Wnt proteins secreted dorsally and the ventrally expressed patterning molecule sonic hedgehog (SHH) produce gradients that progressively specify cell classes. To generate cells with a more posterior identity, such as those of the midbrain or spinal cord, activation of retinoic acid (RA) or fibroblast growth factor (FGF) pathways is required (Danjo et al. 2011, Eiraku et al. 2008, Elkabetz et al. 2008, Peljto & Wichterle 2011). In the absence of exogenous morphogens, neural induction follows a default program toward anterior telencephalic fates, a feature that is recapitulated in vitro (Espuny-Camacho et al. 2013, Kamiya et al. 2011, Paşca et al. 2011, Tropepe et al. 2001, Watanabe et al. 2005).

Cortical Inhibitory Neurons

Cortical interneurons comprise a diverse population of GABAergic neurons thought to play an indispensable role in sculpting the developmental assembly and output of cortical circuits (Le Magueresse & Monyer 2013, Owens & Kriegstein 2002). In the rodent, these cells have been classified into subgroups based on morphological, molecular, and biophysical features. These properties include the expression of calcium-binding proteins or peptide hormones [parvalbumin (PV), calretinin, somatostatin (SST), etc.]; the developmental expression of transcription factors downstream of SHH induction of Nkx2.1 (such as Lhx6); patterns of axonal elaboration; synaptic function; and electrophysiological features (Ascoli et al. 2008, DeFelipe et al. 2013). In marked contrast to excitatory cortical neurons, GABAergic cortical interneurons are generated in the ventral forebrain and traverse long distances to populate the cortex (Wonders & Anderson 2006). The VZ and SVZ of the medial and caudal ganglionic eminences (MGE and CGE, respectively) give rise to most cortical interneuron subtypes. Within these regions, transplantation experiments, in vivo birth dating, and in utero fate mapping have revealed a precise temporal and spatial segregation of subclasses of interneurons with different physiological properties and molecular identities.

Changes in cortical inhibitory interneuron number and function have been repeatedly associated with psychiatric disease (Akbarian & Huang 2006, Chao et al. 2010, Lewis et al. 2012). Generating human cortical interneurons in vitro has thus been a long-standing goal in neural stem cell biology. Previous work has described the in vitro generation of mouse cortical interneurons from ES cells, using an Lhx6::eGFP bacterial artificial chromosome (BAC) transgenic cell line to enrich for MGE-derived PV+ and SST+ cells (Maroof et al. 2010). Using a similar approach, Maroof et al. (2013) and Nicholas et al. (2013) used an Nkx2.1::GFP hES cell line to recapitulate several aspects of the development and maturation of human MGE-derived cortical interneurons in vitro. Both groups employ small-molecule approaches to induce ventral telencephalic fates. Inhibition of the WNT pathway was coupled with dual inhibition of SMAD signaling (Chambers et al. 2009) to enhance forebrain identity, and upon SHH exposure, both studies report efficient generation of FOXG1+/Nkx2.1+ MGE progenitors. Following enrichment for Nkx2.1::GFP, progenitors were differentiated using an astrocyte coculture platform, or in vivo through transplantation into the rodent brain. By characterizing firing properties, synaptic activity, migration, morphology, and gene expression, both studies demonstrate that these human MGE-like progenitors functionally mature into cortical interneurons.

These studies present an essential step forward in our ability to generate interneurons from human pluripotent stem cells. Moving forward, examining the full complement of single-cell gene expression in these populations, as well as more thoroughly reconstructing the dendritic and axonal morphology of these cells, will be important. Additionally, as with many in vitro directed differentiation methods, these approaches result in heterogeneous neuronal populations. Modifying the timing and duration of SHH exposure can induce different ventral telencephalic fates (Maroof et al. 2013). Optimizing these parameters and devising strategies to enrich for specific cell types will thus be essential to obtain pure populations of cortical interneurons. Finally, of particular interest, hES-derived cortical interneurons take a remarkably long time to functionally mature, a feature that is also true in the normal embryonic brain. Even several months after transplantation, Nicholas et al. report relatively few PV$^+$ fast-spiking cells, suggesting that there is still work to be done to induce complete maturation of hES-derived cortical interneurons.

Cortical Excitatory Neurons

Cortical excitatory neurons are born in the dorsal forebrain, arising from actively dividing NPCs called radial glia (Noctor et al. 2002). Within the VZ, radial glia integrate cell intrinsic and extrinsic signals, undergoing characteristic modes of cell division during neurogenesis (Farkas & Huttner 2008, Mione et al. 1997, Noctor et al. 2004). Deeper-layer neurons are generated first, followed by the birth of upper-layer neurons (Leone et al. 2008). In humans and nonhuman primates, studies have described an enlarged proliferative zone called the outer SVZ (oSVZ). This structure is home to newly identified outer radial glia (oRG) progenitor cells, which may underlie some aspects of cortical expansion in the primate lineage (Betizeau et al. 2013, Fietz et al. 2010, Hansen et al. 2010, LaMonica et al. 2012).

The combinatorial expression of lineage-specific transcription factors has been used to delineate the laminar identity of individual classes of excitatory cortical projection neurons (Alcamo et al. 2008; Arlotta et al. 2005; Britanova et al. 2008; Chen et al. 2005, 2008; Cubelos et al. 2010; Greig et al. 2013; Hevner et al. 2001; Lai et al. 2008; Leone et al. 2008; Molyneaux et al. 2005, 2009). For example, expression of the transcription factor CTIP2 is required for the specification of subcortically projecting neurons, whereas SATB2-expressing cells are thought to project callosally to the contralateral hemisphere (Alcamo et al. 2008, Chen et al. 2008). Similarly, dendritic structure and projection neuron firing properties have been assessed as additional means to define pyramidal neuron identity (Chen et al. 2008, Spruston 2008).

The strategies that have been used to generate excitatory neurons from hES cells encompass aggregate-based and adherent systems (**Figure 2**) (Eiraku et al. 2008, Espuny-Camacho et al. 2013, Lancaster et al. 2013, Li et al. 2009, Paşca et al. 2011, Shi et al. 2012b). Forebrain identity is likely a default state for neuronal specification of pluripotent human stem cells, and existing protocols yield neurons with a rostral identity without exogenous morphogen application (in hES/iPSC cultures, inhibitors of SHH signaling like cyclopamine are not required). Employing the SFEBq (serum-free floating culture of embryoid body-like aggregate with quick reaggregation) method, the Sasai group has reported the production of human excitatory neurons from self-organized neuroepithelial structures (Eiraku et al. 2008). Similarly, a modified embryoid body–based approach has also yielded cortical excitatory neurons (Li et al. 2009). Although both aggregate approaches clearly drive cortical excitatory neuron development, a major question remains about the lack of upper layer neuron production in this setting, as defined by the expression of lineage-specific transcription factors. An expanded assessment of function, particularly in the context of in vivo integration of SFEBq-derived human neurons, is also essential for future research.

Several adherent culture systems have also been developed to generate excitatory neurons. Paşca et al. describe the generation of deeper-layer and upper-layer cortical excitatory neurons from hES and iPS cells, as assessed by single-cell gene expression, immunostaining for markers of laminar identity, and population-based microarray analysis (Paşca et al. 2011). A study from the Livesey lab has also used an adherent culture approach to describe sequential generation of functional cortical excitatory neurons from pluripotent stem cells comparable to what is seen in vivo (Shi et al. 2012b). More recently, building on earlier work in mouse ES cells, a report from Espuny-Camacho et al. (2013) comprehensively characterizes hES-derived cortical excitatory neurons along a number of axes encompassing gene expression, morphology, electrophysiological function, and in vivo hodology. Cortical neurons of all layers were generated sequentially and transplanted human cells progressively matured in vivo, forming functional synapses with endogenous rodent cells.

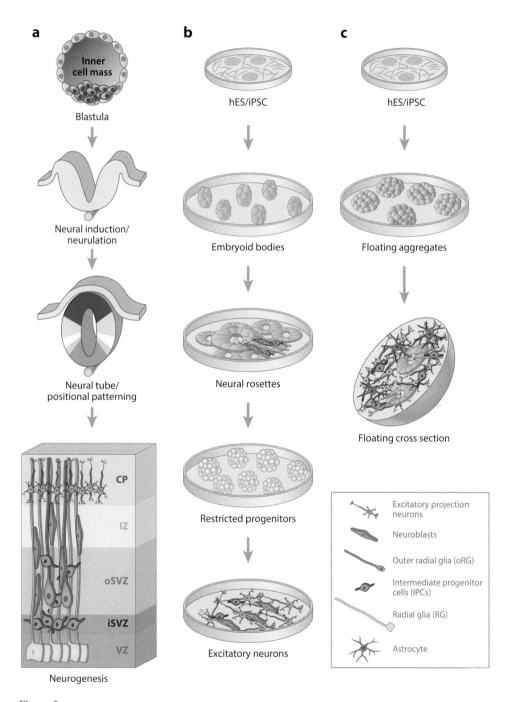

a

Inner cell mass

Blastula

↓

Neural induction/
neurulation

↓

Neural tube/
positional patterning

↓

CP

IZ

oSVZ

iSVZ

VZ

Neurogenesis

b

hES/iPSC

↓

Embryoid bodies

↓

Neural rosettes

↓

Restricted progenitors

↓

Excitatory neurons

c

hES/iPSC

↓

Floating aggregates

↓

Floating cross section

Excitatory projection
neurons

Neuroblasts

Outer radial glia (oRG)

Intermediate progenitor
cells (IPCs)

Radial glia (RG)

Astrocyte

Figure 2

Approaches for the generation of excitatory cortical projection neurons. The developmental events underlying the generation of cortical excitatory neurons in vivo (*a*) are recapitulated in vitro using adherent cultures (*b*) or suspended aggregate methods (*c*). Abbreviations: CP, cortical plate; hES, human embryonic stem cells; iPSC, induced pluripotent stem cell; iSVZ, inner SVZ; IZ, intermediate zone; oSVZ, outer SVZ; SVZ, subventricular zone; VZ, ventricular zone.

This study also did not involve the addition of exogenous morphogens during differentiation, reinforcing the idea of a default specification program favoring the production of neurons with a dorsal forebrain identity.

Despite these advances, a number of important considerations remain. In the context of modeling diseases, it is crucial to develop paradigms that allow for more controlled directed differentiation into specific projection neuron subtypes. The use of single-cell gene expression platforms will be important for identifying unique transcriptional signatures that define subpopulations of neurons. Additionally, there remains a paucity of studies exploring area specification of hES-derived cortical neurons, a feature of particular interest when considering potential cell-replacement strategies. Finally, building on the aggregate methods discussed earlier, a recent study by Lancaster et al. (2013) has reported a novel approach for generating three-dimensional "organoids." This method does not inhibit SMAD signaling during induction, and thus the organoids contain both neural and nonneural tissue. The authors report the generation of cortical projection neurons from NPCs residing in an oSVZ-like region, as well as the generation of GABAergic interneurons. This approach also does not recapitulate the full repertoire of laminar-specific neuronal subtypes seen in vivo. Nevertheless, it presents a step forward in aggregate-based methods, and future work aimed at a complete characterization of the neurons generated using this approach will be of great interest.

Midbrain Dopamine Neurons and Other Neuronal Subtypes

Dopamine neurons in the substantia nigra pars compacta of the midbrain, which are progressively lost in Parkinson's disease (PD), represent a population of neurons that have long been studied in the context of cell-replacement therapies. These neurons are born and patterned in the ventral midbrain through the combined actions of SHH and FGF8 (Crossley et al. 1996, Hynes et al. 1995, Ye et al. 1998). In vitro mouse ES protocols used these known developmental programs to generate progenitors with a midbrain-hindbrain identity and subsequently to obtain functional dopaminergic neurons (Barberi et al. 2003, Kawasaki et al. 2000, Lee et al. 2000, Tabar et al. 2008). Numerous studies have optimized and translated initial reports in the mouse to primate stem cells and hES/iPS cells (Fasano et al. 2010, Perrier et al. 2004, Roy et al. 2006, Soldner et al. 2009). In the most comprehensive study to date, Kriks et al. (2011) coupled Wnt signaling activation with the ventralizing actions of SHH to direct ES cells to become floor-plate midbrain neural progenitors. These progenitors generate enriched populations of functional midbrain dopamine neurons. Using extensive gene expression analysis, electrophysiological characterization, and biochemical assessment, as well as in vivo transplantation in three separate animal models of PD, this work demonstrates bona fide midbrain dopaminergic neuron identity.

Nonetheless, the midbrain contains multiple populations of dopaminergic neurons. Currently available methods are still unable to enrich for individual types of dopamine neurons, an important issue for disease modeling and potential in vivo applications. Recent work has devised reporter strategies to isolate fluorescence activated cell sorting (FACS)-purified hES-derived dopamine neurons, comparable to what has been done with cortical interneurons (Ganat et al. 2012). Combining such approaches with comparative single-cell transcriptomics may identify unique lineage markers that could be used to select for subtypes of dopaminergic cells.

Several other classes of disease-relevant neuronal cells have been generated in vitro from hES cells. These include motor neurons (Amoroso et al. 2013, Dimos et al. 2008, Lee et al. 2007b), striatal medium spiny neurons (Aubry et al. 2008, Carri et al. 2013), nociceptive neurons (Chambers et al. 2012) and other neural crest derivatives (Lee et al. 2007a), and cerebellar neurons (Erceg et al. 2012). Moving forward, thorough characterization of these cell types will be essential for in vitro modeling of neuropsychiatric disorders and for consideration of human-specific features that emerged in these cells during the course of primate evolution.

HUMAN-SPECIFIC ASPECTS IN THE GENERATION OF NEURONS

The ability to obtain diverse neuronal lineages from human pluripotent cells constitutes an unprecedented opportunity to investigate cellular and molecular features that are unique to the human and primate lineage. Furthermore, the generation of patient-specific neurons may shed light on the pathophysiology of neuropsychiatric disorders, for which rodent models fail to display face validity.

Corticogenesis has undergone dramatic changes in primates (Lui et al. 2011). An expanded pool of progenitor cells (Betizeau et al. 2013, Hansen et al. 2010, LaMonica et al. 2012) residing in an enlarged proliferative zone is thought to account for a significantly expanded neocortex (Fietz et al. 2010). As many as 21% of genes expressed in the brain display differential expression patterns in human cortex as compared with rodents (Zeng et al. 2012). Some of these genes are disrupted in disorders of social cognition. For instance, nitric oxide synthase 1 (NOS1) is transiently coexpressed, in a species-dependent manner, with fragile X mental retardation 1 protein (FMRP) in subcortically projecting neurons of Broca's area and the orofacial motor areas (Kwan et al. 2012). Other genes associated with corticogenesis have undergone multiple duplications during more recent phases of human evolution, and as is the case for *SRGAP2*, these genetic events have introduced novel human-specific functions that allow, for example, neoteny during spine maturation (Charrier et al. 2012).

Ramón y Cajal hypothesized that the expansion and complexity of the human cortex can be explained by the presence of more classes of neurons than found in other mammals (Ramón y Cajal 1899). Although this hypothesis has since been challenged, evidence coming primarily from postmortem studies indicates that human neurons and glia display unique features. Human neurons exhibit morphological adaptations, such as increased dendritic complexity, larger and denser dendritic spines (Benavides-Piccione et al. 2002), or larger and more specialized astrocytes (Oberheim et al. 2009). The human cortex also has a higher density of glia than do nonhuman anthropoids (Sherwood et al. 2006), and certain GABAergic cortical interneurons (DeFelipe et al. 2006) have distinct morphologies in primates. Indirect evidence has also suggested unique energetic requirements in human neurons as compared with other nonhuman primates (Fu et al. 2011). Particular subsets of cortical cells have also undergone more dramatic adaptations in primates. For instance, multiple reports have described cortical neurons in humans that express tyrosine hydroxylase (TH$^+$), the rate-limiting enzyme in catecholamine synthesis (Gaspar et al. 1987, Hornung et al. 1989). While rare cortical TH$^+$ inhibitory neurons have been described in rodents, only half of these neurons are GABAergic in the human cortex (Trottier et al. 1989). The function of these catecholaminergic neurons in humans remains unknown. Nonetheless, abnormalities in the expression of TH in patient-derived cortical neurons with a genetically defined form of autism spectrum disorders have been reported in an iPSC-based model (Paşca et al. 2011). This cellular phenotype was unique to human cortical neurons and was not observed in an animal model for this disease. The von Economo neurons (VEN) represent another specialized neuronal cell type displaying species specificity and a strong disease association. VEN are bipolar subcerebral projection neurons localized in layer 5b that express FEZF2 and CTIP2 (Cobos & Seeley 2013). These large-volume neurons are not just morphologically different, but also discretely localized to the anterior cingulate and frontoinsular cortices in large-brained social mammals (Allman et al. 2011). Moreover, these neurons are considered especially vulnerable to degeneration (Seeley et al. 2006) and may play a role in the development of autism spectrum disorders and schizophrenia (Allman et al. 2005, Brüne et al. 2010, Santos et al. 2011).

Upper- and lower-layer excitatory cortical projection neurons, as well as cortical interneurons, have been generated in vitro from human pluripotent stem cells (Espuny-Camacho et al. 2013,

Maroof et al. 2013, Nicholas et al. 2013, Paşca et al. 2011, Shi et al. 2012b). Monitoring and manipulating early stages of corticogenesis in vitro could allow for the dissection of developmental programs underlying primate cortical enlargement and may reveal cellular mechanisms giving rise to the relative expansion of upper cortical layers in humans (DeFelipe 2011). To date, reprogrammed cells have also been generated from nonhuman primates, and their transcriptional signature has uncovered several human-specific cellular adaptations (Marchetto et al. 2013). This study sets the stage for direct comparison of in vitro neural differentiation potential, and corticogenesis in particular, across a number of primate species. Using rigorous assays and careful characterization of individual cell types, these experiments are likely to uncover key cellular phenotypes that contribute to the unique evolutionary adaptations of the human cortex. Moreover, the ability to repeatedly recapitulate neural development in a controlled setting starting from the same clone of pluripotent cells would allow for the interrogation of the relative contribution of stochasticity in human cortical development. For instance, using this platform we could finally examine cells from individuals carrying copy number variants with variable penetrance for neuropsychiatric diseases (e.g., deletions of 1q21.1, 3q19, 16p11.2, 16p13.1). In doing so, we could establish whether penetrance is associated with developmental noise and thereafter identify available mechanisms for compensation. Finally, although some physiological properties of human neurons have been previously examined in surgically dissected human tissue (Molnar et al. 2008), the ability to generate in vitro large quantities of diverse subtypes of functional neurons from pluripotent cells will allow for unique opportunities to probe functional adaptations in human neuronal development and dysfunction.

USING HUMAN-DERIVED NEURAL CELLS TO STUDY NEUROPSYCHIATRIC DISORDERS

Animal models have significantly contributed to our understanding of human pathophysiology. Although a number of endophenotypes for psychiatric disorders can be quantified in animals (Meyer-Lindenberg & Weinberger 2006), rodent and even primate models fall short of recapitulating complex psychiatric phenomena, such as psychosis. Additionally, the success rate for translating drugs effective in rodents is disproportionately low (Rice 2012, van der Worp et al. 2010). Together, these challenges have been impeding progress in gaining mechanistic insights into neuropsychiatric disorders.

Human reprogramming technologies were introduced only recently, but their application to human disease modeling has been abundant. A PubMed search using the terms "induced pluripotent stem cells" and "disease" reveals almost 1,600 articles. To date, in vitro models of brain diseases encompass both developmental and degenerative disorders and are based on the directed differentiation of neurons. Although many of these investigations have attempted to validate findings from rodent models or pathological hallmarks identified in postmortem findings, a significant proportion of studies have used patient-derived neural cells to gain novel insights into neuropsychiatric disease (see **Table 1**). More importantly, in a few instances, either through large unbiased screens or educated guesses, new therapeutic targets or chemical compounds for specific disorders have been identified.

One of the earliest examples of in vitro disease modeling investigated spinal muscular atrophy (Ebert et al. 2009). This severe neurodegenerative disease of infancy is characterized by selective loss of lower motor neurons and is caused primarily by mutations that reduce the level of the survival motor neuron 1 (SMN1) protein. Floating spheres of iPSCs were patterned with RA and SHH. Following several additional weeks of culture in adherent monolayer conditions, motor neuron identity was probed by immunostaining for several transcription factors and the cholinergic enzyme choline acetyltransferase (ChAT). Notably, the proportion of motor neurons in patient cultures was reduced over time, as a result of either impairment in differentiation or decreased neuronal survival. To verify the effects of drugs in patient cells, the authors used a cellular endophenotype—the number of SMN protein aggregates. Previous work has shown that aggregate number is inversely correlated with disease severity (Coovert et al. 1997). Treatment with valproic acid or tobramycin, previously shown to elevate levels of SMN, had a similar effect in patient-derived neurons. Human motor neurons have also been generated from patients with the rapidly progressing neurodegenerative disease amyotrophic lateral sclerosis (ALS). Neurons generated from patients with familial variants of ALS carrying mutations in *Tar DNA binding protein-43* (*TDP-43*) display characteristic cytosolic aggregates and have markedly decreased survival in longitudinal studies (Bilican et al. 2012, Egawa et al. 2012). Transgenic mouse ES-derived motor neurons have also been used to screen drugs for increased cell survival. Identified drugs were subsequently tested in motor neurons derived from ALS patients and strongly improved their survival (Yang et al. 2013). In complementary studies, Serio et al. (2013) generated astrocytes from patients with *TDP-43* mutations. This approach allowed for the interrogation of cell-autonomous effects, an idea explored previously in mouse models of this disorder. In this setting, mutant astrocytes did not affect the survival of cocultured neurons, differentiating familial *TDP-43* ALS from another form of ALS caused by *superoxide dismutase 1* (*SOD1*) mutations (Marchetto et al. 2008).

Gaining access to multiple subtypes of functional neurons from a particular set of patients can facilitate the understanding of tissue-specific effects of a particular genetic event. This notion is illustrated by a series of iPSC studies of familial dysautonomia, a rare but fatal disease caused by a point mutation in the *IκB kinase complex-associated protein* (*IKBKAP*) and characterized by the degeneration of sensory and autonomic neurons (Lee et al. 2009). Following directed differentiation of iPSCs toward five distinct cell lineages, as rigorously defined by FACS purification using lineage-specific surface epitopes, a tissue-specific splicing defect in *IKBKAP* was uncovered. This defect was particularly severe in neural crest precursors, validating the clinical phenotype of the disease as a neurocristopathy. Moreover, differentiation of these precursors revealed impairments in neurogenesis and migration in vitro. In a subsequent paper, the same group used a screening platform with 6,912 small-molecule compounds to test for rescue of *IKBKAP* expression in patient-derived neural crest cells (Lee et al. 2012). This study not only demonstrated that large-scale drug screening is feasible using human neurons, but also revealed that modulation of α2 adrenergic pathways represents a novel therapeutic target for this neuropathy.

In thinking about late-onset human models for neurodegenerative disorders, a major source of concern is the time scale needed to observe any relevant cellular phenotypes in an in vitro setting. One way to overcome this limitation is to use environmental triggers or overload the signaling pathways known to contribute to neurodegeneration. For instance, midbrain dopaminergic neurons generated from patients carrying a high penetrance PD mutation are more sensitive to cell death after exposure to hydrogen peroxide, MG-132, and 6-hydroxy-dopamine (Nguyen et al. 2011). Cortical neurons generated from PD patients display proteotoxic phenotypes as well, and yeast screens can be leveraged to find conserved modalities to reverse these abnormalities (Chung et al. 2013). Another solution for tackling age-related neurodegenerative diseases is to use accelerated models for disease. For instance, Shi et al. (2012a) utilized hES cells carrying a chromosome 21 trisomy as a model for Alzheimer's dementia. The *amyloid precursor protein (APP)* gene, whose cleavage generates the hallmark pathogenic amyloid-beta (Aβ) peptide, is harbored on chromosome 21, and its triplication is the likely reason behind early onset of Alzheimer's in patients with Down syndrome. Cortical neurons from ES cells carrying the extra chromosome, generated using an extensively characterized protocol (Shi et al. 2012b), produce higher levels of Aβ peptides,

Table 1 Summary of some cellular models for neuropsychiatric disorders

Diseases	Genetic events	Derived cell type	Cell-type characterizations	Cellular phenotypes	Publications
Rett syndrome	Missense and nonsense mutations in *MECP2*	Neuronal progenitor cells	Immunostaining	Increased susceptibility for L1 retrotranspositions	Muotri et al. 2010
Familial dysautonomia	Point mutation in *IKBKAP*	Neural crest precursors	Cell sorting, transcriptome analysis, immunostaining	Tissue-specific mis-splicing of *IKBKAP*; defects in neuronal differentiation and migration	Lee et al. 2009, 2012
Timothy syndrome	Point mutation (G406R) in *CACNA1C*	Cortical glutamatergic neurons	Multiplex single-cell analysis, transcriptome analysis, immunostaining, electrophysiology	Ectopic expression of tyrosine hydroxylase (TH) in neurons, defect in the generation of cortical projection neurons; activity-dependent dendritic retraction	Pasca et al. 2011, Krey et al. 2011
Down syndrome	Chromosome 21 trisomy	Cortical glutamatergic neurons	Immunostaining, gene expression, transplantation on rodent slices, electrophysiology	Accumulation of insoluble intra- and extracellular amyloid aggregates; hyperphosphorylated *tau* protein in cell bodies and dendrites	Shi et al. 2012a
Huntington's disease	CAG expansion in *HTT*	Medium spiny neurons	Immunostaining, transcriptome analysis, electrophysiology	Increased cell death over time in culture and after trophic factor withdrawal; abnormal calcium homeostasis upon glutamate stimulation	HD iPSC Consort. 2012
Rett syndrome	Missense and nonsense mutations in *MECP2*	Glutamatergic neurons, GABAergic neurons	Immunostaining, electrophysiology	Reduced spine and synaptic density; smaller soma size	Marchetto et al. 2010

Atypical Rett syndrome	Mutations in *CDKL5*	Glutamatergic neurons, GABAergic neurons	Immunostaining	Aberrant spine structure; reduced number of synapses	Ricciardi et al. 2012
Parkinson's disease	Point mutation in *LRRK2*	Midbrain dopaminergic neurons	Immunostaining, transcriptome analysis, electrophysiology, neurotransmitter release	Neurons highly susceptible to cell death	Nguyen et al. 2011
Amyotrophic lateral sclerosis	Mutations in *TDP-43, SOD1*	Motor neurons	Immunostaining, cell sorting, transcriptome analysis, electrophysiology	Cytosolic aggregates; shorter neurites; decreased survival	Egawa et al. 2012, Bilican et al. 2012, Yang et al. 2013
Spinal muscular atrophy	Mutations in *SMN1*	Motor neurons	Immunostaining	Normal neurogenic potential but reduction in the number of motor neurons over time	Ebert et al. 2009

Abbreviations: *CACNA1C*, calcium channel, voltage-dependent, L type, alpha 1C subunit; *CDKL5*, cyclin-dependent kinase-like 5; *HTT*, huntingtin; *IKBKAP*, inhibitor of kappa light polypeptide gene enhancer in B-cells, kinase complex-associated protein; *LRRK2*, leucine-rich repeat kinase 2; *MECP2*, methyl CpG binding protein 2; *SMN1*, survival of motor neuron 1, telomeric; *SOD1*, superoxide dismutase 1; *TDP-43* (*TARDBP*), TAR DNA binding protein.

secrete more pathogenic Aβ42 peptide, and display abnormally distributed phosphorylated Tau protein. The development of Alzheimer's disease pathological hallmarks in vitro was accelerated in Down syndrome patients, developing over months rather than years. Subsequent studies have demonstrated that cellular phenotypes for both familial and sporadic Alzheimer's dementia can be developed in vitro (Kondo et al. 2013).

The ability to recapitulate critical stages of neural development in vitro allows for the close examination of psychiatric disorders of a developmental origin, such as autism spectrum disorders and schizophrenias. Timothy syndrome (TS) is caused by a point mutation in the voltage-gated calcium channel encoded by the *CACNA1C* gene and is one of the most penetrant forms of autism spectrum disorders (Splawski et al. 2004). Live imaging analysis of cortical neurons and cardiomyocytes derived from these patients revealed defects in calcium signaling (Paşca et al. 2011, Yazawa et al. 2011). Moreover, this gain-of-function mutation was associated with an activity-dependent retraction of dendrites in a mouse model, as well as in iPSC-derived cortical neurons (Krey et al. 2013). Cellular phenotypes in patient cells were rescued with an atypical blocker of the channel but not with classical L-type calcium channel blockers. Unexpectedly, the TS mutation in *CACNA1C* was also associated with defects in corticogenesis, as assessed by single-cell gene expression analysis of patient-derived neurons. Specifically, neuronal cultures from TS patients displayed a reduced production of putative callosally projecting cortical neurons, along with a concomitant increase in the proportion of subcerebral projection neurons. Rett syndrome, caused primarily by a mutation in the X-linked gene *methyl CpG binding protein-2 (MeCP2)* is another example of a neurodevelopmental disorder from which patient-derived neurons have been generated (Marchetto et al. 2010). Glutamatergic and GABAergic neurons from patients carrying multiple mutations showed fewer synaptic contacts, reduced spine density, and a smaller soma size, as well as abnormal electrophysiological properties. Both genetic and pharmacological rescue were demonstrated in neurons differentiated from subjects with Rett syndrome. Some of these observed cellular phenotypes, such as the reduction in the number of excitatory synaptic contacts, have been replicated by another research group (Cheung et al. 2011) and have also been observed in the context of mutations in other X-linked genes, such as *cyclin-dependent kinase-like 5 (CDKL5)*, which cause a Rett-like syndrome (Ricciardi et al. 2012). Considering the level of developmental noise, such multilevel validations are essential for the field. More recent efforts have also made possible the examination of neurons from patients with developmental synaptopathies, such as Phelan-McDermid syndrome (PMD). PMD-derived neurons, carrying a deletion of the protein SHANK3, display defects in excitatory, but not inhibitory, synaptic transmission (Shcheglovitov et al. 2013).

It is compelling to see how interesting basic science findings can be validated in a disease context with iPSC, as is the case with a recent study published by the Gage group. Previous work demonstrated that L1 retrotranspositions are more abundant in the brain, and more specifically in neural precursors, than in other somatic tissue (Coufal et al. 2009, Muotri et al. 2005). Using early neural precursors derived from patients with Rett syndrome, Muotri et al. (2010) further showed that disruptions of *MECP2* greatly influence the frequency of L1 retrotransposons in these cells. This tissue-specific phenomenon could result in an increased rate of somatic mutations in patients with Rett syndrome and alludes to a novel level of complexity that could contribute to neurodevelopmental disorders.

One early aim for deriving human neurons in vitro was to utilize them for cell-replacement therapy. Perhaps the most sustained efforts in this regard have been made toward generating specific populations of midbrain dopaminergic neurons to be used in PD (Kriks et al. 2011). Advances have also been made in the directed differentiation of other neuronal cell types implicated in neurodegenerative disease. For instance, enriched populations of DARPP32-expressing forebrain

GABAergic neurons, but not spinal GABAergic neurons, can successfully correct motor defects when transplanted into the striatum of quinolinic acid–lesioned mice (Ma et al. 2012). In a similar vein, striatal pathology associated with Huntington's disease was also recently recapitulated in vitro by the HD iPSC Consortium (2012). The severity of one of the cellular phenotypes could be correlated with the number of CAG repeats, a feature known to dictate the age of onset of symptoms in these patients.

Although cellular models for neuropsychiatric disorders were initially met with skepticism, a number of well-controlled studies to date indicate that we can surmount the variability associated with reprogramming and differentiation in vitro and, more importantly, that these models can be utilized as a reliable platform for understanding disease pathogenesis. A major goal of this approach is the ability to run large-scale drug screening and perhaps even in vitro clinical trials for rare disorders for which sufficient numbers of patients may not be available. It is becoming increasingly clear that drug responses within specific psychiatric conditions are quite variable. This observation has paved the way for another feasible application of iPSC technology: the development of iPSC-based assays that reliably predict drug responses in individuals. It remains to be seen whether large-scale, multidimensional cellular phenotyping in neurons from patients with idiopathic schizophrenia or autism spectrum disorders will yield novel operational parameters to improve our Kraepelian view of these disorders.

SUMMARY POINTS

1. Developments in stem cell biology have enabled the derivation of pluripotent stem cells from human somatic cells, including from patients with neuropsychiatric diseases.

2. Leveraging knowledge from developmental neurobiology, directed differentiation approaches have made possible the generation of specific neuronal subtypes in vitro, such as cortical excitatory and inhibitory neurons, midbrain dopaminergic neurons, spinal motor neurons, and neural crest derivatives.

3. The ability to recapitulate in vitro early stages of corticogenesis presents a unique opportunity to examine primate- and human-specific aspects of brain development.

4. The use of patient-derived cells has already identified disease-relevant cellular phenotypes. These advances pave the way for running large-scale drug screening on functional patient-derived neurons and may contribute to a paradigm shift in our understanding of psychiatric disorders.

FUTURE DIRECTIONS

1. There is an urgent need to create guidelines and to standardize protocols for the generation, characterization, and maintenance of iPSC. It is also essential to comprehensively characterize various neural differentiation protocols using state-of-the-art technologies, such as single-cell gene expression and RNA sequencing platforms, and to subsequently make standard operating procedures available to the larger scientific community.

2. A comprehensive exploration of the functional properties and transcriptional signatures of forebrain neurons derived from multiple human and nonhuman primate species will provide novel evolutionary insights and guide the study of disorders of social cognition.

3. A major missing element in our understanding of in vitro cellular reprogramming and human neural differentiation is a detailed mapping of epigenetics, imprinting, and X-inactivation phenomena occurring during these processes. This information will be essential for us to reliably address, in in vitro cellular models, neuropsychiatric conditions caused by disruptions in genes governed by such biological events.

4. Genetic manipulation of pluripotent cells should become the norm for iPSC studies to demonstrate causality in a more defined manner, most notably with the inclusion of control experiments utilizing CRISPR (clustered regularly interspaced short palindromic repeats) and TALEN (transcription activator-like effector nucleases)-modified iPSC lines or genetic rescue experiments.

5. Further studies are required to establish the actual age of neurons generated in vitro and to concretely identify their corresponding in vivo human developmental stage. In addition, better approaches for aging human neurons in the dish should be developed.

6. Integrated models of neuropsychiatric disease addressing the question of cell-type-specific defects should be built to address complexities related to the intimate multicellular milieu existing in the brain. Specifically, what is the relative contribution of astrocytes, neurons, endothelial cells, microglia, or oligodendrocytes in modulating a specific cellular phenotype? It is in this setting, in particular, that the requirement for careful assessment of neuronal identity becomes apparent.

7. The field needs to create better infrastructure for sharing iPSC clones between various laboratories and institutions and for collecting clinical details from patients whose cells are reprogrammed for subsequent cellular phenotype-clinical correlation studies.

8. Studies of neuropsychiatric disorders should be highly integrative, using complementary rodent disease models, human cellular models, and postmortem tissue.

9. With appropriate resources, studies on neurons derived from large populations of patients with idiopathic forms of psychiatric disease (i.e., hundreds to thousands of subjects) should be conducted to identify if, with larger statistical power, we can dissect out disease subtypes, predict drug responses, or make clinical prognoses on the basis of cellular endophenotypes.

10. Human cellular models of disease are likely to reveal numerous cellular phenotypes for a given disorder. Some of these abnormalities will be core pathophysiological processes, whereas others will be homeostatic compensatory events or in vitro artifacts. We must develop novel approaches to determine the nature of identified cellular abnormalities so that we can efficiently direct therapeutic targeting.

11. The ultimate proof for the therapeutic potential of patient-derived neurons will come from demonstrating that a drug identified as correcting cellular phenotypes in vitro can result in clinical improvement in patients with a specific disorder. This outcome could arise from high-throughput in vitro screening with FDA-approved drugs and drug repurposing, or by running clinical trials in the dish for rare disorders for which the number of drugs to test surpasses the number of available patients.

DISCLOSURE STATEMENT

The authors are not aware of any affiliations, memberships, funding, or financial holdings that might be perceived as affecting the objectivity of this review.

LITERATURE CITED

Akbarian S, Huang HS. 2006. Molecular and cellular mechanisms of altered GAD1/GAD67 expression in schizophrenia and related disorders. *Brain Res. Rev.* 52:293–304

Alcamo EA, Chirivella L, Dautzenberg M, Dobreva G, Farinas I, et al. 2008. Satb2 regulates callosal projection neuron identity in the developing cerebral cortex. *Neuron* 57:364–77

Allman JM, Tetreault NA, Hakeem AY, Manaye KF, Semendeferi K, et al. 2011. The von Economo neurons in the frontoinsular and anterior cingulate cortex. *Ann. N. Y. Acad. Sci.* 1225:59–71

Allman JM, Watson KK, Tetreault NA, Hakeem AY. 2005. Intuition and autism: a possible role for Von Economo neurons. *Trends Cogn. Sci.* 9:367–73

Altmann CR, Brivanlou AH. 2001. Neural patterning in the vertebrate embryo. *Int. Rev. Cytol.* 203:447–82

Amoroso MW, Croft GF, Williams DJ, O'Keeffe S, Carrasco MA, et al. 2013. Accelerated high-yield generation of limb-innervating motor neurons from human stem cells. *J. Neurosci.* 33:574–86

Anokye-Danso F, Trivedi CM, Juhr D, Gupta M, Cui Z, et al. 2011. Highly efficient miRNA-mediated reprogramming of mouse and human somatic cells to pluripotency. *Cell Stem Cell* 8:376–88

Arlotta P, Molyneaux BJ, Chen J, Inoue J, Kominami R, Macklis JD. 2005. Neuronal subtype-specific genes that control corticospinal motor neuron development in vivo. *Neuron* 45:207–21

Ascoli GA, Alonso-Nanclares L, Anderson SA, Barrionuevo G, Benavides-Piccione R, et al. 2008. Petilla terminology: nomenclature of features of GABAergic interneurons of the cerebral cortex. *Nat. Rev.* 9:557–68

Aubry L, Bugi A, Lefort N, Rousseau F, Peschanski M, Perrier AL. 2008. Striatal progenitors derived from human ES cells mature into DARPP32 neurons in vitro and in quinolinic acid-lesioned rats. *Proc. Natl. Acad. Sci. USA* 105:16707–12

Barberi T, Klivenyi P, Calingasan NY, Lee H, Kawamata H, et al. 2003. Neural subtype specification of fertilization and nuclear transfer embryonic stem cells and application in parkinsonian mice. *Nat. Biotechnol.* 21:1200–7

Benavides-Piccione R, Ballesteros-Yañez I, DeFelipe J, Yuste R. 2002. Cortical area and species differences in dendritic spine morphology. *J. Neurocytol.* 31:337–46

Betizeau M, Cortay V, Patti D, Pfister S, Gautier E, et al. 2013. Precursor diversity and complexity of lineage relationships in the outer subventricular zone of the primate. *Neuron* 80:442–57

Bilican B, Serio A, Barmada SJ, Nishimura AL, Sullivan GJ, et al. 2012. Mutant induced pluripotent stem cell lines recapitulate aspects of TDP-43 proteinopathies and reveal cell-specific vulnerability. *Proc. Natl. Acad. Sci. USA* 109:5803–8

Brain Work. Group. 2013. *Advisory Committee to the NIH Director: Interim Report*. Bethesda, MD: Natl. Inst. Health. **http://www.nih.gov/science/brain/09162013-Interim%20Report_Final%20Composite. pdf**

Britanova O, de Juan Romero C, Cheung A, Kwan KY, Schwark M, et al. 2008. Satb2 is a postmitotic determinant for upper-layer neuron specification in the neocortex. *Neuron* 57:378–92

Brüne M, Schöbel A, Karau R, Benali A, Faustmann PM, et al. 2010. Von Economo neuron density in the anterior cingulate cortex is reduced in early onset schizophrenia. *Acta Neuropathol.* 119:771–78

Caiazzo M, Dell'Anno MT, Dvoretskova E, Lazarevic D, Taverna S, et al. 2011. Direct generation of functional dopaminergic neurons from mouse and human fibroblasts. *Nature* 476:224–27

Carri AD, Onorati M, Lelos MJ, Castiglioni V, Faedo A, et al. 2013. Developmentally coordinated extrinsic signals drive human pluripotent stem cell differentiation toward authentic DARPP-32+ medium-sized spiny neurons. *Development* 140:301–12

Chambers SM, Fasano CA, Papapetrou EP, Tomishima M, Sadelain M, Studer L. 2009. Highly efficient neural conversion of human ES and iPS cells by dual inhibition of SMAD signaling. *Nat. Biotechnol.* 27:275–80

Chambers SM, Qi Y, Mica Y, Lee G, Zhang XJ, et al. 2012. Combined small-molecule inhibition accelerates developmental timing and converts human pluripotent stem cells into nociceptors. *Nat. Biotechnol.* 30:715–20

Chao H-T, Chen H, Samaco RC, Xue M, Chahrour M, et al. 2010. Dysfunction in GABA signalling mediates autism-like stereotypies and Rett syndrome phenotypes. *Nature* 468:263–69

Charrier C, Joshi K, Coutinho-Budd J, Kim J-E, Lambert N, et al. 2012. Inhibition of SRGAP2 function by its human-specific paralogs induces neoteny during spine maturation. *Cell* 149:923–35

Chen B, Schaevitz LR, McConnell SK. 2005. Fezl regulates the differentiation and axon targeting of layer 5 subcortical projection neurons in cerebral cortex. *Proc. Natl. Acad. Sci. USA* 102:17184–89

Chen B, Wang SS, Hattox AM, Rayburn H, Nelson SB, McConnell SK. 2008. The Fezf2-Ctip2 genetic pathway regulates the fate choice of subcortical projection neurons in the developing cerebral cortex. *Proc. Natl. Acad. Sci. USA* 105:11382–87

Cheung AY, Horvath LM, Grafodatskaya D, Pasceri P, Weksberg R, et al. 2011. Isolation of MECP2-null Rett Syndrome patient hiPS cells and isogenic controls through X-chromosome inactivation. *Hum. Mol. Genet.* 20:2103–15

Chung CY, Khurana V, Auluck PK, Tardiff DF, Mazzulli JR, et al. 2013. Identification and rescue of α-synuclein toxicity in Parkinson patient–derived neurons. *Science* 342:983–87

Cobos I, Seeley WW. 2013. Human von Economo neurons express transcription factors associated with layer V subcerebral projection neurons. *Cereb. Cortex.* In press

Coovert DD, Le TT, McAndrew PE, Strasswimmer J, Crawford TO, et al. 1997. The survival motor neuron protein in spinal muscular atrophy. *Hum. Mol. Genet.* 6:1205–14

Coufal NG, Garcia-Perez JL, Peng GE, Yeo GW, Mu Y, et al. 2009. L1 retrotransposition in human neural progenitor cells. *Nature* 460:1127–31

Crossley PH, Martinez S, Martin GR. 1996. Midbrain development induced by FGF8 in the chick embryo. *Nature* 380:66–68

Cubelos B, Sebastian-Serrano A, Beccari L, Calcagnotto ME, Cisneros E, et al. 2010. *Cux1* and *Cux2* regulate dendritic branching, spine morphology, and synapses of the upper layer neurons of the cortex. *Neuron* 66:523–35

Danjo T, Eiraku M, Muguruma K, Watanabe K, Kawada M, et al. 2011. Subregional specification of embryonic stem cell-derived ventral telencephalic tissues by timed and combinatory treatment with extrinsic signals. *J. Neurosci.* 31:1919–33

DeFelipe J. 2011. The evolution of the brain, the human nature of cortical circuits, and intellectual creativity. *Front. Neuroanat.* 5:29

DeFelipe J, Ballesteros-Yañez I, Inda MC, Muñoz A. 2006. Double-bouquet cells in the monkey and human cerebral cortex with special reference to areas 17 and 18. *Prog. Brain Res.* 154:15–32

DeFelipe J, López-Cruz PL, Benavides-Piccione R, Bielza C, Larrañaga P, et al. 2013. New insights into the classification and nomenclature of cortical GABAergic interneurons. *Nat. Rev.* 14:202–16

Dimos JT, Rodolfa KT, Niakan KK, Weisenthal LM, Mitsumoto H, et al. 2008. Induced pluripotent stem cells generated from patients with ALS can be differentiated into motor neurons. *Science* 321:1218–21

Ebert AD, Yu J, Rose FF Jr, Mattis VB, Lorson CL, et al. 2009. Induced pluripotent stem cells from a spinal muscular atrophy patient. *Nature* 457:277–80

Eberwine J, Lovatt D, Buckley P, Dueck H, Francis C, et al. 2012. Quantitative biology of single neurons. *J. R. Soc. Interface* 9:3165–83

Egawa N, Kitaoka S, Tsukita K, Naitoh M, Takahashi K, et al. 2012. Drug screening for ALS using patient-specific induced pluripotent stem cells. *Sci. Transl. Med.* 4:145ra04

Eiraku M, Watanabe K, Matsuo-Takasaki M, Kawada M, Yonemura S, et al. 2008. Self-organized formation of polarized cortical tissues from ESCs and its active manipulation by extrinsic signals. *Cell Stem Cell* 3:519–32

Elkabetz Y, Panagiotakos G, Al Shamy G, Socci ND, Tabar V, Studer L. 2008. Human ES cell-derived neural rosettes reveal a functionally distinct early neural stem cell stage. *Genes Dev.* 22:152–65

Erceg S, Lukovic D, Moreno-Manzano V, Stojkovic M, Bhattacharya SS. 2012. Derivation of cerebellar neurons from human pluripotent stem cells. *Curr. Protoc. Stem Cell Biol.* 20:1H.5–10

Espuny-Camacho I, Michelsen KA, Gall D, Linaro D, Hasche A, et al. 2013. Pyramidal neurons derived from human pluripotent stem cells integrate efficiently into mouse brain circuits in vivo. *Neuron* 77:440–56

Falk S, Sommer L. 2009. Stage- and area-specific control of stem cells in the developing nervous system. *Curr. Opin. Genet. Dev.* 19:454–60

Farkas LM, Huttner WB. 2008. The cell biology of neural stem and progenitor cells and its significance for their proliferation versus differentiation during mammalian brain development. *Curr. Opin. Cell Biol.* 20:707–15

Fasano CA, Chambers SM, Lee G, Tomishima MJ, Studer L. 2010. Efficient derivation of functional floor plate tissue from human embryonic stem cells. *Cell Stem Cell* 6:336–47

Fietz SA, Kelava I, Vogt J, Wilsch-Bräuninger M, Stenzel D, et al. 2010. OSVZ progenitors of human and ferret neocortex are epithelial-like and expand by integrin signaling. *Nat. Neurosci.* 13:690–99

Fu X, Giavalisco P, Liu X, Catchpole G, Fu N, et al. 2011. Rapid metabolic evolution in human prefrontal cortex. *Proc. Natl. Acad. Sci. USA* 108:6181–86

Ganat YM, Calder EL, Kriks S, Nelander J, Tu EY, et al. 2012. Identification of embryonic stem cell-derived midbrain dopaminergic neurons for engraftment. *J. Clin. Investig.* 122:2928–39

Gaspar P, Berger B, Febvret A, Vigny A, Krieger-Poulet M, Borri-Voltattorni C. 1987. Tyrosine hydroxylase-immunoreactive neurons in the human cerebral cortex: a novel catecholaminergic group? *Neurosci. Lett.* 80:257–62

Greig LC, Woodworth MB, Galazo MJ, Padmanabhan H, Macklis JD. 2013. Molecular logic of neocortical projection neuron specification, development and diversity. *Nat. Rev. Neurosci.* 14:755–69

Hansen DV, Lui JH, Parker PRL, Kriegstein AR. 2010. Neurogenic radial glia in the outer subventricular zone of human neocortex. *Nature* 464:554–61

HD iPSC Consort. 2012. Induced pluripotent stem cells from patients with Huntington's disease show CAG-repeat-expansion-associated phenotypes. *Cell Stem Cell* 11:264–78

Hevner RF, Shi L, Justice N, Hsueh Y, Sheng M, et al. 2001. Tbr1 regulates differentiation of the preplate and layer 6. *Neuron* 29:353–66

Hobert O. 2011. Regulation of terminal differentiation programs in the nervous system. *Annu. Rev. Cell Dev. Biol.* 27:681–96

Hornung JP, Törk I, De Tribolet N. 1989. Morphology of tyrosine hydroxylase-immunoreactive neurons in the human cerebral cortex. *Exp. Brain Res.* 76:12–20

Hynes M, Porter JA, Chiang C, Chang D, Tessier-Lavigne M, et al. 1995. Induction of midbrain dopaminergic neurons by Sonic hedgehog. *Neuron* 15:35–44

Kamiya D, Banno S, Sasai N, Ohgushi M, Inomata H, et al. 2011. Intrinsic transition of embryonic stem-cell differentiation into neural progenitors. *Nature* 470:503–9

Kawasaki H, Mizuseki K, Nishikawa S, Kaneko S, Kuwana Y, et al. 2000. Induction of midbrain dopaminergic neurons from ES cells by stromal cell-derived inducing activity. *Neuron* 28:31–40

Kondo T, Asai M, Tsukita K, Kutoku Y, Ohsawa Y, et al. 2013. Modeling Alzheimer's disease with iPSCs reveals stress phenotypes associated with intracellular Aβ and differential drug responsiveness. *Cell Stem Cell* 12:487–96

Krey JF, Paşca SP, Shcheglovitov A, Yazawa M, Schwemberger R, et al. 2013. Timothy syndrome is associated with activity-dependent dendritic retraction in rodent and human neurons. *Nat. Neurosci.* 16:201–9

Kriks S, Shim JW, Piao J, Ganat YM, Wakeman DR, et al. 2011. Dopamine neurons derived from human ES cells efficiently engraft in animal models of Parkinson's disease. *Nature* 480:547–51

Kwan KY, Lam MM, Johnson MB, Dube U, Shim S, et al. 2012. Species-dependent posttranscriptional regulation of NOS1 by FMRP in the developing cerebral cortex. *Cell* 149:899–911

Lai T, Jabaudon D, Molyneaux BJ, Azim E, Arlotta P, et al. 2008. SOX5 controls the sequential generation of distinct corticofugal neuron subtypes. *Neuron* 57:232–47

LaMonica BE, Lui JH, Wang X, Kriegstein AR. 2012. OSVZ progenitors in the human cortex: an updated perspective on neurodevelopmental disease. *Curr. Opin. Neurobiol.* 22:747–53

Lancaster MA, Renner M, Martin CA, Wenzel D, Bicknell LS, et al. 2013. Cerebral organoids model human brain development and microcephaly. *Nature* 501:373–79

Le Dréau G, Martí E. 2012. Dorsal-ventral patterning of the neural tube: a tale of three signals. *Dev. Neurobiol.* 72:1471–81

Le Magueresse C, Monyer H. 2013. GABAergic interneurons shape the functional maturation of the cortex. *Neuron* 77:388–405

Lee G, Kim H, Elkabetz Y, Al Shamy G, Panagiotakos G, et al. 2007a. Isolation and directed differentiation of neural crest stem cells derived from human embryonic stem cells. *Nat. Biotechnol.* 25:1468–75

Lee G, Papapetrou EP, Kim H, Chambers SM, Tomishima MJ, et al. 2009. Modelling pathogenesis and treatment of familial dysautonomia using patient-specific iPSCs. *Nature* 461:402–6

Lee G, Ramirez CN, Kim H, Zeltner N, Liu B, et al. 2012. Large-scale screening using familial dysautonomia induced pluripotent stem cells identifies compounds that rescue IKBKAP expression. *Nat. Biotechnol.* 30:1244–48

Lee H, Shamy GA, Elkabetz Y, Schofield CM, Harrsion NL, et al. 2007b. Directed differentiation and transplantation of human embryonic stem cell-derived motoneurons. *Stem Cells* 25:1931–39

Lee S-H, Lumelsky N, Studer L, Auerbach JM, McKay RD. 2000. Efficient generation of midbrain and hindbrain neurons from mouse embryonic stem cells. *Nat. Biotechnol.* 18:675–79

Leone DP, Srinivasan K, Chen B, Alcamo E, McConnell SK. 2008. The determination of projection neuron identity in the developing cerebral cortex. *Curr. Opin. Neurobiol.* 18:28–35

Lewis DA, Curley AA, Glausier JR, Volk DW. 2012. Cortical parvalbumin interneurons and cognitive dysfunction in schizophrenia. *Trends Neurosci.* 35:57–67

Li XJ, Zhang X, Johnson MA, Wang ZB, Lavaute T, Zhang SC. 2009. Coordination of sonic hedgehog and Wnt signaling determines ventral and dorsal telencephalic neuron types from human embryonic stem cells. *Development* 136:4055–63

Lu B, Jan L, Jan Y-N. 2000. Control of cell divisions in the nervous system: symmetry and asymmetry. *Annu. Rev. Neurosci.* 23:531–56

Lui JH, Hansen DV, Kriegstein AR. 2011. Development and evolution of the human neocortex. *Cell* 146:18–36

Ma L, Hu B, Liu Y, Vermilyea SC, Liu H, et al. 2012. Human embryonic stem cell-derived GABA neurons correct locomotion deficits in quinolinic acid-lesioned mice. *Cell Stem Cell* 10:455–64

Mackler SA, Brooks BP, Eberwine JH. 1992. Stimulus-induced coordinate changes in mRNA abundance in single postsynaptic hippocampal CA1 neurons. *Neuron* 9:539–48

Marchetto MC, Carromeu C, Acab A, Yu D, Yeo GW, et al. 2010. A model for neural development and treatment of Rett syndrome using human induced pluripotent stem cells. *Cell* 143:527–39

Marchetto MC, Muotri AR, Mu Y, Smith AM, Cezar GG, Gage FH. 2008. Non-cell-autonomous effect of human SOD1 G37R astrocytes on motor neurons derived from human embryonic stem cells. *Cell Stem Cell* 3:649–57

Marchetto MC, Narvaiza I, Denli AM, Benner C, Lazzarini TA, et al. 2013. Differential L1 regulation in pluripotent stem cells of humans and apes. *Nature* 503:525–29

Maroof AM, Brown K, Shi SH, Studer L, Anderson SA. 2010. Prospective isolation of cortical interneuron precursors from mouse embryonic stem cells. *J. Neurosci.* 30:4667–75

Maroof AM, Keros S, Tyson JA, Ying SW, Ganat YM, et al. 2013. Directed differentiation and functional maturation of cortical interneurons from human embryonic stem cells. *Cell Stem Cell* 12:559–72

Meyer-Lindenberg A, Weinberger DR. 2006. Intermediate phenotypes and genetic mechanisms of psychiatric disorders. *Nat. Rev. Neurosci.* 7:818–27

Migliore M, Shepherd GM. 2005. Opinion: an integrated approach to classifying neuronal phenotypes. *Nat. Rev. Neurosci.* 6:810–18

Mione MC, Cavanagh JF, Harris B, Parnavelas JG. 1997. Cell fate specification and symmetrical/asymmetrical divisions in the developing cerebral cortex. *J. Neurosci.* 17:2018–29

Molnár G, Oláh S, Komlósi G, Füle M, Szabadics J, et al. 2008. Complex events initiated by individual spikes in the human cerebral cortex. *PLOS Biol.* 6(9):e222

Molyneaux BJ, Arlotta P, Fame RM, MacDonald JL, MacQuarrie KL, Macklis JD. 2009. Novel subtype-specific genes identify distinct subpopulations of callosal projection neurons. *J. Neurosci.* 29:12343–54

Molyneaux BJ, Arlotta P, Hirata T, Hibi M, Macklis JD. 2005. Fezl is required for the birth and specification of corticospinal motor neurons. *Neuron* 47:817–31

Muotri AR, Chu VT, Marchetto MCN, Deng W, Moran JV, Gage FH. 2005. Somatic mosaicism in neuronal precursor cells mediated by L1 retrotransposition. *Nature* 435:903–10

Muotri AR, Marchetto MC, Coufal NG, Oefner R, Yeo G, et al. 2010. L1 retrotransposition in neurons is modulated by MeCP2. *Nature* 468:443–46

Nguyen HN, Byers B, Cord B, Shcheglovitov A, Byrne J, et al. 2011. LRRK2 mutant iPSC-derived DA neurons demonstrate increased susceptibility to oxidative stress. *Cell Stem Cell* 8:267–80

Nicholas CR, Chen J, Tang Y, Southwell DG, Chalmers N, et al. 2013. Functional maturation of hPSC-derived forebrain interneurons requires an extended timeline and mimics human neural development. *Cell Stem Cell* 12:573–86

Noctor SC, Flint AC, Weissman TA, Wong WS, Clinton BK, Kriegstein AR. 2002. Dividing precursor cells of the embryonic cortical ventricular zone have morphological and molecular characteristics of radial glia. *J. Neurosci.* 22:3161–73

Noctor SC, Martínez-Cerdeño V, Ivic L, Kriegstein AR. 2004. Cortical neurons arise in symmetric and asymmetric division zones and migrate through specific phases. *Nat. Neurosci.* 7:136–44

Oberheim NA, Takano T, Han X, He W, Lin JH, et al. 2009. Uniquely hominid features of adult human astrocytes. *J. Neurosci.* 29:3276–87

Okano H, Temple S. 2009. Cell types to order: temporal specification of CNS stem cells. *Curr. Opin. Neurobiol.* 19:112–19

Onder TT, Kara N, Cherry A, Sinha AU, Zhu N, et al. 2012. Chromatin-modifying enzymes as modulators of reprogramming. *Nature* 483:598–602

Owens DF, Kriegstein AR. 2002. Is there more to GABA than synaptic inhibition? *Nat. Rev. Neurosci.* 3:715–27

Pang ZP, Yang N, Vierbuchen T, Ostermeier A, Fuentes DR, et al. 2011. Induction of human neuronal cells by defined transcription factors. *Nature* 476:220–23

Park IH, Zhao R, West JA, Yabuuchi A, Huo H, et al. 2008. Reprogramming of human somatic cells to pluripotency with defined factors. *Nature* 451:141–46

Paşca SP, Portmann T, Voineagu I, Yazawa M, Shcheglovitov A, et al. 2011. Using iPSC-derived neurons to uncover cellular phenotypes associated with Timothy syndrome. *Nat. Med.* 17:1657–62

Peljto M, Wichterle H. 2011. Programming embryonic stem cells to neuronal subtypes. *Curr. Opin. Neurobiol.* 21:43–51

Perrier AL, Tabar V, Barberi T, Rubio ME, Bruses J, et al. 2004. Derivation of midbrain dopamine neurons from human embryonic stem cells. *Proc. Natl. Acad. Sci. USA* 101:12543–48

Qiu S, Luo S, Evgrafov O, Li R, Schroth GP, et al. 2012. Single-neuron RNA-Seq: technical feasibility and reproducibility. *Front. Genet.* 3:124

Rais Y, Zviran A, Geula S, Gafni O, Chomsky E, et al. 2013. Deterministic direct reprogramming of somatic cells to pluripotency. *Nature* 502:65–70

Ramón y Cajal S. 1899. *La textura del sistema nerviosa del hombre y los vertebrados.* Madrid: Imprenta y Librería de Nicolás Moya

Ricciardi S, Ungaro F, Hambrock M, Rademacher N, Stefanelli G, et al. 2012. CDKL5 ensures excitatory synapse stability by reinforcing NGL-1-PSD95 interaction in the postsynaptic compartment and is impaired in patient iPSC-derived neurons. *Nat. Cell Biol.* 14:911–23

Rice J. 2012. Animal models: not close enough. *Nature* 484:S9

Roy NS, Cleren C, Singh SK, Yang L, Beal MF, Goldman SA. 2006. Functional engraftment of human ES cell-derived dopaminergic neurons enriched by coculture with telomerase-immortalized midbrain astrocytes. *Nat. Med.* 12:1259–68

Santos M, Uppal N, Butti C, Wicinski B, Schmeidler J, et al. 2011. Von Economo neurons in autism: a stereologic study of the frontoinsular cortex in children. *Brain Res.* 1380:206–17

Seeley WW, Carlin DA, Allman JM, Macedo MN, Bush C, et al. 2006. Early frontotemporal dementia targets neurons unique to apes and humans. *Ann. Neurol.* 60:660–67

Serio A, Bilican B, Barmada SJ, Ando DM, Zhao C, et al. 2013. Astrocyte pathology and the absence of non-cell autonomy in an induced pluripotent stem cell model of TDP-43 proteinopathy. *Proc. Natl. Acad. Sci. USA* 110:4697–702

Shcheglovitov A, Shcheglovitova O, Yazawa M, Portmann T, Shu R, et al. 2013. SHANK3 and IGF1 restore synaptic deficits in neurons from 22q13 deletion syndrome patients. *Nature* 503:267–71

Sherwood CC, Stimpson CD, Raghanti MA, Wildman DE, Uddin M, et al. 2006. Evolution of increased glia-neuron ratios in the human frontal cortex. *Proc. Natl. Acad. Sci. USA* 103:13606–11

Shi Y, Kirwan P, Smith J, MacLean G, Orkin SH, Livesey FJ. 2012a. A human stem cell model of early Alzheimer's disease pathology in Down syndrome. *Sci. Transl. Med.* 4:124ra29

Shi Y, Kirwan P, Smith J, Robinson HPC, Livesey FJ. 2012b. Human cerebral cortex development from pluripotent stem cells to functional excitatory synapses. *Nat. Neurosci.* 15:477–86, S1

Soldner F, Hockemeyer D, Beard C, Gao Q, Bell GW, et al. 2009. Parkinson's disease patient-derived induced pluripotent stem cells free of viral reprogramming factors. *Cell* 136:964–77

Son EY, Ichida JK, Wainger BJ, Toma JS, Rafuse VF, et al. 2011. Conversion of mouse and human fibroblasts into functional spinal motor neurons. *Cell Stem Cell* 9:205–18

Splawski I, Timothy KW, Sharpe LM, Decher N, Kumar P, et al. 2004. Ca(V)1.2 calcium channel dysfunction causes a multisystem disorder including arrhythmia and autism. *Cell* 119:19–31

Spruston N. 2008. Pyramidal neurons: dendritic structure and synaptic integration. *Nat. Rev. Neurosci.* 9:206–21

Srinivasan K, Leone DP, Bateson RK, Dobreva G, Kohwi Y, et al. 2012. A network of genetic repression and derepression specifies projection fates in the developing neocortex. *Proc. Natl. Acad. Sci. USA* 109:19071–78

Tabar V, Tomishima M, Panagiotakos G, Wakayama S, Menon J, et al. 2008. Therapeutic cloning in individual parkinsonian mice. *Nat. Med.* 14:379–81

Takahashi K, Tanabe K, Ohnuki M, Narita M, Ichisaka T, et al. 2007. Induction of pluripotent stem cells from adult human fibroblasts by defined factors. *Cell* 131:861–72

Tiberi L, Vanderhaeghen P, van den Ameele J. 2012. Cortical neurogenesis and morphogens: diversity of cues, sources and functions. *Curr. Opin. Cell Biol.* 24:269–76

Toledo-Rodriguez M, Blumenfeld B, Wu C, Luo J, Attali B, et al. 2004. Correlation maps allow neuronal electrical properties to be predicted from single-cell gene expression profiles in rat neocortex. *Cereb. Cortex* 14:1310–27

Tropepe V, Hitoshi S, Sirard C, Mak TW, Rossant J, van der Kooy D. 2001. Direct neural fate specification from embryonic stem cells: a primitive mammalian neural stem cell stage acquired through a default mechanism. *Neuron* 30:65–78

Trottier S, Geffard M, Evrard B. 1989. Co-localization of tyrosine hydroxylase and GABA immunoreactivities in human cortical neurons. *Neurosci. Lett.* 106:76–82

van der Worp HB, Howells DW, Sena ES, Porritt MJ, Rewell S, et al. 2010. Can animal models of disease reliably inform human studies? *PLOS Med.* 7:e1000245

Warren L, Manos PD, Ahfeldt T, Loh YH, Li H, et al. 2010. Highly efficient reprogramming to pluripotency and directed differentiation of human cells with synthetic modified mRNA. *Cell Stem Cell* 7:618–30

Watanabe K, Kamiya D, Nishiyama A, Katayama T, Nozaki S, et al. 2005. Directed differentiation of telencephalic precursors from embryonic stem cells. *Nat. Neurosci.* 8:288–96

Wonders CP, Anderson SA. 2006. The origin and specification of cortical interneurons. *Nat. Rev. Neurosci.* 7:687–96

Yang YM, Gupta SK, Kim KJ, Powers BE, Cerqueira A, et al. 2013. A small molecule screen in stem-cell-derived motor neurons identifies a kinase inhibitor as a candidate therapeutic for ALS. *Cell Stem Cell* 12:713–26

Yazawa M, Hsueh B, Jia X, Paşca AM, Bernstein JA, et al. 2011. Using induced pluripotent stem cells to investigate cardiac phenotypes in Timothy syndrome. *Nature* 471:230–34

Ye W, Shimamura K, Rubenstein JL, Hynes MA, Rosenthal A. 1998. FGF and Shh signals control dopaminergic and serotonergic cell fate in the anterior neural plate. *Cell* 93:755–66

Yoo AS, Sun AX, Li L, Shcheglovitov A, Portmann T, et al. 2011. MicroRNA-mediated conversion of human fibroblasts to neurons. *Nature* 476:228–31

Yoshioka N, Gros E, Li H-R, Kumar S, Deacon DC, et al. 2013. Efficient generation of human iPSCs by a synthetic self-replicative RNA. *Cell Stem Cell* 13:246–54

Yu J, Vodyanik MA, Smuga-Otto K, Antosiewicz-Bourget J, Frane JL, et al. 2007. Induced pluripotent stem cell lines derived from human somatic cells. *Science* 318:1917–20

Zeng H, Shen EH, Hohmann JG, Oh SW, Bernard A, et al. 2012. Large-scale cellular-resolution gene profiling in human neocortex reveals species-specific molecular signatures. *Cell* 149:483–96

Zhang Y, Pak C, Han Y, Ahlenius H, Zhang Z, et al. 2013. Rapid single-step induction of functional neurons from human pluripotent stem cells. *Neuron* 78:785–98

Neuropeptidergic Control of Sleep and Wakefulness

Constance Richter,[1] Ian G. Woods,[2]
and Alexander F. Schier[1]

[1] Department of Molecular and Cellular Biology, Center for Brain Science, Division of Sleep Biology, Harvard University, Cambridge, Massachusetts 02138; email: crichter@fas.harvard.edu, schier@fas.harvard.edu

[2] Department of Biology, Ithaca College, Ithaca, New York 14850; email: iwoods@ithaca.edu

Annu. Rev. Neurosci. 2014. 37:503–31

The *Annual Review of Neuroscience* is online at neuro.annualreviews.org

This article's doi:
10.1146/annurev-neuro-062111-150447

Keywords

REM, NREM, hypocretin, MCH, feeding, stress, local sleep

Abstract

Sleep and wake are fundamental behavioral states whose molecular regulation remains mysterious. Brain states and body functions change dramatically between sleep and wake, are regulated by circadian and homeostatic processes, and depend on the nutritional and emotional condition of the animal. Sleep-wake transitions require the coordination of several brain regions and engage multiple neurochemical systems, including neuropeptides. Neuropeptides serve two main functions in sleep-wake regulation. First, they represent physiological states such as energy level or stress in response to environmental and internal stimuli. Second, neuropeptides excite or inhibit their target neurons to induce, stabilize, or switch between sleep-wake states. Thus, neuropeptides integrate physiological subsystems such as circadian time, previous neuron usage, energy homeostasis, and stress and growth status to generate appropriate sleep-wake behaviors. We review the roles of more than 20 neuropeptides in sleep and wake to lay the foundation for future studies uncovering the mechanisms that underlie the initiation, maintenance, and exit of sleep and wake states.

Contents

INTRODUCTION

Three behavioral criteria define sleep: inactivity, rapid reversibility, and reduced responsiveness to external stimuli (reviewed in Campbell & Tobler 1984, Allada & Siegel 2008). Lower responsiveness distinguishes sleep from inactive wakefulness and indicates an increased arousal threshold, often measured as a lengthened reaction time upon stimulation (reviewed in Allada & Siegel 2008). In mammals, recordings of cortical activity—known as electroencephalograms (EEGs)—complement these behavioral criteria (reviewed in Brown et al. 2012). During non–rapid eye movement sleep (NREM sleep), cortical neurons fire synchronously, producing high-amplitude slow waves [also known as delta waves or depicted as slow-wave activity (SWA)]. In rapid eye movement (REM) sleep, asynchronous cortical activity generates small fast waves, a pattern paradoxically similar to that of waking brains [paradoxical sleep (PS)]. During NREM sleep, muscles partially relax and completely lose their tone during REM sleep. Thus, wake and sleep behaviors consist of substates that can be distinguished by EEG, muscle tone, and behavior (reviewed in Harris & Thiele 2011, Brown et al. 2012).

Anatomy of Sleep-Wake Regulation

Two processes regulate sleep and wakefulness: circadian rhythms and homeostatic drive. Sleep-wake behaviors often oscillate with the rhythm of day and night, whereas sleep length and depth correlate with the amount of preceding wakefulness (reviewed in Brown et al. 2012). These regulatory processes act on specific brain areas, including four major centers that activate the cortex during wake (**Figure 1**) (reviewed in Lin et al. 2011). Cholinergic and monoaminergic nuclei in the brain stem (*a*) activate the posterior hypothalamus (*b*) which then activates the cortex. The brain stem nuclei and the hypothalamus also activate cholinergic neurons in the basal forebrain (*c*) and neurons in the thalamus (*d*) (reviewed in Saper et al. 2010), both of which can activate the cortex.

Additionally, arousal nuclei inhibit sleep-active neurons in the ventral lateral preoptic (VLPO) area (reviewed in Saper et al. 2010). Conversely, neurons in the VLPO send inhibitory γ-aminobutyric acid (GABA)-ergic and galaninergic projections back to arousal centers in the hypothalamus and the brain stem. This mutual inhibition of wake- and sleep-promoting neurons generates a flip-flop switch that is thought to mediate the sharp transitions between sleep and wake (reviewed in Saper et al. 2010). A second switch regulates NREM and REM-sleep transitions through mutual inhibition between REM-activating and REM-suppressing neurons (**Figure 1**) (reviewed in España & Scammell 2011, McCarley 2011).

Finally, the suprachiasmatic nucleus (SCN) generates circadian rhythmicity by translating day–night information from the retina into transcriptional and translational feedback loops of clock genes (**Figure 2**) (reviewed in Welsh et al. 2010, Colwell 2011). Homeostatic sleep regulation relies on sleep-promoting molecules, such as adenosine, which accumulate during wake (reviewed in Brown et al. 2012). Thus the coordinated alternating activity of several brain nuclei and transmitter systems regulates sleep-wake states.

Figure 1

Sleep-wake regulation. (*a*) Wake: Brain stem arousal nuclei (*pink*) containing ACh, DA, 5-HT, or NA activate the thalamus, hypothalamus, spinal cord motor neurons, and the basal forebrain, and inhibit the ventrolateral preoptic area (GABA, galanin); hypothalamic arousal centers [*pink*: HA; *dark purple*: Hcrt] activate the cortex and arousal-related regions in the basal forebrain and brain stem; the thalamus activates the cortex. (*b*) NREM sleep: hypothalamic preoptic area nuclei (*dark blue*), containing GABA and galanin, inhibit brain stem and hypothalamic arousal nuclei; endogenous sleep regulatory substances [adenosine and NO] inhibit basal forebrain arousal nuclei, hypocretin neurons, and TMN neurons; and adenosine activates VLPO neurons. (*c*) REM sleep: REM-active brain stem nuclei, including LDT/PPT/SLD/PC and containing ACh, Glu, or GABA, promote activity in the basal forebrain and cortex and induce muscle atonia and rapid eye movements; hypothalamic neurons, containing MCH, promote REM sleep by suppressing REM-inhibitory brain centers, including vlPAG/LPT/DR/LC. NREM–REM switch: During NREM sleep, serotonergic DR and noradrenergic LC neurons inhibit LDT/PPT neurons. During REM sleep DR/LC neurons become silent, enabling the cholinergic LDT/PPT neurons to generate the hallmarks of REM sleep, including rapid eye movements, EEG activation, and muscle atonia. This reciprocal activity between REM-on (LDT/PPT) and REM-off (DR/LC) neurons drives the cycling between REM and NREM sleep episodes. Additionally, GABAergic neurons participate in the mutual inhibition of REM-activating and REM-suppressing neurons. During REM sleep, SLD/PC neurons use ascending and descending projections to activate the cortex and to promote muscle atonia. During NREM sleep, vlPAG/LPT neurons inhibit SLD/PC neurons. Recently, a pair of novel sleep-wake centers were identified in the brain stem: The glutamatergic MPB in the dorsal pontine tegmentum regulates arousal (Fuller et al. 2011), and the GABAergic PZ in the pontomedullary junction promotes sleep (Anaclet et al. 2012). Figure based on information from the following reviews: Saper et al. (2010), España & Scammell (2011), McCarley (2011), Monti et al. (2013), and Urade et al. (2011). Arrowheads in all figures indicate the effect on target structures and not the nature of the synaptic contact. For exact positions of nuclei, please refer to the primary literature and the Allen Mouse Brain Atlas (Lein et al. 2007, website: ©2014 Allen Institute for Brain Science). Abbreviations: 5-HT, serotonin; ACh, acetylcholine; A, adenosine; DA, dopamine; EEG, electroencephalogram; GABA, γ-amino-butyric acid; Glu, glutamine; HA, histamine; Hcrt, hypocretin; MCH, melanin-concentrating hormone; NA, noradrenaline; NO, nitric oxide.

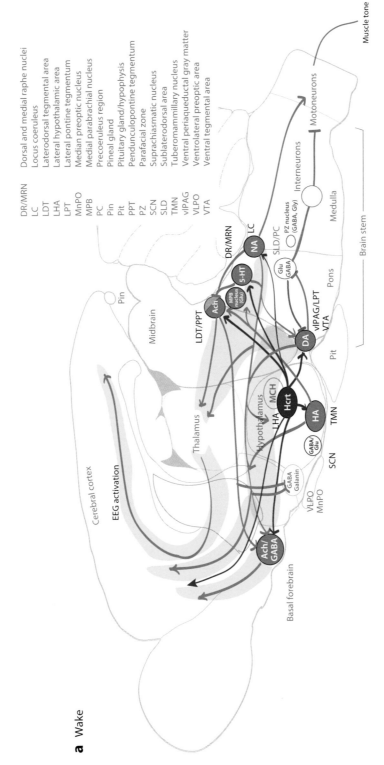

a Wake

DR/MRN	Dorsal and medial raphe nuclei
LC	Locus coeruleus
LDT	Laterodorsal tegmental area
LHA	Lateral hypothalamic area
LPT	Lateral pontine tegmentum
MnPO	Median preoptic nucleus
MPB	Medial parabrachial nucleus
PC	Precoeruleus region
Pin	Pineal gland
Pit	Pituitary gland/hypophysis
PPT	Pedunculopontine tegmentum
PZ	Parafacial zone
SCN	Suprachiasmatic nucleus
SLD	Sublaterodorsal area
TMN	Tuberomammillary nucleus
vlPAG	Ventral periaqueductal gray matter
VLPO	Ventrolateral preoptic area
VTA	Ventral tegmental area

Figure 1

(*Continued*)

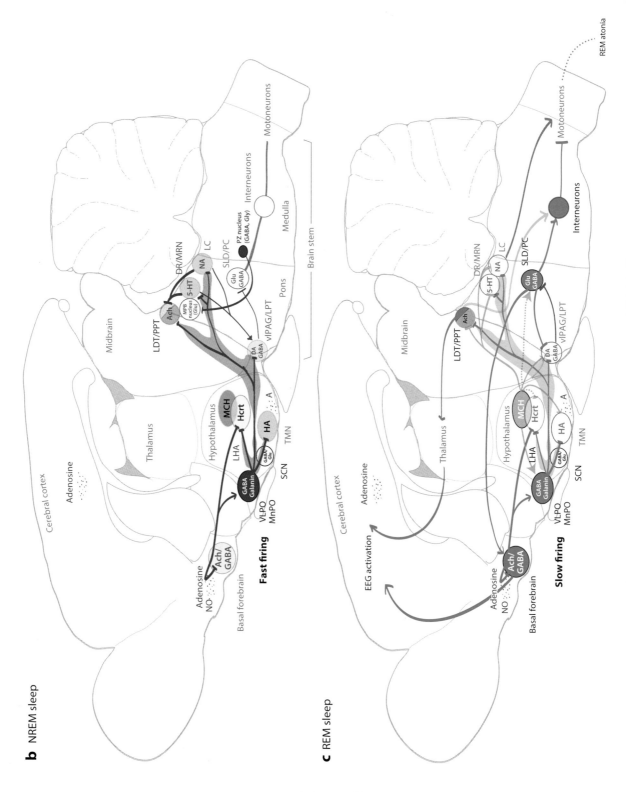

b NREM sleep

c REM sleep

Circadian regulation

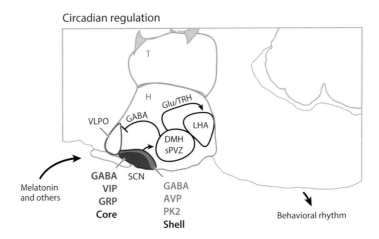

Figure 2

Peptide regulation of circadian rhythm. Circadian regulation: The SCN relays its rhythmic activity via the DMH and the sPVZ to the LHA and the VLPO to promote wake and inhibit sleep, respectively. Input to the SCN originates within the retina, thalamus, and arousal centers and also includes melatonin and other secreted molecules. Figure based on information from the following reviews: Chou et al. (2003), Colwell (2011), Welsh et al. (2010). Abbreviations: AVP, arginine-vasopressin; GRP, gastrin releasing peptide; PK2, prokineticin 2; TRH, thyrotropin releasing hormone; VIP, vasoactive intestinal polypeptide. For anatomic abbreviations, see **Figures 1** and **3**.

Neuropeptide Biogenesis and Signaling

In the mammalian brain, almost 70 different secreted neuropeptides (reviewed in Burbach 2011) complement neurotransmitters and modulate excitability of their targets. After their release, most neuropeptides bind to G protein–coupled receptors (GPCRs) to alter membrane excitability, transcription, and synaptogenesis (reviewed in Ludwig & Leng 2006), enabling neuropeptides to regulate a wide range of behaviors, including sleep-wake states. In contrast with neurotransmitters, neuropeptides are directly encoded by the genome, facilitating their study through genetic approaches.

Neuropeptide activity can be regulated at several levels. After transcription and translation, neuropeptides are processed by peptidases and convertases to generate mature neuropeptides of 3–40 amino acids in length, and these mature peptides often require additional modification for full activity (reviewed in Hook et al. 2008). Neuropeptides can be released from large dense core vesicles along the entire axon, so varying control of their release site(s) can provide another layer of regulation. Neuropeptide release is generally slower and requires stronger neuron activation as compared with the release of neurotransmitters (reviewed in Burbach 2011, Hokfelt 2000). Control of neuropeptide transport and clearance provides an additional layer of regulation. Neuropeptides have half-lives of up to 20 min, allowing them to signal to distant targets, whereas neurotransmitters are cleared from the synapse in milliseconds (reviewed in Lester et al. 1994, Ludwig & Leng 2006). Finally, because neurons often contain several neuropeptides and neurotransmitters, a neuropeptide may have different roles in different neuron populations, depending on its anatomical and biochemical context (reviewed in Bargmann 2012, Schöne & Burdakov 2012, van den Pol 2012).

NEUROPEPTIDERGIC REGULATION OF SLEEP-WAKE CENTERS

Here, we review the roles of more than 20 neuropeptides in sleep-wake regulation (**Table 1**). We focus on signals whose sleep-wake functions have not been discussed extensively and refer

Table 1 Neuropeptides and peptide hormones involved in sleep-wake regulation

Protein name (mouse gene symbol)	Suggested sleep-wake role	Wake	NREM sleep	REM sleep	Selected references	Other roles	Receptors (mouse gene symbol)
Adrenocorticotropic hormone (*Pomc*)	Promotes wake	+	NR	NR	Chastrette et al. 1990, Clow et al. 2010	Stress response, Inhibition of feeding, Motivation and reward, Pain	*Mc1r, Mc2r (main), Mc3r, Mc4r, Mc5r*
Brain-derived neurotrophic factor (*Bdnf*)	May promote NREM sleep, Potentially inhibits wake	(−)	(+)	NR	Tononi & Cirelli 2012	Inhibition of feeding, Synaptic plasticity	*Trkb, Ngfr*
Cholecystokinin (*Cck*)	Promotes wake and NREM sleep	+	+	NR	Obal & Krueger 2003	Inhibition of feeding, Pain, Anxiety	*Cckar, Cckbr*
Cocaine- and amphetamine-regulated transcript (*Cartpt*)	Promotes wake	+	−	−	Keating et al. 2010	Inhibition of feeding, Anxiety, Antidepressant, Conditioned place preference	NA
Corticotropin-releasing hormone (*Crh*)	Promotes wake, Modulates REM sleep	+	−	+/−	Kimura et al. 2010, Romanowski et al. 2010	Anxiety, Depression, Stress	*Crhr1, Crhr2*
Cortistatin (*Cort*)	Promotes NREM sleep, Inhibits REM sleep	NR	+	−	de Lecea 2008	Inhibition of growth hormone release, Memory and learning, Pain suppression	*Sstr1-5, Ghsru*
Dynorphin (*Pdyn*)	Potentially increases NREM sleep	NR	(+)	NR	Greco et al. 2008	Opioid peptide, Promotion of feeding	*Oprk1 (KOR)*
Endomorphin 1/EM1 (NA)	May promote wake	(+)	−	NR	Greco et al. 2008	Pain	*Oprm1*
Epidermal growth factor (*Egf*)	Suppresses locomotion in response to circadian cues	(−)/Timing	Timing	NR	Kushikata et al. 1998, Kramer et al. 2001	Growth stimulation	*Egfr*

(Continued)

Table 1 *(Continued)*

Protein name (mouse gene symbol)	Suggested sleep-wake role	Wake	NREM sleep	REM sleep	Selected references	Other roles	Receptors mouse gene symbol
Galanin (*Gal*)	May promote sleep	−	(+)	(+)	Pieribone et al. 1995, Murck et al. 2004, Woods et al. 2014	Pain perception Anxiety Nerve regeneration and neural stem cell proliferation	*Galr1, Galr2, Galr3*
Ghrelin (*Ghrl*)	Promotes wake	+	−	−	García-García et al. 2014	Promotion of feeding Counteracting leptin	*Ghsra*
Growth hormone (*Gh*)	May inhibit NREM sleep Promotes REM sleep	NR	(−)	+	Obal & Krueger 2003, Steiger 2007	Cell growth	*Ghr*
Growth-hormone-releasing hormone (*Ghrh*)	Promotes NREM sleep	NR	+	NR	Obal & Krueger 2003, Steiger 2007	Stimulation of growth hormone release	*Ghrhr*
Hypocretin (*Hcrt*)	Consolidates wake Inhibits REM sleep	+	−	−	Sakurai 2007, Carter et al. 2013	Promotion of feeding Thermoregulation Mood Reward Energy homeostasis Sensing of external and internal environment	*Hcrtr1, Hcrtr2*
Interleukin 1 beta (*Il-1β*)	May promote sleep	−	+	NR	Jewett & Krueger 2012	Inflammatory cytokine Synaptic plasticity	*Il1r1, Il1r2*
Leptin (*Lep*)	May promote wake Promotes NREM sleep May inhibit REM sleep	(−)	+	(−)	Sinton et al. 1999, Laposky et al. 2006, Leininger 2011	Inhibition of feeding Promotion of energy expenditure Reproduction Thermogenesis Synaptic plasticity Neuroprotection	*Lepr*
Melanin-concentrating hormone (*Pmch*)	Promotes sleep	−	+	+	Jego et al. 2013, Konadhode et al. 2013, Monti et al. 2013	Promotion of feeding Memory formation	*Mcbr1 (SLC-1)*

Neuropeptide	Effect				References	Function	Receptors
Melanocyte-stimulating hormones (*Pomc*)	May promote NREM sleep	NR	(+)	NR	Chastrette et al. 1990	Stress regulation Inhibition of feeding Motivation Pain Reward	*Mc1r, Mc3r, Mc4r, Mc5r*
Neuromedin S (*Nms*)	Promotes lethargus in *C. elegans*	(−)	(+)	NR	Nelson et al. 2013	Circadian rhythm Feeding	*Nmur1, Nmur2*
Neuropeptide B (*Npb*)	Promotes NREM sleep	−	+	NR	Hirashima et al. 2011	Feeding Pain sensation	*Npbwr1* (NPBWR2, not in rodents)
Neuropeptide S (*Nps*)	Promotes wake	+	(−)	(−)	Brown et al. 2012	Locomotion Anxiety Pain	*Npsr1*
Neuropeptide Y (*Npy*)	Downregulates CNS excitability Modulates wake and NREM sleep	−/+	+/−	NR	Dyzma et al. 2010	Promotion of feeding Anxiety Epilepsy Addiction Reproduction Immune regulation Neuroprotection	*Npy1r, Npy2r, Npy4r, Npy5r, Npy6r*
Neurotensin (*Nts*)	Promotes wake Inhibits NREM sleep May promote REM sleep	+	−	(+)	Cape et al. 2000, Fitzpatrick et al. 2012, Furutani et al. 2013	Anxiety Depression Increase in stress response	*Ntsr1, Ntsr2*
Nociceptin (*Pnoc*)	May inhibit wake Modulates REM sleep	(−)	NR	(−/+)	Devine et al. 1996, Xie et al. 2008; Rizzi et al. 2011	Pain sensing	*Oprl1*
Obestatin (*Gbrl*)	Promotes NREM sleep	NR	+	NR	García-García et al. 2014	Counteracting ghrelin	*Gpr39*/NA

(*Continued*)

Table 1 (*Continued*)

Protein name (mouse gene symbol)	Suggested sleep-wake role	Wake	NREM sleep	REM sleep	Selected references	Other roles	Receptors mouse gene symbol
Pituitary adenylyl cyclase-activating polypeptide (*Adcyap1*)	May promote wake Promotes REM sleep	(+)	NR	+	Ahnaou et al. 1999	Circadian rhythms Feeding Stress Memory Pain	*Adcyap1r1* (*PAC1*), *Vipr1* (*VPAC1*), *Vipr2* (*VPAC2*)
Prolactin (*Prl*)	Promotes REM sleep	NR	NR	+	García-García et al. 2009	Stress response Reduction in stress response during lactation	*Prlr*
Somatostatin (*Sst*)	Promotes wake and REM sleep Inhibits NREM sleep	+	−	+	Obal & Krueger 2003, Steiger 2007	Inhibition of growth hormone release	*Sstr1-5*
Substance P/ Tachykinin 1 (*Tac1*)	May promote sleep	−	(+)	(+)	Zhang et al. 2004	Pain Neurogenic inflammation Muscle contractility	*Tacr1-3*
Transforming growth factor alpha (*Tgfa*)	Suppresses locomotion in response to circadian cues	(−)/Timing	Timing	NR	Kramer et al. 2001	Expression in cancer Stimulates proliferation	*Egfr*
Tumor necrosis factor alpha (*Tnf*)	May promote sleep	−	+	(+)	Jewett & Krueger 2012	Inflammatory cytokine synaptic scaling Hippocampal neurogenesis	*Tnfrsf1a, Tnfrsf1b*
Vasoactive intestinal peptide (*Vip*)	Promotes REM sleep	−	NR	+	Riou et al. 1982, Hu et al. 2011	Inhibition of feeding Circadian rhythm	*Vipr1* (*VPAC1*), *Vipr2* (*VPAC2*)

Abbreviations: +, promotes; −, inhibits; NA, not applicable; NR, no role/not known; NREM, non–rapid eye movement; REM, rapid eye movement.

the reader interested in the detailed roles of hypocretin/orexin to the accompanying review by Gao & Horvath (2014) in this volume. Several themes will become apparent in this section. First, there does not appear to be one master regulatory neuropeptide devoted solely to regulating sleep and wakefulness. Even hypocretin, which has a discrete expression pattern and strong effects on sleep and wakefulness, has several additional roles, from feeding to reward, and regulates multiple brain regions. Thus, that neuropeptides have multiple roles makes it impossible for investigators to assign one definitive function. Second, at the molecular level, the same prepropeptide can generate multiple mature peptides with different potencies or functions, and several receptors and receptor isoforms can recognize a given neuropeptide. Third, at the cellular level, the spatial and temporal expressions of signals and receptors are diverse and complex, and several neuropeptides, classic neurotransmitters, and receptors can be coexpressed in the same neuron. Fourth, at the circuit level, neuronal connectivity is highly complex; many brain regions regulate each other at any given time. Fifth, at the physiological level, phenotypic effects can be masked by overlapping roles between related or even unrelated signals and by compensatory feedback mechanisms that maintain homeostasis. Perhaps most important, the effects of neuropeptidergic signaling are extremely context dependent on the behavioral state of the animal, the state of the environment, and the experience of the animal. All these effects undermine attempts at simple classification. Thus, the pictures we paint in the following sections are necessarily incomplete but should provide an overview of the field and help lay the foundation for future studies.

Hypocretin Promotes Wakefulness and Suppresses REM Sleep

We lack molecular understanding of most sleep disorders despite their prevalence (Colten & Altevogt 2006). One exception is narcolepsy, characterized by daytime sleepiness, the premature onset of REM sleep, fragmented nighttime sleep, and cataplexy, the emotionally triggered sudden loss of muscle tone while conscious (reviewed in Sakurai 2007, Sehgal & Mignot 2011).

In humans, narcolepsy results from loss of hypocretin neurons (reviewed in Sakurai 2007, Carter et al. 2013, Gao & Horvath 2014). In rodents, hypocretin promotes wakefulness, suppresses REM sleep, and is important to maintain stable sleep-wake states (reviewed in Gao & Horvath 2014). In zebrafish, a model which complements mammalian models by providing a combination of genetic tractability, high-throughput behavioral analyses, and conserved neuroanatomy, hypocretin increases arousal and inhibits rest (Prober et al. 2006; reviewed in Chiu & Prober 2013).

From their location in the hypothalamus, hypocretin neurons, which are most active during wake, project to the cortex and major arousal centers (**Supplemental Figure 1**) (reviewed in Sakurai 2007, Gao & Horvath 2014). By enhancing activity of arousal nuclei, hypocretin neurons stabilize wake. In addition, hypocretin indirectly represses NREM- and REM-promoting regions (reviewed in Sakurai 2007, Sakurai & Mieda 2011). Hypocretin loss could therefore disinhibit REM nuclei, such as the sublaterodorsal and precoeruleus (SLD/PC) and potentially the laterodorsal and pedunculopontine tegmental nuclei (LDT/PPT) neurons, and thereby sensitize them to triggers of REM sleep (Burgess et al. 2013, Oishi et al. 2013).

Hypocretin neurons integrate cues from several physiological systems and thereby respond to changes in energy, mood, and stress (reviewed in Gao & Horvath 2014). Potentially harmful conditions such as fasting or stress excite hypocretin neurons via state-specific signals such as ghrelin or arginine vasopressin (AVP). Conversely, increases in glucose can context-dependently inhibit hypocretin neurons (Venner et al. 2011). In addition, hypocretin neurons receive inputs from stress systems and indirect inputs via the dorsomedial hypothalamus (DMH) from the circadian system (reviewed in Sakurai 2007). Thus hypocretin neurons integrate metabolic, emotional, and circadian signals to match wake levels to environmental needs.

Galanin: A Potential Sleep Promoter?

Some data propose that galanin decreases arousal. Galanin mRNA is found within GABA-positive VLPO neurons (Gaus et al. 2002). The VLPO sends inhibitory projections to the tuberomammillary nucleus (TMN) and to other arousal systems in the brain stem, including the dorsal raphe (DR) and the locus coeruleus (LC) (Sherin et al. 1998). Some of these target areas, the LC for example, express the three galanin receptors: GALR1, GALR2, and GALR3 (O'Donnell et al. 1999, Mennicken et al. 2002). Furthermore, galanin reduces activity of LC neurons in slice preparations (Seutin et al. 1989, Pieribone et al. 1995). In addition, intravenous galanin administration increases REM sleep duration in young men (Murck et al. 2004). In zebrafish, overexpression of *galanin* decreases spontaneous locomotor activity and reduces responsiveness to sensory stimuli (Woods et al. 2014). Altogether, these data suggest that galanin has sedating functions.

Melanin-Concentrating Hormone Counteracts Hypocretin and Promotes REM Sleep

The location and projections of melanin-concentrating hormone (MCH) neurons are remarkably similar to those of hypocretin neurons (reviewed in Monti et al. 2013). MCH neurons target nuclei that promote arousal and REM sleep, including the LC/DR, LDT/PPT, and SLD (reviewed in Torterolo et al. 2011, Monti et al. 2013). However, hypocretin and MCH reside in separate neuron populations, and in contrast to hypocretin neurons, MCH neurons are sleep-active, especially during REM sleep (reviewed in Torterolo et al. 2011). Furthermore, hypocretin and MCH neurons regulate each other: Hypocretin excites MCH neurons, whereas MCH signaling has an inhibitory effect on hypocretin neurons (reviewed in Monti et al. 2013).

Functional evidence supports a role for MCH in sleep. For example, MCH knockout mice exhibit less NREM sleep, are more active, and are more sensitive to wake-promoting stimuli (Willie et al. 2008). MCH injections into the SLD increase REM sleep (reviewed in Torterolo et al. 2011), whereas injections into VLPO neurons increase NREM sleep (Benedetto et al. 2013). Recent optogenetic studies demonstrated that 24 h activation of MCH neurons increased both NREM and REM sleep (Konadhode et al. 2013), and acute activation and inhibition showed that MCH neurons maintain REM sleep by inhibiting arousal nuclei (e.g., TMN) (Jego et al. 2013). Thus MCH may promote sleep and may act as the sleep-active counterpart to the hypocretin and other arousal systems.

Paradoxically, loss of the MCH receptor MCHR1 increases REM sleep, especially after sleep deprivation (Adamantidis et al. 2008), suggesting that MCHR1 promotes wakefulness. This finding is surprising considering the similar effects of receptor and ligand mutations on energy balance and activity, but some explanations have been proposed. For example, differences in genetic background can have major effects on sleep (Tafti 2007). Alternatively, MCHR1 may mediate autoinhibition of MCH neurons directly or indirectly (Chee et al. 2013). Loss of such an inhibitory feedback loop would allow MCH neurons to release other inhibitory cotransmitters, such as GABA, which could promote sleep (Adamantidis et al. 2008; reviewed in Torterolo et al. 2011).

Dynorphin Modulates Hypocretin Neurons and May Affect Sleep

The endogenous opioid prodynorphin is another candidate regulator of sleep, proposed largely on the basis of its coexpression with hypocretin in lateral hypothalamic area (LHA) neurons (Chou et al. 2001). Infusion of dynorphin A(1–17) into the VLPO increases NREM sleep by activating the κ opioid receptor (KOR) (Greco et al. 2008), potentially reflecting its endogenous release from the parabrachial subnucleus onto VLPO neurons (Khachaturian et al. 1982, Greco et al. 2008). In contrast, systemic infusion of KOR agonists into rats reduces EEG power spectra, indicating CNS activation, yet inhibits movement, potentially via KOR-mediated blockade of dopamine release (Mulder et al. 1984, Coltro Campi & Clarke 1995). To our knowledge, no explicit sleep-wake phenotype has been reported for dynorphin or for KOR knockout mice; instead, these mice exhibit a complex anxiety phenotype that depends on genetic background (Kastenberger et al. 2012).

Thus the physiological significance of dynorphin coexpression in hypocretin neurons is unclear (Chou et al. 2001), although dynorphin and hypocretin can act synergistically to enhance TMN activity: Hypocretin directly activates TMN neurons, and dynorphin inhibits GABAergic inputs from the VLPO (Eriksson et al. 2004). Dynorphin also inhibits hypocretin neurons for short time periods (Li & van den Pol 2006, Williams & Behn 2011). Dynorphin regulation of sleep and arousal may therefore strongly depend on the neuronal context.

PACAP Regulates REM Sleep and Wake

Sleep-wake regulation by the *Adcyap1*-encoded pituitary adenylate cyclase-activating polypeptide (PACAP) is complex owing to its additional functions in anxiety, locomotion, and circadian entrainment (reviewed in Vaudry et al. 2009). Nevertheless, some studies suggest PACAP promotes REM sleep. For example, injection of PACAP into REM nuclei increases REM-sleep duration (Ahnaou et al. 1999). In addition, PACAP is expressed in the SLD, also known as the subcoeruleus area, a REM-active nucleus that blocks muscle tone during REM sleep (Ahnaou et al. 2006). Furthermore, knockout of either PACAP or PAC1-R increases locomotor activity (Hashimoto et al. 2001, Otto et al. 2001), supporting PACAP's sedating role. However, PACAP may also increase arousal. For example, intracerebroventricular (icv) and systemic injections of PACAP into rats enhance activities such as walking and grooming (Masuo et al. 1995). These discrepancies may arise from different genetic backgrounds or differences in the dose or target area of injections. For example, PACAP may promote REM activity of SLD neurons but could also influence other wake-promoting targets, e.g., the LC (Ahnaou et al. 2006).

Functions for PACAP in sleep-wake regulation are further complicated by its involvement in stress- and anxiety-related behaviors (reviewed in Hashimoto et al. 2011). PACAP may simultaneously regulate multiple behaviors via independent action upon discrete brain regions. In zebrafish, for example, genetic overexpression of a PACAP paralog (*adcyap1b*) increased sensory responsiveness without affecting overall levels of locomotor activity (Woods et al. 2014). Studies of PACAP may thus lead to confounding results if induced behaviors influence each other. Dissection of the multiple functions of PACAP will require inhibition or activation of PACAP signaling in specific brain regions.

TGF Alpha and EGF Mediate Circadian Influences on Locomotor Activity

Transforming growth factor alpha (TGF-α) and epidermal growth factor (EGF) signal through receptor protein kinases, in contrast with typical neuropeptides, which signal via GPCRs. Nevertheless, TGF-α and EGF can originate from and act on neurons and can regulate sleep-wake behavior (Kushikata et al. 1998, Kramer et al. 2001, Snodgrass-Belt et al. 2005, Gilbert & Davis 2009). For example, infusion of TGF-α or EGF into the brains of hamsters decreases locomotion, fragments sleep-wake behaviors, and shifts the timing of sleep-wake states. Although overall sleep and wake durations are unaffected, infused hamsters change their usual circadian rhythm to an ultradian rhythm, with 5–6 sleep-wake cycles in 24 h (Kramer et al. 2001, Snodgrass-Belt et al. 2005). Conversely, mice with reduced EGF receptor function are hyperactive and less sensitive to light-mediated inhibition of activity (Kramer et al. 2001), although this phenotype may vary considerably (Mrosovsky et al. 2005). The expression pattern of TGF-α supports a role in sleep regulation. For example, TGF-α expression in the SCN varies diurnally and decreases during the night, when hamsters are active (Kramer et al. 2001). Collectively, these observations suggest that TGF-α and EGF have conserved roles in suppressing locomotor activity in coordination with the circadian rhythm.

ENERGY HOMEOSTASIS AND SLEEP

Arousal and metabolic state are often inversely correlated: Hunger heightens arousal and satiety promotes rest (Antin et al. 1975, Borbély 1977). This notion suggests cross talk between regulatory regions for feeding and arousal behaviors, potentially via common neuronal pathways (reviewed in Sternson 2013). Indeed, centers regulating both behaviors connect anatomically, reside in the same brain regions (hypothalamus and brain stem), and employ overlapping neuron and transmitter systems (**Figure 3a**) (reviewed in Smith & Ferguson 2008). The arousal peptide hypocretin, for example, is also a primary regulator of appetite and feeding (reviewed in Gao & Horvath 2014).

Hypocretin neurons respond to feeding cues such as changing glucose concentrations or the satiety hormone leptin (Yamanaka et al. 2003, Venner et al. 2011; reviewed in Gao & Horvath 2014). Hypocretin also indirectly influences feeding by modulating the mesolimbic dopamine system (reviewed in Tsujino & Sakurai 2013) and directly promotes feeding by activating neuropeptide Y (NPY)/agouti-related protein (AgRP) neurons and inhibiting proopiomelanocortin (POMC)/cocaine and amphetamine-regulated transcript (CART) neurons (reviewed in Adamantidis & de Lecea 2009). Similarly, neuropeptides associated with feeding can modulate arousal systems.

Leptin Can Promote Sleep

Leptin is expressed in the periphery, primarily by fat cells (reviewed in Harvey 2007), whereas its receptor LepRb is expressed in several brain nuclei (Scott et al. 2009). Leptin's primary role is to promote satiety and stimulate energy expenditure (reviewed in Leinninger 2011), but several observations indicate that leptin also regulates sleep and wake. Expression of the leptin receptor

---

Figure 3

(a) Energy homeostasis. Here, we focus on one example feeding circuit, originating in the arcuate nucleus of the hypothalamus (ARC). The ARC is a well-studied integrator of signals relevant to energy homeostasis. Peripheral signals (e.g. leptin, ghrelin, CCK) and central signals from gastrointestinal, gustatory, and arousal regions are integrated within the ARC, which in turn influences locomotion, energy expenditure, and arousal. Feedback among hypothalamic circuits is shown in the square inserts (*frontal views flattened into one plane*). Leptin inhibits feeding through differential regulation of gene expression: leptin enhances POMC expression but lowers AgRP expression (reviewed in Varela & Horvath 2012). LepRb expression in other brain regions indicates the distributed regulation of feeding (reviewed in Sternson 2013). Hypocretin neurons promote feeding and directly influence the Arc. In addition, hypocretin promotes locomotion. MCH neurons promote feeding but inhibit locomotion. Other aspects of feeding are relevant for arousal, including reward and emotional circuits but are beyond the scope of this review. (b) Awakening and stress signaling via HPA axis activation: The hypothalamic stress circuit induces central and peripheral changes that contribute to wake. Centrally, stress-responsive PVH neurons activate the NTS to promote energy expenditure and locomotion. Peripherally, PVH neurons stimulate the pituitary gland via CRH to release ACTH, which induces corticosterone secretion from the adrenal glands. Corticosterone elevates blood pressure and mobilizes energy resources. The stress circuitry is also employed during awakening, which is under circadian control by the SCN/DMH. Awakening is supported by the LHA, which promotes arousal via the LC. Abbreviations: ACTH, adrenocorticotropic hormone; AgRP, agouti-related protein; CART, cocaine and amphetamine-regulated transcript; CCK, cholecystokinin; CRH, corticotropin-releasing hormone; HPA, hypothalamic-pituitary-adrenal; LepRb, leptin receptor; NPY, neuropeptide Y; POMC, proopiomelanocortin.

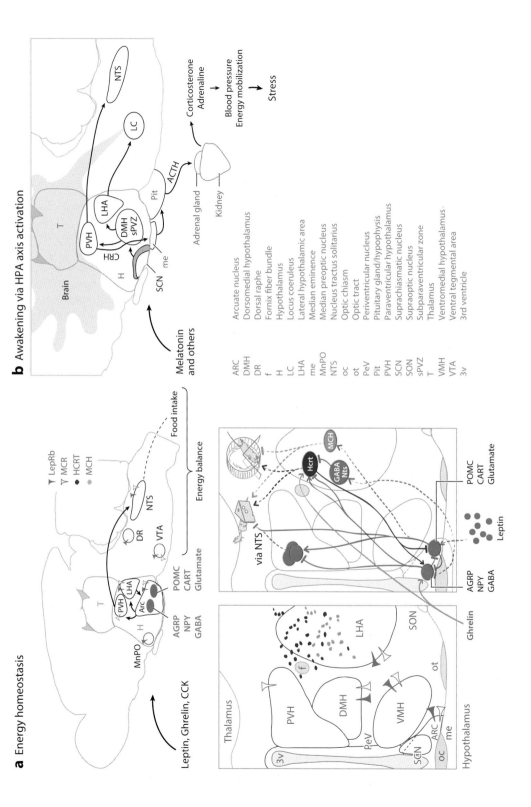

a Energy homeostasis

Food intake

Energy balance

Leptin, Ghrelin, CCK

POMC
CART
Glutamate

AGRP
NPY
GABA

Y LepRb
Y MCR
● HCRT
● MCH

b Awakening via HPA axis activation

Stress

Blood pressure
Energy mobilization

Corticosterone
Adrenaline

Adrenal gland

Kidney

ACTH

Melatonin
and others

ARC	Arcuate nucleus
DMH	Dorsomedial hypothalamus
DR	Dorsal raphe
f	Fornix fiber bundle
H	Hypothalamus
LC	Locus coeruleus
LHA	Lateral hypothalamic area
me	Median eminence
MnPO	Median preoptic nucleus
NTS	Nucleus tractus solitarius
oc	Optic chiasm
ot	Optic tract
PeV	Periventricular nucleus
Pit	Pituitary gland/hypophysis
PVH	Paraventricular hypothalamus
SCN	Suprachiasmatic nucleus
SON	Supraoptic nucleus
sPVZ	Subparaventricular zone
T	Thalamus
VMH	Ventromedial hypothalamus
VTA	Ventral tegmental area
3v	3rd ventricle

LepRb overlaps with several arousal and sleep nuclei (Elmquist et al. 1998, Scott et al. 2009, Patterson et al. 2011). In the LHA, a subset of leptin-responsive GABAergic neurons projects to hypocretin neurons (reviewed in Leinninger 2011), suggesting that leptin-responsive neurons inhibit hypocretin neurons and thereby lower wakefulness during satiety (Yamanaka et al. 2003, Venner et al. 2011). Mice lacking leptin signaling sleep more but exhibit fragmented sleep and wake (Laposky et al. 2006). Accordingly, exogenous leptin promotes sleep: Systemic leptin increases slow wave and REM sleep (Sinton et al. 1999). These sleep-promoting effects are abolished during fasting, when leptin levels decrease (Sinton et al. 1999). Thus, leptin may induce sleep only above a certain threshold, triggering sleep only when energy resources are high.

Ghrelin, Cholecystokinin, and Neuropeptide Y Have Diverse Roles in Sleep-Wake Regulation

Ghrelin targets brain systems that promote feeding (reviewed in Kojima & Kangawa 2010, García-García et al. 2014). It is mainly produced in the stomach and lowly expressed in the brain (reviewed in Furness et al. 2011). Several lines of evidence support a role for ghrelin in sleep-wake regulation. Injections of ghrelin promote wakefulness and suppress NREM and REM sleep in rats (Szentirmai 2012). Conversely, the ghrelin receptor GHS-R1a, which is expressed in arousal nuclei such as the DR, LDT, and SCN (Zigman et al. 2006), is necessary for the heightened arousal behaviors normally observed upon fasting or introduction to novel environments (Esposito et al. 2012). In addition, ghrelin axons can excite hypocretin neurons (reviewed in Kageyama et al. 2010), which further supports a role for ghrelin in enhancing arousal.

Paradoxically, ghrelin also promotes sleep. Mice lacking preproghrelin have slightly increased wake and REM sleep and fragmented NREM sleep, indicating preproghrelin may promote NREM sleep (Szentirmai et al. 2007). Furthermore, systemic injection of ghrelin increases NREM sleep in young men but not in women (reviewed in García-García et al. 2014). How can one reconcile sleep- and wake-promoting effects of preproghrelin? Preproghrelin also encodes obestatin, which promotes NREM sleep upon injection (Szentirmai & Krueger 2006), and loss of obestatin could increase wakefulness in preproghrelin knockout mice. Alternatively, long-term loss of preproghrelin may activate compensatory mechanisms.

The cholecystokinin (CCK) precursor produces several brain–gut peptides that regulate satiety and also sleep and wakefulness (reviewed in Obal & Krueger 2003). For example, systemic CCK-8S increases NREM sleep (reviewed in Obal & Krueger 2003), and inhibition of the CCKA receptor (CCKAR/CCK1R) blocks the prolonged sleep induced by refeeding after fasting (Shemyakin & Kapás 2001), thus suggesting that CCK promotes sleep.

Paradoxically, CCKAR knockout mice display reduced locomotor activity and no sleep phenotype (Sei et al. 1999), and systemic injections of CCK4 elicit anxiety-like behaviors in humans and rodents (Li et al. 2013), which potentially indicates higher arousal. Similarly, locomotor activity is strikingly increased upon overexpression of *cck* in zebrafish larvae (Woods et al. 2014). Furthermore, the CCK-8S peptide directly activates hypocretin neurons (Tsujino et al. 2005), which express the CCKAR (Dalal et al. 2013), supporting CCK's wake-promoting role.

To reconcile these contradictory observations, investigators have suggested that peripheral and central sources of CCK may affect behavior differently (reviewed in Obal & Krueger 2003). For example, sleep-promotion effects of CCK may result from peripheral activation of the vagal nerve by CCK, whereas central CCK may instead excite hypocretin neurons and counterbalance the postfeeding increases in sleep drive.

NPY is an important regulator of feeding and energy expenditure, but it also affects other behaviors, including sleep (Erickson et al. 1996; reviewed in Dyzma et al. 2010). For example, NPY can reduce motor activity in rats and can promote sleep-like states in certain genetic backgrounds and shorten sleep latency in young men (reviewed in Dyzma et al. 2010). Furthermore, NPY inhibits hypocretin neurons directly, potentially through the NPY receptors NPY1R and NPY2R (Fu et al. 2004), which are lowly expressed in hypocretin neurons (Dalal et al. 2013). Thus NPY may promote sleep.

In apparent contradiction, central injection of NPY promotes wake in mice if injected at light onset (reviewed in Dyzma et al. 2010), when mice begin to sleep and increase their VLPO activity. This wake-promoting effect may be explained by inhibition of VLPO neurons, which express NPY1R (Kishi et al. 2005), indicating context-dependent activity of NPY neurons and their targets. Given at sleep onset, for example, high NPY may promote wake by inhibiting VLPO neurons, whereas during wake, high NPY may be sedating by inhibiting hypocretin neurons.

CART May Promote Wakefulness

Cocaine and amphetamine-regulated transcript (CART) may enhance arousal. For example, intracerebroventricular CART injection promotes wakefulness in rats (Keating et al. 2010), and larval zebrafish overexpressing *cart* respond more strongly to sensory stimuli (Woods et al. 2014). In addition, CART knockout mice react less to arousing stimuli, such as cocaine or a novel environment (reviewed in Moffett et al. 2006). On the basis of its expression, CART may affect regulation of arousal at multiple levels. For example, in the olfactory bulb and retina, CART may modulate sensory inputs; in mood- and reward-associated nuclei, CART could function in motivation; and in arousal-associated nuclei [e.g., LHA, LC, DR, or periaqueductal gray matter (PAG)], CART could directly regulate arousal (Koylu et al. 1997, 1998). The latter possibility is favorable because CART is coexpressed with several sleep-wake-related neuropeptides in the hypothalamus, including MCH, dynorphin, and neurotensin (Broberger 1999, Elias et al. 2001, Hanriot et al. 2007). CART also increases anxiety-like behaviors, which may also influence arousal behaviors such as locomotion (Chaki et al. 2003). In the future, it will be important to identify the CART receptor, which is still unknown.

STRESS AND SLEEP

Molecules and neuroanatomical structures relevant to stress also affect sleep-wake regulation. Stress signals such as pain or fear activate the hypothalamic-pituitary-adrenal (HPA) axis, which releases hormones that elevate energy metabolism and blood pressure, thereby potentially stabilizing wake states (reviewed in Kalsbeek et al. 2010). Stress also influences REM sleep. In particular, acute anxiogenic stress increases wake in rats and lowers REM sleep, whereas repeated stress reduces wake and increases REM sleep (O'Malley et al. 2013).

Corticotropin-Releasing Hormone Regulates REM Sleep

Corticotropin-releasing hormone (CRH) initiates the stress response (reviewed in Kalsbeek et al. 2010). From the hypothalamus, CRH neurons project to their targets (**Figure 3b**), including the LDT and PPT, two REM-sleep nuclei that express the receptor CRHR1 (Sauvage & Steckler 2001). Functional relevance of these connections is supported by the observation that the neonatal peak of REM sleep requires CRHR1 (Feng et al. 2007), and long-term overexpression of CRH in mice promotes REM sleep (Kimura et al. 2010). In contrast, acute injections of CRH promote wake at the expense of REM and NREM sleep (Romanowski et al. 2010). Acute effects of CRH on wake and NREM sleep require CRHR1, which is expressed in hypocretin and LC neurons (Sauvage & Steckler 2001, Winsky-Sommerer et al. 2004). However, CRH-mediated suppression of REM sleep and HPA-axis activation persist in CNS-specific CRHR1 knockout mice, which suggests that CRH may have additional roles. We can therefore distinguish acute and chronic CRH effects: Short-term CRH induces wake and suppresses NREM sleep, potentially via hypocretin and LC activation, whereas long-term CRH increases REM sleep, potentially through activation of LDT/PPT.

Prolactin May Promote REM Sleep

The stress-induced peptide prolactin potentially regulates REM sleep (reviewed in García-García et al. 2009). For example, circadian changes in serum prolactin peak during REM sleep (reviewed in Gan & Quinton 2010); prolactin injections into rodent and cat brains increase REM sleep, whereas mice without prolactin have reduced REM sleep (reviewed in García-García et al. 2009).

Sleep-wake nuclei such as the DR and LC are targets of prolactin neurons in the periventricular nucleus and in the LHA (Siaud et al. 1989; Paut-Pagano et al. 1993). Prolactin is therefore an interesting candidate for REM-sleep regulation, particularly in response to stress.

GROWTH, REPAIR, DEFENSE PEPTIDES, AND SLEEP

The somatotropic axis regulates primarily animal size and metabolism. However, members of this pathway, including growth hormone (GH), growth-hormone-releasing hormone (GHRH), and somatostatin (SST), also regulate sleep-wake behaviors.

GH and GHRH Promote Sleep

Both GH and GHRH can promote sleep centrally. For example, systemic and icv administrations of GHRH enhance NREM sleep (reviewed in Steiger 2007), potentially via direct activation of VLPO and median preoptic nucleus (MnPO) neurons (Peterfi et al. 2010). Similarly, systemic injection of GH increases REM sleep (reviewed in Obal & Krueger 2003), perhaps by affecting DR and LC neurons, which express the GH receptor (reviewed in Hallberg & Nyberg 2012). Loss-of-function studies also support sleep-promoting roles for these peptides. Inhibition of endogenous GHRH prevents spontaneous sleep, and disruption of GHRH signaling in dwarf rats decreases REM and NREM sleep (reviewed in Obal & Krueger 2003). In addition, inhibition of GHRH signaling reduces SWA during NREM sleep, possibly via desynchronization of cortical neurons (Liao et al. 2010; reviewed in Obal et al. 2003). Similarly, loss of GH function decreases REM sleep in rats (Peterfi et al. 2006; reviewed in Obal & Krueger 2003). Expression of GHRH also supports a sleep-promoting role for this peptide: GHRH transcription increases upon sleep deprivation and follows circadian patterns, peaking during NREM sleep (reviewed in Obal & Krueger 2003).

Somatostatin and Cortistatin Have Opposite Actions on Sleep

Both SST and its analog octreotide inhibit sleep, potentially by inhibiting GHRH neurons via the somatostatin receptor 2 (SST2) (reviewed in Obal & Krueger 2003). However, SST also promotes REM sleep in rats (reviewed in Obal & Krueger 2003), suggesting potential GHRH-independent functions for this peptide. Cortistatin (CST) is structurally similar to SST and inhibits GH release and depresses neurons (reviewed in de Lecea 2008, Martel et al. 2012). Icv infusion of CST, however, lowers locomotion, and increases NREM sleep, potentially by enhancing synchrony of cortical neurons (reviewed in de Lecea 2008). Expression of CST is consistent with this function:

mRNA levels peak before sleep onset and increase during sleep deprivation (reviewed in de Lecea 2008).

Brain-Derived Neurotrophic Factor Modulates NREM Sleep

A current hypothesis suggests sleep may renormalize synaptic connections, and increased synaptic strength may correlate with subsequent stronger SWA (reviewed in Tononi & Cirelli 2012). As brain-derived neurotrophic factor (BDNF) mediates activity-dependent synaptic plasticity, among its many functions, and BDNF levels peak after extended wakefulness, BDNF may link plasticity and SWA of NREM sleep (reviewed in Tononi & Cirelli 2012, Bachmann et al. 2012). Cortical injection of BDNF transiently increases subsequent SWA, indicating deeper NREM sleep (reviewed in Tononi & Cirelli 2012). These EEG changes were unilaterally induced, suggesting that BDNF induces local sleep, similar to GHRH and Cst. Conversely, reduced neuron activity-dependent BDNF secretion in humans, caused by heterozygosity for a single-nucleotide polymorphism (SNP) in the BDNF gene, decreased SWA during sleep (Bachmann et al. 2012); likewise, mice without BDNF function develop stress-induced hyperactivity, indicating that BDNF may be sedating (Rios et al. 2001). Taken together, these data suggest a role for BDNF in promoting NREM-sleep depth.

TNF and IL-1β Promote Sleep

Increased sleepiness during illnesses, such as influenza and bacterial infection, suggests that the immune system influences arousal state (reviewed in Zielinski & Krueger 2011). Accordingly, cytokines such as tumor necrosis factor (TNF) and interleukin 1 beta (IL-1β) can enter the brain directly during an immune response and, among other things, induce sleep-like states (reviewed in Majde & Krueger 2005). Expression of these cytokines suggests they may also regulate sleep under normal conditions. For example, their expression varies diurnally; levels of IL-1β increase centrally during sleep deprivation; and glial cells release TNF and IL-1β in response to ATP, which accumulates extracellularly during wake (reviewed in Jewett & Krueger 2012). Both IL-1β and TNF promote NREM sleep, potentially by inhibiting wake-active DR or LC neurons, activating preoptic sleep-active neurons respectively, and stimulating release of sleep-inducing substances such as prostaglandin D2 (PGD2), adenosine, and GHRH (reviewed in Obal & Krueger 2003, Huang et al. 2007, Krueger et al. 2008, Urade & Hayaishi 2011, Jewett & Krueger 2012). Loss-of-function studies support this sleep-promoting function. For example, disruption of TNF signaling reduces REM and NREM sleep and lowers SWA locally (Kapás et al. 2008; reviewed in Jewett & Krueger 2012). In addition, TNF can change neuronal excitability by altering the distribution of AMPA (α-amino-3-hydroxy-5-methyl-4-isoxazolepropionic acid) and adenosine receptors at the cell surface (reviewed in McCoy & Tansey 2008, Jewett & Krueger 2012). Moreover, unilateral injection of TNF enhances IL-1β and SWA locally in the injected hemisphere (reviewed in Krueger 2008). Thus these cytokines induce sleep locally by altering cortical network states and promoting the release of sleep-inducing substances (reviewed in Jewett & Krueger 2012) and globally by influencing sleep-wake regulatory brain centers.

CONCLUSIONS AND PROSPECTS

The studies discussed here reveal how numerous neuropeptides regulate sleep-wake behaviors, and these data establish tantalizing connections with other behavioral states such as hunger, stress, anxiety, and infection. Despite this progress, it is still unclear how neuropeptides and their receptors generate stable behavioral states and transitions. Some of these challenges arise from differences in experimental settings. For example, the site, dose, and timing of neuropeptide administration can cause varying effects, and diverse assays such as EEG, locomotion, and feeding are difficult to compare. More standardized manipulations and assays may help to clarify some of the current controversies. Moreover, it is unclear whether all the major neuropeptidergic regulators of sleep have been discovered. It therefore remains unclear whether some of the currently described effects are mediated by additional signals.

Several technical advances promise to overcome these limitations: Genetic and optical methods are so sophisticated that one can now genetically manipulate signals and receptors in specific neurons in combination with optical methods to change the activity of neuron subsets. Standardized behavioral assays and careful measurements of neural activity will provide a better understanding of neuropeptide networks and will facilitate the generation of computational models for sleep-wake regulation.

If the insights gained from simpler neuromodularity systems are any indication, our analyses of sleep-wake states will continue to be extraordinarily diverse and complex. For example, classic studies of the crustacean stomatogastric ganglion have revealed how a circuit consisting of only ~30 neurons can be regulated by more than one dozen neuromodulators to generate many different outputs (reviewed in Marder 2012). The impressive progress in this system has been due to a combination of connectivity mapping, activity mapping, molecular studies, and computational modeling and provides a blueprint for the analysis of sleep-wake states; however, the complexity of this seemingly simple circuit is humbling when considering the much-more-complicated mammalian brains and behaviors.

The recent emergence of nonmammalian model systems in sleep research also promises novel insights. Systems such as *Drosophila*, *Caenorhabditis elegans*, and zebrafish have sleep-like states (Shaw et al. 2000, Cho & Sternberg 2014; reviewed in Crocker & Sehgal 2010, Chiu & Prober 2013, Nelson & Raizen 2013, Rihel & Schier 2013). For example, both *Drosophila* and zebrafish display the fundamental behavioral properties associated with sleep, including reduced locomotion and decreased arousal, which are regulated by circadian and homeostatic mechanisms. Moreover, neurotransmitters that regulate mammalian sleep also affect *Drosophila* and zebrafish sleep-like states. For example, a large-scale drug screen in zebrafish uncovered that many neuromodulators induce sleep-wake phenotypes in zebrafish, similar to phenotypes observed in mammals, ranging from the noradrenaline, serotonin, dopamine, GABA, glutamate, histamine, adenosine, and melatonin systems to numerous modulators of NF-κB, which is a likely integrator of cytokine, prostaglandin, and adenosine signals (Rihel et al. 2010).

Neuropeptidergic regulation is also conserved. For example, overexpression of hypocretin in larval zebrafish leads to increased wakefulness at the expense of rest, and adult zebrafish with mutations in the hypocretin receptor exhibit sleep fragmentation (reviewed in Chiu & Prober 2013). In vivo measurements of hypocretin neural activity reveal that these hypothalamic neurons are maximally active during episodes of spontaneous locomotor activity and are inactive during rest (reviewed in Chiu & Prober 2013), consistent with results obtained in mammals (reviewed in Gao & Horvath 2014). These studies reveal that key aspects of the neuropeptidergic regulation of sleep-wake states are preserved in nonmammalian model systems.

Many nonmammalian model systems also have the advantage that large numbers of animals can be analyzed simultaneously in well-defined and reproducible conditions. For example, in one screen of nearly 6,000 small molecules, several hundred structures were identified that alter the locomotor behavior of larval zebrafish (Rihel et al. 2010). The large data sets generated by such screens can be used to organize compounds by their multidimensional phenotypic output using clustering algorithms (Rihel et al. 2010). Importantly, the parallel comparison of different molecules in standardized settings generates behavioral profiles that organize compounds into clusters of similar phenotypes and can identify novel relationships between behavioral modulators. The same logic has been recently applied to the neuropeptidergic regulation of arousal in zebrafish by quantifying spontaneous locomotor behaviors and responsiveness to sensory stimuli (Woods et al. 2014). Phenotypic clustering revealed both shared and divergent features of neuropeptidergic functions and revealed surprising relationships between hypocretin and calcitonin gene-related peptide, between galanin and nociceptin, and between CART and PACAP. These assays can now be used in genetic screens to identify novel regulators of sleep and wakefulness.

Model systems with translucent bodies and fewer neurons (302 in *C. elegans*; ~100,000 in zebrafish larva) also facilitate the analysis and manipulation of circuit activity. For example, a recent study dissected the generation of roaming and dwelling states in *C. elegans* using genetic and optical tools and demonstrated that serotonin signaling and its associated circuitry lead to dwelling states, whereas neuropeptidergic pigment-dispersing factor (PDF) signaling generates roaming states (Flavell et al. 2013). These studies begin to outline how the entry into long-lasting but reversible behavioral states can be triggered by activation of distinct signaling pathways and neurons (reviewed in Schier 2013).

Traditional model systems will help to identify molecular and neuronal regulators of sleep and wakefulness, but they cannot explain the remarkable diversity of sleep-related behaviors in different animals (reviewed in Siegel 2011, Tobler 2011). Although sleep/wake states across species can be generalized behaviorally by activity levels, rapid transitions, and changes in arousal threshold, the particulars of sleep behaviors in different animals can vary widely (reviewed in Siegel 2011). For example, sleep-wake periods are strikingly diverse across different species, ranging from 3–4 h per day in elephants to 18–20 h per day in opossums (reviewed in Siegel 2005). Circadian regulation of sleep can also vary. For example, most lab rat strains are diurnal, whereas the unstriped Nile grass rat is nocturnal (Martinez et al. 2002). These behavioral differences likely reflect adaptations to diverse lifestyles and habitats (reviewed in Siegel 2009), but it is unclear how these variations are generated. Neuropeptides may contribute to the diversity of sleep and wake behaviors, similar to the generation of morphological diversity by changes in the expression of conserved developmental regulators (reviewed in Carroll 2008).

In addition, sleep behaviors can also differ dramatically within single organisms. For example, Finnish bats are nocturnal from June to September but shift to diurnal cycles in spring (reviewed in Saper et al. 2005), and sleep length decreases over the human lifetime. Analogously, disease states can induce specific changes in sleep regulation. For example, in Alzheimer's patients, who often suffer from fragmented sleep, increased sleep-wake fragmentation correlates with lower levels of hypocretin (Friedman et al. 2007; reviewed in Zeitzer 2013). Conversely, sleep-wake changes influence disease states. For instance, sleep disorders often accompany psychological conditions. Insomnia, for example, increases by fivefold the risk of developing a depression (Colten & Altevogt 2006). Furthermore, sleep deprivation correlates with inflammation, increased levels of Alzheimer's-associated amyloid beta, and impaired immune system function (Zager et al. 2007, Kang et al. 2009; reviewed in Krueger 2008). Additionally, some sleep-associated neuropeptides are neuroprotective, including galanin, dynorphin, and leptin (Hobson et al. 2008, Wang et al. 2012; reviewed in Signore et al. 2008). Thus, the study of neuropeptides is important not only to solve the fundamental mystery of sleep-wake regulation, but also to understand the contributions of sleep and wakefulness to health and disease.

DISCLOSURE STATEMENT

The authors are not aware of any affiliations, memberships, funding, or financial holdings that might be perceived as affecting the objectivity of this review.

ACKNOWLEDGMENTS

We thank J. Rihel, A. Adamantidis, D. Burdakov, C. Chavkin, C. Cirelli, C. Colwell, F. Davis, J. Deussing, J. Fahrenkrug, K. Fitzpatrick, M. Hirasawa, T. Hökfelt, G. Keating, M. Kimura, J. Krueger, D. Lambert, H. Landolt, G. Leinninger, M. López, J. Lu, E. Maratos-Flier, D. McGinty, H. Münzberg, M. Myers, K. Ressler, T. Sakurai, T. Scammell, P. Shiromani, I. Soltesz, E. Szentirmai, R. Szymusiak, G. Tononi, F. Turek, A. van den Pol, C. Viollet, C. Weitz, and M. Zielinski for invaluable comments on the manuscript. We apologize that space limits precluded the discussion of the roles of additional neuropeptides and the citation of all primary literature. Our research has been supported by the American Cancer Society (I.G.W.; PF-07-262-01-DDC), EMBO (ALTF 607-2012) (C.R.), the Life Sciences Research Foundation (Simons Foundation Fellow) (C.R.), the Harvard Stem Cell Institute (A.F.S.), the McKnight Endowment Fund for Neuroscience (A.F.S.), and the NIH (A.F.S.; R01HL109525, R21NS071598).

LITERATURE CITED

Adamantidis A, de Lecea L. 2009. The hypocretins as sensors for metabolism and arousal. *J. Physiol.* 587:33–40

Adamantidis A, Salvert D, Goutagny R, Lakaye B, Gervasoni D, et al. 2008. Sleep architecture of the melanin-concentrating hormone receptor 1-knockout mice. *Eur. J. Neurosci.* 27:1793–800

Ahnaou A, Basille M, Gonzalez B, Vaudry H, Hamon M, et al. 1999. Long-term enhancement of REM sleep by the pituitary adenylyl cyclase-activating polypeptide (PACAP) in the pontine reticular formation of the rat. *Eur. J. Neurosci.* 11:4051–58

Ahnaou A, Yon L, Arluison M, Vaudry H, Hannibal J, et al. 2006. Immunocytochemical distribution of VIP and PACAP in the rat brain stem: implications for REM sleep physiology. *Ann. N. Y. Acad. Sci.* 1070:135–42

Allada R, Siegel JM. 2008. Unearthing the phylogenetic roots of sleep. *Curr. Biol.* 18:R670–79

Anaclet C, Lin JS, Vetrivelan R, Krenzer M, Vong L, et al. 2012. Identification and characterization of a sleep-active cell group in the rostral medullary brainstem. *J. Neurosci.* 32:17970–76

Antin J, Gibbs J, Holt J, Young RC, Smith GP. 1975. Cholecystokinin elicits the complete behavioral sequence of satiety in rats. *J. Comp. Physiol. Psychol.* 89:784–90

Bachmann V, Klein C, Bodenmann S, Schäfer N, Berger W, et al. 2012. The BDNF Val66Met polymorphism modulates sleep intensity: EEG frequency- and state-specificity. *Sleep* 35:335–44

Bargmann CI. 2012. Beyond the connectome: how neuromodulators shape neural circuits. *Bioessays* 34:458–65

Benedetto L, Rodriguez-Servetti Z, Lagos P, D'Almeida V, Monti JM, Torterolo P. 2013. Microinjection of melanin concentrating hormone into the lateral preoptic area promotes non-REM sleep in the rat. *Peptides* 39:11–15

Borbély AA. 1977. Sleep in the rat during food deprivation and subsequent restitution of food. *Brain Res.* 124:457–71

Broberger C. 1999. Hypothalamic cocaine- and amphetamine-regulated transcript (CART) neurons: histochemical relationship to thyrotropin-releasing hormone, melanin-concentrating hormone, orexin/hypocretin and neuropeptide Y. *Brain Res.* 848:101–13

Brown RE, Basheer R, McKenna JT, Strecker RE, McCarley RW. 2012. Control of sleep and wakefulness. *Physiol. Rev.* 92:1087–187

Burbach JPH. 2011. What are neuropeptides? In *Neuropeptides: Methods and Protocols*, ed. A Merighi, pp. 1–36. New York: Springer Sci.-Bus. Media

Burgess CR, Oishi Y, Mochizuki T, Peever JH, Scammell TE. 2013. Amygdala lesions reduce cataplexy in orexin knock-out mice. *J. Neurosci.* 33:9734–42

Campbell SS, Tobler I. 1984. Animal sleep: a review of sleep duration across phylogeny. *Neurosci. Biobehav. Rev.* 8:269–300

Cape EG, Manns ID, Alonso A, Beaudet A, Jones BE. 2000. Neurotensin-induced bursting of cholinergic basal forebrain neurons promotes gamma and theta cortical activity together with waking and paradoxical sleep. *J. Neurosci.* 20:8452–61

Carroll SB. 2008. Evo-devo and an expanding evolutionary synthesis: a genetic theory of morphological evolution. *Cell* 134:25–36

Carter ME, de Lecea L, Adamantidis A. 2013. Functional wiring of hypocretin and LC-NE neurons: implications for arousal. *Front. Behav. Neurosci.* 7:43

Chaki S, Kawashima N, Suzuki Y, Shimazaki T, Okuyama S. 2003. Cocaine- and amphetamine-regulated transcript peptide produces anxiety-like behavior in rodents. *Eur. J. Pharmacol.* 464:49–54

Chastrette N, Cespuglio R, Jouvet M. 1990. Proopiomelanocortin (POMC)-derived peptides and sleep in the rat. Part 1—Hypnogenic properties of ACTH derivatives. *Neuropeptides* 15:61–74

Chee MJ, Pissios P, Maratos-Flier E. 2013. Neurochemical characterization of neurons expressing melanin-concentrating hormone receptor 1 in the mouse hypothalamus. *J. Comp. Neurol.* 521:2208–34

Chiu CN, Prober DA. 2013. Regulation of zebrafish sleep and arousal states: current and prospective approaches. *Front. Neural Circuits* 7:58

Cho JY, Sternberg PW. 2014. Multilevel modulation of a sensory motor circuit during *C. elegans* sleep and arousal. *Cell* 156:249–60

Chou TC, Lee CE, Lu J, Elmquist JK, Hara J, et al. 2001. Orexin (hypocretin) neurons contain dynorphin. *J. Neurosci.* 21:RC168

Chou TC, Scammell TE, Gooley JJ, Gaus SE, Saper CB, Lu J. 2003. Critical role of dorsomedial hypothalamic nucleus in a wide range of behavioral circadian rhythms. *J. Neurosci.* 23:10691–702

Clow A, Hucklebridge F, Stalder T, Evans P, Thorn L. 2010. The cortisol awakening response: more than a measure of HPA axis function. *Neurosci. Biobehav. Rev.* 35:97–103

Colten HR, Altevogt BM. 2006. *Sleep Disorders and Sleep Deprivation: An Unmet Public Health Problem.* Washington, DC: Natl. Acad. Press

Coltro Campi C, Clarke GD. 1995. Effects of highly selective κ-opioid agonists on EEG power spectra and behavioural correlates in conscious rats. *Pharmacol. Biochem. Behav.* 51:611–16

Colwell CS. 2011. Linking neural activity and molecular oscillations in the SCN. *Nat. Rev. Neurosci.* 12:553–69

Crocker A, Sehgal A. 2010. Genetic analysis of sleep. *Genes Dev.* 24:1220–35

Dalal J, Roh JH, Maloney SE, Akuffo A, Shah S, et al. 2013. Translational profiling of hypocretin neurons identifies candidate molecules for sleep regulation. *Genes Dev.* 27:565–78

de Lecea L. 2008. Cortistatin—functions in the central nervous system. *Mol. Cell Endocrinol.* 286:88–95

Devine DP, Taylor L, Reinscheid RK, Monsma FJJ, Civelli O, Akil H. 1996. Rats rapidly develop tolerance to the locomotor-inhibiting effects of the novel neuropeptide orphanin FQ. *Neurochem. Res.* 21:1387–96

Dyzma M, Boudjeltia KZ, Faraut B, Kerkhofs M. 2010. Neuropeptide Y and sleep. *Sleep Med. Rev.* 14:161–65

Elias CF, Lee CE, Kelly JF, Ahima RS, Kuhar M, et al. 2001. Characterization of CART neurons in the rat and human hypothalamus. *J. Comp. Neurol.* 432:1–19

Elmquist JK, Bjørbaek C, Ahima RS, Flier JS, Saper CB. 1998. Distributions of leptin receptor mRNA isoforms in the rat brain. *J. Comp. Neurol.* 395:535–47

Erickson JC, Clegg KE, Palmiter RD. 1996. Sensitivity to leptin and susceptibility to seizures of mice lacking neuropeptide Y. *Nature* 381:415–21

Eriksson KS, Sergeeva OA, Selbach O, Haas HL. 2004. Orexin (hypocretin)/dynorphin neurons control GABAergic inputs to tuberomammillary neurons. *Eur. J. Neurosci.* 19:1278–84

España RA, Scammell TE. 2011. Sleep neurobiology from a clinical perspective. *Sleep* 34:845–58

Esposito M, Pellinen J, Kapás L, Szentirmai É. 2012. Impaired wake-promoting mechanisms in ghrelin receptor-deficient mice. *Eur. J. Neurosci.* 35:233–43

Feng P, Liu X, Vurbic D, Fan H, Wang S. 2007. Neonatal REM sleep is regulated by corticotropin releasing factor. *Behav. Brain Res.* 182:95–102

Fitzpatrick K, Winrow CJ, Gotter AL, Millstein J, Arbuzova J, et al. 2012. Altered sleep and affect in the neurotensin receptor 1 knockout mouse. *Sleep* 35:949–56

Flavell SW, Pokala N, Macosko EZ, Albrecht DR, Larsch J, Bargmann CI. 2013. Serotonin and the neuropeptide PDF initiate and extend opposing behavioral states in *C. elegans*. *Cell* 154:1023–35

Friedman LF, Zeitzer JM, Lin L, Hoff D, Mignot E, et al. 2007. In Alzheimer disease, increased wake fragmentation found in those with lower hypocretin-1. *Neurology* 68:793–94

Fu LY, Acuna-Goycolea C, van den Pol AN. 2004. Neuropeptide Y inhibits hypocretin/orexin neurons by multiple presynaptic and postsynaptic mechanisms: tonic depression of the hypothalamic arousal system. *J. Neurosci.* 24:8741–51

Fuller PM, Sherman D, Pedersen NP, Saper CB, Lu J. 2011. Reassessment of the structural basis of the ascending arousal system. *J. Comp. Neurol.* 519:933–56

Furness JB, Hunne B, Matsuda N, Yin L, Russo D, et al. 2011. Investigation of the presence of ghrelin in the central nervous system of the rat and mouse. *Neuroscience* 193:1–9

Furutani N, Hondo M, Kageyama H, Tsujino N, Mieda M, et al. 2013. Neurotensin co-expressed in orexin-producing neurons in the lateral hypothalamus plays an important role in regulation of sleep/wakefulness states. *PLoS ONE* 8:e62391

Gan EH, Quinton R. 2010. Physiological significance of the rhythmic secretion of hypothalamic and pituitary hormones. *Prog. Brain Res.* 181:111–26

Gao X-B, Horvath T. 2014. Function and dysfunction of hypocretin/orexin: an energetics point of view. *Annu. Rev. Neurosci.* 37:101–16

García-García F, Acosta-Peña E, Venebra-Muñoz A, Murillo-Rodríguez E. 2009. Sleep-inducing factors. *CNS Neurol. Disord. Drug Targets* 8:235–44

García-García F, Juárez-Aguilar E, Santiago-García J, Cardinali DP. 2014. Ghrelin and its interactions with growth hormone, leptin and orexins: implications for the sleep-wake cycle and metabolism. *Sleep Med. Rev.* 18:89–97

Gaus SE, Strecker RE, Tate BA, Parker RA, Saper CB. 2002. Ventrolateral preoptic nucleus contains sleep-active, galaninergic neurons in multiple mammalian species. *Neuroscience* 115:285–94

Gilbert J, Davis FC. 2009. Behavioral effects of systemic transforming growth factor-alpha in Syrian hamsters. *Behav. Brain Res.* 198:440–48

Greco MA, Fuller PM, Jhou TC, Martin-Schild S, Zadina JE, et al. 2008. Opioidergic projections to sleep-active neurons in the ventrolateral preoptic nucleus. *Brain Res.* 1245:96–107

Hallberg M, Nyberg F. 2012. Growth hormone receptors in the brain and their potential as therapeutic targets in central nervous system disorders. *Open Endocrinol. J.* 6:27–33

Hanriot L, Camargo N, Courau AC, Leger L, Luppi PH, Peyron C. 2007. Characterization of the melanin-concentrating hormone neurons activated during paradoxical sleep hypersomnia in rats. *J. Comp. Neurol.* 505:147–57

Harris KD, Thiele A. 2011. Cortical state and attention. *Nat. Rev. Neurosci.* 12:509–23

Harvey J. 2007. Leptin: a diverse regulator of neuronal function. *J. Neurochem.* 100:307–13

Hashimoto H, Shintani N, Tanaka K, Mori W, Hirose M, et al. 2001. Altered psychomotor behaviors in mice lacking pituitary adenylate cyclase-activating polypeptide (PACAP). *Proc. Natl. Acad. Sci. USA* 98:13355–60

Hashimoto H, Shintani N, Tanida M, Hayata A, Hashimoto R, Baba A. 2011. PACAP is implicated in the stress axes. *Curr. Pharm. Des.* 17:985–89

Hirashima N, Tsunematsu T, Ichiki K, Tanaka H, Kilduff TS, Yamanaka A. 2011. Neuropeptide B induces slow wave sleep in mice. *Sleep* 34:1–37

Hobson SA, Bacon A, Elliot-Hunt CR, Holmes FE, Kerr NC, et al. 2008. Galanin acts as a trophic factor to the central and peripheral nervous systems. *Cell. Mol. Life Sci.* 65:1806–12

Hokfelt T, Broberger C, Xu ZQ, Sergeyev V, Ubink R, Diez M. 2000. Neuropeptides—an overview. *Neuropharmacology* 39:1337–56

Hook V, Funkelstein L, Lu D, Bark S, Wegrzyn J, Hwang SR. 2008. Proteases for processing proneuropeptides into peptide neurotransmitters and hormones. *Annu. Rev. Pharmacol. Toxicol.* 48:393–423

Hu WP, Li JD, Colwell CS, Zhou QY. 2011. Decreased REM sleep and altered circadian sleep regulation in mice lacking vasoactive intestinal polypeptide. *Sleep* 34:49–56

Huang ZL, Urade Y, Hayaishi O. 2007. Prostaglandins and adenosine in the regulation of sleep and wakefulness. *Curr. Opin. Pharmacol.* 7:33–38

Jego S, Glasgow SD, Herrera CG, Ekstrand M, Reed SJ, et al. 2013. Optogenetic identification of a rapid eye movement sleep modulatory circuit in the hypothalamus. *Nat. Neurosci.* 16:1637–43

Jewett KA, Krueger JM. 2012. Humoral sleep regulation; interleukin-1 and tumor necrosis factor. *Vitam. Horm.* 89:241–57

Kageyama H, Takenoya F, Shiba K, Shioda S. 2010. Neuronal circuits involving ghrelin in the hypothalamus-mediated regulation of feeding. *Neuropeptides* 44:133–38

Kalsbeek A, Yi CX, la Fleur SE, Buijs RM, Fliers E. 2010. Suprachiasmatic nucleus and autonomic nervous system influences on awakening from sleep. *Int. Rev. Neurobiol.* 93:91–107

Kang JE, Lim MM, Bateman RJ, Lee JJ, Smyth LP, et al. 2009. Amyloid-β dynamics are regulated by orexin and the sleep-wake cycle. *Science* 326:1005–7

Kapás L, Bohnet SG, Traynor TR, Majde JA, Szentirmai E, et al. 2008. Spontaneous and influenza virus-induced sleep are altered in TNF-α double-receptor deficient mice. *J. Appl. Physiol.* 105:1187–98

Kastenberger I, Lutsch C, Herzog H, Schwarzer C. 2012. Influence of sex and genetic background on anxiety-related and stress-induced behaviour of prodynorphin-deficient mice. *PLoS ONE* 7:e34251

Keating GL, Kuhar MJ, Bliwise DL, Rye DB. 2010. Wake promoting effects of cocaine and amphetamine-regulated transcript (CART). *Neuropeptides* 44:241–46

Khachaturian H, Watson SJ, Lewis ME, Coy D, Goldstein A, Akil H. 1982. Dynorphin immunocytochemistry in the rat central nervous system. *Peptides* 3:941–54

Kimura M, Muller-Preuss P, Lu A, Wiesner E, Flachskamm C, et al. 2010. Conditional corticotropin-releasing hormone overexpression in the mouse forebrain enhances rapid eye movement sleep. *Mol. Psychiatry* 15:154–65

Kishi T, Aschkenasi CJ, Choi BJ, Lopez ME, Lee CE, et al. 2005. Neuropeptide Y Y1 receptor mRNA in rodent brain: distribution and colocalization with melanocortin-4 receptor. *J. Comp. Neurol.* 482:217–43

Kojima M, Kangawa K. 2010. Ghrelin: more than endogenous growth hormone secretagogue. *Ann. N. Y. Acad. Sci.* 1200:140–48

Konadhode RR, Pelluru D, Blanco-Centurion C, Zayachkivsky A, Liu M, et al. 2013. Optogenetic stimulation of MCH neurons increases sleep. *J. Neurosci.* 33:10257–63

Koylu EO, Couceyro PR, Lambert PD, Kuhar MJ. 1998. Cocaine- and amphetamine-regulated transcript peptide immunohistochemical localization in the rat brain. *J. Comp. Neurol.* 391:115–32

Koylu EO, Couceyro PR, Lambert PD, Ling NC, DeSouza EB, Kuhar MJ. 1997. Immunohistochemical localization of novel CART peptides in rat hypothalamus, pituitary and adrenal gland. *J. Neuroendocrinol.* 9:823–33

Kramer A, Yang FC, Snodgrass P, Li X, Scammell TE, et al. 2001. Regulation of daily locomotor activity and sleep by hypothalamic EGF receptor signaling. *Science* 294:2511–15

Krueger JM. 2008. The role of cytokines in sleep regulation. *Curr. Pharm. Des.* 14:3408–16

Krueger JM, Rector DM, Roy S, Van Dongen HPA, Belenky G, Panksepp J. 2008. Sleep as a fundamental property of neuronal assemblies. *Nat. Rev. Neurosci.* 9:910–19

Kryger MH, Roth T, Dement WC, Siegel JM, eds. 2011. *Principles and Practice of Sleep Medicine*. Philadelphia, PA: Saunders/Elsevier. 5th ed.

Kushikata T, Fang J, Chen Z, Wang Y, Krueger JM. 1998. Epidermal growth factor enhances spontaneous sleep in rabbits. *Am. J. Physiol.* 275:R509–14

Laposky AD, Shelton J, Bass J, Dugovic C, Perrino N, Turek FW. 2006. Altered sleep regulation in leptin-deficient mice. *Am. J. Physiol. Regul. Integr. Comp. Physiol.* 290:R894–903

Lein ES, Hawrylycz MJ, Ao N, Ayres M, Bensinger A, et al. Genome-wide atlas of gene expression in the adult mouse brain. *Nature* 445:168–76

Leinninger GM. 2011. Lateral thinking about leptin: a review of leptin action via the lateral hypothalamus. *Physiol. Behav.* 104:572–81

Lester HA, Mager S, Quick MW, Corey JL. 1994. Permeation properties of neurotransmitter transporters. *Annu. Rev. Pharmacol. Toxicol.* 34:219–49

Li H, Ohta H, Izumi H, Matsuda Y, Seki M, et al. 2013. Behavioral and cortical EEG evaluations confirm the roles of both CCKA and CCKB receptors in mouse CCK-induced anxiety. *Behav. Brain Res.* 237:325–32

Li Y, van den Pol AN. 2006. Differential target-dependent actions of coexpressed inhibitory dynorphin and excitatory hypocretin/orexin neuropeptides. *J. Neurosci.* 26:13037–47

Liao F, Taishi P, Churchill L, Urza MJ, Krueger JM. 2010. Localized suppression of cortical growth hormone-releasing hormone receptors state-specifically attenuates electroencephalographic delta waves. *J. Neurosci.* 30:4151–59

Lin JS, Anaclet C, Sergeeva OA, Haas HL. 2011. The waking brain: an update. *Cell. Mol. Life Sci.* 68:2499–512

Ludwig M, Leng G. 2006. Dendritic peptide release and peptide-dependent behaviours. *Nat. Rev. Neurosci.* 7:126–36

Majde JA, Krueger JM. 2005. Links between the innate immune system and sleep. *J. Allergy Clin. Immunol.* 116:1188–98

Marder E. 2012. Neuromodulation of neuronal circuits: back to the future. *Neuron* 76:1–11

Martel G, Dutar P, Epelbaum J, Viollet C. 2012. Somatostatinergic systems: an update on brain functions in normal and pathological aging. *Front. Endocrinol.* 3:154

Martínez GS, Smale L, Nunez AA. 2002. Diurnal and nocturnal rodents show rhythms in orexinergic neurons. *Brain Res.* 955:1–7

Masuo Y, Noguchi J, Morita S, Matsumoto Y. 1995. Effects of intracerebroventricular administration of pituitary adenylate cyclase-activating polypeptide (PACAP) on the motor activity and reserpine-induced hypothermia in murines. *Brain Res.* 700:219–26

McCarley RW. 2011. Neurobiology of REM sleep. *Handb. Clin. Neurol.* 98:151–71

McCoy MK, Tansey MG. 2008. TNF signaling inhibition in the CNS: implications for normal brain function and neurodegenerative disease. *J. Neuroinflamm.* 5:45

Mennicken F, Hoffert C, Pelletier M, Ahmad S, O'Donnell D. 2002. Restricted distribution of galanin receptor 3 (GalR3) mRNA in the adult rat central nervous system. *J. Chem. Neuroanat.* 24:257–68

Moffett M, Stanek L, Harley J, Rogge G, Asnicar M, et al. 2006. Studies of cocaine- and amphetamine-regulated transcript (CART) knockout mice. *Peptides* 27:2037–45

Monti JM, Torterolo P, Lagos P. 2013. Melanin-concentrating hormone control of sleep-wake behavior. *Sleep Med. Rev.* 17:293–98

Mrosovsky N, Redlin U, Roberts RB, Threadgill DW. 2005. Masking in waved-2 mice: EGF receptor control of locomotion questioned. *Chronobiol. Int.* 22:963–74

Mulder AH, Wardeh G, Hogenboom F, Frankhuyzen AL. 1984. Kappa- and delta-opioid receptor agonists differentially inhibit striatal dopamine and acetylcholine release. *Nature* 308:278–80

Murck H, Held K, Ziegenbein M, Künzel H, Holsboer F, Steiger A. 2004. Intravenous administration of the neuropeptide galanin has fast antidepressant efficacy and affects the sleep EEG. *Psychoneuroendocrinology* 29:1205–11

Nelson MD, Raizen DM. 2013. A sleep state during *C. elegans* development. *Curr. Opin. Neurobiol.* 23:824–30

Nelson MD, Trojanowski NF, George-Raizen JB, Smith CJ, Yu CC, et al. 2013. The neuropeptide NLP-22 regulates a sleep-like state in *Caenorhabditis elegans*. *Nat. Commun.* 4:2846

O'Malley MW, Fishman RL, Ciraulo DA, Datta S. 2013. Effect of five-consecutive-day exposure to an anxiogenic stressor on sleep-wake activity in rats. *Front. Neurol.* 4:15

Obal F Jr, Alt J, Taishi P, Gardi J, Krueger JM. 2003. Sleep in mice with nonfunctional growth hormone-releasing hormone receptors. *Am. J. Physiol. Regul. Integr. Comp. Physiol.* 284:R131–39

Obal F Jr, Krueger JM. 2003. Biochemical regulation of non-rapid-eye-movement sleep. *Front. Biosci.* 8:d520–50

O'Donnell D, Ahmad S, Wahlestedt C, Walker P. 1999. Expression of the novel galanin receptor subtype GALR2 in the adult rat CNS: distinct distribution from GALR1. *J. Comp. Neurol.* 409:469–81

Oishi Y, Williams RH, Agostinelli L, Arrigoni E, Fuller PM, et al. 2013. Role of the medial prefrontal cortex in cataplexy. *J. Neurosci.* 33:9743–51

Otto C, Martin M, Wolfer DP, Lipp HP, Maldonado R, Schütz G. 2001. Altered emotional behavior in PACAP-type-I-receptor-deficient mice. *Brain Res. Mol. Brain Res.* 92:78–84

Patterson CM, Leshan RL, Jones JC, Myers MGJ. 2011. Molecular mapping of mouse brain regions innervated by leptin receptor-expressing cells. *Brain Res.* 1378:18–28

Paut-Pagano L, Roky R, Valatx JL, Kitahama K, Jouvet M. 1993. Anatomical distribution of prolactin-like immunoreactivity in the rat brain. *Neuroendocrinology* 58:682–95

Peterfi Z, McGinty D, Sarai E, Szymusiak R. 2010. Growth hormone-releasing hormone activates sleep regulatory neurons of the rat preoptic hypothalamus. *Am. J. Physiol. Regul. Integr. Comp. Physiol.* 298:R147–56

Peterfi Z, Obal F Jr, Taishi P, Gardi J, Kacsoh B, et al. 2006. Sleep in spontaneous dwarf rats. *Brain Res.* 1108:133–46

Pieribone VA, Xu ZQ, Zhang X, Grillner S, Bartfai T, Hökfelt T. 1995. Galanin induces a hyperpolarization of norepinephrine-containing locus coeruleus neurons in the brainstem slice. *Neuroscience* 64:861–74

Prober DA, Rihel J, Onah AA, Sung RJ, Schier AF. 2006. Hypocretin/orexin overexpression induces an insomnia-like phenotype in zebrafish. *J. Neurosci.* 26:13400–10

Rihel J, Prober DA, Arvanites A, Lam K, Zimmerman S, et al. 2010. Zebrafish behavioral profiling links drugs to biological targets and rest/wake regulation. *Science* 327:348–51

Rihel J, Schier AF. 2013. Sites of action of sleep and wake drugs: insights from model organisms. *Curr. Opin. Neurobiol.* 23:831–40

Rios M, Fan G, Fekete C, Kelly J, Bates B, et al. 2001. Conditional deletion of brain-derived neurotrophic factor in the postnatal brain leads to obesity and hyperactivity. *Mol. Endocrinol.* 15:1748–57

Riou F, Cespuglio R, Jouvet M. 1982. Endogenous peptides and sleep in the rat: III. The hypnogenic properties of vasoactive intestinal polypeptide. *Neuropeptides* 2:265–77

Rizzi A, Molinari S, Marti M, Marzola G, Calo' G. 2011. Nociceptin/orphanin FQ receptor knockout rats: in vitro and in vivo studies. *Neuropharmacology* 60:572–79

Romanowski CP, Fenzl T, Flachskamm C, Wurst W, Holsboer F, et al. 2010. Central deficiency of corticotropin-releasing hormone receptor type 1 (CRH-R1) abolishes effects of CRH on NREM but not on REM sleep in mice. *Sleep* 33:427–36

Sakurai T. 2007. The neural circuit of orexin (hypocretin): maintaining sleep and wakefulness. *Nat. Rev. Neurosci.* 8:171–81

Sakurai T, Mieda M. 2011. Connectomics of orexin-producing neurons: interface of systems of emotion, energy homeostasis and arousal. *Trends Pharmacol. Sci.* 32:451–62

Saper CB, Fuller PM, Pedersen NP, Lu J, Scammell TE. 2010. Sleep state switching. *Neuron* 68:1023–42

Saper CB, Lu J, Chou TC, Gooley J. 2005. The hypothalamic integrator for circadian rhythms. *Trends Neurosci.* 28:152–57

Sauvage M, Steckler T. 2001. Detection of corticotropin-releasing hormone receptor 1 immunoreactivity in cholinergic, dopaminergic and noradrenergic neurons of the murine basal forebrain and brainstem nuclei—potential implication for arousal and attention. *Neuroscience* 104:643–52

Schier AF. 2013. Should I stay or should I go: neuromodulators of behavioral states. *Cell* 154:955–56

Schöne C, Burdakov D. 2012. Glutamate and GABA as rapid effectors of hypothalamic "peptidergic" neurons. *Front. Behav. Neurosci.* 6:81

Scott MM, Lachey JL, Sternson SM, Lee CE, Elias CF, et al. 2009. Leptin targets in the mouse brain. *J. Comp. Neurol.* 514:518–32

Sehgal A, Mignot E. 2011. Genetics of sleep and sleep disorders. *Cell* 146:194–207

Sei M, Sei H, Shima K. 1999. Spontaneous activity, sleep, and body temperature in rats lacking the CCK-A receptor. *Physiol. Behav.* 68:25–29

Seutin V, Verbanck P, Massotte L, Dresse A. 1989. Galanin decreases the activity of locus coeruleus neurons in vitro. *Eur. J. Pharmacol.* 164:373–76

Shaw PJ, Cirelli C, Greenspan RJ, Tononi G. 2000. Correlates of sleep and waking in *Drosophila melanogaster*. *Science* 287:1834–37

Shemyakin A, Kapás L. 2001. L-364,718, a cholecystokinin-A receptor antagonist, suppresses feeding-induced sleep in rats. *Am. J. Physiol. Regul. Integr. Comp. Physiol.* 280:R1420–26

Sherin JE, Elmquist JK, Torrealba F, Saper CB. 1998. Innervation of histaminergic tuberomammillary neurons by GABAergic and galaninergic neurons in the ventrolateral preoptic nucleus of the rat. *J. Neurosci.* 18:4705–21

Siaud P, Manzoni O, Balmefrezol M, Barbanel G, Assenmacher I, Alonso G. 1989. The organization of prolactin-like-immunoreactive neurons in the rat central nervous system. Light- and electron-microscopic immunocytochemical studies. *Cell Tissue Res.* 255:107–15

Siegel JM. 2005. Clues to the functions of mammalian sleep. *Nature* 437:1264–71

Siegel JM. 2009. Sleep viewed as a state of adaptive inactivity. *Nat. Rev. Neurosci.* 10:747–53

Siegel JM. 2011. Sleep in animals: a state of adaptive inactivity. See Kryger et al. 2011, pp. 126–38

Signore AP, Zhang F, Weng Z, Gao Y, Chen J. 2008. Leptin neuroprotection in the CNS: mechanisms and therapeutic potentials. *J. Neurochem.* 106:1977–90

Sinton CM, Fitch TE, Gershenfeld HK. 1999. The effects of leptin on REM sleep and slow wave delta in rats are reversed by food deprivation. *J. Sleep Res.* 8:197–203

Smith PM, Ferguson AV. 2008. Neurophysiology of hunger and satiety. *Dev. Disabil. Res. Rev.* 14:96–104

Snodgrass-Belt P, Gilbert JL, Davis FC. 2005. Central administration of transforming growth factor-alpha and neuregulin-1 suppress active behaviors and cause weight loss in hamsters. *Brain Res.* 1038:171–82

Steiger A. 2007. Neurochemical regulation of sleep. *J. Psychiatr. Res.* 41:537–52

Sternson SM. 2013. Hypothalamic survival circuits: blueprints for purposive behaviors. *Neuron* 77:810–24

Szentirmai E. 2012. Central but not systemic administration of ghrelin induces wakefulness in mice. *PLoS ONE* 7:e41172

Szentirmai E, Kapás L, Sun Y, Smith RG, Krueger JM. 2007. Spontaneous sleep and homeostatic sleep regulation in ghrelin knockout mice. *Am. J. Physiol. Regul. Integr. Comp. Physiol.* 293:R510–17

Szentirmai E, Krueger JM. 2006. Obestatin alters sleep in rats. *Neurosci. Lett.* 404:222–26

Tafti M. 2007. Quantitative genetics of sleep in inbred mice. *Dialogues Clin. Neurosci.* 9:273–78

Tobler I. 2011. Phylogeny of sleep regulation. See Kryger et al. 2011, pp. 76–91

Tononi G, Cirelli C. 2012. Time to be SHY? Some comments on sleep and synaptic homeostasis. *Neural Plast.* 2012:415250

Torterolo P, Lagos P, Monti JM. 2011. Melanin-concentrating hormone: a new sleep factor? *Front. Neurol.* 2:14

Tsujino N, Sakurai T. 2013. Role of orexin in modulating arousal, feeding, and motivation. *Front. Behav. Neurosci.* 7:28

Tsujino N, Yamanaka A, Ichiki K, Muraki Y, Kilduff TS, et al. 2005. Cholecystokinin activates orexin/hypocretin neurons through the cholecystokinin A receptor. *J. Neurosci.* 25:7459–69

Urade Y, Hayaishi O. 2011. Prostaglandin D2 and sleep/wake regulation. *Sleep Med. Rev.* 15:411–18

van den Pol AN. 2012. Neuropeptide transmission in brain circuits. *Neuron* 76:98–115

Varela L, Horvath TL. 2012. Leptin and insulin pathways in POMC and AgRP neurons that modulate energy balance and glucose homeostasis. *EMBO Rep.* 13:1079–86

Vaudry D, Falluel-Morel A, Bourgault S, Basille M, Burel D, et al. 2009. Pituitary adenylate cyclase-activating polypeptide and its receptors: 20 years after the discovery. *Pharmacol. Rev.* 61:283–357

Venner A, Karnani MM, Gonzalez JA, Jensen LT, Fugger L, Burdakov D. 2011. Orexin neurons as conditional glucosensors: paradoxical regulation of sugar sensing by intracellular fuels. *J. Physiol.* 589:5701–8

Wang Q, Shin EJ, Nguyen XK, Li Q, Bach JH, et al. 2012. Endogenous dynorphin protects against neurotoxin-elicited nigrostriatal dopaminergic neuron damage and motor deficits in mice. *J. Neuroinflamm.* 9:124

Welsh DK, Takahashi JS, Kay SA. 2010. Suprachiasmatic nucleus: cell autonomy and network properties. *Annu. Rev. Physiol.* 72:551–77

Williams KS, Behn CG. 2011. Dynamic interactions between orexin and dynorphin may delay onset of functional orexin effects: a modeling study. *J. Biol. Rhythms* 26:171–81

Willie JT, Sinton CM, Maratos-Flier E, Yanagisawa M. 2008. Abnormal response of melanin-concentrating hormone deficient mice to fasting: hyperactivity and rapid eye movement sleep suppression. *Neuroscience* 156:819–29

Winsky-Sommerer R, Yamanaka A, Diano S, Borok E, Roberts AJ, et al. 2004. Interaction between the corticotropin-releasing factor system and hypocretins (orexins): a novel circuit mediating stress response. *J. Neurosci.* 24:11439–48

Woods IG, Schoppik D, Shi V, Zimmerman S, Coleman HA, et al. 2014. Neuropeptidergic signaling partitions arousal behaviors in zebrafish. *J. Neurosci.* 34:3142–60

Xie X, Wisor JP, Hara J, Crowder TL, LeWinter R, et al. 2008. Hypocretin/orexin and nociceptin/orphanin FQ coordinately regulate analgesia in a mouse model of stress-induced analgesia. *J. Clin. Invest.* 118:2471–81

Yamanaka A, Beuckmann CT, Willie JT, Hara J, Tsujino N, et al. 2003. Hypothalamic orexin neurons regulate arousal according to energy balance in mice. *Neuron* 38:701–13

Zager A, Andersen ML, Ruiz FS, Antunes IB, Tufik S. 2007. Effects of acute and chronic sleep loss on immune modulation of rats. *Am. J. Physiol. Regul. Integr. Comp. Physiol.* 293:R504–9

Zeitzer JM. 2013. Control of sleep and wakefulness in health and disease. *Prog. Mol. Biol. Transl. Sci.* 119:137–54

Zhang G, Wang L, Liu H, Zhang J. 2004. Substance P promotes sleep in the ventrolateral preoptic area of rats. *Brain Res.* 1028:225–32

Zielinski MR, Krueger JM. 2011. Sleep and innate immunity. *Front. Biosci.* 3:632–42

Zigman JM, Jones JE, Lee CE, Saper CB, Elmquist JK. 2006. Expression of ghrelin receptor mRNA in the rat and the mouse brain. *J. Comp. Neurol.* 494:528–48

Cumulative Indexes

Contributing Authors, Volumes 28–37

Dawson TM, 28:57–84
Dawson VL, 28:57–84
Dayan P, 32:95–126
De Biasi M, 34:105–30
de Bono M, 28:451–501
Dehaene S, 32:185–208
Deisseroth K, 34:389–412
Dellovade T, 29:539–63
Devineni AV, 36:121–38
De Vos KJ, 31:151–73
Dhaka A, 29:135–61
Di H, 37:457–78
Dib-Hajj SD, 33:325–47
Dillon C, 28:25–55
DiMauro S, 31:91–123
Dölen G, 35:417–43
Dolmetsch RE, 37:479–501
Donoghue JP, 32:249–66
Doupe AJ, 36:489–517
Dudai Y, 35:227–47
Dum RP, 32:413–34
During MJ, 35:331–45

E

Eatock RA, 34:501–34
Ehlers MD, 29:325–62
Eichenbaum H, 30:123–52
Eickhoff SB, 37:409–34
Eijkelkamp N, 36:519–46
El-Amraoui A, 35:509–28
Emes RD, 35:111–31
Emoto K, 30:399–423

F

Fancy SPJ, 34:21–43
Fekete DM, 36:361–81
Feldheim D, 36:51–77
Feldman DE, 32:33–55
Feller MB, 31:479–509
Feng G, 35:49–71
Fenno L, 34:389–412
Fernald RD, 35:133–51
Field GD, 30:1–30
Fields HL, 30:289–316
Fiez JA, 32:413–34
Fishell G, 34:535–67
Flames N, 34:153–84
Flavell SW, 31:563–90
Forbes CE, 33:299–324
Fotowat H, 34:1–19

Fox PT, 37:409–34
Franklin RJM, 34:21–43
Freeman MR, 33:245–67
Fregni F, 28:377–401
Freund TF, 35:529–58
Friedrich RW, 36:383–402
Fries P, 32:209–24
Frigon A, 34:413–40
Fu Y-H, 36:25–50
Fusi S, 33:173–202

G

Gabbiani F, 34:1–19
Gage FH, 29:77–103
Gallant JL, 29:477–505
Gandhi NJ, 34:205–31
Ganguli S, 35:485–508
Gao Q, 30:367–98
Gao X-B, 37:101–16
Garner CC, 28:251–74
Gerfen CR, 34:441–66
Ghosh KK, 32:435–506
Ginty DD, 28:191–222
Gitlin JD, 30:317–37
Glebova NO, 28:191–222
Goda Y, 28:25–55
Gohl DM, 37:307–27
Gold JI, 30:535–74
Goldberg ME, 33:1–21
Gosseries O, 37:457–78
Grafman J, 33:299–324
Grant SGN, 35:111–31
Graybiel AM, 31:359–87
Graziano M, 29:105–34
Green CB, 35:445–62
Green CS, 35:391–416
Greenberg ME, 31:563–90
Greene NDE, 37:221–42
Grierson AJ, 31:151–73
Groves AK, 36:361–81
Grueber WB, 36:547–68
Guntupalli JS, 37:435–56

H

Haag J, 33:49–70
Haddad R, 37:363–85
Haelterman NA, 37:137–59
Hagoort P, 37:347–62
Han V, 31:1–25
Hardy J, 37:79–100

Hariri AR, 32:225–47
Harrington DL, 36:313–36
Harris KM, 31:47–67
Hatsopoulos NG, 32:249–66
Hatten ME, 28:89–108
Häusser M, 28:503–32
Haxby JV, 37:435–56
He Z, 34:131–52
Heberlein U, 36:121–38
Heintz N, 28:89–108
Henderson CE, 33:409–40
Henriques DYP, 34:309–31
Hikosaka O, 37:289–306
Hillman EMC, 37:161–81
Hines M, 34:69–88
Hjelmstad GO, 30:289–316
Ho ETW, 32:435–506
Hobert O, 34:153–84
Holtzheimer PE, 34:289–307
Holtzman DM, 31:175–93
Horton JC, 28:303–26
Horvath T, 37:101–16
Horvath TL, 30:367–98
Howell GR, 35:153–79
Huang W, 37:17–38
Huang ZJ, 36:183–215
Hübener M, 35:309–30
Huberman AD, 31:479–509
Huys QJM, 32:95–126
Hyman SE, 29:565–98

I

Indefrey P, 37:347–62
Inostroza M, 36:79–102
Isa T, 35:559–78

J

Jacobson SG, 33:441–72
Jadzinsky PD, 36:403–28
Jaiswal M, 37:137–59
Jan Y-N, 30:399–423
Jazayeri M, 37:205–20
Johansen-Berg H, 32:75–94
John SWM, 35:153–79
Jung MW, 35:287–308

K

Kalaska JF, 33:269–98
Kanning KC, 33:409–40
Kanold PO, 33:23–48

Kaplan A, 33:409–40
Katnani HA, 34:205–31
Katona I, 35:529–58
Kelley MW, 29:363–86
Kelsch W, 33:131–49
Kennedy MJ, 29:325–62
Kevrekidis IG, 28:533–63
Kiecker C, 35:347–67
Kiehn O, 29:279–306
Kim HF, 37:289–306
Kim MD, 30:399–423
King-Casas B, 29:417–48
Knierim JJ, 35:267–86
Knudsen EI, 30:57–78
Koleske AJ, 33:349–78
Kourtzi Z, 34:45–67
Krakauer JW, 33:89–108
Krauzlis RJ, 36:165–82
Kreitzer AC, 32:127–47;
  37:117–35
Kriegstein A, 32:149–84
Kristan WB Jr, 31:271–94
Kropff E, 31:69–89
Kuhl PK, 31:511–34
Kullmann DM, 33:151–72
Kuwabara T, 29:77–103
Kwan KY, 30:339–65

L

Laird AR, 37:409–34
Lancaster JL, 37:409–34
Larkum ME, 36:1–24
Laureys S, 37:457–78
LeDoux JE, 32:289–313
Lee D, 35:287–308
Lee YS, 33:221–43
Lemon RN, 31:195–218
Lempert KM, 37:263–87
Leopold DA, 35:91–109
Levelt CN, 35:309–30
Lin Y-C, 33:349–78
Lingnau A, 37:1–15
Liu K, 34:131–52
Livingstone MS, 31:411–37
Lois C, 33:131–49
London M, 28:503–32
Love J, 30:451–74
Lovejoy LP, 36:165–82
Luhmann HJ, 33:23–48
Lumsden A, 35:347–67
Luo L, 28:127–56

M

Ma WJ, 37:205–20
Madsen E, 30:317–37
Mainen ZF, 29:163–201
Major G, 36:1–24
Malenka RC, 29:565–98
Marder E, 37:329–46
Margolis EB, 30:289–316
Maricq AV, 28:451–501
Mason CA, 31:295–315
Mayberg HS, 34:289–307
McAllister AK, 30:425–50
McCandliss BD, 30:475–503
McInnes RR, 33:441–72
McLaughlin T, 28:327–55
McMahon SB, 29:507–38
Meck WH, 36:313–36
Medendorp WP, 34:309–31
Merabet LB, 28:377–401
Merchant H, 36:313–36
Miesenböck G, 28:533–63
Miller CCJ, 31:151–73
Miller EK, 33:203–19
Ming G-l, 28:223–50;
  37:243–62
Mink JW, 29:229–57
Mintun MA, 29:449–76
Mohawk JA, 35:445–62
Montague PR, 29:417–48
Montcouquiol M, 29:363–86
Moore DJ, 28:57–84
Moore T, 36:451–66
Mori K, 34:467–99
Moser EI, 31:69–89
Moser M-B, 31:69–89
Mukamel EA, 32:435–506
Muotri AR, 29:77–103
Murray EA, 30:99–122
Murthy VN, 34:233–58

N

Nave K-A, 31:247–69, 535–61
Naya Y, 37:39–53
Nelson AB, 37:117–35
Nestler EJ, 29:565–98
Nickells RW, 35:153–79
Nicola SM, 30:289–316
Nieder A, 32:185–208
Nienborg H, 35:463–83
Noudoost B, 36:451–66

O

O'Brien RJ, 34:185–204
O'Donnell M, 32:383–412
O'Leary DDM, 28:127–56,
  327–55
O'Leary T, 37:329–46
Oppenheim RW, 29:1–35
Orban GA, 34:361–88
Orban T, 36:139–64
Orr HT, 30:575–621

P

Padoa-Schioppa C, 34:333–59
Palczewski K, 36:139–64
Panagiotakos G, 37:479–501
Park KK, 34:131–52
Parrish JZ, 30:399–423
Paşca SP, 37:479–501
Pascual-Leone A, 28:377–401
Patapoutian A, 29:135–61
Peça J, 35:49–71
Perlmutter JS, 29:229–57
Petersen CCH, 37:183–203
Petersen SE, 35:73–89
Petit C, 35:509–28
Petros TJ, 31:295–315
Pezet S, 29:507–38
Phelps EA, 37:263–87
Polleux F, 32:347–81
Poo C, 37:363–85
Poo M-m, 35:181–201
Posner MI, 35:73–89
Pouget A, 35:391–416
Ptácek LJ, 36:25–50
Purdon PL, 34:601–28

Q

Quick K, 36:519–46

R

Raichle ME, 29:449–76
Rajan K, 28:357–76
Ranganath C, 30:123–52
Ranum LPW, 29:259–77
Rebsam A, 31:295–315
Reiff DF, 33:49–70
Richter C, 37:503–31
Rivera-Gaxiola M, 31:511–34

Robbins TW, 32:267–87
Rodrigues SM, 32:289–313
Roelfsema PR, 29:203–27
Romanski LM, 32:315–46
Romer JT, 29:539–63
Roska B, 36:467–88
Roskies AL, 33:109–30
Rossignol S, 34:413–40
Roth BL, 37:387–407
Rowitch DH, 34:21–43
Rubin LL, 29:539–63
Rudy B, 34:535–67
Rushworth MFS, 32:75–94
Russell JF, 36:25–50

S

Sabatini BL, 30:79–97
Safieddine S, 35:509–28
Sahakian BJ, 32:57–74
Sahani M, 36:337–59
Sahel J-A, 36:467–88
Sakai K, 31:219–45
Sakano H, 34:467–99
Saksida LM, 30:99–122
Salinas PC, 31:339–58
Salzman CD, 33:173–202
Sandoval H, 37:137–59
Sapolsky RM, 32:289–313
Sawtell NB, 31:1–25
Schafer RJ, 36:451–66
Scheiffele P, 33:473–507
Schier AF, 37:503–31
Schiller J, 36:1–24
Schlaggar BL, 30:475–503
Schnitzer MJ, 32:435–506
Scholz J, 32:1–32
Schon EA, 31:91–123
Schrater P, 35:391–416
Schultz W, 30:259–88
Seger CA, 33:203–19
Seo H, 35:287–308
Shadlen MN, 30:535–74
Shadmehr R, 33:89–108
Shapiro L, 30:451–74
Sharpee TO, 36:103–20
Shen K, 33:473–507
Shenoy KV, 36:337–59
Shilyansky C, 33:221–43
Shruti S, 37:329–46
Shulman GL, 34:569–99

Shulman JM, 37:137–59
Silies M, 37:307–27
Silva AJ, 33:221–43
Sim S, 33:131–49
Sincich LC, 28:303–26
Singer T, 35:1–23
Sleegers K, 33:71–88
Smith MA, 33:89–108
Sokol-Hessner P, 37:263–87
Sommer MA, 31:317–38
Sompolinsky H, 35:485–508
Song H, 28:223–50; 37:243–62
Songer JE, 34:501–34
Soto I, 35:153–79
Squire LR, 34:259–88
Squire RF, 36:451–66
Stephan AH, 35:369–89
Sternson SM, 37:387–407
Stevens B, 35:369–89
Stocker RF, 30:505–33
Strick PL, 32:413–34
Strickland S, 30:209–33
Strnad L, 37:1–15
Sun W, 29:1–35
Surmeier DJ, 34:441–66
Suzuki WA, 37:39–53

T

Takahashi JS, 35:445–62
Takahashi KA, 29:37–75
Tam SJ, 33:379–408
Tan L, 37:79–100
Taube JS, 30:181–207
Tedeschi A, 34:131–52
Thiele A, 36:271–94
Thompson PM, 28:1–23
Ting JT, 35:49–71
Toga AW, 28:1–23
Trapp BD, 31:247–69, 535–61
Trussell LO, 35:249–65
Tsao DY, 31:411–37
Turrigiano G, 34:89–103

U

Uchida N, 37:363–85

V

Van Broeckhoven C, 33:71–88

Van Dort CJ, 34:601–28
Vargas ME, 30:153–79
Viswanath V, 29:135–61
Vogels TP, 28:357–76
Vollrath MA, 30:339–65
Vosshall LB, 30:505–33

W

Waites CL, 28:251–74
Wallis JD, 30:31–56
Wang X-J, 35:203–25
Watts RJ, 33:379–408
Waxman SG, 33:325–47
West AB, 28:57–84
Wilson RI, 29:163–201;
  36:217–41
Wilt BA, 32:435–506
Wixted JT, 34:259–88
Wong PC, 34:185–204
Wood JN, 36:519–46
Woods IG, 37:503–31
Woolf CJ, 32:1–32
Wright AF, 33:441–72
Wu MC-K, 29:477–505
Wurtz RH, 31:317–38

Y

Yamamoto A, 37:55–77
Yamamoto S, 37:289–306
Yasuda M, 37:289–306
Yeo G, 29:77–103
Yizhar O, 34:389–412
Yonelinas AR, 30:123–52
Yoon WH, 37:137–59
Yu J-T, 37:79–100
Yu W-M, 30:209–33
Yue Z, 37:55–77
Yuste R, 36:429–49

Z

Zeng H, 36:183–215
Zénon A, 36:165–82
Zhang K, 35:267–86
Zhang KD, 36:361–81
Zipursky SL, 36:547–68
Zoghbi HY, 30:575–621
Zou Y, 31:339–58

Article Titles, Volumes 28–37